PRECISION AGRICULTURE BASICS

2ND EDITION

PRECISION AGRICULTURE BASICS

2ND EDITION

Edited by David E. Clay, Skye Brugler & Katsutoshi Mizuta

ASA American Society of Agronomy
CSSA Crop Science Society of America
SSSA Soil Science Society of America

WILEY

Copublication by American Society of Agronomy, Inc., Crop Science Society of America, Inc., Soil Science Society of America, Inc. and John Wiley & Sons, Inc.

Editorial Correspondence:
American Society of Agronomy, Inc.
Crop Science Society of America, Inc.
Soil Science Society of America, Inc.

5585 Guilford Road, Madison, WI 53711-58011, USA
agronomy.org - crops.org - soils.org

Registered Office:
John Wiley & Sons, Inc., 111 River Street, Hoboken, NJ 07030, USA

For details of our global editorial offices, customer services, and more information about Wiley products, visit us at www.wiley.com.

Wiley also publishes its books in a variety of electronic formats and by print-on-demand. Some content that appears in standard print versions of this book may not be available in other formats.

Library of Congress Cataloging-in-Publication Data Applied for:

Paperback: 9780891184089

Cover Design: Wiley
Cover Image: JamesBrey/Getty Images

This book was typeset in 11/13 Palatino LT Std by Straive™.

Table of Contents

Editors

Dr. David E. Clay is the editor-in-chief for the American Society of Agronomy, Distinguished Professor Soil Biogeochemistry, and the South Dakota Corn Council Endowed Chair of Precision Soybean. In addition, he is a Fellow of the American Society of Agronomy and a former President of the American Society of Agronomy (2022) and former Editor for the Agronomy Journal. His research goal is to develop and test sustainable agricultural management systems that enhance environmental quality, maintain rural economies, and sustainably produce food while improving soil health. He teaches Introductory Soil Science, Environmental Soil Chemistry, and the Use of Sensors in Agriculture.

Skye Brugler M.S is a Research Associate and PhD candidate at South Dakota State University in the Department of Agronomy, Horticulture, and Plant Science. Her research involves 4R management strategies and creating system-based approaches to nitrogen management through a combination of soil science, microbiology, and plant physiology. Her research goals include improving soil health and maintaining rural economies through practices that produce more food with less resources. She is passionate about creating materials for students, farmers, and professionals to improve agricultural outreach and education.

Dr. Katsutoshi Mizuta is an Assistant Professor of Computational, Agricultural and Environmental Sciences at the University of Kentucky, Department of Plant and Soil Sciences. He is currently serving as a guest editor for the Frontiers in Soil Science Journal and the Crop Science Journal. His research integrates precision agriculture and pedo-econometrics to develop data-driven technologies using proximal and remote sensing data with artificial intelligence, for the rapid, cost-effective, and accurate characterization of soil and crop health as well as quantitative assessment of agroecological functions and management efficiency. He teaches Precision Management for Agriculture and Natural Resource Conservation.

Contributors

Chapter 1

David E. Clay
Department of Agronomy
Horticulture, and Plant Science
South Dakota State University

Skye Brugler
Department of Agronomy
Horticulture, and Plant Science
South Dakota State University

Katsutoshi Mizuta
Department of Plant and Soil Sciences
University of Kentucky

Chapter 2

Maaz Gardezi
Virginia Polytechnic Institute and State University

Bhavna Joshi
Virginia Polytechnic Institute and State University

Chapter 3

Harkirat Kaur
Department of Plant and Agroecosystem Sciences
University of Wisconsin-Madison
Madison, WI

Luan Oliveira
Department of Horticulture
University of Georgia.

Simerjeet Virk
Department of Biosystems Engineering
Auburn University
Alabama

David E. Clay
Department of Agronomy
Horticulture, and Plant Science
South Dakota State University

Ali Mirzakhani Nafchi
Department of Agronomy
Horticulture, and Plant Science
South Dakota State University

Chapter 4

Ali Mirzakhani Nafchi
South Dakota State University
Ajay Sharda, Kansas State University

David E. Clay
Department of Agronomy
Horticulture, and Plant Science
South Dakota State University

Simerjeet Virk
Biosystems Engineering
Auburn University, Alabama

Chapter 5

Sharon A. Clay
Department of Agronomy
Horticulture, and Plant Science
South Dakota State University
Brookings, SD.

Bruce J. Erickson
Department of Agronomy
Purdue University, Indiana

Newell R. Kitchen
Agronomic, Geologic, and Environmental Services
Columbia, Missouri

Chapter 6

Katsutoshi Mizuta
Department of Plant and Soil Sciences
University of Kentucky.

Skye Brugler
Department of Agronomy, Horticulture, and Plant Science
South Dakota State University

Srinadh Kodali
Department of Agronomy
Horticulture, and Plant Science
South Dakota State University

Ahmed Harb Rabia
Department of Agricultural and Biosystems Engineering
North Dakota State University
and
Damanhour University
Department of Natural Resources and Agricultural Engineering
Egypt

Xi Zhang
Department of Biosystems Engineering and Soil Science
University of Tennessee,

David E. Clay
Department of Agronomy
Horticulture, and Plant Science
South Dakota State University

Chapter 7

Timothy Stombaugh
Biosystems and Agricultural Engineering
University of Kentucky

Won Suk Lee
Department of Agricultural and Biological Engineering
University of Florida

Ali Mirzakhani Nafchi
Departments of Agronomy
Horticulture and Plant Science, and Agricultural & Biosystem Engineering
South Dakota State University

Kent Shannon
Division of Plant Science and Technology
University of Missouri

Hasan Mirzakhaninafchi
Department of Agricultural & Biosystem Engineering
South Dakota State University

Chapter 8

Liang Liang
Department of Geography
University of Kentucky

Katsutoshi Mizuta
Department of Plant and Soil Sciences
University of Kentucky

Indunil P. Dharmasiri
Department of Management
Virginia Polytechnic Institute and State University

Leonardo M. Bastos
Department of Crop and Soil Sciences
University of Georgia

Rob Proulx
Department of Agricultural and Biosystems Engineering
North Dakota State University

Chapter 9

Leonardo M. Bastos
Department of Crop and Soil Sciences
University of Georgia

Katsutoshi Mizuta
Department of Plant and Soil Sciences
University of Kentucky

Rob Proulx
Department of Agricultural and Biosystems Engineering
North Dakota State University.

Chapter 10

Carrie Werkmeister
Department of Agronomy
Horticulture, and Plant Science
South Dakota State University

Douglas Malo
Department of Agronomy
Horticulture, and Plant Science
South Dakota State University

Chapter 11

Robin Gebbers
Agromechatronics Department
Leibniz Institute for Agricultural Engineering and Bioeconomy (ATB)
Potsdam, Germany

Katsutoshi Mizuta
Department of Plant and Soil Sciences
University of Kentucky

Viacheslav Adamchuk
Bioresource Engineering Department
McGill University, Quebec, Canada

Chapter 12

Asim Biswas
School of Environmental Sciences
University of Guelph
Ontario, Canada

Robin Gebbers
Agromechatronics Department
Leibniz Institute for Agricultural Engineering and Bioeconomy (ATB)
Potsdam, Germany

Viacheslav Adamchuk
Bioresource Engineering Department
McGill University
Quebec, Canada

Gaurav Jha
Department of Agronomy
Kansas State University

Katsutoshi Mizuta
Department of Plant and Soil Sciences
University of Kentucky

Slawomir Kurpaska
Department of Bioprocess Engineering
Energy, and Automation
University of Agriculture
Krokow, Poland

Maciek Gliniak
University of Minnesota

Ce Yang
Department of Bioproducts and Biosystems Engineering
University of Minnesota

Naoya Kadota
School of Environmental Sciences
University of Guelph, Ontario, Canada

Hiteshkumar Bhogilal Vasava
School of Environmental Sciences
University of Guelph
Ontario, Canada

Chapter 13

Ce Yang,
Department of Bioproducts Engineering
University of Minnesota

Alireza Sanaeifar
Department of Bioproducts Engineering
University of Minnesota

Lang Qiao,
Department of Bioproducts Engineering
University of Minnesota

Sravanthi Bachina
Precision Agriculture Technology
Fort Hays Tech North Central, Kansas

Asim Biswas
School of Environmental Sciences
University of Guelph, Ontario, Canada

Robin Gebbers
Agromechatronics Department
Leibniz Institute for Agricultural Engineering and Bioeconomy (ATB)
Potsdam, Germany

Viacheslav Adamchuk
Bioresource Engineering Department
McGill University
Quebec, Canada

Gaurav Jha
Department of Agronomy
Kansas State University

Katsutoshi Mizuta
Department of Plant and Soil Sciences
University of Kentucky

Slawomir Kurpaska
Department of Bioprocess Engineering
Energy, and Automation
University of Agriculture, Krokow, Poland

Maciek Gliniak
Department of Bioprocess Engineering
Energy, and Automation
University of Agriculture
Krokow, Poland

Dharani Suresh Babu
Vegetable Crops Reserach Unit
USDA-ARS, Madison
WS, and Plant and Agroecosystem Sciences
University of Wisconsin

Chapter 14

Richard Ferguson
University of Nebraska-Lincoln

Brian Wardlow
University of Nebraska-Lincoln

Jackson Stansell
Sentinel Ag

Gaurav Jha
Kansas State University

Leonardo Bastos
Department of Crop and Soil Sciences
University of Georgia

Guillermo Balboa
University of Nebraska-Lincoln

Sheetal Kumari
Virginia Polytechnic Institute and State University

Chapter 15

Lorena Lacerda
Department of Crop and Soil Sciences
University of Georgia

Meetpal Kukal
Department of Soil and Water Systems
University of Idaho

Chapter 16

Meetpal S. Kukal
Department of Soil and Water Systems
University of Idaho

Emily Bedwell
Department of Soil and Water Systems
University of Idaho

Jasmine Neupane
Division of Plant Science and Technology
University of Missouri

Unius Arinaitwe
South Dakota State University

Chapter 17

Ivan Grijalva
Department of Entomology
Louisiana State University

Brian McCornack
Department of Entomology
Kansas State University

Sharon A. Clay
Department of Agronomy
Horticulture and Plant Science
South Dakota State University

Chapter 18

J. Anita Dille
Department of Agronomy
Kansas State University

Rodrigo Werle
Department of Plant and Agroecosystem Sciences
University of Wisconsin-Madison

Thomas R. Butts
Department of Botany and Plant Pathology
Purdue University
Indiana

Lynn Sosnoskie
School of Integrated Plant Science-Horticulture Section
Cornell University, New York

Thierry Besançon
Philip E. Marucci Center for Blueberry and Cranberry Research and Extension
Rutgers University, New Jersey

Chapter 19

Febina Mathew
Department of Plant Pathology
North Dakota State University

Harikat Kaur
Department of Plant and Agroecosystem Sciences
University of Wisconsin–Madison

Milsha George
Department of Plant Pathology
North Dakota State University

Karthika Mohan
Department of Plant Pathology
North Dakota State University

Taofeek Mukaila
Department of Plant Pathology
North Dakota State University

Nitha Rafi
Department of Plant Pathology
North Dakota State University

Sharon A. Clay
Department of Agronomy
Horticulture and Plant Science
South Dakota State University

Chapter 20

John P. Fulton
Department of Food
Agriculture, and Biological Engineering
The Ohio State University

Elizabeth Hawkins
Department of Extension
The Ohio State University

Kent Shannon
Division of Plant Science and Technology
Agricultural Systems Technology University of Missouri

Sunoj Shajahan
Department of Agricultural and Biological Engineering
University of Illinois Urbana-Champaign

Simerjeet Virk
Department of Biosystems Engineering
Auburn University, Alabama

Chapter 21

Peter M. Kyveryga
Department of Agronomy
Iowa State University

Matthew E. Caroll
Iowa Soybean Association
Research Center for Farming Innovation, Iowa

Takashi S. T. Tanaka
Department of Agroecology
Faculty of Technical Sciences
Aarhus University, Slagelse, Denmark

Chapter 22

Ali Mirzakhani Nafchi
Departments of Agronomy
Horticulture and Plant Science, and Agricultural & Biosystem Engineering
South Dakota State University

John P. Fulton
Department of Food
Agriculture, and Biological Engineering
The Ohio State University

Chapter 23

Tong Wang
Ness School of Management and Economics
South Dakota State University

Akinsola Oyebanji
Ness School of Management and Economics
South Dakota State University

David E. Clay
Department of Agronomy
Horticulture, and Plant Science
South Dakota State University

Chapter 24

Xi Zhang
Department of Biosystems Engineering and Soil Science
The University of Tennessee

Kristen Veum
Cropping Systems and Water Quality Research Unit
USDA-ARS
University of Missouri

Yuchuan Fan
Department of Soil, Water, and Ecosystem Sciences
University of Florida Institute of Food and Agricultural Sciences

Vander Luis Novaes Nunes
Department of Agronomy
Horticulture, and Plant Science
South Dakota State University

Wei Dai
Department of Biosystems Engineering and Soil Science
The University of Tennessee

Skye Brugler
Department of Agronomy
Horticulture, and Plant Science
South Dakota State University

David E. Clay
Department of Agronomy
Horticulture, and Plant Science
South Dakota State University

Chapter 25

Jasmine Neupane
Division of Plant Science and Technology
University of Missouri-Columbia

Gretchen Sassenrath
Department of Agronomy
Kansas State University

Deepak R. Joshi
Department of Agronomy
Kansas State University

Jorge Delgado
USDA-ARS Soil Management and Sugar Beet Research Unit
Fort Collins, Colorado

Rojina Kafle
Department of Agronomy
Horticulture, and Plant Science
South Dakota State University

Dharani Suresh Babu
Vegetable Crops Research Unit
USDA-ARS, Madison
WS and Plant and Agroecosystem Sciences
University of Wisconsin

Chapter 26

Ali Mirzakhani Nafchi
Department of Agronomy
Horticulture, and Plant Science
South Dakota State University

Douglas Prairie
Department of Agronomy
Horticulture, and Plant Science
South Dakota State University

Todd Trooien
Department of Agronomy
Horticulture, and Plant Science
South Dakota State University

Nicholas Uilk
Department of Agronomy
Horticulture, and Plant Science
South Dakota State University

Xufei Yang
Department of Agronomy
Horticulture, and Plant Science
South Dakota State University

Mazhar Sher
Department of Agronomy,
Horticulture, and Plant Science
South Dakota State University

Karishma Kumari
Department of Agronomy
Horticulture, and Plant Science
South Dakota State University

Ahmed Abdalla
Department of Agronomy
Horticulture, and Plant Science
South Dakota State University

Hasan Mirzakhaninafchi
University of Georgia

Chapter 27

Deepak R. Joshi
Department of Agronomy
Kansas State University

Skye Brugler
Department of Agronomy
Horticulture and Plant Science
South Dakota State University

David Mulla
Department of Soil
Water, and Climate
University of Minnesota

Jitender Rathore
School of Plant and Environmental Sciences
Virginia Polytechnic Institute and State University

Ahmed Harb Rabia
Department of Agricultural and Biosystems Engineering
North Dakota State University

Pappu Yadav
Department of Agricultural and Biosystems Engineering
South Dakota State University

Chapter 28

Heidi Sieverding
Civil and Environmental Engineering Department
South Dakota School of Mines and Technology

Nowrin Mostafa Tuly
Department of Agronomy
Horticulture, and Plant Science
South Dakota State University

Skye Brugler
Department of Agronomy
Horticulture, and Plant Science
South Dakota State University

James J. Stone
Civil and Environmental Engineering Department
South Dakota School of Mines and Technology

Chapter 29

Unius Arinaitwe
Department of Agronomy
Horticulture, and Plant Science
South Dakota State University

Thandiwe Nleya
Department of Agronomy
Horticulture, and Plant Science
South Dakota State University

Dwarika Bhattarai
Department of Agronomy
Horticulture, and Plant Science
South Dakota State University

Rojina Kafle
Department of Agronomy
Horticulture, and Plant Science
South Dakota State University

Shailesh Pandit
Department of Agronomy
Horticulture, and Plant Science
South Dakota State University

Janet Moriles Miller
Department of Agronomy
Horticulture, and Plant Science
South Dakota State University

Abraham Hangamaisho
Department of Plant Pathology
Microbiology and Biotechnology
North Dakota State University

Rafael Otto
Department of Soil Science
University of São Paulo,
Brazil

Guilherme Martineli Sanches
Department of Soil Science
Escola Superior de Agricultura Luiz de Queiroz
da Universidade de São Paulo
Brazil

Prakriti Sharma
Department of Plant Pathology
Kansas State University

Varshith Kommineni
Department of Agronomy
Horticulture, and Plant Science
South Dakota State University

Chapter 30

Katsutoshi Mizuta
Department of Plant and Soil Sciences
University of Kentucky

Skye Brugler
Department of Agronomy
Horticulture, and Plant Science
South Dakota State University

Sravanthi Bachina
Department of Agronomy
Horticulture, and Plant Science
South Dakota State University

Sarah E. LaRose
Agricultural Sciences Education and
Communication Department
Purdue University
Indiana

Nick Schiltz
Department of Agricultural Sciences
River Community College
Minnesota

Chapter 31

Liujun (Johnny) Li
Department of Soil and Water Systems and
Department of Computer Science
University of Idaho

Sravanthi Bachina
Department of Precision Agriculture Technology
Fort Hays Technical Northwest
Kansas

Viacheslav Adamchuk
Department of Bioresource Engineering
McGill University
Quebec, Canada

David Bullock
Department of Agricultural and Consumer
Economics
University of Illinois, Urbana-Champaign

Qianqian Du
Department of Agricultural and Consumer
Economics
University of Illinois, Urbana-Champaign

Ivan Grijalva
Department of Entomology
Louisiana State University

Saurav Upadhyaya
Department of Entomology
Louisiana State University

Brian McCornack
Department of Entomology
Kansas State University

Brennan Lewis
Department of Agronomy
Horticulture and Plant Science
South Dakota State University

Chapter 32

Brennan Lewis
Department of Agronomy
Horticulture, and Plant Science
South Dakota State University

Dwarika Bhattarai
Department of Agronomy
Horticulture, and Plant Science
South Dakota State University

Skye Brugler
Department of Agronomy
Horticulture, and Plant Science
South Dakota State University

Shailesh Pandit
Department of Agronomy
Horticulture, and Plant Science
South Dakota State University

Nowrin Mostafa Tuly
Department of Agronomy
Horticulture, and Plant Science
South Dakota State University

Janet Moriles-Miller
Department of Agronomy
Horticulture, and Plant Science
South Dakota State University

Sravanthi Bachina,
Precision Agriculture Technology
Fort Hays Tech North Central, Kansas

Norah Ochonye
Department of Agronomy
Horticulture, and Plant Science
South Dakota State University

Varshith Kommineni
Department of Agronomy
Horticulture, and Plant Science
South Dakota State University

Rojina Kafle
Department of Agronomy
Horticulture, and Plant Science
South Dakota State University

Graig Reicks
Department of Agronomy
Horticulture, and Plant Science
South Dakota State University

Deepak Joshi
Department of Agronomy
Kansas State University

Student Reviewers and Mentors

Fifty-four undergraduate and graduate students from universities across the US and globally reviewed this textbook. These students were supported in their review by experienced editors from the American Society of Agronomy (ASA), Crop Science Society of America (CSSA), and Soil Science Society of America (SSSA). We are grateful to the students and mentors for helping to ensure this book is both educational and engaging.

Student Reviewers

Ahmed Abdalla, Akanksha Singh, Amrinder Jakhar, Ana Gabriela Morales Ona, Ananya Gairola, Beata Matuszek, Benjamin Amoah, Bhavna Joshi, Bhawana Acharya, Binita Ghimire, Blake Toennies, Bruno Paulus Scheffer, Caitlin Terry, Chloe Yi-Luo Cho, Claire Spickermann, Egondu Umeobi, Ejezie Chisom, Eman Eldeeb, Emily Marsh, Emmanuel G. Cooper, Eric Youngerman, Hasan Mirzakhaninafchi, Isaac A. Oluwatimilehin, Israel Akinlabi, Jean Fritz Saint Preux, Jeferson Prass Pimentel, Jitender Rathore, Karem Meza, Karishma Kumari, Khuma Bhusal, Lora Mohanty, Maria Carolina de Carvalho Rocha Souza, Nazir Adam Sharifi, Nishadini Widanagamage, Nitish Joshi, Amarachi Norah Ochonye, Nowrin Mostafa Tuly, Ogbonnaya Bright, Riya Jain, Sagar Regmi, Sandesh Poudel, Saurav Ranabhat, Shailesh Pandit, Sheetal Kumari, Sourajit Dey, Sravanthi Bachina, Srinadh Kodali, Udit Debangshi, Ujjwal Sigdel, Usman Mohammed, Varshith Kommineni, Victor Ugwuegbu, Victoria Kwamboka, and Vipin Kumar.

About the Companion Website

This book is accompanied by a companion website containing videos, student questions and other instructional materials.

https://www.wiley.com/go/precisionagriculture2e

To access the website materials, click on the link under Related Resources on the book's page on wiley.com. You can also find this by searching www.wiley.com for the ISBN 9780891184089.

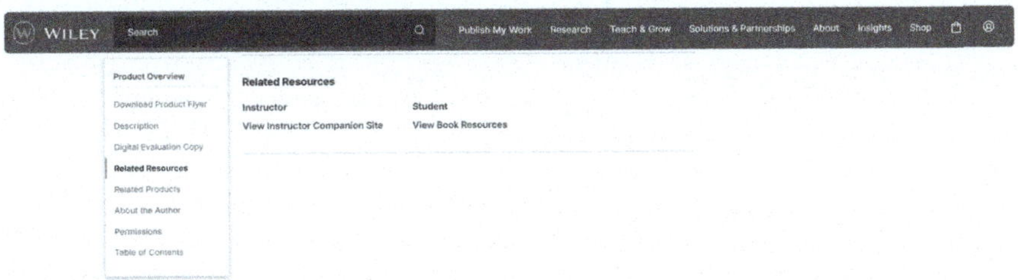

An Introduction to *Precision Agriculture Basics* 1

David E. Clay, Skye Brugler, and Katsutoshi Mizuta

Chapter Purpose

This book is an update of the first edition of *Precision Agriculture Basics* (Shannon et al., 2018). It is also based, in part on *Practical Mathematics for Precision Farming* (Clay et al., 2017a).

This chapter provides an overview of precision farming and how new scientific discoveries in molecular biology, machine learning, and sensing technologies can change our understanding of the practices used on managed lands. This book contains 32 written chapters and numerous videos featuring farmers and leading precision agriculture experts. The book is designed for self-learning, integration into the classroom, and extension workshops. Each chapter provides questions and answers that can be used for student learning.

What Is Precision Farming?

Pierre Robert, often regarded as the father of precision farming, actively promoted the idea of "soil specific crop management" during the early 1990s (Mulla & Khosla, 2016). Since then, precision agriculture has been updated. According to the International Society of Precision Agriculture, it is currently defined as a management strategy that gathers, processes and analyzes temporal, spatial and individual plant and animal data and combines it with other information to support management decisions according to estimated variability for improved resource use efficiency, productivity, quality, profitability and sustainability of agricultural production (January, 2024, International Society of Precision Agriculture; https://www.ispag.org/).

Precision agriculture can be defined in many ways, but at its core, it focuses on managing the variability found within a field. Since the development of agriculture, farmers and land managers have recognized the land's natural variability and have used the best available knowledge to create systems that meet the food, economic, and environmental security needs of individual communities. For example, in North America, Native Americans intercropped corn, bean, and squash to increase food production, whereas in Europe, many farmers reduced the need for fallow by using the Norfolk rotation. This rotation provided nitrogen (N) through N fixation, as well as land for grazing and food production. These systems improved food security by increasing both the quantity and quality of food produced.

Today, we are seeing similar yield increases. For example, in South Dakota (USA), corn and soybean yields increased annually by 2.76 bu/ac (155 kg/ha^2) and 0.65 bu/ac (43 kg/ha), respectively, from 2000 to 2020 (Joshi et al., 2025). Similar to the impact of new technologies on food security during the seventeenth century in Europe, we believe that future yields can also be improved by using precision farming tools such as global positioning systems (GPS), molecular biology, geographic information systems (GIS), yield monitors, variable rate mapping, robotics, machine learning, and remote and proximal sensing.

These technologies are increasingly used across disciplines such as agronomy, horticulture, entomology, plant pathology, animal science, range science, soil science, fisheries and marine sciences, biology, sociology, economics, and environmental science. We believe that the full integration of applied and basic tools into precision farming will further enhance food, energy, and economic and environmental security and that ongoing research in these areas will continue to increase precision farming opportunities. With the expanding understanding and use of technologies, we are beginning to learn how to measure, interpret, and manage, plants' phenotypic and molecular responses to abiotic and biotic stress (Brugler, 2026; Clay et al., 2009; Hansen et al., 2013; Morales et al., 2012).

As agriculture continues to embrace digitization and incorporate advanced technology, new professional opportunities will be created, and these professionals will need new skills and abilities to accurately fill their roles (Clay et al., 2024; Erickson et al., 2018). This book is designed to help farmers, land managers, and industry leaders learn and develop new skills and abilities to take advantage of the upcoming opportunities in agriculture.

The first edition of *Precision Agriculture Basics*, published in 2018 by the American Society of Agronomy, the Crop Science Society of America, and the Soil Science Society of America, equipped readers with some of these skills. Now, years later, the field of precision agriculture has evolved rapidly, and we have learned that agricultural systems are much more complicated than anyone envisioned 100 years ago. For example, advancements in proximal and remote sensing technologies have improved our ability to understand soil variability and create site-specific prescription fertilizer maps. Additionally, physiological research into plant stress responses has improved our understanding of the multiple factors that can influence yield (Clay et al., 2009; Hansen et al., 2013). This new knowledge can be integrated into novel algorithms using artificial intelligence (AI) and machine learning; topics that were not covered in the original publication.

Therefore, this second edition is not a replacement for the first edition but rather an expansion. This edition builds on the original by updating all 15 chapters and adding 17 new chapters that explore emerging advancements in precision agriculture. These include innovations in remote and proximal sensing technologies, applications of machine learning and AI, and dedicated chapters on soil health, precision conservation, and life cycle analysis or assessment (LCA). This edition also integrates insights from nearly 100 agriculture professionals from universities and industry in both the United States and around the world. Additionally, more than 50 students from these universities volunteered as reviewers, helping to ensure that the chapters are clear, comprehensible, and valuable both for experienced professionals in precision agriculture and for those just beginning their journey.

Characterizing and Managing System Variability (Chapters 4–6)

Two variables driving precision agriculture are spatial variability and temporal variability. Spatial variability refers to how a measured parameter varies from one location to another, as well as by depth, whereas temporal variability refers to how the measured parameter, such as a plant response, changes from one point in time to another. For example, soil water content is often higher in footslope soils than in summit soils (spatial variability), and some soil test values are often lower in the fall than in the spring (temporal variability; Clay et al., 2017b).

Chapter 4 (*Advances in Variable Rate Field Equipment Technologies*), Chapter 5 (*Identifying Understanding and Managing Variability*), and

Chapter 6 *Understanding Soil Variability and Sampling Strategies for Site-specific Nutrient Management*.

(*Soil Variability and Fertility Management*) examine the factors that contribute to variability in crop production systems and the role precision agriculture plays in managing productivity, profitability, and sustainability through data-driven decision-making techniques. Some of these techniques include the use of soil sampling designs, sensor technologies, variable rate technologies, and the Web Soil Survey (Chapter 10) as tools to characterize variability. Additional discussions include the influence of biotic and abiotic factors—including weather, soil properties, and management practices—on plant growth and yield, as well as the challenges of collecting representative soil samples to develop optimal site-specific nutrient management plans.

Locating and Processing Spatial Information (Chapters 7–9)

In precision farming, the ability to locate and process spatial information is essential. This information can be collected by many different approaches, including unmanned aerial vehicles (UAVs) and satellites, which collect remote sensing data, as well as proximal sensors that measure soil characteristics. The ability to collect and process this information has created an opportunity for farmers and land managers to better target treatments to specific problems. Chapter 7 (*Global Navigation Satellite System and Other Global Navigation Satellite Systems for Precision Agriculture*).

Satellite-based Positioning Systems for Precision Agriculture provides an overview of how to locate sampling points using a Global Navigation Satellite System (GNSS).

The use of GIS is discussed in Chapters 8 and 9. Chapter 8 (*Fundamentals of Geographic Information Systems*) provides an introduction to GIS, whereas Chapter 9 (*GIS Applications in Precision Agriculture*) covers the use of GIS in precision agriculture. This second edition, highlights (a) the latest applications of GNSS error corrections and GIS workflows for acquiring, storing, and visualizing geospatial data; (b) how to create field boundaries; and (c) the use of programming for advanced data manipulation and map creation to support better decision-making.

Using Soil Maps as a Starting Point for Understanding Variability (Chapter 10)

The Web Soil Survey, operated by the USDA, is a platform of maps and information that can be integrated into a GIS database containing soil and plant information. The platform has many different data layers, including soil cation exchange capacity, bulk density, drainage, and land suitability for agriculture, planning, and conservation. This information can be used for a variety of purposes, such as management zone delineation, site-specific fertilizer applications, soil sampling maps, and estimating soil organic carbon (Nyaupane et al., 2024).

Remote and Proximal Sensing (Chapters 11–13, 15, and 16)

This section discusses topics ranging from an introduction to proximal and remote sensing to the use of soil and plant sensing. The first chapter in this section is a discussion of the general principles of proximal sensing (Chapter 11). This chapter is followed by discussions on the use of proximal soil sensing for agriculture (Chapter 12), proximal plant sensing (Chapter 13), remote sensing for site-specific crop management (Chapter 14), remote sensing for water stress management (Chapter 15), and precision irrigation technology and water management (Chapter 16).

Measuring Insects, Weeds, and Diseases (Chapters 17–19)

Farmers in the United States and worldwide use a variety of strategies to reduce yield losses from pests, weeds, and diseases. For example, insect management often involves integrated pest management techniques designed to target insect-related problems site-specifically, using comprehensive information about insect life cycles and their interactions with the environment (Chapter 17). Integrated pest management strategies discussed include selecting crop varieties, carefully planning planting dates, practices such as crop rotation and tillage, and strategically using chemical control materials.

Chapter 18 (*Targeted Weed Management Using Precision Agriculture*) discusses specific problems associated with precision weed management and provides a conceptual model for distinguishing weeds from crop plants. Chapter 18 discusses the tools and techniques required for accurate identification, sampling, and data analysis. For example, the chapter proposes an approach that uses spectral differences to separate weeds from soil (green–brown) and weeds from crop plants (green–green). Chapter 19 (Plant Disease Identification and Management Using Remote Sensing) discusses how technologies such as UAVs, satellite imagery, and machine learning can enhance disease diagnosis and yield loss estimation.

Using Yield Monitor Data in On-farm Trials and Economic Analysis (Chapters 20–23)

This section begins with a discussion of yield monitors and how to collect accurate information (Chapter 20). The discussion of yield monitors is followed by the use of yield monitors in on-farm trials (Chapter 21), the importance of archiving the collected information (Chapter 22), and the evaluation of the economic returns of precision agriculture practices using tools such as partial budgeting, cost–benefit analysis, net present value, internal rate of return, and the payback period (Chapter 23).

Soil Health and Precision Conservation (Chapters 24 and 25)

A major strength of precision farming is its ability to identify areas where soil health practices can improve agricultural efficiency (Chapter 24) and to assess whether precision conservation strategies can enhance profitability and environmental sustainability (Chapter 25). Chapter 24 (*Soil Health in Precision Agriculture*) provides an overview of (a) soil health and its importance; (b) the physical, chemical, and biological properties of soil associated with soil health; (c) the spatial variability of soil health indicators; and (d) the link between soil health and precision farming.

Chapter 25 (*Precision Conservation: A Pathway to Improved Sustainability*) discusses the various components of precision conservation, including a framework for building a precision conservation plan, along with four case studies that demonstrate the use of the framework at different scales and explore the intersection of precision agriculture and environmental quality protection.

Using Advancements in Control Systems and Computing to Improve Precision Farming (Chapters 26 and 27)

Chapter 26 (*From Control Systems to Variable Rate Seeding, Pesticide, Chemical Applications and Irrigation*) introduces students to the basics of control systems, electronics, and robotics and discusses how these technologies can be used in site-specific precision nutrient and pest management strategies.

Chapter 27 (*Machine Learning and Artificial Intelligence-driven Solutions in Precision Agriculture*) provides (a) an introduction to machine learning, which has been transforming precision agriculture; and (b) an overview of the strengths and weaknesses of different machine learning algorithms. We believe that a basic understanding of machine learning is necessary to prepare for the future of agriculture.

Understanding and Making Decisions About the Whole System

Chapter 28 (*An Introduction to Using Life Cycle Analysis (LCA) as an Agricultural Sustainability Assessment Tool*). introduces LCA, a method used to calculate the impact of a product or process. The terms *analysis* and *assessment* are used interchangeably within LCAs, which are typically used to give consumers a better understanding of the long-term benefits and damages to nature and people as a result of making and using products. LCAs are used globally by a wide variety of businesses when making decisions about which product(s) to purchase. In addition, LCA results can impact national and international trade, including manufacturing firms (e.g., automotive, electronics, chemicals), retail companies, fuel and energy industries, and agriculture.

Chapter 29 (*Precision Agriculture Basics for Major Agronomic Crops Worldwide*) s) discusses how knowledge of an agricultural system, such as soil variability and environmental conditions in a region, can be used to make the most efficient crop choices and increase food security and profit for the farmer. This chapter discusses precision farming opportunities for many crops, including corn, rice, wheat, oat, barley, millet, sorghum, teff, soybean, cotton, sunflower, rapeseed/canola, potato, chickpea, lentil, field pea, common bean, sugarcane, banana, and grape.

Instructor's Guide and Training Materials (Chapters 30–32 and glossary)

Chapter 30 (*Instructor's Guide to Teaching with* Precision Agriculture Basics) is intended to assist instructors in effectively incorporating the textbook into their curriculum. It also provides an overview of course planning and instructional strategies to enhance student engagement, adapt content to various academic levels, and improve learning outcomes through the use of supplementary teaching tools.

Chapter 31 (*Precision Agriculture Datasets*) provides soil and plant datasets that can be used in workshops and classroom exercises, whereas Chapter 32 (*Soil and Plant Calculations*) offers examples of processing soil tests, along with plant and soil variability information that can be used in decision-making.

As supplementary materials to the second edition, more than 100 newly developed educational videos featuring over 20 contributors from around the world were included to support and enhance learning in precision agriculture.

Summary

Today, farmers can monitor their fields in real time, enabling faster and more informed responses to emerging issues. In parallel, trends in crop production are shifting as farmers gain the ability to make more informed decisions about the crops they grow. As a result of these advancements, the agricultural industry will continue to invest in precision agriculture. However, employers have identified gaps in the knowledge, skills, and abilities of individuals seeking an agriculture-focused career or those interested in integrating precision agriculture into a small business (Erickson et al., 2018). Therefore, the goal of this updated textbook is to address and fill these gaps across various agriculture-focused careers, helping to continue the advancement of precision agriculture into the future.

ACKNOWLEDGMENTS

This document was funded in part by the USDA National Institute of Food and Agriculture under the Multi-State Hatch project (accession number 7008110), Kentucky Corn Growers Association, Kentucky Small Grain Grower Association, University of Kentucky, National Science Foundation (grant numbers 2202706 and 2026431), North Dakota State University, USDA-ARS under agreement number 58-6064-3-011 (FAR0037049), South Dakota State University, USDA Natural Resources Conservation Service Conservation Innovation Grants (NR233A750011G019), USDA Agriculture and Food Research Initiative, American Coalition for Ethanol, GEVO, South Dakota Corn Utilization Council, and the South Dakota Soybean Research and Promotion Council. The use of trade names is for the convenience of the reader and does not imply endorsement by the authors.

REFERENCES

Brugler, S. (2026). *Building a rapid approach to nitrogen stress: integrating molecular biology into management*. [Doctoral dissertation, South Dakota State University (in progress)].

Clay, D. E., Brugler, S., & Joshi, B. (2024). Will artificial intelligence and machine learning change agriculture: A special issue. *Agronomy Journal*, *116*(3), 791–794. https://doi.org/10.1002/agj2.21555.

Clay, D. E., Clay, S. A., & Bruggeman, S. A. (2017a). *Practical mathematics for precision farming*. ASA, CSSA, SSSA.

Clay, D. E., Robinson, C., & DeSutter, M. (2017b). Chapter 6: Soil sampling and understanding soil test results for precision farming. In D. E. Clay, S. A. Clay, & S. A. Bruggeman (Eds.), *Practical mathematics for precision farming* (pp. 105–121). ASA, CSSA, SSSA.

Clay, S. A., Clay, D. E., Horvath, D., Pullis, J., Carlson, C. G., Hansen, S., & Reicks, G. (2009).

Corn responses to competition: Growth alteration vs. yield limiting factors. *Agronomy Journal*, *101*(6), 1522–1529. https://doi.org/10.2134/agronj2008.0213x.

Erickson, B., Fausti, S., Clay, D., & Clay, S. (2018). Knowledge, skills, and abilities in the precision agriculture workforce: An industry survey. *Natural Sciences Education*, *47*(1), 1–11. https://doi.org/10.4195/nse2018.04.0010.

Joshi, D. R., Clay, D. E., Alverson, R., Clay, S. A., Westhoff, S., Johnson, J. M. F., Wang, T., & Sieverding, H. (2025). Tillage intensity reductions when combined with yield increases may slow carbon saturation in the central United States. *Scientific Reports*, *15*, 10697. https://doi.org/10.1038/s41598-025-95388-x.

Hansen, S., Clay, S. A., Clay, D. E., Carlson, C. G., Reicks, G., Jarachi, Y., & Horvath, D. (2013). Landscape features impacts on soil available water, corn biomass, and gene expression during the late vegetative growth stage. *The Plant Genome*, *6*(2), 1–9. https://doi.org/10.3835/plantgenome2012.11.0029.

Morales, J., Hansen, S., Horvath, D. P., Reicks, G., Clay, D. E., & Clay, S. A. (2012). Microarray and growth analyses identify differences and similarities of early corn response to weeds, shade, and nitrogen stress. *Weed Science*, *60*(2), 158–166. https://doi.org/10.1614/WS-D-11-00090.1.

Mulla, D., & Khosla, R. (2016). Historical evolution and recent advances in precision farming. In R. Lal & B. A. Stewart (Eds.), *Soil-specific farming: Precision agriculture* (pp. 1–36). CRC Press.

Nyaupane, K., Mishra, U., Tao, F., Yeo, K., Riley, W. J., Hoffman, F. M., & Gautam, S. (2024). Observational benchmarks inform representation of soil organic carbon dynamics in land surface models. *Biogeosciences*, *21*(22), 5173–5183. https://doi.org/10.5194/bg-21-5173-2024.

Shannon, D. K., Clay, D. E., & Kitchen, N. R. (2018). *Precision agriculture basics*. ASA, CSSA, SSSA.

Adoption, Barriers and Pathways Forward in Precision Agriculture Technologies

2

Bhavna Joshi and Maaz Gardezi

Chapter Purpose

The adoption of precision agriculture technologies involves the use of innovative tools by farmers who are willing to experiment with emerging, and sometimes untested, methods. One approach to reducing adoption barriers is to test the technology prior to making large financial investments (an approach discussed in Chapter 23). This chapter explores the evolution and adoption of precision agriculture technologies, highlighting key milestones and recent trends, common barriers to adoption, strategies for smallholder integration, and future directions. The chapter also emphasizes the importance of trust, co-design, and interdisciplinary collaboration in shaping the future of precision agriculture.

Introduction

Agricultural innovations have transformed the way humans cultivate and manage land. For example, around 12,000 years ago, humans switched from living as hunter–gatherers to building small communities and growing crops. The early farmers, either by rigorous testing or random change, slowly modified their production system. One change that occurred in the sixteenth century was the discovery and adoption of the Norfolk four-course rotation system, which involved rotating four different crops—wheat, turnip, barley, and clover—in succession over four years to maintain soil fertility. The change reduced the need to fallow land, which means leaving the soil to rest without planting any crops. This new method increased yields and improved food security, highlighting that change is a critical component of agriculture. Today, farmers are resolving how to integrate new data-processing techniques into their farm's operation through precision agriculture. This change is not easy because it requires learning new skills and adopting new technologies.

Emerging trends in digitalization in agriculture are driving technological modernization in food and agriculture systems. Precision agriculture, which includes data-driven agricultural technologies with artificial intelligence (AI), machine learning (ML), and real-time environmental sensing, has created unprecedented opportunities to produce food without adversely affecting the environment and climate. Since the turn of the century, precision agriculture has shifted farming away from more "hands-on," manual approaches toward more interdependent human–technology partnerships. By employing data-based agricultural technologies, precision agriculture allows farm managers to generate site-specific farm recommendations that have the potential to improve farm profitability while reducing the environmental impact of agriculture (Cisternas et al., 2020; Lowenberg-DeBoer & Erickson, 2019).

Precision agriculture aims to harness the capacity of big data, in situ sensors, and data analytics to provide farmers with recommendations about where and when to plant, seed, spray, graze, and harvest. The basic premise of precision agriculture is to allow farmers to produce more crops and livestock on less land, with fewer inputs and a smaller environmental footprint. Precision agriculture is largely made possible through the processing of big data or high-volume, high-velocity, multisource agriculture data, such as crop yield data, digital soil information, reflectance information, and elevation data. With help from satellites, drones, sensors, and ML algorithms, precision agriculture technologies enable farmers to integrate information technology into their farm machinery and management. For example, Taylor (2023) suggested that the integration of data-based technologies with farmer knowledge can streamline farm management, reduce economic risks, and improve food security.

At the same time, however, precision agriculture has the potential to reshape farm work, with less hands-on management and more data-driven analysis. New skills are required to create, interpret, and establish recommendations. Preparing this workforce presents associated challenges, including high training and equipment costs, poor broadband internet availability, data ownership, and trust issues with new technologies. Precision agriculture can influence people's expectations of farming. In the past, farmers manually scouted their fields for pests and diseases. But today, AI-driven algorithms are being used to process remote sensing images into field prescriptions (Clay et al., 2024; Joshi et al., 2023). In the future, farmers might only need to know how to convert remote sensing data into maps that show areas where the crops face problems, such as abiotic stressor maps for water scarcity and biotic stressor maps for the presence of pests and diseases. This can help save time and make farming more efficient.

To use data-intensive tools and technologies involved in precision agriculture, farmers may need additional skills and support as they navigate the adoption process (Brugler et al., 2024; Erickson et al., 2018). Two of the biggest challenges to the adoption of precision agriculture technology are (a) the high cost of training and purchasing equipment and (b) the lack of knowledge and technical skills to use this technology. Higher costs can disproportionately affect smaller operations that are already vulnerable to financial risk. Currently, many precision tools, such as variable rate fertilizer applicators, are more affordable and practical for large-scale, conventional farms that grow commodity crops and have the capital to invest in advanced machinery. This creates a risk that precision agriculture could widen existing social disparities within rural communities. Differences in income, age, or digital knowledge and skills may also result in unequal distribution of benefits associated with the adoption of precision agriculture technology among farmers. As a result, smaller farms or farmers with less technical training may be left behind.

This chapter provides a brief overview of the factors that contribute to farmers' adoption of precision agriculture in the United States and elsewhere. This overview is not intended to be exhaustive, but to provide a starting point for discussions. By identifying the barriers to adopting these technologies and how they can be overcome, we aim to propose some recommendations that may be important for the future.

Development of Precision Agriculture Globally

These subsections will discuss the development and growth of precision agriculture globally. Different parts of the world use and think of precision agriculture differently. This is important to consider when learning about precision agriculture because it is much more than just an advanced set of tools and technologies. Precision agriculture not only helps farmers make site-specific recommendations, but it also plays a key role in addressing global food security and reducing environmental harms. This is why it is essential to understand how precision agriculture has developed over time and how it is interpreted and applied in different parts of the world under diverse conditions and challenges.

The Growth of Precision Agriculture Technologies and Techniques

The development of precision farming first took shape in the Western world, particularly in North America and Europe, during the 1980s and '90s. This was the first time global positioning systems (GPS; Chapter 7) and yield monitors were introduced to farmers (Fountas et al., 2024). These technologies were used to produce yield maps and images that started to help farmers monitor and manage their farm (McFadden et al., 2023). Soon thereafter, precision soil sampling techniques, such as grid sampling (Chapter 6), were combined with yield maps (Chapter 20) to determine the physical and chemical properties of the soil and its effectiveness for enhancing crop yield. Combining yield and soil nutrient maps enabled the customization of fertilizer and pesticide applications based on the specific needs of the soil and crop growth. The precision agriculture infrastructure was further improved by the proliferation of data-driven guidance systems for tractors, along with aerial and space-based platforms equipped with sensors that measured electromagnetic radiation.

In more recent years, the fourth agricultural revolution, or "Agriculture 4.0," has become more prominent in farming. This revolution in agriculture is driven by other digital technologies, such as AI, big data, virtual reality, Internet of Things (IoT), quantum computing, blockchain, and robotics. These technologies, when combined, have the potential to increase crop productivity and reduce the impact of agriculture on the environment (Fountas et al., 2024).

Outside of North America, Europe, and Australia in countries of the Global South, also called the "Majority World," precision agriculture solutions followed a different trajectory. In these countries, the development of precision agriculture has been shaped by resource constraints, smaller farm sizes, and different social, economic, and environmental conditions. Ownership of mobile devices in these regions is lower than the global average, and internet access is even scarcer, creating a major barrier for the spread of digital farming technologies (Mehrabi et al., 2020). For example, the share of farming households with internet access is only about 31% in India, 21% in Pakistan, 14% in Bolivia, and as low as 5% in Haiti. Similarly, many Sub-Saharan African countries still lack access to mobile connectivity and electricity (GSMA, 2025; Ritchie et al., 2019), which makes adopting advanced precision agriculture technologies difficult. In such situations, farmers often rely on more accessible solutions, such as low-cost sensors and mobile applications that can work offline or with limited connectivity, to support their farming activities.

Despite the above challenges, precision agriculture has slowly managed to influence farming practices in the Majority World. Its adoption is majorly supported by government initiatives, nongovernmental organizations (NGOs), and international development agencies. These groups play an important role in providing resources, training, and tools to farmers to encourage them to improve their crop productivity more sustainably. As of 2025, major financial support for precision agriculture development in the Global South comes from investments by private venture capital and public–private partnerships initiated by international donors, such as the World Bank, the United States Agency for International Development, the African Development Bank, and the Gates Foundation. These investments often encourage the development of precision agriculture tools in partnership with local farmers, using their knowledge of farming practices, potential adoption barriers, and resource availability. This helps improve acceptance of the technology in the region. For instance, in East Africa, a mobile application called *AfriScout* is widely used by pastoralists to locate fresh pasture and water for their livestock within the surrounding region (Bessenecker, 2025). An NGO called Project Concern International locally developed this application by integrating the local community's knowledge of grazing areas with satellite imagery. The application can operate over 3G networks and allows easy, affordable access to maps showing crop health and water availability in real time. Other mobile applications that are known to provide support to the farmers are *WeFarm*, in East Africa; *Ama Krush*, in India; and *mFarmer*, in Singapore.

In addition to the organizations mentioned above, major corporate players, such as Yara International, IBM, and Microsoft, are developing innovative solutions to enhance soil and crop diagnostics through these applications and field-sensor-based tools. These applications can combine in-depth farmer profiles with transaction, weather, satellite, unmanned aerial vehicles (UAVs), and field/machinery sensor data, and some can also use AI to generate highly tailored, dynamic agronomic and financial advice regarding irrigation, fertilization, pest control, and disease detection. Yara's application, *Atfarm*, was designed in 2018 to help farmers optimize nutrient use and improve their yield. This tool uses satellite imagery to monitor crop growth and generate nitrogen application maps to assist farmers in decision-making. So far, this application has been successful in improving wheat yields in Brazil and corn productivity in Mexico and increasing farm profit in Ireland. Similarly, IBM Research (Brazil) is developing a paper device called *AgroPad,* which uses AI technology to check soil and water health in real time (FAO, 2018). When a drop of water or soil is placed on the paper, the microchip inside it performs a chemical analysis for pH, nitrogen dioxide, magnesium, and chlorine and shows the results to the farmer on their smartphone. In this way, this innovation not only saves time by eliminating the need to send soil samples to a lab but also gives smallholder farmers an affordable way to perform instant nutrient analysis. Other private enterprises, such as Upande Limited, Twiga, Ujuzi Kilimo, Tulaa, AGIN, and Apollo Agriculture, are focusing on tailoring their technologies to smaller plots. These tools are often affordable, easy to set up, and simple to monitor. For example, Ujuzi Kilimo, a Kenyan agricultural technology startup, has developed a handheld IoT-based sensor that rapidly and accurately tests soil health. This sensor is water-resistant and operates on batteries. It is particularly useful for Kenyan smallholder farmers, who make up about 70% of the Kenyan rural population, as it provides them with affordable, personalized recommendations to improve yields (Colquhoun, 2024).

Global Adoption Trends

The adoption of precision agriculture technologies varies widely across the globe, with some innovations being embraced more rapidly than others (Franzen & Mulla, 2015). In regions such as North America and parts of Europe, their uptake is relatively high, especially for machinery-based tools like GPS-guided tractors. In the United States, for instance, GPS-enabled sprayer booms saw an increase in adoption from 39% to 53% between 2011 and 2013, while the yield monitors on combine harvesters had adoption rates ranging from 60% to 80% by 2016 (Nowak, 2021). By contrast, in much of Africa, South Asia, and Latin America, adoption has been slower and much more selective (Lowenberg-DeBoer & Erickson, 2019). Since most farms in these regions are small and medium scale, the use of motorized mechanization is limited. As the size of the farm increases, the possibility of using more advanced technology increases. Sasao and Shibusawa (2000) showed that farms <1 ha (<3 ac) usually rely on manual work, farms between 1 and 10 ha might use some automated machinery and sensors, and farms between 10 and 50 ha are better suited for GPS-based technologies. Overall, due to a paucity of data from the Majority World countries, it is difficult to statistically determine the adoption rates of different precision agriculture technologies within these regions (Lowenberg-DeBoer & Erickson, 2019).

Another factor that contributes to uneven adoption of precision agriculture technologies is the difference in local geography and environment. These differences shape the kind of tools and practices that are effective in optimizing crop growth (Barnes et al., 2019). For example, in the hilly areas of Nepal, most farms are less than half a hectare in size and have limited irrigation. Because of that, projects like Sustainable Intensification of Mixed Methods, which aim to increase agricultural productivity without harming the environment, are useful in helping farmers by introducing them to tools such as two-wheeler tractors and climate-smart irrigation that are appropriate to the local environment (Shrestha & Fry, 2022). In contrast, the relatively flat terrain in the eastern part of South Dakota allows for large swaths of land to be farmed at once. This makes technologies such as variable rate technology (VRT), auto-steer, and guidance systems in tractors more useful (Franzen & Mulla, 2015; Nowak, 2021). Similarly, semi-arid or arid regions often opt for irrigation-based farming and are therefore more likely to adopt precision irrigation technology than those growing crops in wetter regions (Castle et al., 2016). Recently, the use of imagery data has become popular with UAVs equipped with Real-Time Kinematic positioning and ML-based image recognition software showing a strong potential to improve resource allocation by enhancing pest detection (Ammann et al., 2022; Gabriel & Gandorfer, 2023). To provide digital agriculture access to smallholder farmers in the Sichuan Province of China, local cooperatives were established. These cooperatives make collective decisions and share the risks and benefits with each other. Similarly, farmers in North China overcame the farm size-based restrictions by using remote sensing and satellite data to create large-scale management zones (Cammarano et al., 2020).

Barriers to Adoption of Precision Agriculture Technologies

Despite many advancements in precision agriculture worldwide, the regions with less mechanized agricultural practices continue to lag behind high-income nations in adopting smart farming technologies (Lowenberg-DeBoer & Erickson, 2019; Pathak et al., 2019). This is a concern because agriculture is deeply connected across borders through global markets, supply chains, and shared environmental challenges. Uneven adoption of agricultural technologies can increase global production gaps, creating trade imbalances. It also limits the ability of farmers in the vulnerable regions to adapt to climate change, threatening global food security and stability.

Internet Accessibility

The United Nations has recognized that accessibility to precision agriculture technologies depends not only on their availability but also on users' ability to afford and effectively use them (UNDP, 2021). While these challenges are prevalent worldwide, they are exacerbated by the digital divide among nations, which limits some users' access to the internet, and in turn, their ability to benefit from these technologies (Protopop & Shanoyan, 2016). Globally, about 95% of the population without internet access lives in the low- and middle-income countries, such as those in Sub-Saharan Africa and South Asia (Shanahan & Bahia, 2024). Within these countries, rural residents are about 25% less likely than urban residents to use mobile internet devices, and their average download speeds remain below 20 Mbps, compared with nearly 100 Mbps in high-income countries. This results in lower adoption rates of precision agricultural technologies in these countries than in countries such as the United States, Germany, the United Kingdom, Australia, and Denmark (Barnes et al., 2019; Lee et al., 2021; Pathak et al., 2019).

Some of the ways farmers in the Majority World deal with internet issues include adopting low-cost sensors, use of cellphones, web-based platforms, and social media platforms that are accessible in a no-internet or low bandwidth internet environment (Choruma et al., 2024). In Tanzania, the introduction of cost-effective radio and SMS-based platforms improved farmers' knowledge of legume-based sustainable agriculture intensification, which in turn was reflected in better sustainable farming practices (Silvestri et al., 2021). Similarly, in Senegal, a group of farmers who wanted to shift from traditional millet and corn crops to higher-value onion cultivation faced resource constraints (Modi, 2022). To address this, they collectively financed a shared solar power system to operate irrigation motors, thereby avoiding the higher cost of battery-based alternatives. Expenses for electricity were distributed among the farmers according to their individual consumption.

This subsection highlighted the barriers that farmers from the Majority World face in adopting precision agriculture technologies. The next section explores the challenges faced by smallholder farmers and the strategies they use to overcome them.

Smallholder Farmers

Smallholder farmers face a range of difficulties that tend to boil down to less time and capital to train employees and learn new technologies. This is a significant barrier to adoption for many smallholding farmers globally. On a positive note, there are examples where smallholder farmers managed to integrate low-cost precision agricultural technologies into their farming practices based on their specific farm requirements. For example, in a Vietnamese case study, local paddy farmers shifted their traditional farming practices, which included flooding or manual monitoring of the field for irrigation needs, to a precision practice that alternates wetting and drying of the paddy field. IoT-based technology, combined with soil sensors and cloud computing, was used to improve irrigation management. The system could operate water pumps in the field, thereby reducing labor and time requirements. The impact of this system was a 13%–20% reduction in water consumption. This approach, using precision agriculture technology, reduced the expensive water consumption caused by paddy cultivation and also reduced resulting methane emissions (Choudhary & Fock, 2020). In another case study, this one in Chongzhou County in China, smallholder farmers digitized their farming practices by attaining a large farm scale by forming cooperatives (Xie et al., 2021). These cooperatives outsourced their agricultural production to the service providers who mechanized and digitized farming on a larger scale.

Overall, smallholding farmers often face disadvantages because many precision agriculture technologies are designed with large-scale farms in mind. However, targeted innovations and supportive strategies offer opportunities to reduce these barriers and enhance the adoption of precision agriculture.

Recommendations for Future Precision Agriculture Technology Development and Adoption

The adoption of precision agriculture is influenced by more than just the perception and decision-making abilities of the farmers. It is shaped by a broader farming ecosystem that includes factors such as farmers' sociodemographic characteristics (like age, education, and income), affordability and availability of technology, quality of farm resources, support from a farm advisor network, geospatial characteristics of land, market conditions, and national or regional policies that promote or restrict digital innovations (Straub, 2009). Similarly, the diffusion of precision agriculture technologies does not occur on its own and requires that researchers, government agencies, private corporations, and farming communities work together (Geels & Schot, 2007).

Users of new technologies exert substantial influence on the diffusion of innovations while simultaneously being influenced by the dynamic nature of these innovations. As scientific advancements and technological progress unfold, they catalyze transformative changes in the existing applications and demands of users of these technologies (Franzen & Mulla, 2015). These advancements have shaped the nature and function of a variety of precision agriculture tools, such as soil sensors, GPS-based farm equipment, and yield monitors. This ongoing interaction shapes the ecosystem of technological evolution, encompassing social practices and beliefs, the nature of jobs, expectations from innovations and experts, environmental impact, and government policies. Consequently, to comprehensively understand the adoption of precision agriculture technologies, it is crucial to understand the surrounding social, economic, technological, and organizational systems that surround the developments (Robinson et al., 2025). In this section, we propose several recommendations for technology developers and policymakers to improve precision agriculture adoption.

Looking Beyond the Adoption of Precision Agriculture Technologies

To address barriers to adoption, it is crucial to understand how farmers interact with precision agriculture technologies. Adoption is often conceptualized in dichotomous terms—adoption versus non-adoption—yet in practice it is a more complex process (Feder et al., 1985). For example, a farmer with no prior experience may rely on the experience of peers and advisors when deciding whether to purchase a technology, a process that can lead to either adoption or rejection. The choice of technology depends on the farm-specific needs, budget constraints, technical skills, availability, and a farmer's long- and short-term goals. Even after adoption, farmers may withdraw or abandon a technology if its costs outweigh its benefits. In other cases, farmers also adapt or modify technologies to better align with their conditions and practices (Higgins et al., 2023). Therefore, there are ample ways in which farmers engage with technologies. Future research should move beyond adoption metrics to examine the full life cycle of precision agriculture technologies—including withdrawal, rejection, modification, and phase out—and explore how farmers adapt and repurpose these technologies, not only as adopters but as active co-innovators. Some of the key dimensions that advance the discussion of looking beyond adoption are presented in the next subsections.

Building Trust

The adoption of precision agriculture technologies requires trust between farmers and the technology, especially if they have not used it or witnessed its use. Increasingly, these new technologies—such as predictive models, blockchain, autonomous farming vehicles, targeted spray systems, drones, and decision support systems—are becoming data intensive and often rely on AI and ML to create algorithms that create recommendations. However, many farmers do not trust these algorithms, as they do not understand how they work. Moreover, the use of such technologies may require sharing farm-level data, which some farmers view as sensitive information. Therefore, farmers' trust in precision agriculture technologies is contingent on many factors, including the return on investment and the privacy and ownership of the collected data.

With increasing digitalization of farming, concerns about data collection, ownership, privacy, and regulation are growing (Brugler et al., 2024). Because of these concerns, along with the lack of clarity and transparency about the usage of their farm data, farmers often lose trust in data-driven technologies (Wiseman et al., 2019). To answer questions about data-driven precision agriculture technology, farmers seek information from crop advisors, nutritionists, agronomists, and other farmers (McBride & Daberkow, 2003). However, these advisory groups are often ill-equipped to handle these queries (Erickson et al., 2018). This diminishes trust in the available workforce for agronomic recommendations and increases dependence on the private sector to disseminate knowledge pertaining to precision farming. Therefore, for building trust in precision agriculture technologies, the training of the workforce in the technical and social aspects of these technologies is essential (Brugler et al., 2024). This approach will enhance communication and facilitate strategic investment in precision agriculture technologies that align with the user's requirements rather than being influenced and imposed by the market.

Co-designing Precision Agriculture Technologies with Users (Farmers)

Co-designing precision agriculture technologies with potential users is one approach to reducing adoption barriers. It is a method for innovating "responsibly" by connecting "the practice of research and innovation in the present to the future that it promises and helps bring about" (Stilgoe, 2013, pg. xii). It uses a participatory approach to imagine, identify, and solve a problem by including the ideas, interests, and perspectives of multiple stakeholders relevant to the problem (Tzachor et al., 2022). It focuses on the socioeconomic context of the end-users, where they integrate their experiences, knowledge, practices, and future farming goals in product development (Robinson et al., 2025). Therefore, co-designing of agricultural technology (Ag-Tech) is a farmer-centric problem-solving approach (Burch & Legun, 2021; Meynard et al., 2012).

Currently, precision agriculture technologies are employed on 27% of farms across the United States, with over 50% adoption in the leading grain-producing states (Abbott, 2024). As the farm size increases, the adoption rate of these technologies, particularly auto-steering, yield monitors, yield maps, and soil maps, also increases (McFadden & Lim, 2024). To encourage adoption by small farmers, precision agriculture technologies should be made affordable and suitable to their farming needs (Santos Valle & Kienzle, 2020). Additionally, these technologies should be simple to use and should not require a major time commitment by farmers for data collection and analysis (Abbott, 2024). However, it is recognized that the perspectives of innovation designers and scientists in North America about "good data" or "good technology," or about being a "good farmer" in general, favor commodity crops and large-scale farms (Bronson, 2019). For example, a dominant view among the designers is that a "good farmer" is someone whose major goal is to maximize profits to acquire more land and machinery to dominate nature (Foureaux & Daum, 2025). However, this idea overlooks the identity of a farmer as a family person, a nature lover, or a community steward who is constrained by a budget and likes to manage the farm personally. Excluding smallholder farmers from the picture of the so-called "ideal farmer" is a risk to global food security, as nearly 90% of farmers worldwide operate on a small scale. Therefore, the new precision agriculture technologies should be designed with these farmers' needs and goals in mind, so that future precision agriculture technologies can reflect their realities and be more likely to be used. For instance, innovations should favor the crops that they grow rather than compelling them to transform their production system into a corporate-centered model. The innovations should thus be farmer- and environment-centric rather than vice versa. We believe that by integrating farmers into the design process, their role can be transformed from passive data collectors into verifiers of data-driven models.

Precision Agriculture Governance

In the context of precision agriculture technologies, the primary challenges to governance include inequitable access, inadequate data regulations, environmental concerns, and social implications for farmers and seasonal laborers. Although these technologies possess the potential to revolutionize farming practices, without inclusive governance frameworks, the benefits of precision agriculture technologies may be disproportionately accrued to large farms. The limited financial resources, technical expertise, and knowledge possessed by smallholder farmers, coupled with inadequate infrastructure, including internet accessibility, significantly contribute to the digital divide between smallholder and commercial farms (Barnes et al., 2019). Technologies like automated machinery, GPS-enabled tractors, and AI-based harvesters can change the skill requirements for farm labor (Rotz et al., 2019).

A major challenge for policymakers is to strike a balance between the goals of environmental sustainability, access and training required for using precision agriculture technologies, building suitable digital infrastructure, and fair and transparent data regulation. In addition, precision agriculture technologies should be rooted in transparency and accountability (Gardezi et al., 2023). This means that Ag-Tech companies should make clear and accessible information about how the technology works available to farmers. For example, they should explain how data from sensors or drones will be collected and used, who will own it, and what risks may be involved in the process. By ensuring inclusive adoption of precision agriculture, safeguarding farmers' privacy rights through data regulation, providing access to training and technologies, and fostering effective communication and awareness about the technology, trust in the adoption of precision agriculture technologies can be established (Wiseman et al., 2019).

The Need for Interdisciplinary Studies on Precision Agriculture Innovations and Adoption

The adoption of precision agriculture technologies involves a complex interplay between social, economic, technological, and environmental factors (Lemay & Boggs, 2024). An interdisciplinary research approach is required to comprehensively understand the interaction between these factors. This means that researchers from diverse disciplines, such as sociology, economics, geology, animal science, agronomy, genetics, environmental quality, computer science, engineering, statistics, and design, need to work together. The focus of such academic collaboration must be farmer-centric and environment-friendly, and it should generate evidence to guide and train the future workforce.

Future research on the adoption of precision agriculture technologies should broaden its scope beyond the predictability models relevant to the profits and investments of farmers (Straub, 2009) to include the following questions (Rose et al., 2023):

1. How do farmers engage with these technologies on a day-to-day basis?

2. How is farmer engagement influenced by technology upgrades and societal priorities?

3. How can precision agriculture accessibility and affordability be improved for the small farmer?

4. What roles can farm advisors, peers, and the government play in facilitating a smooth transition to the emerging precision agriculture innovations?

All these questions belong to different arenas of disciplinary expertise. Therefore, to get a holistic view of the factors affecting the adoption of precision agriculture technologies, an interdisciplinary approach to collaboration is essential (Ryan et al., 2023).

Conclusion

To conclude, this chapter discussed different aspects of precision agriculture adoption, starting from its early evolution to its spread across other countries, the barriers to its adoption, and the steps needed for improving the adoption of these technologies. These steps include building trust in technologies, engaging farmers as co-designers of farm innovations, and strengthening governance around precision agriculture and AI to ensure the fair use of data. The chapter also illustrated different ways in which the smallholder farmers are using precision agriculture technologies to boost crop production and manage environmental risks. Precision agriculture offers great promise because of the efficiency and accuracy it provides. However, more work is needed to make it affordable and accessible to farmers across different regions and countries. For global food security and climate resilience, future precision tools must reflect the needs of farmers from diverse geographies, contexts, and socioeconomic backgrounds. This inclusivity will support wider adoption and help scale Ag-Tech innovations.

Study Questions

1. How has agricultural innovation changed productivity?
2. What are the potential data sources in precision agriculture?
3. What are the barriers to precision agriculture adoption?
4. Why is privacy important to farmers?
5. Why are interdisciplinary studies important?
6. What is co-creation?

ACKNOWLEDGMENTS

This paper was partially funded by National Science Foundation grant number 2202706 and 2026431, South Dakota State University, the USDA National Institute of Food and Agriculture under Award No. 2023-67023-40216, the USDA Natural Resources Conservation Service Conservation Innovation Grants (NRCS-CIG; NR233A750011G019), the USDA National Institute of Food and Agriculture (NIFA; Multistate Project S1090) Program under (7008110), American Coalition for Ethanol, GEVO, the South Dakota Corn Utilization Council and the South Dakota Soybean Research and Promotion Council. The use of trade names is for the convenience of the reader and does not imply endorsement by the authors.

REFERENCES AND ADDITIONAL INFORMATION

Abbott, C. (2024, December 11). Two thirds of large crop farms use precision agriculture technology, says report. *Successful Farming*. https://www.agriculture.com/two-thirds-of-large-crop-farms-use-precision-agriculture-technology-says-report-8759483.

Ammann, J., Umstätter, C., & El Benni, N. (2022). The adoption of precision agriculture enabling technologies in Swiss outdoor vegetable production: A Delphi study. *Precision Agriculture*, *23*(4), 1354–1374. https://doi.org/10.1007/s11119-022-09889-0.

Barnes, A. P., Soto, I., Eory, V., Beck, B., Balafoutis, A., Sánchez, B., Vangeyte, J., Fountas, S., van der Wal, T., & Gómez-Barbero, M. (2019). Exploring the adoption of precision agricultural technologies: A cross regional study of EU farmers. *Land Use Policy*, *80*, 163–174. https://doi.org/10.1016/j.landusepol.2018.10.004.

Bessenecker, C. (2025). *Indigenous practices meet mobile technology* [UBS]. Sustainability and Social Impact. https://www.ubs.com/global/en/sustainability-impact/social-impact-and-philanthropy/globalvisionaries/gv/2021/afriscout.html.

Bronson, K. (2019). Looking through a responsible innovation lens at uneven engagements with digital farming. *NJAS: Wageningen Journal of Life Sciences*, *90–91*(1), 1–6. https://doi.org/10.1016/j.njas.2019.03.001.

Brugler, S., Gardezi, M., Dadkhah, A., Rizzo, D. M., Zia, A., & Clay, S. A. (2024). Improving decision support systems with machine learning: Identifying barriers to adoption. *Agronomy Journal, 116*(3), 1229–1236. https://doi.org/10.1002/agj2.21432.

Burch, K. A., & Legun, K. (2021). Overcoming barriers to including agricultural workers in the co-design of new AgTech: Lessons from a COVID-19-present world. *Culture, Agriculture, Food and Environment, 43*(2), 147–160. https://doi.org/10.1111/cuag.12277.

Cammarano, D., Zha, H., Wilson, L., Li, Y., Batchelor, W. D., & Miao, Y. (2020). A remote sensing-based approach to management zone delineation in small scale farming systems. *Agronomy, 10*(11), 1767. https://doi.org/10.3390/agronomy10111767.

Castle, M. H., Lubben, B. D., & Luck, J. D. (2016). Factors influencing the adoption of precision agriculture technologies by Nebraska producers. *UNL Digital Commons, 49*. http://digitalcommons.unl.edu/ageconworkpap/49.

Choruma, D. J., Dirwai, T. L., Mutenje, M. J., Mustafa, M., Chimonyo, V. G. P., Jacobs-Mata, I., & Mabhaudhi, T. (2024). Digitalisation in agriculture: A scoping review of technologies in practice, challenges, and opportunities for smallholder farmers in sub-Saharan Africa. *Journal of Agriculture and Food Research, 18*, 101286. https://doi.org/10.1016/j.jafr.2024.101286.

Choudhary, V., & Fock, K. (2020, April 23). Precision agriculture for smallholder farmers in Vietnam: How the Internet of Things helps smallholder paddy farmers use water more efficiently. *World Bank Blogs*. https://blogs.worldbank.org/en/eastasiapacific/precision-agriculture-smallholder-farmers-vietnam-how-internet-things-helps.

Cisternas, I., Velásquez, I., Caro, A., & Rodríguez, A. (2020). Systematic literature review of implementations of precision agriculture. *Computers and Electronics in Agriculture, 176*, 105626. https://doi.org/10.1016/j.compag.2020.105626.

Clay, D. E., Brugler, S., & Joshi, B. (2024). Will artificial intelligence and machine learning change agriculture: A special issue. *Agronomy Journal, 116*(3), 791–794. https://doi.org/10.1002/agj2.21555.

Colquhoun, A. (2024, November 29). *Boosting soil health in Kenya: UjuziKilimo's IoT solution*. GSMA. https://www.gsma.com/solutions-and-impact/connectivity-for-good/mobile-for-development/blog/boosting-soil-health-in-kenya-ujuzikilimos-iot-solution.

Erickson, B., Fausti, S., Clay, D., & Clay, S. (2018). Knowledge, skills, and abilities in the precision agriculture workforce: An industry survey. *Natural Sciences Education, 47*(1), 180010. https://doi.org/10.4195/nse2018.04.0010.

FAO. (2018, November 9). *AgroPad: A real-time on location and chemical analysis of the soil or water using AI*. https://www.fao.org/e-agriculture/news/agropad-real-time-location-and-chemical-nalysis-soil-or-water-using-ai.

Feder, G., Just, R. E., & Zilberman, D. (1985). Adoption of agricultural innovations in developing countries: A survey. *Economic Development and Cultural Change, 33*(2), 255–298. https://doi.org/10.1086/451461.

Fountas, S., Espejo-García, B., Kasimati, A., Gemtou, M., Panoutsopoulos, H., & Anastasiou, E. (2024). Agriculture 5.0: Cutting-edge technologies, trends, and challenges. *IT Professional, 26*(1), 40–47. https://doi.org/10.1109/MITP.2024.3358972.

Foureaux, S., & Daum, T. (2025). "But don't think it is a game": Agricultural videogames and "good farming.". *Journal of Rural Studies, 117*, 103686. https://doi.org/10.1016/j.jrurstud.2025.103686.

Franzen, D., & Mulla, D. (2015). A history of precision agriculture. In Q. Zhang (Ed.), *Precision agriculture technology for crop farming* (1st ed., pp. 1–19). CRC Press. https://doi.org/10.1201/b19336-1.

Gabriel, A., & Gandorfer, M. (2023). Adoption of digital technologies in agriculture—An inventory in a European small-scale farming region. *Precision Agriculture, 24*(1), 68–91. https://doi.org/10.1007/s11119-022-09931-1.

Gardezi, M., Joshi, B., Rizzo, D. M., Ryan, M., Prutzer, E., Brugler, S., & Dadkhah, A. (2023). Artificial intelligence in farming: Challenges and opportunities for building trust. *Agronomy Journal, 116*(3), 1217–1228. https://doi.org/10.1002/agj2.21353.

Geels, F. W., & Schot, J. (2007). Typology of sociotechnical transition pathways. *Research Policy, 36*(3), 399–417. https://doi.org/10.1016/j.respol.2007.01.003.

GSMA. (2025). *The GSMA Mobile Connectivity Index*. https://www.mobileconnectivityindex.com.

Higgins, V., van der Velden, D., Bechtet, N., Bryant, M., Battersby, J., Belle, M., & Klerkx, L. (2023). Deliberative assembling: Tinkering and farmer agency in precision agriculture implementation. *Journal of Rural Studies, 100*, 103023. https://doi.org/10.1016/j.jrurstud.2023.103023.

Joshi, D. R., Clay, S. A., Sharma, P., Rekabdarkolaee, H. M., Kharel, T., Risso, D. M., Thapa, R., & Clay, D. E.

(2023). Artificial intelligence and satellite-based remote sensing can be used to predict in-season soybean (*Glycine max*) yield. *Agronomy Journal, 116*(3), 917–930. https://doi.org/10.1002/agj2.21473.

Lee, C.-L., Strong, R., & Dooley, K. E. (2021). Analyzing precision agriculture adoption across the globe: A systematic review of scholarship from 1999–2020. *Sustainability, 13*(18), 10295. https://doi.org/10.3390/su131810295.

Lemay, M. A., & Boggs, J. (2024). Determinants of adoption of automation and robotics technology in the agriculture sector—A mixed methods, narrative, interpretive knowledge synthesis. *PLOS Sustainability and Transformation, 3*(11), e0000110. https://doi.org/10.1371/journal.pstr.0000110.

Lowenberg-DeBoer, J., & Erickson, B. (2019). Setting the record straight on precision agriculture adoption. *Agronomy Journal, 111*(4), 1552–1569. https://doi.org/10.2134/agronj2018.12.0779.

McBride, W. D., & Daberkow, S. G. (2003). Information and the adoption of precision farming technologies. *Journal of Agribusiness, 21*(1), 21–38. https://doi.org/10.22004/ag.econ.14671.

McFadden, J., & Lim, K. (2024, December 10). *Precision agriculture use increases with farm size and varies widely by technology*. USDA Economic Research Service. https://www.ers.usda.gov/data-products/charts-of-note/chart-detail?chartId=110550.

McFadden, J., Njuki, E., & Griffin, T. (2023). *Precision agriculture in the digital era: Recent adoption on U.S farms*. US Department of Agriculture, Economic Information Bulletin No. EIB-248. Economic Research Service.

Mehrabi, Z., McDowell, M. J., Ricciardi, V., Levers, C., Martinez, J. D., Mehrabi, N., Wittman, H., Ramankutty, N., & Jarvis, A. (2020). The global divide in data-driven farming. *Nature Sustainability*, 4(2), 154–160. https://doi.org/10.1038/s41893-020-00631-0.

Meynard, J.-M., Dedieu, B., & Bos, A. P. (2012). Re-design and co-design of farming systems. An overview of methods and practices. In I. Darnhofer, D. Gibbon, & B. Dedieu (Eds.), *Farming systems research into the 21st century: The new dynamic* (pp. 405–429). Springer Netherlands. https://doi.org/10.1007/978-94-007-4503-2_18.

Modi, V. (2022). Solar for small-holder farmers made affordable through innovative co-design and financing—lessons learnt. *Solar Compass, 3–4*, 100027. https://doi.org/10.1016/j.solcom.2022.100027.

Nowak, B. (2021). Precision agriculture: Where do we stand? A review of the adoption of precision agriculture technologies on field crops farms in developed countries. *Agricultural Research, 10*(4), 515–522. https://doi.org/10.1007/s40003-021-00539-x.

Pathak, H. S., Brown, P., & Best, T. (2019). A systematic literature review of the factors affecting the precision agriculture adoption process. *Precision Agriculture, 20*(6), 1292–1316. https://doi.org/10.1007/s11119-019-09653-x.

Protopop, I., & Shanoyan, A. (2016). Big data and smallholder farmers: Big data applications in the agri-food supply chain in developing countries. *International Food and Agribusiness Management Review, 19*(A), 173–190.

Ritchie, H., Rosado, P., & Roser, M. (2019, September 20). *Access to energy*. Our World in Data. https://ourworldindata.org/energy-access.

Robinson, D. K. R., Borrás, S., & Boon, W. P. C. (2025). Reimagining diffusion: Eight building blocks for reconceptualising the generalisation of innovation into society. *Technological Forecasting and Social Change, 213*, 123844. https://doi.org/10.1016/j.techfore.2024.123844.

Rose, D. C., Barkemeyer, A., de Boon, A., Price, C., & Roche, D. (2023). The old, the new, or the old made new? Everyday counter-narratives of the so-called fourth agricultural revolution. *Agriculture and Human Values, 40*(2), 423–439. https://doi.org/10.1007/s10460-022-10374-7.

Rotz, S., Gravely, E., Mosby, I., Duncan, E., Finnis, E., Horgan, M., LeBlanc, J., Martin, R., Neufeld, H. T., Nixon, A., Pant, L., Shalla, V., & Fraser, E. (2019). Automated pastures and the digital divide: How agricultural technologies are shaping labour and rural communities. *Journal of Rural Studies, 68*, 112–122. https://doi.org/10.1016/j.jrurstud.2019.01.023.

Ryan, M., Isakhanyan, G., & Tekinerdogan, B. (2023). An interdisciplinary approach to artificial intelligence in agriculture. *NJAS: Impact in Agricultural and Life Sciences, 95*(1), 2168568. https://doi.org/10.1080/27685241.2023.2168568.

Santos Valle, S., & Kienzle, J. (2020). *Agriculture 4.0 – Agricultural robotics and automated equipment for sustainable crop production* (No. Vol 24; Integrated Crop Management, p. 40). FAO. https://openknowledge.fao.org/server/api/core/bitstreams/14fc8bf7-fdeb-4c7c-a2ed-b2d59118a70b/content.

Sasao, A., & Shibusawa, S. (2000). Prospects and strategies for precision farming in Japan. *Japan Agricultural Research Quarterly*, *34*(4), 233–238. https://www.jircas.go.jp/sites/default/files/publication/jarq/34-4-233-238_0.pdf.

Shanahan, M., & Bahia, K. (2024). *The state of mobile internet connectivity—Key findings 2024* (p. 7). GSMA 2024. https://www.gsma.com/r/wp-content/uploads/2024/10/The-State-of-Mobile-Internet-Connectivity-Report-Key-Findings-2024.pdf.

Shrestha, N., & Fry, C. (2022, August 22). *Supporting Nepal's hill farmers to enhance their livelihoods through mixed farming*. International Water Management Institute (IWMI). https://www.iwmi.org/2022/08/supporting-nepals-hill-farmers-to-enhance-their-livelihoods-through-mixed-farming.

Silvestri, S., Richard, M., Edward, B., Dharmesh, G., & Dannie, R. (2021). Going digital in agriculture: How radio and SMS can scale-up smallholder participation in legume-based sustainable agricultural intensification practices and technologies in Tanzania. *International Journal of Agricultural Sustainability*, *19*(5–6), 583–594. https://doi.org/10.1080/14735903.2020.1750796.

Stilgoe, J. (2013). Foreword: Why responsible innovation? In R. Owen, J. Bessant, & M. Heintz (Eds.), *Responsible Innovation* (pp. 364). John Wiley & Sons, Incorporated.

Straub, E. T. (2009). Understanding technology adoption: Theory and future directions for informal learning. *Review of Educational Research*, *79*(2), 625–649. https://doi.org/10.3102/0034654308325896.

Taylor, J. A. (2023). Precision agriculture. In M. J. Goss & M. Oliver (Eds.), *Encyclopedia of soils in the environment* (pp. 710–725). Elsevier. https://doi.org/10.1016/B978-0-12-822974-3.00261-5.

Tzachor, A., Devare, M., King, B., Avin, S., & Ó hÉigeartaigh, S. (2022). Responsible artificial intelligence in agriculture requires systemic understanding of risks and externalities. *Nature Machine Intelligence*, *4*(2), 104–109. https://doi.org/10.1038/s42256-022-00440-4.

UNDP. (2021). *Precision agriculture for smallholder farmers* (p. 80). United Nations Development Programme. https://www.undp.org/publications/precision-agriculture-smallholder-farmers.

Wiseman, L., Sanderson, J., Zhang, A., & Jakku, E. (2019). Farmers and their data: An examination of farmers' reluctance to share their data through the lens of the laws impacting smart farming. *NJAS—Wageningen Journal of Life Sciences*, *90–91*(1), 1–10. https://doi.org/10.1016/j.njas.2019.04.007.

Xie, L., Luo, B., & Zhong, W. (2021). How are smallholder farmers involved in digital agriculture in developing countries: A case study from China. *Land*, *10*(3), 245. https://doi.org/10.3390/land10030245.

Discovery, Adoption, and Training for Precision Agriculture

3

Harkirat Kaur, Luan Oliveira, Simerjeet Virk, David E. Clay, and Ali Mirzakhani Nafci

Purpose

This chapter was not included in the first edition of *Precision Agriculture Basics*.

In January 2024, the International Society of Precision Agriculture (https://www.ispag.org/) defined precision agriculture as, a management strategy that gathers, processes and analyzes temporal, spatial and individual plant and animal data and combines it with other information to support management decisions according to estimated variability for improved resource use efficiency, productivity, quality, profitability and sustainability of agricultural production.

Meeting this definition means that to manage variability effectively, resources should be placed where they are most needed. The goals of this chapter are to discuss the discovery and development of precision agriculture and to identify the educational requirements and training resources needed for students and the current workforce.

The Interdisciplinary Evolution of Precision Agriculture

In 1936, Alan Turing discovered how to replace manual calculations with a machine, producing the foundational ideas for modern digital computation. Turing's discoveries were followed by significant advancements, including the launch of NASA's Landsat 1 satellite in 1972, the launch of the first Global Positioning System (GPS) satellite in 1978, the release of ArcView 1.0 in 1991, and the development of artificial intelligence (AI) in the 1980s. During this same period, there were simultaneous breakthroughs in molecular biology that set the stage for genomics-driven plant breeding and the ability to better understand how plants respond to biotic and abiotic stress (Hansen et al., 2013). Genomic information was later used by researchers such as Norman Borlaug, an agricultural scientist and plant pathologist known as the Father of the Green Revolution, to develop dwarf wheat, which increased food production and reduced the risk of famine across the globe. He received the Nobel Prize for this in 1970.

These technological advances led to the first International Conference on Precision Agriculture, organized by the University of Minnesota, the American Society of Agronomy, the Crop Science Society of America, and the Soil Science Society of America in 1992, as well as the release of the Yield Monitor by Ag Leader in 1992 and the launch of the Web Soil Survey by the USDA Natural Resources Conservation Service (USDA-NRCS) in 2005. At the time, we did not understand how important these discoveries

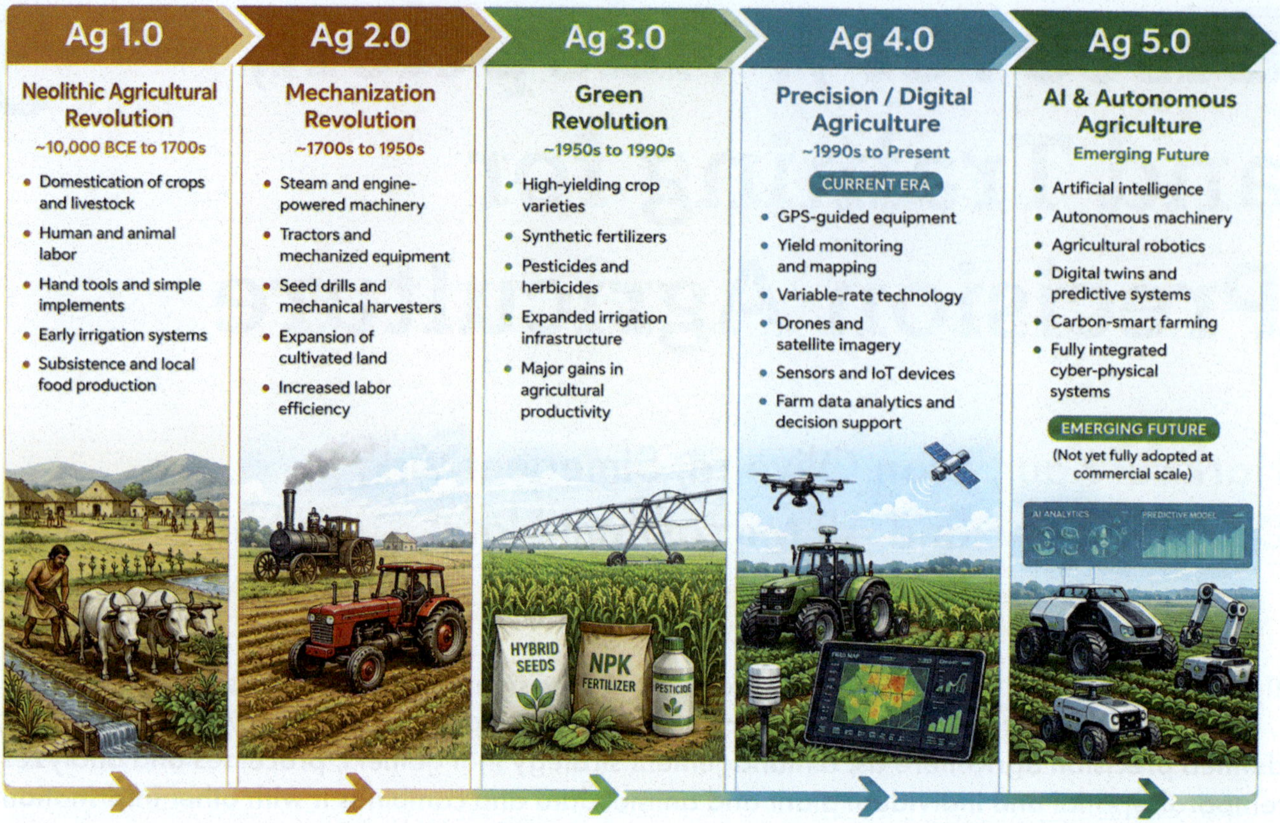

Figure 3.1. Timescale of agricultural discoveries from Agriculture 1.0 to Agriculture 5.0. (Adapted from Tisdell & Svizzero, (2018) and Bowles & Choi (2019), created using OpenAI image generation tools, prompted by Harkirat Kaur).

would be for implementing precision agriculture, but by the time of this publication, it is clear that each invention was an important step toward modern precision agriculture (Figure 3.1).

The Computer Revolution

Computing started with people conducting calculations, and it evolved into computers that could perform simple, and then extremely complex, calculations very quickly. These computers started out as room-sized machines requiring specialized operators and have become pocket-sized smartphones that can easily download and use a variety of applications.

These applications can use artificial Intelligence and machine learning, as computing technologies, combined with data that can be collected from, shared on, and stored on the internet, have created opportunities to help farmers with tasks such as increasing the precision of planting, fertilizing, spraying, and harvesting. The goal of precision agriculture in the twenty-first century is to fully integrate these technologies into our production and management systems.

GPS and Global Navigation Satellite Systems Integration

Originally developed for military navigation in 1973, GPS became fully operational in 1993 (Chapter 7). When selective availability was turned off in 2000, location accuracy improved significantly, dropping from around 100 to 20 m. Subsequent advances included Satellite-Based Augmentation Systems (SBAS), the Wide Area Augmentation System (WAAS) and European Geostationary Navigation Overlay System (EGNOS), which provided free, satellite-broadcast corrections and improved positioning accuracy to the submeter level, effectively bridging the gap between autonomous GPS and more advanced correction methods. Technologies such as Real-Time Kinematic (RTK) and Real-Time eXtended (RTX) further refined positioning to centimeter-level accuracy. Real-Time Kinematic base stations allow high precision within a localized network, while RTX provides global correction without dependence on base stations. For precision agriculture, these improvements have been transformative and are now embedded into all modern machinery including tractors planters, sprayers, and harvesters.

These technologies reduced uncertainty and increased the precision with which a farmer can manage the variability of their field by creating point-specific, detailed management maps. These changes led to precise planting, spraying, and fertilizer applications. These practices have further evolved into strategies such as variable rate applications for seeding rates and fertilizers to increase profitability and reduce overapplication of nutrients.

Geographic Data Processing and GIS

GPS and Global Navigation Satellite Systems (GNSS) data are fed into Geographic Information Systems (GIS), which serve as the backbone of spatial data analysis and management in precision farming (Chapters 8 and 9). Geographic Information Systems enable the integration, visualization, and analysis of spatial and temporal datasets, facilitating site-specific decision-making. Originally developed in the 1950s through geostatistics for the mining and petroleum industries, these tools now help generate prescription maps for nutrients, irrigation, and pest management in agriculture. The GPS interface with yield monitors, remote sensors, and satellite data to analyze variability and optimize inputs. Machine learning models further enhance this analysis by identifying patterns and generating adaptive recommendations based on both historical and real-time data.

Weaving Technology and Knowledge into Precision Farming

Since the 1980s, researchers and industry innovators have been working to integrate emerging technologies with agronomic knowledge. Dr. Pierre Robert of the University of Minnesota laid the foundation for precision agriculture in 1983. Around the same time, the sustainable agriculture movement gave rise to the Low-Input Sustainable Agriculture (LISA) initiative in 1989, which emphasized resource optimization—a principle that remains central to precision agriculture today. The 1990s marked a turning point with the adoption of yield monitors and GPS-enabled spatial management. Geographic Information Systems integration followed, providing tools to manage variability with greater granularity. By the mid-2000s, the gradual but steady adoption of precision agriculture tools began accelerating, fueled by public and private investment in agricultural research and development. Data from Lowenberg-DeBoer and Erickson (2019) shows widespread adoption, with yield monitors installed on 68% of U.S. cropland and 29% of acreage receiving variable-rate fertilizer applications. In the 2010s and afterward, AI, cloud computing, and robotics began reshaping agricultural operations. These tools now drive automation, enhance sustainability, and enable real-time optimization of inputs, reducing waste and maximizing profitability.

Integrating yield measurement tools with molecular biology offers an opportunity to better manage variability across the landscape. In many fields, both non-biological (abiotic)– and biological (biotic)–driven stresses result in lower yields. For example, many farmers have observed that yields can be drastically lower in upland environments than in lowland ones. The factors for identifying why yields are lower in upland than lowland soils can be determined by using molecular biology techniques that quantify gene expression in plants under different treatment stressors (Sun et al., 2022). For example, Hansen et al. (2013) evaluated the gene expression of corn plants in upland summit area soils compared with those in lowland footslope soils within the same field. They found that gene expression related to drought stress and tolerance in corn, even within the same field, was higher in summit areas than in footslope areas. This study also found that plants in summit areas had a reduced ability to take up nutrients and to respond to wounding by pests. This suggests that adding nutrients to these areas may not improve production or be economically optimal unless additional water is supplied as well. This example demonstrates how precision agriculture research can inform and change the way farmers manage their fields.

Unlocking the Full Potential of Precision Agriculture: Training the Existing Workforce

The adoption of precision agriculture technologies involves the use of innovative tools by farmers willing to experiment with emerging, and sometimes untested, methods. It involves weighing the strengths and weaknesses of the practices (Chavas & Nauges, 2020), considering the financial implications, and assessing the feasibility of implementation with the current number of trained professionals available to provide guidance (Cochrane, 1979; Feder & Umali, 1993). The decision to adopt (or not to adopt) technologies is complex, and challenges vary globally. This section focuses on adoption through the framework of extension and outreach. For more details on the adoption of these technologies and the complex challenges farmers face worldwide, see Chapter 2.

Despite technological progress, a persistent gap remains between innovation and on-farm implementation. Many growers and professionals lack the training needed to successfully deploy and interpret precision agriculture information and making these tools significantly less effective in management decision-making. To fully harness the benefits of modern digital agriculture, a coordinated effort is required to educate and empower the existing workforce, especially in underserved or resource-limited areas. Providing appropriate training is essential to unlocking the full potential of precision agriculture technologies worldwide. In addition, farming efficiency is important to consider, but not all practices improve efficiency (Daberkow & McBride, 2003).

Extension and Outreach Reduce Limitations and Provide Guidance

Extension professionals and outreach programs are key in identifying low-cost, scalable precision agriculture prescriptions. Extension professionals can identify precision agriculture technologies or practices that work for farmers in their local region to improve the adoption of precision agriculture among small-scale family farms (Lowenberg-DeBoer & Erickson, 2019).

However, the ability to adopt precision agriculture technologies often varies between small and large farms. Larger operations typically have greater financial flexibility, enabling them to absorb the higher upfront costs and risks associated with testing new technologies, whereas smaller, family-owned farms often operate with tighter profit margins and limited access to capital, making technology adoption a more consequential decision. As a result, unsuccessful investments may pose disproportionately high financial risk to smaller farmers.

Farmers need to see the financial benefits of adopting precision agriculture technologies because profit margins are often narrow, and these practices may be contrary to their classical training (Kendall et al., 2022). Extension services act as a third party to observe farmers' operations and help identify these financial benefits, often through the adoption of precision agriculture technologies. Additionally, extension programs can provide both formal and informal training to farmers interested in specific topics.

Tools Used by Extension Professionals

Farmers and agricultural professionals often need clear, accessible materials and information that break down complex precision agriculture concepts into practical applications. A combination of printed resources, digital tools, case studies, and targeted consulting has proven effective in bridging the gap between research and practice (Lybbert & Sumner, 2012). Successful training materials must deliver information through multiple channels to all farmers (Eastwood et al., 2017).

A commonly used educational tool in precision agriculture is the development of short educational handouts. These handouts are helpful because they offer step-by-step guides to various precision agriculture practices and technologies, including variable rate applications, soil fertility mapping, and remote sensing. Farmers benefit from simple, easy-to-follow materials that enable them to interpret nutrient maps, analyze yield data and cost–benefit scenarios, and create customized precision agriculture adoption plans. Factsheets focusing on calibration techniques for precision equipment, best management practices for resource efficiency, and strategies for integrating precision agriculture tools into traditional farming methods have been instrumental in outreach efforts. Beyond printed materials, the internet has transformed precision agriculture education for farmers and their advisors by offering online courses, interactive decision-support tools, mobile applications, YouTube tutorials, webinars, and virtual field days that provide real-time insights into specific needs (Chivenge et al., 2022; Ravikumar & Gopu, 2016; Tey & Brindal, 2012), ranging from fixing equipment to using remote sensing in crop monitoring.

Modern precision agriculture tools, such as Internet of Things (IoT), have become widely used to enhance real-time monitoring of soil conditions, automate irrigation schedules to enhance real-time monitoring of plants, soil, and environmental conditions, as well as improve overall farm management efficiency. However, the full potential of IoT in precision agriculture can only be realized through targeted training programs that teach and train farmers on how to interpret and act upon sensor-generated data (Fastellini & Schillaci, 2020). As a result, Extension programs have begun focusing on how the IoT can be used in precision agriculture applications, prioritizing education on data-driven decision-making and providing hands-on experience with integrated sensor technology (Khanna & Kaur, 2019; Wolfert et al., 2017).

In France, Matese and Di Gennaro (2015) assisted growers trained in integrating sensors with the IoT to use multispectral imaging, GPS-guided equipment, and soil conductivity mapping in their vineyards to optimize yield quality and sustainability. The ability to take advantage of these technologies may require one-on-one training that enhances engagement and enables practitioners to fully implement the practices (Howard, 2002). As precision agriculture continues to evolve, land-grant universities and their Extension services will remain vital in bridging the gap between scientific research and on-farm implementation.

One of the most important foundational steps in training the current and future workforce is providing hands-on experience with various precision agriculture tools and technologies, geospatial analysis, and data-driven decision-making (Vermunt et al., 2022). To address this, workforce development programs should collaborate with universities to develop educational materials that translate complex precision agriculture concepts into simple-to-implement practices and offer hands-on workshops, field trials, and internships with industry partners (Khanna & Kaur, 2019). For example, collaborative efforts among Michigan State University, Purdue University, Kansas State University, Penn State, and the University of Wisconsin-Madison developed region-specific educational bulletins that help improve corn and soybean production management (Pierpaoli et al., 2013).

With the rise of digital learning, video-based outreach and online conversations have emerged as powerful tools for precision agriculture education. Farmers increasingly turn to online platforms—including YouTube, university Extension websites, and social media—to access educational content. Studies suggest that videos improve knowledge retention and provide effective visual demonstrations, making complex concepts and technologies—such as UAV-based crop scouting, GPS-guided machinery, and automated irrigation systems—more accessible and easier to understand. To maximize effectiveness, video-based educational efforts should align with critical periods in agricultural decision-making. In addition, keeping videos concise, providing clear visuals and step-by-step instructions, and including farmer testimonials that highlight localized content enhance viewer engagement while retaining educational value (Coggins et al., 2022; Prokopy et al., 2017). Moreover, as precision agriculture continues to evolve, educators must continually update, refine, and modify the training materials.

Improving the Effectiveness of Extension

The current and future agricultural workforce often struggles to connect complex precision agriculture concepts with real-world decision-making. Strengthening these connections is needed to improve training effectiveness (Finger et al., 2019). A promising approach to improving precision agriculture training is to develop interactive farm management exercises that simulate real-world decision-making scenarios. For example, exercises can be structured by immersing students in a simulated farming environment (Wachenheim & Hanson, 2024). In one simulation, Dakota Levy, a recent agronomy graduate, collaborated with his parents on their 2,400-acre farm in southeastern North Dakota. The family cultivates soybean, corn, wheat, and sunflower. After attending a workshop on precision agriculture technologies hosted by North Dakota State University Extension, Dakota and his father, Arnold, became interested in implementing precision agriculture technologies to enhance their farming operations. The workshop featured experts discussing various precision agriculture technology services, including UAV imagery analysis, soil testing, variable rate fertilization, and seeding. Inspired by the potential benefits, Dakota proposed using partial budgeting to assess the economic viability of adopting a specific precision agriculture technology bundle on the farm.

By incorporating new and emerging technologies into these workshops, such as variable rate fertilizer equipment and how to analyze and interpret remote sensing data, farmers and agronomists gain hands-on experience in applying precision techniques, helping them identify what is optimal for their operations to increase profitability and sustainability (Khanna & Kaur, 2019). A framework that can be used in extension workshops to allow farmers to simulate scenarios includes working through the following four steps:

1. Identify the farmer's long-t erm goals and how adopting certain precision agriculture tools or practices can help achieve these goals.
2. Obtain prior data (e.g., remote sensing, Web Soil Survey, yield, soil test results, farmer knowledge) from the site and determine additional information that is needed (i.e., soil or tissue sampling, economic/data analysis, technology purchases, or additional expert guidance that may be needed).
3. Scale the implementation plan up from a small area of the field to a larger area by using on-farm strip trial research concepts (Chapter 21).
4. Collect the data from these smaller experiments and determine whether adjustments are needed.

To further engage farmers and agronomists, it is essential to investigate existing data and possible management options, including hands-on experience in accessing and using data from soil sampling and field trials (Chapters 6 and 21), yield monitors (Chapter 20), remote sensing technologies and software (Chapters 8, 9,14, and 15) and the Web Soil Survey (Chapter 10). Learning to analyze data empowers farmers and agronomists to make decisions that are optimal for their soil texture, organic matter content, drainage, and water-holding capacity and determine the best management strategies for each growing season. For example, overlaying yield data on an elevation map can provide a farmer with valuable insights into why some areas of the field have lower or higher yields than others (Figure 3.2).

Beyond outreach to farmers and agronomists, extension professionals must turn their attention to classrooms that are teaching the next generation of farmers. These classrooms should introduce students to these technologies and data early on and challenge them to formulate management strategies and recommendations based on factors they will encounter on their own farms, such as hybrid selection, soil conditions, and moisture availability. Case studies of the success of variable rate seeding in different growing regions can help students understand the practical advantages of precision agriculture–based planting strategies. Similarly, students should be engaged in activities that demonstrate the economic impact of delayed planting due to weather constraints using historical weather data and crop yield response models. These exercises provide insight into mitigation strategies, such as adjusting hybrid maturity, modifying fertilization plans, or adopting alternative crop rotations (Mulla, 2013).

Another integral component of precision agriculture, specifically in undergraduate education, involves learning how to determine the economic feasibility of precision technologies. By participating in exercises where students use real data from GPS-guided planters, yield monitors, and

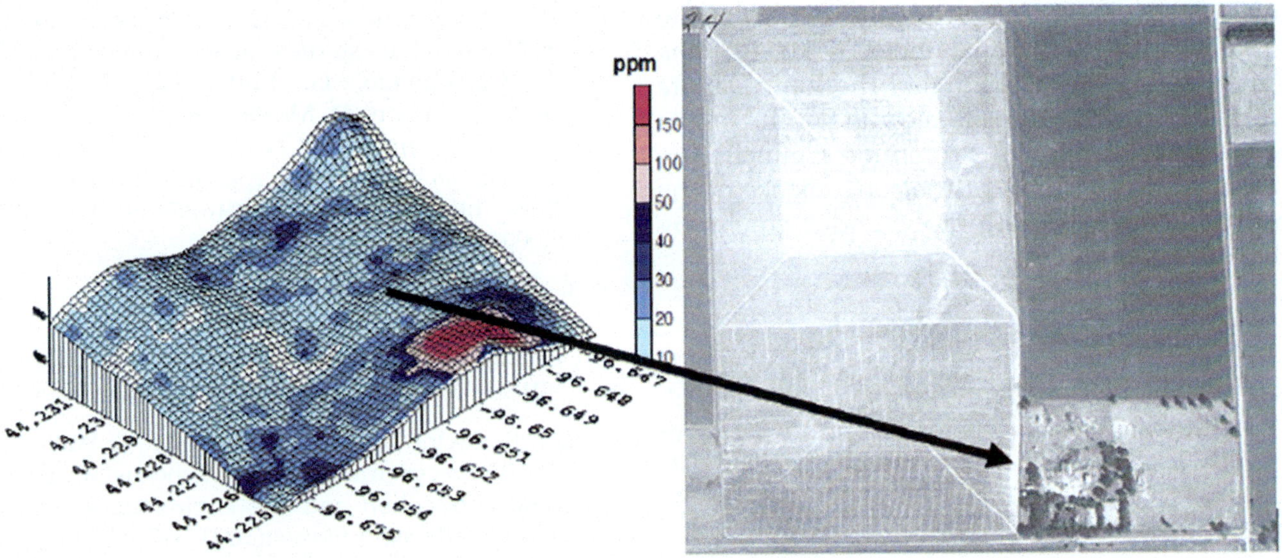

Figure 3.2. Soil test phosphorus (P) concentration and a historical image collected from the study site. The high soil test P level affecting current yields is associated with a homestead that was located at the site 50 years ago. (Courtesy D. Clay, South Dakota State University).

UAVs or drones, both with and without AI models. This can help them understand the importance of economic assessments and how to knowledgeably implement the decision-making power of AI applications (Kutter et al., 2011; Wolfert et al., 2017).

Ensuring that precision agriculture education is accessible and engaging for undergraduates requires a dynamic approach that combines simulation-based learning, real-world data analysis, and hands-on problem-solving exercises. As precision agriculture continues to evolve, extension and outreach professionals must create learning experiences that teach the current generation while also preparing the next generation to face the challenges in agriculture (Pierpaoli et al., 2013).

Conclusion

This chapter provided insights into developing effective teaching and outreach programs for farmers. The key factors identified for the successful development of training materials were (a) that the training information addressed the specific questions of the practitioner; (b) that farmers and agricultural professionals often require clear, accessible materials that break down complex PA concepts into practical applications; (c) that a combination of printed resources, digital tools, case studies, and targeted consulting may be required; and (d) that successful training materials must focus on delivering information through multiple channels.

Study Questions

1. You are an extension professional in South Dakota tasked with creating a workshop to help farmers understand how to use Web Soil Survey data for decision-making. Discuss how you would create and implement this workshop.

2. Nitrogen-rate recommendations using the corn nitrogen rate calculator. Instructions: You are an extension professional in Iowa, and you are tasked with creating a workshop that helps farmers calculate the optimal nitrogen application rate for their field. Discuss how you would create and implement this workshop.

ACKNOWLEDGMENTS

Support for this document was provided by the Precision Farming Systems community in the American Society of Agronomy, South Dakota State University, USDA Natural Resources Conservation Service Conservation Innovation Grants (NR233A750011G019), USDA National

Institute of Food and Agriculture (Multistate Project S1090) Program under (7008110), the American Coalition for Ethanol, the South Dakota Corn Utilization Council, and the South Dakota Soybean Research and Promotion Council. The use of trade names is for the convenience of the reader and does not imply endorsement by the authors.

REFERENCES

Bowles, S., & Choi, J. K. (2019). The Neolithic agricultural revolution and the origins of private property. *Journal of Political Economy*, *127*(5), 2186–2228. https://www.journals.uchicago.edu/doi/full/10.1086/701789.

Chavas, J.-P., & Nauges, C. (2020). Uncertainty, learning, and technology adoption in agriculture. *Applied Economic Perspectives and Policy*, *42*(1), 42–53. https://doi.org/10.1002/aepp.13003.

Chivenge, P., Zingore, S., Ezui, K. S., Njoroge, S., Bunquin, M. A., Dobermann, A., & Saito, K. (2022). Progress in research on site-specific nutrient management for smallholder farmers in sub-Saharan Africa. *Field Crops Research*, *281*, 108503. https://doi.org/10.1016/j.fcr.2022.108503.

Cochrane, W. W. (1979). *The development of American agriculture: A historical analysis*. University of Minnesota Press.

Coggins, S., McCampbell, M., Sharma, A., Sharma, R., Haefele, S. M., Karki, E., Hetherington, J., Smith, J., & Brown, B. (2022). How have smallholder farmers used digital extension tools? Developer and user voices from Sub-Saharan Africa, South Asia and Southeast Asia. *Global Food Security*, *32*, 100577. https://doi.org/10.1016/j.gfs.2021.100577.

Daberkow, S. G., & McBride, W. D. (2003). Farm and operator characteristics affecting the awareness and adoption of precision agriculture technologies in the US. *Precision Agriculture*, *4*(2), 163–177. https://doi.org/10.1023/A:1024557205871.

Eastwood, C., Klerkx, L., & Nettle, R. (2017). Dynamics and distribution of public and private research and extension roles for technological innovation and diffusion: Case studies of the implementation and adaptation of precision farming technologies. *Journal of rural studies*, *49*, 1–12. https://doi.org/10.1016/j.jrurstud.2016.11.008.

Fastellini, G., & Schillaci, C. (2020). Precision farming and IoT case studies across the world. In A. Castrignanò, G. Buttafuoco, R. Khosla, A. M. Mouazen, D. Moshou, & O. Naud (Eds.), *Agricultural Internet of Things and decision support for precision smart farming* (pp. 331–415). Academic Press. https://doi.org/10.1016/B978-0-12-818373-1.00007-X.

Feder, G., & Umali, D. L. (1993). The adoption of agricultural innovations: A review. *Technological Forecasting and Social Change*, *43*(3–4), 215–239. https://doi.org/10.1016/0040-1625(93)90053-A.

Finger, R., Swinton, S. M., El Benni, N., & Walter, A. (2019). Precision farming at the nexus of agricultural production and the environment. *Annual Review of Resource Economics*, *11*, 313–335. https://doi.org/10.1146/annurev-resource-100518-093929.

Hansen, S., Clay, S. A., Clay, D. E., Carlson, C. G., Reicks, G., Jarachi, Y., & Horvath, D. (2013). Landscape features impacts on soil available water, corn biomass, and gene expression during the late vegetative growth stage. *The Plant Genome*, *6*(2), 1–9. https://doi.org/10.3835/plantgenome2012.11.0029.

Howard, W. (2002). *Four case studies in precision agriculture*. *California Polytechnic State University*. https://digitalcommons.calpoly.edu/agb_fac/146/.

Kendall, H., Clark, B., Li, W., Jin, S., Jones, G. D., Chen, J., Taylor, J., Li, Z., & Frewer, L. J. (2022). Precision agriculture technology adoption: A qualitative study of small-scale commercial "family farms" located in the North China Plain. *Precision Agriculture*, *23*(1), 319–351. https://doi.org/10.1007/s11119-021-09839-2.

Khanna, A., & Kaur, S. (2019). Evolution of Internet of Things (IoT) and its significant impact in the field of precision agriculture. *Computers and Electronics in Agriculture*, *157*, 218–231. https://doi.org/10.1016/j.compag.2018.12.039.

Kutter, T., Tiemann, S., Siebert, R., & Fountas, S. (2011). The role of communication and cooperation in the adoption of precision farming. *Precision Agriculture*, *12*(1), 2–17. https://doi.org/10.1007/s11119-009-9150-0.

Lowenberg-DeBoer, J., & Erickson, B. (2019). Setting the record straight on precision agriculture adoption. *Agronomy Journal*, *111*(4), 1552–1569. http://doi.org/10.2134/agronj2018.12.0779.

Lybbert, T. J., & Sumner, D. A. (2012). Agricultural technologies for climate change in developing countries: Policy options for innovation and technology diffusion. *Food Policy*, *37*(1), 114–123. https://ideas.repec.org/a/eee/jfpoli/v37y2012i1p114-123.html.

Matese, A., & Di Gennaro, S. F. (2015). Technology in precision viticulture: A state-of-the-art review. *International Journal of Wine Research*, *7*, 69–81. https://doi.org/10.2147/IJWR.S69405.

Mulla, D. J. (2013). Twenty five years of remote sensing in precision agriculture: Key advances and remaining knowledge gaps. *Biosystems Engineering*, *114*(4), 358–371. https://doi.org/10.1016/j.biosystemseng.2012.08.009.

Pierpaoli, E., Carli, G., Pignatti, E., & Canavari, M. (2013). Drivers of precision agriculture technologies adoption: A literature review. *Procedia Technology*, *8*, 61–69. http://doi.org/10.1016/j.protcy.2013.11.010.

Prokopy, L. S., Bartels, W.-L., Burniske, G., & Power, R. (2017). Agricultural extension and climate change communication. In H. von Storch (Ed.), *Oxford Research Encyclopedia of Climate Science*. Oxford University Press. https://doi.org/10.1093/acrefore/9780190228620.013.429.

Ravikumar, R.R., & Gopu, J. (2016). An over view of the implementation of precision farming projects in Tamil Nadu, India. Munich Personal RePEc Archive Paper #7374. https://www.researchgatenet/publication/347438748_An_Over_View_of_the_Implementation_of_Precision_Farming_Projects_in_Tamil_Nadu_India.

Sun, Y., Shang, L., Zhu, Q.-H., Fan, L., & Guo, L. (2022). Twenty years of plant genome sequencing achievements and challenges. *Trends in Plant Science*, *27*(4), P391–P401. https://doi.org/10.1016/j.tplants.2021.10.006.

Tey, Y. S., & Brindal, M. (2012). Factors influencing the adoption of precision agricultural technologies: A review for policy implications. *Precision Agriculture*, *13*(6), 713–730. https://link.springer.com/article/10.1007/s11119-012-9273-6.

Tisdell, C. A., & Svizzero, S. (2018). The agricultural revolution, childe's theory of economic development as outlined in man makes himself, and contemporary economic theories. *History of Economics Review*, *71*(1), 55–72. https://doi.org/10.1080/10370196.2019.1598252.

Vermunt, D. A., Wojtynia, N., Hekkert, M. P., Van Dijk, J., Verburg, R., Verweij, P. A., Wassen, M., & Runhaar, H. (2022). Five mechanisms blocking the transition towards 'nature-inclusive' agriculture: A systemic analysis of Dutch dairy farming. *Agricultural Systems*, *195*, 103280. https://doi.org/10.1016/j.agsy.2021.103280.

Wachenheim, C., & Hanson, E. (2024). Adopting precision input management: A teaching case study. *Applied Economics Teaching Resources*, *6*(1), 1–6. https://doi.org/10.22004/ag.econ.341812.

Wolfert, S., Ge, L., Verdouw, C., & Bogaardt, M.-J. (2017). Big Data in smart farming—A review. *Agricultural Systems*, *153*, 69–80. https://doi.org/10.1016/j.agsy.2017.01.023.

Advances in Variable Rate Field Equipment Technologies

4

Ali Mirzakhani Nafchi, Ajay Sharda, David E. Clay, and Simerjeet Virk

Purpose

Variable rate equipment is evolving toward autonomous, sensor-dense platforms that use machine learning algorithms to develop agronomic recommendations on managing nutrients, seeding rates, and pests across the field. However, the measurement of variability through various forms of data collection and the analysis of the data through machine learning are beyond the scope of this chapter and are covered in many of the following chapters of this book. In general, variable rate equipment will follow directions to the best of its ability. To create appropriate directions, it is important to understand the basic components of precision farming equipment. This chapter discusses opportunities for variable rate equipment; recent options for variable seed application, pesticides, and fertilizers; recent advances in variable rate equipment; the basic components of variable rate equipment; and future research needs, including plant-scale actuation, per-plant phenotyping, interoperable agronomic knowledge graphs, and resilient autonomy in low-connectivity environments. This is a revision and update of the first edition original chapter by Sharda et al. (2018).

Opportunities for Variable Rate Equipment

Providing food for a rapidly growing world population can be accomplished only by increasing the amount of cultivated land, increasing genetic potential, improving management, or a combination of all three. Due to the physical and biochemical constraints associated with increasing the amount of cultivated land or increasing genetic yield potential, it is unlikely that any one approach can double food production by 2050. For example, between 1950 and 2010, U.S. corn yields increased at approximately 93.5 kg/ha per year (1.5 bu/ac per year). This yield gain can be attributed to several factors, including improved genetics, increased fertilizer use, and better soil management. Extrapolating that yield gain rate to 2010 to 2050 would result in a corn grain yield enhancement of 3.8 Mg/ha (60 bu/ac), representing a 43% increase in corn yields—falling short of the goal to double production. Precision agriculture can be used to reduce the projected shortfall in food production relative to population growth. The power of precision farming lies in its ability to increase sustainable intensification on currently farmed land, thereby reducing the need to convert grasslands or forests into annual crop production systems.

Variable rate technologies are systems that can take in information about the variability of the landscape and use it to vary and match the appropriate inputs across a field. For example, in a South Dakota no-tillage field, soybean yields ranged from less than 2 Mg/ha (30 bu/ac) to 4 Mg/ha (60 bu/ac; Figure 4.1). This variability was primarily attributed to water stress, which changed the amounts of water and fertilizer input needed in different topographic areas of the field. In the past, this field would have been managed based on the field's average yield, average pest population, and average soil nutrient information rather than its inherent variability. Precision agriculture provides technology to manage areas with various yield-limiting factors differently by shifting from field-average to plant-level prescriptions generated from high-resolution sampling, which may also include machine learning algorithms.

In many production systems, weather extremes have limited plant growth (Clay & Trooien, 2017). With today's technologies, growers can monitor how extreme weather affects crop growth using sensor networks connected to the Internet of Things. For example, they can use soil moisture sensors, satellite evapotranspiration products, and unmanned aircraft system (UAS) thermal imagery to localize water stress at sub-field scales. Differential amounts of water and associated stress can impact many factors, including the nitrogen (N) mineralization rate, weed distribution, and the plants' ability to resist pests and take up nutrients (Hansen et al., 2013; Kim et al., 2008). These stress maps can be used to enhance variable rate prescriptions, leading to improved agricultural efficiency. For example, if water stress reduces the ability of the plant to control diseases or utilize nutrients, then these areas could be targeted for corrective treatments (Hansen et al., 2013; Kim et al., 2008). Theoretically, targeting treatments to critical areas should improve treatment efficiency, close the gap between observed yield and the plants' genetic potential, reduce input waste, and enhance economic returns.

Using Maps to Create Variable Rate Treatments

Precision farming relies on information and science-based recommendations to create effective treatment options. In 2025, high temporal and spatial resolution data can be obtained from yield monitors, satellite constellations such as PlanetScope imagery, Sentinel imagery, unmanned aerial vehicle (UAV) imagery, and other remote-sensing products, soil laboratories, in-field sensors, and weather stations. Machine learning or process-based algorithms can be used to process this information into site-specific recommendations. These recommendations can then be implemented by

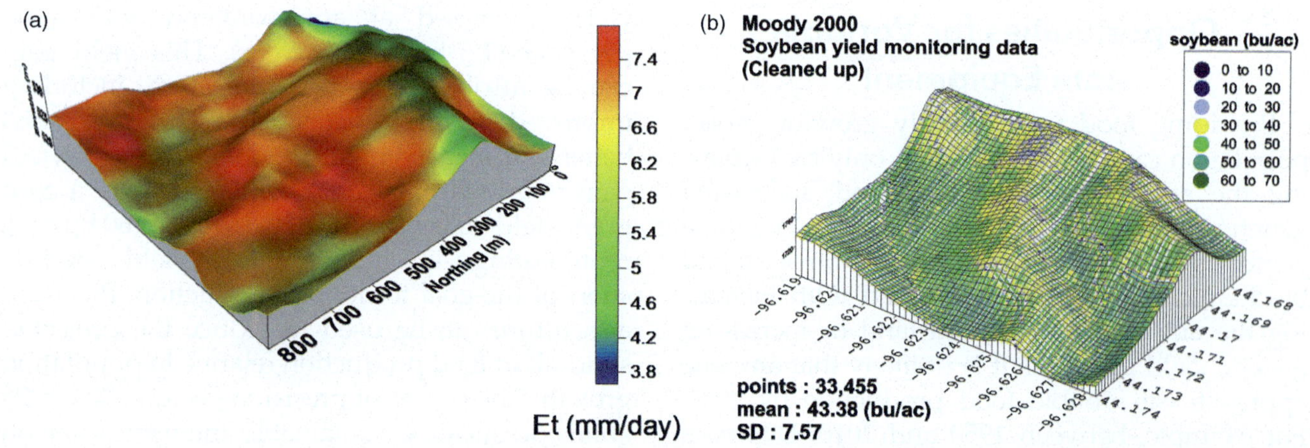

Figure 4.1. An example evapotranspiration (ET) map is shown on the left, with a corresponding yield map from the same field shown on the right (Mishra et al., 2008 / with permission of Taylor & Francis Group). Area with low yields (yellow in right image) were water stressed and had low yields, typically in the ridge or shoulder areas of the field.

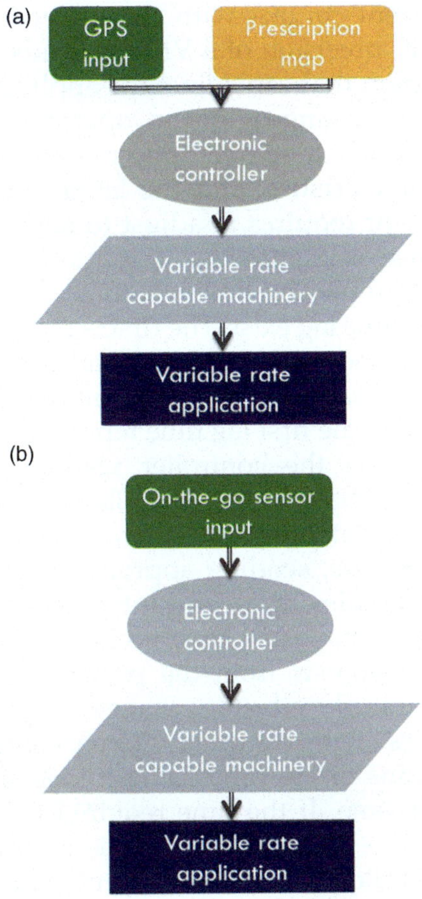

Figure 4.2. Flow charts and control methodologies for (a) map-based and (b) sensor-based strategies.

variable rate equipment (Figure 4.2). However, the ability of the variable rate equipment to solve the problem depends on the accuracy of the data and the associated algorithm. For example, Figure 4.3 shows that Planet's Normalized Difference Vegetation Index (NDVI) can be identified using remote sensing data collected by a space-based satellite and a UAV. Based on this information, precision prescriptions can be created and implemented using appropriate equipment and site-specific N-rate applications.

Variable Rate Applicator Components

In precision agriculture, the goal of variable rate application (VRA) is to apply the desired amount of a product at predetermined locations within a field. The agronomics involved in determining the desired prescription for a given field is important, but the performance of the equipment relies on the user's understanding of the technology. For example, when entering a heavier soil zone, the sprayer needs approximately 1–2 s to reach the higher gallons per acre (GPA or liters per hectare), and the look-ahead adjusts for this lag so that the correct rate lands in the right place. Knowing this can ensure the

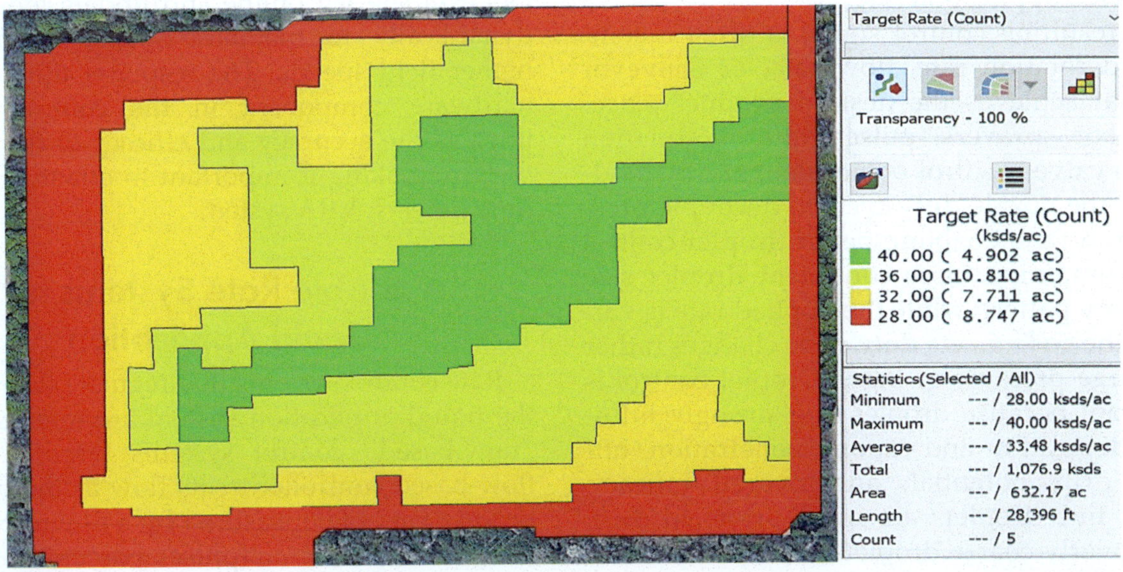

Figure 4.3. Preset map-based variable rate control strategies (prescription/map-based approach with management zones and assigned a target application rate for each zone). The machine uses GPS to apply the mapped rate as it moves through the field. (Courtesy of Ali Mirzakhani Nafchi).

correct rate is administered each time without error. It is important for agronomists, farmers, and applicators to ensure that the prescription map or sensor algorithm used is appropriate for the product and the application rates. In modern systems, electronic rate control systems are designed to automatically adjust instantaneous application rates in response to changes in machine and field operating parameters. This chapter provides a general overview of the electronic control systems used in variable rate technologies; however, for more detailed information, see Chapter 26.

Typically, machine operating parameters, such as ground speed, application width, and site-specific target application rates, are used to calculate the rate at each point within the field. This might mean that an entire spray boom, an entire dry spread, a spray boom broken into smaller sections, a group of planter row units driven by a common drive, an individual planter row unit, or an individual nozzle is adjusted in real time on the basis of the provided prescription map. This can vary from one machine to another depending on its capabilities. The main job of the electronic control unit (ECU) is to combine the prescription map or sensor algorithm rate with the machine's real-time kinematic (RTK) positioning system, Inertial Measurement Unit (IMU)–derived orientation, and speed to compute the required control signal that matches the desired application rate.

The resulting control signal is often a commanded rotation rate, flow rate, or conveyor rate that achieves the desired product placement. For sprayers, pulse-width modulation (PWM) valves control power delivery by modulating the nozzle duty cycle to decouple pressure from rate, enabling better droplet control and maintaining more consistent droplet size and spray pattern while the applied rate is varied on the go (e.g., via duty cycle changes rather than large pressure swings). Droplet control is important because droplet size strongly influences deposition and canopy penetration, off-target drift potential, and overall efficacy. Overly fine droplets are more prone to drift, while overly coarse droplets can reduce coverage and increase the likelihood of runoff or insufficient contact with the target. The accuracy and precision of a VRA depends on a control system that can adjust the application rates based on changes in the operator's driving style, the machine operating parameters, product characteristics, and the length of time the equipment requires to adjust to new directions (response or lag time).

For example, in a flow rate control routine, two types of time lag exist. The first is the control system response time, and the second is the required time for dynamic stabilization due to system configuration. The first lag time represents the interval between the controller reading a control command based on the Global Positioning System (GPS) vehicle location, reading the current flow rate, sending appropriate control signals, and achieving a stable intended system flow rate. The second lag time occurs as the vehicle system readjusts from the present state to the newly introduced step change and emerging dynamics. Rate controllers often contend with both forms of lag time using a look-ahead distance or time. If the time required to achieve a rate change is known or can be approximated, the controller can use the current vehicle speed, bearing, and location to initiate rate changes so that the commanded rate is changed before the vehicle reaches the transition zone while ensuring the resulting application placement is as desired or prescribed. Auto-calibrating controllers estimate lag online and update look-ahead dynamically to maintain spatial accuracy at higher field speeds. The response time of each hardware component in the control system impacts the accuracy and efficacy of the overall system, making it important to quantify the lag time for each VRA system.

Variable Rate Systems for Liquid Application

Rate control systems that regulate flow rate in the liquid application system are referred to as "flow-based" control systems (Figure 4.4). A flow-based controller treats flow as the primary process variable, measured by a flowmeter with a calibrated K-factor (pulses per unit volume), and drives a proportional valve (or PWM-actuated

Figure 4.4. A sensor-based, on-the-go approach (mounted sensors; e.g., crop or canopy sensors) with an in-cab controller estimate field conditions in real time and continuously adjust the application rates without a pre-made prescription map.

valve) using a closed-loop Proportional, Integral, and Derivative (PID) algorithm (proportional gain [Kp], integral gain [Ki], derivative gain [Kd]) to maintain the target set point. Current systems are typically ISOBUS (ISO 11783) compatible, allowing for a universal language between the equipment and the software. The systems are also integrated with Global Navigation Satellite Systems (GNSS; including RTK) for speed and position, enabling geo-aware rate control. Flow-based systems are commonly used because it is easier to manage system flow than pressure. This is because volumetric flow is approximately linear with the valve position after valve characterization, whereas pressure–flow relationships follow the orifice equation ($Q \propto \sqrt{\Delta p}$), complicating direct pressure control. To set up a flow-based system, the operator programs the flow meter and regulates the valve calibration numbers (VCNs) in the controller as instructed by the manufacturer. The flowmeter K-factor (or calibration number) maps pulses to volume, while the VCN defines the valve gain, deadband, and direction to ensure a stable loop response. Once the controller is programmed, the operator rarely needs to alter the recommended VCNs. The flow meter calibration number establishes the system flow rate, whereas the regulating valve VCN defines the expected response of the control valve motor during system flow rate management. The alternative to a flow-based control system is to control system pressure. This alternative is known as a "pressure-based" control system. Pressure-based systems regulate boom line pressure with feedback from pressure transducers; nozzle flow that varies with $\sqrt{\Delta P}$, so droplet size and rate are indirectly controlled and must be compensated for by speed changes. See Chapter 26 for a discussion of both pressure- and flow-based control of liquid application (Franzen, 2018).

The rate controller for a liquid application system needs to compute the flow rate as the amount of product per unit of time (in gallons or liters per minute) to issue the proper command signal. Controllers typically compute a per-nozzle set point from the target application rate and vehicle motion data, then aggregate these into section or boom commands while enforcing valve and pump constraints. For each nozzle, the controller calculates the desired flow based on the desired application rate (in gallons per acre or liters per hectare), the current vehicle speed (in miles or kilometers per hour), and the application width (in feet or meters) of the equipment. A common nozzle flow (in gallons per minute that is converted to liter per minute by multiplying by 3.785) shows the computation for nozzle flow rate performed by the controller:

$$\frac{\text{flow}}{\text{nozzle}}\left(\frac{\text{gal}}{\text{min}}\right)=$$
$$\left(\text{Desired Rate}\left(\frac{\text{gal}}{\text{acre}}\right)\times\text{Speed}\left(\frac{\text{mile}}{\text{hour}}\right)\times\text{Applicator Width}\,(\text{ft})\right)\times\left(\frac{1\,\text{hr}}{60\,\text{min}}\times\frac{5{,}280\,\text{ft}}{\text{mile}}\times\frac{\text{acre}}{43{,}560\,\text{ft}^2}\right) \quad (4.1)$$

To determine the total flow rate (Equation 4.2) from the tank, multiply the flow rate per nozzle from Equation 4.1 by the number of nozzles at this rate. This universal equation is useful when the flow rate needs to be calculated for individual nozzles, a group of nozzles on a boom section, or across an entire boom, as well as for PWM duty cycle limits, and pump capacity.

$$\text{Total flow rate}(\text{GPM}) = \text{Per}-\text{nozzle GPM} \times \text{Number of active nozzles} \quad (4.2)$$

Automatic Section Control Technology

Many variable rate control systems employ automatic section control (ASC) technology. Automatic section control operates as a supervisory layer that computes ON/OFF or PWM duty cycle commands with look-ahead distance ($L = V \times T$ to compensate for vehicle speed, V, and actuation latency, T). Automatic section control turns the entire boom (Figure 4.5), individual nozzles, or control sections ON or OFF as needed during field operations. Turn compensation adjusts the duty cycle based on the local boom tip velocity to maintain the target GPA on curves. Most commonly, ASC is used to control arrays of application points (e.g., row units, nozzles), with the entire array divided into smaller subsections that are controlled independently. In sprayers, ASC is achieved using shutoff valves on boom sections or solenoids on nozzle bodies. For dry fertilizer spreaders, proportional PWM valves are used, whereas planters use row-unit motors or clutches on the main shaft. The controller automatically disables the control section(s) in areas previously applied with crop input (seed, fertilizer, or chemical) or in no-application areas, based on coverage maps stored in the controller.

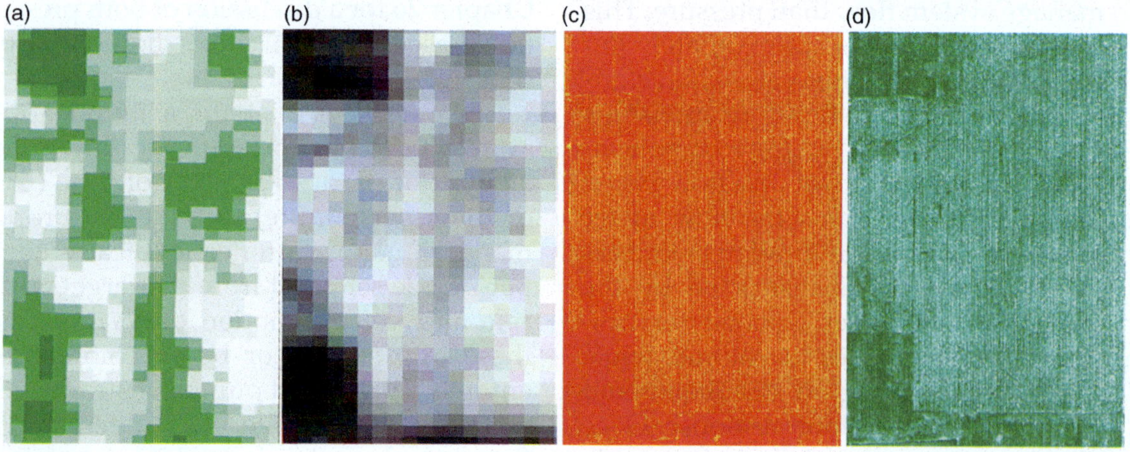

Figure 4.5. Comparison of field-scale NDVI information derived from remote sensing at different platforms. From left to right: (a) satellite-derived NDVI; (b) satellite image used for NDVI calculation; (c) UAV-derived NDVI; (d) UAV NDVI/vigor map highlighting within-field variability for site-specific N-rate prescriptions. (Courtesy of Ali Mirzakhani Nafchi).

Blockchain-anchored as-applied logs and explainable artificial intelligence (AI) summaries provide end-to-end traceability. Automatic section control provide cost savings and improve environmental stewardship by reducing overapplication and unwanted application in buffer strips or waterways. Although many section control units allow the operator to manually shut off individual sections, an ASC rate controller can control sections more effectively than many human operators. This allows for minimizing overlaps and missed areas while maximizing operator and machine productivity. Autonomy-ready controllers negotiate tasks and dynamically adapt section granularity for optimal throughput.

Rate controllers equipped with ASC track the location of each section with respect to field boundaries, keep-out zones, and previously applied areas as the application equipment moves through the field. The controller automatically manages each section's ON and OFF states and controls the flow rate for each section to match the rate defined in the prescription map. The controller also stores the ON/OFF state and application rate for each section to generate a map of the actual application in the field. This is known as the "as-applied" map and may deviate from the prescription map when a boom section partially overlaps two different management zones. The system application rate is managed through a closed-loop feedback control scheme (classical PID with feed-forward from ground speed) as described in Chapter 26.

Automatic Section Control on Sprayers

The product flow to the boom is controlled using a boom shutoff valve (Figures 4.4 and 4.5). These valves allow the operator to turn the sprayer ON and OFF when turning equipment around at the end of the field, an area often referred to as the headlands. Most agricultural sprayers have booms split into smaller sections, allowing independent ON and OFF control of these sections across the boom. This setup allows the operator to manually turn OFF a portion of the boom, rather than the entire boom, using a switch box located in the tractor cab. Partial boom width may be required at times during field operations. Therefore, turning sections OFF can reduce overapplication of the product. In the United States, two- and three-way boom shutoff valves are popular on agricultural sprayers. Pulse-width modulated nozzle valves are also used to decouple flow from pressure, enabling droplet size control while maintaining the appropriate rate.

The control system commonly uses a GPS or GNSS to provide speed measurements. The system uses an inline flow meter to measure the flow rate (in gallons or liters per minute), providing feedback to the rate controller, which adjusts

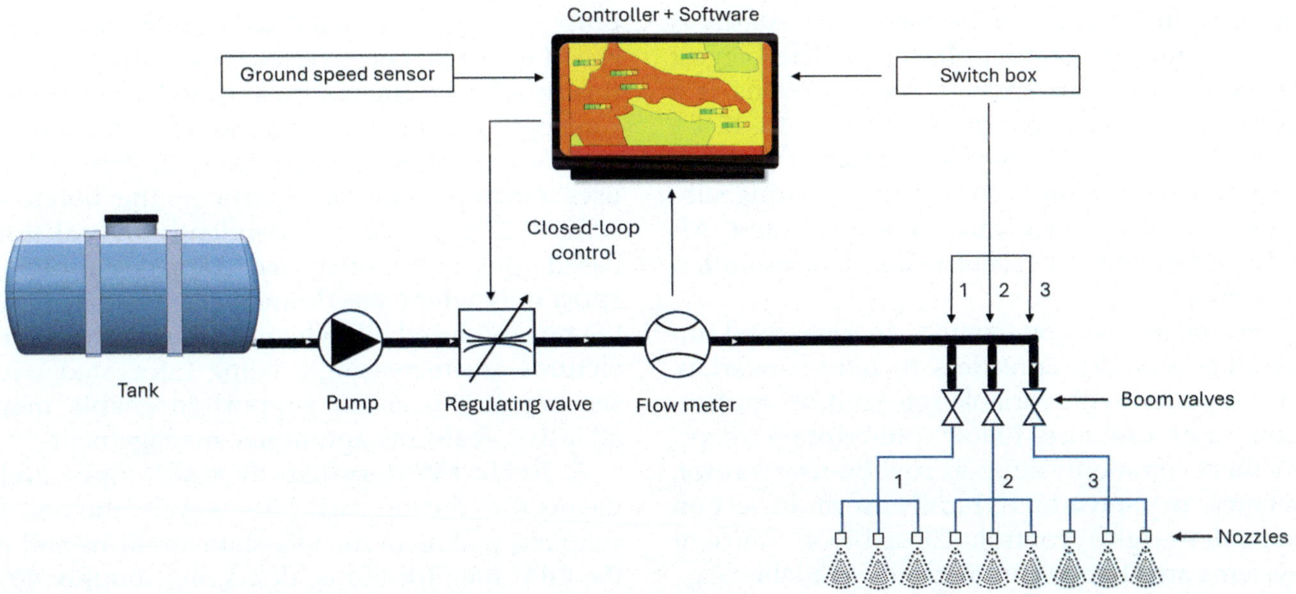

Figure 4.6. Schematic diagram of a typical microprocessor-based rate control system.

the regulating valve to achieve the target system flow. The target system flow is a function of ground speed (in miles or kilometers per hour), application width (in feet or meters), and rate (in gallons per acre or liters per hectare) set in the controller by the operator. Further, a control system uses feedback from the system flow meter and sends control commands to the regulating valve to adjust the system flow. The two-way boom valve is simply an ON/OFF boom valve that either allows product flow (ON state) or no-flow (OFF state) to the boom section (Figure 4.6). When the two-way boom valve is shut off, the product volume intended for that boom section temporarily remains in the system as the control system regulates flow to the new target value. A sprayer setup with two-way boom shutoff valves use flow feedback and a regulating valve to adjust flow and spray width.

The benefits of section control include reductions in areas where chemicals are under- and overapplied. Underapplication of chemicals may not be sufficient to effectively control pests (Hoehne & Brumett, 1982), resulting in potential yield loss and the development of pest resistance. Overapplication of chemicals increases input costs and the risk of environmental damage (Grisso et al., 1989; Miller & Smith, 1992). Automatic section control can potentially save 15.2 to 17.5% of nutrients and pesticide applied to each field (Luck et al., 2010). An additional benefit is a reduction in the frequency of loading inputs into the machine because less material is wasted during spray applications. The biggest benefit of ASC for sprayer boom sections is a decrease in overapplication from 6.2 to 12.4% (Luck et al., 2010), which ultimately reduces costs. Further savings can be achieved using sub-meter accuracy, fast valve response, and AI-assisted boundary detection using high-resolution base maps.

Automatic section control is also used on planters and dry spreaders to reduce overlap. When paired with variable rate control, application rates can also follow prescription maps. Planters commonly support row-by-row control, whereas dry spreaders typically use multi-section (e.g., left/right) control. Closed-loop nutrient systems are already commercially available (e.g., flow-based feedback for liquid systems and scale or weigh-bar feedback for dry spreaders). These control architectures are being integrated with sensing and decision support to enable more adaptive, real-time agronomic management.

Pulse-Width Modulation Technology

A PWM system uses solenoid valves mounted on individual nozzle bodies (normally closed, two way valves driven by a solid-state controller) to provide automatic proportional flow compensation based on the sprayer's speed and desired application rate via closed-loop rate control on a CAN/ISOBUS network. The solenoid valve is an electromagnet that opens and closes the nozzle flow via a plunger held in the closed position by a spring, exhibiting millisecond-scale actuation and release times under rated pressure. Pulse-width modulation is covered in general in Chapter 26. Figure 4.6 shows the nozzles ON and OFF states at two different duty cycles, with a greater duty cycle resulting in a higher average volumetric flow. In practice, the duty cycles are fast enough that the spray deposition is spread evenly, as if the lower duty cycle setting were simply applied by a nozzle with a lower flow rate at a given pressure (pattern pulsation is minimized by phase-offsetting adjacent nozzles and maintaining constant manifold pressure). Commercially available PWM systems operate at frequencies between 10 and 30 Hz, which means the nozzles are turning ON and OFF 10–30 times per second. Modern systems also incorporate individual nozzle turn compensation and dual-pulse strategies to stabilize droplet size across speed changes. For a system operating at 10 Hz, this provides a 100-ms cycle time to operate the solenoids mounted on each nozzle. An example is the John Deere See and Spray system, which uses nozzle-by-nozzle control in the boom to actuate each nozzle independently in real time based on camera-detected targets, applying spray only where weeds are present and keeping the nozzle closed elsewhere. These control architectures are increasingly being integrated with sensing and decision support to enable more adaptive, real-time agronomic management.

A 10-Hz PWM system at a 40% duty cycle means that during each 100-ms cycle, the nozzle solenoid will be in the ON state for 40 ms and in the OFF state for 60 ms, delivering roughly 40% of the full-open flow at the same pressure differential (ΔP). The same system at an 80% duty cycle means the nozzle will be in the ON state for 80 ms (Figure 4.6). Therefore, an 80% duty cycle would release twice as much product in 100 ms compared with a 40% duty cycle.

Pulse-width modulated systems maintain a constant desired application pressure irrespective of the number of nozzles in the ON or OFF states during different duty cycles (Mangus et al., 2016) by decoupling pressure regulation [pump/relief or pump Variable Frequency drive(VFD)] from metering (valve duty cycle; Sharda et al., 2010, 2011, 2013). Through PWM systems, liquid application at a constant pressure helps maintain uniform nozzle-to-nozzle overlap and droplet size. Nozzles for PWM systems on a boom are configured so that adjacent nozzles are often phase-shifted rather than strictly alternating ON and OFF, reducing pattern pulsing. At higher duty cycles, multiple nozzles may be ON simultaneously.

Pulse-width modulated nozzle technology enables very fine-grained as-applied maps, with application rates for each nozzle being recorded and logged individually. Another important advantage of the PWM system is its ability to instantly turn individual nozzles ON or OFF. This is possible because the valve at each nozzle actuates quickly, holding the product in place when the pressure is closed. As a result, when the solenoid valve is energized, the nozzle applies the product at the desired pressure and then quickly closes. This eliminates or greatly reduces product drainage when the valve is in the OFF state.

Pulse-width modulated systems often use flow control and feedback components to regulate the application rate. The change in flow rate is required to maintain the target application pressure and to provide sufficient product in the plumbing system to complement the changing duty cycle. End users should consult equipment manufacturers and carefully program the flow control valves mounted on the pumps to achieve the desired flow control response.

A Universal Terminal or sprayer control display provides a user interface to import prescription maps, monitor spray coverage, receive signals from the flow meter, send signals to the PWM system, implement control commands to the flow control valve, and record as-applied data. Bennur and Taylor (2010) evaluated commercially available rate controllers using PWM with fixed-orifice nozzles and fast-close section valves with variable-orifice nozzles. Under simulated ground speed changes and map-based variable rate technology inputs, step–response times ranged from 0.5 to 2.1 s. They recommended establishing configuration-specific response times to optimize real-time, sensor-based VRA. Separately, Yang (2001) reported that side dress VRA stabilized within 1–2 s, with mean application rate errors of 2.5–5.8% across years and fertilizers.

Variable Seeding Rate Systems

Like variable rate liquid application systems, variable rate seeding can be implemented with varying levels of granularity, ranging from whole-planter to individual row-unit control. Before hydraulic drives were available, most planter seed meters maintained the target seeding rate using a ground drive that engaged with the soil and turned at a speed relative to the planter's ground speed. Seeding rates were adjusted from field to field or crop to crop by changing the seed disc (which alters the number of seeds per revolution) and/or by changing the gear ratio between the driving and driven shafts. On mechanical ground-drive planters, one (or a few) ground-contact wheels drive a common shaft for the row units, so the meters do not adjust independently, resulting in higher populations on inside rows and lower populations on outside rows. True curve compensation can be achieved with individual mechanical, hydraulic, or electric row drives that set the meter rpm for each row independently. The primary drawbacks of a mechanical ground drive are its many moving parts that require regular replacement and the need for manual seeding rate adjustments.

With the introduction of hydraulic drive seed meters and, later, electric drive meters, the seeding rate for different sections or individual row units on planters can be controlled precisely (Nafchi, 2025). Initially, most growers were still interested in using a constant seeding rate across an entire field or farm for a given crop. Thus, maintaining the proper seed meter rotational speed at varying ground speeds requires adjusting the meter speed according to the measured ground speed. Equation 4.3 can be used to calculate the required seed meter rotational speed for a given seed population prescription, ground speed, and seed meter plate. This equation was originally used by controllers to maintain the proper meter speed for a fixed population while accounting for varying ground speed. Today, this same equation is commonly used to calculate the appropriate meter speed for varying ground speeds and populations, including swath control and curve compensation, in which inner and outer rows receive different rpm set points:

$$\text{speed}\left(\frac{\text{revolution}}{\text{min}}\right)=\left(\frac{\text{Population}\left(\frac{\text{Seeds}}{\text{acre}}\right)\times\text{Speed}\left(\frac{\text{mile}}{\text{hr}}\right)\times\text{Row Spacing}(\text{ft})}{\text{Seed Slots}\left(\frac{\text{seeds}}{\text{revolution}}\right)}\right)\times\left(\frac{\text{hour}}{60\,\text{min}}\times\frac{5{,}280\,\text{ft}}{\text{mile}}\right) \tag{4.3}$$

The earliest hydraulic drive planters used a single hydraulic motor with fixed-displacement or load-sensing pumps, PWM-modulated proportional valves, and mechanical chain or sprocket reduction across the toolbar to drive the entire planter. Newer systems have incorporated multiple hydraulic motors to drive sections of the planter, similar to boom-section control valves in liquid application systems. While it is technically possible to drive each row unit on a planter with an individual hydraulic motor, the cost of proportional electrohydraulic valves required for each row unit and the higher service burden compared with electric architecture can make this approach prohibitively expensive. A low-cost option to enable individual row shutoff is to use a clutch system on each row unit that can disengage the drive shaft, reducing or eliminating planting overlap. However, it does not offer variable rate capabilities and is now largely considered a legacy approach on premium planters. Most hydraulic systems are unsuited for planting contoured or terraced fields because each hydraulic drive powering a planter section must operate at the same speed, whereas row units at the ends of the toolbar travel at different speeds during turns (modern section controllers mitigate this limitation but do not eliminate it). Thus, Equation 4.3 can be solved simultaneously for each row unit on a planter, enabling turn speed compensation at the individual row level, making it an option on many planters. Many retrofit packages are commercially available to provide hydraulic planter rate control, enabling older planters to perform variable rate planting with section control. A major drawback of hydraulic drives when compared with electric drives is they can have greater maintenance needs.

In addition, electric drives are generally less expensive than hydraulic drives; an example of an electric drive seed meter is shown in Figure 4.7. In these systems, the seed plate is either mounted directly to the electronic motor shaft or is driven by internal gear teeth around the perimeter of the seed plate that mesh with the electric motor. A row control module (RCM) maintains the desired seed plate speed to match the target seeding rate by using real-time ground speed data for each individual row unit and seed-tube sensors to monitor singulation and spacing quality in real time. Because each row has its own seed meter drive, individual row control can be implemented for each row unit without significant additional complexity or tradeoff, allowing for higher travel speeds while maintaining placement quality. Contour farming and point-row shutoff functionalities can increase productivity by reducing the over-planted area of a field, especially in one with an irregular shape or terraces (Velandia et al., 2013).

Several commercially available variable rate seeding options exist, most of which use hydraulic or electric drives. Hydraulic drives operate at high rotational speeds with speed reduction at each seed meter, whereas electronic drives operate at much lower rotational speeds. For example, Horsch's Maestro planter (Figure 4.7a) features an electronic drive seed metering system with a seed plate directly driven by an electric motor and a planetary gearbox assembly (Horsch Maschinen GmbH) that includes integrated curve compensation. The Precision Planting vSet meter (Precision Planting; Figure 4.7b) is an electronically driven seed meter often paired with vDrive and automated downforce. Also on the market is John Deere's ExactEmerge Row Unit, which fea-

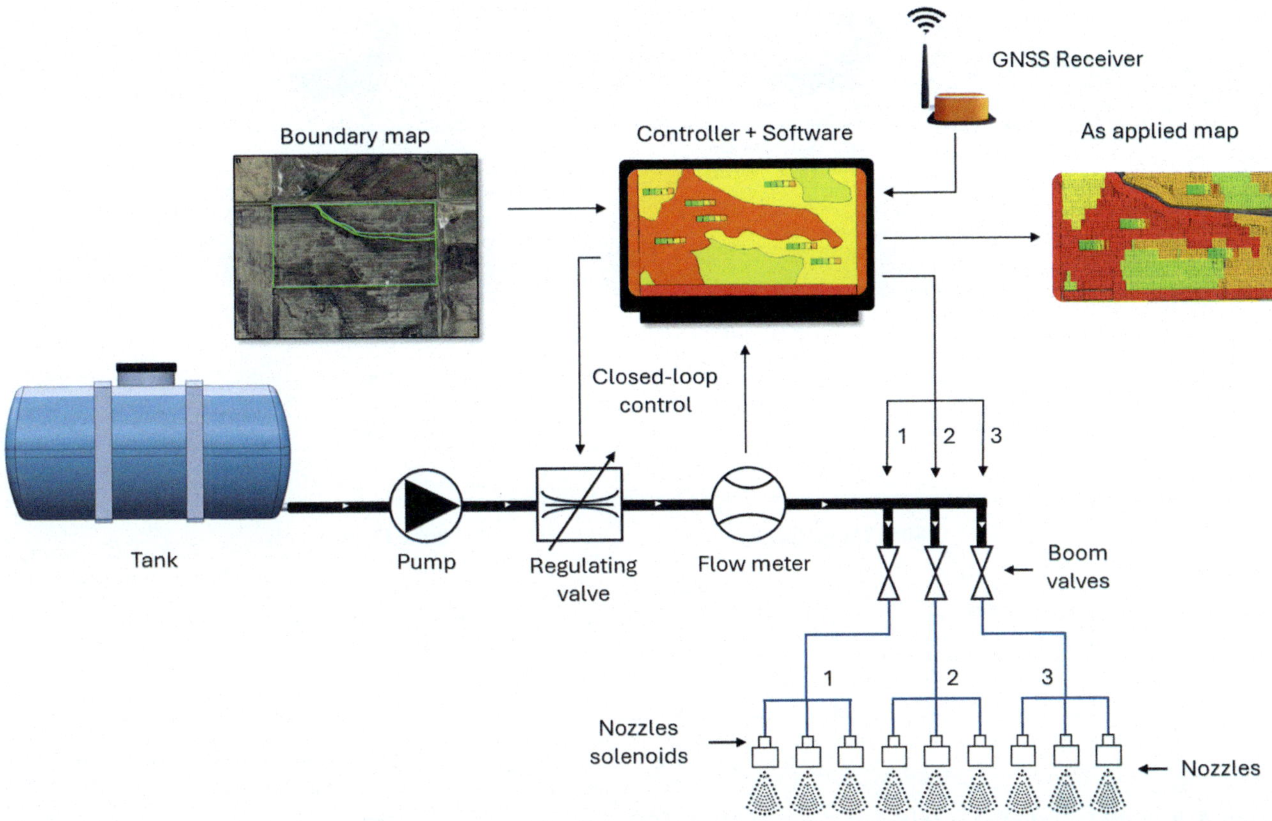

Figure 4.7. Overview of the typical components and setup for a sprayer equipped with ASC. (a) An electronic drive seed metering system with a seed plate directly driven by an electric motor and a planetary gearbox assembly, including integrated curve compensation; (b) An electronically driven seed meter often paired with vDrive and automated downforce.

tures an electronically driven seed meter and an electronically driven seed belt–brush delivery system (Deere & Company). In contrast to the gearmotor approach used by previous manufacturers, Kinze Manufacturing's seed meters use stepper motors, eliminating the need for gear reduction and allowing finer control of rotational angle (Kinze Manufacturing). Other seed meter planting systems are available from many leading equipment manufacturers and will likely converge on open, secure ISOBUS/AgIO data layers to enable cross-brand interoperability.

Multi-Hybrid/Cultivar Technology

In the multi-cultivar planter, two or more cultivars can be planted in a single field at variable seeding rates. The cultivars are stored in separate holding boxes and metered by either a single or multiple electric motor metering units into a single seed tube (Figure 4.8). During planting, the RCM receives feedback on ground speed, the desired cultivar for the location, and the target seeding rate. The RCM uses this information to achieve the desired seeding rate. When switching from one hybrid to another, the planter ECU synchronizes turning one meter OFF and another ON, or switches between the two seed hoppers, ensuring there are no gaps left in the row. It is important to place the correct cultivar in the appropriate seed tanks so that target cultivars are planted in the intended area.

Variable Rate Systems for Dry Fertilizer Application

As with liquid application systems, accurate dry fertilizer metering and rate control are critical for applying the appropriate fertilizer rate. The operator typically sets the application rate, swath width, product density, and gate area, whereas a controller maintains the product flow rate during field application. The product flow rate is controlled by monitoring the conveyor belt or chain speed using a rotary encoder that reads the

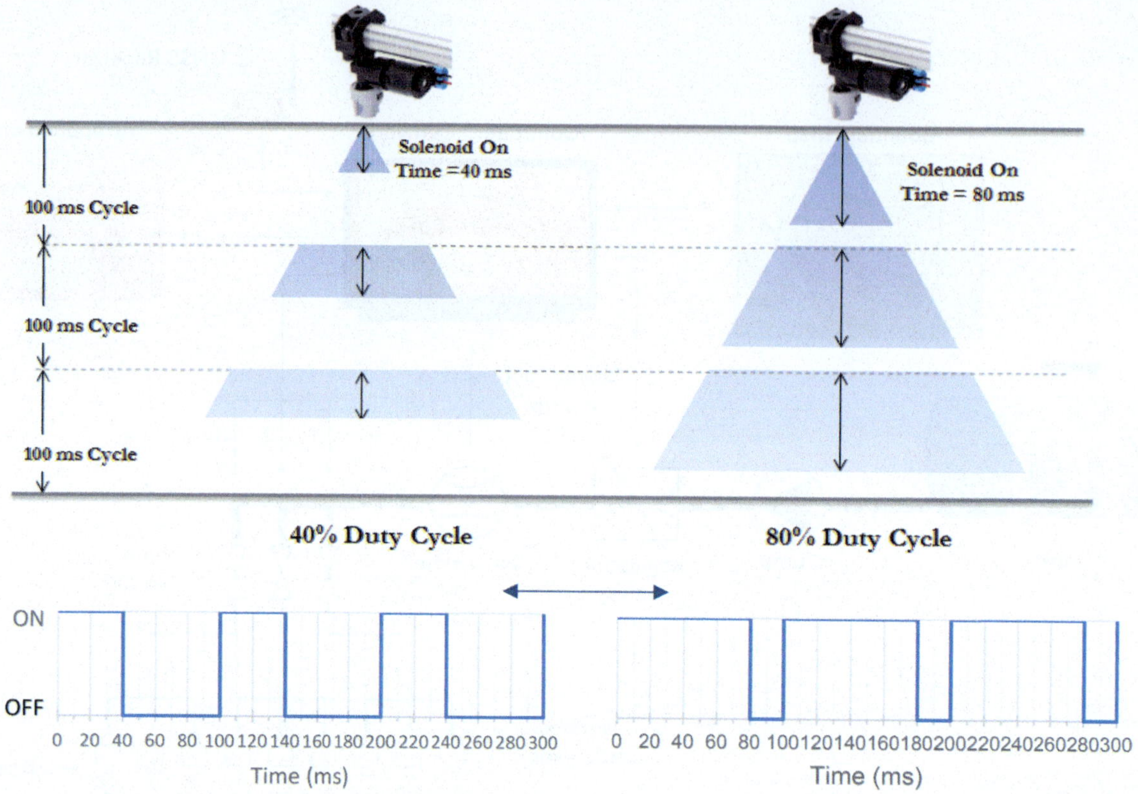

Figure 4.8. PWM control of a solenoid valve at 10 Hz (100-ms period). Left: 40% duty → 40-ms ON per cycle; Right: 80% duty → 80-ms ON per cycle. Bottom traces show the gate signal; upper graphics illustrate the resulting ON time per 100-ms cycle.

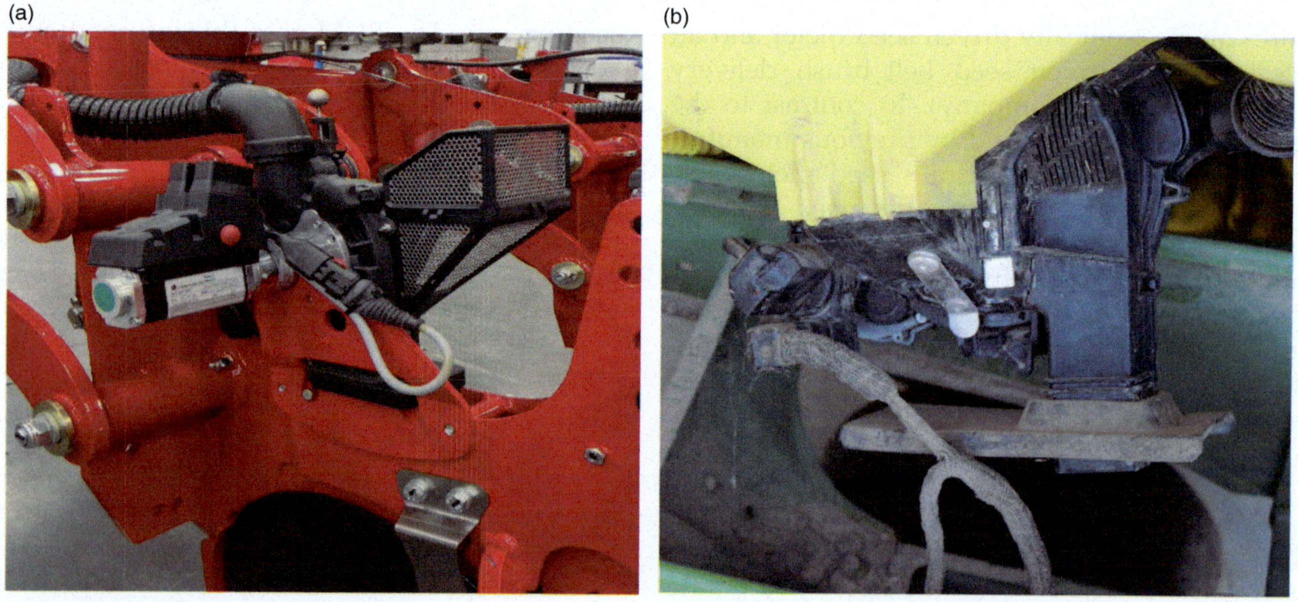

Figure 4.9. Examples of electronic drive seed metering systems: (a) Horsch Maestro electric seed plate drive with planetary gearbox and curve compensation; (b) Precision Planting vSet electronically driven seed meter.

rotational shaft's position and generates an output signal. The conveyor speed must be calibrated against material flow to ensure accurate control of the product flow rate. The encoder provides real-time conveyor speed feedback to the ECU, which actuates a proportional PWM valve to adjust the conveyor speed and achieve the target application rate (Figure 4.9). For hydraulic drives, valve duty cycle maps (percentage of PWM [PWM%]) are correlated to conveyor rpm through calibration curves (in kilograms per minute vs. rpm), and the controller updates set points using ground speed input (radar or GNSS). In pneumatic systems, the metered material enters a dedicated tube where a fanned airflow carries it to the respective outlets or nozzles. In spinner spreaders, a flow divider distributes the product that falls onto the spinning discs to ensure uniform spatial distribution. Bulk density is a critical parameter typically entered into the rate controller or field computer. Although bulk density is often considered constant during operation, it should be measured for each load and the controller settings updated accordingly. Optional mass-flow sensing or load-cell (weigh) feedback can further close the loop by continuously correcting the controller's effective calibration factor—often represented as an effective bulk density term—in real time. Pneumatic spreaders, used for row-crop fertilization and, in some cases, granular herbicide applications, commonly provide independent left/right metering or drive control for the corresponding boom sections in dry product applications. Modern spinner-disc spreaders can adjust the effective swath by changing the disc's rotational speed (rpm) and by using pattern deflectors or limiters to narrow the spread or restrict edge throw. Equation 4.3 is typically used by the rate controller to calculate and update the conveyor speed for dry material applications. Note that the equation depends on target rate and ground speed, which may vary due to operator actions or prescription variability in the field.

$$\text{Belt Speed}\left(\frac{\text{ft}}{\text{min}}\right)=\left(\frac{\text{Target Rate}\left(\frac{\text{lb}}{\text{acre}}\right)\times\text{Speed}\left(\frac{\text{mile}}{\text{hr}}\right)\times\text{Width}(\text{ft})}{\text{Bulk Density}\left(\frac{\text{lb}}{\text{ft}^2}\right)\times\text{Coverage Area}\left(\text{ft}^2\right)}\right)$$

$$\times\left(\left(\frac{5{,}280\,\text{ft}}{\text{mile}}\right)\times\left(\frac{\text{acre}}{43{,}560\,\text{ft}^2}\right)\times\left(\frac{\text{hr}}{60\,\text{min}}\right)\right) \tag{4.4}$$

Variable Rate Irrigation

Variable rate irrigation (VRI) is a technology that adjusts water application across a field to align with variability in soil, crop, or environmental conditions (Figures 4.13, 4.14). This technology enables the application of varying depths of irrigation water across crop fields. In control systems terms, VRI treats the field as a set of georeferenced "control elements" and modulates the application rate as a function of position (x, y) and time (t). Additional information on the agronomics of irrigation scheduling is available in Chapters 15 and 16. The primary goal of these systems is to increase water-use efficiency (I_{wue}) by preventing over- or underwatering of the crop throughout the growing season. Different VRI options exist, typically depending on the manufacturer, but systems are generally grouped into two categories: sector control or zone control.

Sector control VRI allows the pivot to change speed as it rotates during irrigation application and is often referred to as speed control. Speed modulation adjusts the instantaneous application rate (in millimeters per hour) through the angular velocity, ω, of the pivot, producing annular "slices" where depth is proportional to dwell time. Different systems may operate with varying increments of control zones ranging from 1° to 10° of pivot rotation over the entire pivot rotation range (90° for corner systems, up to 360° for center pivot systems). The result is a center pivot field with pie-shaped slices representing different watering zones (Figure 4.10).

Zone control VRI systems use valves to independently control water flow to individual sprinklers or zones for precise irrigation. These technologies increase the capacity to vary irrigation depths along the pivot lateral by controlling

Figure 4.10. An example multi-hybrid planter row unit with row-level electronic control and pneumatic seed delivery components that support switching between hybrids.

the output of the sprinklers themselves. The pivot's longitudinal axis is partitioned into hydraulic or electrical sections with independent actuation. Zone VRI systems are generally subdivided into two categories, where sprinklers can be controlled in banks of multiple sprinklers (Figure 4.11a) or by an individual sprinkler (Figure 4.11b), allowing for more precise irrigation management zone development across the field. In general, sprinkler flow is controlled by PWM solenoid valves. Pulse-width modulated control typically operates at a fixed carrier frequency while varying duty cycle to achieve proportional discharge; optional flowmeters enable closed-loop control by correcting for pressure variations along the lateral. Pulsing the valves ON and OFF at a set frequency and duty cycle enables proportional control of water.

Successful implementation of a VRI system will likely require a prescription map and continuous management of the pivot throughout the irrigation season to meet crop water needs. Technological advancements in hardware, software, and communication systems provide the tools needed to successfully manage and apply prescriptions to irrigated fields. However, a major limitation lies in assessing geospatial datasets and developing prescription maps to address the numerous factors that affect yield variability and plant available water.

Unmanned Aerial Vehicles for Implementing Precision Agriculture

Unmanned aerial vehicles provide rapid, high-resolution, and cost-effective spatial and decision-making data for VRAs. Beyond remote sensing, UAVs are increasingly used in direct application tasks, such as spraying and spreading, and also play a role in logistical support for autonomous agricultural systems (Behjati et al., 2024; Schutte et al., 2025).

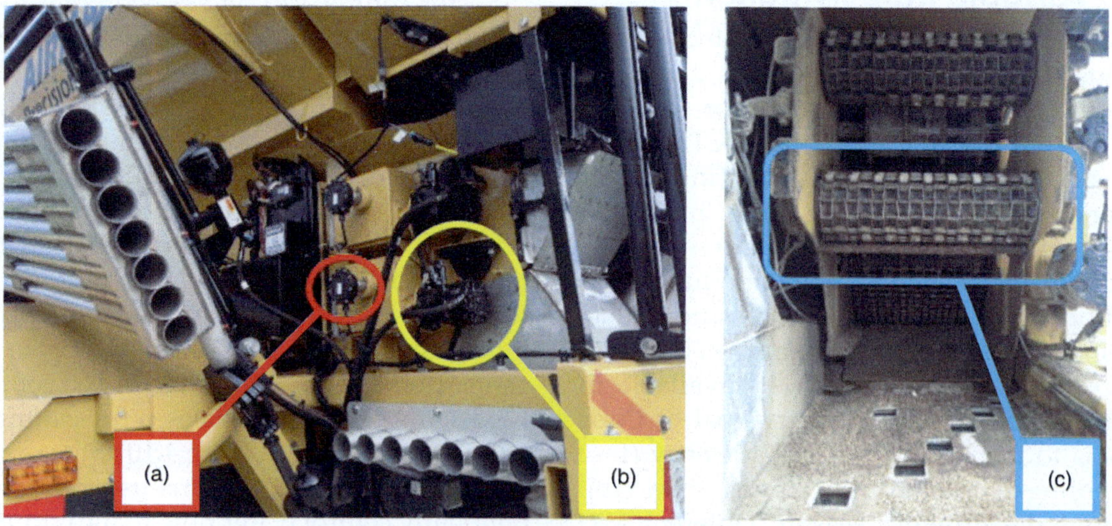

Figure 4.11. Metering control mechanism with (a) encoder (within red circle) and (b) proportional PWM (within yellow circle) valve (left), and (c) apron chain (within blue circle) for metering product for a pneumatic spreader (right).

Some UAVs can apply variable rates of cover crop seeds, pesticides, herbicides, and fertilizers depending on within-field variability. For example, seeding cover crops into standing cash crops reduces logistical challenges. It enables timely operations in conditions where existing crop size, soil compaction, or field accessibility pose concerns. However, limitations include restricted payload capacity, and flight duration can limit large-scale operations. Scouting UAVs can support logistics in variable rate management beyond direct application by collecting data on field variability, crop growth stages, logistical needs, and simulated operational bottlenecks to inform decision-making for operators and autonomous systems (Nafchi et al., 2024). Unmanned aerial vehicles can also deliver the spatial intelligence needed for robotic applicators to operate efficiently. Unmanned aerial vehicle–based prescription maps can guide robotic sprayers to treat only infected areas or direct autonomous spreaders to target specific nutrient-deficient zones. Below is an example of using N-Rich Biosensors and UAVs to improve nitrogen use efficiency (NUE). The Nitrogen-Rich Biosensor Spot (NRBS) method is a field-scale, geospatial decision-support workflow that places georeferenced, supra-optimal "N-rich spots" within management zones as in-field references for variable rate N application. At planting, each spot receives a small supplemental N dose; as the crop develops, multispectral satellite or UAS imagery (including red edge and near-infrared bands that enable NDVI and Normalized Difference Red Edge (NDRE), canopy cover, and chlorophyll proxies) quantifies differences in response between spots and surrounding crops (Figure 4.12). A positive change in index or change in biomass flags N deficiency in that zone, while no change indicates non-limiting N. Algorithms combine these contrasts with soil, yield, electrical conductivity (EC), topography, and weather data to generate zone-specific N-rate prescriptions, which are exported as controller-ready shapefiles. Using the crop itself as an in situ biosensor under real-time conditions, NRBS enhances NUE and profit while reducing environmental risk.

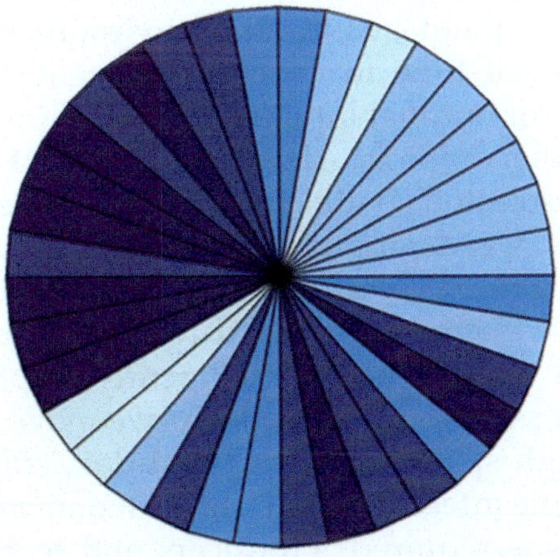

Figure 4.12. Potential sector (or speed) control irrigation application depth control areas (differing shades of blue) at 10° increments.

The planter can establish these proposed spots during planting using coated fertilizer to prevent leaching. These NRBSs can also be established

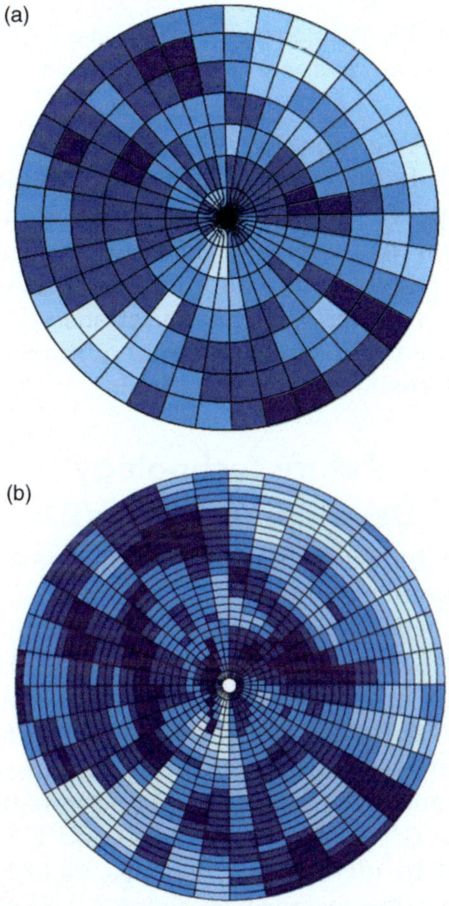

Figure 4.13. (a) Potential zone pivot irrigation application depth control areas (differing shades of blue) at 10° increments with banks or sections of sprinkler controlled along the pivot lateral; (b) potential zone pivot irrigation application depth control areas (differing shades of blue) at 10° increments with individual sprinkler control along the pivot lateral.

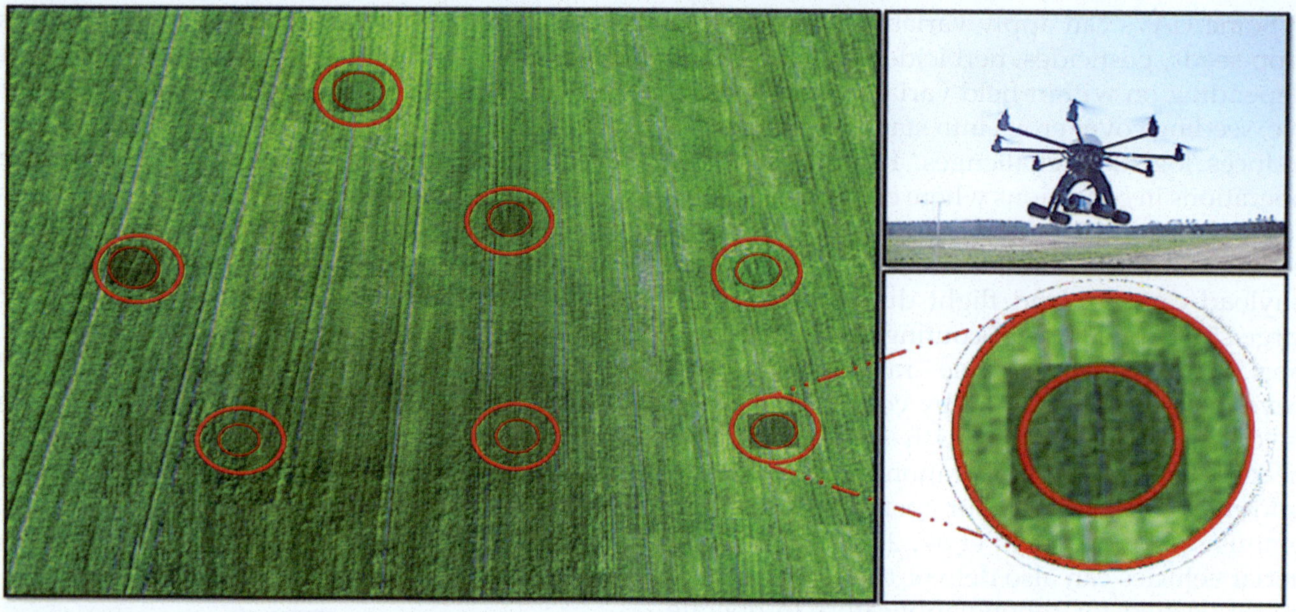

Figure 4.14. Establishing the NRBSs in a field and monitoring the plants in the NRBSs and around these spots to determine the growth rate and in-season plants' N response. (Courtesy of Ali Mirzakhani Nafchi).

with a UAV equipped with a spreader or sprayer to apply fertilizer (only for the N-rich bio reference spots). The NRBS method is a novel approach that assesses wheat and corn crops' N response to dynamic conditions, such as weather patterns and topographic variations. This innovative method has the potential to improve crop nutrient management nationwide.

Future Technology Developments for VRA

In the future, technological advancements will likely continue improving the precision and accuracy of VRA equipment, with machine learning and AI assisting in the decision-making process. For instance, John Deere's See and Spray uses advanced AI and edge computing technology to make real-time decisions for site-specific management. In addition, ground robotic technologies (Upadhyay et al., 2024) and UAVs will be used to identify and manage yield-limiting problems. For instance, Precision AI, a Canadian startup, is developing aerial spraying technology that can scan and spray weeds from the sky. As we advance toward very high spatial and temporal data collection, automatic integration of data with analytics systems is also required. Producers may benefit from locally based prediction models that allow managers to assess the potential advantages of various management strategies. Recent advances in technology have focused on increasing the size and operational speed of equipment, as well as improving control systems. Other work continues to automate or augment human operators using swarms of smaller machines instead of creating ever larger ones. Much of the focus has been on real-time visualization of machine performance; real-time feedback to operators for in-field adjustments to machine setup; improved sensing, feedback, and control; and intelligent control systems for varying field and operating scenarios. Future technologies are needed to assess soil health, soil moisture, plant health, and environmental factors in real time to enable intelligent machine control for optimal input applications.

Another rapidly developing aspect is the high-resolution spatial and temporal assessment of pests, weeds, input uptake, and crop growth parameters. Newer sensing, application, and control systems are needed for both aerial and ground platforms to intelligently apply crop inputs based on plant needs. As we move toward very high spatial and temporal data collection, automatic integration of data with data analytics systems is required. Producers and technology users would greatly benefit from global prediction models that allow producers to input local information and use global analytic models to develop input management strategies. Future agriculture will also require small, unmanned aerial and ground systems that can automatically scout, collect data on crops and fields, and develop application recommendations with optimal site-specific input application rates. Overall, there is a need for complete integration of information sources, analytics, and application technologies so producers can use them to maximize yields with optimal inputs and minimal environmental impact for sustainable agricultural production.

Future Picture of 2050 UAVs in Agriculture

By 2050, UAVs are expected to evolve from current supportive tools into UAVs functioning as integrated autonomous aerial platforms that work with tractors, ground robots, and decision support systems serving as collaborative robotic partners in agriculture (Nafchi et al., 2024). Unmanned aerial vehicle swarms enabled by AI and machine-to-machine communication (Tsolakis et al., 2022) will likely become a standard feature of farm operations capable of continuous monitoring, as well as on-demand human supervision and intervention.

Conclusion

As precision agriculture advances toward autonomous platforms that use machine learning algorithms to generate agronomic recommendations for input applications across the field, it can further enhance food production by shifting from field-average decisions to plant-level prescriptions based on high-resolution sampling. Aerial imagery can provide high temporal and spatial resolution data, which can be processed into site-specific recommendations and implemented using variable rate equipment. Automatic section control manages sections or individual nozzles, reduces overlap and unwanted application, and records an as-applied map.

ACKNOWLEDGMENTS

Support for this document was provided by the Precision Farming Systems community in the American Society of Agronomy, South Dakota State University, Kansas State University, USDA Natural Resources Conservation Service Conservation Innovation Grants (NR233A750011G019), USDA National Institute of Food and Agriculture under the Multi-State Hatch project (project number: KY006151-7008110-S1090), American Coalition for Ethanol, the South Dakota Corn Utilization Council, and the South Dakota Soybean Research and Promotion Council. The use of trade names is for the convenience of the reader and does not imply endorsement by the authors.

Study Questions

1. Explain how variable cultivar seeding is accomplished.
2. There are two types of lag time in a flow rate control routine. (1) What are they, and (2) how can variable rate applicators control for them?
3. How can automatic section control (ASC) in variable application systems improve efficiency?
4. How does pulse-width modulation (PWM) control spray application rates?
5. Why must constant pressure be maintained in a PWM system?
6. Why are flow-based control systems easier to manage than pressure-based control systems in liquid applications?
7. What are some differences between the electric drive seed meter and the hydraulic drive system for variable rate seeding? How can they help improve performance?
8. How does the controller in a hydraulic-driven dry fertilizer application system use conveyor speed calibration to maintain the target application rate?
9. What is the main difference between sector control and zone control when using a variable rate irrigation system?
10. How does having either too much or too little water influence yield, and how can variable rate application help with this?
11. What role can machine learning play in variable rate agriculture?
12. Why is response or lag time critical in variable rate application systems?

13. How does automatic section control (ASC) help reduce environmental impact?
14. What is the difference between flow- and pressure-based control in liquid application?
15. Why are electric drive seed meters better suited for contour farming than hydraulic or ground-drive systems?
16. How does pulse-width modulation (PWM) improve droplet size control in spraying?
17. What are the two main types of variable rate irrigation (VRI), and how do they differ?
18. How do UAVs enhance N management using the Nitrogen-Rich Biosensor Spot (NRBS) method?
19. What limitations do UAV-based direct application systems have?
20. What role are UAV swarms expected to play in agriculture by 2050?

REFERENCES

Behjati, M., Shamshiri, R. R., & Hameed, I. A. (2024). Agricultural mobile robots for plant health assessment and drought stress detection. *IntechOpen*. https://doi.org/10.5772/intechopen.115219.

Bennur, P. J., & Taylor, R. K. (2010). Evaluating the response time of a rate controller used with a sensor-based, variable rate application system. *Applied Engineering in Agriculture*, *26*(6), 1069–1075. https://doi.org/10.13031/2013.35903.

Clay, D. E., Kitchen, N. R., Byamukama, E., & Bruggeman, S. A. (2017a). Chapter 7: Calculations supporting management zones. In D. E. Clay, S. A. Clay, & S. A. Bruggeman (Eds.), *Practical mathematics and agronomy for precision farming* (pp. 123–135). ASA, CSSA, SSSA.

Clay, D.E., and T.P. Trooien. 2017. Chapter 8: Understanding soil water and yield variability in precision farming. Clay, D.E., S.A. Clay, and S. Bruggeman (eds), Practical Mathematics and Precision Farming. American Society of Agronomy.

Franzen, A. (2018). In D. K. Shannon, D. R. Clay, & N. R. Kitchen (Eds.), *Electronics and control systems*. Precision Agriculture Basics, ASA,CSSA.SSSAS.

Grisso, R. D., Dickey, E. C., & Schulze, L. D. (1989). The cost of misapplication of herbicides. *Applied Engineering in Agriculture*, *5*(3), 344–347.

Hansen, S., Clay, S. A., Clay, D. E., Carlson, C. G., Reicks, G., Jarachi, J., & Horvath, D. (2013). Landscape features impacts on soil available water, corn biomass, and gene expression during the late vegetative growth stage. *The Plant Genome*, *6*(2), 1–9. https://doi.org/10.3835/plantgenome2012.11.0029.

Kim, K.-I., Clay, D. E., Carlson, C. G., Clay, S. A., & Trooien, T. (2008). Do synergistic relationships between nitrogen and water influence the ability of corn to use nitrogen derived from fertilizer and soil? *Agronomy Journal*, *100*(3), 551–556. https://doi.org/10.2134/agronj2007.0064.

Luck, J. D., Pitla, S. K., Shearer, S. A., Mueller, T. G., Dillon, C. R., Fulton, J. P., & Higgins, S. F. (2010). Potential for pesticide and nutrient savings via map-based automatic boom section control of spray nozzles. *Computers and Electronics in Agriculture*, *70*(1), 19–26. https://doi.org/10.1016/j.compag.2009.08.003.

Miller, M. S., & Smith, D. B. (1992). A review of application error for sprayers. *Transactions of the American Society of Agricultural and Biological Engineers*, *35*(3), 787–791. https://doi.org/10.13031/2013.28663.

Mishra, U., Clay, D., Trooien, T., Dalsted, K., Malo, D., & Carlson, C. G. (2008). Assessing the value of using a remote sensing-based evapotranspiration map in site-specific management. *Journal of Plant Nutrition*, *31*(7), 1188–1202. https://doi.org/10.1080/01904160802134491.

Nafchi, A. M. *Multiple cover crop mixed-seed/mixed-type fertilizer precision planter*. U.S. Patent Application 18/776,858 filed January 23, 2025.

Nafchi, A. M., Abdalla, A., & Azad, B. *360 plant image capturing system and related methods*. U.S. Patent Application 18/528,161 filed June 6, 2024.

Schütte, T., Dworak, V., Shamshiri, R. R., & Weltzien, C. (2025). Cooperative lane mapping for autonomous mobile robots in horticulture by aerial photogrammetry. *Computers and Electronics in Agriculture*, *237*(Pt A), 110515. https://doi.org/10.1016/j.compag.2025.110515.

Sharda, A., Franzen, A., Clay, D. E., & Luck, J. D. (2018). Chapter 11: Precision variable equipment. In D. K. Shannon, D. E. Clay, & N. R. Kitchen (Eds.), *Precision agriculture basics* (pp. 155–168). ASA, CSSA, SSSA.

Sharda, A., Fulton, J. P., McDonald, T. P., & Brodbeck, C. P. (2011). Real-time nozzle flow uniformity when using automatic section control on agricultural sprayers. *Computers and Electronics in Agriculture, 79*(2), 169–179. https://doi.org/10.1016/j.compag.2011.09.006.

Sharda, A., Fulton, J. P., McDonald, T. P., Zech, W. C., Darr, M. J., & Brodbeck, C. J. (2010). Real-time pressure and flow dynamics due to boom section and individual nozzle control on agricultural sprayers. *Transactions of the American Society of Agricultural and Biological Engineers, 53*(5), 1363–1371. https://doi.org/10.13031/2013.34891.

Sharda, A., Luck, J. D., Fulton, J. P., McDonald, T. P., & Shearer, S. A. (2013). Field application uniformity and accuracy of two rate control systems with automatic section capabilities on agricultural sprayers. *Precision Agriculture, 14*(3), 307–322. https://doi.org/10.1007/s11119-012-9296-z.

Tsolakis, N., Bechtsis, D., Vasileiadis, G., Menexes, I., & Bochtis, D. D. (2022). Sustainability in the digital farming era: A cyber-physical analysis approach for drone applications in agriculture 4.0. In D. D. Bochtis, S. Pearson, M. Lampridi, V. Marinoudi, & P. M. Pardalos (Eds.), *Information and communication technologies for agriculture—theme IV: Actions* (pp. 29–53). Springer International Publishing.

Upadhyay, A., Zhang, Y., Koparan, C., Rai, N., Howatt, K., Bajwa, S., & Sun, X. (2024). Advances in ground robotic technologies for site-specific weed management in precision agriculture: A review. *Computers and Electronics in Agriculture, 225,* 109363. https://doi.org/10.1016/j.compag.2024.109363.

Velandia, M., Buschermohle, M., Larson, J. A., Thompson, N. M., & Jernigan, B. M. (2013). The economics of automatic section control technology for planters: A case study of Middle and West Tennessee farms. *Computers and Electronics in Agriculture, 95,* 1–10. https://doi.org/10.1016/j.compag.2013.03.006.

Yang, C. (2001). A variable rate applicator for controlling rates of two liquid fertilizers. *Applied Engineering in Agriculture, 17*(3), 409–417. https://doi.org/10.13031/2013.6203.

Identifying, Understanding, and Managing Variability

5

Sharon A. Clay, Bruce J. Erickson, and Newell R. Kitchen

Chapter Purpose

This is a revision of the chapter, Understanding and Identifying Variability, Kitchen and Clay (2018). It contains (a) discussion of the different sources of variability typically found in crop production systems, (b) descriptions and examples of both natural and anthropogenic (human-made) variation, and (c) discussions of how weather, soil, and management, alone, or in combination, create variability. Students will learn that by identifying, measuring, and understanding the causes of variability, positive outcomes from precision farming practices can improve field management profitability and sustainability.

Introduction

Mechanization and economies of scale often led to a one-size-fits-all, uniform approach to farm management. However, today, the availability of digital technologies provides the opportunity to switch from whole-field management to within-field management. The first step in this switch is to understand the root causes and the extent of within-field variability. If little variation exists, the varying management will have a minimal ability to increase efficiency and profitability in the system. However, if variability is high, there is great potential for improvement. This improvement is achieved by developing strategies and management decisions that account for and capitalize on the variability of the field.

Because the ability to improve efficiency and profitability depends on the amount of variability in a field, it is essential to recognize spatial and temporal variability in a location and understand how this variability impacts management decisions. For example, variability occurs with pests, yields, elevation, soil water, and nutrient availability, but it will also shape management decisions, including crop variety selection, crop rotations, land consolidation, and livestock confinement locations. These management choices can, in turn, alter the land itself, resulting in further variability over time. Understanding this relationship between variability and management decisions is key to precision farming. Managing uniformly in the presence of variability is illogical. However, keeping variability in mind as management decisions are made can reduce operational and input costs, improving productivity and sustainable environmental outcomes.

Types of Variability

Spatial Variability

Spatial variability refers to quantitative or qualitative differences among locations. Some spatial variations across the landscape are readily observable—note the aerial photograph in Figure 5.1 that shows differences in vegetation and soils. These differences could be attributed to crops planted, landscape position, soil organic matter content, soil moisture, and soil texture. All of these factors can impact productivity. However, sometimes variations are not obvious. For example, even in the flattest of landscapes, such as the Red River Valley fields of North Dakota, small variations (an inch or less) in topography can result in differences in soil parent material, drainage patterns, denitrification rates, salt accumulation, and other properties that greatly influence agronomic production decisions (Franzen, 2007).

Spatial variability can also result from past management (Figure 5.1). For example, a large field today may have been consolidated from two or more smaller fields that were under different management practices in the past. These smaller fields may have grown different crops, been under different tillage practices, or received different amounts of manure or fertilizer. However, the legacy effect of these differences may still be evident today, and for decades to come. Recognizing the differences in properties across the landscape may aid in the precision management of these subfield areas.

Spatial variation in single or multiple features can be represented on a map. For example, a topographical map shows the spatial variation in elevation, a quantitative feature. A field map showing different hybrids planted in different areas within a field is an example of a qualitative feature. The ease of measuring and mapping spatial variability increased exponentially when position coordinates could be determined quickly and inexpensively using Global Positioning System and Global Navigation Satellite Systems (GPS/GNSS) technologies (Chapter 7). When crops, pests, or soil parameters are measured and recorded in association with GPS coordinates, this is georeferencing (see Chapters 8 and 9).

Georeferenced data can be processed using geospatial analysis software programs, such as Geographic Information Systems (GIS), which are designed to analyze, manage, and create maps based on spatial relationships. These maps can then be used to make management decisions

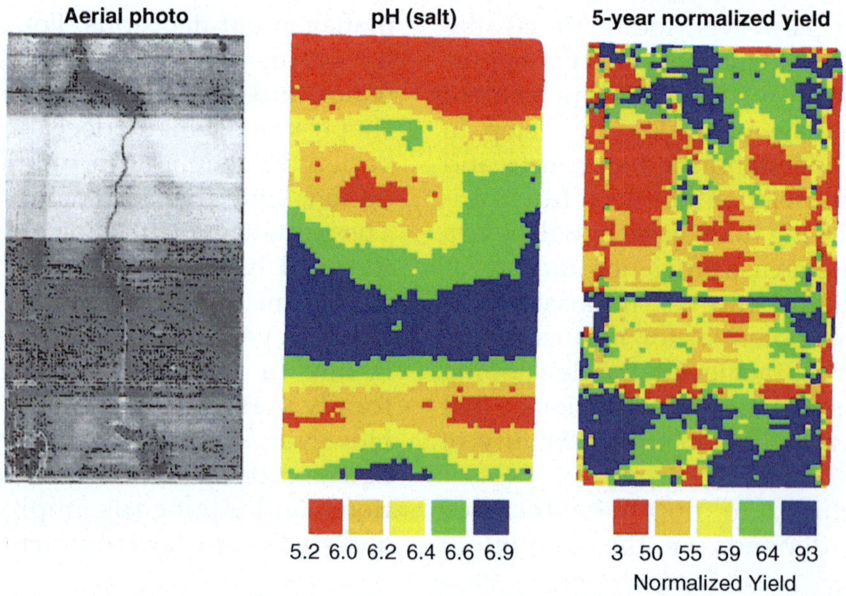

Figure 5.1. Aerial photograph, soil pH, and 5-year grain yield for a field located in central Missouri. (Courtesy USDA-ARS).

and, in some cases, loaded into farm machinery to carry out specific tasks (e.g., managing fertilizer input, pesticide rates, differential seeding rates) at specific locations.

Temporal Variability

Temporal variability refers to changes in the features observed over time. Changes may occur over a short time frame, days to weeks (e.g., plant growth, nitrification of applied fertilizer, pest infestations, soil water content, carbon sequestration, gully formation after a single large rainfall event). Additionally, temporal changes occur naturally over time, ranging from years to decades (e.g., soil test values, crops grown).

Temporal variability in agricultural production systems can be observed across many different biophysical objects and processes, including plants, animals, insects, microbes, soil, weather patterns, soil nutrients, pesticides, and more. Without an understanding of how these processes vary over time, it is not possible to anticipate future events. For example, soil test phosphorus and potassium values may be lower in the fall than in the spring, which could lead to overapplication if only fall values were considered for fertilizer recommendations. Another example is weed and insect pests that may not be detected one day but have severe infestations a few days later.

The rate of temporal change depends on the specific factor being tracked. For example, hourly measurements of humidity and temperature may be needed to examine disease or insect progression, whereas calculating growing degree day (GDD) accumulation (a temporal measure of heat units) from planting or from the start of the season within a single season can be used to monitor day-to-day plant growth, pest emergence, and predict disease outbreaks. Measurements recorded over many years for rainfall, temperature, snowpack, wind, and hail damage can also be used to identify optimal planting dates and select crop varieties.

Physical, Biological, and Management-Induced Variability

Physical factors of variability are non-living, abiotic influences that impact crop growth and productivity. These include weather, water, temperature, sunlight, wind speed, and soil properties (although soil also plays a substantial biological role that can be dramatically altered by management). Some factors may have a positive influence on crops, such as warm temperatures and optimal rainfall at the right times during the growing season. However, extreme weather conditions, such as freezing or extreme heat, scouring winds, and extremes in water (e.g., flooding, hail, snow, drought), can reduce yields. The occurrence of some abiotic stresses is predictable, such as freezing conditions in winter or drought in arid environments. Other abiotic stresses may be unpredictable, such as very heavy rainfall events (e.g., 100-year rainfalls) or extended months or years with very little rainfall (e.g., the Dust Bowl in the 1930s in the U.S. Great Plains). Managing unpredictable stress is much more difficult than managing predictable stress.

Biological factors that contribute to field variability are those caused by living organisms. Agricultural pests are unwanted organisms that reduce crop yields, crop quality, or are detrimental to the agricultural operation. These include pathogens (viruses, fungi, and bacteria), unwanted plants (weeds), insects, nematodes, and vertebrate animals (ground squirrels, mice, birds, deer, etc.).

On the other hand, beneficial biological factors that improve plant growth include ladybugs that eat aphids and scavenger beetles that feed on harmful insects. Soils are full of microbes that naturally vary in space and time, and many of these are critical and beneficial to plant growth, affecting nutrient availability and interactions with pests. These include nitrogen-fixing and denitrifying bacteria, fungi that aid phosphorus availability to plants, and compounds released by plants that may influence weed seed dormancy and germination. Because biotic factors are living organisms, they, along with the crop plant, are affected by many abiotic stresses, including temperature and water availability.

Management-induced variability refers to the human-derived practices that result in long-lasting changes to the landscape (Figure 5.1). An example is soil erosion, where tillage can increase susceptibility to erosion. Other examples are crop rotations and the application of fertilizers, manures, soil amendments, and irrigation water.

Drivers of Variability

Weather as Driver of Variability

Weather is defined as the atmospheric conditions at a specific location over a brief period, including temperature, precipitation, wind, cloud cover, and air pressure, among other factors. Weather can vary in time and space, and it is influenced by local topography. For instance, on a calm evening, a field with hills and valleys can experience noticeable temperature differences. Cool air may settle into valleys, and with it bring higher humidity, dampness, and dew, while the hillsides and hilltops may remain warmer and drier. Fields adjacent to large bodies of water, such as near the seashore or the Great Lakes, can experience the moderating effect of being close to those water bodies. For example, these fields may not be as cold in the winter, nor as hot in the summer, as fields farther inland. This buffering effect may make these landscapes better suited for growing high-value crops such as grapes, fruit trees, and berries.

Across a region, long-term average weather (i.e., climate) influences the temporal variability of crop production. For example, the length of the growing season (e.g., using 100-day corn hybrids in northern climates vs. 120-day corn hybrids in more southern climates). Additionally, planting dates and seeding rates will differ across latitudes, and daily weather conditions affect a wide range of operational decisions and risks each year. These decisions and risks may include

- The likelihood of being able to do field work with suitable soil conditions
- Timing and feasibility of chemical applications (e.g., fertilizers and pesticides)
- Fire risk during harvest operations
- Potential for storm damage (e.g., wind, hail; Figure 5.2)
- Soil erosion caused by wind and/or water (Figure 5.3).

Water as a Driver of Variability

The amount of water available to plants is seldom ideal. In many cases, crop fields can be too dry—usually due to a lack of rainfall, but also to low humidity, intense sunlight, wind, and soils not capable of holding water, such as sands. Excess water is also a common concern. This can be due to abundant rainfall, as well as to fields that lie so flat that excess surface water takes a long time to drain, evaporate, or both. Other factors include soil layers such as hardpans (compacted layers that limit root penetration and

Figure 5.2. Damage from wind and hail can vary greatly in space and time. The corn in the foreground has been affected by wind more than the corn farther back in the field. (Courtesy of Pioneer, https://www.pioneer.com/us/agronomy/root-lodging-in-corn.html)

Figure 5.3. United States 1930s Dust Bowl. (USDA-ARS).

USING WEATHER FOR MODELING VARIABILITY

Models have been developed that use weather parameters to predict the likelihood of crop stress and pest growth. For example, some aphid infestations in northern climates may result from southern aphids being carried northward by jet stream currents and deposited on northern host plant species (MacRae et al., 2011). If the jet stream currents are moving in a different direction, are too forceful, or are too cold, the aphids will not survive this transport mechanism. Therefore, the risk of aphid outbreaks in the north can be predicted by examining aphid populations in the southern states and the upper atmospheric wind currents and temperatures. Likewise, temperature alone, or combined with soil water conditions, can be used to predict weed emergence and growth rates (Forcella, 1998; Forcella et al., 2000). These predictions, in turn, can be used to schedule mechanical or chemical weed control applications.

water movement), dense till (glacially deposited soils), and fragipans (naturally occurring, brittle subsurface horizons) that slow soil drainage. These factors commonly vary across fields and among fields.

Field relief, also called topography, and soil characteristics can sometimes introduce water variability within and among fields. For example, a 2-in. (5-cm) rainfall event in a field with hills and valleys may result in differences in water infiltration. Upland areas may have less infiltration, perhaps an inch (2.5 cm) of water going into the soil due to runoff, compared with lower-elevation areas, where water may pool, and perhaps 3 in. (7.6 cm) could infiltrate. Similarly, surface and subsurface field drainage are seldom uniform across a field, and small changes in elevation can have a huge impact on soil water. Soils with small elevation changes and high clay content, such as the Red River Valley fields of North Dakota, commonly install tile to address the variability. In other areas, such as the Mississippi Delta region of the South, field leveling techniques combined with drainage ditches are common ways to help remove excess water. These management strategies create their own variation in the system.

Even with modern irrigation equipment and techniques, irrigation is often not evenly applied due to equipment problems or calibration issues. Older irrigation methods, such as furrow irrigation, lack control and precision. When irrigation is underapplied, crops can suffer. When irrigation is overapplied, excessive watering can reduce germination and increase the risk of disease or nutrient deficiencies. However, uniform applications are not ideal for many fields, as soil water-holding capacity can be highly variable within a field. In these situations, the variable application of irrigation water is warranted (see Chapters 15 and 16).

Soil as a Driver of Variability

Unless extreme weather events occur, soil texture (the relative amount of sand, silt, and clay) at a specific location is considered a permanent property. However, texture can vary widely across landscape positions, soil types, and soil depths. Soil texture influences multiple soil properties, including water-holding capacity, drainage, fertility, biological activity, and more.

Soil organic carbon (SOC) is not permanent at any given location. It is influenced by weather (soil temperature and moisture), as well as tillage intensity, crop rotation, sampling depth, and residue amount and type. Generally, removing biomass (such as silage or hay harvest or baling cornstalks) or increasing tillage intensity reduces SOC. SOC is important because it is related to water infiltration rate, aggregate stability, and plant-available water.

Soil is often considered one of the most complex and least-explored frontiers in agriculture. In particular, it is difficult to explore subsurface layers. The subsurface layers of the soil can be observed during earthwork for road or building construction, by digging pits to view the soil profile, or by using a soil probe to collect samples from different soil horizons (Chapters 6 and 10). Inspection of the soil profile shows that the soil can be separated into layers based on color, structure, texture, and other properties. These layers are called soil horizons. Inspection of the soil profile also shows that the soil properties may change rapidly or slowly with distance, depending on the soil property of interest (Schepers & Meisinger, 1994).

The most ideal representation of spatial soil differences is through a continuum map showing gradual changes. One of the most promising approaches to measure the soil continuum is the use of in-field sensing techniques. Since sensing measurements can often be obtained close together, they provide an efficient means to characterize spatial variability within fields. Due to the high spatial density of the collected information, mobile sensors can also improve the overall accuracy of the measurement process, even if point accuracy is lower than that of laboratory analysis (Sudduth et al., 1997). Various approaches for mobile sensing of key physical and chemical soil properties are available with commercial systems (Adamchuk et al., 2004; Sudduth et al., 1997). As geophysical sensors have developed, applications for their use in precision agriculture purposes have also been discovered and are discussed in detail in

SOIL AS A CONTINUUM

Soil properties have been mapped for decades using mapping units that can be accessed through the USDA's Web Soil Survey (Chapter 10). Figure 5.4 shows a composite of the thousands of soils mapped in the United States. Every map unit is made up of the soils for which it is named, plus some minor components that may not fit the description. In addition, soil properties across the landscape are nearly always a continuum and do not change abruptly at the boundary between one soil mapping unit and another. A continuum can be described as a continuous sequence in which adjacent elements or properties are not perceptibly different from each other, though with increasing distance, can be quite distinct. Furthermore, some of the properties are not static and may have changed with time. Specific changes in a field may include pH (due to the type of nitrogen fertilizer or lime applications), soil organic matter amounts (due to tillage or residue management), or soil electrical conductivity (due to adding tile drainage or salt deposition from subsurface sediments). The USDA soil maps were originally on paper but are now digitized and available online. Information on the Web Soil Survey is available in Chapter 10.

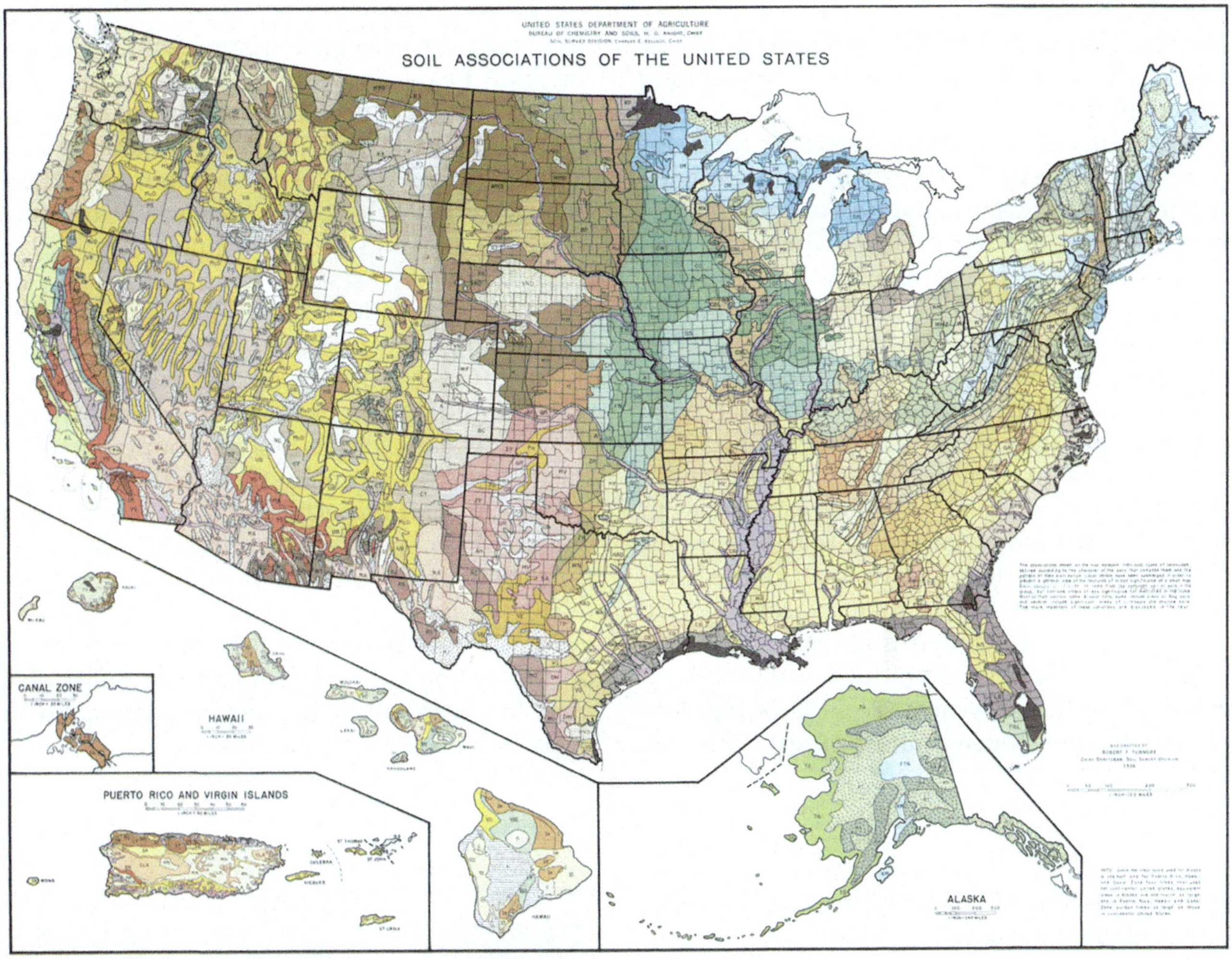

Figure 5.4. Soil associations of the United States, which are aggregates of similarly mapped soils. From Brevik E. C., & Hartemink A. E. Soil maps of the United States of America. Soil Sci Soc Am J. 2013; 77(4): 1117–1132. (https://doi.org/10.2136/sssaj2012.0390)

Chapters 11–13. Nevertheless, even with all these tools, soil by its very nature is difficult to quantify, characterize, and represent. It is important to acknowledge that soil properties change gradually across the landscape in time and space, rather than abruptly at defined boundaries. Managing crop production systems, where biophysical processes vary across temporal and spatial scales, underscores the need for soil representations that reflect the soil's continuous nature.

Soil Erosion as a Driver of Variability

Tilling the soil, and the erosion that commonly follows, can introduce soil variability. Soil that took thousands of years to form can quickly degrade. The intensity and severity of this degradation are driven primarily by physical factors (e.g., rainfall intensity and duration for water erosion, wind speed and duration for wind erosion) and within-field, site-specific conditions (e.g., slope length, steepness, aspect, and crop management practices such as tillage and types

SOIL EROSION

The Dust Bowl is a well-known event in U.S. history that demonstrates the devastating impact of topsoil loss from soil erosion on crop productivity. In the North American Midwest during the 18th and 19th centuries, grasslands and forest lands were converted to annual cropping systems, leaving a lasting impression on the landscape (Clay et al., 2017). Over time, farm mechanization and the intensification of agriculture, such as aggressive land clearing and tillage, became the norm. For the first time, extensive quantities of flat grasslands were plowed and put into grain production. This rapid expansion of agriculture was then accelerated by World War I, causing grain prices to soar. This sustained intensification of agriculture had an immediate and dramatic increase in soil erosion, which was then exacerbated by years of extreme rainfall and drought in the 1920s and 1930s, severely reducing the ability to grow crops. In wet years, severe flooding, loss of nutrient-rich topsoil, and property damage occurred. In drought years, wind erosion stripped fields of the topsoil (see Figure. 5.4.) and deposited it across the continent and into the ocean. By the 1930s, streams were filled with sediment, and much of the topsoil needed for crop growth was lost, with the subsoil clay even exposed in some areas (Bennett, 1939). The impact was devastating for those relying on crop production to make a living, with many abandoning their farming operations for metropolitan areas to survive, destroying many rural communities in the process. An erosion survey in 1934 disclosed that 25% of cropped acres in this region had lost more than 75% of the original topsoil, and it is estimated that this resulted in a yield decline of greater than 50%. This historic event demonstrates the importance of protecting topsoil from erosion. Additionally, events such as this can leave lasting impressions on the soil, resulting in topsoil variability that must be considered when making management decisions almost a century later.

of tillage). As a result, soil variation within fields is magnified by erosion.

While conservation measures that began in the 1930s have reduced erosion overall, water and wind erosion still occur. For example, a blinding dust storm was documented in Delhi, India, in 2013, where visibility dropped to near zero. In California during March 2016 and in Illinois in 2023, dust storms triggered multi-vehicle pileups on highways.

An example of how erosion can magnify the within-field variability is shown for a 90-acre field located in Missouri (Figure 5.5; Lerch et al., 2005). For this field, topsoil is defined as the soil thickness to the depth of a claypan horizon. The map on the left is the estimated topsoil depth prior to 1800. Nearby fields that have never been cropped were used to develop this representation. The center map is a topsoil depth map measured in 1999 using soil electrical conductivity sensing (see Chapter 12). The map on the right shows the net loss of topsoil (the difference between the first two maps). Note how the

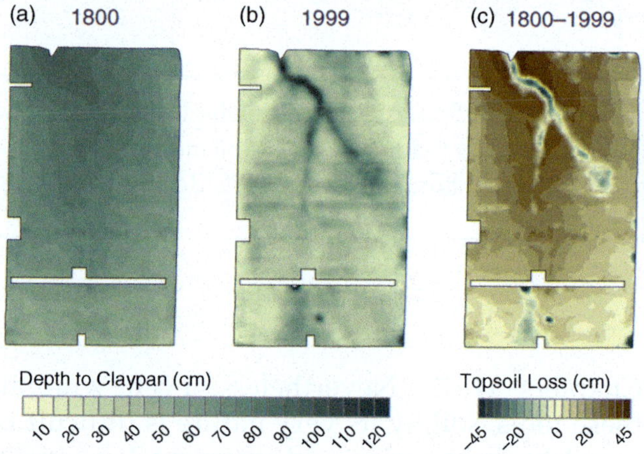

Figure 5.5. Impact of erosion on within-field variability. (a) A map of the estimated topsoil depth in 1800, prior to tillage. (b) A map of topsoil depth measured in 1999 with the aid of soil electrical conductivity sensing. (c) A map of the net loss of topsoil (difference between the first two maps). The "depth to claypan" is a measure of the thickness of the topsoil. This Missouri field provides an example of how erosion has resulted in both loss of productivity and increased within-field variability.

field was much more variable in 1999 than in 1800. Most of the field has lost surface soil; only a few areas have increased in topsoil. On average, this field was calculated to have lost 6.7 in. (17 cm) of soil since cultivation began 120 years ago. Seventeen centimeters of soil depth is equivalent to losing 7 tons of soil per acre per year (15.6 Mg/ha or 15.6 metric tons/ha). Yield mapping in this field shows that grain production is the least where topsoil loss from erosion is greatest (Kitchen et al., 2005). Erosion has resulted in both lost crop productivity and increased within-field variability.

Soil Compaction as a Driver of Variability

Soil compaction from field equipment operations is a concern in many crop production fields. When soil is compacted, pore space is reduced, and other physical and chemical soil properties (e.g., water content, air or water permeability, strength) are affected so that root development and crop growth are reduced. Generally, the depth of compaction varies from 4 to 24 in. (10–60 cm), but the effect on crop growth is usually more pronounced in the surface 6 in. (~15 cm). An increase in soil compaction results in higher soil bulk density (the relative amount of solids as opposed to pore space in a given volume). The ability of roots to penetrate the soil decreases with increasing bulk densities, which in turn can reduce water and nutrient availability, slow plant growth by restricting root penetration, and decrease water infiltration and air movement (Allmaras et al., 1988). This in turn can cause reduced seed germination and emergence, crop establishment, and yields (Soane & Van Ouwerkerk, 1994). Furthermore, when crops are stunted by compaction, unused nutrients from fertilizer or manure applications become more vulnerable to off-field movement into ground and surface waters.

Soil water content is usually the most important factor influencing soil compaction (Soane & Van Ouwerkerk, 1994). Wet soil is very vulnerable to compaction. The degree of compaction created by tillage and heavy machinery traffic on wet soils is often a function of soil texture (e.g., sand, silt, and clay particle proportions), initial soil density, soil structure, and organic matter content. Because these properties vary within fields, the degree of compaction within fields also varies (Jung et al., 2010). Avoiding field work on wet fields, as well as using controlled trafficking that restricts machinery to designated pathways, can help limit soil compaction.

Soil Fertility and Management as a Driver of Variability

From a soil fertility perspective, the greatest drivers of variability within a field are (a) soil pH, often due to lime applications and (b) nutrient fertilizer applications, the use of nitrogen, phosphorus, potassium, and others to reduce nutrient deficiencies. Research shows that interactions among the soil-forming factors, discussed in detail in Chapter 6, and prior management impact current nutrient variability (Figure 5.6).

As seen in a time sequence of aerial images (Figure 5.6, top), this field, like many others, was historically divided and managed as smaller fields. The boundaries of those smaller fields changed, and small fields were combined to form larger fields. Although many years have passed, the smaller field boundaries and locations of farm homes are evident in the soil fertility maps derived from an intensive grid-soil sampling from 1999 (Figure 5.6, bottom). Abrupt changes in soil fertility going from north to south (up and down in the figure image) approximately line up with boundaries of the historic smaller fields, indicating that these smaller fields were not managed in the same way. For example, lime was applied to all the subfields within the landscape shown, but one large area, seen as red on the soil pH map, was evidently limed more than the rest of the field. The higher pH observed on the very southern edge of the field is attributed to lime dust coming off the gravel road that borders the field on that side. Furthermore, the highest soil test phosphorus in the field (red areas) aligns with the farm home locations and with where farm animals were kept in corrals and barns.

Variability can also result from uneven application of animal manure. Because of convenience, manure was often spread closer to barns and animal lots. As such, plant-available nutrients like phosphorus and potassium, as well as organic matter content, can be highly variable in these fields. This example demonstrates that soils exhibit a "memory effect" (or legacy effect) relative to human activities. Effects of past management can persist and affect field properties for many years. This variability shows how difficult it would be to

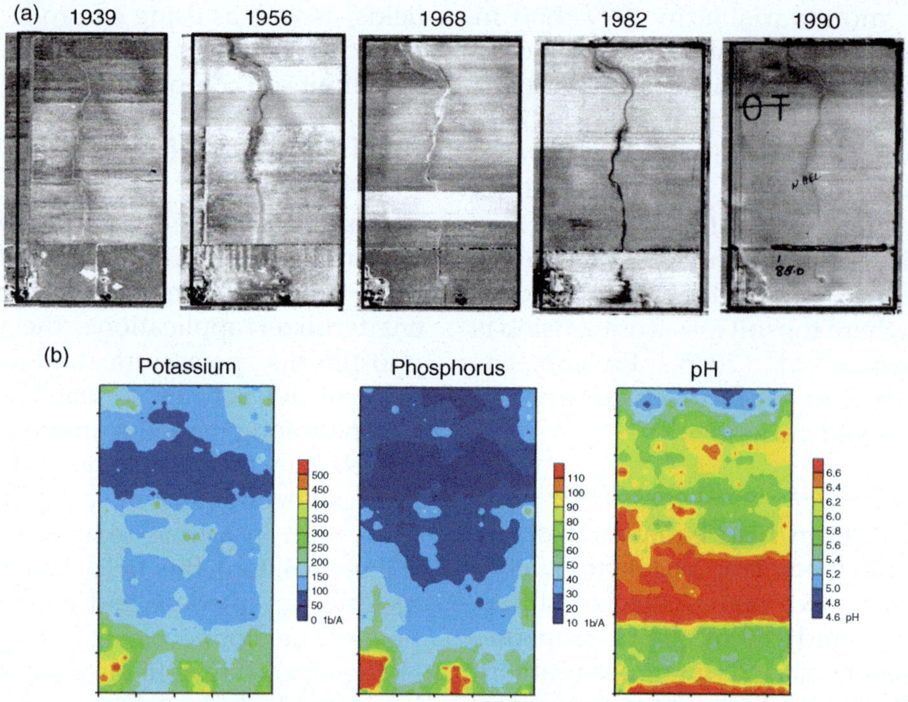

Figure 5.6. (a) A time sequence of aerial images of a Missouri field over a 50-year period. (b) Soil fertility maps of the same field obtained from intensive grid-soil maps in 1999. Much of the variation in soil fertility for the field arises from historic anthropogenic activities.

develop a nutrient management plan based on a single composite sample or even several samples collected and analyzed separately. Another source of management variability occurs when the application equipment unevenly applies fertilizer materials. This variability complicates soil sampling and can increase yield variability (Figure 5.7).

Figure 5.7. Areal image of a field showing a physical response to uneven fertilizer application (dark strip indicates more, or an overlap, of applied fertilizer).

Pests as a Driver of Variability

The spread of pests (insects, weeds, and diseases) is not uniform in most crop fields. One potential reason for the patchiness is that different species thrive under different environments. Some plant species (e.g., stinkgrass, field bindweed, sandbur) grow better in dry soils, whereas others (fall panicum, common cocklebur) typically do better in moist soils (Figure 5.8). Species shifts can also occur due to annual weather patterns. Diseases, such as root rots, are found most often in wetter soils, whereas dry soils provide a more favorable habitat for nematodes. Rootworms' impact on crops is often worse within fields, as the adults may avoid field edges when laying eggs. In contrast, deer and rodent damage are often more common along field edges.

Management practices can also increase pest variability (Chapters 17–19). For example, mechanical and chemical weed control methods are seldom 100% effective; therefore, surviving weeds may, over time, adapt to these control methods. Seeds produced by these surviving

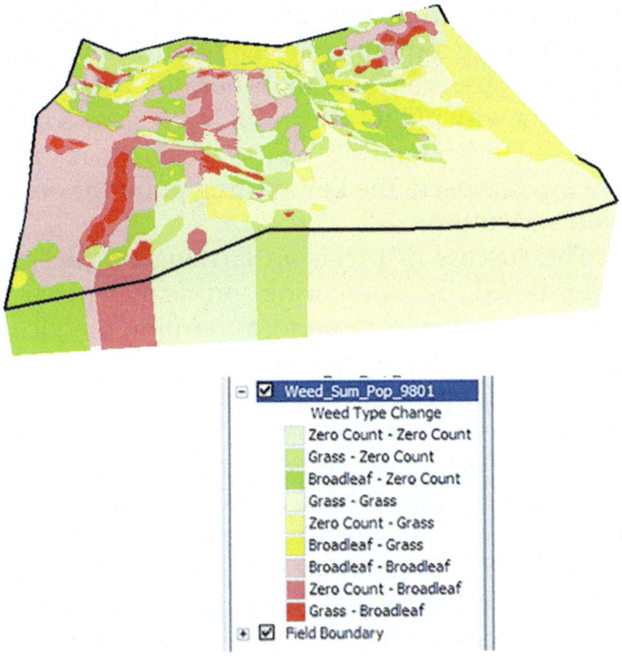

Figure 5.8. Changes in weed type in a 160-acre (64.7 ha) field intensively mapped in 1998 and 2001 (mapped as no weed, grass, or broadleaf species present in each year). The legend indicates the species shift between sampling dates. The darkest red areas were mapped originally as grass species and in the subsequent sample as broadleaf areas (Clay, unpublished data).

plants are generally localized but, if unchecked, can spread throughout a field. As such, weed stress in crops is typically not uniform within fields (Figure 5.8). In addition, "one year's seeding is seven years' weeding" is a saying often used to describe the consequences of poor weed management. Unlike crop seeds that are bred to have uniform germination, the seeds of most weed species have inherent dormancy, resulting in staggered germination. For example, not all weed seeds germinate at the same time (accounting for the differences observed in weed flushes throughout a season), and not all seeds produced the previous year germinate the following year. In addition, new research is revealing that some weeds alter soil microbial populations, ultimately enhancing or impairing their own growth (i.e., plant–soil microbial feedback loops; Lou et al., 2014). While the magnitude of change may not be a short-term concern, the legacy effects may hamper future crop production.

The Complexity of Field Variability

Some may wonder why precision farming seems complex. The need to work with a complex system is the result of attempting to manage an interaction between the genetics, environment, and management (G × E × M). Added complexity arises when spatial and temporal variations are included within the Environment component of this expression (E), or when management varies across the field.

Narratively, this expression may be stated as follows: Plant growth and productivity is the result of crop type and hybrid and/or variety planted (G), influenced by unique environments defined by time and location combinations (e.g., soil texture, soil organic matter, landscape slope and aspect, rainfall amount and intensity, temperature, radiation; E_{ST}), but modified by crop management practices (e.g., tillage, fertilization, planting date, weed control, cover crops; M). The number of unique plant growth outcomes when combinations of spatial and temporal factors are included becomes astronomical. This is why the ability to predict yield and make optimal management decisions are so complex. Precision farming explores and addresses the yield-impacting interactions of the $G \times E_{ST} \times M$ function.

Sustainable Production

A guiding principle of sustainable agricultural production systems is the ability to meet society's food, feed, fiber, and fuel needs today without compromising the needs of future generations. While farmers are concerned about the environment, their primary concern is often economic, since their farms represent their livelihood. When crop grain yields fail to provide revenue greater than the costs, such land is called "marginal." This often occurs in only some portions of the field (Massey et al., 2008) and may include eroded hillsides, poorly drained soils, and salt-affected soils. Once identified, alternative land uses should be identified for these locations (Mudgal et al., 2010). As such, improved site-specific management changes to these marginal soils should simultaneously address both economic and environmental risks.

Quantifying and Addressing Variability

In the past, farmers kept mental or written notes about their perceptions and observations regarding yield-limiting factors; these factors were often described in different locations of each field they managed. Associated with this information is what cultivar was planted, how much and what types of fertilizer were applied, weed patches, and more. Over time, farmers began to recognize which areas were the most productive and how to invest their resources. Today, farmers have more tools than ever to identify and address these variations. Increases in the amount of information available for analysis may feel overwhelming and complicated. The key is to ultimately quantify and address the aspects of variability in a field, with precision farming often boiling down to tailoring management to match spatial and temporal variability. Quantification of variability can be highly technical, using geospatial analysis techniques (Chapters 8 and 9) and the latest algorithms (Chapter 27) to develop and predict techniques that will increase production using data such as the following:

- Combine yield monitor data
- Reflectance data collected by unmanned aerial vehicles, airplanes, and satellites
- Soil elevation, apparent electrical conductivity, and other soil sensing
- Sprayer boom height sensor and weed detection
- Soil characteristics reported by the Web Soil Survey
- Field-specific weather reports.

However, precision farming can also be low-tech (Chapters 2 and 3): walking the field and understanding the variability through living and working with the land for generations. Humans have always been aware of the importance of recognizing the variability in the landscape to grow food. If the weather were the same each year and crop fields were uniform, one would expect uniform yields across all the tillable acres, and there may be no need to vary management practices. However, even in the past, before the ability to collect large amounts of environmental data, farmers observed and took note of where and possibly why plants grew differently across their fields. Understanding this variability is, and always has been, the key starting point for precision agriculture.

The success of precision farming practices is determined by comparing precision management technologies to uniform farming practices. For example, changing nitrogen rates or seeding rates on the go can be compared with uniform application or seeding rates. As crop genetics improve, drought-tolerant varieties or more disease-resistant varieties may be planted in problem areas. As precision agriculture becomes more widely adopted, the new standard for judging the performance of newer precision agriculture strategies will be against the older, less precise methods. In other words, today's precision farming is likely to be the standard against which tomorrow's precision farming will be compared, quantifying and addressing variability as the key factor.

Conclusions

Managing variability to reduce costs and sustain greater economic yield and field health (i.e., improved environmental services) is an overarching principle of precision farming. The specific intensity of precision management (by square foot, acre, soil type; by field; or by the minute, day, or month) will be dictated by the skill and knowledge of the land manager, the variability of the land base, and the cost and return of various management techniques. This book introduces many concepts in precision farming, but it starts with understanding the sources and magnitude of variability. These variability basics are foundational for precision farming, and they aid in appreciating the many different and diverse practice options that can be considered when addressing variability. In addition to understanding variability, knowledge of how to manage this is needed. An example of this is the deliberate decision to use or not to use variable-rate technology based on economic or agronomic benefits. Precision farming is not the use of increasingly complex technology but, rather, the application of knowledge to solve complex problems. Precision farming may or may not use technology to implement the potential solution on the farm.

New information investigating variability is continuously reported from scientists worldwide. This information, combined with new sensors and application technology, advancements in moving and managing data, improved crop genetics, enhanced methods of data analysis and visualization, and new tools such as artificial intelligence, will continue to advance our understanding of variability. However, these must be tempered with the wisdom of biological systems and implemented with agronomic knowledge and skills—precision farming is fully dependent on our understanding of how crops grow and their growing environments.

Study Questions

1. Describe how soil erosion creates variability in crop grain yield.
2. What technology enabled inexpensive and rapid mapping of spatial variability?
3. What does it mean for some forms of variability to be derived from an anthropogenic source?
4. Provide an example of how pests create variability and then identify a management practice you might use to address that variability.
5. What does the term "E_{ST}" in the expression $G \times E_{ST} \times M$ represent?
6. Name two examples of abiotic factors that can cause within-field variability.
7. Name two examples of biotic factors that can cause within-field variability.
8. What types of agronomic knowledge are needed to optimize precision farming possibilities?

ACKNOWLEDGMENTS

Support for this document was provided by the Precision Farming Systems community in the American Society of Agronomy, South Dakota State University, USDA Natural Resources Conservation Service Conservation Innovation Grants (NR233A750011G019), American Coalition for Ethanol, GEVO, the South Dakota Corn Utilization Council, and the South Dakota Soybean Research and Promotion Council. The use of trade names is for the convenience of the reader and does not imply endorsement by the authors.

REFERENCES

Adamchuk, V. I., Hummel, J. W., Morgan, M. T., & Upadhyaya, S. K. (2004). On-the-go soil sensors for precision agriculture. *Computer Electronics in Agriculture*, *44*(1), 71–91. https://doi.org/10.1016/j.compag.2004.03.002.

Allmaras, R. R., Kraft, J. M., & Miller, D. E. (1988). Effects of soil compaction and incorporated crop residue on root health. *Annual Review of Phytopathology*, *26*, 219–243. https://doi.org/10.1146/annurev.py.26.090188.001251.

Bennett, H. H. (1939). Central prairie and eastern timbered border region. In *Soil Conservation* (pp. 685–714). McGraw-Hill. https://doi.org/10.1097/00010694-194007000-00019.

Clay, D. E., Clay, S. A., DeSutter, T., & Reese, C. (2017). *From plows, horses, and harnesses to precision technologies in the North American Great Plains. Oxford research encyclopedia of environmental science*, Oxford University Press. https://doi.org/10.1093/acrefore/9780199389414.013.196.

Forcella, F. (1998). Real-time assessment of seed dormancy and seedling growth for weed management. *Seed Science Research*, *8*(2), 201–209. https://doi.org/10.1017/S0960258500004116.

Forcella, F., Benech-Arnold, R. L., Sanchez, R., & Ghersa, C. M. (2000). Modeling seedling emergence. *Field Crops Research*, *67*(2), 123–139. https://doi.org/10.1016/S0378-4290(00)00088-5.

Franzen, D. W. (2007). Nitrogen management in sugar beet using remote sensing and GIS. In F. J. Pierce & D. E. Clay (Eds.), *GIS applications in agriculture* (pp. 35–48). CRC Press. https://doi.org/10.1201/9781420007718.

Jung, K.-Y., Kitchen, N. R., Sudduth, K. A., Lee, K.-S., & Chung, S.-O. (2010). Soil compaction varies by crop management system over a claypan soil landscape. *Soil and Tillage Research*, *107*(1), 1–10. https://doi.org/10.1016/j.still.2009.12.007.

Kitchen, N. R., & Clay, S. A. (2018). Understanding and identifying variability. In D. K. Shannon, D. E. Clay, & N. R. Kitchen (Eds.), *Precision agriculture basics* (pp. 13–24). ASA, CSSA, SSSA.

Kitchen, N. R., Sudduth, K. A., Myers, D. B., Massey, R. E., Sadler, E. J., Lerch, R. N., Hummel, J. W., & Palm, H. L. (2005). Development of a conservation-oriented precision agriculture system: Crop production assessment and plan implementation. *Journal of Soil and Water Conservation*, *60*(6), 421–430. https://doi.org/10.1080/00224561.2005.12435835.

Lerch, R. N., Kitchen, N. R., Kremer, R. J., Donald, W. W., Alberts, E. . E., Sadler, E. J., Sudduth, K. A., Myers, D. B., & Ghidey, F. (2005). Development of a conservation-oriented precision agriculture system: Water and soil quality assessment. *Journal of Soil Water Conservation*, *60*(6), 411–421. https://doi.org/10.1080/00224561.2005.12435834.

Lou, Y., Clay, S. A., Davis, A. S., Dille, A., Felix, J., Ramirez, A. H. M., Sprague, C. L., & Yannarell, A. C. (2014). An affinity-effect relationship for microbial com- munities in plant–soil feedback loops. *Microbial Ecology*, *67*(4), 866–876. https://doi.org/10.1007/s00248-013-0349-2.

MacRae, I., Carroll, M., & Zhu., M. (2011). Site-specific management of green peach aphid, *Myzus periscae* (Sulzer). In S. A. Clay (Ed.), *GIS applications in agriculture* (Vol. 3: *Invasive species*, pp. 167–189). CRC Press.

Massey, R. E., Myers, D. B., Kitchen, N. R., & Sudduth, K. A. (2008). Profitability maps as an input for site-specific management decision making. *Agronomy Journal*, *100*(1), 52–59. https://doi.org/10.2134/agronj2007.0057.

Mudgal, A., Baffaut, C., Anderson, S. H., Sadler, E. J., & Thompson, A. L. (2010). APEX model assessment of variable landscapes on runoff and dissolved herbicides. *Transactions of the ASABE*, *53*(4), 1047–1058. https://doi.org/10.13031/2013.32595.

Schepers, J. S., & Meisinger, J. J. (1994). Field indicators of nitrogen mineralization. In J. J. Havlin & J. S. Jacobsen (Eds.), *Soil testing: Prospects for improving nutrient recommendations* (Vol. *40*, pp. 31–47). Soil Science Society America. Spec. Publ. SSSA. https://doi.org/10.2136/sssaspecpub40.c3.

Soane, B. D., & van Ouwerkerk, C. (1994). *Soil compaction in crop production* (Vol. *11*). Elsevier Science. https://www.sciencedirect.com/bookseries/developments-in-agricultural-engineering/vol/11.

Sudduth, K. A., Hummel, J. W., & Birrell, S. J. (1997). Sensors for site-specific management. In F. J. Pierce & E. J. Sadler (Eds.), *The state of site-specific management for agriculture* (pp. 183–210). ASA, CSSA, SSSA. https://doi.org/10.2134/1997.stateofsitespecific.c10.

Understanding Soil Variability and Sampling Strategies for Site-specific Nutrient Management

6

Katsutoshi Mizuta, Skye Brugler, Srinadh Kodali, Ahmed Harb Rabia, Xi Zhang, and David E. Clay

Chapter Purpose

One of the aims in precision agriculture is to optimize fertilizer applications to enhance productivity and sustainability. Analyzing the optimal application rate requires an accurate estimate of the soil's ability to supply nutrients to the growing crop. This is typically achieved by collecting and analyzing soil samples. The purpose of this chapter is to introduce the spatial and temporal variability of soil properties and how and why they change across a field. It also emphasizes the importance of selecting appropriate soil sampling strategies to support site-specific management decisions.

Introduction

Soil properties within agricultural fields can vary both spatially and temporally. This variability, much like that observed in cropping systems discussed in Chapter 5, arises from a combination of natural processes and human interventions. Soil-forming factors, such as climate, organisms, topography, parent material, and time, interact with management practices to create complex patterns of variation. As a result, collecting soil samples from a single location is rarely sufficient for developing accurate fertilizer recommendations. Instead, multiple samples must be collected across the field to capture the full spectrum of variability.

To ensure that nutrient management practices are both effective and sustainable, soil sampling strategies should align with the established stewardship framework commonly known as 4R (also called 4Rplus or 5R), which emphasizes the importance of applying the right source of nutrients at the right rate, time, and place. Additionally, the right manner is sometimes included, which is why the terms 4Rplus or 5R are occasionally used. The last R may also involve conservation practices, such as cover cropping, reduced tillage, or soil health management. Designing effective soil sampling protocols is challenging due to the complex interactions between natural and anthropogenic

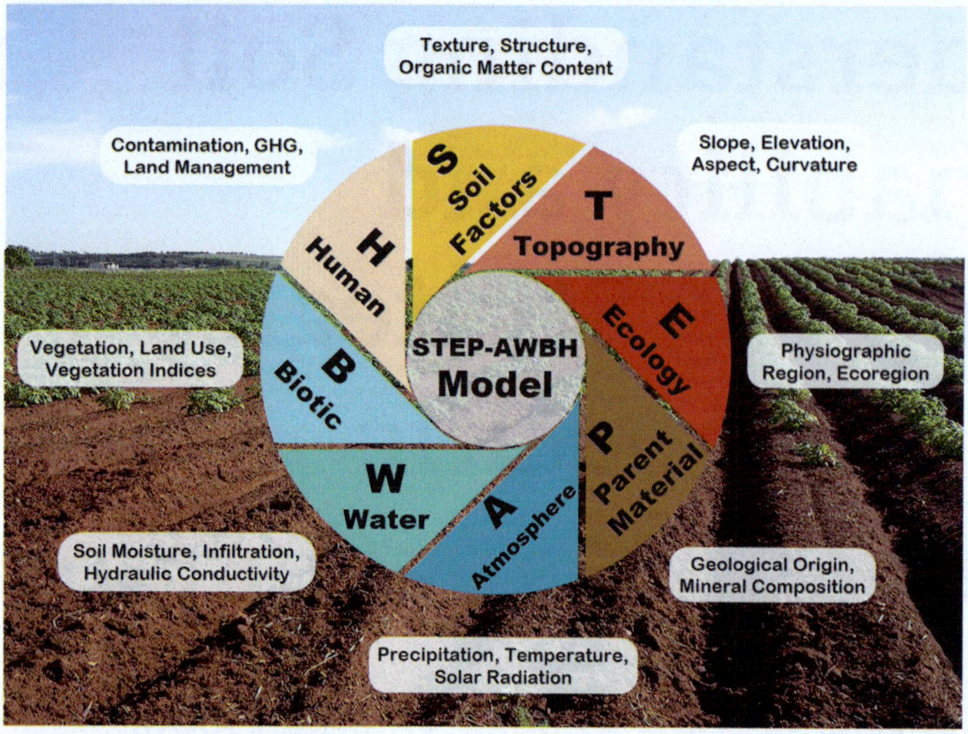

Figure 6.1. Sources of within-field variability as categorized by the factorial STEP-AWBH model. The text in boxes with a white background show examples for each category. (Modified based on Grunwald et al., 2011; Courtesy Ahmed Harb Rabia.)

(human-induced) factors (Franzen, 2023). These interactions can be better addressed using modeling frameworks, such as factorial and spatiotemporal models (Grunwald, 2021). One such approach to creating efficient sampling protocols is to use a model such as the "STEP-AWBH" (phonetically "step-up") model (Grunwald et al., 2011), a space–time framework that categorizes variability into multiple distinct factors (Figure 6.1). This model integrates soil (S), topographic (T), ecological (E), parent material (P), atmospheric (A), water (W), biotic (B), and human-induced (H) factors to identify within-field variation and aid in site-specific nutrient management. By examining these factors individually and in combination, researchers and practitioners can better understand within-field variability and develop more precise nutrient management strategies. Figure 6.1 illustrates the STEP-AWBH model, showing examples of how each factor contributes to soil variability. In the next section, although we have other similar models (e.g., SCOPAN [McBratney et al., 2003]), we will simplify the sources of soil variability and discuss how those factors can affect nutrient availability and crop growth.

Sources of Soil Variability Induced by Intrinsic and Human Factors

Soil nutrient content does not remain constant across a field spatially or over time. Instead, it fluctuates due to a combination of intrinsic factors related to natural soil formation and human-induced factors stemming from agricultural management. Intrinsic variability in soil properties arises from the five soil-forming factors identified by Hans Jenny—climate, organisms, relief or topography, parent material, and time (Jenny, 1941)—and human factors (e.g., management practices). Practices such as tillage, irrigation, fertilization, and land leveling can alter soil structure, nutrient distribution, and water dynamics (Grunwald et al., 2011). These factors influence the types of minerals present in the soil, the rate at which nutrients are released into the soil solution, the availability of these nutrients for plant uptake, and the movement of nutrients within the soil profile.

Climate

Among the soil-forming factors, climate is particularly dynamic, influencing soil properties through both short-term weather events and long-term climatic trends. Extreme weather events include, but are not limited to, high winds that exceed 100 mi/hr (160 km/hr), droughts, hailstorms, flooding, hurricanes, tornadoes, and ocean inundation. The magnitude of impact caused by these events depends on geographic location, intensity, frequency, and duration. For example, a heavy rainstorm may result in significant runoff, causing nutrient loss or deposition at different locations within a field depending on topography and soil characteristics. In contrast, gentle rainfall may enhance nutrient infiltration and retention. Seasonal variations in precipitation and temperature also play a role, affecting microbial activity, organic matter decomposition, and nutrient cycling. Drought conditions can reduce soil moisture and slow nutrient uptake, while temperature extremes can influence the rate of chemical reactions in the soil.

Organisms

Organisms, including plants, animals, and microorganisms, play a crucial role in soil formation and fertility, and their activity contributes significantly to within-field variability. Plants contribute biomass that, upon decomposition, becomes soil organic matter. Earthworms, through their burrowing, enhance soil aeration, water infiltration, and nutrient availability for plants. Plant roots interact with soil biota, altering the distribution of nutrients through exudation and rhizosphere microbial activity. Microbial communities contribute to essential processes, such as nitrogen fixation, phosphorus solubilization, and carbon cycling, leading to spatial differences in soil fertility through their burrowing. The complexity of these interactions is amplified by the influence of climate at both the macro and micro scales. For example, a region of a field with favorable environmental conditions may support a higher abundance of earthworms or microbial populations, leading to improved soil structure, organic matter decomposition, and nutrient cycling in that area. This also results in spatial differences in soil fertility and structure. By maintaining living roots in the soil, minimizing soil disturbances, and following the 4R management principles, we can help support the resilience of soil organisms and improve soil health under variable environmental conditions.

Relief or Topography

Topographical features, such as slope and elevation, affect water movement, erosion, and soil depth, thereby affecting nutrient availability across a field. Yield responses often differ between hilltops and low-lying areas (Franzen, 2018). In summit and shoulder positions, plants may suffer from water scarcity, while those in footslope and toeslope positions may be challenged by excess moisture. Runoff from elevated areas can strip away topsoil and essential nutrients, reducing fertility, whereas lower landscape positions tend to accumulate sediments and nutrients, resulting in higher organic matter content and improved fertility. Topographic variation also affects solar radiation exposure and moisture retention. Slopes that receive less direct sunlight, such as northern slopes in the Northern Hemisphere, tend to retain more soil moisture due to slower evaporation rates. These differences contribute to spatial variability in nutrient dynamics and plant growth. The integration of topographic data into nutrient management has been greatly enhanced by the use of Digital Elevation Models, which can be downloaded and used with other forms of spatial data to show and analyze the impact of variations in topographic features on soil water and nutrient availability.

Parent Materials

Parent materials are the geological substrates from which soils develop, and they can be broadly classified into residual and transported types. Residual parent materials form in place through the weathering of bedrock, which may include igneous rocks like granite and basalt, sedimentary rocks like limestone and sandstone, or metamorphic rocks like schist and gneiss. In contrast, transported parent materials have been relocated by natural forces, such as water (alluvium), gravity (colluvium), wind (aeolian deposits like loess), and ice (glacial till and outwash). The nature of the parent material influences soil mineralogy, texture, and nutrient retention and the soil's capacity to supply nutrients to plants. For instance, sandy soils derived from coarse parent materials, such as sandstone or glacial outwash, are more prone to mid-season drought due to their lower water-holding capacity, while finer-textured soils with parent materials such as shale or basalt retain moisture and nutrients more effectively.

The mineral composition also affects the availability of soil nutrients for plants. For example, in Kentucky's Bluegrass region of the United States, agricultural productivity has historically benefited from limestone, which naturally supplies phosphorus (Ragland et al., 2025). However, not all mineral-derived nutrients are readily available in all soil types due to differences in clay mineral composition. Certain clay minerals, such as vermiculite, that expand in wet conditions, contribute to the soil's high cation exchange capacity (CEC) but can also cause some nutrients, such as potassium and ammonium, to become temporarily unavailable by collapsing and trapping them between clay sheets in dry conditions. Understanding the characteristics of parent materials is therefore essential for interpreting soil fertility and designing effective nutrient management strategies.

Time

Soil formation is a relatively slow process that contributes to long-term spatial and temporal variability in soil properties. Because pedogenic processes, such as weathering, organic matter accumulation, and horizon development, occur over centuries, differences in soil age and development lead to variation across landscapes. Evans et al. (2019) reported that in arable systems in the United Kingdom, it can take between 104 and 385 years to form just 1 cm of topsoil. Additionally, the nutrients available in the soil take time to weather from rock and become available for plants.

Harvesting can remove some of these nutrients from the soil, which must then be replenished since crops use the nutrients from the soil to grow and develop. Historically, fallow periods between cropping cycles were thought to restore nutrients through natural processes, such as the decomposition of plant residues. Nutrients are not only extracted from the soil through the harvested grain, but they are also stored in non-harvested biomass, such as roots, stalks, and leaves. These materials return nutrients to the soil upon decomposition, although this process takes time (Joshi et al., 2024). In either case, the time required for organic residues to decompose and return nutrients to the soil adds another temporal dimension to soil variability, as nutrient availability fluctuates with cropping cycles and decomposition rates. In modern systems, nutrients are returned to the soil through fertilizer applications, supplemented by the in-field decomposition of non-harvested biomass from a rotation of different cash crops or cover crops. In either case, the time required for organic residues to decompose and return nutrients to the soil adds another temporal dimension to soil variability, as nutrient availability fluctuates with cropping cycles and decomposition rates.

Additionally, soil properties can vary significantly over short periods due to seasonal changes. For example, Omer et al. (2018) found that 15 of 21 soil parameters measured at the surface (0–15 cm) in arid systems varied significantly depending on the timing of sampling within a year. These parameters include wet aggregate stability, mean weight diameter of dry aggregates, dry aggregates smaller than 0.25 mm, available water capacity, soil organic matter, permanganate oxidizable carbon, soil bulk density, pH, nitrate-nitrogen, extractable potassium and phosphorus, sodium, sodium adsorption ratio, and micronutrients (zinc and copper). Such findings underscore the importance of accounting for temporal variability when collecting soil samples for site-specific management, as soil development and nutrient dynamics can differ even within small geographic areas.

Human-induced Variability

Human activities have long influenced soil properties, and understanding these impacts is essential for adopting and implementing precision nutrient management practices (Clay et al., 1997, 2002). Past and present human activities and land-use decisions influence the distribution of nutrients across a landscape, soil erodibility, and overall productivity. For instance, agricultural practices, such as tillage, irrigation, and fertilizer application, can modify soil structure, organic matter content, and nutrient levels, resulting in localized differences within a field. These patterned differences, referred to as "field systematic variability," become more pronounced when high rates of fertilizer or manure are applied, as areas with differing texture, drainage, or organic matter content retain and cycle nutrients at different rates. Such variability can also differ in magnitude across macro- and micronutrients and may further influence their interactions within the soil system. For example, nitrogen fertilization can lower soil pH, which in turn affects the solubility and uptake of micronutrients, such as iron, zinc, and manganese. Similarly, over time, large quantities of manure can leave an effect on the soil known as legacy phosphorus. This can be observed especially in field areas that held livestock during the era of homesteading, where a lasting high phosphorus concentration is left in that area of soil that must be sampled separately from the rest of the field (Figure 6.2).

Human activity may also impact the physical accessibility of nutrients to plants. Soil compaction is often caused by heavy machinery and repeated traffic. This compaction can restrict root growth, limiting nutrient availability for plants. Root penetration becomes increasingly difficult when compaction, represented by soil bulk density, exceeds 1.80 g/cm^3 in sandy soils, 1.65 g/cm^3 in silty soils, and 1.47 g/cm^3 in clayey soils. Although it is common in the United States to assume that the soil mass at a 6-in. depth in 1 ac of cropland equals approximately 2 million pounds, compaction is rarely uniform across a field, making it difficult to sample and monitor. Additionally, repeated tillage at the same depth, particularly under wet conditions, can cause compaction and lead to the formation of dense layers known as tillage pans (Raney et al., 1955). These pans reduce water infiltration, increase erosion, and strip away topsoil and surface

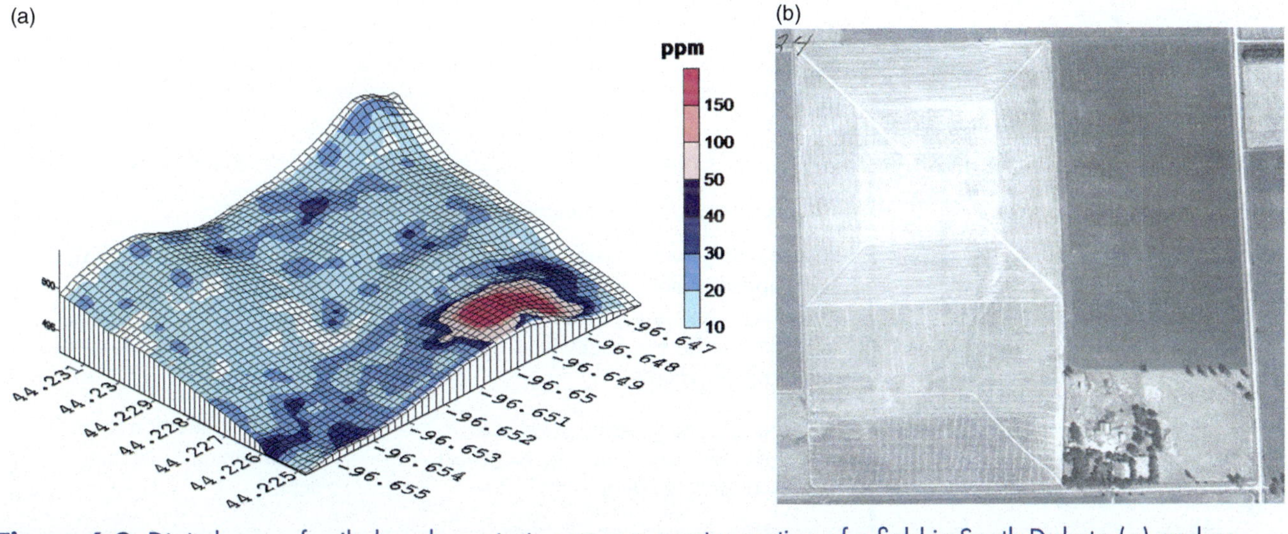

Figure 6.2. Digital map of soil phosphorus tests across a quarter section of a field in South Dakota (a) and an aerial photograph of the same site taken in the 1950s (b). Between the time of the aerial image and the 1995 soil sampling, the original homestead was removed, and the land was consolidated into a single field planted with corn or soybeans. (Courtesy South Dakota State University).

nutrients. Their presence contributes to uneven crop development and introduces significant variability in soil properties across agricultural landscapes.

Soil Sampling Methods

Developing an effective soil sampling protocol requires careful consideration of both intrinsic and human-induced variability. Due to the diverse conditions encountered in different fields, a one-size-fits-all sampling strategy is neither available nor suitable. Instead, the selection of appropriate sampling methods must be based on aspects that include, but are not limited to, the following:

1. The identification of the soil sampling goals and objectives (e.g., identifying optimal lime, fertilizer, and irrigation rates; assessing nutrient leaching rates; evaluating spatial and temporal changes in carbon stocks)
2. The importance of intrinsic and human-induced variation
3. The prior management practices (e.g., crop rotation, cover cropping, no-tillage, fertilizer additions)
4. The equipment, money, and labor available for sampling
5. The size and time constraints for sampling
6. The location of soil testing laboratories and prices for soil analysis
7. How and when fertilizers will be applied in the future
8. The collaborating partners (e.g., partnering crop consultants, fertilizer retailers, county agents).

The goal of soil sampling can vary widely, ranging from determining optimal lime or fertilizer rates to assessing nutrient leaching, evaluating changes in carbon stocks, or meeting regulatory standards for conservation practices. Soil sampling goals can influence the depth, frequency, and density of sampling, as well as the types of soil properties to be analyzed, whether physical, chemical, and/or biological (Lyons et al., 2023; Figure 6.3). For example, if the goal is to calculate the lime recommendation rate for a uniformly managed field over the years, a lower sampling density may suffice. In contrast, fields managed under the USDA Natural Resources Conservation Service (NRCS) Nutrient Management (Ac.) (590) Conservation Practice Standard (Best Nutrient Management Practices) may require more intensive sampling to qualify for subsidies, with procedures and sample counts varying by state (Crouse et al., 2014; Lyons et al., 2023).

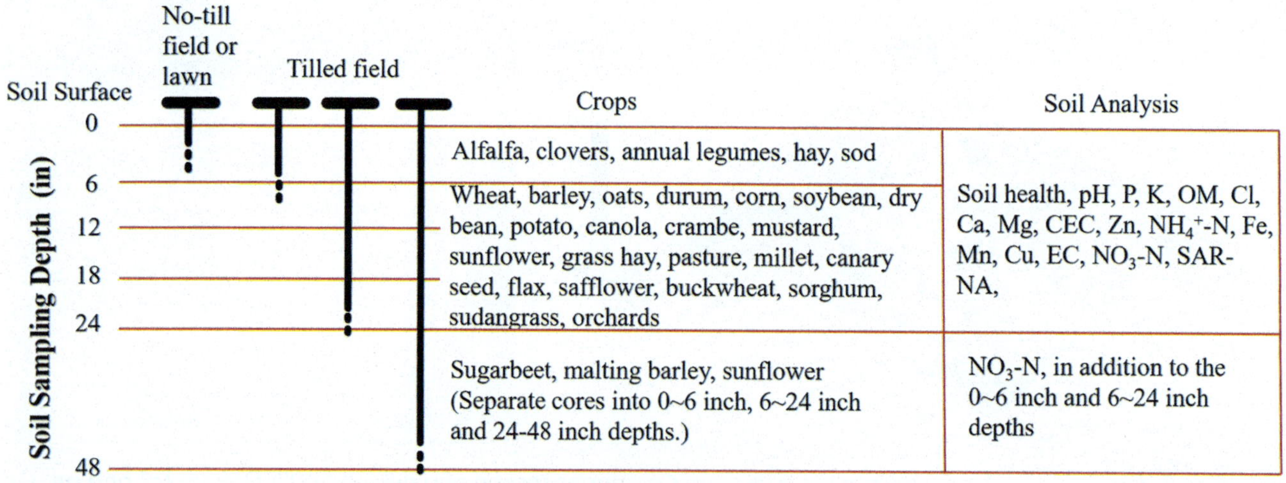

Figure 6.3. Recommended soil sampling depths (modified based on Franzen, 2023). The dotted line indicates that sampling depth may be locally adjusted based on crop type, sampling objectives (soil analysis), and management practices, such as tillage. (Courtesy Skye Brugler and Katsutoshi Mizuta).

Random Sampling

Random sampling is one of the foundational approaches in soil sampling, where each location within a field has an equal probability of being selected. This method includes composite sampling, a common approach in which soil cores are collected and combined. Random sampling uses two primary types of composite sampling: (a) whole-field composite sampling, and (b) individual location composite sampling.

Whole-field composite sampling involves collecting a single sample composed of 15–20 soil cores randomly distributed across the entire field or a designated portion of it. This simple approach provides a general estimate of soil fertility and has been widely used by farmers for decades. However, it does not capture within-field variability and therefore has limited ability to support site-specific nutrient management. In contrast, individual location composite sampling entails collecting multiple samples from specific areas within the field (blue squares in Figure 6.4). Each composite sample typically consists of five to eight cores taken within a radius of approximately 3 m. These cores are thoroughly mixed and submitted for laboratory analysis. Depending on the protocol, the exact coordinates of sampling points may or may not be recorded. This method is most suitable for fields with minimal variability and no recent history of fertilizer or manure application as outlined in Table 6.1. However, users may miss critical zones in a field unless enough random points are taken to capture the full variability of soil properties. Note that when collecting random samples, field borders and any livestock enclosures from old homesteads should be avoided, as these locations may not represent broader field conditions.

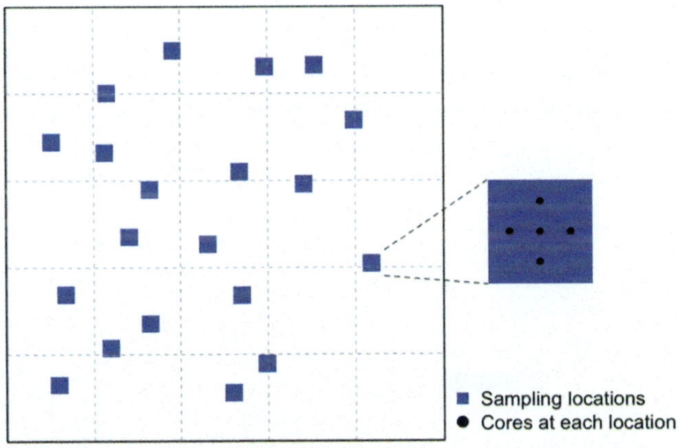

Figure 6.4. Illustration of random soil sampling across a 500 m × 500 m field. Blue squares indicate 20 randomly selected sampling locations. Dashed gray lines represent a 1-ha reference grid (100 m × 100 m) used to visualize spatial distribution. The zoomed-in panel shows a single sampling location, where five soil cores (black dots) are collected within a 3-m radius and composited into a single representative sample. (Courtesy Srinadh Kodali).

The pros and cons of random sampling may be best understood through a more familiar analogy. Imagine you have a large pizza with different toppings spread unevenly, pepperoni on one side, mushrooms in the center, and olives near the edges (Figure 6.5a). In random sampling, you close your eyes and randomly pick spots to taste. You might end up sampling only the cheese parts or get multiple bites from the pepperoni side, missing the mushrooms and olives entirely (Figure 6.5b). As a result, you might conclude that the entire pizza is mostly pepperoni because that's what you experienced with the sample (Figure 6.5c). In the same way, random soil sampling can produce a misleading picture of field conditions since it may overlook important variability in nutrient levels or soil properties.

Grid Sampling

Grid sampling offers a more structured alternative to random sampling. In this method, the field is overlaid with a systematic grid, resembling a chessboard, where samples are collected at evenly spaced intervals (Table 6.1). The most common grid size used by farmers is 1 ha (approximately 2.5 ac), which corresponds to the minimum map unit size defined by the USDA *Soil Survey Manual* (Soil Science Division Staff, 2017). (See Chapter 10 for information on the Web Soil Survey.) At each grid point, a composite sample is collected from the appropriate depth, thoroughly mixed, and analyzed (Figure 6.6). This method is particularly useful in fields where high rates of fertilizer or manure have been applied, or no logical management zones exist (Franzen & Peck, 1995; Mallarino & Wittry, 2004).

For example, in a field where site-specific nutrient applications have not previously been implemented, a screening grid of 2.5 ac (See Chapter 9) can provide a baseline understanding

Table 6.1. Comparative Overview of Soil Sampling Techniques. Note that categories such as low, medium, and high are relative and may vary depending on user perception and field-specific conditions.

Sampling strategy	Description	Statistical bias	Preparation intensity (e.g., acquisition of necessary data)	Labor intensity in field	Cost for labor/ equipment
Simple random sampling	Collect samples at random locations within the field.	High	Low	Low (generally low sample size requirement)	Low
Grid sampling	Take samples at regular intervals, with centroids arranged in a grid pattern.	Moderate	Moderate	Moderate (a general design requires a large sample size)	High
Staggered start grid pattern	Collect samples using a grid pattern with staggered starting points.	Low to moderate	Moderate	Moderate to high	Moderate to high
Systematic unaligned grid	Use randomized sampling locations within each grid cell.	Low to moderate	Moderate to high	High	Moderate
Zone/stratified random sampling	Divide the field into zones and randomly sample within each zone.	Moderate	High	Low to moderate	Low to moderate
Conditioned Latin hypercube sampling	Use statistical clustering methods to determine sampling locations to ensure samples represent the variability of soil and covariates across the field.	Very low	High	High	High

Note that categories such as low, medium, and high are relative and may vary depending on user perception and field-specific conditions.

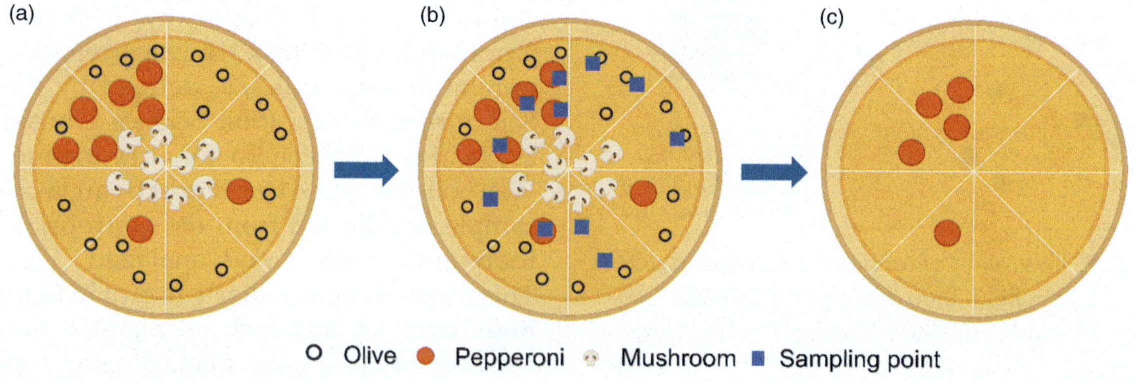

Figure 6.5. Conceptual representation of random sampling using a pizza analogy. (a) The pizza symbolizes a field with unevenly distributed soil properties (e.g., olives, pepperoni, mushrooms). (b) Random sampling points (blue squares) may overrepresent or miss certain features. (c) Post analysis, the field is evaluated for spatial variability to inform site-specific management decisions.

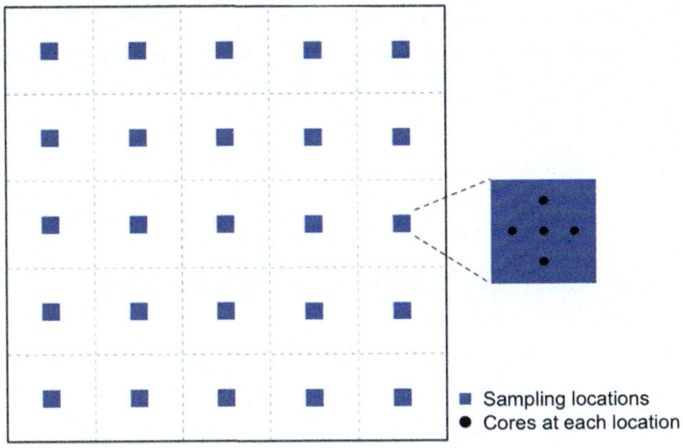

Figure 6.6. Grid sampling approach illustrated with blue squares marking sampling locations placed at grid centers (or grid intersections) to ensure systematic field coverage. (Courtesy Srinadh Kodali).

of soil fertility for nonmobile nutrients, such as phosphorus and potassium. If the field has been fertilized sufficiently over time (the "build soil test" approach), most samples may fall into the high fertility category for these nutrients, justifying continued use of this sampling method. However, if soil test results reveal a wide range of nutrient levels, increasing the sampling intensity might be necessary, dividing the field into smaller grids to cover more sampling area. Once spatial patterns for phosphorus and potassium are established, future sampling efforts can be adjusted to focus on other nutrients. Despite its advantages, grid sampling may still miss important variability if sampling points fall between zones of rapid soil change. Franzen and Peck (1995) provide additional information on the limitations and applications of grid sampling.

Back to our pizza analogy, if you slice the pizza into a grid, like a chessboard (Figure 6.7b), take a bite exactly from the center of each square. This gives you good overall coverage and you will be able to sample all parts of the pizza. However, you might miss small surprises, like olives clustered along the edges of slices or mushrooms at the center (Figure 6.7c). Grid sampling provides structured, uniform coverage but may overlook finer soil details that fall between sampling points. It can also be an expensive and inefficient method if the soil variability is relatively low.

Staggered Grid Start

A variation of grid sampling known as the staggered grid start balances efficiency with improved data quality by introducing a random offset to each row of sampling points (Table 6.1). Instead of sampling at fixed locations, such as the center or corner of each grid cell, this method shifts the starting point of each row, creating a zig-zag pattern (Figure 6.8). This approach reduces alignment bias by avoiding systematic errors in a single direction and is particularly effective in irregular landscapes where subtle soil variability may be missed by traditional grid sampling. Multiple soil cores are collected at each location to ensure that the samples accurately represent local conditions.

Back to our pizza analogy, if you wanted to taste a pizza using a staggered grid sampling pattern, you would shift every other row so that

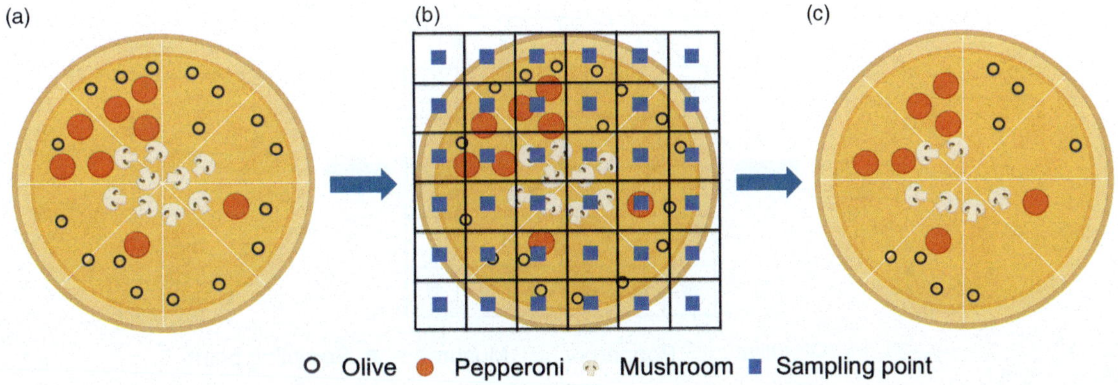

Figure 6.7. Grid sampling visualized using a pizza analogy. Toppings represent soil variability, and evenly spaced sampling points provide structured coverage. Features between grid points may be missed, highlighting potential alignment bias. (Courtesy Srinadh Kodali).

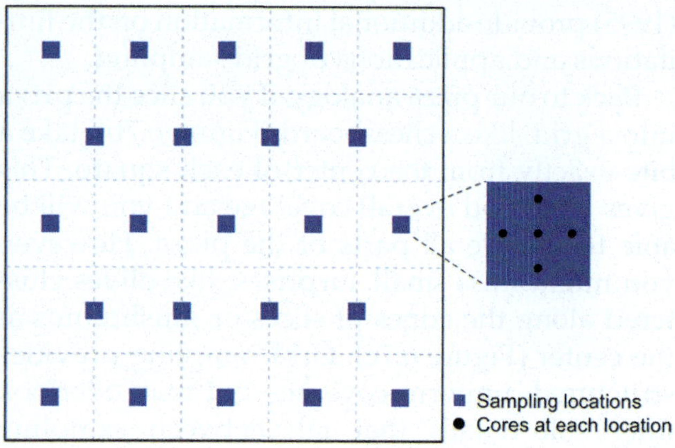

Figure 6.8. Illustration of the staggered grid sampling approach. Note that sampling points (blue squares) are offset between rows to reduce alignment bias, and multiple soil cores (black dots) are collected at each location to better represent local variability. (Courtesy Srinadh Kodali).

your bites fall in a zig-zag pattern across the pizza (Figure 6.9). This is different from random or grid sampling, in which you slice the pizza into perfect squares and sample from the center of each. Instead, this pattern helps you sample areas that might otherwise be missed, like pepperoni that tends to fall between the slices or sauce swirls near the edges. Similarly, in soil sampling, a staggered grid reduces the chance of aligning with repeating field features, helping capture more representative variability across the field. One caveat of this approach is that it accounts for bias in one direction (horizontal or vertical) but may still leave bias in the other direction. Therefore, the decision to use the method should consider the direction in which variability needs to be captured in the field.

Systematic Unaligned Grid

Another refinement of grid sampling is the systematic unaligned grid (Figure 6.10). In this method, the field is divided into grid cells; however, instead of sampling at a fixed point within each cell, a single point is randomly selected. This random offset avoids coinciding with repeating spatial patterns and helps mitigate bias in both horizontal and vertical directions, making it particularly useful in fields with streaks of over- or underfertilization or in cases where multiple fields are managed as a single unit (Table 6.1).

In our pizza analogy, if you were to use a systematic unaligned grid method to get the most representative taste of the pizza, you would still divide the pizza into squares (Figure 6.11b). However, instead of always taking a bite from the center of each slice as you would in random and grid sampling, you would randomly choose a spot within each square to taste (Figure 6.11). This way, every part of the pizza is covered, but you avoid a systematic pattern and bias. This way, you are more likely to catch scattered toppings like stray olives or uneven pepperoni. Similarly, in the field, this sampling strategy ensures full spatial coverage while reducing the risk of bias from repetitive patterns.

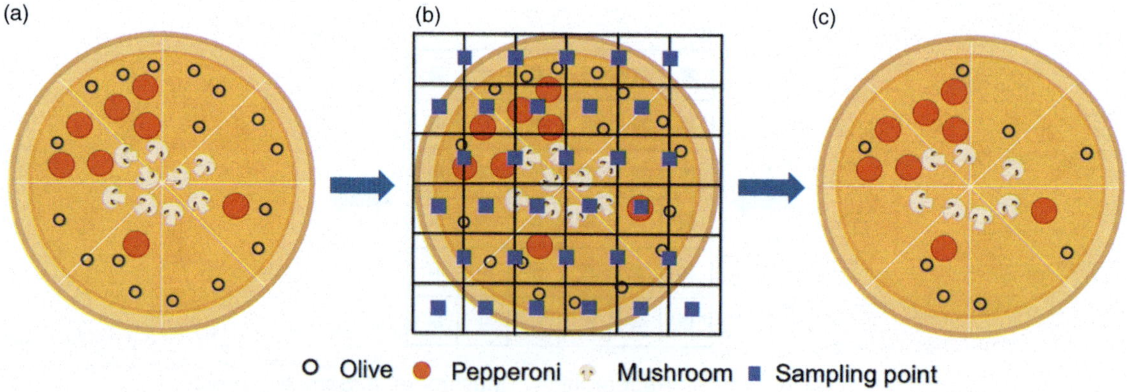

Figure 6.9. Staggered grid sampling depicted using a pizza analogy. Offset sampling points capture variability more effectively by avoiding alignment with repeating patterns, such as toppings distributed between slices. (Courtesy Srinadh Kodali).

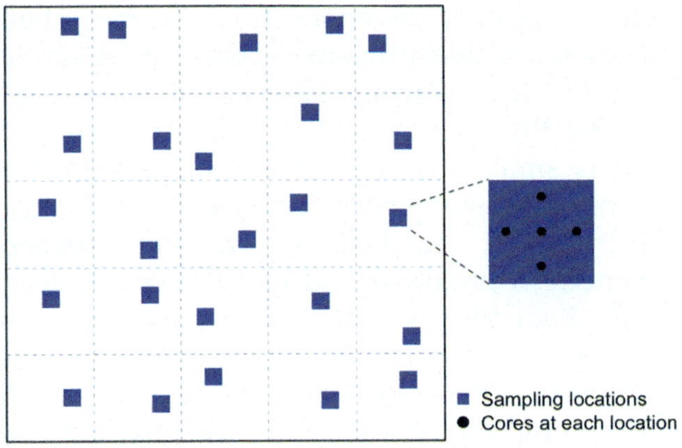

Figure 6.10. Systematic unaligned grid sampling approach. Sampling points (blue squares) are randomly shifted within each grid cell to avoid alignment with spatial patterns, enhancing the detection of variability. (Courtesy Srinadh Kodali).

Zone/Stratified Random Sampling

Zone sampling, also known as stratified random sampling, is a targeted soil sampling strategy that divides a field into distinct zones based on underlying spatial variability. These zones, or strata, are typically delineated using factors such as topography, soil type, historical management practices, or yield patterns (Fridgen et al., 2004). Within each zone, soil samples are collected randomly, under the assumption that nutrient levels are randomly distributed within the zone (Figure 6.12). Zones can be based on topography, parent material, or previous management. This method is particularly useful for producers who seek spatially explicit soil information but prefer a more cost-effective alternative to grid sampling. Zone sampling offers a balance between precision and affordability. Soil survey boundaries published by the NRCS and accessible via the Web Soil Survey are commonly used to identify initial zone boundaries. To delineate and verify management zones, Whelan and McBratney (2003) suggest a five-step process that includes the following:

1. Gathering relevant data (e.g., soil, environment, and yield) layers
2. Partitioning the data into distinct zones or classes
3. Determining the optimal number of zones
4. Creating the management zone map
5. Analyzing the variation of the chosen parameters within each zone.

Back to our pizza analogy, if we want to get the most representative taste of our pizza using zone sampling and we already know the pizza has three zones of toppings: pepperoni on the left, mushrooms in the center, and olives mainly near the crust (Figure 6.13a). Instead of randomly guessing where to taste, you would need to deliberately take bites from each of the topping zones (Figure 6.13b). This gives you a more complete and accurate sense of how the whole pizza tastes (Figure 6.13c). Similarly, in zone-based soil

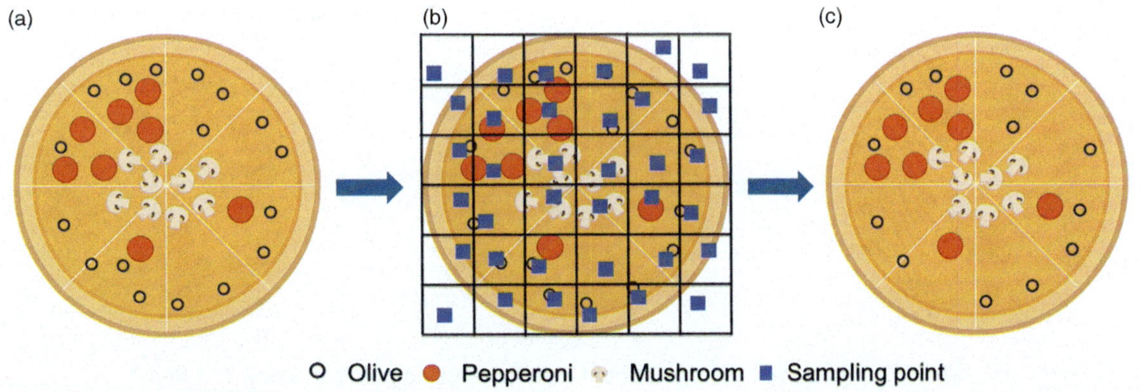

Figure 6.11. Systematic unaligned grid sampling illustrated using a pizza analogy: Varied toppings represent different soil properties, and systematic grid points ensure more even coverage but may still miss key features between sampling locations. (Courtesy Srinadh Kodali).

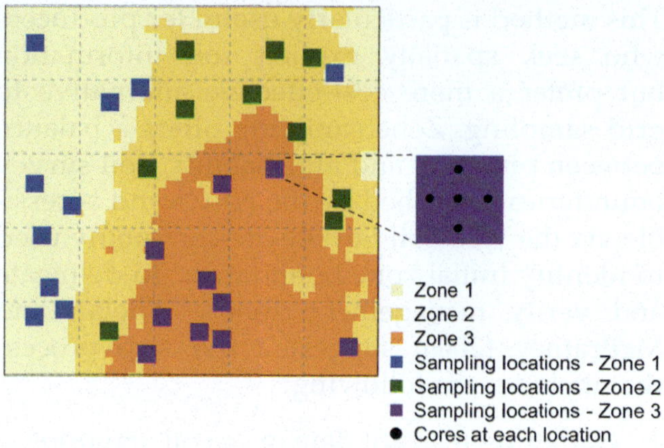

Figure 6.12. The field is divided into three management zones (shaded in yellow-orange) based on interpolated soil conductivity. Within each zone, sampling points (colored squares) are randomly selected, spaced to avoid clustering, and constrained within field boundaries. The zoomed-in view shows five soil cores (black dots) collected within a small radius and combined into one sample. (Courtesy Srinadh Kodali).

sampling, you target areas that are known to behave differently, ensuring your sampling captures most of the variability across the field.

While management zone sampling can be more efficient, less expensive, and less time consuming, there are some challenges that users must be aware of before using the method (Nawar et al., 2017). These challenges include (a) collecting the appropriate information that limits yields to identify the zones, (b) measuring these factors at a scale sufficient to characterize spatial variation, (c) ensuring that variation is greater between the zones while variation within each zone is small, and (d) identifying an optimal number of management zones that is both cost-effective and practical for growers. Recent advances in proximal and remote sensing technology have the potential to assist in data collection for determining the optimal number and location of zones (see Chapters 11–15 for more information on remote and proximal sensing technologies).

When constructing a management zone map, integrating multiple years of yield data can help avoid and limit some of these challenges by ensuring the spatial patterns analyzed are stable and consistent in normal years (Blackmore, 2000; Miao et al., 2018). Regardless of whether a field has a history of a single crop or diverse crop rotations, the same general procedure can be followed. For instance, a field in continuous wheat might have varying yields across years, so standardizing such yield data can improve zone identification. Standardization involves converting yield values (bushels per acre or megagrams per hectare) into relative yield by dividing each yield value at each location within a field by the maximum yield observed in the entire field. This process is repeated for each year's yield data. Based on the standardized data, the mean values and coefficients of variation for each yield cell are determined by dividing the standard deviation

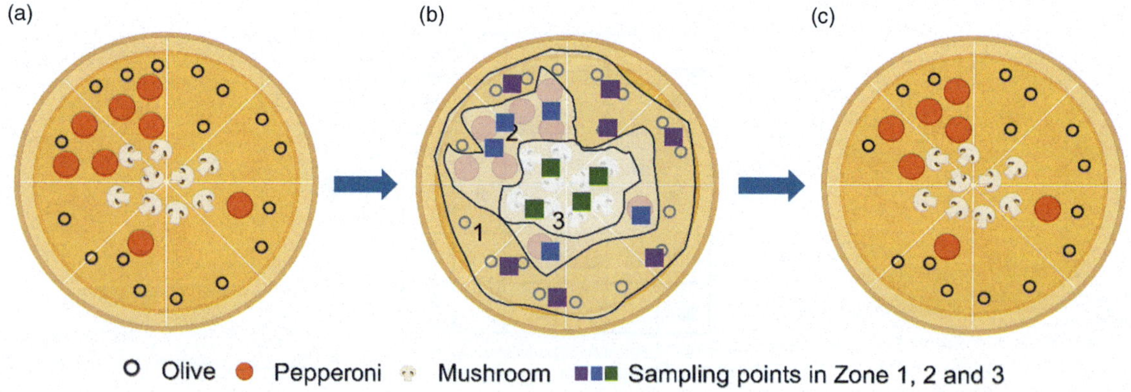

Figure 6.13. Zone-based stratified random sampling illustrated using a pizza analogy: Toppings are unevenly distributed across the pizza, so it is first divided into zones. Random sampling points are then chosen within each zone to capture the variability of toppings, ensuring that all topping types are represented in the overall sample. (Courtesy Srinadh Kodali).

of each year's yield data by the average. The resulting data can be used to classify the field into zones characterized by high yield with low variability or low yield with low variability.

Management zone maps may also be based on elevation, soil nutrient levels, crop reflectance, soil electrical conductivity (EC), and soil classification (Lacerda et al., 2022). Interpolation maps of these elements can be produced and superimposed geographically to aid in zone delineation as discussed in Chapters 8 and 9. Zone sampling is particularly effective for testing soil nitrate when fertilizer recommendations depend on residual nitrate levels, which are influenced by water movement and crop productivity (Clay et al., 2017; Morris et al., 2018).

The number of zones selected typically ranges from three to five, although there is no universal standard. The optimal number depends on the spatial complexity of the field. To ensure the effectiveness of zone-based sampling, it is essential to minimize within-zone variability. Metrics such as normalized classification entropy and the fuzziness performance index can be used to evaluate zone consistency. Tools, such as the "smart-map" Quantum Geographic Information System (QGIS) plug-in, which employs fuzzy *k*-means or support vector machine algorithms (Pereira et al., 2022), and proprietary software like SMS Advanced, are available to assist crop consultants in creating management zones.

Ultimately, both the map developer and the end user must understand the delineation and purpose of the management zones created to move beyond uniform application strategies. In some cases, small-scale variability within zones may necessitate further subdivision. Targeted sampling within well-defined zones enhances the accuracy of nutrient recommendations, leading to more efficient input use and improved crop performance.

Other Sampling Approaches

Beyond the commonly used methods, several advanced soil sampling strategies have been developed to address specific research and management needs (Allen et al., 2010; Lawrence et al., 2020; Nagel et al., 2024; Rodrigues et al., 2025). Among these, the conditioned Latin hypercube sampling method stands out as a sophisticated form of stratified random sampling that incorporates ancillary spatial data. This approach selects sampling locations to maximize the representation of variability in both soil properties and environmental covariates, which may be continuous or categorical (Minasny & McBratney, 2010). Although this method offers low statistical bias and high representativeness, it is computationally intensive and requires substantial resources to implement, making it more suitable for research applications than for routine field management (Table 6.1).

Soil Sampling Tools and Software

The evolution of soil sampling tools and software has significantly enhanced the precision and efficiency of data collection in agricultural systems. Historically, manual tools, such as hand probes and augers, have been the standard for collecting soil samples at various depths. These instruments are valued for their simplicity and affordability; however, they can be labor-intensive and less effective in large-scale operations or in soils with high compaction or gravel content.

To address these limitations, hydraulic soil sampling systems have been developed and are commonly mounted on or towed by all-terrain vehicles (ATVs), utility task vehicles (UTVs), or trucks. These mechanized systems enable consistent and efficient core collection with minimal physical exertion, even in challenging soil conditions. Figure 6.14 illustrates the range of soil samplers and their depth capabilities, highlighting

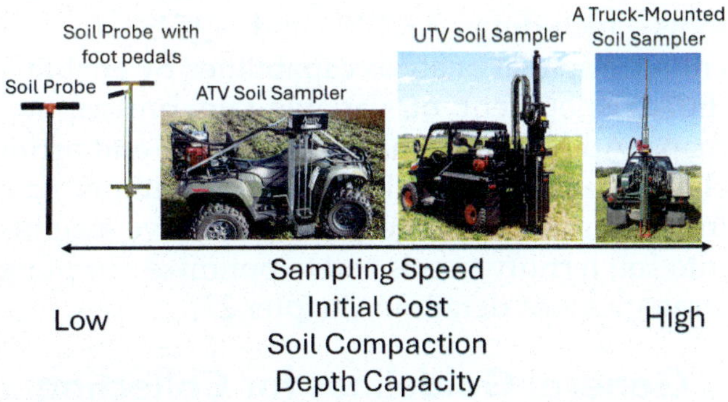

Figure 6.14. Soil sampler chart: features and depth capabilities. Note that heavier equipment can easily and efficiently collect samples from deeper layers and is also capable of collecting shallow samples.ATV refers to all-terrain vehicles; UTV denotes utility task vehicles. (Courtesy Katsutoshi Mizuta).

how heavier equipment can penetrate deeper soil layers. However, this increased capability comes with higher operational costs.

Among truck-mounted probes, two widely used models are the Bull probe, typically mounted on the side of the vehicle, and the Giddings probe, mounted on the rear. The Giddings probe, produced by the Giddings Machine Company in Windsor, Colorado, operates using gas engines or hydraulic motors that power either a coring mechanism or a rotating auger. For bulk density measurements, larger tip sizes may help reduce sample compaction, although the effect may vary depending on soil type. Additionally, using larger tip sizes requires more power to operate effectively.

In addition to physical sampling tools, digital technologies have transformed how soil sampling protocols are designed and executed. The Soil Sampling Planner Optimizer Tool, developed by the University of Nebraska-Lincoln in collaboration with the USDA Agricultural Research Service, provides a web-based platform that simplifies the planning process by optimizing sampling locations based on field variability (https://sspot.scinet.usda.gov).

The Global Navigation Satellite System has allowed the use of sampling devices to become standard in precision agriculture, ensuring spatial accuracy and repeatability. These devices integrate seamlessly with farm management software platforms, such as AgLeader™, Trimble Ag Software™, and the John Deere Operations Center™, which facilitate the mapping of sample locations, tracking of nutrient levels, and coordination with other precision technologies. GIS tools, including ArcGIS and QGIS, further enhance spatial analysis capabilities by enabling the development of variable-rate prescription maps. Additionally, machine learning and artificial intelligence (AI)-based algorithms are increasingly used to generate predictive insights into soil fertility trends and to optimize sampling strategies (see details in Chapter 27).

General Guidelines for Collecting Representative Soil Samples

Collecting representative soil samples is a critical step in accurately characterizing field conditions for precision nutrient management. This process requires careful planning and attention to detail. The following general guidelines should be considered before initiating soil sampling:

1. Make an initial assessment of a field's landscape and note any visible variations (e.g., plant growth, crop yield) caused by soil-forming factors, as well as previous management practices, such as variable-rate fertilizer applications, known sites of homesteading, or livestock confinement.
2. Determine management objectives; for example, a goal might be to determine fertilizer requirements for the upcoming growing season or to evaluate nutrient leaching potential.
3. Choose an appropriate sampling approach (see details in the section "Soil Sampling Methods") and identify the appropriate number of soil samples based on the factors listed above.
4. Identify the expected costs and returns of the chosen sampling approach, including the cost of sample analysis (e.g., the Louisiana State University Agricultural Center Soil Testing Lab charges $10 per sample for a routine soil testing package) and reevaluate the sampling strategy based on the available budget, equipment, and labor.
5. Develop the appropriate protocol(s) for soil sampling to ensure safe and effective operations in the field.

Identifying Variability from Crops, Fertilizers, and Manure

Historical crop rotations, fertilizer applications, and manure additions can significantly influence soil nutrient variability over time. To improve the accuracy and efficiency of soil sampling, it is essential to account for these management-induced changes. One of the key challenges in sampling is determining the appropriate depth, which can vary depending on crop root depth, the specific nutrient or soil property being analyzed, the analytical methods used, and the management system in place, such as conventional tillage versus no-till practices. Pedogenic processes also play a role in determining the sampling depth.

The method of fertilizer application further complicates sampling (Kitchen et al., 1990). For example, banded fertilizer applications can create zones of elevated nutrient concentration that persist for years (Clay et al., 2002). To assist in decision-making, Figure 6.15 presents a step-by-step framework that helps farmers determine whether uniform or variable-rate management is appropriate based on the presence, cause, and actionable potential of within-field yield variability. This decision tree supports the development of sampling strategies that align with the spatial and temporal dynamics of nutrient availability.

Summary

Precision agriculture depends on a comprehensive understanding of soil variability, which arises from both intrinsic factors, such as climate, parent material, and topography, and human-induced influences, including management practices and land use history. Recognizing and quantifying these sources of variability is essential for developing site-specific nutrient recommendations that align with the principles of 4R (4Rplus or 5R) management.

The factorial models, such as the STEP-AWBH model, offer a structured framework for catego-

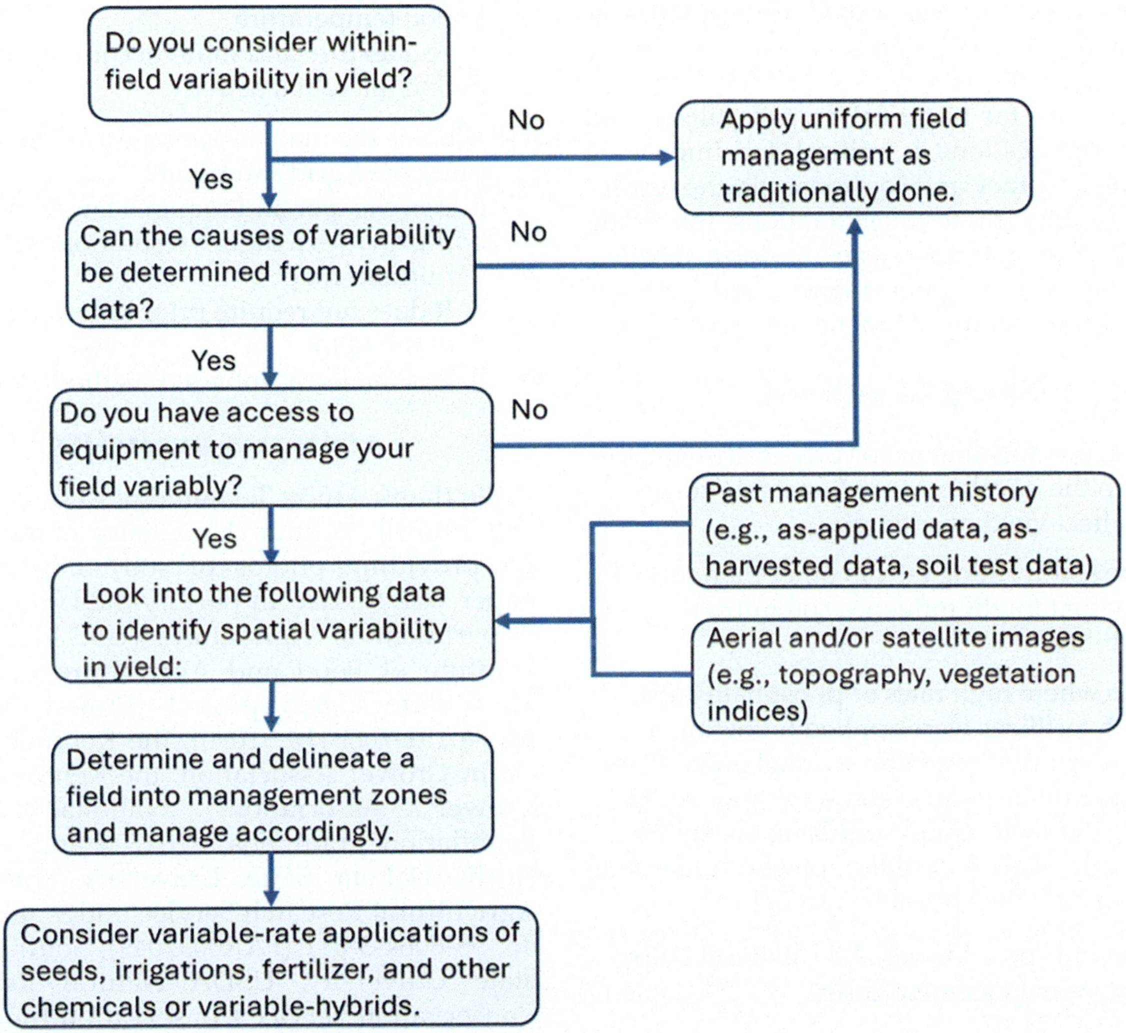

Figure 6.15. Decision tree for selecting appropriate field management strategies based on within-field yield variability. (Modified based on Whelan, 2018; Courtesy Katsutoshi Mizuta).

rizing soil variability into distinct factors that affect soil properties, crop growth, and overall field productivity. This chapter has examined these factors in detail and provided guidance on selecting appropriate soil sampling strategies to support precision nutrient management.

Effective soil sampling is foundational to accurate landscape representation and informed decision-making. The design of sampling protocols must consider a range of factors, including the intended goals, field history, soil type, crop characteristics, available resources, and collaborative partnerships. Advances in soil sampling tools and software have greatly improved the precision and efficiency of data collection. While traditional tools, such as hand probes and augers, remain useful, a Global Navigation Satellite System–enabled devices and AI-driven platforms offer enhanced spatial accuracy and predictive capabilities.

Accounting for historical crop, fertilizer, and manure applications is critical for improving sampling accuracy and confidence in the results. By integrating these considerations into sampling strategies, farmers can optimize resource use, enhance crop performance, and promote sustainable agricultural practices.

Study Questions

1. Which soil-forming factor physically contributes to the development of unstable zones that affect yield over time?
2. Name soil-forming factors other than topography that might influence soil nutrient variability.
3. Fields where high rates of phosphate and potash fertilizer were applied following a "build soil test" program would benefit from a higher intensity grid soil sampling method (more than one sample per 2.5 ac) next time, especially if 90–95% of the soil test values in a field exceed the critical level. (T/F).
4. Name four possible variables that can help delineate management zones.
5. Compare sampling strategies discussed in this chapter and address their pros and cons from the aspects of bias, cost, and feasibility.
6. What are the 4Rs, and what might be considered the 5th R or the "plus" in 4Rplus?
7. Discuss the pros and cons of the common assumption that soil mass per acre at a 6-in. depth is equivalent to 2,000,000 lb (when the actual soil bulk density is 1.25 g/cm^3.
8. Grid sampling is the most cost-effective method for detecting soil nutrient variability in highly heterogeneous fields. (T/F).
9. Soil compaction from heavy machinery can lead to nutrient stratification and reduced root penetration by plants. (T/F).
10. Which soil property is most affected by parent material?
    a. Soil aggregate stability
    b. Soil temperature
    c. Soil texture and mineral composition
    d. Soil microbial activity
11. What is the main advantage of zone sampling over grid sampling?
    a. It is cheaper and requires fewer samples
    b. It provides a higher resolution soil nutrient map
    c. It does not require prior knowledge of the field
    d. It is the most unbiased method available

ACKNOWLEDGMENTS

We thank Amity Technology LLC, AMS Inc., and Riley J. Walton (University of Kentucky) for providing photos of soil samplers. This paper was funded in part by the United States Department of Agriculture (USDA) National Institute of Food and Agriculture under the Multi-State Hatch project [Project Number: KY006151-7008110-S1090], the Kentucky Small Grain Grower Association, the Kentucky Corn Grower Association, National Science Foundation grant nos. 2202706 and 2026431, North Dakota State University, the USDA Agricultural Research Service under agreement no. 58-6064-3-011 (FAR0037049), South Dakota State University, USDA Natural Resources Conservation Service Conservation Innovation Grants (NR233A750011G019), USDA Agriculture and Food Research Initiative, American Coalition for Ethanol, GEVO, the

Kentucky Corn Grower Association, the South Dakota Corn Utilization Council and the South Dakota Soybean Research and Promotion Council. The use of trade names is for the convenience of the reader and does not imply endorsement by the authors.

REFERENCES

Allen, D. E., Pringle, M. J., Page, K. L., & Dalal, R. C. (2010). A review of sampling designs for the measurement of soil organic carbon in Australian grazing lands. *The Rangeland Journal*, *32*(2), 227–246. https://doi.org/10.1071/RJ09043.

Blackmore, S. (2000). The interpretation of trends from multiple yield maps. *Computers and Electronics in Agriculture*, *26*(1), 37–51. https://doi.org/10.1016/S0168-1699(99)00075-7.

Clay, D. E., Carlson, C. G., Brix-Davis, K., Oolman, J., & Berg, B. (1997). Soil sampling strategies for estimating residual nitrogen. *Journal of Production Agriculture*, *10*(3), 446–452. https://doi.org/10.2134/jpa1997.0446.

Clay, D. E., Kitchen, N., Carlson, C. G., Kleinjan, J. L., & Tjentland, W. A. (2002). Collecting representative soil samples for N and P fertilizer recommendations. *Crop Management*, *1*(1), 1–5. https://doi.org/10.1094/CM-2002-1216-01-MA.

Clay, D. E., Robinson, C., & DeSutter, T. M. (2017). Chapter 6: Soil testing and understanding soil testing results for precision farming. In D. E. Clay, S. A. Clay, & S. Bruggeman (Eds.), *Practical Mathematics for Precision Farming*. American Society of Agronomy.

Crouse, K. K., Hardy, D. H., Heckendorn, S., Huluka, G., Joines, D. K., Kissel, D. E., Miller, R., Mitchell, C. C., Moore, K. P., Mylavarapu, R., Oldham, J. L., Provin, T., Savoy, H. J., Sikora, F. J., Sonon, L., Wang, J. J., Warncke, D. D., & Zhang, H. 2014. *Soil test methods from the Southeastern United States*. Southern Cooperative Series Bulletin no. 419, Southern Cooperative. https://aesl.ces.uga.edu/sera6/PUB/MethodsManualFinalSERA6.pdf.

Evans, D. L., Quinton, J. N., Tye, A. M., Rodés, Á., Davies, J. A. C., Mudd, S. M., & Quine, T. A. (2019). Arable soil formation and erosion: A hillslope-based cosmogenic nuclide study in the United Kingdom. *Soil*, *5*, 253–263. https://doi.org/10.5194/soil-5-253-2019.

Franzen, D. 2018. *Developing zone soil sampling maps*. North Dakota State University Extension SF1176-2. https://www.sbreb.org/wp-content/uploads/2018/05/Developing-Zone-Soil-Sampling-Maps.pdf.

Franzen, D. W. 2023. *Soil sampling as a basis for fertilizer application*. North Dakota State University Extension SF990. https://www.ndsu.edu/agriculture/sites/default/files/2023-09/sf990.pdf.

Franzen, D. W., & Peck, T. R. (1995). Field soil sampling density for variable rate fertilization. *Journal of Production Agriculture*, *8*(4), 568–574. https://doi.org/10.2134/jpa1995.0568.

Fridgen, J. J., Kitchen, N. R., Sudduth, K. A., Drummond, S. T., Wiebold, W. J., & Fraisse, C. W. (2004). Management zone analyst (MZA). *Agronomy Journal*, *96*(1), 100–108. https://doi.org/10.2134/agronj2004.1000.

Grunwald, S. 2021. Grand challenges in pedometrics-AI research. *Frontiers in Soil Science*, *1*, 714323. https://doi.org/10.3389/fsoil.2021.714323.

Grunwald, S., Thompson, J. A., & Boettinger, J. L. (2011). Digital soil mapping and modeling at continental scales: Finding solutions for global issues. *Soil Science Society of America Journal*, *75*(4), 1201–1213. https://doi.org/10.2136/sssaj2011.0025.

Jenny, H. (1941). *Factors of soil formation—A system of quantitative pedology*. Dover Publications.

Joshi, D. R., Moriles-Miller, J., Clay, S. A., Reicks, G., Westhoff, S., & Clay, D. E. (2024). Rye cover crop had a nitrogen credit one year after termination in a frigid environment. *Journal of Soil and Water Conservation*, *79*(5), 261–267. https://doi.org/10.2489/jswc.2024.00097.

Kitchen, N. R., Westfall, D. G., & Havlin, J. L. (1990). Soil sampling under no-till banded phosphorus. *Soil Science Society of America Journal*, *54*(6), 1661–1665. https://doi.org/10.2136/sssaj1990.03615995005400060026x.

Lacerda, L., Miao, Y., Mizuta, K., & Stueve, K. 2022. Identifying key factors influencing yield spatial pattern and temporal stability for management zone delineation. *15th International Conference on Precision Agriculture*.

Lawrence, P. G., Roper, W., Morris, T. F., & Guillard, K. (2020). Guiding soil sampling strategies using classical and spatial statistics: A review. *Agronomy Journal*, *112*(1), 493–510. https://doi.org/10.1002/agj2.20048.

Lyons, S. E., Clark, J. D., Osmond, D. L., Parvej, M. R., Pearce, A. W., Slaton, N. A., & Spargo, J. T. (2023). Current status of US soil test phosphorus and potassium recommendations and analytical methods. *Soil Science Society of America Journal*, *87*(4), 985–998. https://doi.org/10.1002/saj2.20536.

Mallarino, A. P., & Wittry, D. J. (2004). Efficacy of grid and zone soil sampling approaches for site-specific assessment of phosphorus, potassium, pH, and organic matter. *Precision Agriculture*, *5*, 131–144. https://doi.org/10.1023/B:PRAG.0000022358.24102.1b.

McBratney, A. B., Mendonça Santos, M. L., & Minasny, B. (2003). On digital soil mapping. *Geoderma*, *117*(1–2), 3–52. https://doi.org/10.1016/S0016-7061(03)00223-4.

Miao, Y., Mulla, D. J., & Robert, P. C. (2018). An integrated approach to site-specific management zone delineation. *Frontiers of Agricultural Science and Engineering*, *5*(4), 432–441. https://doi.org/10.15302/J-FASE-2018230.

Minasny, B., & McBratney, A. B. (2010). Conditioned Latin hypercube sampling for calibrating soil sensor data to soil properties. In R. A. V. Rossel, A. B. McBratney, & B. Minasny (Eds.), *Proximal soil sensing* (pp. 111–119). Springer.

Morris, T. F., Murrell, T. S., Beegle, D. B., Camberato, J. J., Ferguson, R. B., Grove, J., Ketterings, Q., Kyveryga, P. M., Laboski, C. A. M., McGrath, J. M., Meisinger, J. J., Melkonian, J., Moebius-Clune, B. N., Nafziger, E. D., Osmond, D., Sawyer, J. E., Scharf, P. C., Smith, W., Spargo, J. T., . . . Yang, H. (2018). Strengths and limitations of nitrogen rate recommendations for corn and opportunities for improvement. *Agronomy Journal*, *110*(1), 1–37. https://doi.org/10.2134/agronj2017.02.0112.

Nagel, P., Mizuta, K., & Khalsa, S. J. 2024. Towards a standardized protocol and policy for acceptance of innovative soil testing. *IEEE International Geoscience and Remote Sensing Symposium*. https://doi.org/10.1109/IGARSS53475.2024.10642192.

Nawar, S., Corstanje, R., Halcro, G., Mulla, D., & Mouazen, A. M. (2017). Chapter four—Delineation of soil management zones for variable-rate fertilization: A review. In *Advances in agronomy* (Vol. *143*, pp. 175–245). https://doi.org/10.1016/bs.agron.2017.01.003.

Omer, M., Idowu, O. J., Ulery, A. L., VanLeeuwen, D., & Guldan, S. J. (2018). Seasonal changes of soil quality indicators in selected arid cropping systems. *Agriculture*, *8*(8), 124. https://doi.org/10.3390/agriculture8080124.

Pereira, G. W., Valente, D. S. M., de Queiroz, D. M., de Coelho, A. L., F., Costa, M. M., & Grift, T. (2022). *Smart-map*: An open-source QGIS plugin for digital mapping using machine learning techniques and ordinary kriging. *Agronomy*, *12*(6), 1350. https://doi.org/10.3390/agronomy12061350.

Ragland, J., Egli, D. B., Mizuta, K., Greb, S., & Levy, J. E. 2025. *The role of phosphorus in Kentucky agricultural development: A story of the haves and the have–nots*. Extension publications (No. ID-278). University of Kentucky Cooperative Extension. https://publications.mgcafe.uky.edu/id-278.

Raney, W. A., Edminster, T. W., & Allaway, W. H. (1955). Current status of research in soil compaction. *Soil Science Society of America Journal*, *19*(4), 423–428. https://doi.org/10.2136/sssaj1955.03615995001900040008x.

Rodrigues, H., Ceddia, M. B., Vasques, G. M., Grunwald, S., Babaeian, E., & Villela, A. L. O. (2025). autoRA: An algorithm to automatically delineate reference areas—A case study to map soil classes in Bahia, Brazil. *Land*, *14*(3), 604. https://doi.org/10.3390/land14030604.

Soil Science Division Staff. (2017). *Soil survey manual*. Government Printing Office.

Whelan, B. (2018). Site-specific crop management. In A. B. McBratney, B. Minasny, & U. Stockmann (Eds.), *Pedometrics* (pp. 597–622). Springer, Cham. https://doi.org/10.1007/978-3-319-63439-5_20.

Whelan, B. M., & McBratney, A. B. 2003. Definition and interpretation of potential management zones in Australia. *11th Australian Agronomy Conference*, Geelong, Victoria.

7

GPS and Other Global Navigation Satellite Systems for Precision Agriculture

Timothy Stombaugh, Won Suk Lee, Ali Mirzakhani Nafchi, Kent Shannon, and Hasan Mirzakhaninafchi

Chapter Purpose

Over the years, many different positioning technologies have been used to target variability and address yield limitations. Some of these include placing mechanical furrow followers on vehicles, using dead reckoning, burying cables across a field, and placing radio frequency transponders at various locations around the field. Many of these technologies lacked the accuracy necessary for common field applications and were impractical because they required either extensive installation of permanent equipment or systematic transfer of stationary equipment from one field to another. The development of the Global Positioning System and other satellite-based positioning systems, collectively known as Global Navigation Satellite Systems (GNSS), has overcome these limitations by providing real-time position information for vehicles, people, and devices as they travel across a field. Today, GNSS technology makes possible a variety of technologies that are used daily by many farmers globally. These technologies are used to apply variable rates of fertilizer, identify and control pests, scout fields, and harvest crops. These technologies enhance the average farmer's ability to increase efficiency and productivity, improving precision agriculture globally. In this chapter, readers are provided with a basic description of GNSS technology, the fundamental operational principles, common sources of error, and how different levels of accuracy are achieved through differential corrections and advanced computational techniques.

Global Navigation Satellite Systems

The Global Positioning System (GPS) was developed by the U.S. Department of Defense to help determine the location of military assets. When it became fully functional in the early 1990s, it was quickly adopted in agriculture because it could provide reasonably accurate position indication using hardware for spatial management and machine control that could be completely contained on a mobile vehicle. A plethora of machine technologies were quickly developed to use GPS technology for this same purpose.

The realization of GPS inspired the further development of systems by the United States and other nations (Table 7.1). These systems have a wide range of uses and coverage. For example, some of these systems have global coverage, while others target specific geographic regions. Some systems can be used for basic position determination, while others are intended to increase accuracy and/or reliability in other systems. In general, all systems designed to determine location on the Earth with reasonable accuracy operate on very similar principles using similar, if not identical, broadcast frequencies. Many higher-end positioning receivers simultaneously use signals from multiple satellite positioning systems to increase reliability and accuracy. Therefore, to more accurately represent this ever-growing technology, the general term Global Navigation Satellite Systems (GNSS) has been accepted to describe the collective group of satellite-based positioning systems.

Satellite Ranging

Most GNSS satellite systems consist of multiple satellites orbiting the Earth. For example, GPS originally consisted of 24 core satellites in six different orbital planes. In 2011, the constellation was modified to consist of 27 core satellites, and it has since expanded to 31 (Figure 7.1). Its satellites have an orbital period of approximately 12 hr, which means that each satellite circles the Earth twice in a 24-hr period. The constellation was designed so that at least four satellites would be visible from any point on the Earth's surface. As will be discussed later, signals received from at least four satellites are required for complete three-dimensional (3D) position determination. However, the accuracy of the real position can be improved by using signals from more than four satellites simultaneously. Additional information may be obtained by using signals from one of the many satellites orbiting the Earth (Table 7.1).

Table 7.1. Current Functional Satellite-based Positioning Systems

System	Owner
GPS	United States
GLONASS	Russia
Galileo	European Union
BeiDou	China

Each GNSS satellite transmits radio signals carrying digital information used to determine the range, or distance, between the satellite and the receiver. This distance is often referred to as the pseudo range. The signal frequencies of these transmissions vary slightly from system to system but generally fall within the range of 1,100–1,600 MHz. Most satellites broadcast on several different frequencies simultaneously.

There are two fundamental pieces of information broadcast by each satellite: its ephemeris and a timing code. The ephemeris can be described as the satellite's orbital location. The timing code is used by the receiver to calculate the distance to the satellite. The receivers compare an internally generated timing code with the identical timing code sent by the satellite to determine the time delay between the codes, which is the elapsed time for the signal to travel from the satellite to the receiver. These calculations must be very precise. Since radio signals travel at the speed of light (2.99×10^8 m/s or 186,000 mi/s), the transmission time between satellites and receivers is less than 0.07 seconds. The potential for error is great, as a miscalculation of 0.001 seconds would result in a distance error of over 300 km (186 mi).

A GNSS receiver uses the distance measurements between itself and multiple satellites to calculate its location using a mathematical technique called trilateration (Figure 7.2). A precise distance measurement from one satellite will locate the receiver somewhere on a sphere centered on that satellite. A second distance measurement to a second satellite will narrow the position to the intersection of the two spheres, which forms a circle. A third measurement will further narrow the position possibilities to two

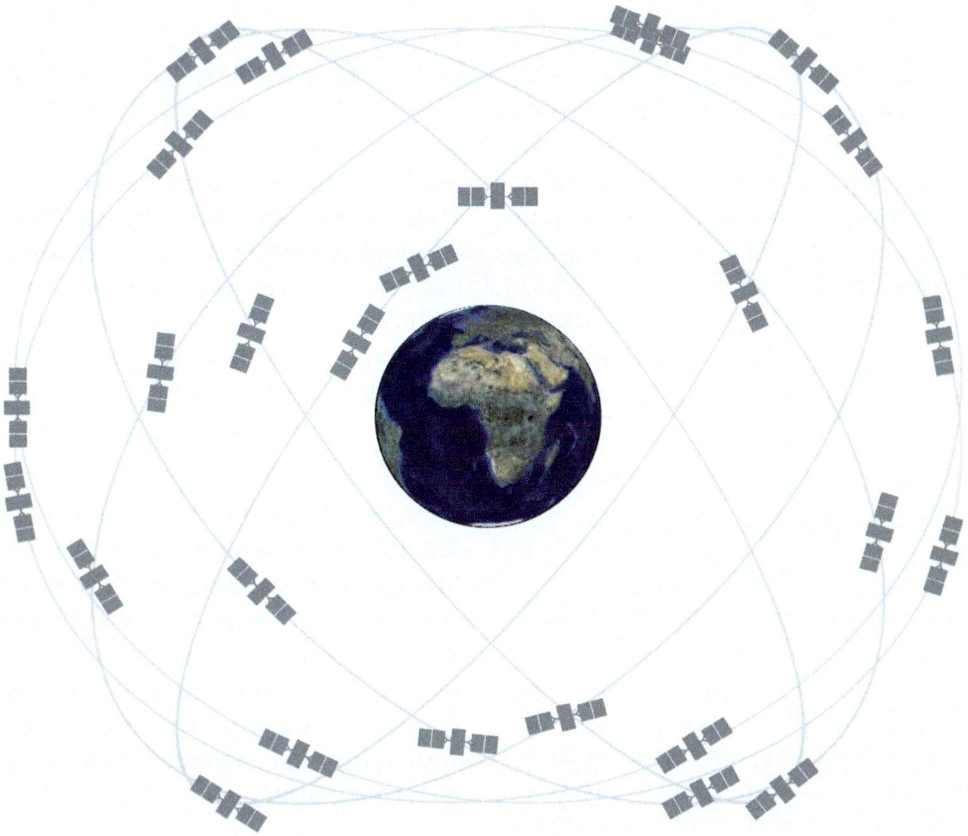

Figure 7.1. Locations of 27 GPS satellite orbital planes.

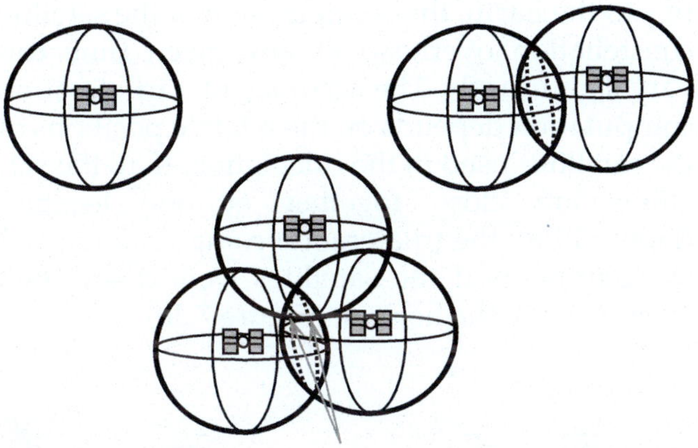

Figure 7.2. Illustration of trilateration position determination using one satellite (left), two satellites (right), and three satellites (bottom).

points (the intersection of a circle and a sphere). It is often possible to deduce which of those two points is the correct point by using an algorithm that might examine the recent position history or the reasonableness of the two solutions (one of which will often be nowhere near the surface of the Earth). Information from the fourth satellite is needed to confidently identify the correct position solution and to precisely synchronize the timing codes generated by the satellites and the receiver. To enhance accuracy, most manufacturers implement hardware and software techniques that use as many satellite signals as possible.

Some GNSS receivers will estimate and report a position if information is available from only three satellites. However, in this case, the receiver may display a message such as "2D Position" along with the position indication, and the user must be aware that the position solution will have limited accuracy, especially in the vertical direction. A common technique for computing a 2D position is to estimate the vertical (altitude) component of the solution based on the Earth's geoidal height, an elevation prediction derived from a mathematical model of the Earth's shape at the receiver's approximate position. This mathematical estimation allows the receiver to compute an estimate of the horizontal position coordinates based on the limited satellite data, but again, the accuracy of the solution may not be acceptable for many applications.

A NOTE ABOUT CELL PHONE POSITIONING

All cellular phones and tablets can determine their location without using GNSS signals. They use a form of trilateration based on the locations of nearby cellular towers, Wi-Fi access points, or Bluetooth devices to estimate the position. The accuracy of these positioning techniques is often no better than 100 m and is not sufficient for precision agriculture. Almost all cell phones also include a GNSS receiver, which provides better position accuracy in areas where satellite signals are available. Most devices have settings that allow the user to adjust the accuracy of position determination, which can impact battery life. These settings essentially turn the different positioning systems on or off. When set to the highest accuracy setting, cellular phones, tablets, and other handheld devices can be used for many precision agriculture tasks such as scouting, soil sampling, and boundary mapping.

Global Navigation Satellite System Errors

The accuracy of the position information received from a GNSS receiver depends on the quality of the receiver and the conditions under which it is used. The primary causes of GNSS position errors include clock imprecision, ephemeris variation, poor satellite configuration, atmospheric interference, and multipath errors. Understanding these errors will help users properly select and use GNSS equipment.

Global Navigation Satellite System Clock Imprecision

Clock imprecision errors result from the limited precision of the physical clocks within the receiver and the lack of exact synchronization between the receiver and satellite clocks. Compensation for clock limitations in receivers is part of the reason why a minimum of four satellite signals is needed to achieve a fully realized 3D position solution. More precise GNSS receivers improve accuracy by using higher-quality clocks and more accurate timing measurements. These enhancements will, of course, increase equipment costs.

Ephemeris Errors

Ephemeris errors, also known as ephemeris variation, are the differences between the actual and predicted locations of satellites. Because the satellites are orbiting within gravitational fields, their positions and movements can be accurately predicted. Nevertheless, due to either miscalculation of the data or variations in the orbital path, there may still be some errors in the broadcast satellite locations. Most GNSS systems use base stations placed strategically at precisely surveyed locations within their coverage areas to monitor the status of individual satellites and update ephemeris information. They will also update information about the health of each satellite, indicating its expected reliability for position determination.

Satellite Configuration

Because most GNSS satellites are moving relative to the Earth, the configuration of the satellite constellation overhead at any given time can vary significantly. The accuracy of a trilateration computation depends on the relative positions of the satellites used in the calculation. If all the satellites are close together at one location (Figure 7.3a), the trilateration computation is not as accurate as if the satellites were distributed evenly above the horizon (Figure 7.3b).

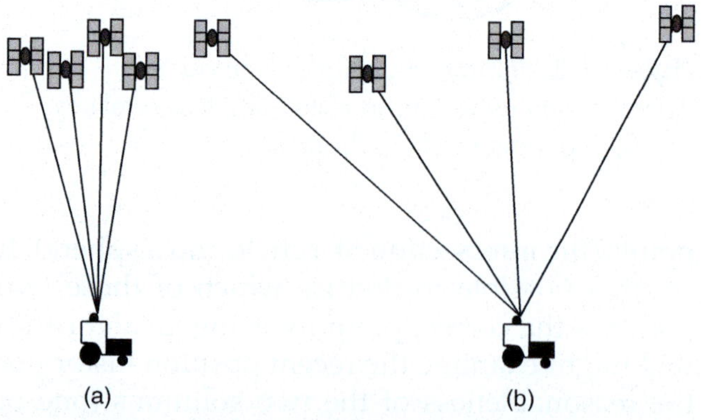

Figure 7.3. Satellite configurations showing (a) poor and (b) good DOP.

The configuration of the satellites at any point in time is quantified by the Dilution of Precision (DOP). Smaller DOP values indicate better accuracy; a DOP value of less than two is considered good. The DOP calculations can be refined to provide more specific indications of the horizontal, vertical, and time dilutions. Some GNSS receivers automatically report DOP values. Other receivers may not report the DOP but instead combine it with other parameters to estimate the accuracy of the position.

Dilution of Precision can be predicted using orbital models, and some users requiring the utmost precision will plan to do field work when the satellites are in the best configuration. Multiple software tools are available to help users plan for variations in satellite constellations. An example output from one software tool (https://gnssplanning.com/#/settings) shows how the number of visible satellites and DOP vary over a 24-hr period on a particular day (Figure 7.4).

Dilution of Precision can be affected by any object that blocks satellite transmissions. Common objects that block radio signals include buildings, metallic objects, and trees. In urban settings, operating between tall buildings would leave only part of the sky visible in the area between the buildings. This is referred to as an urban canyon. In agriculture, operating along tree lines at the edge of fields is a common situation that similarly reduces the visibility of satellites, resulting in poorer accuracy. In general, placing a GNSS antenna at locations where signals can be blocked will reduce accuracy. In precision agriculture, these locations include on a combine cab near a metal grain bin extension panel, or near a metal-clad building or grain bin.

Atmospheric Interference

As radio waves enter and pass through the Earth's atmosphere, they can be bent, or refracted, in much the same way that light is refracted when

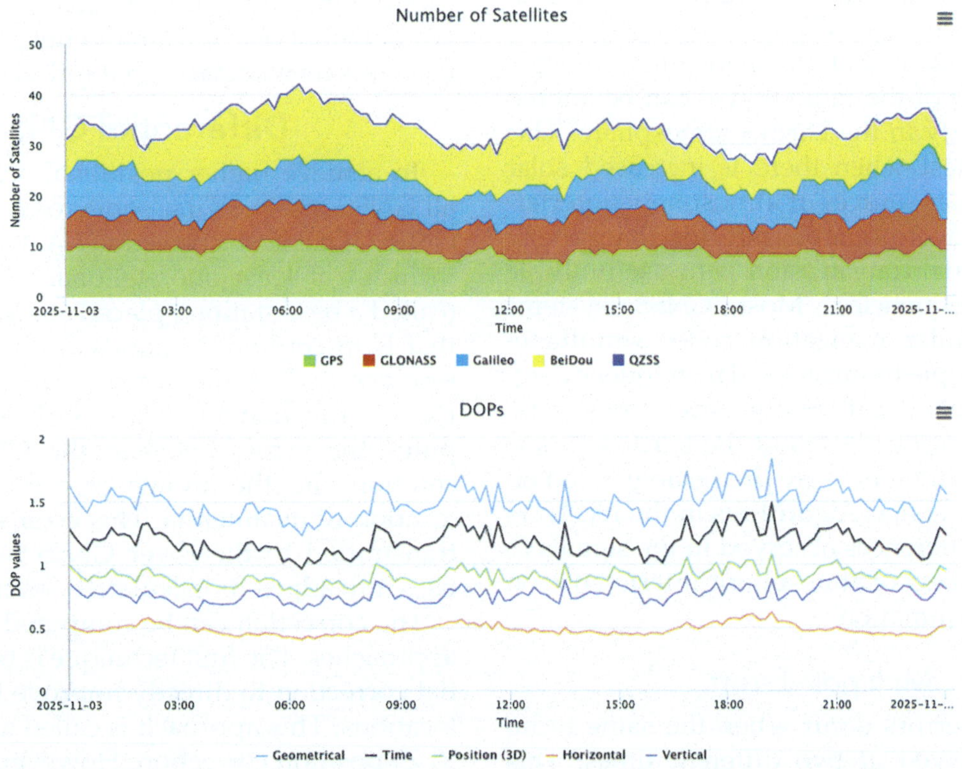

Figure 7.4. Example output from an online GNSS planning calculator showing the expected number of satellites available (top) and various measures of DOP (bottom) for a 24-hr period at a given receiver location using multiple GNSS systems. (Example output from https://gnssplanning.com/#/settings).

Figure 7.5. Illustration of atmospheric effects on radio signals.

it passes through a water surface (Figure 7.5). If the radio waves are bent, error is increased because the signal travels a longer path length from the satellite to the receiver. Atmospheric interference is greater for satellites lower on the horizon than for those directly overhead. To counteract this interference, many receivers ignore or mask satellites located below a certain angle above the horizon. On more sophisticated GNSS receivers, this mask angle may be changeable and can be selected by the user to achieve the most optimal angle. Typical values for a mask angle range from 8 to 14° above the horizon. The trade-off with masking satellites is that the remaining satellites may have a poor configuration, which increases DOP.

A second source of atmospheric interference is that the speed of the radio waves can be altered by ionic activity in the Earth's atmosphere. This error is greatest when there is increased solar activity or other sources of atmospheric interference. One way to compensate for ionospheric effects on signal transmission is to use multiple broadcast frequencies. Most GNSS systems broadcast similar navigation codes simultaneously on multiple frequencies. The frequency signals will behave differently when they pass through the various layers of the Earth's atmosphere. These differences are predictable based on the signal frequency. Algorithms in the receivers can exploit differences observed in the signals to reduce the impact of atmospheric interference on position determination.

Multipath Errors

Multipath errors occur when the same radio signal is received at two different times. This happens when the radio signal bounces off an object (Figure 7.6). For example, a GNSS transmission could be received twice if it comes

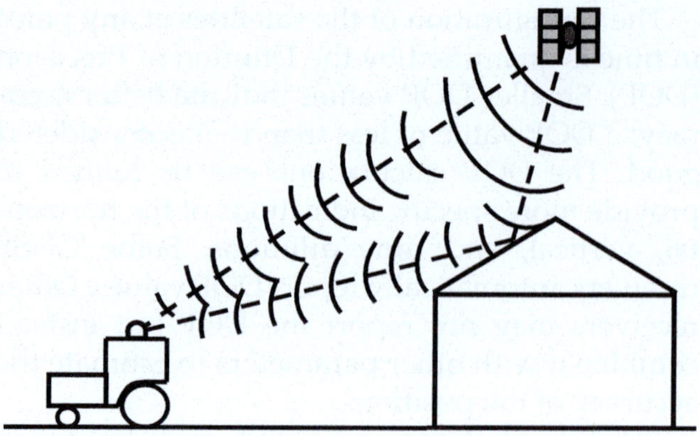

Figure 7.6. Illustration of multipath errors in GNSS transmissions.

directly from the satellite and the signal is also reflected off a building. Multipath errors can cause significant problems for GNSS receivers, especially when operating near metal buildings and bodies of water. Placing the receiver antenna near any metallic object can increase the chances of multipath. Manufacturers can reduce multipath errors by placing choke rings around antennae and by using hardware and software filtering techniques. However, users should be careful when operating in areas where multipath is a potential risk and should not place stationary receivers near surfaces that reflect radio signals.

Differential GNSS

To achieve higher positional accuracy, nearly all GNSS receivers use some form of Differential GNSS (DGNSS) correction. A DGNSS system includes at least one stationary GNSS receiver (called a base station), placed at a known location that is in communication with the user's GNSS receiver (called the rover). An explanation of DGNSS functionality is that the base station computes the errors between the GNSS-indicated solution and the known or precisely surveyed location of its antenna. This error information is transmitted to the rover GNSS receiver, which uses it to adjust its calculated position.

The correction can be computed using several approaches. The first technique is to use differential correction to directly improve the calculated locations. This approach is called a Cartesian (*X*, *Y*, *Z*) position correction. However, the accuracy of Cartesian corrections decreases as the distance from the base station increases. The second approach is to correct the data transmitted by each individual GNSS satellite. Correcting individual satellite signals will lessen the effect of baseline separation between the rover and the base station, but the accuracy still degrades as the distance increases.

Numerous open-source and proprietary computational techniques can be used in DGNSS. One technique used by some manufacturers is called Precise Point Positioning (PPP). A PPP correction signal relies on data from a network of base stations to estimate the clock, ephemeris, and some additional bias errors of each satellite transmission. Using these specific error corrections decreases the effect of baseline separation distance, making the corrections less geographically restricted.

The best accuracy is achieved by using a DGNSS technique called Real-Time Kinematic (RTK) GNSS. In contrast to other systems that use only the code data broadcast by the satellites to determine positions, RTK systems also use information about the carrier frequency of the satellite radio signals. Essentially, the receiver counts the number of carrier signal wavelengths between the satellite and the receiver itself, and the number is subsequently converted into a distance based on the signal frequency. Since most GNSS satellites broadcast on multiple frequencies, some RTK receivers use multiple signals simultaneously. These receivers may be called dual- or multifrequency RTK receivers as opposed to single-frequency receivers. Multifrequency receivers require antennae tuned to multiple frequencies and additional computational capabilities.

One challenge for RTK receivers is the initial determination of the number of wavelengths between the receiver on the rover and the satellite. Because of this, RTK systems often require longer initial acquisition times before they converge to the expected accuracy. Once the initial determination is made, the receivers can maintain a lock and continuously calculate distances. Although technological advances are allowing equipment to become smaller, RTK receivers are traditionally physically larger and require more power than non-RTK systems.

There are two primary methods for computing RTK corrections. The simpler method is called Observation Space Representation. Observation Space Representation systems use multiple reference stations to determine the range errors for each satellite transmission. These errors are incorporated by the rover to improve the accuracy of the trilateration solution. A more sophisticated method is called State Space Representation. State Space Representation computations decompose satellite transmissions into specific components, including satellite clock and ephemeris errors, atmospheric interference effects, and signal biases. State Space Representation reduces the impact of baseline separation between the rover and the base station by enabling a more location-specific estimation of correction information. However, these techniques often require increased computational capabilities in the receivers and may involve proprietary signals and methods, including two-way communication between the rover and the base station network.

One user consideration for any DGNSS system is acquisition time. As mentioned earlier, one challenge for RTK receivers is determining the initial number of carrier wavelengths for each visible satellite. Because of this, RTK systems often require longer initial acquisition times before they converge to a specific location accuracy. Precise Point Positioning and some other DGNSS systems aim to reduce this convergence time, which could reduce accuracy. A user needs to understand how the potential trade-off between accuracy and acquisition time may impact field operations, especially if there is a chance of signal interruption requiring reacquisition.

Accuracy of DGNSS Corrections

There are many differential correction services available for users to choose from. The differences between these services lie in the location and number of base station receivers, calculation algorithms, and correction data delivery method.

Single base station RTK systems are common in agricultural and construction applications, and they are called Local Area DGNSS systems. These systems achieve very high accuracy when the base station is correctly placed and the rover operates relatively close to it. Single base station systems generally rely on land-based radio modems to transfer correction information to the rover.

A growing number of DGNSS correction systems with broader geographic coverage are maintained by private or public entities. Examples of commonly used free systems are the U.S. Wide Area Augmentation System (WAAS) and the Geostationary Navigation Overlay Service (EGNOS), which is maintained by the European Space Agency. WAAS was developed by the Federal Aviation Administration to add redundancy to GPS solutions, providing the reliability necessary for passenger airline operations. These broader coverage DGNSS systems generally consist of an array of independent base stations spaced as widely as possible to minimize equipment costs while still meeting the accuracy requirements. In these systems, the rover generally uses correction information from the nearest base station. The correction signals are usually transferred to the rover either via satellite or cellular communications. Wide Area DGNSS systems provide lower accuracy than single base station RTK systems.

While some base station networks are established and maintained by private entities, all U.S. states have established public RTK base station networks. Most of the base stations in these state networks are registered under the Continuously Operating Reference Station (CORS) program, a base station certification program administered by the National Geodetic Survey. For a GNSS reference station to be registered as a CORS base station, it must adhere to a set of standards governing the construction, location, and maintenance of the station. Most states integrate the CORS stations into networks that provide public access to RTK-level corrections.

Communication of differential correction information between base stations and rovers can be accomplished in several ways. One technique is to establish a terrestrial-based radio frequency (RF) link, as is common with single base station systems. Historically, one of the first broad coverage augmentation services was developed and maintained by the United States Coast Guard (USCG). It consisted of DGNSS reference stations and land-based towers with RF transmitters that broadcast correction information. Some GNSS receivers at that time were manufactured with the capability to receive satellite and USCG radio transmissions. However, this network was only useful near navigable water bodies that were of interest to the USCG. It was used heavily in agriculture but is no longer supported and became obsolete in 2020.

Terrestrial RF links are still common with single base station RTK systems on farms and construction sites where baseline distances are relatively short. They commonly use a lower-powered spread spectrum radio modem set to establish the data link between the base station and rover(s). The challenge with the lower-powered RF links is that many obstacles can disrupt the line-of-sight transmission of radio signals. This problem can be reduced by using higher-powered radios; however, these typically require Federal Communications Commission licensing. Historically, the rules governing this communication technology can change, which can result in equipment becoming obsolete.

To overcome the coverage challenges of terrestrial RF links, many current differential correction systems use communication satellites to relay correction information from the base station network to GNSS receivers. These systems are often referred to as satellite-based augmentation systems (SBAS). Traditionally, SBAS were designed for broader coverage, recognizing that accuracies would be lower than those achievable with single base station systems. Most SBAS satellites are geostationary, meaning that their location does not change relative to the Earth's surface. This means that their coverage will be geographically restricted (Table 7.2). The most common SBAS signals used by GNSS receivers are the U.S-maintained WAAS and the Europe-maintained EGNOS systems. Signals from WAAS and EGNOS satellites are broadcast on the same frequencies as GPS signals; therefore, additional antennae or a tuned receiver are not required. Currently, nearly all GNSS devices, including

Table 7.2. Examples of Satellite-based Augmentation Systems

System	Coverage
WAAS	United States
EGNOS	Western Europe
MSAS	Japan
GAGAN	India
SDCM	Russia
SNAS	China
SACCSA	South/Central America

those in cellular phones and tablets, rely on WAAS and/or EGNOS to improve accuracy. Depending on the hardware capabilities, time of day, and rover location, accuracy as good as 1 m or slightly better is achievable with free SBAS corrections.

Wide area distribution of DGNSS signals can also be accomplished using internet protocols and cellular communications. The most common examples of internet-based distribution are the state-maintained RTK networks. The corrections from state-based RTK networks are usually delivered over the internet using a standardized protocol called Network Transport of Radio Technical Commission for Maritime Services (RTCM) via Internet Protocol (NTRIP). Radio Technical Commission for Maritime Services is a data format protocol for differential correction information, which will be discussed further in Section 7.7.

The NTRIP correction information is streamed from the base station network to a web interface. The rover then accesses the data stream through a cellular web connection. The link can be established with a cell phone tethered to the rover receiver. More commonly, a dedicated cellular modem, either integrated into the GNSS receiver or contained in a separate hardware device, is used to establish the link. These dedicated cellular modems can have more powerful receivers and antennae than cellular phones, resulting in better cellular reception. Cellular-based correction systems rely on the availability of cellular service, which can be a problem in some agricultural areas.

In the past, people used general accuracy levels to categorize receivers into different groups, such as uncorrected, sub-meter, decimeter, or RTK. However, these distinctions have become blurred as more DGNSS systems have been introduced. Uncorrected devices, which typically have accuracies no better than about 3–10 m, have been shown to be useful for basic agricultural tasks such as scouting, soil sampling, and field area determination. Sub-meter receivers, as the name implies, provide accuracy to less than a meter and are useful for mapping, yield monitoring, and even some basic guidance functions with larger machinery. Almost all GNSS devices, including cell phones, use one of the free SBAS systems and provide sub-meter accuracy. Examples of different accuracy types are highlighted in Table 7.3.

A key factor affecting the DGNSS receiver accuracy is the distance between the base station and the rover. This accuracy decreases with increasing distance and is often reported as the degradation rate in parts per million (ppm). An example specification might look like 2 cm + 5 ppm. This receiver would be expected to have a positional accuracy of 2 cm when it is very close to the base station. Degradation of specification refers to the baseline distance, or the distance between the rover and the base station, and it is interpreted as a loss of the specified number of units of accuracy for every million units of separation. In the example above, the degradation specification of 5 ppm is interpreted to mean that the 2-cm fundamental accuracy would degrade

Table 7.3. A Summarization of GNSS Receiver Accuracy based on DGNSS Correction Method and Type of Receiver

Receiver type	Accuracy (typical root mean square error)	Use case	Correction methods
Single-frequency (L1) GNSS	1–3 m	Basic navigation, smartphones, Geographic Information System mapping	SBAS (WAAS, EGNOS)
Dual-frequency (L1/L2, L1/L5) GNSS	0.5–1 m	Precision agriculture, surveying, unmanned aerial vehicles (UAVs)	SBAS, PPP, RTK
RTK GNSS	1–2 cm	High-precision surveying, construction, autonomous vehicles	RTK corrections from base station
PPP GNSS	10–50 cm (3–5 cm with long convergence time)	Guidance systems, scientific research, UAV mapping	Commercial correction services

by 5 cm for every million centimeters, or 10 km, of distance between the rover and base station. Typical RTK systems degrade in accuracy by about 1 cm for every 10 km (1 in. for every 15 mi) of baseline separation.

Though many consumers and manufacturers prefer to use rather general terminology and simplified numeric indicators to describe GNSS accuracy, the quantification and specification of accuracy is a complicated process. There are four critical aspects of GNSS accuracy that impact agricultural operations. The first aspect is the distinction between absolute and relative performance. A receiver with good relative accuracy would exhibit consistent position determination relative to earlier position solutions from the same receiver. That receiver might be very adequate for minimizing overlapping and skipped coverage in a single field operation. If, however, it would be desirable to use multiple machines in the same field or to overlay data from one field operation with another, the lack of absolute accuracy could cause significant spatial offsets between the field operations or map layers. The absolute accuracy of a receiver is a measure of the position's deviation from some established reference frame. A good absolute accuracy would ensure that multiple machines working in the same field would be able to follow the same path and that multiple data layers would overlay correctly without geographic offset.

The second critical aspect of GNSS position quality is the distinction between short-term and long-term accuracy. Because most GNSS satellite constellations are constantly changing, multiple factors, such as satellite constellation changes and ionospheric interference, which occurs when GNSS signals pass through charged particles in the Earth's atmosphere, can cause GNSS position measurements to drift over time. Consequently, position measurements taken over a shorter time interval will tend to be more precise than those collected over extended periods. Because of this, many manufacturers will report a short-term accuracy performance parameter often called "pass-to-pass" accuracy. This is typically based on a 15-min relative accuracy performance, but that time reference can change depending on the application. For many receivers, long-term accuracy is approximately three times worse than short-term pass-to-pass accuracy.

The third critical aspect of GNSS accuracy is the distinction between using static and dynamic GNSS receivers. A relatively simple way to measure accuracy is to record position locations from a GNSS system placed at a known fixed location over a period of time. This static accuracy is not the same as the dynamic accuracy measured when the receiver is moving. Solution algorithms and data filtering tuned to improve static accuracy and stability can introduce positional errors when the receiver is moving. For example, positional filtering in the receiver's solution algorithm could cause the receiver to indicate excessive overshoot during a corner maneuver (Figure 7.7).

The fourth critical aspect of GNSS accuracy is the distinction between horizontal and vertical accuracy. Vertical errors are generally larger than horizontal errors because the trilateration calculations are based on satellites located at high altitudes. Since many agricultural operations do not rely on the vertical component of position, manufacturers often report horizontal and vertical accuracies separately.

Because there are many ways to measure and evaluate GNSS receiver accuracy, it can be difficult to directly compare systems offered by different manufacturers. This inconsistency can lead to misconceptions, as receivers with similar

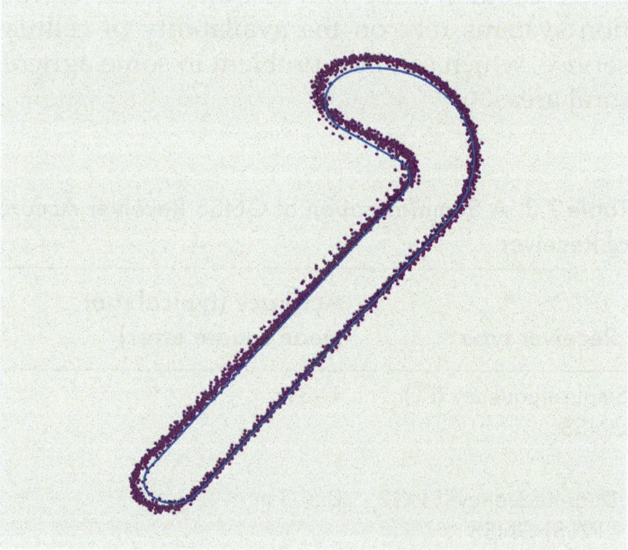

Figure 7.7. Example of dynamic GNSS performance from a receiver traveling clockwise on a serpentine track, showing an incorrect indication of overshoot on corners.

specified accuracies based on different statistical inferences can have quite different performance. van Diggelen (2007), in an article in GPS World magazine, showed the complexities involved in statistical reporting methods and discussed techniques that could be used to compare receivers. Fortunately, standards have been developed, making it easier to compare receivers.

Because satellite constellations are constantly changing, GNSS performance varies over time. The major GNSS constellations nearly repeat every 12 hr and come even closer to a complete repeated cycle every 24 hr. Testing standards account for this by specifying appropriate testing times and durations based on celestial cycles. The location of the receiver can also affect the achievable accuracy. For example, the satellite orbital planes do not cross the polar regions of the Earth. This means that a receiver located closer to the poles will not see satellites in the polar direction. This phenomenon can be illustrated by a plot of satellite paths relative to a point in the Earth's Northern Hemisphere (Figure 7.8). In discussions about satellite configurations and DOP, this lack of satellite dispersion can cause accuracy to degrade at higher latitudes. It also causes larger errors in the north–south direction than the east–west direction. Rather than differentiating this in performance specifications, most testing standards require combining all horizontal errors into a single value.

Global Navigation Satellite System Receiver Development Trends

As with any technology, hardware and software innovations impact device capabilities and costs. Global Navigation Satellite System users desire highly accurate, lightweight systems at low cost. This is especially important for the use of unmanned aerial vehicles (UAVs) in assisting with scouting and chemical application. Precision chemical application or image acquisition with a UAV requires RTK-level precision. The key

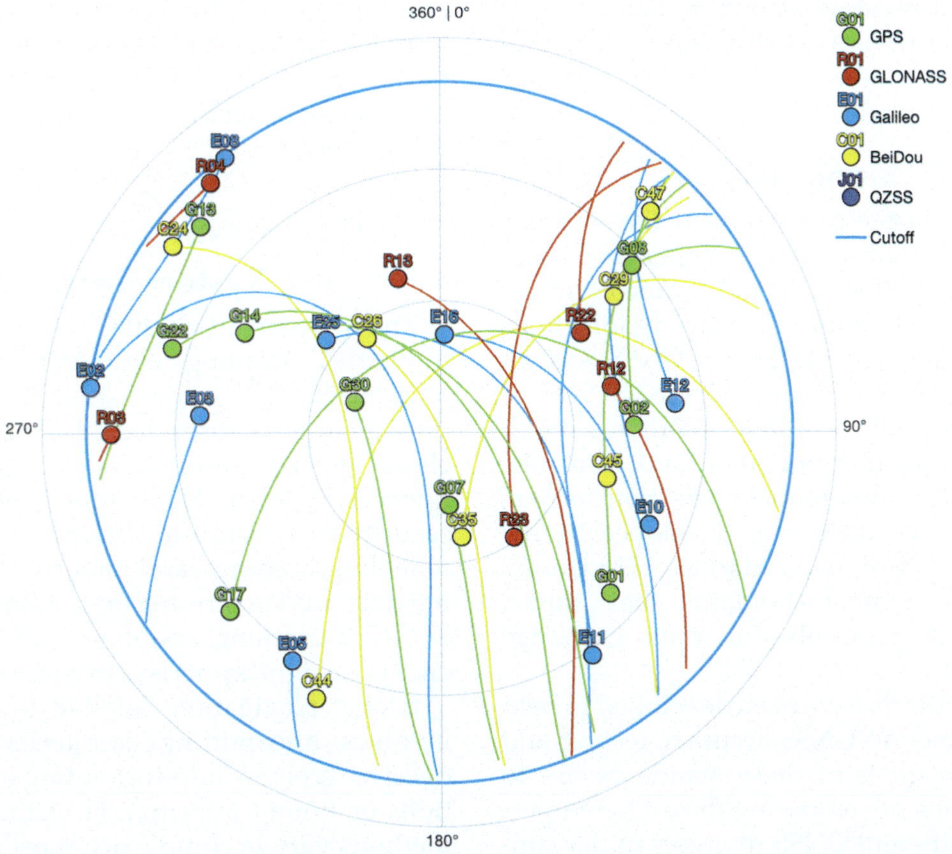

Figure 7.8. Plot of satellite paths on a particular day as viewed vertically from a point on the Northern Hemisphere of the Earth. (Example output from free satellite planning software available at https://gnssplanning.com/#/settings).

technologies that have facilitated the development of lightweight, low-cost GNSS receivers include system-on-chip manufacturing, open-source libraries, more efficient signal processing algorithms, and improved antenna calibration.

There have been many studies evaluating the performance of new low-cost DGNSS receivers for agricultural applications (Hamza et al., 2024, 2025; Jackson et al., 2018; Janos & Kuras, 2021; Janos et al., 2022; Li et al., 2023; Robustelli et al., 2023; Tunini et al., 2022; Valente et al., 2020). Many of these receivers track multiple constellations and have RTK capabilities via NTRIP DGNSS corrections. In general, these studies found that many receivers are acceptable for use in typical precision agriculture applications. However, some receivers performed well only in open areas, which might limit their use in applications such as forestry, where signal obstruction is common. They also reported that other receivers exhibited lower observation quality than expected due to a variety of problems outside the scope of this chapter, such as low carrier-to-noise ratio, multipath effect, cycle slips, phase, pseudo-range noise, and susceptibility to intentional jamming. Additionally, to avoid problems, users should check the manufacturer's specifications carefully when purchasing low-cost receivers.

Standards

Standards are necessary to ensure the accuracy and functionality of satellite systems, as well as the compatibility of signals and protocols across manufacturers. There are several organizations that have created standards for GNSS system functionality. The International GNSS Service (https://igs.org/), for example, maintains standards affecting satellite operation and transmissions. The RTCM differential correction format, used by many receivers, was developed by the RTCM. The National Marine Electronics Association developed standards that impact communication protocols for data exchange between devices.

Several standards have been developed to help bring consistency to GNSS accuracy testing and reporting. The goals of these standards are to provide a means of comparing the expected performance of different GNSS receivers under conditions typically found in agricultural fields. A standard developed by the Institute of Navigation (ION, 1997) outlined detailed procedures for performing static tests and a standardized way to report the findings of those tests. This standard is no longer updated or maintained by ION, but its content served as the basis for a set of international standards maintained by the International Organization for Standardization (ISO) that specify common terminology, test procedures, and reporting methodology for GNSS receiver performance.

An ISO standard related to GNSS accuracy is ISO 17123-8, which specifies field procedures for testing the precision of GNSS field measurement systems used in surveying, building, and industrial applications. This standard defines how to measure GNSS accuracy in RTK mode. It allows users to verify the suitability of their GNSS equipment for static applications.

Two additional ISO standards were developed for dynamic agricultural field operations. ISO 12188-1 outlines accuracy testing and reporting standards for GNSS equipment used in dynamic motions typical of agricultural field operations. This standard clarifies terminology related to absolute and relative accuracy, as well as short-term versus long-term performance. The second part of ISO 12188-2 focuses on the performance of automated steering systems that use GNSS for vehicle location.

Global Navigation Satellite System Security

The widespread use of GNSS in transportation industries, farming equipment, smartphones, and smartwatches suggests that any disruption due to natural causes, technical failures, or malicious attacks could have far-reaching consequences. Global Navigation Satellite System security in agriculture is critical to ensuring the reliability, accuracy, and integrity of the positioning data used in farming operations. Threats such as GNSS jamming, spoofing, and signal interference pose significant risks to agricultural systems.

Global Navigation Satellite System spoofing involves transmitting counterfeit signals that deceive receivers into computing erroneous positions or timing information. Although spoofing methods vary in complexity, most attacks exploit fundamental vulnerabilities in unencrypted, low-power GPS signals.

In simple spoofing, attackers overpower legitimate GPS signals with fake ones at higher power, causing receivers to calculate incorrect positions or timing. In intermediate spoofing, such as "meaconing," attackers capture authentic satellite signals and rebroadcast them after a delay. This causes the receiver's clock and position estimates to drift, leading to navigational errors over time. In "replay" attacks, attackers record genuine GPS signals and play them back, effectively "repeating" past signal data. Receivers then believe the current time or location corresponds to the recorded signals, leading to temporal or positional inaccuracies.

In advanced spoofing attacks, artificial intelligence–driven or complex signal-generation methods closely mimic legitimate GPS signals to synchronize counterfeit signals with real ones, maintaining realistic timing and phase relationships to evade detection. In non-synchronized spoofing, fake signals are transmitted without matching the phase or timing of real satellite signals. Although often easier to detect (due to abrupt positioning errors), non-synchronized spoofing can still be highly effective if the target receivers lack anomaly-detection capabilities.

In synchronized spoofing, counterfeit signals are precisely aligned (in phase and timing) with legitimate satellite signals. This alignment makes detection much more difficult, as the spoofed signals closely mirror the real signals. Synchronized spoofing uses two notable techniques:

1. Selective delay spoofing, which selectively delays the satellite signals to the receiver. This approach results in an inaccurate positional location.
2. Time manipulation spoofing, which alters the satellite timestamp. This misleads receivers about the correct time synchronization. Time manipulation can disrupt navigation, financial transactions, and critical infrastructure operations that depend on GPS time signals.

Coordinated spoofing is a specialized scenario in which multiple spoofing devices or attackers work in concert, often combining techniques such as meaconing, replay, and selective delay, to deceive multiple receivers simultaneously. This orchestration can generate very realistic false positional data across several targets, increasing an attack's overall impact. In a replay-attack scenario, a well-positioned adversary could record authentic satellite transmissions using a high-gain antenna and a software-defined radio. By rebroadcasting these signals a short time later at higher power, the receiver would erroneously lock onto delayed information, causing incremental drift in navigational accuracy over time, all without the user's immediate awareness.

Cybersecurity Challenges in Precision Agriculture

Unmanned aerial vehicles use GNSS to provide precise location and time information for navigation and autonomous mission planning. Unmanned aerial vehicles are particularly vulnerable to spoofing attacks because their GNSS signals are unencrypted. Hybrid GNSS spoofing threats can disrupt UAV operations by overwhelming navigation and communication systems. When GNSS signals are spoofed in even one UAV, corrupted location data can spread across the entire network, disrupting flight coordination and risking mid-air collisions or faulty mission execution.

Autonomous vehicles rely on GNSS for real-time navigation and route optimization. Despite advancements in sensor-fusion technologies, GNSS remains the primary component for accurate positioning. Global Navigation Satellite System spoofing attacks on autonomous vehicles can cause navigation errors by feeding incorrect location data to the vehicle's control system. Such attacks could lead to traffic disruptions and accidents.

To address these sector-wide dangers, researchers and industry practitioners have developed a range of detection techniques, each targeting unique signal anomalies and system discrepancies that can help prevent a spoofing attack.

Summary

GNSS technology is a core precision agriculture technology that provides real-time position and time information to machines and devices as they move across fields. Because of technological advances, the accuracy and reliability of GNSS systems have improved while the size and cost of receivers have decreased.

Study Questions

1. What is the minimum number of satellite signals needed for a GNSS receiver to compute a 3D position? Explain.
2. Use internet resources to explore the status of the GPS constellation and report the total number of satellites currently functional in the constellation, how many have L2C capabilities, and how many broadcast the L5 codes.
3. Use internet resources to explore the CORS stations in your state. Print a state map showing all the locations. For the one closest to your location, find out where it is and when it was established.
4. A farmer wants to purchase an RTK GNSS system for a farm. The base station antenna will be placed on top of a centrally located grain bin. If the accuracy of the RTK GNSS system is 1.5 cm + 4 ppm, what is the accuracy at a field located 35 km away?
5. For a particular date and location, use a constellation modeling tool such as the one found at http://gnssplanning.com to plot the availability of GPS and GNSS satellites and the DOP values. Are there particular times of the day that would be really good or potentially bad for GNSS work?
6. Explain how the absolute and relative accuracy of GNSS receivers would affect their use in machine guidance applications.
7. Explain the difference between short-term and long-term accuracy and how these two types of accuracy relate to GNSS position quality.
8. A farmer has a field that is bordered on one side by a large manufacturing facility in a three-story metal-clad building. Why might their GNSS equipment perform poorly near that edge of the field?
9. A GPS satellite is at an altitude of 20,100 km directly above a receiver. Calculate how many wavelengths of the L1 and L2 GPS frequencies there would be between the satellite and the receiver.
10. A farmer in Canada is getting ready to plant a field that is perfectly square and nearly completely flat using a GNSS-based autosteer system. In which direction (N-S or E-W) should they plant the field to get the best guidance accuracy? Explain your reasoning.

ACKNOWLEDGMENTS

Support for this document was provided by the Precision Farming Systems community in the American Society of Agronomy, South Dakota State University, USDA Natural Resources Conservation Service Conservation Innovation Grants (NR233A750011G019), the American Coalition for Ethanol, the South Dakota Corn Utilization Council, and the South Dakota Soybean Research and Promotion Council. The use of trade names is for the convenience of the reader and does not imply endorsement by the authors.

REFERENCES

Hamza, V., Stopar, B., Sterle, O., & Pavlovčič-Prešeren, P. (2024). Observations and positioning quality of low-cost GNSS receivers: A review. *GPS Solutions*, *28*, 149. https://doi.org/10.1007/s10291-024-01686-8.

Hamza, V., Stopar, B., Sterle, O., & Pavlovčič-Prešeren, P. (2025). Recent advances and applications of low-cost GNSS receivers: A review. *GPS Solutions*, *29*, 56. https://doi.org/10.1007/s10291-025-01815-.x.

Institute of Navigation (ION). (1997). ION STD 101. *Recommended test procedures for GPS receivers*. Institute of Navigation.

Jackson, J., Davis, B., & Gebre-Egziabher, D. (2018). A performance assessment of low-cost RTK GNSS receivers, *2018 IEEE/ION Position, Location and Navigation Symposium (PLANS)* (pp. 642–649), Institute of Navigation. https://doi.org/10.1109/PLANS.2018.8373438.

Janos, D., & Kuras, P. (2021). Evaluation of low-cost GNSS receiver under demanding conditions in RTK network mode. *Sensors*, *21*(16), 5552. https://doi.org/10.3390/s21165552.

Janos, D., Kuras, P., & Ortyl, Ł. (2022). Evaluation of low-cost RTK GNSS receiver in motion under demanding conditions. *Measurement*, *201*, 111647. https://doi.org/10.1016/j.measurement.2022.111647.

Li, X., Gou, H., Lin, X., Shen, Z., Lyu, H., Zhou, Y., Wang, H., & Zhang, Q. (2023). Performance analysis of frequency-mixed PPP-RTK using low-cost GNSS chipset with different antenna configurations. *Satellite Navigation*, *4*, 26. https://doi.org/10.1186/s43020-023-00116-3.

Robustelli, U., Cutugno, M., & Pugliano, G. (2023). Low-cost GNSS and PPP-RTK: Investigating the capabilities of the u-blox ZED-F9P module. *Sensors*, *23*(13), 6074. https://doi.org/10.3390/s23136074.

Tunini, L., Zuliani, D., & Magrin, A. (2022). Applicability of cost-effective GNSS sensors for crustal deformation studies. *Sensors*, *22*(1), 350. https://doi.org/10.3390/s22010350.

Valente, D. S. M., Momin, A., Grift, T., & Hansen, A. (2020). Accuracy and precision evaluation of two low-cost RTK global navigation satellite systems. *Computers and Electronics in Agriculture*, *168*, 105142. https://doi.org/10.1016/j.compag.2019.105142.

van Diggelen, F. (2007). GPS accuracy: Lies, damn lies, and statistics. *GPS World*, *January*, 41–45. https://www.scribd.com/document/669798980/GPS-Accuracy-Lies-Damn-Lies.

Fundamentals of Geographic Information Systems

8

Liang Liang, Katsutoshi Mizuta, Indunil P. Dharmasiri, Leonardo M. Bastos, and Rob Proulx

Chapter Purpose

A Geographic Information System (GIS) is a computer-based program for geospatial data processing. Geospatial data describes objects, fields, and events that are tied to specific geolocations over the surface of the Earth. Users of GIS explore and analyze geospatial data to extract meaningful geographic information for a wide range of applications. This chapter provides a concise introduction to GIS in the context of precision agriculture. It outlines the typical GIS workflow and covers the essentials of spatial data management and cartographic representation of geographic information. This chapter is a revision of Brase (2018), with most materials rewritten and updated. This chapter is not intended to be a comprehensive description of GIS. Additional information is available in Chang (2018), Kurt et al. (2016), Longley et al. (2015), Price (2023), Rano et al. (2022), Toromade and Chiekezie (2024), and Zurmotai (2020).

Introduction

Geographic Information Systems (GIS) are a suite of computer-based programs specialized in processing and visualizing geospatial data. Geospatial data are specifically spatial data georeferenced to locations on the surface of the Earth, while spatial data is used in a more general way. In most contexts, the term "geospatial" may be used interchangeably with "spatial", given that most of the spatial data is for the Earth's surface. For example, almost 80% of all the open data available from U.S. government agencies is geospatial (https://data.gov). More details on the various uses of spatial data can be found in Chapters 4, 5, 6, 9, 10, and 27.

GIS derives its origin from computerized mapping in the 1960s and since then has evolved to become a comprehensive set of computer-based tools that can effectively collect, manage, and analyze geospatial data. Further, GIS can be used to engage users and support decision-making by providing meaningful geographic information. GIS combines standard database functions, such as querying and statistical analysis, with the visualization and geographic analysis capabilities provided by maps. Precision farming requires accurate geospatial data and associated tools for analyzing fertilizer application, planning irrigation, and performing yield assessment and prediction. The synergistic use of GIS, remote sensing, Global Navigation Satellite Systems (GNSS), including Global Positioning System (GPS), with other precision agriculture technologies (e.g., variable-rate application, crop scouting, yield monitoring) provides growers with opportunities to visualize and analyze geospatial data and apply spatially explicit and location-smart management solutions. Similar to other information systems, using GIS requires the following:

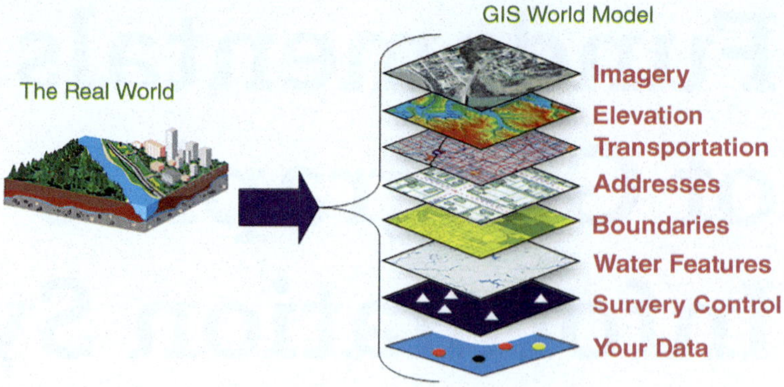

Figure 8.1. Overlaying information within a GIS program. Each data layer represents one type of real-world object. Henrico.us Henrico County GIS, from Brase (2018) / John Wiley & Sons.

1. Software that can handle spatial data, available either commercially (e.g., ArcGIS) or open source (e.g., QGIS)

2. Hardware that resides with data and service providers (e.g., cloud servers) or end users (e.g., computers, tablets, or other mobile devices capable of running the software)

3. People and institutions who use or produce spatial data, who are responsible for setting standards, identifying the needs, and defining the questions to be addressed

4. Spatial data to be analyzed (what makes GIS unique).

GIS has the advantage of presenting the real world with multiple layers of digital information corresponding to varied phenomena of interest within a given area (Figure 8.1). This multilevel representation allows for more targeted queries and analysis of the relationships between the different layers of geographic information. Analysis across layers can improve our understanding of various land systems. In addition, GIS helps link scientific knowledge with practical solutions to meet real-world challenges. For example, combining location information with yield, elevation, and evaporation data allows the user to identify water as the limiting factor in the summit/shoulder area of the field (Figure 8.2, Mishra et al., 2008).

Additionally, in precision agriculture, GIS utilizes satellite imagery, GPS, and field data to map, monitor, and manage agricultural landscapes (Rano et al., 2022; Zurmotai, 2020). GIS helps farmers to precisely locate and map spatial variations in soil characteristics, crop health, pests, and environmental conditions. For instance, farmers can use GIS to identify trends, predict outcomes, and apply strategies that maximize agricultural productivity while minimizing environmental impact. They can also use GIS to answer spatial questions, such as how soil organic matter influences crop yield across a field (Toromade & Chiekezie, 2024).

While GIS has the potential to significantly improve decision-making, and readers may be excited to get started using it, it is important to note that data quality determines the quality of the outcome. For example, poor spatial resolution, outdated layers, misaligned coordinate systems, or inadequate data processing can all lead to low-quality outcomes. Therefore, having sufficient understanding of and skills in spatial data and its analysis are as important as being clear about the question to be answered or the problem to be addressed. A discussion about scientific questions and problems in precision farming is beyond the scope of this chapter but is available in other chapters of this book. The skills one will need to become sufficient in answering questions with GIS can be described in the following four goals:

1. Understand the functions and workflow of GIS and different types of GIS software.

2. Understand the characteristics, advantages, and limitations of different geospatial data.

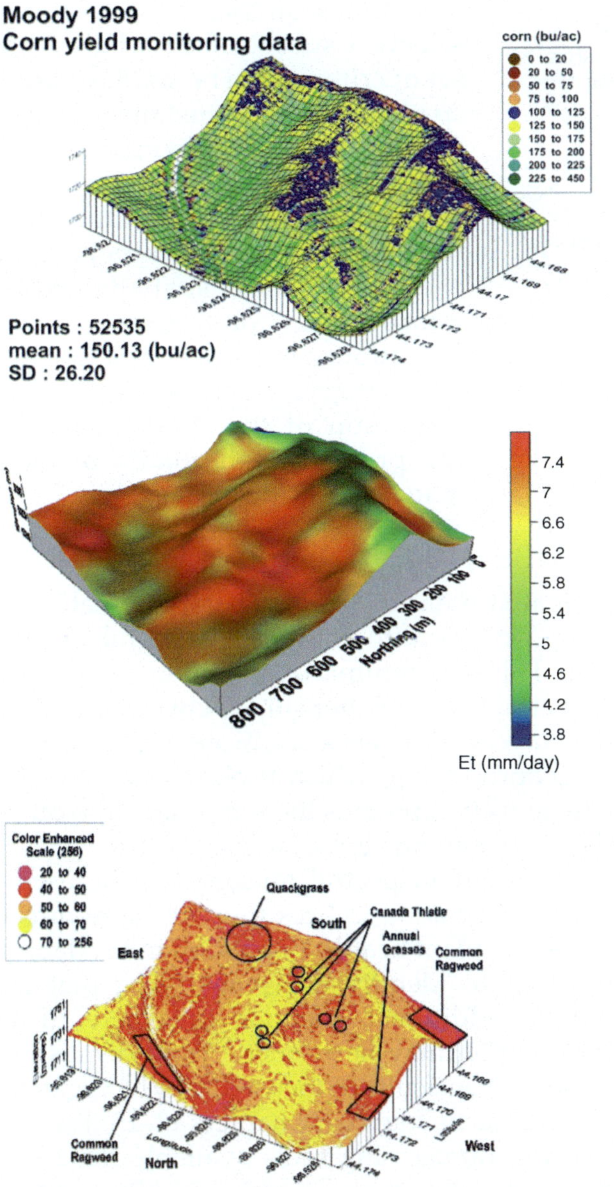

Figure 8.2. The corn yield monitor dataset from 1999, weed distribution on color infrared imagery, and evapotranspiration (Et) data layer from the field. All images are superimposed on an elevation map. Overlaying information helps managers better understand and manage systems. Courtesy South Dakota State University.

3. Learn the basic principles of cartography to use and create maps effectively.

4. Use geoprocessing tools to perform spatial data analysis and transformation.

In the present chapter, the first three goals are addressed in the following sections, while the last will be covered in Chapter 9 of this book.

GIS Functions and Workflow

A typical GIS workflow includes the following steps involving the major functions of GIS (Figure 8.3):

1. Acquire or create geospatial data.

2. Store and manage geospatial data in appropriate formats.

3. Display and visualize geospatial data.

4. Analyze and transform geospatial data.

5. Present and share geographic information with users/targeted audiences.

Data Acquisition

In precision agriculture, geospatial data provides detailed insights that enhance decision-making and efficiency. Geospatial data can be obtained from various agricultural machinery and technologies. For instance, as-applied fertilizer data can be directly extracted from fertilizer applicators, whereas field boundaries can be obtained by walking the field perimeter. Remote sensing data can be collected by unmanned aerial vehicles (UAVs) or spaceborne satellites equipped with high-resolution scanners (Chapters 7, 14, 15). Precision agriculture specialists may also use data from relevant government agencies, such as the U.S. Department of Agriculture Natural Resources Conservation Service Geospatial Data Gateway (https://datagateway.nrcs.usda.gov), the web soil survey (Chapter 10), or historical information maintained by clients. To be used in GIS, all data should be recorded in digital format and georeferenced with explicit locational information.

Data Storage and Management

All spatial data must be stored in appropriate formats to be used in a GIS program. For instance, geographic coordinates (latitude and longitude) of soil sample locations may be recorded into a spreadsheet in comma-separated values (.csv) or text (.txt) format to be imported into a GIS program as point data. For any geospatial data, both the geometric/map component providing the location, size, and shape of spatial features (e.g., point locations) and the attribute component providing the meaning/nature of data (e.g., soil types at

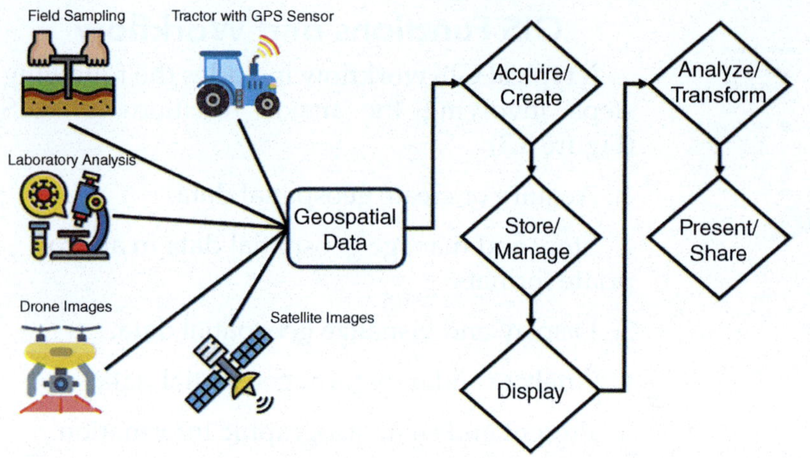

Figure 8.3. A diagram of GIS workflow. Courtesy Katsutoshi Mizuta.

Sometimes queries, where the user selects a subset of data based on a set of criteria, can be used to explore spatial data. For instance, a query might be used to identify locations of soils with >6 cm of plant available water. In precision agriculture, high-resolution imagery of a field can be useful to derive spatially explicit information on crop growth conditions. At the completion of a GIS project, depending on the resolution of the spatial data, GIS can produce high-quality maps for management purposes.

the point locations) are needed (Figure 8.4). Additional information on gathering precision farming data is available in various chapters of this book and differs based on the precision agriculture question being asked. More details on managing and storing this data are included in Chapter 22.

Data Display and Exploration

Displaying geospatial data as maps helps visualize the spatial phenomenon of interest. GIS programs provide functionalities to facilitate the convenient display and viewing of geospatial data. For example, users can "zoom in" and "pan" to make the digital map dynamic and interactive. This initial step allows users to better understand their data.

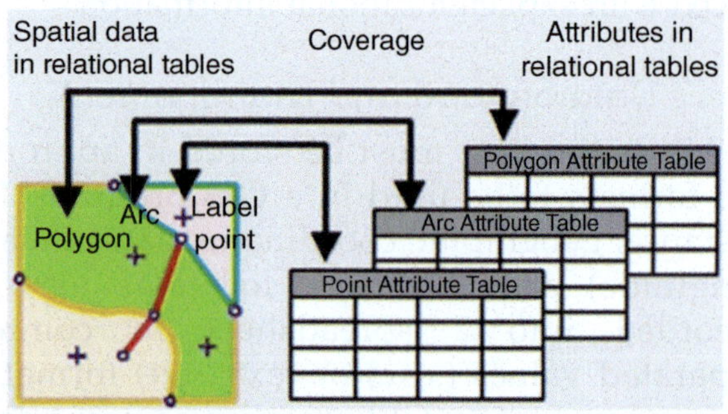

Figure 8.4. Spatial data requires both the geometries and attributes of features to be properly stored and linked together. http://www.geog.ucsb.edu, adapted from Brase (2018).

Data Analysis and Transformation

Spatial data analysis and transformation are often necessary to extract meaningful information. This step summarizes, combines, compares, and establishes relationships between the spatial data layers. Such tasks in GIS are often referred to as geoprocessing, which typically involves specific tools that process the input data into output data containing information of interest. For example, multispectral imagery can be used to derive vegetation indices that are more directly related to plant health. Vegetation index data can be used to identify areas in a field that may require additional attention or where further information may be needed. More details on a variety of agriculture-focused uses of vegetation indices can be found in Chapters 12–15. There are many geoprocessing tools available for accomplishing a wide range of tasks. Specific examples are provided in Chapter 9 of this book.

Presentation and Sharing of Results

The final step in a GIS workflow involves presenting meaningful geographic information to the interested parties. The standard output of GIS work is a map, often accompanied by charts, tables, and explanatory texts. The GIS analyst needs to understand the principles of cartography to effectively display the information. This task involves choosing appropriate coordinate systems, map symbology, and arranging map elements in a visually appealing layout. While paper format maps were the norm in the past, today most geographic information is shared digitally via the internet or cloud-based map services.

GIS Software

Many different companies produce GIS programs (Table 8.1). These programs have different strengths, weaknesses, and costs. Some software programs are designed for agriculture, while others are for general applications. All software discussed in this chapter can be run locally from a PC desktop. But there are also cloud-based GIS programs that run on web servers. One may also classify GIS programs based on whether they are open source and freely available or proprietary and commercial. This introduction focuses on the distinction between general-purpose and agriculture-focused GIS but will also mention other distinctions along the way. Understanding these distinctions will help users choose tools suited to their technical skill level and operational needs in precision agriculture.

ArcGIS and QGIS are examples of general-purpose GIS software that could be used across many applications. ArcGIS is considered one of the most robust GIS software available, as it supports many different data types and can handle multiple spatial data layers. It has hundreds of tools that can be used to manipulate and analyze data layers. QGIS is an alternative free general-purpose GIS software with similar functionalities. The proprietary ArcGIS software requires an annual subscription fee, but it often offers a smoother user experience and includes high-quality training and customer support. Therefore, ArcGIS is widely used in many educational and governmental organizations. On the other hand, there is also a robust community that develops and supports free and open-source programs, such as QGIS (https://www.osgeo.org).

Although these general-purpose GIS programs are highly useful within many analytical contexts, agricultural professionals who work directly with farmers, such as crop consultants or retail agronomists, may likely use an ag-focused GIS, such as SMS AgLeader or FieldView. Ag-focused GIS feature tools that are designed for agricultural applications, providing straightforward solutions to questions that agricultural professionals encounter daily. For example, SMS AgLeader automatically calculates acreages for field polygon features and can easily average several years of yield data with a single click, whereas ArcGIS requires multiple steps to perform similar tasks and may additionally require unit conversions. These ag-focused GIS programs are proprietary and require annual subscription fees.

The inclusion of tools focused on agricultural applications generally makes ag-focused GIS simpler to use than general-purpose GIS. Among the varied ag-focused GIS programs, their analytical capacities differ. For example, the SMS AgLeader interface is more similar to a general-purpose GIS, and its large number of analytical tools makes it a robust option for data analysis and transformation. In contrast, FieldView contains a limited number of analytical tools compared with SMS AgLeader. Given that using more comprehensive programs often requires a steeper learning curve, agricultural professionals may choose relatively simple, more user-friendly GIS programs as long as the available functionalities meet their specific needs.

Additionally, most GIS, whether general-purpose or ag-focused, feature a graphical user interface (GUI) that lets users point, click, or tap their way through menus, tools, and options to interact with the software. The functions and tools available via GUI are limited, and when no tool is available for a specific analysis, users may need to develop their own tools using a programming language such as Python or R (demonstrated in Chapter 9). A programming language can be used to develop customized tools within a GIS program or perform spatial data analysis independently. Programming-based GIS analysis is not only free but also offers versatile analytical capabilities and makes performing repetitive data analysis tasks highly efficient. Certainly, the major hurdle is the steep learning curve for users who lack computer programming training. But knowing this possibility, users can seek professional support should the need to develop customized tools arise.

Spatial Data

Spatial (or geospatial) data describes the location and characteristics of spatial features, such as roads, land parcels, and soil sampling points. The locations of spatial features must be

Table 8.1. The comparison of features of six different GIS program options

Software	QGIS	ArcGIS Pro	SMS AgLeader	FieldView	R	Python
Category	General	General	Ag-Focused	Ag-Focused	General	General
Cost	Free	$700/yr (Creator)	$995/yr (Basic)	$499/yr	Free	Free
Open-source	Yes	No	No	No	Yes	Yes
Add-in/customizable	Yes	Yes/limited	Yes/limited	Yes/limited	Yes	Yes
Desktop vs. cloud	Desktop	Desktop & Cloud	Desktop	Cloud	Desktop	Desktop & Cloud
Graphical user interface (GUI) vs. programming (PGM)	GUI & PGM	GUI & PGM	GUI	GUI	PGM	PGM
Knowledge of expertise	Medium	Medium	Low	Low	High	High
Mobile access	Yes (QField)* Asynchronous	Yes (Field Maps) *Synchronous	Yes	Yes	No	No
Converting data recorded by agricultural machinery to Shapefile/Raster/Text	No	No	Yes (read & write)	Yes (read only)	No	No
Additional categories	Multi-platform, extensive plugins, community support, regular updates, data analysis	Comprehensive tools, integration with ESRI products, advanced analysis, high-quality support, regular updates	Precision farming tools, data management, equipment integration, technical support, regular updates	Real-time data, mobile access, user-friendly, integration with equipment, data analysis	Statistical analysis, extensive libraries, open-source, customizable, community support	Versatile, extensive libraries, integration with other tools, open-source, community support

* QField is designed for offline-first field data collection, with data stored locally on the device and synchronized later through manual transfer or cloud services such as QFieldCloud. ArcGIS Field Maps supports real-time synchronization of collected data to ArcGIS Online when connectivity is available. When operating offline, data collected in Field Maps are stored locally and uploaded automatically once an internet connection is restored.

quantified using geographic references, such as latitude and longitude coordinates or street addresses. In addition to locations, spatial data must carry information about the characteristics of spatial features (such as road names or areas of a land parcel), which are stored as attributes in tables. The geometries of spatial data are described in GIS using two separate models, namely, vector and raster data. Figure 8.5 shows examples of both vector and raster datasets for the same field.

Vector Data

A vector data model is feature- or object-based, and the term object refers to real-world physical things or events. The term feature refers to the representation of a discrete object on a map. For example, in agriculture, a field boundary is an important object because it defines the outer borders of a field. Each boundary object would be represented as a polygon feature that depicts the field's shape. Another example is soil samples, represented as point features corresponding to their locations. All real-world objects can be mapped as features in GIS following the vector data model.

Vector Data Geometry

The vector data model has three geometric data types: points, lines, and polygons (Table 8.2). Points are the simplest features, represented as locations. Lines are created by a series of connected points, which include the beginning and ending points (called nodes) and intermediate points (called vertices). A polygon is created by a series of connected points, with the nodes coinciding, forming an enclosed space that distinguishes the inside from the outside of the polygon. A field boundary is an example of a two-dimensional (2D) object being represented as a polygon (Figure 8.6). Depending on the scale, wind turbines could be represented as points, and roadways as lines.

Table 8.2. Different vector data geometric types

Map symbol type	Definition and use
Point	A single coordinate pair (x, y) is used to represent a discrete location in space, such as a soil sample or a GPS marker.
Line	A one-dimensional sequence of connected points, used to represent a linear feature, such as a fence or a transect.
Polygon	A two-dimensional shape formed by enclosed lines, used to represent an area feature, such as a field boundary or a land parcel.

Figure 8.5. A map created using QGIS showing both vector and raster data in a field. The vector data includes field boundaries (red polygon), soil sampling locations (brown points), and irrigation (black line).

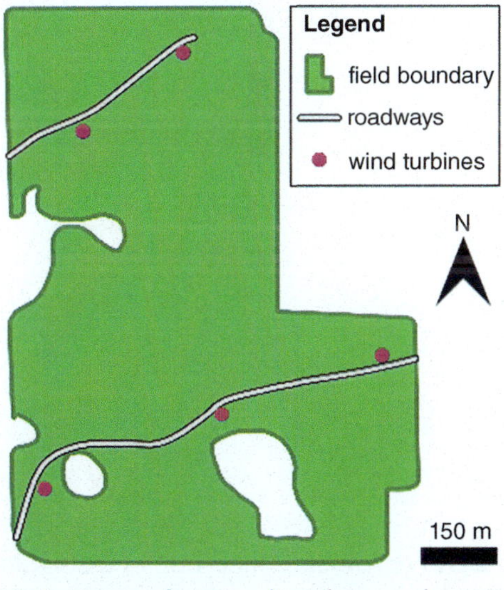

Figure 8.6. Types of vector data that can be used in GIS. Vector data consists of points, lines, and polygons. This field map uses point data to represent the locations of wind turbines, lines to represent the length and location of roadways, and a polygon to represent the area and location of the field boundary. Courtesy Rob Proulx.

Vector features are organized in the hierarchical structure of *Feature–Feature Class–Feature Dataset*. Using the previous example in agriculture, a field boundary is a single polygon feature on a map. When multiple field boundaries are mapped in the same data layer, they form a polygon feature class. In addition to field boundaries, we have a point feature class when there is more than one feature in a field. For example, multiple wind turbines form a point feature class. Multiple roadway lines in a map form a line feature class. All three feature classes of different geometry types are represented in the same spatial reference and comprise a feature dataset for that study area. The hierarchical structure of *Feature–Feature Class–Feature Dataset* shows us how diverse types of discrete objects are effectively represented and organized in GIS using the vector data model.

Vector Data Attributes

All vector data have attribute tables that define the characteristics of spatial features. These attributes are information for objects in the real world, represented by features on a map. For a feature class, attributes are organized in a matrix or a spreadsheet with rows and columns. Each row within the matrix stores a record of characteristics or information of a specific feature. The columns for different attributes are commonly referred to as fields. A cell is the intersection of a column and a row, representing one piece of attribute data for a specific feature. For instance, in the following example (Figure 8.7), the feature class of Georgia counties contains multiple polygons of county boundaries. These county polygons correspond to the records in the attribute table. Each county has a unique row (a record)

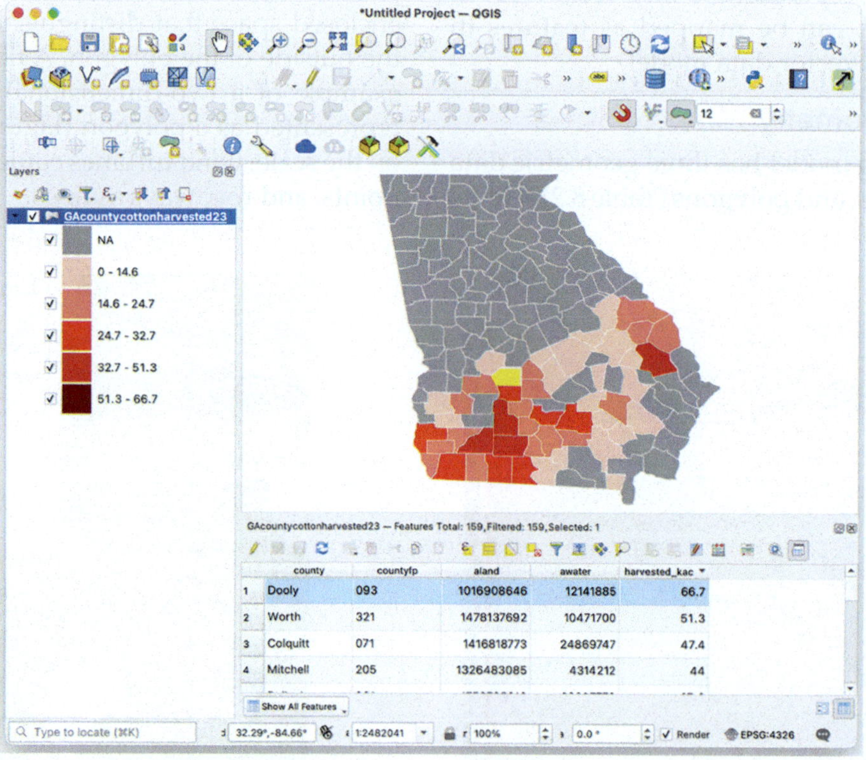

Figure 8.7. Feature class map and its associated attribute table produced by QGIS. In the map, each polygon feature represents a county in Georgia. Counties are colored according to the feature *harvested_kac*, representing the cotton harvested area in 2023. The map and attributes are linked; therefore, when one county record is highlighted in the attribute table, the corresponding polygon feature in the map is also highlighted. The highlighted record is for Dooly County, which had the largest cotton harvested area in 2023. Note that while *harvested_kac* is an "attribute" or "field," 66.7 is an "attribute value." Courtesy Leonardo M. Bastos.

with values in multiple columns (fields). The attribute fields (columns) are county name (*county*), county code (*countyfp*), land area (*aland*), water area (*awater*), and cotton harvested area in thousand acres in 2023 (*harvested_kac*; obtained from USDA-NASS, 2025).

Vector Data Formats

Some typical vector file formats include shapefile, geodatabase feature classes, Keyhole Markup Language (KML), and Geographic JavaScript Object Notation (GeoJSON). Shapefile is the most common vector data format and stores its geometry and attributes in separate files (i.e., .shp files and .dbf files). Viewed using File Explorer, a shapefile contains multiple files with the same name but different extensions. Therefore, when transferring shapefiles, all associated files must be included. Local management of shapefiles using a GIS catalog tool is recommended. When sharing shapefiles, a good approach is to first compress all the files into a single package. Files associated with feature classes in a geodatabase are not accessible/comprehensible using File Explorer. In addition to managing the geodatabases using a dedicated GIS, one must share the entire folder (ideally compressed to a Zip file) to ensure all necessary files are included. Other vector data formats include KML and GeoJSON. KML was developed for use with Google Earth, and GeoJSON was developed for web compatibility. These two formats store vector data in single files, making sharing easier and more straightforward.

Raster Data

Continuous data in GIS is typically represented using a raster data model. Raster data format stores data in an array of cells (a.k.a. pixels) arranged by rows and columns in a grid (Figure 8.8). Each grid cell represents a specific area in the real world with a specific size. Each grid cell is georeferenced to its location and contains a cell value that corresponds to an attribute of interest, such as elevation or precipitation. Typically, a raster layer can only have a single attribute represented as cell values. To incorporate multiple attributes, multiple raster layers would be needed. Another important characteristic is that attribute tables for raster data are available only if the cell values are integers. This

15	18	0	6	10	8	14
14	17	16	7	11	9	17
13	16	32	8	12	10	20
12	17	48	9	13	11	23
11	18	31	10	14	12	26
10	19	18	11	15	13	29
9	15	6	12	16	14	32

Figure 8.8. A raster comprises an array of cells or pixels. Each cell or pixel has a value corresponding to an attribute of interest. Courtesy South Dakota State University.

is because the floating-point data describing continuous phenomena over space can take on infinite potential values, making a table representation impossible.

Properties of Raster Data

The spatial extent of a raster is defined by its top, bottom, left, and right coordinates or minimum and maximum x and y values. The relative position of the cells is defined by their respective row and column numbers, usually counted from the upper-left corner of the raster grid. The cell size is defined as the side length of square-shaped cells in the real world. For example, a grid cell size of 1 m means that each cell represents an area of 1×1 m. The attribute value is often a continuous value that represents elevation, yield, nutrients, or distance to an object. Given that these values cover an entire land area continuously, they are best represented with a raster data model. The other advantage of raster is that each grid cell in a layer is aligned with the respective grid cells in other layers, allowing more convenient multilayer analysis.

Cell size, also known as spatial resolution in the context of remote sensing, is the most fundamental characteristic of raster data. In a case when the cell size is relatively large, a cell value can represent a mixture of the properties of interest within the area covered by the cell. For instance, when land cover type, such as forests, is assigned to a cell with a size of 30×30 m, this only indicates that most of the area within the cell is forests, but does not necessarily rule out the possible presence of other land cover types. Smaller cell sizes (finer spatial resolutions) provide more detail and greater precision in the features identified (Figure 8.9), but at the same time, they also increase the data volume and computing resources required. Hence, the appropriate cell size for the raster must be chosen to match the spatial scale of the phenomenon of interest.

Further, cell values may have different precisions, referred to as pixel depth, cell depth, or bit depth. Raster cells that have 1-bit pixel depth can only have two possible values (0 and 1), whereas cells that have 8-bit pixel depth (2^8) can have values ranging from 0 to 255. The pixel depth describes the number of levels available for cell values and is referred to as radiometric resolution in remote sensing. For example, the older Landsat MSS has 6-bit depth that can resolve 64 (2^6) different levels of brightness, whereas Landsat TM has 8-bit depth that provide 256 (2^8) different levels. The newest Landsat OLI has 12-bit depth that can provide 4096 (2^{12}) brightness values.

In the context of remote sensing, imagery raster data also has different spectral and temporal resolutions. Spectral resolution refers to the number, the width, and the nominal positions (e.g., red, near-infrared) of the spectral bands. For example, the latest aerial photos usually come with four bands covering the blue, green, red, and near-infrared spectral intervals. Temporal resolution describes how often the data is collected. For example, Landsat data typically has a temporal resolution of 16 days. Different types of resolutions (spatial, radiometric, spectral, and temporal) describe the precision with which the remotely sensed raster data represent the real world. Remote sensing is a major source of raster data and can provide regular and comprehensive information for precision agriculture. More details about remote sensing can be found in Chapters 14 and 15 in this book.

Types of Raster Data

According to the data type stored in a raster grid, raster data can be classified as thematic, continuous, or imagery. Thematic data often represents categorical information, such as land use, land cover types, soil types, and crop types. Thematic raster data usually comes with attribute tables that contain the corresponding categorical values. The second type of raster data is continuous, representing surfaces or phenomena that vary smoothly across space. Examples of continuous raster data include digital elevation models (DEMs), gridded temperature, and gridded precipitation data. Depending on whether continuous values were stored with or without a decimal point (i.e., as integers or floating-point values), continuous raster layers may or may not have attribute tables associated with them. The last type of raster layer is imagery. Imagery raster types include satellite images, air photos, and scanned maps. Given that all digital images have a limited bit depth, they all come with attribute tables associated with them. Imagery raster properties are described in more detail in the section "Properties of Raster Data". Note that digital images taken by cell phones are stored in a raster format. However, because these images are not georeferenced, they cannot be mapped in a GIS.

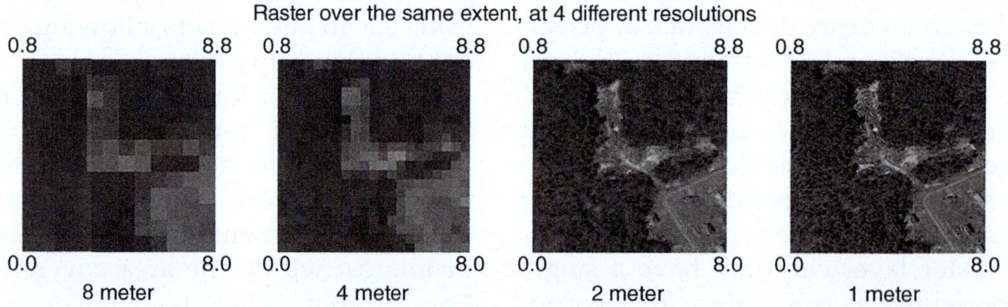

Figure 8.9. Examples of image raster data with different cell sizes/spatial resolutions. When viewed closely, you may see the individual pixels. A raster at the same extent (geographic coverage) with more pixels will have a higher resolution (toward the right in the figure). A raster with a lower resolution is stretched over the same extent with fewer pixels, and it will appear blurrier (toward the left in the figure). National Ecological Observatory Network https://www.neonscience.org/resources/learning-hub, under the Creative Commons Attribution 2.0 (CC BY 2.0).

Raster File Formats

Common raster formats are GeoTIFF (tif, .tiff), ERDAS Imagine IMG (.img), ESRI Grid, and ASCII Grid (.asc). GeoTIFF is an extension of the Tagged Image File Format (TIFF) and embeds spatial reference information into the raster file. GeoTIFF is a widely used format for storing remote sensing imagery and other typical raster data types. GeoTIFF is compatible with most GIS and image processing programs. The IMG format is native to ERDAS Imagine and can be directly accessed by ArcGIS and QGIS. ESRI Grid is native to ESRI ArcGIS programs. However, given its lack of compatibility with other programs, ESRI Grid has been replaced by GeoTIFF as the default raster export format in ArcGIS. The ASCII Grid, developed by ESRI, is a simple raster format that can be read directly using a text editor. While it is open source and program independent, the ASCII Grid requires more storage space and processing power than the binary GeoTIFF and lacks the capability to support more complex metadata and multilayer raster datasets.

Some specialized file formats that support data compression are available for storing large and complex raster datasets. These file formats include Multi-resolution Seamless Image Database (MrSID) and Joint Photographic Experts Group 2000 (JPEG 2000). MrSID employs an effective wavelet compression method and efficiently stores multiple layers of raster data with varied spatial resolutions for the same image in a single file. While proprietary, MrSID is compatible with most GIS programs, including ArcGIS, ERDAS Imagine, and QGIS. Government agencies often use MrSID format to provide access to multi-resolution imagery. JPEG 2000 is an improved version of the original JPEG format that uses wavelet compression with high-quality large images. Most GIS programs can handle JPEG 2000, but its application extends more into other industries, such as digital cinema and medical imaging.

Different GIS programs have varying levels of compatibility with their respective raster data file formats. The compatibility includes the ability to read, write, and analyze. For optimized data integrity and functionality, it is recommended to use the native file format of a particular GIS program for analysis tasks. Most programs can read GeoTIFF, making it an ideal raster file format for work across programs. GeoTIFF is also a good choice for long-term data storage.

Data Integration

Using raster and vector data in the same project is common. For example, data collected for a precision farming project may contain a range of vector and raster data formats (Table 8.3). A range of data formats results from different field data types (e.g., soil, climate, vegetation) using various formats (Figure 8.1). While spatially discrete features (e.g., soil samples) are represented using vector format, spatially continuous features (e.g., temperatures) are represented using raster format. If necessary, vector data can be converted to raster format via rasterization or vice versa via vectorization. Most GIS programs provide the relevant tools to handle format conversions. However, conversions may result in some data loss. More commonly, the different layers serve different purposes in their original formats and are used in an integrated fashion. For instance, field polygons may be used to extract zonal statistics from a raster layer containing vegetation indices. Soil samples of a given property (e.g., water, nitrogen content) may also be used to generate continuous surfaces through interpolation. All data layers for a given project or study area must be georeferenced and placed in the same spatial reference to ensure proper display and analysis.

Table 8.3. Examples of data types for a precision agriculture project

Data layers	Data type
Field boundary	Vector polygon
Soil type	Vector polygon
Soil samples	Vector points
Pest scouting	Vector points, lines, and polygon
As-applied input layers and yield monitor data	Vector points, polygons, and raster
Input prescription layer	Vector polygons and raster
Field image by a drone, plane, or satellite	Imagery raster
Vegetation indices	Continuous raster
Digital elevation models	Continuous raster

In addition to the typical file formats used exclusively for storing vector or raster data, file formats that can store both data types are also available. Examples include ArcGIS's geodatabase and QGIS's GeoPackage. Either case supports storing vector and raster data in a single location, making spatial data management more convenient. The primary advantage of using "all-inclusive" file types is that raster and vector data from the same study area can be stored in the same location. The data layers in a geodatabase must be stored in the same spatial reference (coordinate system), making interlayer operations and analysis significantly easier. Note that a geodatabase is proprietary, and its full function is only supported in ArcGIS, while GeoPackage is open source and more interoperable with different programs. QGIS (through Geospatial Data Abstraction Library [GDAL]) can read and write vectors to ArcGIS file geodatabase (one type of geodatabase) and read rasters from it. It cannot, however, write rasters to a file geodatabase.

Attribute Data

In GIS, attribute data comes in different types, including text, numeric, date, and binary large object (BLOB). The user needs to select the attribute data types that correspond to the nature and properties of data collected when creating new fields in an attribute table. Text data provides qualitative information on spatial features and may be defined with varied field lengths in an attribute table. Numeric data provides quantitative information and may be stored with varied precisions, such as short integer (5 digits/2 bytes), long integer (10 digits/4 bytes), single-precision floating point (7 significant digits/4 bytes), and double-precision floating point (15 significant digits/8 bytes). An appropriate choice of numeric data type should accommodate the precision of original data values and, at the same time, optimize data storage and processing efficiency. For example, a field identifier (ID) may be stored in short integer format, while county Federal Information Processing Standards (FIPS) codes may be better stored in a long integer format because the possible values may exceed the range allowed for short integers. The BLOB type of data is less commonly used but can store a large amount of binary data in a single field. Binary data may include annotations, images, or multimedia files. There is an alternative way of classifying attribute data types in relation to map symbology, which is covered in the section "Data Types and Map Symbols".

Multiple attribute tables in GIS are managed using a relational database structure. Within a table, different spatial features are identified with unique IDs. Attribute information, in additional tables for the same spatial features, may be linked to the first table using a common ID field (a key). For example, the attribute table of a polygon feature class representing Iowa county boundaries contains unique ID fields for county names and FIPS. Now, if a user wants to see how corn yields differ across Iowa by county, a nonspatial attribute table (such as from a spreadsheet) containing information about corn yield in each county can be linked to the polygon layer's attribute table using the common field of county name or FIPS. Then, the spatial variation of corn yield by county in the state can be presented on a map.

In GIS, tables can be linked using *join* and/or *relate*. A *join* operation temporarily appends two tables together to produce one unified table. The *relate* operation links the tables without putting them together. A *join* operation is limited to one-to-one and many-to-one relationship cardinalities between records of the two tables when a unique record from the second table can be joined to each record in the first table. As for more complex relationship cardinalities (Figure 8.10), i.e., one-to-many or many-to-many, a *join* operation cannot be carried out without compromising the data structure. For instance, a field may stretch over multiple soil types, forming a one-to-many relationship. When attempting to combine the attributes of a field feature class and soil types into one table, each field record is duplicated to accommodate multiple soil types, creating a large flat file that is difficult to manage. In such a case, a *relate* operation should be used instead, as it accommodates all relationship cardinalities without affecting the structure of the tables.

Making a Map

Much of the work in GIS involves making a map to display the spatial data. To place spatial features on a map, an understanding of cartography is needed. First, all spatial features on the Earth (geospatial features) are located using a pair of geographic coordinates (i.e., latitude and

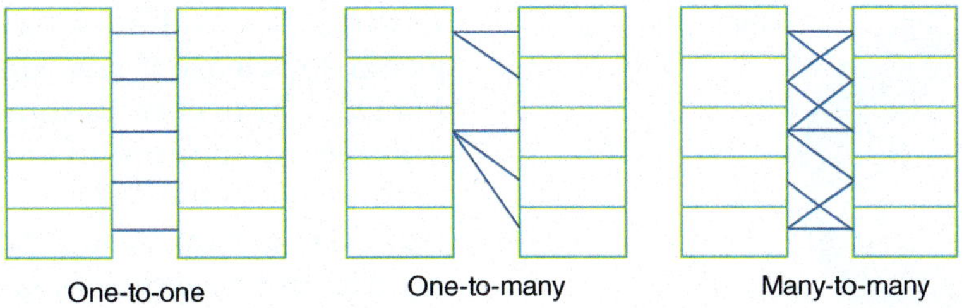

Figure 8.10. Diagrams showing different relationship cardinalities used for the database in a GIS. Given the same underlying structure, the many-to-one cardinality is handled as the one-to-many cardinality in reverse. Courtesy University of Kentucky.

longitude). A third coordinate (altitude) is typically not required to locate things on the surface of the Earth but is necessary for aviation or bathymetry applications. Given that the Earth is spherical (more accurately, spheroidal) while our perception of the land is flat, all geospatial features of interest that exist in the 3D space must be transformed into a 2D plane to be drawn on a map. This 3D–2D transformation is called map projection. Geographic coordinate systems (GCS) and projections determine the coordinate reference system (CRS) through which things on a map are related to their true locations. In GIS, a CRS is also referred to as a spatial reference.

Second, a GIS user needs to understand the basics of map symbology and map design principles. Some questions regarding map symbology include the following: What symbology type and color scheme are suitable for showing different crop varieties across multiple fields of a farm? What map type is ideal for showing gradients of soil properties? The decision should be based on the nature of the variable of interest and the purpose of the map. Questions in relation to map design include the following:

1. What scale should the map have?
2. What map elements should be included?
3. What map layout is appropriate according to the study area shape?

A properly designed map will help users effectively convey geographic information to target audiences.

Geographic Coordinate Systems

A geographic coordinate system (GCS) consists of a 3D geographic grid that is composed of parallels (lines of equal latitude) and meridians (lines of equal longitude). The parallels are numbered northward or southward relative to the Equator as the origin (0° latitude), while the meridians are numbered eastward or westward relative to the Prime Meridian (which runs through Greenwich, England) as the origin (0° longitude), dividing the Earth's surface into different hemispheres (Figure 8.11). Note that a given location's latitude and longitude values are angular distances to the Equator and the Prime Meridian, respectively (Figure 8.12). Latitudinal values range from 90°S to 90°N, while longitudinal values range from 180°W to 180°E. Latitude and longitude values may be expressed in degrees, minutes, and seconds with direction designations (i.e., N/S, W/E) or decimal degrees (DD) for more efficient computation. Coordinates in DD designate the directions using signs: northern latitudes are positive, while southern latitudes are negative; similarly, eastern longitudes are positive, and western longitudes are negative.

Given that the Earth is not a perfect sphere but an oblate spheroid (flattened at the poles), a GCS requires a horizontal datum (a reference framework) to be used to ensure an accurate

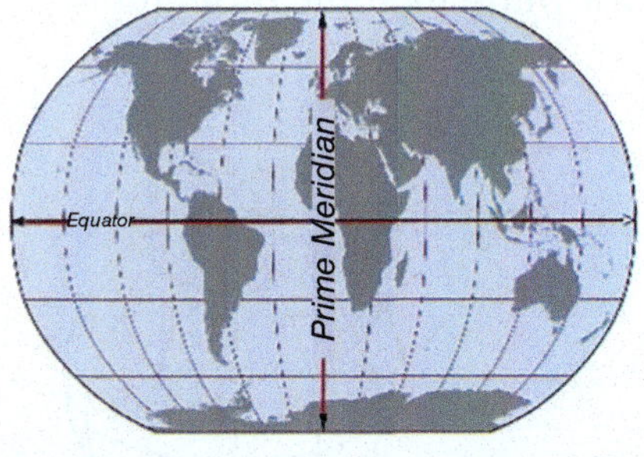

Figure 8.11. The diagram represents the Geographic Coordinate System with two origin lines: the Equator dividing the Northern Hemisphere and Southern Hemisphere and the Prime Meridian dividing the Eastern Hemisphere from the Western Hemisphere. Kmf164 / Wikimedia / Public Domain.

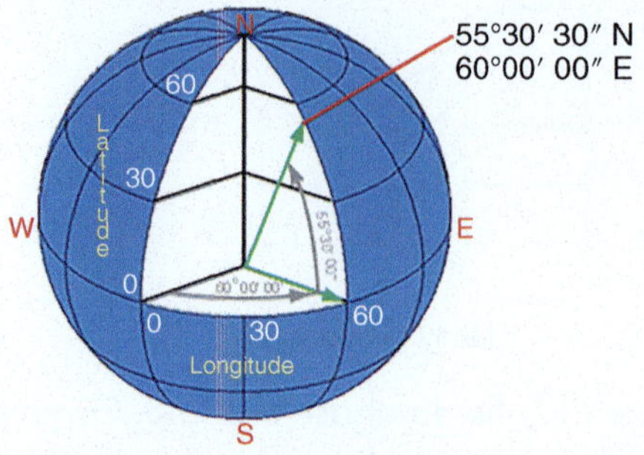

Figure 8.12. To calculate a latitude value, one must determine the angle between the line extending from the Equator to the center of the Earth and the line extending from the location to the center of the Earth. Latitude and longitude represent the angular distances from the lines of origin. The example point is located at a 60° angle east of the Prime Meridian. The latitude is 55 degrees, 30 minutes, and 30 seconds north of the Equator (55°30′30″ N). Hunter College, CUNY /Geographic coordinate systems/ www.geo.hunter.cuny.edu/last accessed on october 10, 2025.

and consistent representation of locations on Earth. A vertical datum is also needed for accurately measuring elevations, but we focus on the horizontal datum in this introduction. A horizontal datum is set up by "pinning" a *spheroid* to an *origin* to ensure alignment with Earth's physical surface. For example, the historical North American Datum of 1927 (NAD27) is a datum based on the Clarke 1866 spheroid and has an origin at Meades Ranch, Kansas. The older NAD27 has been replaced by the North American Datum of 1983 (NAD83), which is based on a more accurate GRS80 spheroid with its origin set at the Earth's center. NAD83 is fixed to the North American plate and does not account for plate tectonic movements. As of this writing, the National Geodetic Survey is projected to release a new datum that will replace NAD83. The new datum will have improved accuracy in mapping U.S. territories. For global applications, the World Geodetic System 1984 (WGS84) datum should be used. WGS84 is geocentric and based on a slightly refined WGS84 spheroid. In addition, WGS84 accounts for plate motions and provides more accurate and consistent coordinates across continents. WGS84 is the datum used for the GPS, and most GPS receivers purchased in the United States are factory-set to WGS84. Other GNSS use datums that are virtually identical to WGS84. It should be noted that the GCS based on WGS84 datum is referred to by the same name. Therefore, WGS84 may refer to the spheroid, the datum, or the GCS, depending on the specific context.

Map Projections

A map projection transforms a 3D Earth surface into a 2D flat map, or converts a GCS to a projected coordinate system (PCS). The procedure of map projection is illustrated by a developable (flattenable) geometric surface (e.g., a cone, a cylinder, or a plane) wrapped around a reference globe (Figure 8.13). A light source (usually placed at the center of the reference globe but varies depending on the projection design) projects the features on the surface of the reference globe to the geometric surface. Then, the geometric surface, in the case of a cone or cylinder, can be cut and developed into a 2D map. Conic or cylindrical projections are more commonly used for maps covering inhabited continents, while a planar projection is typically used for mapping polar regions.

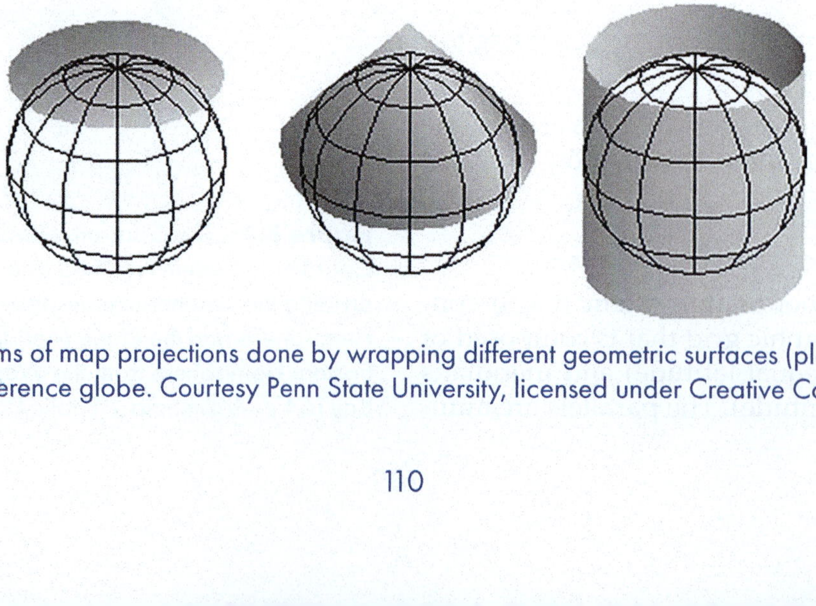

Figure 8.13. Diagrams of map projections done by wrapping different geometric surfaces (plane, cone, or cylinder) around a reference globe. Courtesy Penn State University, licensed under Creative Commons CC BY-NC-SA 4.0.

More detailed descriptions of map projections are beyond the scope of this chapter. However, the readers should realize that map projections cannot be done without producing geometric distortions. It is important to recognize that different map projections preserve certain geometric properties while distorting others. For example, an equal-area or equivalent projection preserves areas but allows distortion of shapes, while a conformal projection preserves shapes but distorts areas. An equal-area projection, such as the Albers Equal-Area Conic projection, is often used for national-scale mapping of natural resources in the United States.

Projected Coordinate Systems

Map projections convert a GCS to different PCSs. The unit of the PCSs is linear and uses meters or feet as opposed to angles of latitude and longitude. To increase measurement accuracy and minimize geometric distortions, multiple locally customized projections are used to cover extensive regions. A widely used PCS in precision agriculture is the Universal Transverse Mercator (UTM) Grid System (Figure 8.14). The UTM system divides the Earth's surface between 84°N and 80°S into 60 zones of 6° wide each. Within each zone, a customized transverse Mercator projection (one that uses a cylinder that wraps around the reference globe horizontally) is used to achieve the highest possible accuracy. The UTM achieves a minimum accuracy of 1 part in 2500, meaning that a distance measured over 2500 m would be accurate within a meter of the true distance. The UTM zones are labeled with numbers and directions, such as UTM zone 14N, which covers central Texas.

A GIS user needs to select the appropriate UTM zone for a particular region of interest. Obviously, the UTM projection system is designed mainly for regional-scale applications rather than national-scale mapping since large countries often span multiple UTM zones. Note that the PCSs typically move the origin of the coordinate system to the lower-left corner of each zone to avoid negative coordinate values. Thus, the coordinates of the center of the zone (the origin) are referred to as false easting (x coordinate) and false northing (y coordinate).

In the United States, the State Plane Coordinate System (SPCS) provides a higher accuracy of 1 part in 10,000. The SPCS uses a transverse Mercator projection for areas (states) elongated in the north–south direction and a Lambert conformal conic projection for areas elongated in the east–west direction. Depending on the shape of each state, one of the two projection types is used. For states with larger areas, multiple zones are projected using a chosen projection. For example, Iowa is divided into two zones (North and

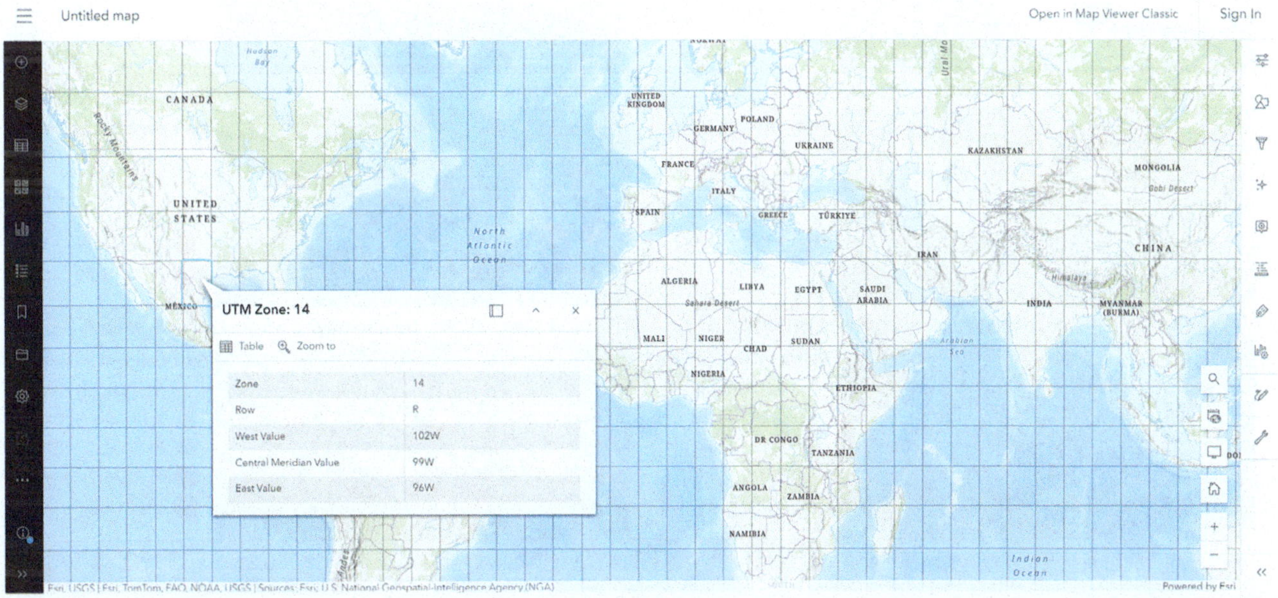

Figure 8.14. A screenshot of a world map of Universal Transverse Mercator (UTM) grids illustrating how the Earth is divided into different zones/grids. Each grid functions with its own coordinate system. ESRI, U.S. National Geospatial-Intelligence Agency.

South), and a Lambert conformal conic projection is applied to each zone.

Different countries have developed similar grid systems for their specific mapping and surveying needs. For instance, the Gauss-Krüger projection (like UTM but with narrower zones) is often used for specifically divided zones in many countries outside the United States. GIS is designed to convert data created using one PCS to another. Therefore, it is important to understand the types of projections and use GIS to perform the projections or conversions as needed.

Data Types and Map Symbols

Making an effective map involves choosing the right symbols. Cartographic symbols vary by shape, size, and color. Shapes and sizes are mainly used for vector data. Different shapes can be used to represent qualitative differences in features. For instance, different point symbols can be used to denote sampling locations for various soil or crop properties, and different sizes of the same symbol type can be used to represent quantitative differences of a variable of interest (e.g., soil nitrogen content). For line features, different line thicknesses can indicate variation in a measured quantity, such as river discharge or traffic volume. Polygon features representing fields or administrative units do not vary in shape or size, so colors must be used to differentiate them qualitatively or quantitatively. For example, different colors can be used to demonstrate varied crop types or soil properties of fields. In addition, aggregated data for polygon features, such as population or crop yield for different counties, can be represented using a single symbol type (e.g., a circle) with varying sizes. This approach is sometimes preferred, given that the fields or administrative units vary in size, which may either increase or reduce the visual impact of colors. For a raster dataset, color is the default visual variable, given the fixed cell size.

For a given map, a specific map symbology must be chosen according to the attribute data type being mapped (Figure 8.15, Table 8.4). Categorical data without ranking (nominal data), such as land use and land cover types or soil survey map units, can be mapped using unique colors (Figure 8.15a). Categorical data with

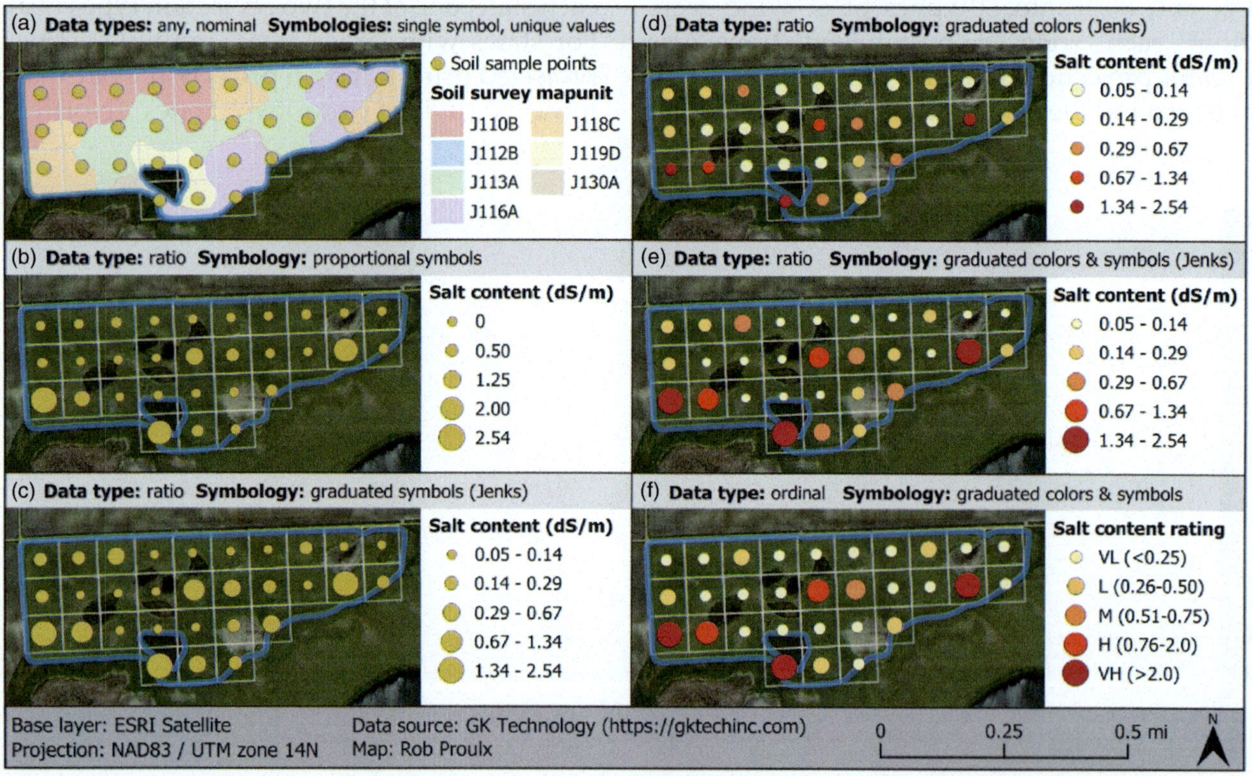

Figure 8.15. Examples of map symbol types and their corresponding attribute data types, as applied to soil test laboratory results from soil samples collected in a grid pattern within a 106-acre field. Courtesy Rob Proulx.

Table 8.4. Commonly used map symbol types and their corresponding attribute data types

Map symbol type	Description	Applicable data type
Single symbol	Single symbol/color to all features, to show only their locations and geometries	All data types
Unique values	Unique colors to distinct categories of features to highlight their qualitative differences	Nominal data
Graduated colors	A range of gradually transitioning colors to show quantitative differences of features in classes	Ordinal, interval, and ratio data
Graduated symbols	Quantitative differences of features in classes are shown in varied symbol sizes	Ordinal, interval, and ratio data
Proportional symbols	Quantitative differences in features are represented by unclassed and proportionally sized symbols	Ratio data
Dot density	Point samples within polygons with the density of points corresponding to values of a quantitative attribute of the polygons	Ratio data

meaningful orders or rankings (ordinal data), such as drought stress levels in crops or soil test interpretation ratings (i.e., low, medium, and high), can be represented using a graduated color symbology (Figure 8.15f). Graduated colors are based on a color ramp that shows quantitative transitions. For example, drought stress levels may be represented using a color ramp from light brown to dark brown.

Graduated colors are also commonly used for continuous data with consistent intervals between values, including interval and ratio data (Figure 8.15d,e). The difference between interval and ratio data is that ratio data has a true zero point and allows for meaningful divisions between values. In contrast, interval data only has consistent intervals between values, with no true zero. Ratio data in agriculture includes crop yield, land area, rainfall amount, or fertilizer application rates. Examples of interval data include temperatures (measured in Celsius or Fahrenheit), pH levels, or vegetation indices derived from remote sensing. Normalized difference vegetation index (NDVI) values are interval data, given the presence of negative values and the lack of a true zero. A GIS user must remember that the use of interval data is limited in mathematical computations in spatial data analysis.

It should be noted that graduated colors or graduated symbols (with size variations to indicate quantity change in aggregated data, such as Figure 8.15c) are based on data classifications that break down the attribute values into manageable numbers of classes. The choice of a classification method usually makes a visible difference on the map. Jenks natural breaks classification method (Figure 8.15c–e) is a widely employed approach that maximizes relative differences among groups in the data and works well for most data distributions. The user can also customize the breaks of the classes or use a statistical method.

Although graduated symbology can represent relative differences in any quantitative data (ordinal, interval, or ratio), a proportional symbology is needed to map the absolute values of ratio data. Proportional symbols (e.g., Figure 8.15b) and dot density maps are such examples. Please note that these map symbology types use unclassified data. For example, a proportional symbol map showing different crop yields from the various counties of a state would have the symbol sizes drawn in proportion to the actual yield values. Or, on a dot density map showing population densities across the country, each dot would represent a fixed number of people.

Map Scale

A map scale shows the relationship between a distance on the map and the corresponding true distance in the real world. It can be precisely expressed using a representative fraction (RF), written as one over a large number (1/x) or one

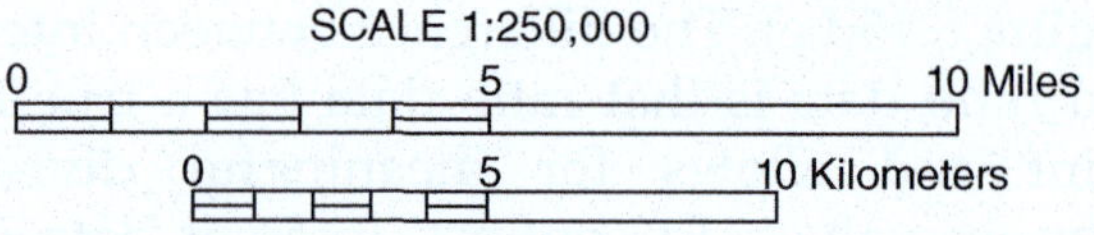

Figure 8.16. Examples of map scales that can be used in a map. Wikimedia Commons, licensed under Creative Commons CC0 1.0.

to a large number (1:x). For example, an RF 1:24,000 means a 1-unit distance on the map equals a 24,000-unit distance on the ground, irrespective of the unit chosen. One can also use verbal/written scales by simply stating that 1 in. on the map equals 2000 ft on the ground. The most common map scale format is a graphic or bar scale, which uses a line segment with labeled divisions to show their actual length on the ground (Figure 8.16).

When dynamically adding a scale bar to a map, it is often better to use a fixed-division-value (adjust width) fitting strategy rather than a fixed-width (adjust value) strategy. This is because the former can ensure that desirable rounded values (e.g., 100 mi or 50 m) are used instead of any random number when the map scale is adjusted. Finally, a GIS user should know the convention of describing the relative size of map scales. By convention, whether a map scale is large or small is based on its RF value. Therefore, a large-scale map (with a large RF value, e.g., 1:2000) covers a smaller area with more details, while a small-scale map (with a small RF value, e.g., 1:200,000) covers a larger area with fewer details.

Map Layout

In addition to map scale, a map should also have a title, a legend, a north arrow, neat lines, a scale bar, data sources, optional annotations, and an inset location map if the area is not well known. These map elements are also referred to as map surrounds, as they are typically placed around the map body to help quantify and attribute meanings to the map. A well-made map requires careful consideration of the styles, sizes, and placements of these elements. The following are some general recommendations for laying out these map elements.

In arranging map elements, a GIS user should pay attention to the *focus*, *order*, and *balance* of a map layout. The map body should always be the *focus* of a map; therefore, the largest space should always be given to the map body, with no other map elements drawing too much of the reader's attention away. When choosing a layout, the shape of the map body should be considered. For example, when mapping a state with a west–east orientation, like Kentucky, a landscape layout should be chosen; likewise, for mapping a state with a north–south orientation, like California, a portrait layout is more suitable. This maximizes the area occupied by the map body in the map space. In addition, the map body should be centered at the visual center of the map space, which is about 5% higher than the geometric center of the page.

The arrangement of map elements should also create a visual *order* or hierarchy. This hierarchy should first draw the reader's attention to the map body; next to the second-tier elements, such as the map title and the legend; and finally to the third-tier elements, including the remaining elements, such as a north arrow, a scale bar, and annotations. The second-tier elements define the content of the map, and the third-tier elements provide additional and necessary details and spatial context. An oversized north arrow or scale bar, for example, often breaks the rule of order and distracts readers from capturing the main content of the map. In addition, a map with *balance* in mind places the map elements evenly across the map space around the map body, so no part of the map space is too blank or too crowded. The GIS user needs to consider the shape of the map body and place the map elements strategically to occupy the surrounding space in a balanced way. One can find detailed demonstrations of these principles in a comprehensive GIS or cartography text (Chang, 2018).

Adding a location map can be very helpful in mapping an area that is not well known. A location map is typically on a smaller map scale and covers a larger area that contains well-known geographic features. For example, when mapping a field in Georgia, USA, the main body of the map should be the field, while a smaller location map may include the continental United

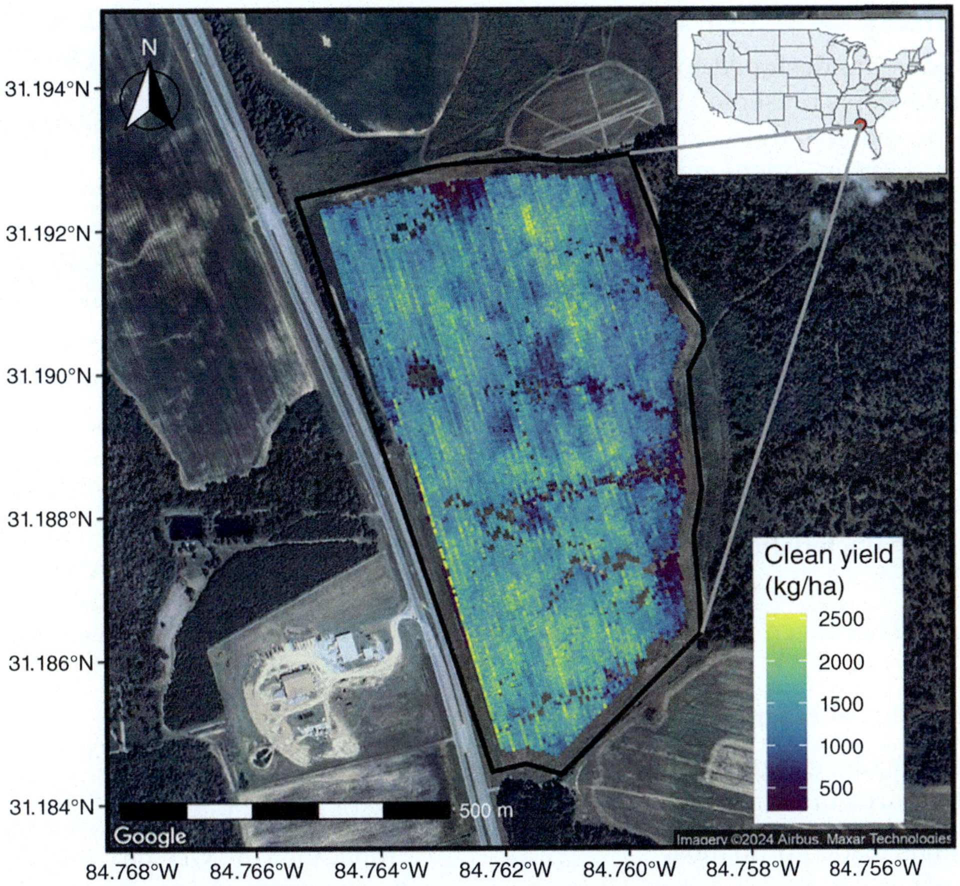

Figure 8.17. A sample study area map designed with focus, order, and balance in mind. The study area map is from a cotton field in Georgia, USA, depicting cleaned yield monitor data. Courtesy Leonardo M. Bastos.

States to show where the field is within a familiar geographic context (Figure 8.17). An inset location map is more reader-friendly, but when space is limited, graticule lines of latitude and longitude would be helpful. A north arrow is the most basic geographic reference element. When geographic coordinate graticules are provided, a north arrow becomes optional. The example in Figure 8.17 provides all three types of geographic reference elements for demonstration purposes. Finally, using a rectangle-shaped neat line or bounding box is always recommended to confine all map elements and hold the map in place.

Summary

This chapter provides a concise introduction to the workflow of GIS, types of software, geospatial data, and the essentials of cartography. The subjects are covered with the students of precision agriculture who do not have prior GIS training in mind. The chapter was not designed to replace detailed training on the use of GIS in precision agriculture. Users who would like to learn more about GIS can find recommended textbooks in the "References and Additional Resources" section below. This chapter has not covered spatial data analysis, which is available in Chapter 9 of this book.

ACKNOWLEDGMENTS

Partial support for this document was provided by South Dakota Corn, the South Dakota Soybean Research and Promotion Council, South Dakota State University, and the USDA National Institute of Food and Agriculture through the Multistate Project S1090 Program under grant number 7008110. The use of trade names is for the convenience of the reader and does not imply endorsement by the authors.

Study Questions

These questions are designed to help readers review the key concepts presented in the chapter. Most answers are found in the respective sections of the chapter. Certain open-ended questions have answers that are user-specific.

1. Describe GIS functions and workflow.
2. What are the different types of GIS software?
3. Imagine you are tasked with creating a GIS-based decision support system for a precision farming operation. Which GIS programs would you choose to work with and why?
4. Describe the characteristics of raster and vector data.
5. What is a good approach for sharing shapefiles?
6. How is attribute data managed in GIS?
7. Define and describe geographic coordinate systems and horizontal datum.
8. What datum is loaded onto most GPS systems?
9. What datum is commonly used in the GNSS in your country?
10. What are projected coordinate systems? Can you give some examples and describe them? Which system is commonly used in precision agriculture?
11. What are some of the typical map symbology types and attribute data types with respect to measurement scales for all data types (i.e., nominal, ordinal, interval, and ratio data)?
12. What are the essential elements to include when making a map?

REFERENCES AND ADDITIONAL RESOURCES

Brase, T. (2018). Basics of geographic information system. Chapter 4. In D. K. Shannon, D. E. Clay, & N. R. Kitchen (Eds.), *Precision agriculture basics*. ASA, CSSA, SSSA.

Chang, K.-T. (2018). *Introduction to geographic information systems* (9th ed.). McGraw-Hill Education.

Kurt, M. G., Smith, J. R., & Pirelli, L. (2016). *Mastering QGIS* (2nd ed.). Packt Publishing Ltd.

Longley, P. A., Goodchild, M. F., Maguire, D. J., & Rhind, D. W. (2015). *Geographic information science and systems* (4th ed.). John Wiley & Sons.

Mishra, U., Clay, D., Trooien, T., Dalsted, K., Malo, D., & Carlson, C. G. (2008). Assessing the value of using a remote sensing-based evapotranspiration map in site-specific management. *Journal of Plant Nutrition*, *31*(7), 1188–1202. https://doi.org/10.1080/01904160802134491.

Price, M. H. (2023). *Mastering ArcGIS Pro* (2nd ed.). McGraw Hill.

Rano, S. H., Afroz, M., & Rahman, M. M. (2022). Application of GIS on monitoring agricultural insect pests: A review. *Journal of Food and Agriculture*, *3(1)*, 19–23. http://doi.org/10.26480/rfna.01.2022.19.23.

Toromade, A. S., & Chiekezie, N. R. (2024). GIS-driven agriculture: Pioneering precision farming and promoting sustainable agricultural practices. *World Journal of Advanced Science and Technology*, *6*(1), 57–72. https://doi.org/10.53346/wjast.2024.6.1.0047.

USDA-NASS. (2025) United States Department of Agriculture National Agricultural Statistical Service.

Zurmotai, N. H. (2020). GIS, remote sensing and GPS: Their activity, integration and fieldwork. *International Journal of Applied Research*, *6(9)*, 328–332. https://www.allresearchjournal.com/archives/2020/vol6issue9/PartE/6-7-86-415.pdf.

GIS Applications to Precision Agriculture

9

Leonardo M. Bastos, Katsutoshi Mizuta, and Rob Proulx

Chapter Purpose

This chapter builds on Chapter 8 (*Fundamentals of Geographic Information Systems*). It covers various Geographic Information Systems (GIS) operations, including collecting georeferenced points, creating field boundaries, and generating a vector grid for soil sampling. The chapter also discusses yield data standardization, soil property interpolation, terrain analysis, cropping imagery, and extracting pixel values. Additionally, the chapter introduces GIS programming using popular languages like R, highlighting vector and raster operations, and demonstrating how to create static and interactive maps. All examples in this chapter were created using Quantum GIS (QGIS; version 3.38.0), a free, open-source GIS software.

Every precision agriculture analysis follows a similar cycle that involves (a) data collection, (b) data processing, (c) data analysis, (d) implementation, (e) assessment, and (f) adaptation (Figure 9.1). While some analyses may be simpler and not include all these steps (such as simply visually assessing satellite imagery), many precision agriculture practices involve these steps in, for example, creating management zones that are used to implement a variable rate application of a certain input. This cycle will be put into perspective as we navigate multiple GIS applications in precision agriculture in this chapter.

Exploring Common Vector Operations in Precision Agriculture

Collecting Georeferenced Points

In precision agriculture, one of the first steps is to collect georeferenced information. This information can be used to answer a variety of questions related to the spatial or temporal variability of the field. For example, the data can be used to (a) create variable fertilizer application maps, (b) determine the effectiveness of imposed treatments, and (c) create variable profitability maps. The term georeferenced means that the spatial location of a measured value is known. Say, for example, the nitrate-N concentration (a measured value) at a specific point in space (the exact location of measurement) is 27 ppm. One approach to georeferencing sampling points like that is to use a Global Navigation Satellite System/Global Positioning System (GNSS/GPS) receiver to identify the sampling locations or use previously georeferenced information from a satellite (Chapter 7). For example, if the coordinates on a satellite image are known, then by aligning the sampling locations on the image, the geographic coordinates can be extracted from the image. This process is called digitizing. Digitizing is the action of drawing a spatial feature into a map for further use.

Transforming Tabular Data to Vector

Once the data is collected, it is important to store it in a format that can be used for future reference. It is common to store spatial data as comma-delimited (.csv) or text (.txt) values within a spreadsheet file containing the spatial coordinates and individual values of the measurement of interest. In this form, datasets can be used to determine the means, standard deviation, medians, and confidence intervals. However, to use these files in a Geographic Information System (GIS) program, they need to be converted to vector geospatial files. In Quantum GIS (QGIS), the tool required for this operation is called "Create points layer from table." This tool converts the x, y location from a table, csv, or txt formats into vector objects, normally of type point (Figure 9.2).

Creating Field Boundaries

Field boundaries represent the geospatial extent of a particular area. A field boundary can be used (a) to identify what is and is not included in the study area, (b) as the starting layer of a grid, and (c) to pull open-source data into the dataset. Field boundaries can be created by collecting location information with a GPS as you walk or drive along the field boundary. Additionally, a boundary can be created using a previously collected georeferenced layer as a guide in digitizing. This is typically conducted using a satellite imagery basemap (Figure 9.3).

In QGIS, the QuickMapServices plug-in provides access to satellite imagery basemaps, such as Google Satellite Hybrid and ESRI Satellite. A step-by-step process for identifying boundaries from these data layers is beyond the scope of this chapter and is available in "Lesson 5.1: Creating a New Vector Dataset" in the *QGIS Training Manual* (https://docs.qgis.org/3.40/en/docs/training_manual).

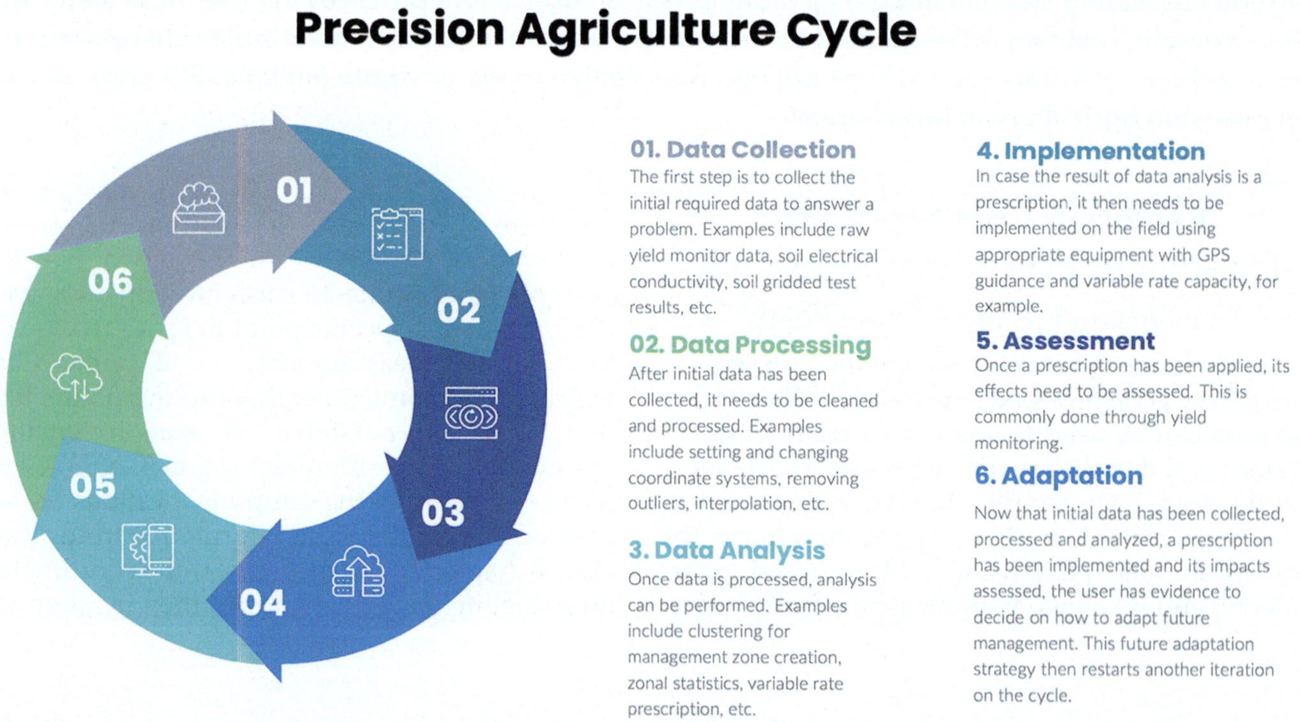

Figure 9.1. Precision agriculture cycle, including six steps ranging from data collection to adaptation. (Courtesy of Leo Bastos).

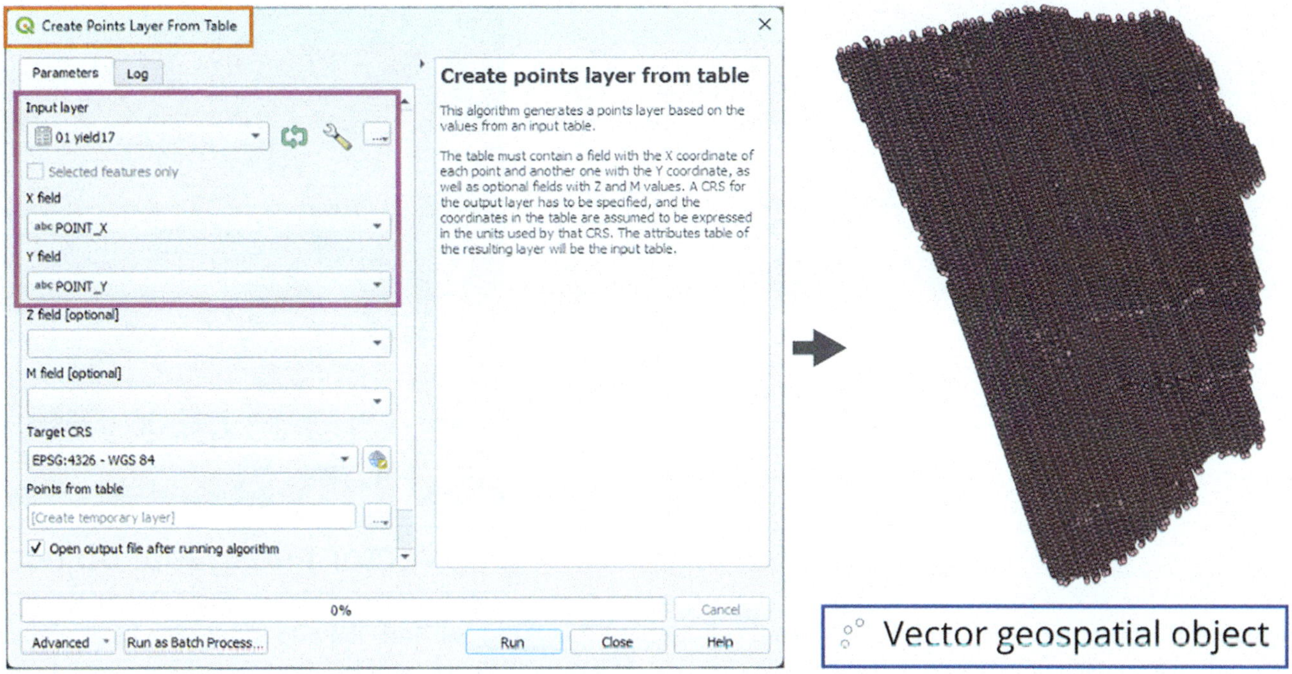

Figure 9.2. Transforming tabular data into geospatial data in Quantum Geographic Information System (QGIS). (Courtesy of Leo Bastos and Rob Proulx).

Figure 9.3. Using a satellite image basemap to identify the field boundaries. The blue digitized line is the boundary. (Courtesy of Leo Bastos).

Creating a Grid for Sampling a Field

A grid can be created using the field boundary and the desired size of the grid cells. A grid is a continuous collection of adjacent cells defined by length and width. For additional detail on grid cells, see Chapter 8. In QGIS, grids are created using the "Create grid" tool (Figure 9.4) or the "Create Fishnet" tool in ArcGIS Pro. The creation of a grid relies on distance and should be referenced to a distance-based measurement system (like the Universal Transverse Mercator [UTM] Grid System). Maps based on latitude and longitude should be referenced to an appropriate datum (like WGS84). The grid cell size depends on what is being measured and often ranges from 0.5 to 2.5 ha. There are trade-offs when selecting a grid cell size. In general, reducing the grid cell increases the cost and the level of detail.

Extracting Sampling Point Locations from Grid Cells

Once the sampling grid is created, it is important to determine from where exactly the samples will be collected. Three commonly used approaches to collect spatial information from a grid cell include

1. Sampling the center of each grid cell
2. Sampling at the corners of each cell
3. Sampling from a random location within each cell.

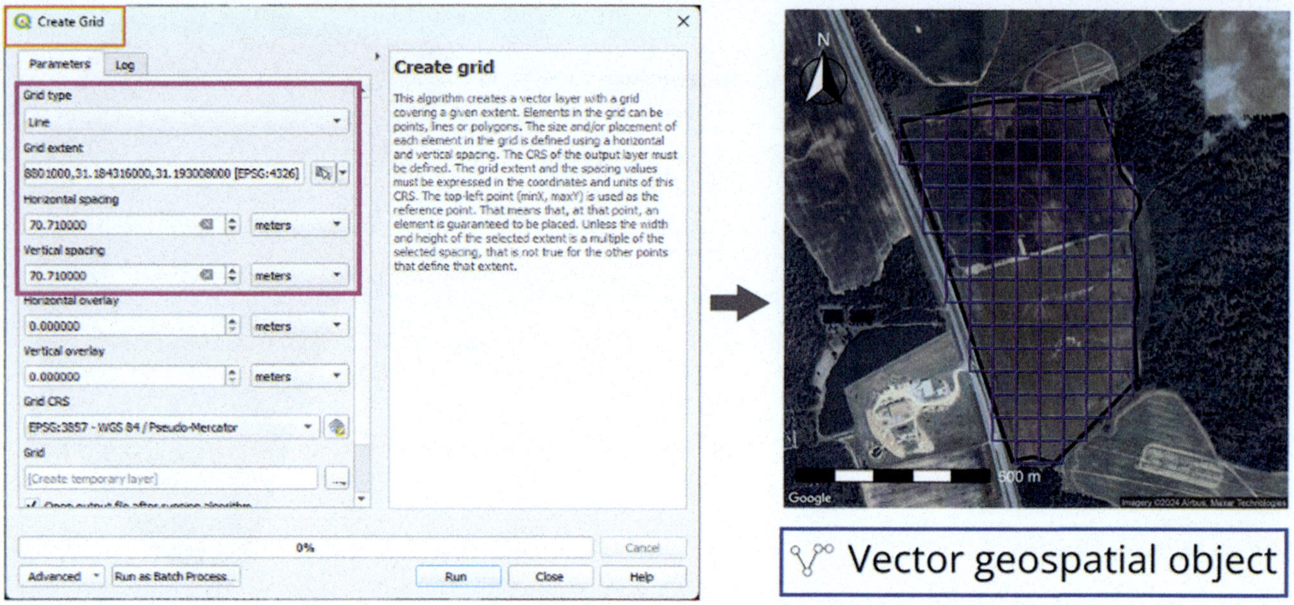

Figure 9.4. Using the create grid tool to build a 0.5-ha (i.e., 70 × 70 m) cell grid. The grid is shown in blue, and it is overlaid on top of the field boundary and satellite imagery basemap. (Courtesy of Leo Bastos and Rob Proulx).

More details on the collection of spatial data are discussed in Chapter 6. Regardless of the approach, each sampling location should consist of approximately 10 subsamples combined into a composite sample representing that point. Using the cell center or corner as sampling locations are common approaches, but those approaches may be biased because the distance between sampling points remains constant, potentially missing variations within the grid and limiting some interpolation methods such as Kriging. Kriging is a spatial interpolation technique used to estimate values at unknown locations based on known values at nearby locations. Sampling bias with a grid cell approach can be reduced by randomly identifying sampling locations. In QGIS, random locations are identified using an operation called "Random points in polygon." The operation requires a polygonized grid (a sampling grid made out of polygons) as input and creates a vector geospatial point object as output (Figure 9.5).

Buffering

Buffering creates a zone of specified distance around a geographic feature. For example, a 10-m wide buffer on either side of a road would result in a 20-m wide corridor that follows the road. Buffering is often used to obtain a more representative value of a given area. For example, when extracting a vegetation index at specific point locations, all the values around that point are averaged to provide a more representative value for the area around each point (Figure 9.6, left panel). In contrast, a negative buffer can be used on yield monitor data near the field boundary to reduce errors resulting from maneuvering, partial swaths, and changes in velocity (Figure 9.6, right panel).

Yield Monitor Data Cleaning

The process of collecting and cleaning yield monitor data is outside the scope of this chapter (see Chapter 20 for more detail on yield monitor data). The present chapter focuses heavily on the use of yield monitor data for spatial analysis. Raw yield monitor data can suffer from multiple sources of error, including flow delay, partial swaths, abrupt velocity changes, positional GPS inaccuracies, and others (see Chapter 20). Yield data points that originated under error-prone conditions need to be assessed and, if necessary, removed before using yield data for the assessment, modification, or implementation of crop management practices. Yield Editor (Sudduth et al., 2012, 2019) is a popular software used for yield monitor data evaluation and cleaning. It includes 12 different filters to identify and potentially remove erroneous data points. The

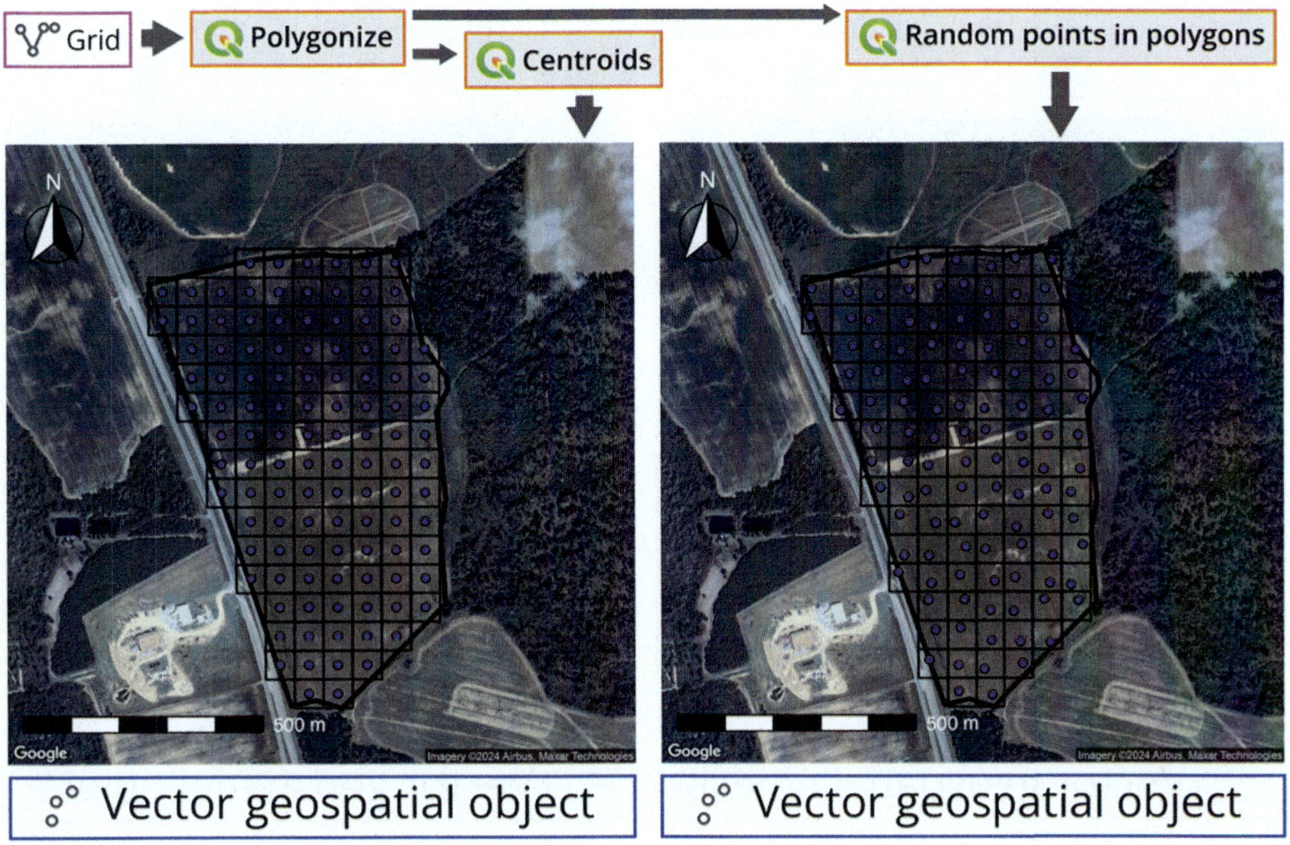

Figure 9.5. Centroid (left panel) and random (right panel) points in blue extracted from a 0.5-ha grid in black overlaid on top of the field boundary in black and satellite imagery basemap. (Courtesy of Leo Bastos and Rob Proulx).

software is free and was developed for Windows operating systems. An example of implementing some of the filters available in Yield Editor is demonstrated below, depicting raw yield data from a cotton field in 2017 (Figure 9.7, left panel), along with its cleaned version (Figure 9.7, right panel). For further discussions about yield monitor cleaning, see Chapter 20.

Spatial Joins

To improve our understanding of a production system, it is often necessary to combine data layers from different sources into a single feature dataset. In GIS, this process is called "Join." Two general types of spatial joins include (a) joining a vector layer with nonspatial data and (b) joining two or more vector layers.

Joining Nonspatial Data

To join spatial and nonspatial data layers, a common attribute is needed to establish the connection. For example, when downloading a spatial layer of interest from an open-source dataset such as the Web Soil Survey (Chapter 10), a unique key for each polygon or map unit, called "Mapunit Key" is included. This will accompany a Tabular directory, with a text file named "muaggatt." This contains the name of each map unit in addition to map unit keys as "MUKEY." By matching the Mapunit Key in the polygon map layer with the MUKEY in the muaggatt text file, you can see the map unit name for each map unit in the spatial data layer (Figure 9.8). Such integration enriches the spatial analysis with additional attributes from the nonspatial dataset, facilitating more comprehensive and informed decision-making. The detailed procedure to work with the Soil Survey Geographic Database (SSURGO) by the U.S. Department of Agriculture Natural Resources Conservation Service can be found in the SSURGO Portal User Guide (https://www.nrcs.usda.gov).

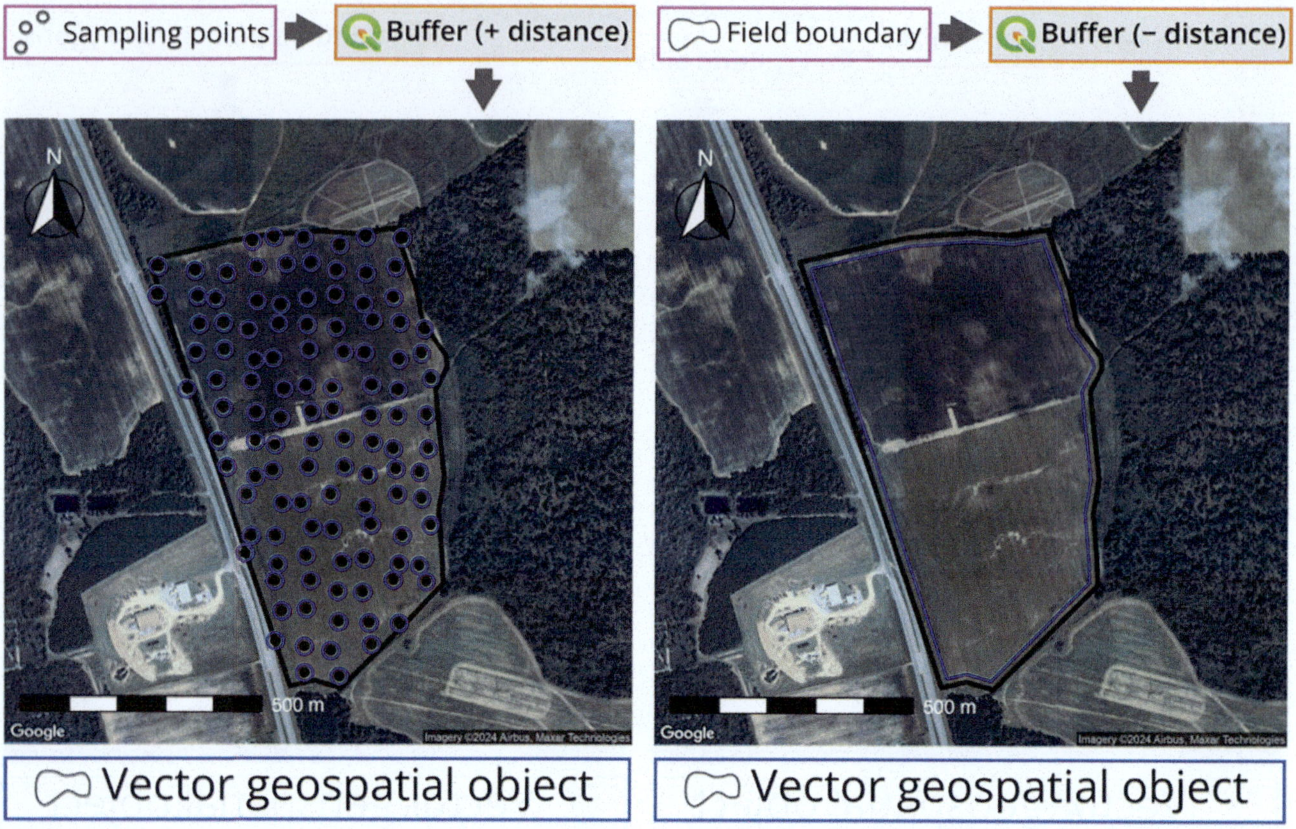

Figure 9.6. Positive buffer in blue around each sampling point in black (left panel), and negative buffer in blue inside the field boundary in black (right panel), overlaid on satellite imagery basemap. (Courtesy of Leo Bastos and Rob Proulx).

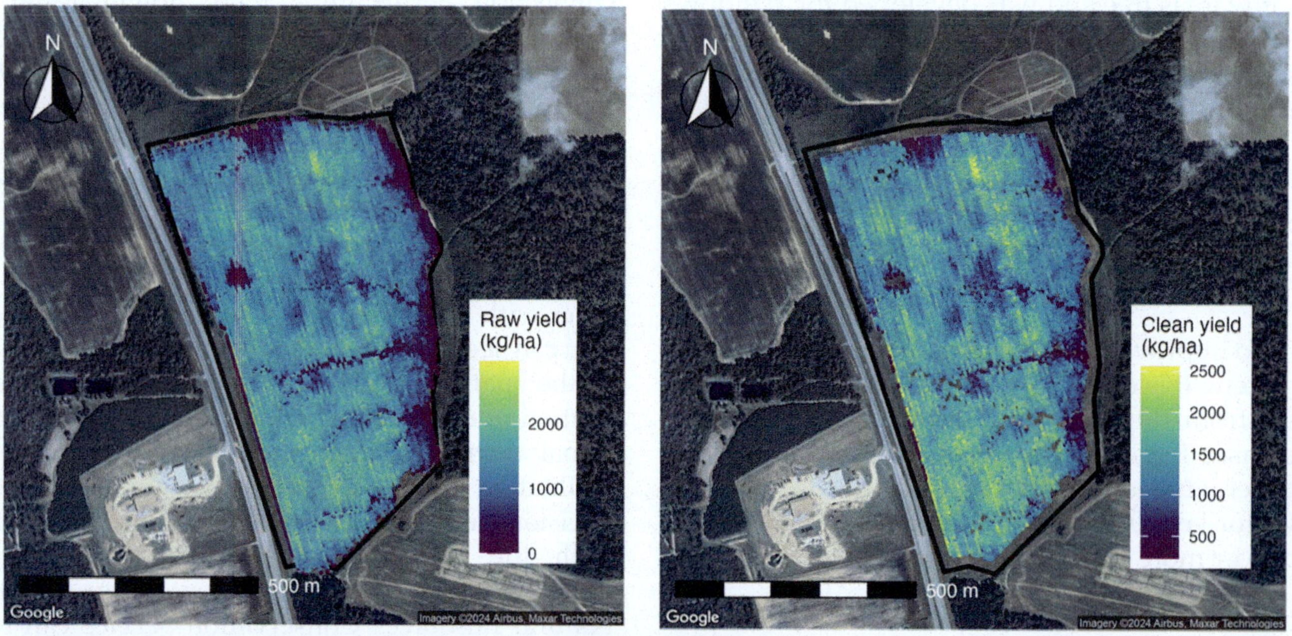

Figure 9.7. Raw yield monitor data from a cotton crop in 2017 (left panel) and its cleaned version after applying six filters available on Yield Editor software (right panel). (Courtesy of Leo Bastos).

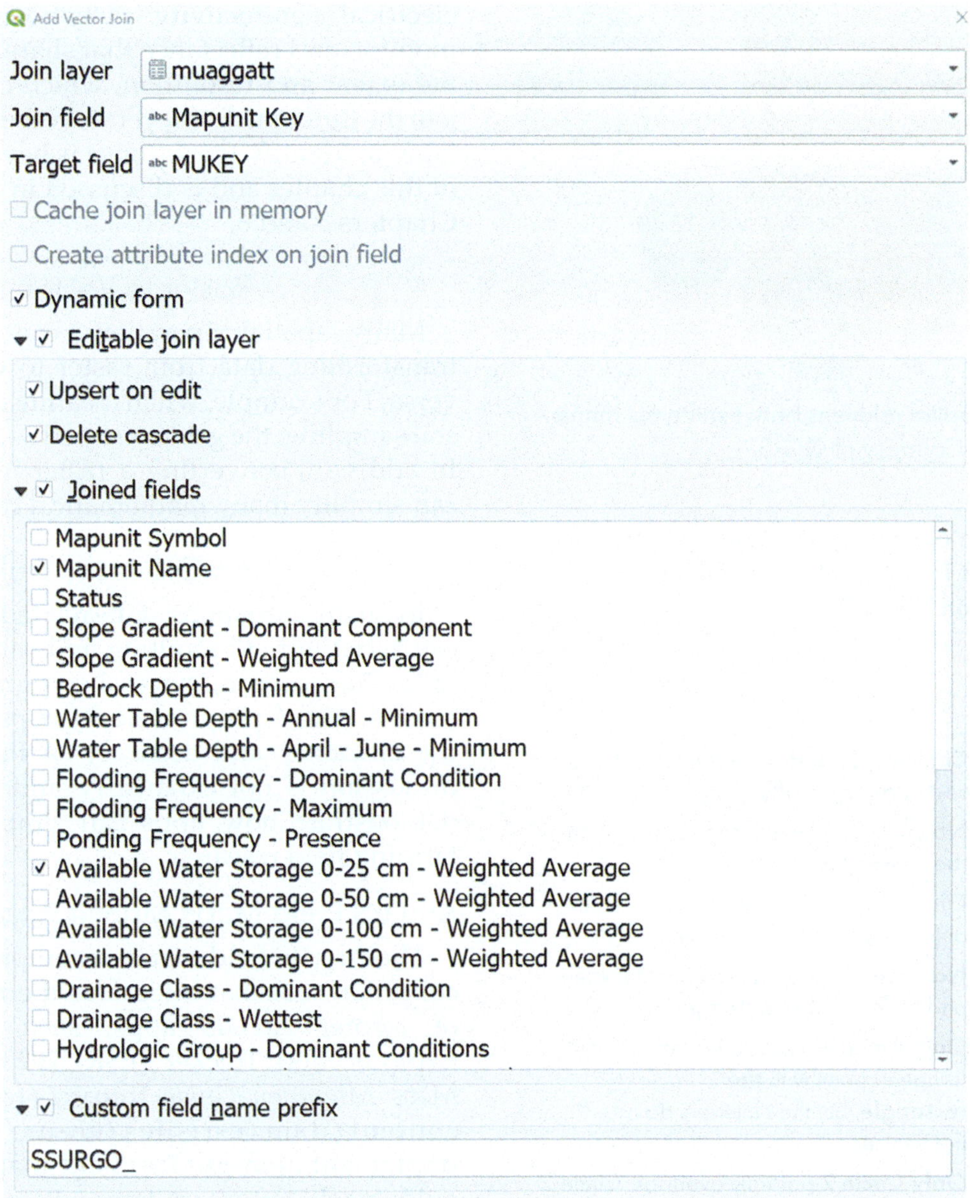

Figure 9.8. Vector Join Function in Quantum Geographic Information System (QGIS). (Courtesy of Katsutoshi Mizuta).

Joining Two Vector Layers Together

While individual geospatial data layers can be useful, it may be possible to conduct more sophisticated analyses by joining them with associated information. When joining vector layers, there are two types of joins that can be used: (a) symmetrical joins and (b) asymmetrical joins. Symmetrical joins are used when all layers being joined share the same extent and resolution. A symmetrical join would be used when creating management zones. This is because this type of join is used when information across different layers needs to match spatially. An asymmetrical join is used when the layers being joined do not fully overlap and the topological relationship (overlap, difference, complement, etc.) is the analysis of interest. For either join type, all layers being joined must share the same coordinate reference system (CRS). Understanding the spatial relationships between different geometric features is crucial when joining vector layers. The figure and table above demonstrate how different spatial operators can be used to identify the relationships between polygons (circles and a rectangle; Figure 9.9, Table 9.1).

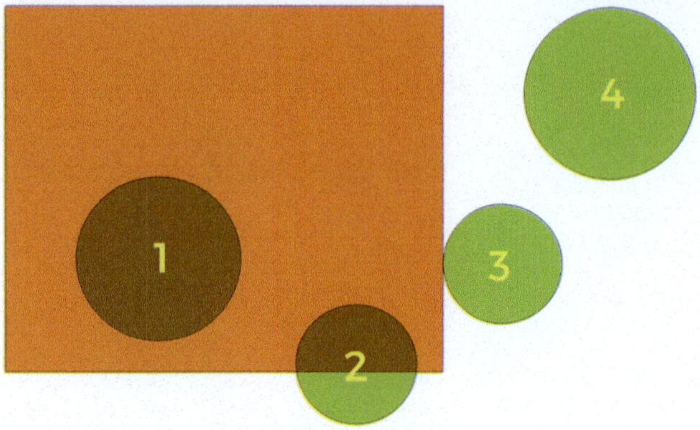

Figure 9.9. Spatial relations between layers (https://docs.qgis.org). Courtesy of QGIS.

Table 9.1. Join Names, Query Type, and Resulting Layer Based on Topological Relationships Demonstrated on Figure 9.8.

Join name (predicate)	Query	Result
Intersect	**Circles 1, 2, and 3** intersect **the rectangle spatially**.	Circles 1, 2, 3
Contain	**No circle** completely contains **the rectangle**.	None
Disjoint	**Only Circle 4** does not share any space with **the rectangle**.	Circle 4
Equal	**No circle** is exactly the same size and shape as **the rectangle**.	None
Touch	**Only Circle 3** shares a boundary (tangent) point with **the rectangle**, but their interiors do not intersect.	Circle 3
Overlap	**Only Circle 2** partially overlaps **the rectangle**.	Circle 2
Are within	**Only Circle 1** is entirely contained within **the rectangle**.	Circle 1
Cross	**No circles** "cross" **the rectangle**.	None

Management Zone Creation (Clustering)

Management zones can be made by using the join operations discussed above. These zones are parts of a field that share similarities across one or more spatial data layers. For example, a field that is separated into different yield or soil nutrient zones. Management zones are commonly based on measurements such as soil apparent electrical conductivity, elevation, slope, yield monitor, and reflectance that share the same spatial extent and resolution, which can be used to join the different data layers together (Figure 9.10). How to identify these zones is beyond the scope of this chapter and is discussed in more detail in Chapters 5 and 6.

Raster to Vector

Many spatial operations can benefit from transforming data from raster to vector, or vice versa. For example, when creating a grid to guide soil sampling, the grid is created as a raster object. In addition, converting a raster to a vector file can simplify many mathematical calculations.

Using GIS in the Field

In many situations, GIS-related mobile apps can assist in field scouting and sampling. In the following section, several programs are discussed. These programs and apps can enhance the efficiency and accuracy of fieldwork. This discussion is not exhaustive and is selected to demonstrate how apps can simplify precision farming field research.

ArcGIS Field Maps Mobile Application

The ArcGIS field map is a comprehensive data collection tool that integrates the functionalities of former ArcGIS apps, such as Collector, Explorer, and Tracker (ESRI 2023 *ArcGIS Field Maps Migration Guide* [https://www.esri.com/content/dam/esrisites/en-us/media/pdf/implementation-guides/field-maps-migration-guide.pdf]). Future updates will incorporate features from ArcGIS Workforce and ArcGIS Navigator, consolidating them into a single app. Field Maps is a free, essential app for viewing maps, collecting data, sharing locations, and maintaining situational awareness in the field.

A key ability of ArcGIS Field Maps is that it can create field boundary maps in ArcGIS Pro, which can then be viewed using a mobile app. The app will show the user's location and collect GPS points as the user walks through the field, creating a field boundary map. The app also has a useful navigation function that allows a user to identify the direction to the next sampling point or the next flagging location. It can also be used to measure distances between rows and count seedling density in the field. Global

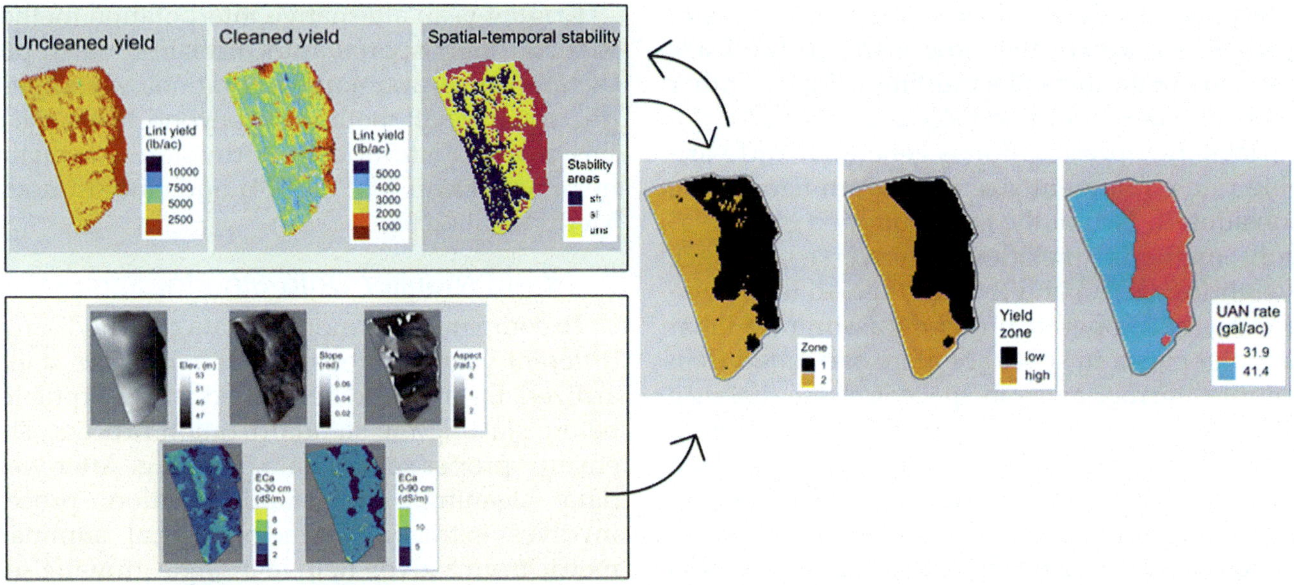

Figure 9.10. Management zone creation workflow includes interpolated cleaned yield data, terrain properties, and soil apparent electrical conductivity. These layers were clustered using the algorithm *k*-means and further smoothed to represent two zones of different yield potential (low and high). Different zones then received a different prescription of in-season nitrogen fertilizer application in the form of urea-ammonium nitrate (UAN) rates. (Courtesy of Leo Bastos).

Positioning System information for ground control points can also be collected and uploaded to unmanned aerial vehicle (UAV) flight plans to collect even more spatial and temporal data from the field.

QField Mobile Application

QField is an asynchronous app that meets the basic needs for data collection, visualization, and map markup. Maps generated by QGIS can be exported with a plug-in ("QField Sync") and viewed in QField by copying and pasting them into a smartphone app. A user can create points or draw lines on a device in a field and update the field maps by importing the updated files back into files to be accessed by QGIS. Both Field Maps and QField can capture photos, videos, and audio recordings in the field and integrate with a map on a PC.

Trimble Mobile Manager Mobile Application

Trimble Mobile Manager enables the configuration and management of Trimble GNSS receivers in the field. This system supports real-time positioning correction services and integrates with satellite systems to enhance data accuracy.

SoilWeb Mobile Application

SoilWeb is a tool that provides soil information based on the user's current location. It does not allow users to choose a location for soil information but offers access to SSURGO data, which can also be accessed via the Web Soil Survey in a web browser. The information includes soil properties, topography, vegetation, climate, forestry, productivity, drainage, and parent materials. SoilWeb can be used to identify management zones based on soil classification. More detail about the Web Soil Survey can be found in Chapter 10.

Exploring Common Raster Operations in Precision Agriculture

Interpolation of Point Data

Interpolation is a general term for a mathematical technique that estimates values at points between sampling locations. It is used because it would be extremely difficult, time-consuming, and expensive to sample at every point in the field. Interpolation requires the point vector layer containing the data (the data to be interpolated), as well as a raster grid upon which the interpolated values will be estimated. The raster grid cell size can be the same as the grid cell size used to create the soil sampling grid, but other grid cell sizes could be selected, especially if a variable rate prescription is generated and variable rate equipment size and swath need to be considered. Two of the most common interpolation methods include (a) inverse distance weighting (IDW) and (b) Kriging. All interpolation methods operate under the assumption that samples closer in space are more correlated than samples farther away in space. Inverse distance weighting is a deterministic interpolation method that predicts the value at an unknown location based on the values of neighboring sample locations. Inverse distance weighting assigns specific or fixed (deterministic) weights that decrease with increasing distance. An example of using IDW on soil potassium (K) levels based on gridded soil sampling data is shown on Figure 9.11.

Kriging is an alternative interpolation method that considers spatial autocorrelation when predicting values at unknown locations. A complete discussion of all methods is beyond the scope of this chapter; additional information is available in Li and Heap (2011, 2014) and Robinson and Metternicht (2006).

Yield Monitor Data Standardization

To compare yield monitor data collected across different years or crops, the data must be standardized. Data should be cleaned and interpolated before starting the standardization process. This ensures proper spatial data handling. After yield data cleaning, the standardization process involves extracting some statistical summary metric from a given field in a single growing season and expressing all yield values relative to this metric (yield at a given location divided by the summary metric yield). Commonly used summary metrics in standardization are the field

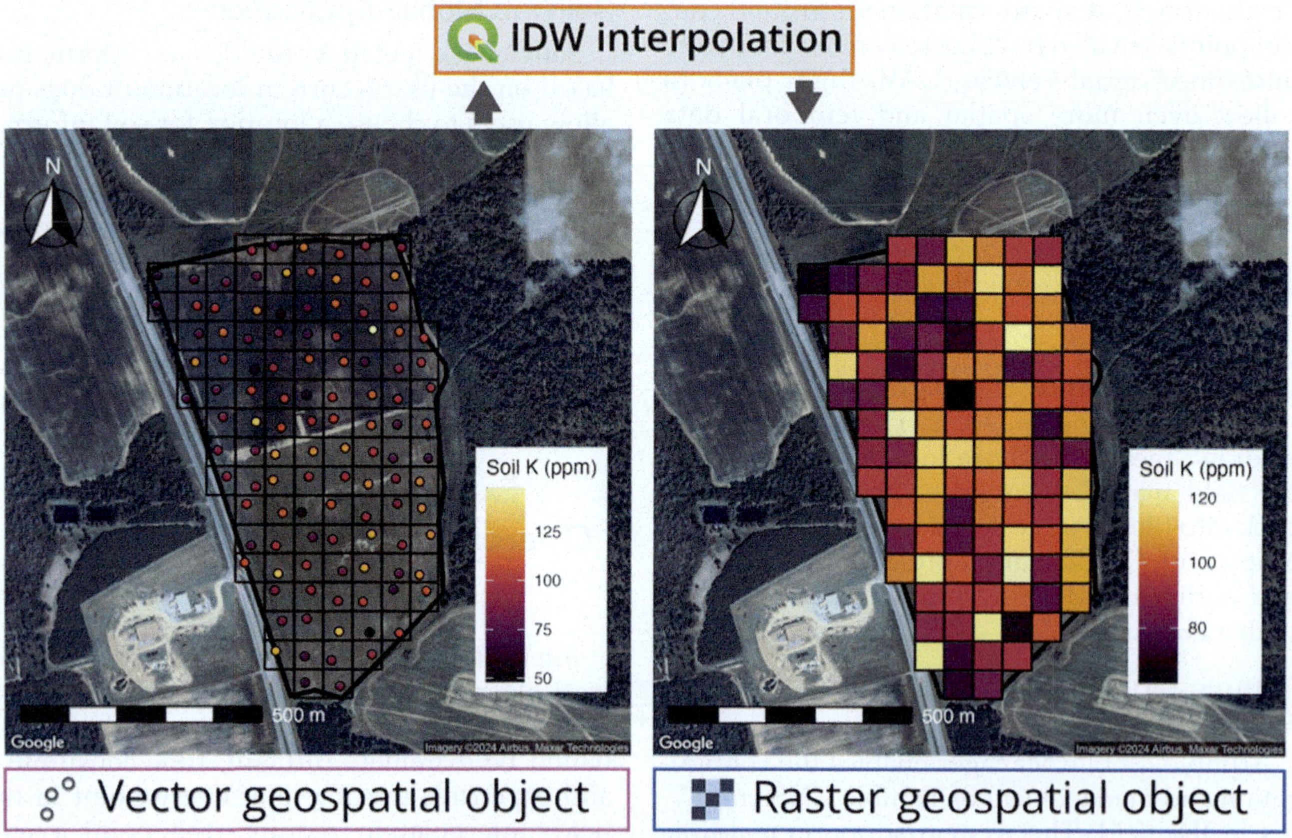

Figure 9.11. Point soil potassium (K) data (left panel) is interpolated using inverse distance weighting (IDW), resulting in a smooth estimation of the entire field (right panel). (Courtesy of Leo Bastos and Rob Proulx).

mean, median, or maximum yield value. If using the mean or median value, then standardized yield values will be centered around 1 (or 100%), with yield values lower than the mean (or median) being below 1, and yield values greater than the mean (or median) being above 1. The choice between using the mean or median depends on the yield data's statistical distribution: If the distribution is skewed, then the median is a better metric because it is less affected by outliers. If using the maximum value, then the standardized yield values will range from 0 to 1. Figure 9.12 shows an example of yield data standardized by the median field-level yield value.

Elevation and Secondary Terrain Variables

Elevation data in point vector format can be collected every time a GPS-equipped tractor or equipment travels through a field. Elevation data in a raster format can be obtained using aerial spectral imagery and/or radar or lidar. If elevation data are collected in point vector format, then interpolation will be required to transform them into a continuous raster digital elevation model (DEM). If elevation data are collected in raster format, then no further processing is required. Once elevation data are in raster format (i.e., DEM), slope, aspect, flow direction, topographical position index, terrain ruggedness index, and topographic wetness index can be calculated.

Spectral Bands and Vegetation Indices

Spectral imagery collected from sensors on board drones, airplanes, satellites, or other airborne platforms is commonly generated in a raster format and contains one or more layers associated with different bands. Bands represent sections of the electromagnetic spectrum; the most commonly available bands are the blue, green, red, and near-infrared bands. See Chapters 13–15 for additional information on remote and proximal sensing using these bands.

Masking

Masking is a technique used with raster data in which pixels with certain values are ignored or removed before further analysis can be performed. This is common in crop imagery collected with drones, where the spatial resolution is very fine (<5 cm), and different surfaces, such as bare soil, must be identified and separated. For example, a researcher may want to classify pixels as either plant or soil, with the goal of masking soil pixels before calculating the average Normalized Difference Vegetation Index (NDVI) at the plot level. This ensures only plant pixels are used to calculate the plot mean. For

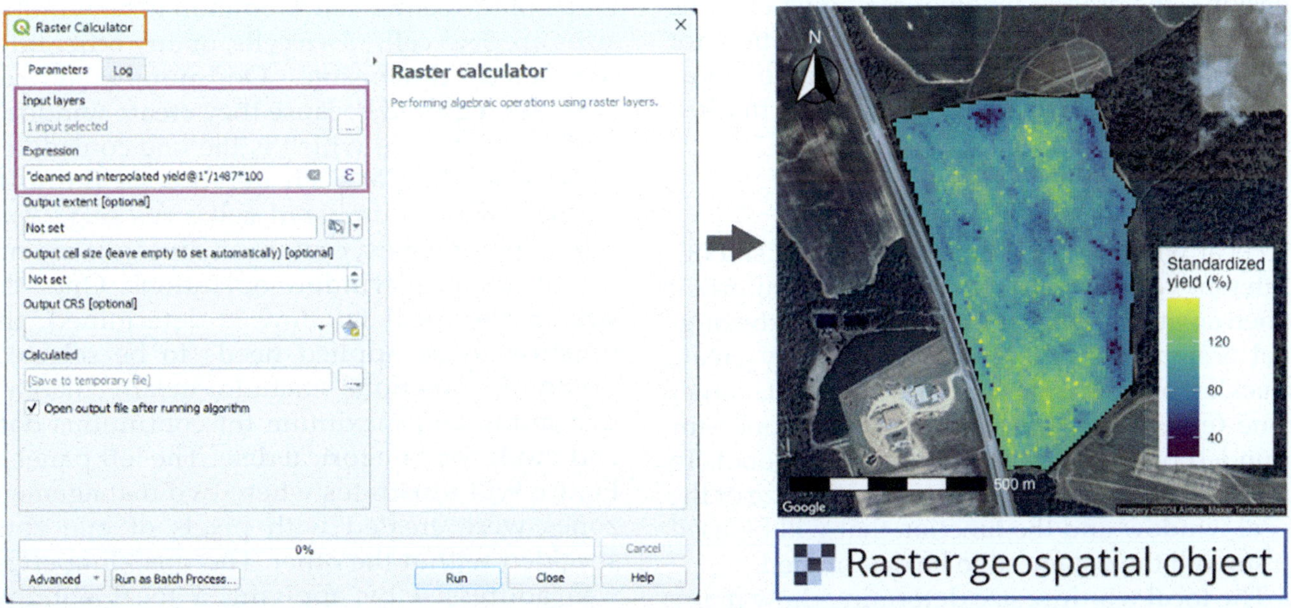

Figure 9.12. Cleaned, interpolated, and standardized yield data for a cotton field in 2017. (Courtesy of Leo Bastos and Rob Proulx).

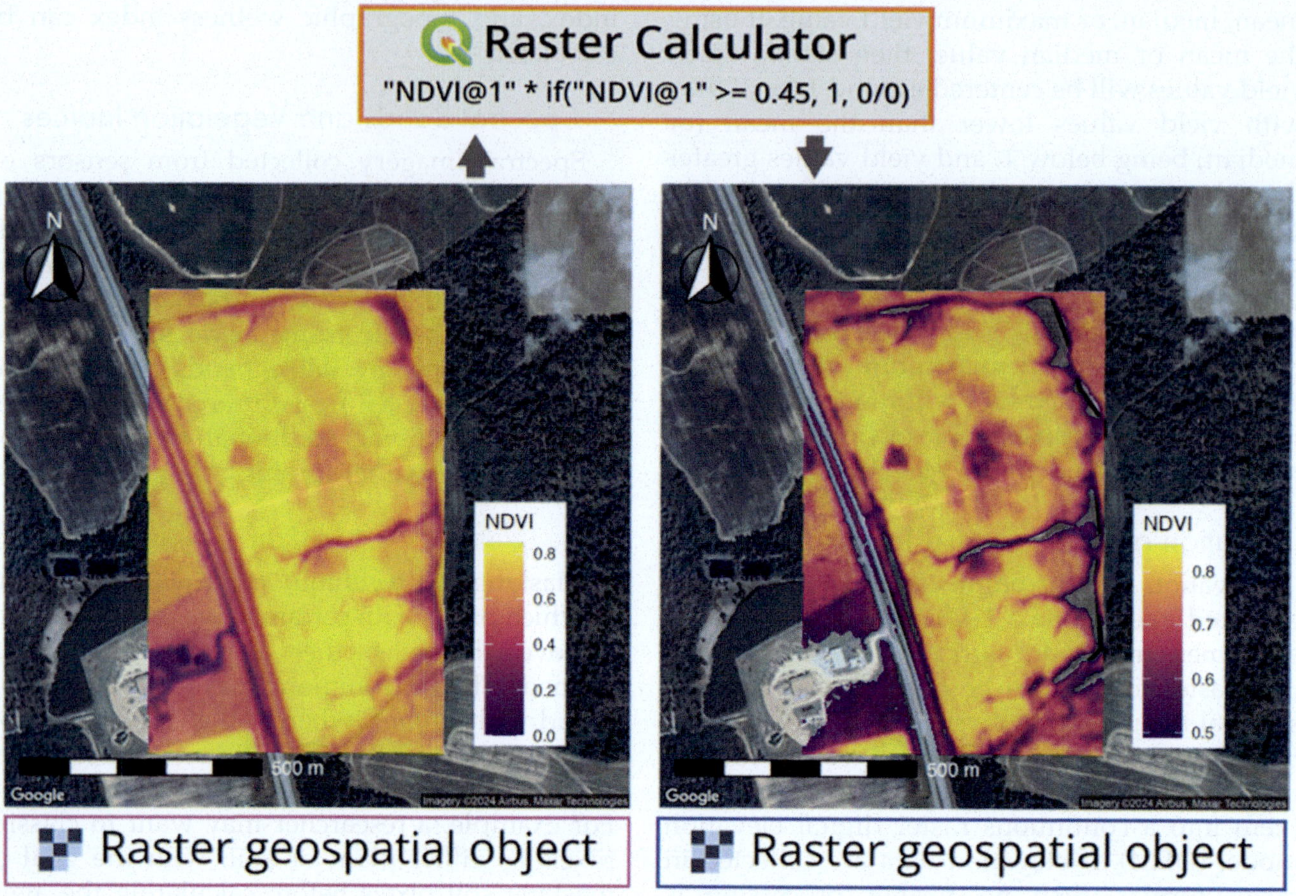

Figure 9.13. Normalized difference vegetation index (NDVI) remote sensing data from PlanetScope satellite in mid-July during the 2017 cotton growing season. The chart of the left was not masked and the chart on the right is masked where pixels with NDVI <0.45 were removed. (Courtesy of Leo Bastos and Rob Proulx).

masking, the process involves removing only the undesired pixels based on their classification, for example, all pixels with an NDVI value below 0.45 were classified as soil and masked (Figure 9.13).

Focal Window Operations

Focal window operations are used to smooth different values or classes. This task is important when creating management zones, since the original assignment of zones to pixels may create speckles of one zone within a different, larger zone (e.g., a few cells of a low-yield zone surrounded by many cells of a high-yield zone). To perform a focal window operation, the size of the focal window and the function that will be used to smooth the data need to be determined.

The focal window size determines how many cells will be considered when deciding the value of the center cell. Common focal window sizes are 3 × 3 cells, 5 × 5 cells, or any other odd-numbered dimension. Odd-numbered cell sizes are required because they create windows with a center cell, which is the one considered to be smoothed at each iteration of the focal window. Even-numbered cell sizes (e.g., 2 × 2 cells) do not have a center cell and thus are not useful for focal window operations. Once the size of the focal window is determined, the function to be applied needs to be selected. Common functions include mean, median, minimum, and maximum for continuous data and mode for categorical data. The left panel of Figure 9.14 illustrates where two management zones were created with pixels of one zone speckled within the other. The speckling effect was addressed by applying a focal window operation with a focal window size of 5 × 5 cells

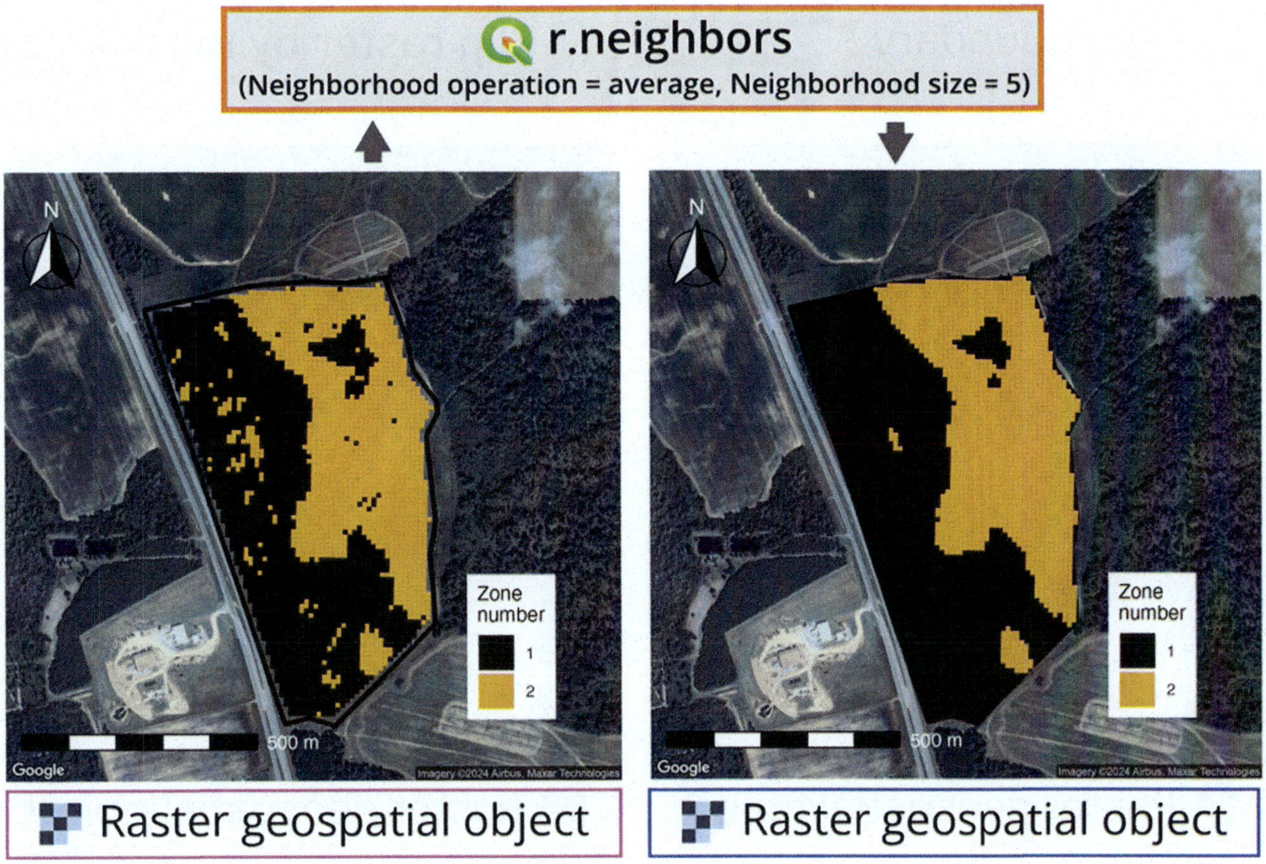

Figure 9.14. Unsmoothed (left panel) and smoothed (right panel) management zone cell assignment based on a 5 × 5 cells focal window size and mode function. (Courtesy of Leo Bastos and Rob Proulx).

and applying the mode function (Figure 9.14, right panel).

Vector–Raster Operations

Some operations require vector and raster data layers. The goal of this subsection is to illustrate common vector–raster operations used in precision agriculture. For the applications described below, the vector and raster layers must share a common CRS.

Cropping Imagery with Field Boundaries

Imagery (raster data) is usually collected over an area larger than the target area to ensure full coverage. Cropping the imagery to include only pixels within a field boundary helps ensure that only necessary data is retained for analysis. To crop imagery to a field boundary, the imagery and the field boundaries are needed. Figure 9.15 presents an example of cropping.

Extracting Pixel Values at Specific Point Locations

Extracting pixel values from specific point locations is a useful task to connect georeferenced sampling locations on the ground to their corresponding imagery data. To extract pixel values at specific point locations, imagery and ground-collected information are needed. Figure 9.16 shows an example where the location of the gridded soil sampling points previously identified for this field was used to extract NDVI data from satellite imagery.

Zonal Statistics

Zonal statistics are used to extract statistical summary metrics (e.g., minimum, median, mean, maximum) of pixel values within a polygon. Some examples where summary statistics are extracted include management zone boundaries,

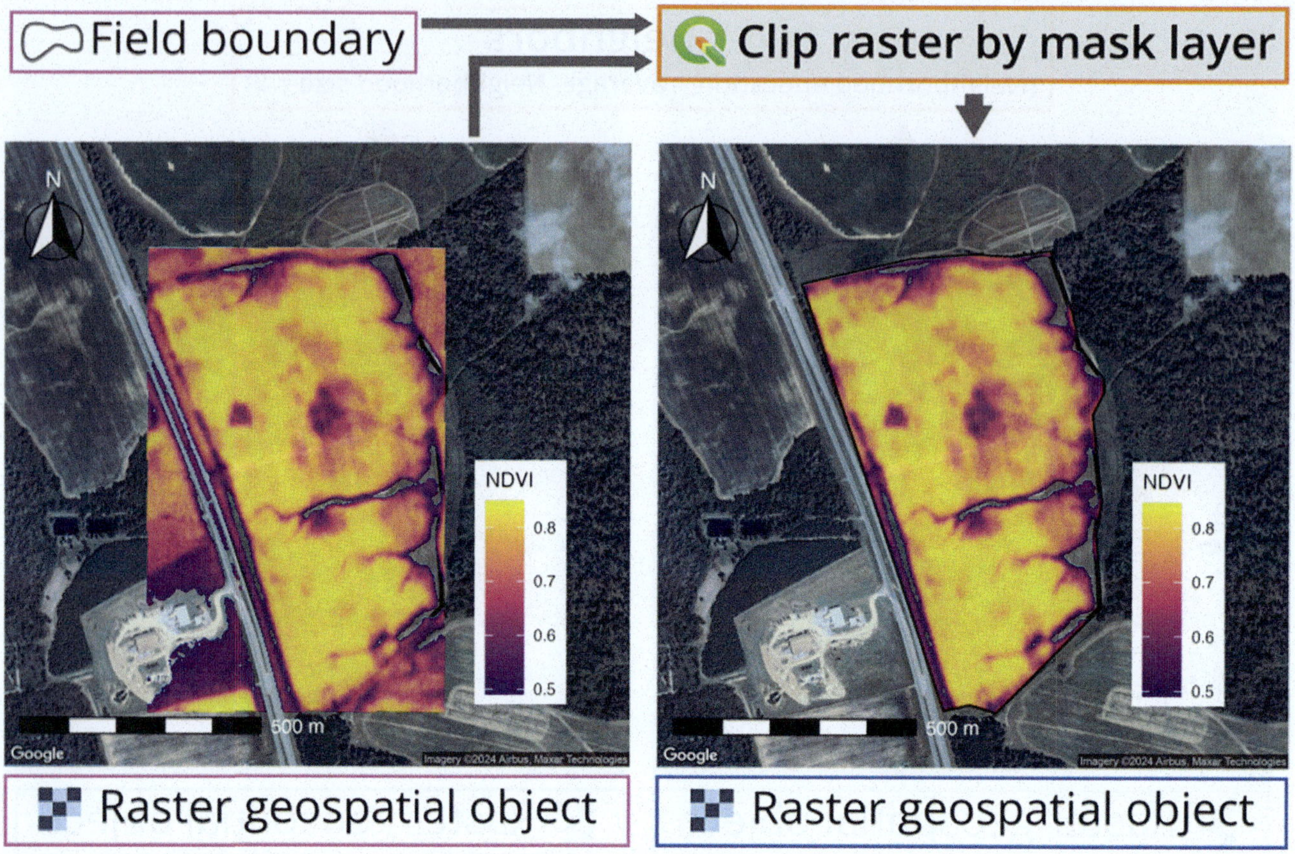

Figure 9.15. Normalized difference vegetation index (NDVI) remote sensing data from PlanetScope satellite in mid-July during the cotton growing season in 2017, cropped to the field boundary. (Courtesy of Leo Bastos and Rob Proulx).

block boundaries, plot boundaries, and a buffered area around a point. One example of this application is averaging subplot data to connect it to plot-level yield data. To compute zonal statistics, the required layers are (a) at least one imagery raster layer and (b) a polygon vector layer containing the boundaries for each "zone." Figure 9.17 depicts the extraction of the mean NDVI value for each of the two management zones identified for this field.

Net Profit Analysis

A complete net profit analysis should incorporate fixed costs, variable costs, depreciation, and other aspects to obtain a full picture of revenue and costs. However, to simplify the problem, partial profit analysis (revenue minus costs) is often determined. This analysis may only consider the return on investment for a single operation, where partial revenue is calculated. In this example, a single fixed rate of nitrogen fertilizer was applied, and therefore, partial costs were the same across the field. In more complex examples where input rates change in a variable rate application, the partial costs would also vary in space (Figure 9.18).

Nutrient Use Efficiency

Nutrient use efficiency metrics are useful in understanding how much of the applied nutrient was used by a crop. In general, higher efficiencies are associated with lower losses to the environment (McLellan et al., 2018). Many nutrient use efficiency metrics exist and are used in agriculture (Fixen et al., 2015). One of the simplest and most common metrics applied to commercial fields is called partial factor productivity (PFP). The PFP value is computed as yield divided by fertilizer application rate. Figure 9.19 provides an example for PFP of nitrogen fertilizer in a cotton crop, where cotton lint yield (in kilograms per hectare) data collected with a cotton picker

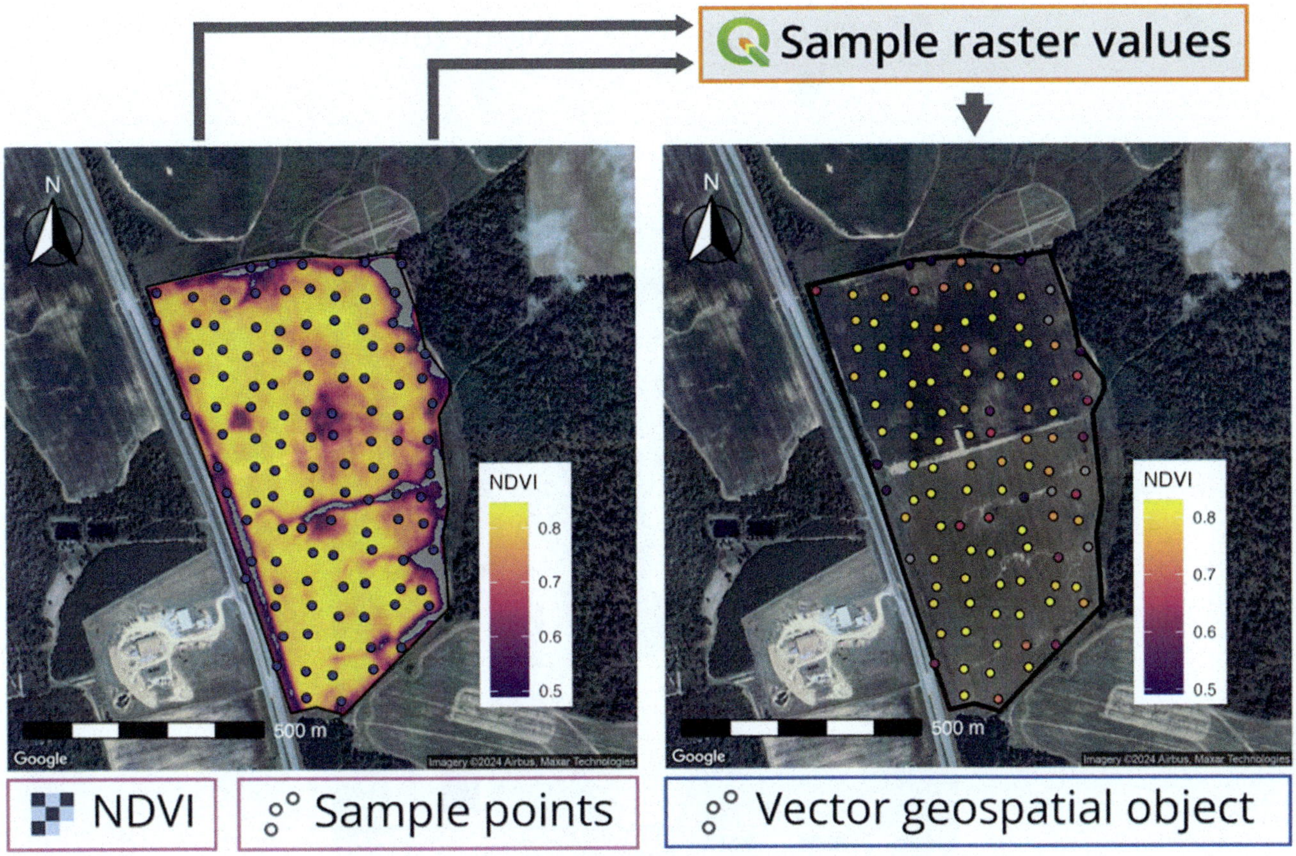

Figure 9.16. Normalized difference vegetation index (NDVI) remote sensing data from the PlanetScope satellite in mid-July during the cotton growing season in 2017 overlaid with the positions of soil sampling points (left panel), and the NDVI data extracted at each point location (right panel). (Courtesy of Leo Bastos and Rob Proulx).

equipped with a yield monitor was divided by nitrogen rate (in kilograms per hectare). In this example, yield varies spatially while the nitrogen rate is applied at a fixed rate of 90 kg N/ha (80.5 lb N/acre)

Using Programming Languages as GIS

Most users learn, use, and interact with GIS software through a graphical user interface (GUI), such as QGIS, ArcGIS, and SMS AgLeader. Graphical user interfaces, discussed further in Chapter 8, are a user-friendly way to interact with any software, where menus, tools, and options are accessed using a computer mouse and keyboard via point-and-click, drag-and-drop, selection, etc. While this way of interacting with GIS software is easy to learn and user-friendly, it is largely manual and difficult to reproduce. Imagine you need to run an analytical workflow that uses 10 different GIS tools, and you need to run each tool on 100 different fields. That is a total of 1,000 GIS operations! This situation is time-consuming and prone to human error, as the user can make mistakes. One way to implement a less time-consuming and easily reproducible workflow is to avoid manual GIS operations and use a programming language instead.

Why Programming Languages Are a Popular Choice

In contrast to GUIs, programming languages use code and scripts to interact with geospatial and non-geospatial data. Two of the most used programming languages in GIS and precision agriculture are R and Python. Both languages are free and open source and are increasingly being taught and used in precision agriculture. Two key benefits of using a programming language are time savings and reproducibility. While learning a programming language does require an

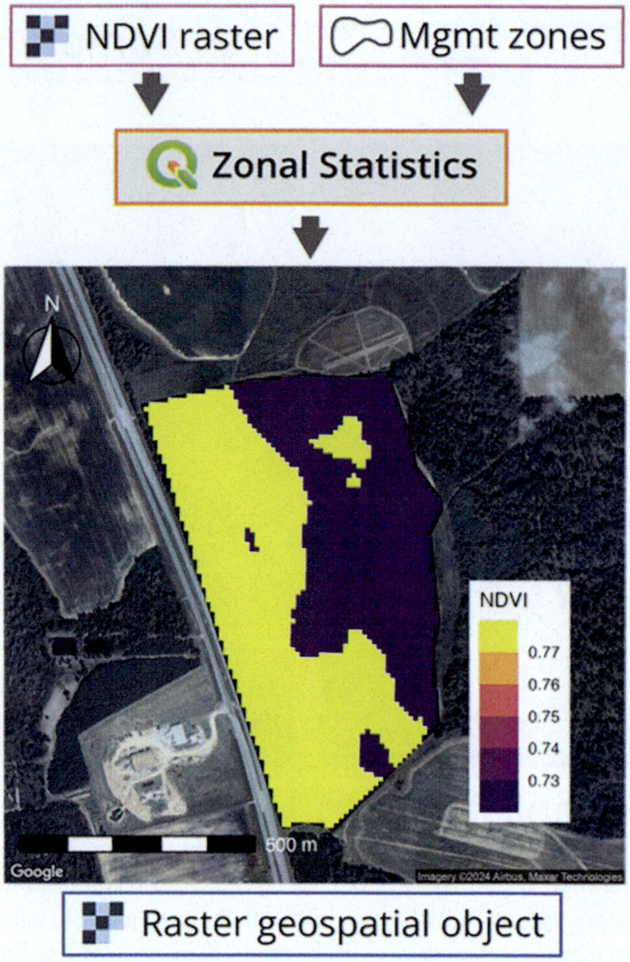

Figure 9.17. Normalized difference vegetation index (NDVI) remote sensing data from the PlanetScope satellite in mid-July during the cotton growing season in 2017 averaged across the two previously identified management zones. (Courtesy of Leo Bastos and Rob Proulx).

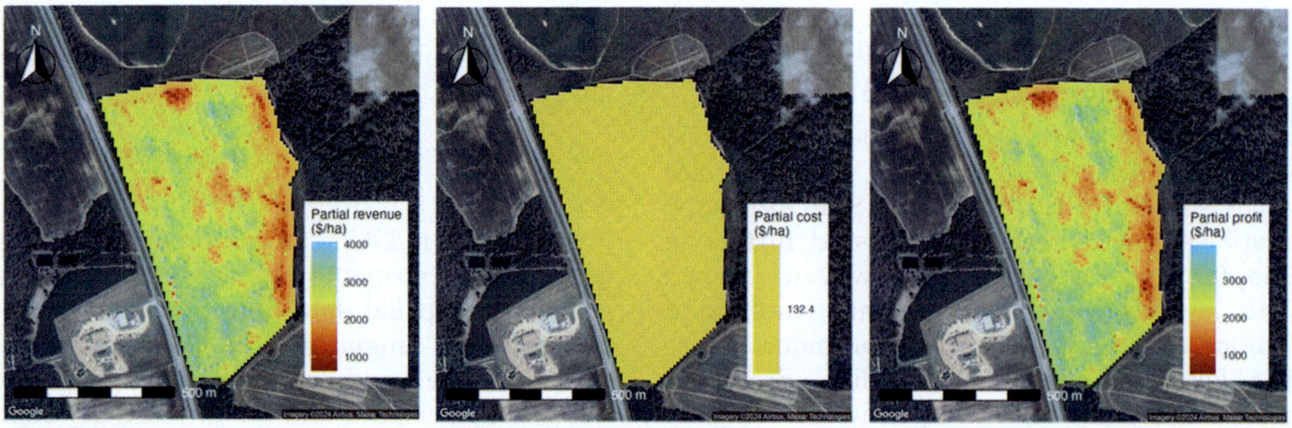

Figure 9.18. Partial revenue map (calculated as cotton lint yield multiplied by a price of $USD 1.87 kg lint or $USD 0.87/lb lint; left panel), partial cost map (calculated as a fixed nitrogen fertilizer rate of 90 kg N/ha or 80 lb N/ac multiplied by a cost of $USD 1.48/kg N or $USD 0.67/lb N; middle panel), and partial profit map (calculated as partial revenue minus partial cost; right panel). (Courtesy of Leo Bastos).

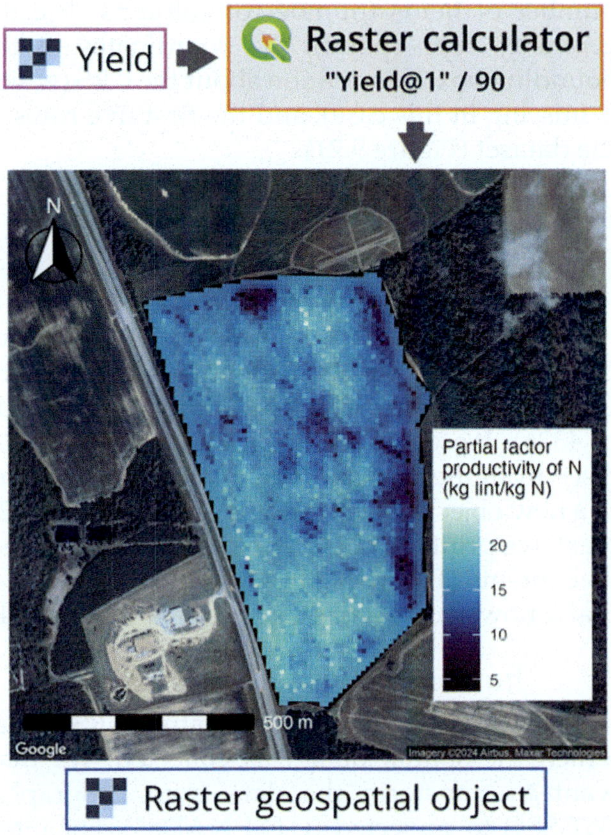

Figure 9.19. Partial factor productivity of nitrogen fertilizer calculated as the ratio between cotton yield (in kg lint/ha) and nitrogen fertilizer rate (in kg N/ha). To convert to lb/ac multiply by 0.89. (Courtesy of Leo Bastos and Rob Proulx).

initial time investment, time savings will accrue over time. That is because once a script is developed, it can be reused for new datasets with only minor adaptations. Another important benefit is reproducibility. An analysis is reproducible when another user can fully replicate your results when given the same starting data and code. Also, as shown below, the code itself serves as documentation of the tools used and the options selected.

In the following discussion, we demonstrate the use of GIS tools using the programming language R. This chapter is not an introduction to R. However, no R knowledge is required since the chapter focuses on understanding the parallels between using a GUI like QGIS and a programming language like R. Beginners to R and RStudio may find the RStudio Education site (https://education.rstudio.com/learn/beginner) helpful for getting started.

Using R as a GIS

R is a free, open-source statistical programming language initially developed in the early 1990s by Ross Ihaca and Robert Gentleman (Ihaka & Gentleman, 1996), two statisticians from the University of Auckland. Since its initial launch, R has grown in adoption and popularity across many scientific fields, including GIS and precision agriculture. Most people use R within an integrated development environment, where RStudio is among the most popular.

In R, it is important to understand four main concepts: (a) functions, (b) function arguments, (c) packages, and (d) objects. A function performs a specific task and is represented by a word (the function name) followed by opening and closing parentheses. For example, if you wanted to calculate the mean of a variable, you would use the function 'mean()'. Inside functions, there are function arguments that are either required information (like the column name where the values for which we want to calculate the mean are stored) or optional information that can change the default behavior of a function (like whether cells containing missing values should be removed before calculating the mean). In our example, assuming we have yield values stored in a variable called 'yield,' and that this variable contains a few missing value cells that we want to remove before calculating the mean, the code for that operation would be 'mean(x = yield, na.rm. = TRUE)'. Notice that we called the function name (mean), provided the name of the variable containing the yield values (x = yield), and asked for missing/not available (na) values to be removed before computing the mean (na.rm. = TRUE). A collection of functions that serve a similar purpose and are bundled together when installed is called a package. The function 'mean()' ships with R as part of the package 'base,' which offers many other functions for calculating statistical summaries and performing other operations in R.

Last, objects in R are named variables that store any type of information the user decides to assign to them for future recall. Going back to our previous example, assuming the result of the call to the mean function on the yield object was the number 250 (bushels per acre). We wanted to save this number for future recall into an object called "yield_mean", we would use the code 'yield_mean <- mean(x = yield, na.rm. = TRUE),' where 'yield_mean' is the name of the object where we are storing the result of the mean operation, and the symbol '<-' is used to say "store the results of the call to my right on the object name to my left" and is referred to as the "assignment symbol." Objects can store any type of information, ranging from a single number, as in our example, to characters, lists, data frames and their various columns, and vector and raster geospatial data, among others.

When installed, R ships with basic functions for statistical analysis and graphics, and is backed up by a large, active user community with many community-driven contributions made in the form of packages. One can think of R as your smartphone, and packages as applications you download and install to perform a specific task. All community-contributed packages in R are available on the Comprehensive R Archive Network (CRAN; https://cran.r-project.org).

In any programming language, there are many ways to perform the same task. R is no different. Whenever possible, our examples use functions and packages that were developed using a shared underlying philosophy, grammar, and data structures found in an overarching collection of packages called the tidyverse (Wickham et al., 2019).

Importing Vector Data Layers into R

In the example in Figure 9.20, we import a geospatial vector layer of type point (yield monitor data) into R. To import a vector data layer, we are going to use the function 'read_sf()' from the package 'sf' (Pebesma, 2018), a very popular package for handling vector spatial data in R. For that, we provide the path where the shapefile is stored, and assign it to an object called 'yield.' We print the object 'yield' to inspect its contents. Notice that we get information about how many features it has (number of points, 31,721), the number of fields (number of columns, 16), the geometry type (point), the coordinates of the bounding box, information about the CRS (which is missing in this case), and the first five rows of the dataset (Figure 9.21).

Assigning Native Coordinate Reference System

Since the CRS is missing, one of the immediate next steps would be to set or assign its native CRS, which is WGS84 (identified through its European Petroleum Survey Group [EPSG] code of 4326). We do that using the function 'st_set_crs()' from the package 'sf,' and assign the result to a new object called 'yield_wgs84' (Figure 9.22). Next, we print the object 'yield_wgs84' to inspect its contents (Figure 9.23). The only difference is that it now shows it has a geodetic CRS (WGS84).

Transforming to a Different CRS

Now that we have set the native CRS under which the spatial data was collected, we may want to transform the data from geographic (WGS84) to projected (UTM NAD83). For more detail on coordinate systems and projections, see Chapter 8. To transform the data, we use the function 'st_transform()' from the package 'sf' by specifying the EPSG code for UTM NAD83 Zone 17 N (6346) and assign the result to a new object called 'yield_utm' (Figure 9.24). Then, we print the object 'yield_utm' to inspect its contents. The only difference is that now it shows that it has a projected CRS (NAD83 2011 UTM Zone 17; Figure 9.25).

Creating a Static Map

Now that our yield data has been imported, set to its native geographic CRS, and transformed to a projected CRS, we can go ahead and plot it for visual assessment. When creating a plot of

```
# Importing the yield shapefile and storing it in a new object called yield
yield <- read_sf("../data/02 processed/yield.shp")

# Printing the first 5 rows of the object yield
print(x = yield,
      n = 5,
      width = 70)
```

Figure 9.20. Example of R code for importing a geospatial vector layer of type point and printing its contents.

```
Simple feature collection with 31721 features and 16 fields
Geometry type: POINT
Dimension:     XY
Bounding box:  xmin: -84.7652 ymin: 31.18432 xmax: -84.7588 ymax: 31.19301
CRS:           NA
# A tibble: 31,721 × 17
    FID Region    Heading Elevation Distance Duration Speed Fuel_Usage
* <dbl> <chr>       <dbl>     <dbl>    <dbl>    <dbl> <dbl>      <dbl>
1     0 First Pi…    343.      172.     5.76        1   4.3       25.4
2     1 First Pi…    343.      173.     7.21        1   4.8       20.9
3     2 First Pi…    343.      173.     7.07        1   4.8       21.4
4     3 First Pi…    343.      173.     7.11        1   4.8       22.6
5     4 First Pi…    343.      173.     7.11        1   4.8       22.6
# i 31,716 more rows
# i 9 more variables: Moisture <dbl>, Mass_Cotto <dbl>, Width <dbl>,
#   Engaged <dbl>, Mass_Yield <dbl>, Wet_Yield <dbl>,
#   Dry_Yield <dbl>, Timestamp <dbl>, geometry <POINT>
```

Figure 9.21. Example of printed output of the object yield.

```
# Setting the native geographic CRS WGS84 on object yield and storing the
result in a new object called yield_wgs84
yield_wgs84 <- st_set_crs(x = yield,
                          value = 4326)
# Printing the first 5 rows of the object yield_wgs84
print(x = yield_wgs84,
      n = 5,
      width = 70)
```

Figure 9.22. Example R code for setting the native coordinate reference system of a vector object and printing its contents.

```
Simple feature collection with 31721 features and 16 fields
Geometry type: POINT
Dimension:     XY
Bounding box:  xmin: -84.7652 ymin: 31.18432 xmax: -84.7588 ymax: 31.19301
Geodetic CRS:  WGS 84
# A tibble: 31,721 × 17
    FID Region    Heading Elevation Distance Duration Speed Fuel_Usage
* <dbl> <chr>       <dbl>     <dbl>    <dbl>    <dbl> <dbl>      <dbl>
1     0 First Pi…    343.      172.     5.76        1   4.3       25.4
2     1 First Pi…    343.      173.     7.21        1   4.8       20.9
3     2 First Pi…    343.      173.     7.07        1   4.8       21.4
4     3 First Pi…    343.      173.     7.11        1   4.8       22.6
5     4 First Pi…    343.      173.     7.11        1   4.8       22.6
# i 31,716 more rows
# i 9 more variables: Moisture <dbl>, Mass_Cotto <dbl>, Width <dbl>,
#   Engaged <dbl>, Mass_Yield <dbl>, Wet_Yield <dbl>,
#   Dry_Yield <dbl>, Timestamp <dbl>, geometry <POINT [°]>
```

Figure 9.23. Example of printed output of the object 'yield_wgs84'.

```
# Transforming from geographic to projected CRS on object yield_wgs84 and
storing the result in a new object called yield_utm
yield_utm <- st_transform(x = yield_wgs84,
                          crs = 6346)
# Printing the first 5 rows of the object yield_utm
print(x = yield_utm,
      n = 5,
      width = 70)
```

Figure 9.24. Example R code for transforming the coordinate reference system of a vector object and printing its contents.

```
Simple feature collection with 31721 features and 16 fields
Geometry type: POINT
Dimension:     XY
Bounding box:  xmin: 141154 ymin: 3456129 xmax: 141756.7 ymax: 3457094
Projected CRS: NAD83(2011) / UTM zone 17N
# A tibble: 31,721 x 17
    FID Region    Heading Elevation Distance Duration Speed Fuel_Usage
* <dbl> <chr>       <dbl>     <dbl>    <dbl>    <dbl> <dbl>      <dbl>
1     0 First Pi…    343.      172.     5.76        1   4.3       25.4
2     1 First Pi…    343.      173.     7.21        1   4.8       20.9
3     2 First Pi…    343.      173.     7.07        1   4.8       21.4
4     3 First Pi…    343.      173.     7.11        1   4.8       22.6
5     4 First Pi…    343.      173.     7.11        1   4.8       22.6
# i 31,716 more rows
# i 9 more variables: Moisture <dbl>, Mass_Cotto <dbl>, Width <dbl>,
#   Engaged <dbl>, Mass_Yield <dbl>, Wet_Yield <dbl>,
#   Dry_Yield <dbl>, Timestamp <dbl>, geometry <POINT [m]>
```

Figure 9.25. Printed output of the object 'yield_utm'.

any sort, including geospatial data, we can create either a static or an interactive plot. For static plots we will use functions from the package 'ggplot2' (Wickham, 2016). For that, we start it with a call to the function 'ggplot()' and sequentially add layers to it using the '+' sign at the end of each line of code. In this case, we will add a vector geospatial layer using the function 'geom_sf(),' since our vector data is of type 'sf,' change the color scale to viridis (colorblind-friendly scale), and change the theme of the plot to more closely match that of a map (Figure 9.26).

Creating an Interactive Map

To create interactive maps, we will use the package 'mapview' (Appelhans et al., 2023), which has a function with the same name. Inside this function, we specify the value for different arguments including the name of the data frame object (x = yield_utm), the column to be used to color the points (zcol = "Dry_Yield"), and change the point size (cex = 0.1), the point outline width for improved appearance (lwd = 0), and a basemap option (map.types = "Esri.World-Imagery"; Figure 9.27).

Summary

This chapter, an extension of Chapter 8, explored how to use geoprocessing tools to perform spatial data analysis and transformation. This was a non-exhaustive set of GIS applications for precision agriculture using multiple data from a cotton field in South Georgia, USA. The chapter divided the applications into two categories: vector or raster data types, and their combined uses. We also covered the use of R, one of the most common statistical programming languages used in GIS, and demonstrated code and output examples for some precision agriculture operations.

```
ggplot () +
  geom_sf(#define data to be used
          data = yield_utm,
          #assign point color to column Dry_Yield
          mapping = aes(color = Dry_Yield),
          #change the size of the points
          size = .5) +
  #change default color scale to viridis
    (colorblind-friendly)
  scale_color_viridis_c() +
  #change the plot theme to match that of a map
  theme_map()
```

Figure 9.26. Example R code for creating a static map and output map.

```
mapview(
        #define data to be used
        x = yield_utm,
        #assign point color to column Dry_Yield
        zcol = "Dry_Yield",
        #change the size of the points
        cex = .1,
        #removing point outline
        lwd = 0,
        #change basemap
        map.types = "Esri.WorldImagery")
```

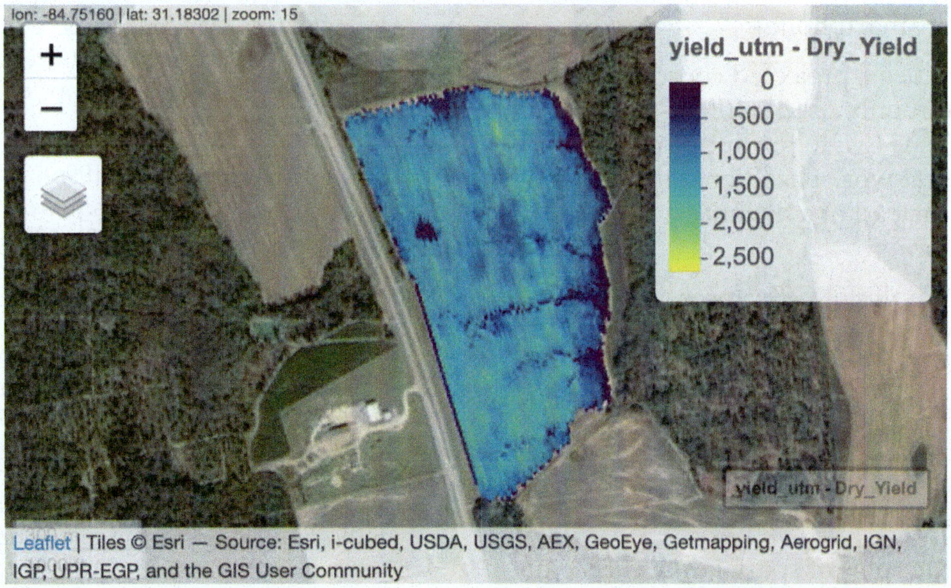

Figure 9.27. Example R code for creating an interactive map and output map.

ACKNOWLEDGMENTS

Partial funding was provided by the U.S. Department of Agriculture National Institute of Food and Agriculture under the Multi-State Hatch project (Project Number: KY006151-7008110-S1090), South Dakota State University, South Dakota Corn, the South Dakota Soybean Research and Promotion Council, and the Kentucky Corn Grower Association. The authors thank Dr. Lorena Lacerda and Dr. George Vellidis for contributing to the datasets used in this chapter. The use of trade names is for the convenience of the reader and does not imply endorsement by the authors.

Study Questions

1. Georeferenced points are essential for creating accurate boundaries in precision agriculture.
2. Vector-based operations in GIS are used to analyze and manipulate raster data.
3. What type of data is typically represented in raster format in GIS?
   A. Points
   B. Lines
   C. Polygons
   D. Continuous surfaces
4. How would you create a workflow for creating a field boundary (polygon) based on GPS locations collected by your mobile device (e.g., smartphone)?
5. Should you use a projected coordinate system commonly used in precision agriculture, such as UTM, to produce maps using functions that work based on latitude and longitude instead of a distance-based system?
6. Several examples of GIS processing tools are provided within this chapter. Find and research two GIS tools not listed and describe their characteristics and functions.
7. List four examples of spatial data characterized as vector data used in precision agriculture.
8. List four examples of spatial data characterized as raster data used in precision agriculture.
9. A grower asks you to perform gridded soil sampling on one of her corn fields. Describe the sequence of steps required to accomplish them, from the initial data needed, tools used, all the way to delivering an interpolated map.
10. A grower asks you to create management zones on one of his cotton fields. List at least five geospatial data layers that could be used for this purpose.
11. Following from the situation in Question 10, if you were using R as GIS, list the functions and their arguments required to import a vector geospatial data, set its native CRS (if missing), and transform its CRS to a different one.
12. Should precision agriculture prioritize environmental sustainability over maximizing crop yields? (Debate)

REFERENCES AND ADDITIONAL RESOURCES

Appelhans, T., Detsch, F., Reudenbach, C., & Woellauer, S. (2023). *mapview: Interactive viewing of spatial data in R*. https://doi.org/10.32614/CRAN.package.mapview.

Fixen, P. E., Brentrup, F., Bruulsema, T., Garcia, F., Norton, R., & Zingore, S. (2015). Nutrient/fertilizer use efficiency: Measurement, current situation and trends. In P. Drechsel, P. Heffer, H. Magen, R. Mikkelsen, & D. Wichelns (Eds.), *Managing water and fertilizer for sustainable agricultural intensification* (pp. 8–38). IFA, IWMI, IPNI, IPI.

Ihaka, R., & Gentleman, R. (1996). R: A language for data analysis and graphics. *Journal of Computational and Graphical Statistics*, *5*(3), 299–314. https://doi.org/10.1080/10618600.1996.10474713.

Li, J., & Heap, A. D. (2011). A review of comparative studies of spatial interpolation methods in environmental sciences: Performance and impact factors. *Ecological Informatics*, *6*(3–4), 228–241. https://doi.org/10.1016/j.ecoinf.2010.12.003.

Li, J., & Heap, A. D. (2014). Spatial interpolation methods applied in the environmental sciences: A review. *Environmental Modelling & Software*, *53*, 173–189. https://doi.org/10.1016/j.envsoft.2013.12.008.

McLellan, E. L., Cassman, K. G., Eagle, A. J., Woodbury, P. B., Sela, S., Tonitto, C., Marjerison, R. D., & van Es, H. M. (2018). The nitrogen balancing act: Tracking the environmental performance of food production. *Bioscience*, *68*(3), 194–203. https://doi.org/10.1093/biosci/bix164.

Pebesma, E. (2018). Simple features for R: Standardized support for spatial vector data. *R Journal*, *10*(1), 439–446. https://doi.org/10.32614/RJ-2018-009.

Robinson, T. P., & Metternicht, G. (2006). Testing the performance of spatial interpolation techniques for mapping soil properties. *Computers and Electronics in Agriculture*, *50*(2), 97–108. https://doi.org/10.1016/j.compag.2005.07.003.

Sudduth, K. A., Drummond, S. T., & Myers, D. B. (2012). *Yield Editor 2.0: Software for automated removal of yield map errors*. Dallas, Texas, July 29–August 1, 2012. https://doi.org/10.13031/2013.41893.

Sudduth, K. A., Drummond, S. T., & Myers, D. B. (2019). *Yield Editor 2.0.7*. https://www.sidalc.net/search/Record/dat-usda-us-article24664038.

Wickham, H. (2016). *ggplot2: Elegant graphics for data analysis*. Springer-Verlag. https://ggplot2.tidyverse.org.

Wickham, H., Averick, M., Bryan, J., Chang, W., McGowan, L. D., François, R., Grolemund, G., Hayes, A., Henry, L., Hester, J., Kuhn, M., Pedersen, T. L., Miller, E., Bache, S. M., Müller, K., Ooms, J., Robinson, D., Seidel, D. P., Spinu, V., . . . Yutani, H. (2019). Welcome to the tidyverse. *Journal of Open Source Software*, *4*(43), 1686. https://doi.org/10.21105/joss.01686.

Web Soil Survey and Soil Health Assessments

10

Carrie Werkmeister and Douglas Malo

Chapter Purpose

The Web Soil Survey (WSS) is an interactive online tool that provides fundamental soils information that can be used for many purposes, ranging from locating sites for landfills to identifying areas suitable for crop production. This chapter focuses on how to use soils data in agricultural land management decisions.

Introduction

To provide better access to soil survey information, the United States Department of Agriculture Natural Resources Conservation Service (USDA NRCS), along with other federal, state, and local agencies, launched the Web Soil Survey (WSS) in 2005. Unlike traditional printed county soil surveys, the WSS provides an updated, interactive tool that makes soil data accessible. This free-to-the-public interactive online platform provides fundamental soils information and access to soils data that can be downloaded, viewed, analyzed, fused with other data layers, or used to create customized reports for agriculture, engineering, land management, and soil health purposes within the United States (Soil Survey Staff et al., 2025; Figure 10.1). The WSS provides a modern alternative to the traditional hardbound county soil survey publications. It delivers information more quickly, offers electronic access to complete and annually updated soil data, and eliminates the high cost of printing. The WSS allows users to download spatial, tabular, and interpretive data for use in geographic information systems (GIS) software or handheld devices, making it possible to view soils information in the field and tailor it to specific locations or investigations. For example, Joshi et al. (2019) used downloaded and fused land capability classification of South Dakota and Nebraska soil mapping units (MUs) with visual land use assessments to assess the risks for converting grassland to cropland. When combined with other data layers or models, data downloaded from the WSS can also be used to identify management zone boundaries, improve fertilizer recommendations, and estimate erosion risks or be used as a benchmark for soil health improvements (Chang et al., 2004).

Accessibility of the WSS

The WSS can be accessed at https://websoilsurvey.nrcs.usda.gov (Figure 10.2). Rather than provide step-by-step instructions for using the WSS, which are available elsewhere (Malo, 2008, 2016a, 2016b; Soil Survey Staff et al., 2025), this chapter highlights the use of the data in agricultural land management decisions and provides a list of abbreviations and acronyms (Tables 10.1 and 10.2). When using the

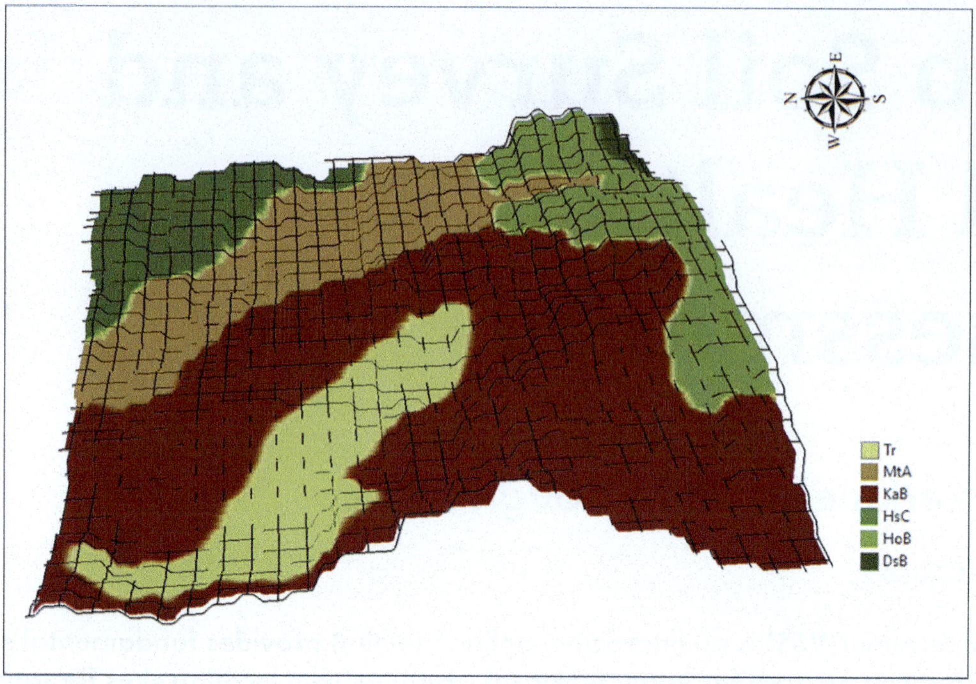

Figure 10.1. Three-dimensional soil series field map generated using data from the WSS. The figure illustrates spatial distribution across a landscape of mapped soil series for a field research site. Note: DsB, Doland loam soil (2–6% slopes); HoB, Houdek clay loam soils (2–6% slopes); HsC, Houdek–Shindler clay loam soils (5–9% slopes); KaB, Kranzburg–Brookings silty clay loam soils (1–6% slopes); MtA, Moody–Trent complex (0–2% slopes); Tr, Trent silty clay loam soils (0–3% slopes). (Courtesy Sravanthi Bachina).

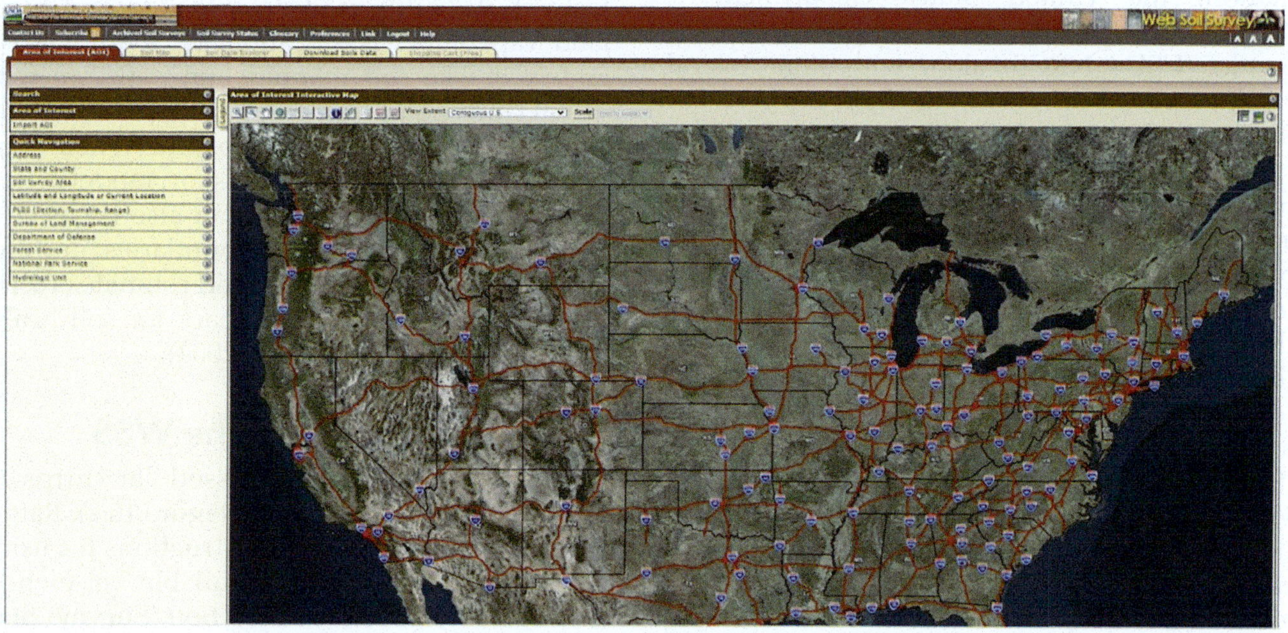

Figure 10.2. Web Soil Survey home page, where the user will begin. https://websoilsurvey.nrcs.usda.gov/app (United states Department of Agriculture / Soil Survey Staff (2025) / Public Domain.

WSS click on the "Start WSS" to engage in the interactive mapping tool platform. One of the first steps is to define your area of interest (AOI). The WSS provides multiple options (e.g., address, state, and county; soil survey area; latitude and longitude; legal land description [section, township, range]; manual selection) for selecting the AOI.

Table 10.1. Abbreviations and acronyms used in this chapter

Item agency/name	Acronym/ abbreviation
Acre	ac
Area of Interest	AOI
Cation exchange capacity	CEC
Crop Productivity Index	CPI
Doland loam soil	DsB
Earth Resource Observation and Science	EROS
Electrical conductivity	EC
Geographic Information System	GIS
Global Positioning System	GPS
Hectare	ha
Houdek clay loam soil	HoB
Houdek–Shindler clay loam soils	HsC
Inorganic carbon	IC
Kellogg Soil Survey Lab	KSSL
Kranzburg–Brookings silty clay loam soils	KaB
Major Land Resource Area	MLRA
Map Unit	MU
Megabyte	MB
Moody–Trent complex	MtA
National Commodity Crop Productivity Index	NCCPI
National Cooperative Soil Survey	NCSS
National Institute of Food and Agriculture	NIFA
Natural Resources Conservation Service	NRCS
Organic matter	OM
Series Extent Explorer	SEE
Sodium Adsorption ratio	SAR
Soil Landscapes of the United States	SOLUS
Soil organic carbon	SOC
Soil organic matter	SOM
Soil Survey Geographic Database	SSURGO
South Dakota State University	SDSU
Trent silty clay loam soil	Tr
Total carbon	TC
United States Department of Agriculture	USDA
United States Department of Interior	USDOI
Web Soil Survey	WSS

Table 10.2. Common applications of the WSS

Precision agriculture mapping
Soil health benchmarking
GIS and remote sensing
Soil fertility—fertilizer rate planning
Lime recommendations
Assist with spray application
Provide knowledge for variable rate technology
Crop suitability
Crop selection and rotation planning
Irrigation suitability
Identify soil types
Compare soils across regions
Evaluate field research sites—support research projects
Assist in technical writing and reporting
Understand soil limitations
Assist farm management decisions
Build digital soil platforms
Modeling crop growth and inputs
Assist with industry certifications
Provide information for collegiate contests
Drainage design and planning
Erosion risk and soil loss
Manure and wastewater application
Site section for buildings and infrastructure
Pasture and rangeland management
Tillage decisions
Conservation compliance
Custom soil productivity reports—used by USDA, loan applications or insurance companies
Monitor field trends or areas over time

Once the AOI is set, the platform generates an interactive map tied to the location. Users can then create numerous types of maps, including soil drainage, soil textures, soil pH, restrictive layers, slopes, and many others. The WSS complements, rather than replaces, classical soil test reports and can be used to fill missing data gaps in experiments. For example, Clay et al. (2012) used bulk density information downloaded from the WSS to determine partial carbon intensity values for South Dakota.

The background data archive of the WSS is the Soil Survey Geographic Database (SSURGO).

The SSURGO contains soil data collected over the last 100 years through field soil survey and sampling and collecting and analyzing soil samples for many physical and chemical properties at scales ranging from 1:12,000 to 1:63,360. The data archived in SSURGO include many soil physical and chemical properties (e.g., plant-available water, electrical conductivity [EC], cation exchange capacity [CEC], frequency of flooding, expected yields, texture, land capability class, and soil organic matter [SOM]).

The WSS is used by producers, engineers, agronomists, government agencies, construction professionals, gardeners, environmentalists, and land use planners for many purposes, including locating sites for landfills, orchards, house construction, roads, and drainage maps, as well as assessing land value for tax and sales purposes. There are a number of approaches to accessing the soil data (Table 10.3; Malo, 2016b). For example, the SoilWeb tool developed by the University of California, Davis is a mobile-friendly app that

Table 10.3. Online WSS sources for soil and natural resources information (Malo, 2008). Web addresses verified May 31, 2025

Name	Information available	Web address
California Soil Resource Lab	Soil survey data, soil web apps	http://casoilresource.lawr.ucdavis.edu
Canada Centre for Mapping and Earth Observation	General remote sensing information and educational materials	https://natural-resources.canada.ca/science-data/science-research/research-centres/canada-centre-mapping-earth-observation
NRCS—Lab Data Mart	Soil physical, chemical, and characterization data	https://www.nrcs.usda.gov/conservation-basics/natural-resource-concerns/soil/lab-data-mart
NRCS—Official Soil Series Descriptions	Detailed official soil series descriptions for soils in the USA	https://www.nrcs.usda.gov/resources/data-and-reports/official-soil-series-descriptions-osd
NRCS—Soil	Home page, soil classification, lab data	http://www.nrcs.usda.gov/wps/portal/nrcs/site/soils/home
NRCS—Soil Quality	Soil quality definitions, assessment, management, resources, and publications	https://www.nrcs.usda.gov/conservation-basics/natural-resource-concerns/soils/soil-health
NRCS—Soil Survey Manual	Soil Survey Manual Publication	https://www.nrcs.usda.gov/resources/guides-and-instructions/soil-survey-manual
NRCS—SOLUS	Soil Landscapes of the United States, soil property maps	https://www.nrcs.usda.gov/resources/data-and-reports/soil-landscapes-of-the-united-states-solus
NRCS—SSURGO/STATSGO2 Metadata	Soil Data Mart, metadata links and reports	https://www.nrcs.usda.gov/resources/data-and-reports/ssurgo/stats2go-metadata
Soil Explorer	Soil series and landscapes	https://soilexplorer.net
Soil Orders	Images of 12 soil orders	http://soils.cals.uidaho.edu/soilorders
Soil Series Explorer (SEE)	Soil series application viewer	https://casoilresource.lawr.ucdavis.edu/see
SoilWeb	Soil survey data compatible with technology	https://casoilresource.lawr.ucdavis.edu/gmap
USDA—National Agricultural Statistics Service	Agricultural statistics for state and county	http://www.nass.usda.gov
USDA—Web Soil Survey (WSS)	Detailed soil survey information	http://websoilsurvey.nrcs.usda.gov/app/HomePage.htm
USGS—The National Map	Various kinds and scales of U.S. maps	http://nationalmap.gov/viewer.html
Web GIS	TerraServer-USA-USA aerial and topographic images of land in the USA	http://www.webgis.com/terraserver.html

allows the end user to link their Global Positioning System (GPS) location to enable quick access to WSS information (California Soil Resource Lab, 2015). There are a few other tools, including GoogleEarth, Series Extent Explorer (SEE), GIS, and Soil Properties Explorer, that can help in accessing and processing data (Tables 10.3 and 10.4; Malo, 2016b).

Table 10.4. Online sources for soil and natural resources information (Malo, 2008). Web addresses verified May 31, 2025

Name	Information available	Web address
Bureau of Land Management, USDOI	Home page, projects and activities	https://www.blm.gov/
California Soil Resource Lab	Soil survey data, soil web apps	http://casoilresource.lawr.ucdavis.edu
Canada Centre for Mapping and Earth Observation	General remote sensing information and educational materials	https://natural-resources.canada.ca/science-data/science-research/research-centres/canada-centre-mapping-earth-observation
Canada Centre for Remote Sensing	General remote sensing information	https://natural-resources.canada.ca/science-data/science-research/research-centres/canada-centre-remote-sensing
Data Gateway	Current agricultural research results, education, and publications	http://cris.nifa.usda.gov
FAO Soil Portal	World Reference for soil classification system	http://www.fao.org/soils-portal/soil-survey/soil-classification/world-reference-base/en
Google Maps/Google Earth	Various maps of the USA in multidimensions	https://earth.google.com/web/
iGrow (SDSU Extension Service)	Production and management information	http://igrow.org
National Information Management and Support System	Agricultural research activities and projects in the state, region, and country	https://nimss.org
National Weather Service—NOAA	Weather data, drought monitoring, current conditions	http://www.weather.gov
NRCS—Field Office Technical Guide	Provide county specific scientific technical and reference information on soil, water, air, plant, and animal conservation	http://www.nrcs.usda.gov/wps/portal/nrcs/main/national/technical/fotg
NRCS—Hydric Soils	Hydric soils information	http://www.nrcs.usda.gov/wps/portal/nrcs/main/soils/use/hydric
NRCS—Kellogg Soil Curvey Laboratory (KSSL)	Source for NCSS soil analytical data	https://www.nrcs.usda.gov/conservation-basics/natural-resource-concerns/soil/kellogg-soil-survey-laboratory-kssl
NRCS—Lab Data Mart	Soil physical, chemical, and characterization data	https://www.nrcs.usda.gov/conservation-basics/natural-resource-concerns/soil/lab-data-mart
NRCS—Major Land Resource Area (MLRA)	Physiography, geology, climate, water resources, soils, biological resources, and kinds of land use	https://www.nrcs.usda.gov/resources/data-and-reports/major-land-resource-area-mlra
NRCS—National Cooperative Soil Survey Soil Characterization Data	Lab Data	https://ncsslabdatamart.sc.egov.usda.gov

(Continued)

Table 10.4. (Continued)

Name	Information available	Web address
NRCS—National Information and Centers	National NRCS Centers (e.g., Water, Climate, Soil Survey, Agroforestry, and others)	https://www.nrcs.usda.gov/national-information-and-centers
NRCS—National Resources Inventory (NRI)	Statistical survey, reports, data	https://www.nrcs.usda.gov/nri
NRCS—National Soil Survey Handbook	Technical guide for soil survey projects and activities	https://www.nrcs.usda.gov/resources/guides-and-instructions/national-soil-survey-handbook
NRCS—National Water and Climate Center	Climate and water conservation planning information	http://www.wcc.nrcs.usda.gov
NRCS—Official Soil Series Descriptions	Detailed official soil series descriptions for soils in the USA	https://www.nrcs.usda. gov/resources/data-and-reports/official-soil-series-descriptions-osd
NRCS—Range Resources	Procedures for the inventory, analysis, treatment, and management of grazing land resources	http://www.nrcs.usda.gov/wps/portal/nrcs/detailfull/national/landuse/rangepasture/?cid=stelprdb1043084
NRCS—Soil	Home page, soil classification, lab data	http://www.nrcs.usda.gov/wps/portal/nrcs/site/soils/home
NRCS—Soil Health	Soil health/quality definition, assessment, management, resources, and publications	http://www.nrcs.usda.gov/wps /portal/nrcs/main/soils/health/
NRCS—Soil Quality	Soil quality definitions, assessment, management, resources, and publications	https://www.nrcs.usda.gov/conservation-basics/natural-resource-concerns/soils/soil-health
NRCS—Soil Survey Manual	Soil Survey Manual Publication	https://www.nrcs.usda.gov/resources/guides-and-instructions/soil-survey-manual
NRCS—SOLUS	Soil Landscapes of the United States, soil property maps	https://www.nrcs.usda.gov/resources/data-and-reports/soil-landscapes-of-the-united-states-solus
NRCS—SSURGO/STATSGO2 Metadata	Soil Data Mart, metadata links and reports	https://www.nrcs.usda.gov/resources/data-and-reports/ssurgo/stats2go-metadata
NRCS—State Offices/Centers	Service Center locator and contact information	https://www.farmers.gov/working-with-us/service-center-locator
NRCS—Technical References	Website for manuals, technical guides, and references used by NRCS	https://www.nrcs.usda.gov/resources/guides-and-instructions/technical-references-for-soils
Soil Explorer	Soil series and landscapes	https://soilexplorer.net
Soil Orders	Images of 12 soil orders	http://soils.cals.uidaho.edu/soilorders
Soil Series Explorer (SEE)	Soil series application viewer	https://casoilresource.lawr.ucdavis.edu/see
SoilWeb	Soil survey data compatible with technology	https://casoilresource.lawr.ucdavis.edu/gmap
U.S. Census Bureau	Source of population, housing, economic, and geographic data by state, town, county, or ZIP Code area	https://data.census.gov
USDA—Agricultural Research Service	Home page, research results, and projects	http://www.ars.usda.gov/main/main.htm
USDA—Forest Service	Home page	https://www.fs.usda.gov
USDA—National Agricultural Statistics Service	Agricultural statistics for state and county	http://www.nass.usda.gov

Table 10.4. (Continued)

Name	Information available	Web address
USDA—National Institute of Food and Agriculture (NIFA)	USDA- NIFA Home page and agricultural research information	https://www.nifa.usda.gov
USDA—Site Risk Water Quality Tool (SRWQ Tool)	Water quality tool, evaluate nutrient loss	https://www.nrcs.usda.gov/resources/tech-tools/site-risk-for-water-quality-tool-srwq-tool
USDA—Web Soil Survey (WSS)	Detailed soil survey information	http://websoilsurvey.nrcs.usda.gov/app/HomePage.htm
USGS—Earth Resource Observation and Science (EROS) Center	Home page, satellite and aerial images, research projects and programs	http://eros.usgs.gov
USGS—The National Map	Various kinds and scales of U.S. maps	http://nationalmap.gov/viewer.html
USGS—Upper Midwest Aerospace Consortium	USGS remote sensing applications and projects	https://pubs.usgs.gov/publication/70159667
Web GIS	TerraServer-USA aerial and topographic images of land in the USA	http://www.webgis.com/terraserver.html

Data Availability

A map can be developed that displays the soils across the landscape of a defined AOI. Associated with the map are key descriptors for the MUs and symbols. The MUs group together soils with similar characteristics. Soil MUs can be classified as consociation, complexes, associations, and undifferentiated. Consociations are areas where the dominant soil type occupies most of the area. Complexes are where two or more components are intertwined. Associations are where two or more components are present but are more widely separated than in a complex. Undifferentiated MUs are areas where there are two or more soil types, but they are not well enough defined to be mapped as separate entities. Each MU contains information about (a) the dominant soil series with description and classification; (b) the precipitation, temperature, and frost-free period; (c) the elevation, slope, drainage class, permeability, depth to water table, available water supply, flooding frequency, land capability classification, ecological site, and forage suitability; and (d) the soil profile relative to texture, drainage characteristics, and chemical properties (e.g., calcium carbonate, sodium adsorption ratio [SAR], gypsum, and salinity).

A MU symbol is the identifier on the soils map and is used as a soil survey reference. Within each MU, there are dominant and minor soils. The dominant and minor soils share many similarities; however, they may differ slightly in soil properties. Landscapes, which are often linked by slopes, are a key factor in soil development, and certain soils tend to be regularly found at specific landscape positions (Figure 10.3). Soil formation and landscape position are closely related; understanding the landscape helps users recognize where specific soils occur and why their characteristics vary across the topography. Differences in elevation, slope, and drainage influence the distribution of soil MU and help explain patterns observed across the landscape. Landscapes will contain soils with different drainage and land capability classes (Soil Survey Staff et al., 2025).

Land capability classes and limitation subclasses provide information on soil fitness for specific land-use activities, identify the major limitations (e.g., erosion, wetness, soil profile feature [including texture, salinity, pH, thickness], and climate), and provide ratings with maps and interpretive descriptions. The WSS enables users, with supportive information, to make decisions about best land management practices for vegetation productivity and land and water management (Soil Survey Staff et al., 2025). For example, water-holding capacity, drainage, and permeability directly affect soil productivity. This information is important in assisting with regenerative

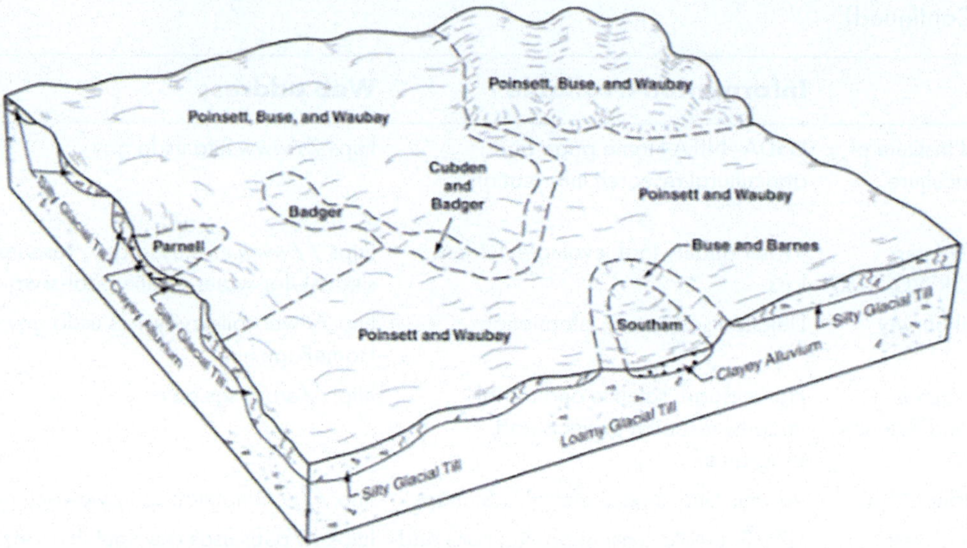

Figure 10.3. Example of MUs across a landscape within Brookings County, SD. (Adapted from Soil Survey Staff (2004)).

agricultural planning, as well as understanding what areas of the landscape have the potential for improvement under a conservation management plan. Information of particular interest includes the following:

1. The National Commodity Crop Productivity Index (NCCPI), which provides a numerical rating (0.01–1.0) based on soil suitability for agricultural production based on soil properties and characteristics, site conditions, and climatic factors. Higher rating numbers identify the potential for producing major commodity crops (e.g., corn, soybeans, wheat, cotton; Soil Survey Staff et al., 2025).

2. The hydric soils rating, which identifies soils that were formed under water saturation, flooding, or ponding conditions long enough to produce anaerobic/reducing conditions.

3. The Soil Landscapes of the United States (SOLUS), which predicts soil physical and chemical properties in both horizontal and vertical dimensions. SOLUS100 provides a 100-m spatial resolution at seven soil depths for 20 soil properties, including soil texture, pH, calcium carbonate equivalent, organic carbon, gypsum, EC, SAR, effective exchange capacity, bulk density, bedrock depth, and 11 additional properties.

4. The Crop Productivity Index (CPI), which ranks soil from 0 to 100 for soil productivity.

5. Range productivity, which is an estimate in pounds per acre that can be expected under favorable, normal, and unfavorable conditions.

6. Irrigated/nonirrigated crop yields that provide average yield per acre for the crop and soil type under consideration, using an estimated yield for each crop along with management choices.

Web Soil Survey information can be used alongside management tools to assist in the planning process. For example, agronomists can use management zones to tailor fertility needs to each field. The WSS is a powerful, comprehensive data tool that aids planners in optimizing crop and range management production by providing knowledge to assist with regenerative agriculture and knowledge about sustainable practices. The WSS platform is informative and valuable for anyone involved in land-based decision-making (Malo, 2016b).

Web Soil Survey Data Limitations

Because the WSS provides an easily accessible database, the risk exists that it will be used for purposes other than what it was designed for.

For example, the WSS does not replace annual soil sampling for fertilizer recommendation. Additionally, the WSS was not intended for use as the primary regulatory tool in site-specific permitting decisions. The WSS is very useful for agricultural planning at a regional level, and it is an invaluable tool for comparing soil properties over large geographical areas.

Reliable, fast internet service is needed for WSS, and at times, the WSS platform can be sluggish and slow. Compatibility issues may arise with certain web browsers. If the dataset exceeds 8 MB, instant online viewing will be unavailable, in which case, the data and/or maps will be emailed to the user.

The mapping resolution for soil delineations in the WSS is limited to approximately 0.8 ha (2 ac). If a user would like greater accuracy for smaller sectors, for instance, for precision farming or highly detailed engineering projects, it is recommended that a soil pedologist update the WSS map. Soil surveys have been mapped at different scales across the country, impacting the level of detail available. Scale generally influences the level of detail shown on the map.

Field observations and scattered data points represent landscape variations and form the foundation for soil surveys. Interpolations and extrapolations (estimations made between or beyond sampled points) are then used to extend the information between soil sampling data points and across the survey area. Soil is highly variable and can change over a distance of just a few meters (yards) within a small geographic area. Soil scientists use interpretations of soil series based on generalized landscape patterns reflected in the MUs to account for variability. Remember that MUs may contain minor soil inclusions with different soil properties. Therefore, caution should be exercised when using WSS data for site-specific land management decisions. Additionally, the WSS MU boundaries may not match current observations due to variable weather patterns, erosion, and changes related to management. Revisions to the WSS data may take years to become publicly available.

Since the WSS data is not continuously updated, real-time georeferenced yield monitoring, nutrient, soil test results, EC data, and in-field in-person assessments should be interfaced with WSS data where possible. The WSS yield estimates are based on general assumptions (e.g., static conditions within the NCCPI and CPI and do not consider how the adoption of regenerative agricultural management or how climate change impacts productivity). The merging of WSS and real-time data improves accuracy, given that changes in weather patterns or management may not be reflected within the current WSS platform information.

Web Soil Survey, Soil Organic Carbon, and Soil Health

Soil health, soil quality, and carbon sequestration are intrinsically linked within the soil ecosystem. This section discusses how these factors can be evaluated using the WSS. For a broader discussion of soil health, see Chapter 24. Soil is a living, breathing, and cyclical system, with SOM serving as a driving force in the ecosystem. Soil organic matter is the food source for many soil organisms. These organisms consume organic matter and convert it into carbon dioxide and humus (stable, dark organic material formed from decomposed plant and animal matter). Soil organic matter and humus help to build soil structure, improve productivity, increase water infiltration, and reduce runoff and erosion (Devine et al., 2014; Lal, 2004; Tariq et al., 2024).

Soil organic matter can be used as a soil health benchmark by comparing the measured SOM value from the field to the value on the WSS (Figure 10.4). If the measured SOM value exceeds the WSS value, it indicates that years of soil health management practices have effectively improved the soil health and quality in this field. Some practices known to increase SOM include planting cover crops and reducing tillage intensity (Clay et al., 2012; Joshi et al., 2025; Liu et al., 2013; Soil Survey Staff et al., 2025; Weil & Brady, 2017). The data in the WSS can also be used to estimate soil carbon stock calculations for landscapes. This calculation requires knowing the SOM, bulk density, soil depth, and rock fraction present in the measurement. Caution should be exercised when the complete carbon account picture is not taken into account. Soils created under glaciation tend to have significant levels of

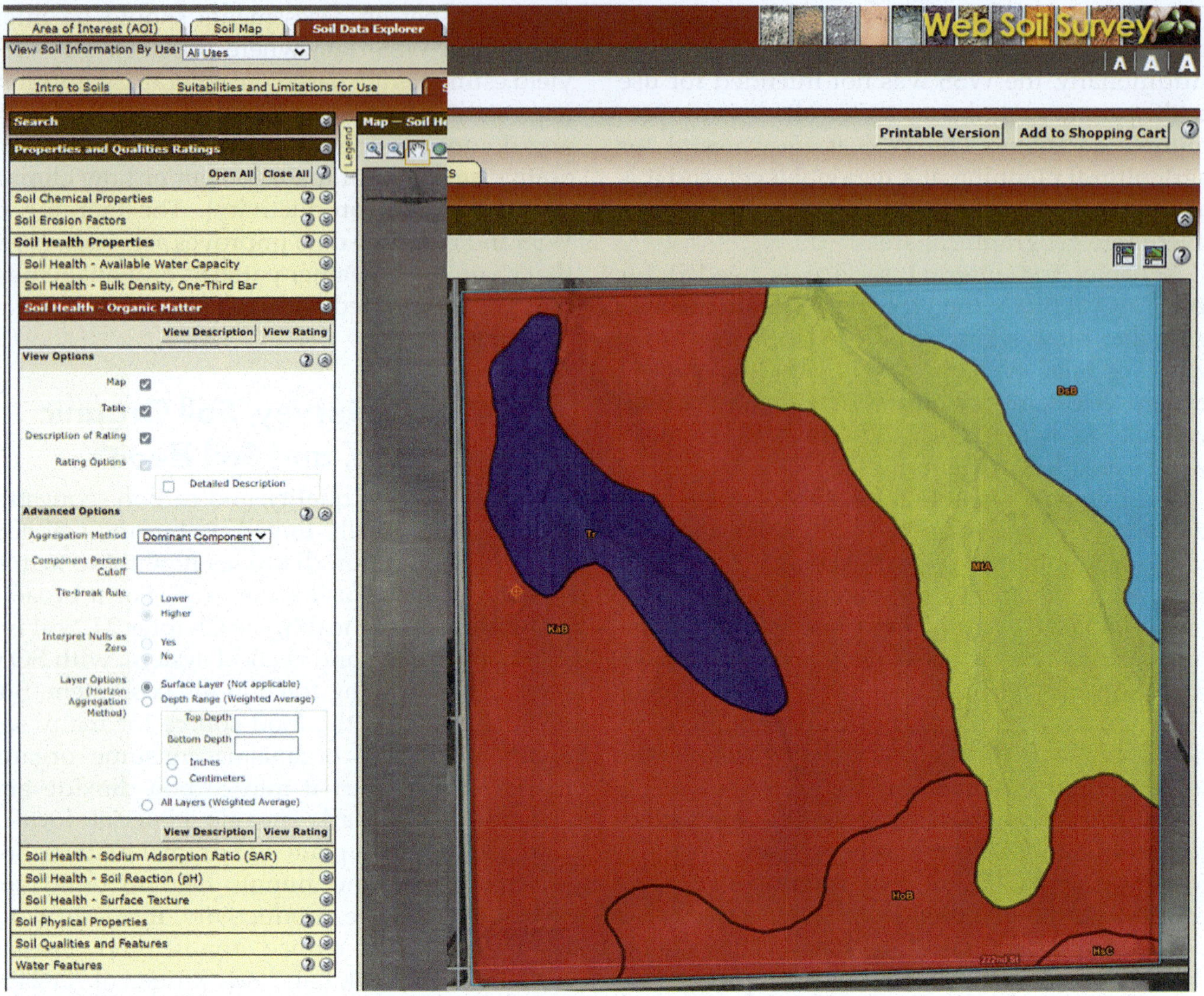

Figure 10.4. Soil health—Organic matter percentage (weight of decomposed plant, animal, and microbial residues ≤2 mm). Note: DsB, teal (5.0%); KaB, HsC, HoB, red (3.0%); MtA, yellow (3.25%); Tr, blue (5.25%). https://websoilsurvey.nrcs.usda.gov/app (Soil Survey Staff et al., 2025).

carbon present as inorganic carbon (IC; e.g., calcium and magnesium carbonates). The amount of IC tends to be overlooked as part of the soil carbon storage capacity (Batool et al., 2024; Lal, 2016; Soil Survey Staff et al., 2004a). Total soil carbon (TC) is present in soil organic carbon (SOC) and IC forms (Soil Survey Staff et al., 2004a; Xu et al., 2019). Soil organic carbon is calculated by subtracting IC from TC, expressed as a percentage, using the following equations:

$$\text{SOC} = \text{SOM} \times 0.58$$

$$\text{SOC}(\%) = \text{TC}(\%) - \text{IC}(\%)$$

$$\text{Soil carbon stock}(\text{Mg}/\text{ha}) = \text{SOC}(\%) \times \text{Soil depth}(\text{cm}) \times \text{Soil bulk density}(\text{g}/\text{cm}^3)$$

For more examples using these equations, see Chapter 32. A comparison between the measured SOC value and the SOC value reported on the WSS can serve as a benchmark, providing clues to the sustainability of current management practices. If a comparison between the currently measured SOC value shows an increase from the benchmark value, this suggests that soil productivity also increased (Joshi et al., 2025). Organic carbon can occur when fresh biomass additions are greater than the maintenance requirement (Lal, 2004; Palm et al., 2014; Schmidt et al., 2011). The SOC maintenance requirement is the amount of fresh biomass needed to maintain the SOC at its current level.

If the soil health assessment indicates a decrease in SOC, it is likely that productivity has also decreased (Hao et al., 2025; Lal, 2016). In areas where SOC has decreased, additional monitoring or a change in management may be necessary (Hao et al., 2025; Ogle et al., 2023). When making this comparison, it is important to consider that most of the data provided by the WSS represent expected values, rather than measured values at a specific location. When conducting a soil health assessment, it is important to take into account that soil carbon levels are dynamic and that the rate at which fresh plant biomass is converted into soil humus is influenced by weather, temperature, precipitation, snow, evapotranspiration, moisture levels, microbial activity, time of year, fertilizer, insecticides, soil texture, and pesticides (Devine et al., 2014; Lal, 2004, 2016; Ogle et al., 2023; Zuber & Villamil, 2016).

Decreases in SOC may increase erosion, which is detrimental to economic and environmental sustainability (DeLaune et al., 2020; Minasny et al., 2017; Schmidt et al., 2011). Erosion moves soil from upland areas to downslope areas, and even into streams or rivers. Erosion reduces productivity by physically detaching and transporting soil particles, resulting in the loss of topsoil and the nutrients contained in the soil (Gregorich et al., 1998; Lal, 2005; Weil et al., 2017).

In summary, the WSS does not replace the soil sampling required to produce fertilizer recommendations; rather, it provides valuable information that can be used to determine the impact of management practices on soil health. Conducting soil health assessments is essential for determining whether your practices are helping you achieve your management and soil health goals. When making this comparison, it is important to consider that most of the data provided by the WSS represent expected values rather than measured values at a specific location. Therefore, it is essential to have WSS predications based on actual data collected from the field. Table 10.3 provides additional online sources for soil and natural resources information, while Chapter 24 offers more information on soil health.

Study Questions

1. What is the Web Soil Survey (WSS), and who provides the materials, knowledge, and information within the WSS?
2. Provide three reasons why a farmer might use the WSS in their farming practice.
3. You are advising a producer to use cover crops within the crop rotation. How could the WSS be used to support your recommendations related to soil suitability, drainage, or erosion risk?
4. Within the WSS, create an area of interest (AOI) from your selected field. See references for step-by-step guidance (Malo, 2008, 2016a, 2016b), as well as Section 10.2.
5. If the soil depth is 0–15 cm, the bulk density is 1.26 g/cm^3, and the SOM is 4%, how much SOC is contained in the surface 15 cm?

ACKNOWLEDGMENTS

Support for this document was provided by the Precision Farming Systems community in the American Society of Agronomy, South Dakota State University, USDA Natural Resources Conservation Service Conservation Innovation Grants (NR233A750011G019), American Coalition for Ethanol, the South Dakota Corn Utilization Council, the South Dakota Soybean Research and Promotion Council, and the USDA National Institute of Food and Agriculture (Multistate Project S1090) Program under (7008110). The use of trade names is for the convenience of the reader and does not imply endorsement by the authors.

REFERENCES

Batool, M., Cihacek, L. J., & Alghamdi, R. (2024). Soil inorganic carbon formation and the sequestration of secondary carbonates in global carbon pools: A review. *Soil Systems*, *8*(1), 15. https://doi.org/10.3390/soilsystems8010015.

California Soil Resource Lab. (2015). *SoilWeb apps*. University of California, Davis. https://casoilresource.lawr.ucdavis.edu.

Chang, J., Clay, D. E., Carlson, C. G., Reese, C. L., Clay, S. A., & Ellsbury, M. M. (2004). Defining yield goals and management zones to minimize yield and nitrogen and phosphorus fertilizer recommendation errors. *Agronomy Journal*, *96*(3), 825–831. https://doi.org/10.2134/agronj2004.0825.

Clay, D. E., Chang, J., Clay, S. A., Stone, J., Gelderman, R. H., Carlson, G. C., Reitsma, K., Jones, M., Janssen, L., & Schumacher, T. (2012). Corn yields and no-tillage affects carbon sequestration and carbon footprints. *Agronomy Journal*, *104*(3), 763–770. https://doi.org/10.2134/agronj2011.0353.

DeLaune, R. D., Reddy, K. R., Richardson, C. J., & Megonigal, J. P. (2020). *Methods in Biogeochemistry of Wetlands*. John Wiley & Sons.

Devine, S., Markewitz, D., Hendrix, P., & Coleman, D. (2014). Soil aggregates and associated organic matter under conventional tillage, no-tillage, and forest succession after three decades. *PLoS One*, *9*(1), e84988. https://doi.org/10.1371/journal.pone.0084988.

Gregorich, E. G., Greer, K. J., Anderson, D. W., & Liang, B. C. (1998). Carbon distribution and losses: Erosion and deposition effects. *Soil and Tillage Research*, *47*(3–4), 291–302. https://doi.org/10.1016/S0167-1987(98)00117-2.

Hao, Y., Mao, J., Bachmann, C. M., Hoffman, F. M., Koren, G., Chen, H., Tian, H., Liu, J., Tao, J., Tang, J., Li, L., Liu, L., Apple, M., Shi, M., Jin, M., Zhu, Q., Kannenberg, S., Shi, X., Zhang, X., . . . Dai, Y. (2025). Soil moisture controls over carbon sequestration and greenhouse gas emissions: A review. *npj Climate and Atmospheric Science*, *8*, 16. https://doi.org/10.1038/s41612-024-00888-8.

Joshi, D. R., Clay, D. E., Alverson, R., Clay, S. A., Westhoff, S., Johnson, J. M. F., Wang, T., & Sieverding, H. (2025). Tillage intensity reductions when combined with yield increases may slow soil carbon saturation in the central United States. *Scientific Reports*, *15*, 10697. https://doi.org/10.1038/s41598-025-95388-x.

Joshi, D. R., Clay, D. E., Smart, A., Clay, S. A., Kharel, T. P., & Mishra, U. (2019). Soil and land-use change sustainability in the Northern Great Plains of the USA. In S. Appiah-Opuku (Ed.), *Land-use change and sustainability*. IntechOpen. https://doi.org/10.5772/intechopen.84781.

Lal, R. (2004). Soil carbon sequestration impacts on global climate change and food security. *Science*, *304*(5677), 1623–1627. https://doi.org/10.1126/science.1097396.

Lal, R. (2005). Soil erosion and carbon dynamics. *Soil and Tillage Research*, *81*(2), 137–142. https://doi.org/10.1016/j.still.2004.09.002.

Lal, R. (2016). Soil health and carbon management. *Food and Energy Security*, *5*(4), 212–222. https://doi.org/10.1002/fes3.96.

Liu, S., Wei, Y., Post, W. M., Cook, R. B., Schaefer, K., & Thornton, M. M. (2013). The Unified North American soil map and its implication on the soil organic carbon stock in North America. *Biogeosciences*, *10*(5), 2915–2930. https://doi.org/10.5194/bg-10-2915-2013.

Malo, D. D. (2008). Online sources of basic soils information. In S. Logsdon, D. Clay, D. Moore, & T. Tsegaye (Eds.), *Soil Science Step-by-Step Field Analysis* (pp. 35–48). ASA, CSSA, SSSA. https://doi.org/10.2136/2008.soilsciencestepbystep.c3.

Malo, D. D. (2016a). Chapter 16: Online soil survey information—Web Soil Survey (WSS). In D. E. Clay, C. G. Carlson, S. A. Clay, & E. Byamukama (Eds.), *iGrow corn: Best management practices*. South Dakota State University. https://extension.sdstate.edu/sites/default/files/2019-09/S-0003-16-Corn.pdf.

Malo, D. D. (2016b). Chapter 17: Online soil survey information—SoilWeb Application (SWA). In D. E. Clay, C. G. Carlson, S. A. Clay, & E. Byamukama (Eds.), *iGrow corn: Best management practices*. South Dakota State University. https://extension.sdstate.edu/sites/default/files/2019-09/S-0003-17-Corn.pdf.

Minasny, B., Malone, B. P., McBratney, A. B., Angers, D. A., Arrouays, D., Chambers, A., Chaplot, V., Chen, Z.-S., Cheng, K., Das, B. S., Field, D. J., Gimona, A., Hedley, C. B., Hong, S. Y., Mandal, B., Marchant, B. P., Martin, M., McConkey, B. G., Mulder, V. L., . . . Winowiecki, L. (2017). Soil carbon 4 per mille. *Geoderma*, *292*, 59–86. https://doi.org/10.1016/j.geoderma.2017.01.002.

Ogle, S. M., Breidt, F. J., Del Grosso, S., Gurung, R., Marx, E., Spencer, S., Williams, S., & Manning, D. (2023). Counterfactual scenarios reveal historical impact of cropland management on soil organic carbon stocks in the United States. *Scientific Reports*, *13*, 14564. https://doi.org/10.1038/s41598-023-41307-x.

Palm, C., Blanco-Canqui, H., DeClerck, F., Gatere, L., & Grace, P. (2014). Conservation agriculture and ecosystem services: An overview. *Agriculture, Ecosystems & Environment*, *187*, 87–105. https://doi.org/10.1016/j.agee.2013.10.010.

Schmidt, M. W. I., Torn, M. S., Abiven, S., Dittmar, T., Guggenberger, G., Janssens, I. A., Kleber, M., Kögel-Knabner, I., Lehmann, J., Manning, D. A. C., Nannipieri, P., Rasse, D. P., Weiner, S., & Trumbore, S. E. (2011). Persistence of soil organic matter as an ecosystem property. *Nature*, *478*, 49–56. https://doi.org/10.1038/nature10386.

Soil Survey Staff, National Resources Conservation Services, United States Department of Agriculture. (2004a). *Soil survey laboratory methods manual*. Soil Survey Investigations Report No. 42 (version 4.0). https://nrcspad.sc.egov.usda.gov/DistributionCenter/pdf.aspx?productID=886.

Soil Survey Staff, National Resources Conservation Services, United States Department of Agriculture. (2004b). *Soil survey of Brookings County, South Dakota*. https://archive.org/details/Brookings_SD2004.

Soil Survey Staff, National Resources Conservation Services, United States Department of Agriculture. (2025). *Web soil survey*. https://websoilsurvey.nrcs.usda.gov/app/WebSoilSurvey.aspx.

Tariq, M., Liu, Y., Rizwan, A., Shoukat, C. A., Aftab, Q., Lu, J., & Zhang, Y. (2024). Impact of elevated CO_2 on soil microbiota: A meta-analytical review of carbon and nitrogen metabolism. *Science of the Total Environment*, *950*, 175354. https://doi.org/10.1016/j.scitotenv.2024.175354.

Weil, R. R., & Brady, N. C. (2017). *The nature and properties of soils* (15th ed.). Pearson Education.

Xu, H., Sieverding, H., Kwon, H., Clay, D., Stewart, C., Johnson, J. M. F., Qin, Z., Karlen, D. L., & Wang, M. (2019). A global meta-analysis of soil organic carbon response to corn stover removal. *GCB Bioenergy*, *11*(10), 1215–1233. https://doi.org/10.1111/gcbb.12631.

Zuber, S. M., & Villamil, M. B. (2016). Meta-analysis approach to assess effect of tillage on microbial biomass and enzyme activities. *Soil Biology and Biochemistry*, *97*, 176–187. https://doi.org/10.1016/j.soilbio.2016.03.011.

Fundamentals of Proximal Soil and Crop Sensing

11

Robin Gebbers, Katsutoshi Mizuta, and Viacheslav Adamchuk

Chapter Purpose

This chapter introduces the core principles of crop and soil proximal sensing and provides an overview of physical, chemical, and biological sensor systems. It lays the foundation for Chapters 12 and 13, which discuss the use of soil and plant proximal sensors. In addition, this chapter discusses the basics of sensor calibration and data fusion.

Definition of Proximal Crop and Soil Sensing

Proximal crop and soil sensors detect physical, chemical, or biological phenomena within 2 m of the target (Viscarra Rossel et al., 2011). The goal of proximal crop and soil sensors is to reduce the cost of collecting information while improving management decisions. However, before using sensors for management decisions, a well-defined relationship between the measured signals and the property of interest must be established. In other words, before using the collected data, data processing is necessary to extract relevant information from sensor readings. The integration of proximal sensing into decision-making in precision agriculture is depicted in Figure 11.1. To meet the needs of precision agriculture, sensor output must not only fit into the decision-making process but also be robust, safe, economical, and easy to handle.

The term sensor generally refers to a device that converts a physical measurement into a signal that is read by an observer or by an instrument (McGrath & Ní Scanaill, 2013). The sensor responds to a stimulus, which is a quantity, property, or condition received by the sensor and converted into an electrical or optical signal. Stimuli can be acoustic, chemical, biological, electric, magnetic, optical, mechanical, radioactive, thermal, among others. A sensor—the sensing element—is only one of many components that ensure the sensing system's functionality and applicability. Sensing systems can be analog, as shown in Figure 11.2a. However, many current proximal sensing systems (a) record data in a digital format and link sensor readings to geospatial coordinates and a timestamp from a Global Positioning System; (b) include power management, ruggedized casing, and signal processing; (c) have data communication capability; and (d) include

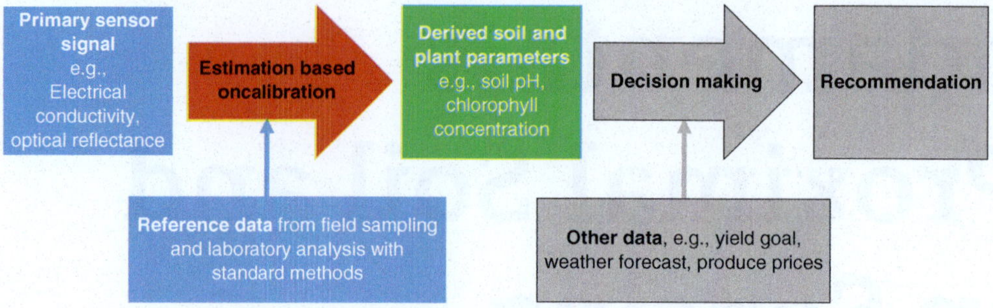

Figure 11.1. Integration of proximal soil and crop sensing into decision-making.

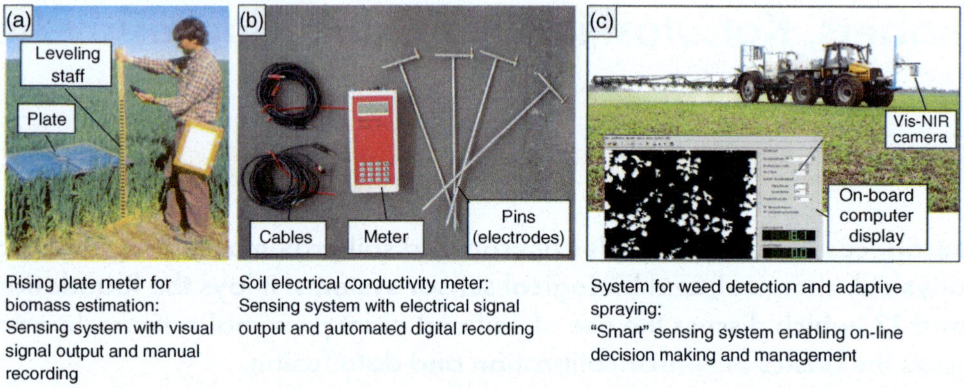

Figure 11.2. Examples of proximal sensing systems demonstrating different levels of automation and integration into site-specific management. (a) Analog sensing with optical indication. (b) Sensing with digital recording. (c) Sensor-based automated decision-making. *Source*: with permission of Robin Gebbers.

sensor calibration accessories, data storage, and a display (Figure 11.2b). In addition, automated mobile proximal sensing systems designed for field-scale mapping or online application control may also contain (a) sensor mounting platforms, (b) automated sampling units, (c) sensor heads, (d) sample preparation equipment, and (e) onboard computing units for real-time data processing and autonomous decision-making (Figure 11.2c). A smartphone is an example of a proximal sampling system; it includes a camera, positioning systems, and other sensors. The reader can explore a smartphone's sensing and data-processing capabilities using the free app phyphox (https://phyphox.org/). Many of the fundamentals described below apply not only to soil and plant sensors but also to other agricultural sensor systems, including weather stations and yield monitors.

Sensor-based Observations

Sensor-based observations have many of the same concerns as soil and plant sampling in that temporal and spatial variation must be considered (Figure 11.3; de Gruijter et al., 2010; Legendre and Legendre, 2012). In sensor-based observations, (a) the physical area or volume and (b) when and how any data processing is performed should be known, and (c) spatial and temporal distances between sampling points and (d) the sampling extent and sampling design should be considered. For example, a red–green–blue (RGB) camera captures images of plants that occupy a specific area at a known time and location (Figures 11.3 and 11.4).

Common strategies to address the spatial and temporal arrangement of sample collection are depicted in Figure 11.5. For proximal sensing, two widely used sampling strategies are

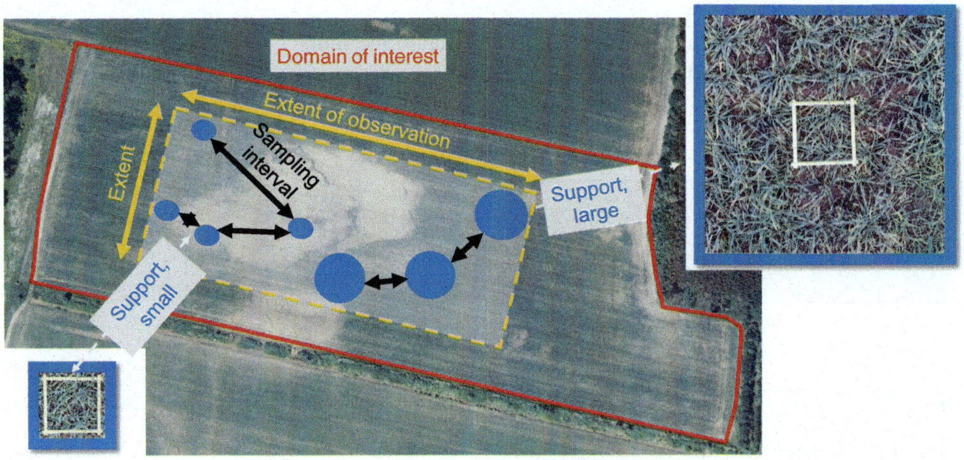

Figure 11.3. Elements of sensor-based spatial sampling and observation. Source: with permission of Robin Gebbers.

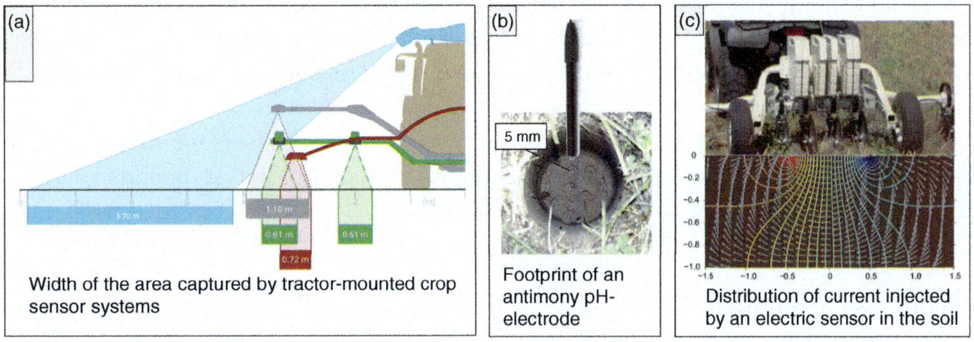

Figure 11.4. Examples of support for proximal sensing systems. Source: (a and b) With permission of Robin Gebbersand (c) Courtesy of Veris technologies.

(a) high-resolution temporal monitoring, often achieved with stationary sensors deployed at a specific site; (b) and spatially dense sampling, facilitated by mobile sensing systems.

Criteria for Proximal Sensor Description

From a practical standpoint, it is logical to differentiate sensor systems based on their target (e.g., crops or soil, nitrogen or phosphate). However, understanding the strengths and limitations of selecting sensor systems, including how to interpret their outputs, requires consideration of additional factors. Sensors can be characterized according to their response to stimuli, mobility, and temporal responsiveness, among other factors (Fraden, 2016; Morris and Langari, 2021). Several of these aspects are discussed below.

First, sensors respond to a stimulus, that is, the quantity, property, or condition received and converted into either an electrical signal, which can be recorded by a digital system, or an optical signal, which can be observed visually (Fraden, 2016; McGrath & Ní Scanaill, 2013). Stimuli, or signals, can be acoustic, chemical, biological, electric, magnetic, optical, mechanical, radioactive, thermal, among others.

Second, mobility refers to the sensor's use for temporal monitoring or spatial mapping. Sensors can be installed at a fixed location for extended periods, ranging from days to years. These sensors are often used to monitor changes in environmental conditions over time. Common examples include soil moisture and temperature sensors buried in the soil. Sensors attached to plants are called wearable plant sensors. Sensors can also collect data while moving across the

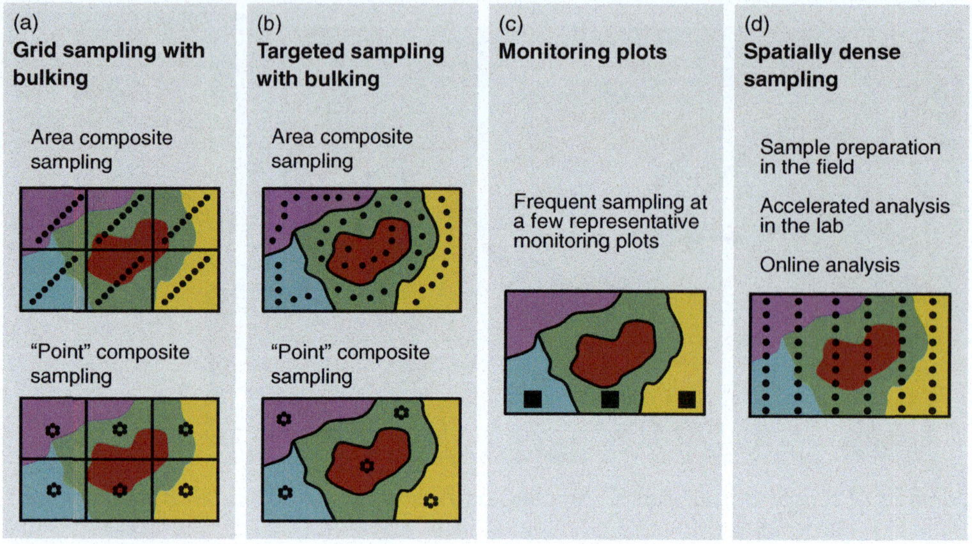

Figure 11.5. Sampling strategies in precision agriculture. (a) Grid sampling with bulking, a strategy that splits the field into equally spaced units, does not use information to guide the sampling. (b) Targeted sampling uses information from ancillary data to select the sampling positions. Strategies A and B use composite (bulk) sampling with subsamples spread out over the zone or focused at one location. (c) Monitoring plots are used for frequent measurements over time, and their positions must be carefully selected to represent the area. (d) Spatially dense sampling is facilitated by different stages of automation, ranging from automated sampling to on-site sample analysis. Source: Schirrmann et al., 2011 / MDPI / CC BY 3.0.

field. These sensors are often called kinematic or on-the-go sensors. Examples include crop sensors, which are frequently mounted on agricultural machinery to measure the Normalized Difference Vegetation Index (NDVI) for fertilizer application, and electromagnetic (EM) induction sensors used for soil mapping. Stop-and-go sensing systems are mobile but require stationary operation for a short period, ranging from seconds to minutes, during measurement. An example is the cone penetrometer, which is pushed vertically into the soil.

Third, temporal responsiveness refers to the time a sensor system requires to perform measurements, process data, support decision-making, and enable applications. Some sensing systems can provide information immediately after rapid data processing and can therefore be used in (near) real-time (online) applications. These systems are sometimes called smart sensors because they have data processing and decision-making embedded within them. For example, smart weed sensors can identify weeds and control the sprayer in real time. If the sensor response is slow or time-consuming, post-processing is necessary, and the resulting data must be used in offline applications. This is currently the case for most mobile soil sensing systems, as their data require cleaning and calibration.

Fourth, the footprint or support refers to the size and shape of the area or volume detected in a single measurement. It forms the basic sampling unit for measurement.

Fifth, proximity refers to how close the sensor is to the target. Some sensors require direct contact, such as tensiometers and penetrometers for soils or "leaf-clip" chlorophyll meters for plants. Examples of sensors that can measure from a distance (proximal or remote) include infrared (IR) temperature sensors, cameras, NDVI sensors for assessing plant stress, and EM induction sensors.

Sixth, samples often need to be conditioned to enable or improve sensor readings. Sample conditioning includes all forms of preparation or modification of a target before measurement. This may include processes such as sample collection, chemical treatment, mechanical alteration, or other forms of physical manipulation. Noninvasive sensors require no conditioning or sample preparation. Optical crop sensors and EM induction sensors are examples that do not require modifying the target. Invasive sensors require sample preparation to ensure accurate measurements. In soil sensing, conditioning may involve flattening the surface, loosening the soil, or excavating the soil.

Seventh, sensing systems can be categorized according to how they obtain the energy needed to generate a stimulus. Passive sensing systems rely on ambient energy sources (e.g., sunlight, electrochemical potential, natural radioactive decay) to detect physical or chemical stimuli from samples. These systems are generally simpler and more energy-efficient, but can be limited by environmental conditions. Active sensing systems generate their own stimuli using internal energy sources such as light, electrical power, or mechanical force—offering greater control and reliability, but requiring higher energy consumption.

Eighth, some sensors may be autonomous and function independently of direct external control. The desirable traits of these sensors include low power consumption and self-calibration. For example, sensors on small field robots must operate efficiently, using limited onboard energy.

Ninth, selectivity refers to a sensor's ability to exclude interference from other stimuli in the sample while detecting the analyte of interest (Bănică, 2012). For example, pH electrodes have relatively high selectivity, whereas soil electrical conductivity sensors are influenced by multiple soil properties, including moisture, texture, salinity, temperature, and bulk density. While high selectivity is preferred for precise measurements, it often entails higher cost, reduced robustness, or decreased responsiveness.

Tenth, the range of detection defines the span within which a sensor can provide reliable and accurate measurements. Operating outside this range may result in degraded performance or even sensor damage.

Eleventh, resolution (e.g., spatial, spectral, radiometric, temporal) is the smallest detectable change in the measured variable that a sensor can reliably identify. Uncertainty reflects the accuracy and precision of a sensor's output and is influenced by the sensor's inherent characteristics and the environmental conditions during measurement. It includes systematic and random errors that must be minimized to ensure reliable data for decision-making. Temporal stability refers to a sensor's ability to maintain consistent performance over time, which can be degraded by factors such as sensor drift, environmental exposure, and wear.

Twelfth, combining multiple sensors into a single platform can provide several benefits. At a weather station, combining a rain gauge with temperature and humidity sensors enhances the usefulness of each sensor, as they provide complementary data. Combining crop reflectance data from an active NDVI sensor with plant temperature information can distinguish between nitrogen (N) stress and water stress.

Thirteenth, when calculating the cost of a sensor system, one should include purchase or rental, licensing, maintenance, operating life, training, sampling and analysis, and data archiving. The benefits of a sensor system may not be immediately apparent, and may require a cost–benefit analysis. For this type of analysis, a longer time span is often necessary to fully capture the sensor's value. Data accessibility and output format are important factors to consider when evaluating cost. While some sensor manufacturers provide unrestricted access to the collected data in a common format, such as Comma-Separated Values (CSV), others encrypt the data and require users to pay a fee to access the information or use proprietary software.

Fourteenth, the sustainability of sensor systems should be taken into account. In other words, users should consider whether stationary sensors can harm the environment after use and how they could be removed from the field.

Overview of Sensing Principles by Stimulus

Sensing principles can be broadly divided into physical and chemical sensors based on the type of stimulus. Physical sensors respond to physical stimuli, such as electrical current, light, heat, and force. Chemical sensors respond to the presence of chemical substances through chemical interactions. Most sensors used in agriculture are physical ones. This is due to their robustness, speed, low cost, ease of use, and additional advantages.

Electromagnetic Radiation

Electromagnetic radiation sensors are commonly used to assess soil moisture, temperature, chlorophyll content, weeds, nutrients, and other parameters. Electromagnetic radiation includes optical radiation and electric charge, as well as radioactivity. It consists of an oscillating EM field that carries momentum and radiant energy. Electromagnetic energy propagates in waves and is classified by wavelength or its inverse, frequency—the number of wave cycles per unit time. The EM radiation spectrum covers a wide range of frequencies, each associated with distinct mechanisms of energy absorption, transmission, and/or reflectance that convey information about the target (Figure 11.6).

Depending on the frequency, the interaction of EM radiation with objects ranges from the macroscopic to the atomic scale. For example, low-frequency (long-wavelength) electrical current moves ions in liquids, whereas high-frequency (short-wavelength) X-ray radiation moves electrons in the orbits of atoms. Thus, EM sensors can be used to detect a wide range of phenomena. Many sensors used for proximal soil and plant sensing evaluate EM stimuli. For example, thermal IR radiation (wavelengths from 3,000 to 14,000 nm) can indicate plant water stress, while near-infrared (NIR) wavelengths (760 to 900 nm) help detect nutrient deficiencies. At wavelengths below 1 nm, high-energy radiation from X-rays and γ-rays—produced by the ejection of electrons from inner atomic orbitals—can provide elemental composition information about soils and plants. The frequency of an EM wave is proportional to its energy, which is expressed in electronvolts (eV). Depending on the discipline, EM sensors and signals are characterized by wavelength, frequency, or energy. When they reach a target, EM radiation may be absorbed, transmitted, reflected, or undergo a combination of these interactions (Figure 11.7).

Absorbed EM energy can generate heat and secondary radiation (e.g., fluorescence) or, if it breaks molecular bonds, produce smaller compounds. Reflection can be either specular or diffuse. Specular reflection occurs when a surface is smooth and behaves like a mirror. Parallel incoming rays are reflected at the same angle. In diffuse reflection, incoming radiation scatters in many directions because of surface roughness. Similarly, transmitted radiation can be diffracted in many directions or refracted at a uniform angle. Diffuse reflection and diffraction are the most common phenomena analyzed in plants and soils.

Electrical Charge

Electrical charge sensors use electric circuits to measure an object's ability to conduct or store electrical charge, become polarized (a separation

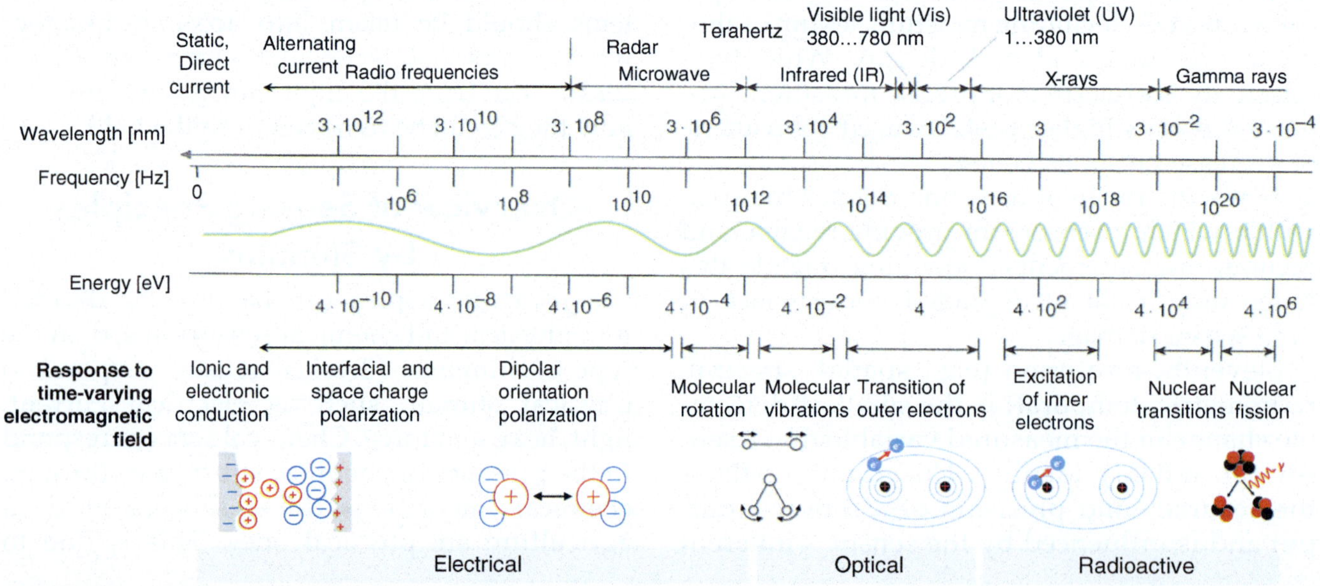

Figure 11.6. Electromagnetic spectrum.

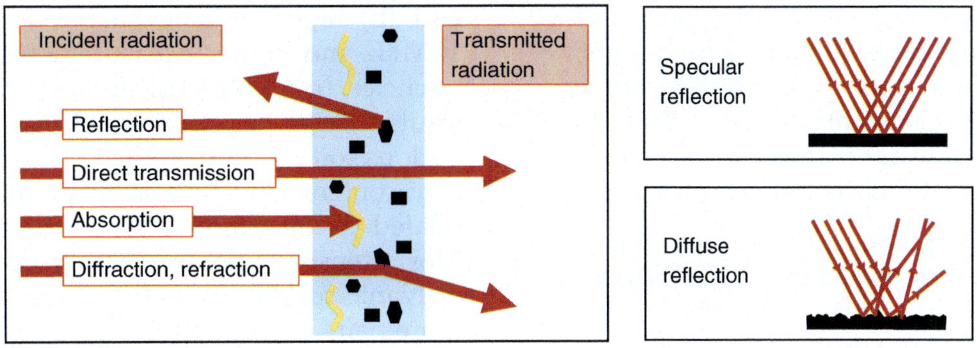

Figure 11.7. Illustration of absorption, reflection, and transmission.

of electric charge), or support the formation of magnetic fields. They are also known as electrical sensors. These sensors are widely used to measure soil water content, salinity, and texture. Their main practical application in plant analysis is moisture sensing in harvest combines and storage, whereas sensing properties such as plant root mass, plant water content, and nutrient status is mainly a matter of research.

Electrical charge sensors assess EM properties from 0 Hz (static, direct current) to over 3×10^{10} Hz or 300 GHz (radar frequencies; Figure 11.6). The response of materials is frequency dependent and can be summarized as sustained flow, space charge polarization, dipolar orientational relaxation, and atomic polarization (Figure 11.8; Everett, 2013; Loewer et al., 2017).

- Sustained flow of an electrical current at low frequencies (including direct current) occurs from 0 to about 300 kHz. This flow is attributed to the free flow of charged particles (e.g., ions) in the bulk solution and at interfaces. Electrical conductivity (EC) is a measure of how well a material conducts electricity. A sustained flow of electrical current can be carried by electrons in metals and semiconductors, and by ions in ionic liquids. In soil and plants, ionic conductivity prevails at low frequencies. In soil, minerals acting as semiconductors can contribute to conductivity, and in rare cases, electronic conduction is even possible. However, the influence of these mechanisms on a sustained electric current throughout the entire soil or plant volume is

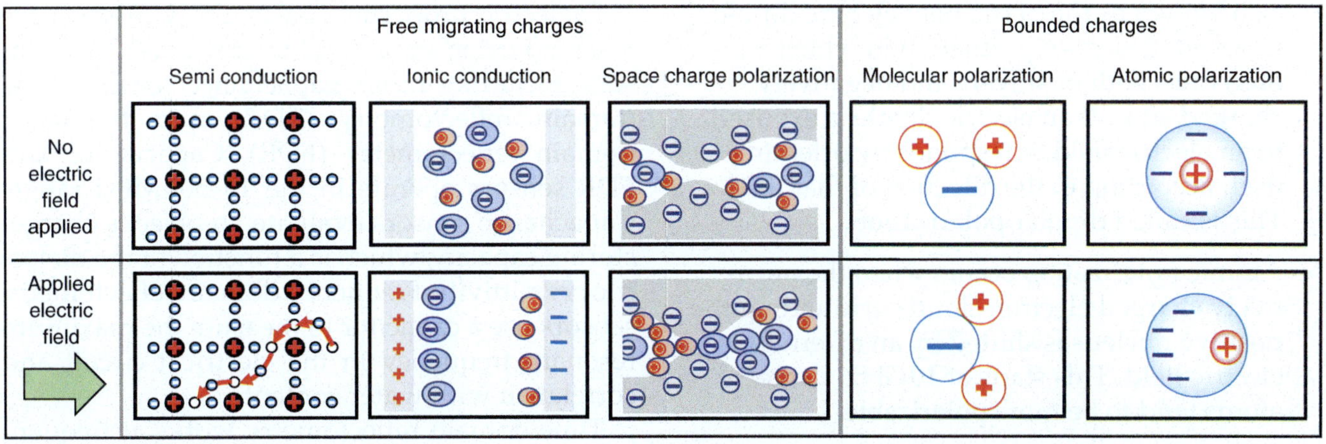

Figure 11.8. Responses to an electrical field (lower row). The upper row shows the situation when no electric field is applied. Free charges, such as electrons in semiconductors and dissolved ions, can migrate at the macroscopic level and contribute to the electrical conduction (left). Bounded charges can respond only at the microscopic and submicroscopic levels by changing the polarization around a charge center (right).

usually minimal (Everett, 2013). Soils and plants are mixtures of many components, arranged in various spatial configurations with distinct electrical responses. The sum of the current transfer by these mixed mechanisms and materials is called bulk EC.

- Space charge polarization occurs when ion movement is restricted by pore constriction and charged interfaces. Electric polarization is the separation of positive and negative electric charges. This effect is reversible when the applied electric field is removed. The return to an equilibrium is called relaxation. Relaxation takes time, and this is a measurable effect in soils and plants that can provide information on chemical properties (ions) and spatial structures (Everett, 2013; Jócsák et al., 2019; Van Haeverbeke et al., 2023). There are a couple of space charge polarization and relaxation mechanisms that occur at different spatial scales and corresponding frequencies. The macroscale polarization in narrow pores (Figure 11.8) corresponds to millihertz frequencies, while the mesoscale dipolar orientational relaxation of physically bound water and the adsorption/desorption rates of ions correspond to megahertz frequencies.
- Dipolar orientational relaxation of free and immobile molecules with electric polarity, such as water molecules, occurs at the nanoscale, and corresponds to frequencies ranging from gigahertz to terahertz. Polarity, also known as the permanent electric dipole moment, describes a situation in which an object has both positively and negatively charged sites. If an electric dipole is exposed to an electric field, it responds by orienting itself according to the direction of that field. This is called dipolar polarization.
- Atomic polarization occurs when the negatively charged electron cloud surrounding the positive nucleus is shifted by an external electric field. This starts at 10 GHz. Larger atoms tend to be more polarizable.

Based on these mechanisms, electric charge methods have traditionally been divided into two groups: (a) low-frequency methods, typically operating at frequencies from 0 to 100 kHz; and (b) high-frequency methods, operating at 10 MHz and higher. Measurements at lower frequencies, up to 100 kHz, are used to measure EC (or its inverse, resistivity). This is correlated with various soil properties, such as moisture, salinity, clay content, and bulk density. In plants, EC is related to ion concentration and water content. Most sensors measure at fixed frequencies. Galvanic-coupled resistivity sensors use one or more pairs of electrodes that need to be in direct physical contact (coupled) with the object. They are often measured at a single frequency within the range of 0–10 Hz. Electromagnetic induction sensors operate at less than 100 kHz, typically between 8 and 20 kHz. They use coils of wound wire to send and receive EM fields.

At higher frequencies, from about 10 MHz to 300 GHz, the ability of a medium to become polarized in response to an electrical field is measured by the dielectric permittivity (dielectric constant). At high frequencies, the dielectric permittivity is the dominant response of the medium, while the electrolytic conductivity due to ions becomes less relevant because ions cannot move significantly in a rapidly changing field. Because water is highly polar and the dominant component of the liquid phase in soils and plants, dielectric permittivity correlates with water content. This property is used to obtain a more accurate moisture assessment by sensors operating at radio and microwave frequencies. However, since sensor costs increase with frequency, moisture sensors are designed to operate at lower or intermediate frequencies.

Dielectric properties, particularly in soils, are assessed using radio- and microwave-based sensors, including radar, capacitance sensors, time domain reflectometry (TDR), and frequency domain reflectometry (FDR). Capacitance and FDR sensors operate in the 10–500 MHz range. Capacitance sensors evaluate an object's charge-storing capacity, which is a function of the dielectric permittivity. Frequency domain reflectometry sensors use a capacitor to measure the maximum resonant frequency in the electrical circuit and correlate it with water content.

Time domain reflectometry works at frequencies of 500 MHz to 1 GHz. They determine permittivity by measuring the travel time of an electrical pulse along metallic rods inserted into the object. Time domain reflectometry does not use a single frequency; instead, it emits a broadband burst (signal). Due to their higher frequency, they are expected to measure water content more accurately than FDR and capacitance sensors. However, even permittivity measurements taken with high-frequency sensors are affected by ion concentration and space charge. Radar operates in a frequency range similar to that of TDR and FDR, from about 1 MHz to 1 GHz. However, radar systems do not use rods to send their signals. Instead, radar instruments use transmitter and receiver antennas and operate from a remote location, without direct contact. The reflection at the interface between materials with contrasting dielectric properties, such as sand and clay or leaves and air, can be measured with radar. This provides information on soil layering or canopy structure. Besides reflection, the attenuation of a radar signal is an interesting parameter because it is related to water content. With lower frequencies, the depth of exploration increases, but spatial resolution decreases.

Instead of working at a single frequency, it is possible to scan through a range of frequencies one by one. This is called spectral induced polarization, electrical impedance, or dielectric spectroscopy in general, and bioimpedance spectroscopy when applied specifically to organisms, including plants.

The frequency range between microwave and optical IR (see the next section) is called the terahertz range. It spans approximately 0.1–10 THz. Compared with other electrical-charge-based sensor technologies, THz technology is a relatively recent development because THz radiation generators have been either unavailable or extremely expensive. Terahertz sensing has gained interest in agricultural research because it can, like radar, penetrate some materials and, like IR radiation, interact with molecules. These are promising features for enabling nondestructive assessment of the chemical composition inside fruits, leaves, or soils. However, THz sensors are currently not used in practice due to their limited penetration depth, high system costs, and lack of operational robustness (Abdul-Munaim and Abdulmunem, 2025).

In summary, the most relevant physical properties detected by EM fields from 0 to 300 GHz include (a) low-frequency EC, measured by resistivity meters and EM induction sensors, usually at frequencies below 100 kHz; and (b) high-frequency dielectric permittivity (dielectric constant, charge storage capacity), measured by capacitance sensors, TDR, FDR, and radar at higher frequencies. These electrical charge sensors are common in stationary and mobile proximal soil sensing but are rarely used for crop assessment in practice. Most sensors for assessing soil water content and salinity are electrical charge sensors.

Optical Radiation

Optical sensors are widely used to characterize soils and plants because they are relatively inexpensive, fast, and capable of detecting a wide range of properties. Among other things, they can assess an object's chemical composition, temperature, and shape. This is used, for instance, in crops to estimate nitrogen supply, biomass, and drought stress or to detect weeds, pests, and infections.

The optical region of the EM spectrum spans frequencies ranging from 1×10^{11} to 1×10^{17} Hz (Figure 11.6). These frequencies correspond to wavelengths between 1 and 1×10^{6} nm (Table 11.1). Within this range, EM waves are commonly called light or photons. When light strikes an object, energy can be absorbed. This can occur only if the energy quantum of the light (corresponding to its wavelength) equals the difference between the material's possible quantized states—a phenomenon known as optical resonance. Two absorption mechanisms are distinguished based on the wavelength of the incident light. The first mechanism is the absorption of IR

Table 11.1. Division of the Optical EM Spectrum (Data from ISO, 2007)

Name of the waveband	Waveband limits	Interaction with matter
UV	1–380 nm	Electron excitation
Vis	380–780 nm	Electron excitation
NIR	0.78–3 µm	Overtone and combined molecule vibration
Mid-infrared (MIR)	3–50 µm	Fundamental molecular vibration
Far-infrared (FIR)	50 µm–1 mm	Fundamental molecular vibration

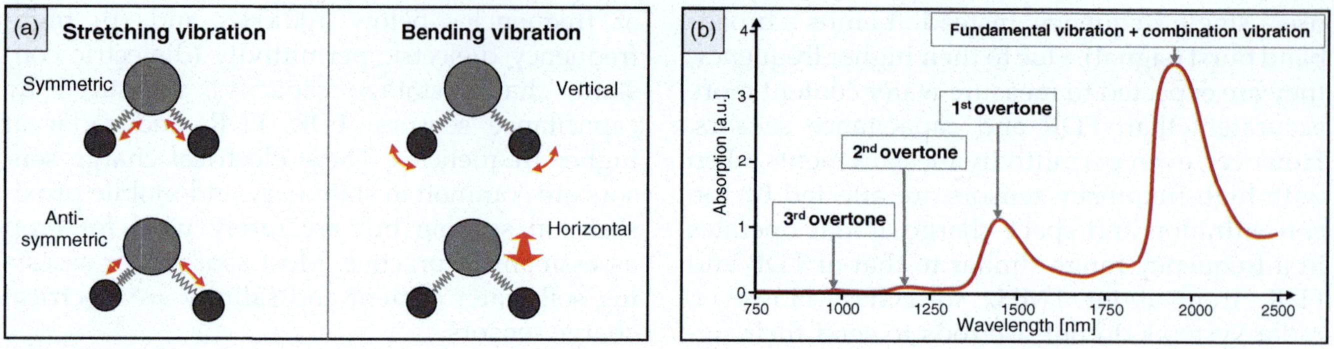

Figure 11.9. Infrared spectroscopy. (a) Absorption mechanisms through molecular vibrations. (b) Fundamental and overtone vibrations of water in the NIR range. Source: Courtesy of SILICANN Systems GmbH.

light (780 nm to 1×10^6 nm) when its frequency corresponds to the vibrational frequency of the molecule (Figure 11.9). The absorbed energy increases the amplitude of molecular vibrations.

The second absorption mechanism involves the absorption of ultraviolet (UV) and visible (Vis) light (10–780 nm) when electrons in the outer orbitals of atoms or molecules move from their ground state to an excited state. The optical range is divided into UV, Vis, and IR regions, each being further subdivided (Table 11.1). The definition and naming of regions in the optical spectrum are not always consistent across disciplines. Compared with the absorption bands in the NIR range, those in the UV-Vis range are much wider. This produces a large overlap, and so spectra from samples containing many substances, such as soils and crops, exhibit less distinct features in the UV-Vis range (Figure 11.10).

Optical sensors operate on the photoelectric effect by which incident light (photons) is converted into a change in electrical current or voltage proportional to the number of photons. These optoelectronic converters are manufactured in various designs, including photodiodes, photoresistors, and photocells. The photons that the converters receive from a sample can result from absorption, reflection, transmission, or emission.

Optical Absorption, Reflection, and Transmission

The primary reflection and absorption mechanisms depend on the resonance between the incident photon energy and the target's available energy levels. Light that is not absorbed is either directly reflected toward the light source or scattered before it reaches the sensor.

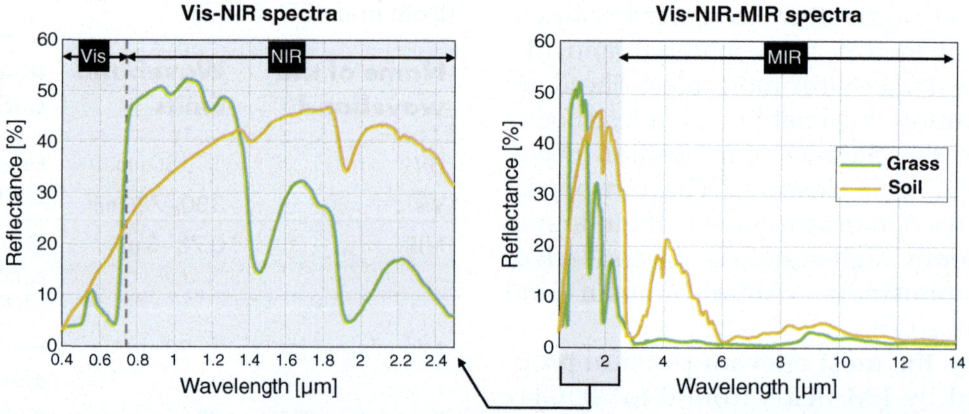

Figure 11.10. Reflectance spectra of grass and soil in the Vis, NIR, and mid-infrared (MIR) range.

A large portion of the incident light striking soils or plants is scattered in multiple directions, resulting in diffuse reflection. In spectral analysis, reflectance is a key parameter defined as the ratio of reflected to incident light. Reflectance values are presented on a scale from 0 to 1, or equivalently, from 0% to 100%. However, not all optical radiation is immediately absorbed or reflected at the surface. Depending on the wavelength and the optical properties of the target, a fraction of light may pass through it. This optical mechanism characterizes plant leaves. An example of reflectance used in precision agriculture is the NDVI, which indicates crop vigor by using reflectance values in the red and NIR ranges.

Optical Emission

Optical emission occurs when an excited sample emits light. This emitted light forms a spectrum that can be analyzed to reveal information about the material's composition or structure as in diffuse reflectance spectroscopy. However, emission spectroscopy can be more sensitive, more selective, or serve as a complement to diffuse reflectance spectroscopy. The main agricultural applications are fluorescence and laser-induced breakdown spectroscopy (LIBS). The fluorescence of chlorophyll is extremely sensitive to plant stress. Laser-induced breakdown spectroscopy is a rapid method for analyzing the elemental composition of soil and plant tissues. The emission lines of elements observed in a LIBS spectrum are very distinct; thus, LIBS provides high selectivity (Figure 11.11).

Modes of Optical Radiation

Optical sensors offer two data analysis modes: (a) spectral analysis, which captures data from different wavebands; and (b) geometric analysis, which captures images or three-dimensional (3D) point clouds. For spectral analysis, the EM energy coming from the sample is split into different energy ranges to produce an EM spectrum. As in remote sensing (Chapters 14 and 15), the different portions, or bands, are identified by their frequency, or wavelength. The bands offer different information about the object. For example, high reflectance in the NIR band indicates that the plant is relatively healthy. Sensors that measure up to 10 bands are classified as multispectral sensors, while those that measure more than 10 bands are called hyperspectral sensors. To generate a light spectrum, an optical element that disperses light is required. There are three main approaches to creating this dispersion (Figure 11.12). The first approach is to separate the bands using optical filters, as done in an RGB camera. Using more narrowband filters can extend this to hyperspectral sensing. The second approach is to use dispersive optics, such as diffraction gratings, which are common in field-deployable Vis-NIR spectrometers. The third approach is to use Fourier transform interferometry, a mathematical tool for analyzing waves. This is done using a scanning mirror without a dispersive element, resulting in a lighter setup and increased sensitivity of the Fourier transform approach. However, the software must perform the dispersion into different wavebands during a

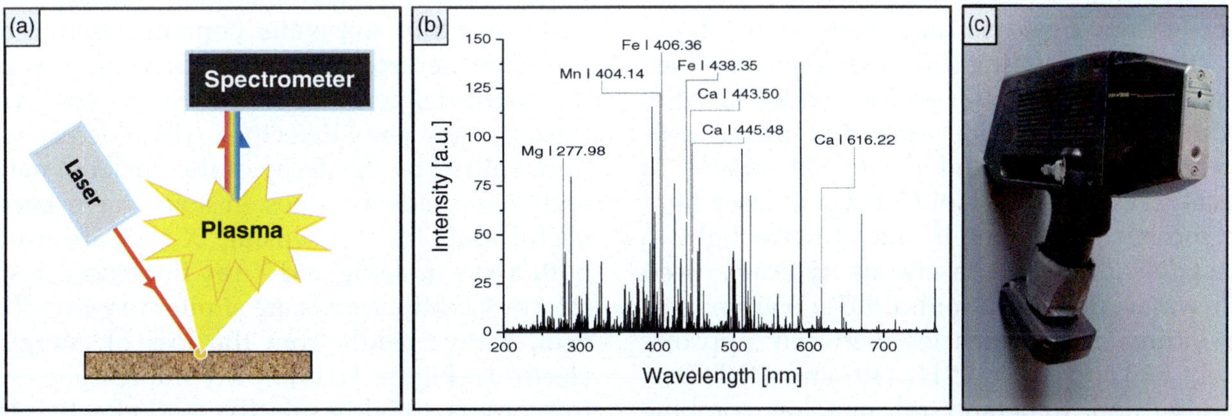

Figure 11.11. Laser-induced breakdown spectroscopy. (a) Measurement principle, (b) example of an emission spectrum from LIBS, (c) example of a handheld LIBS system.

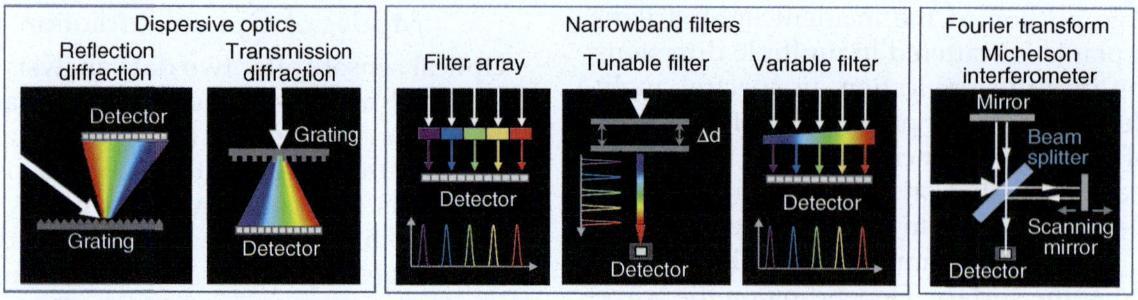

Figure 11.12. Spectral dispersion approaches in miniaturized spectrometers.

post-processing step. These optical dispersion techniques vary in speed, spectral resolution, and cost. Optical filters are relatively inexpensive when only a few wavebands and low spectral resolution are needed. Dispersive optics offer medium spectral resolution but higher speed than Fourier transform spectrometers. Another way to split the light continuum is to use light sources (incident light) with different wavelengths, such as using lamps with filters, light-emitting diodes (LEDs), and lasers.

Geometric analysis is possible when data points are linked to spatial coordinates in 2D or 3D. To generate this data with optical sensors, light reflected from an object is captured as a multi-pixel image in a single shot (camera) or by scanning the surface (laser scanning, line cameras). Three-dimensional data can be generated using stereovision (images from multiple viewpoints) and laser distance measurements (e.g., time-of-flight cameras). If the outlines of objects are extracted and analyzed, this is called shape analysis. If the spatial patterns within objects are analyzed, this is called image texture analysis. Laser technology is fundamental to Light Detection and Ranging (LiDAR) systems, which generate 3D point clouds and are widely used to produce high-resolution elevation maps, estimate plant heights, and create 3D images of objects. LiDAR emits rapid pulses of laser light and measures the time it takes for the light to reflect. Near-infrared lasers are typically used with wavelengths of around 905–1,550 nm, corresponding to frequencies between approximately 1.93 and 3.31×10^{14} Hz (193 and 331 THz).

Multi- and hyperspectral imaging combine spectral and geometric analyses. Due to high costs and sensor complexity, this type of imaging was initially limited to the domain of remote sensing. Nowadays, with the availability of cheaper sensors and faster data processing, interest in proximal hyperspectral sensing is increasing.

Radioactive Radiation

Radioactive radiation consists of high-energy (greater than 10 eV) particles that can ionize atoms and molecules and break chemical bonds. It is classified into α-rays (helium nuclei), β-rays (electrons), γ-rays and X-rays (photons), and *n*-rays (neutrons), among others. Radioactive radiation is generated naturally (by the decay of isotopes), or artificially (e.g., by X-ray tubes). It is valuable in measurement applications because its absorption or emission is highly specific, and certain forms of radiation can penetrate deeper into matter, expanding the effective sensing footprint. However, ionizing radiation poses risks to living organisms depending on its energy and the associated exposure time. Neutrons are specifically absorbed by water; thus, neutron probes, which include a neutron emitter, are highly precise sensors of soil water content. Despite their precision, neutron probes are not commonly used due to the dangers associated with *n*-rays. In contrast, passive γ-ray detectors, which detect radiation emitted by the decay of the nuclei of natural and man-made isotopes in soils, have become useful tools for soil sensing. X-rays are used in both X-ray imaging and X-ray fluorescence spectroscopy (XRF). X-rays are similar to γ-rays; however, X-rays result from the loss of energy by electrons (Figure 11.13). X-ray fluorescence spectroscopy is a highly selective and effective technique for detecting elemental concentrations in soils and crops.

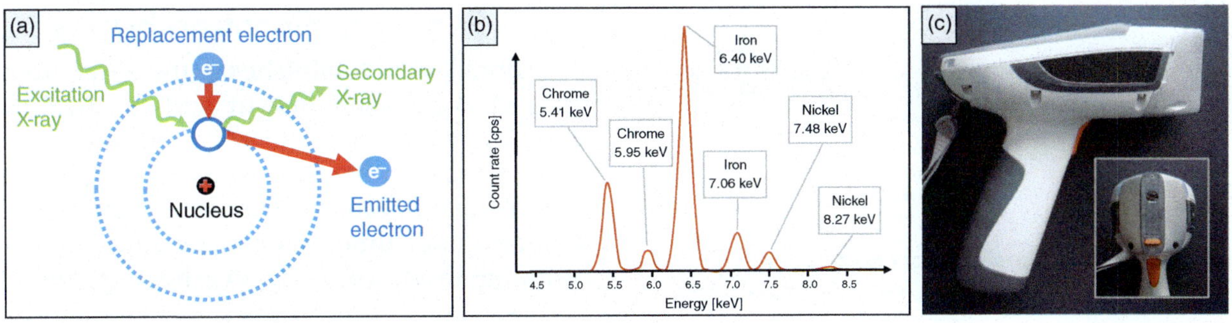

Figure 11.13. X-ray fluorescence spectroscopy. (a) Creation of the secondary (characteristic) X-ray emission after the ejection of an electron from a lower orbit due to its excitation by a primary X-ray from an X-ray tube. (b) X-ray fluorescence spectrum example, (c) example of a handheld XRF system.

Temperature

Temperature represents the average kinetic energy of a substance resulting from the vibrations and collisions of its atoms. The temperature of soil and plants is an indicator of water availability—lower temperatures occur when water is present—and reflects biological activity, which is limited by both low and high temperatures. Temperature sensors can be classified as either contact or noncontact. Contact temperature sensors are used for stationary measurements because of their slow response, good stability, and low cost. These sensors are typical components of soil probes (McInnes, 2002) but can also be parts of leaf sensors (Yu et al., 2016). The most common contact sensors are thermocouples, resistance thermometers, and thermistors (Morris & Langari, 2021). Noncontact temperature sensors measure thermal radiation optically in the thermal IR and microwave bands, ranging from 2 to 14×10^3 nm. Thermal radiation is emitted by a material because of its temperature, which depends on the material's heat. The measurement relies on the photoelectric effect (as described above). Spot sensors are called IR thermometers, whereas imaging devices are called thermal cameras. The temperature measured by thermal radiation sensors depends on both the target's temperature and its thermal emissivity. Emissivity is the ratio of radiation emitted by an object's surface to that emitted by a perfect emitter of radiation (i.e., a black body) at the same temperature. To determine the object's exact temperature using noncontact sensors, its emissivity must be known.

Mechanical Stimulus

Mechanical sensors detect the position, movement (including direction and acceleration), force, mass, pressure, stress, or density of a target. Mechanical soil sensors include soil penetrometers for measuring compaction and tensiometers for assessing soil water potential. Leaf thickness sensors, which measure turgor pressure, and dendrometers, which measure stem circumference are examples of mechanical crop sensors.

Acoustic Stimulus

Acoustic sensors rely on sound waves, which alternately compress and expand a medium (solid, liquid, or gas) at specific frequencies (Fraden, 2016). In soil sensing, microphones record sound as a shank or push probe moves through the soil. The sound is related to soil texture and stone content. The reflection of sound (echo time) can be used to estimate crop height, which is related to biomass and canopy area.

Chemical and Biological Stimuli

Chemical and biological sensors rely on direct, selective chemical interactions between the object and a receptor, also called the sensing element (Figure 11.14). This interaction is then converted into a measurable physical quantity, such as a change in voltage or color. The conversion process is called signal transduction or signaling (Bănică, 2012).

The most common chemical sensor is the pH electrode (Figure 11.15). Its sensitive surface specifically absorbs hydrogen ions, creating an

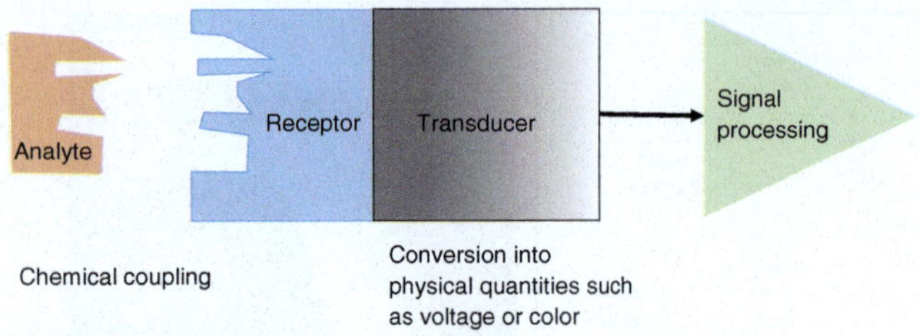

Figure 11.14. Basic principle of chemical and biological sensors. Source: Courtesy of Robin Gebbers.

electrical potential difference between the outer and inner sides proportional to the ions' activity, which is measured by a voltmeter. Another example is the quick nitrate test, which relies on the color change of a chemical agent (a chemical indicator) when it reacts with nitrate.

Biological sensors (biosensors) are a special class of chemical sensors that use biological components—such as tissues, microorganisms, organelles, cell receptors, enzymes, antibodies, and nucleic acids—as recognition elements. The detection can be very specific due to the "key-and-lock principle." Chemical and biological sensors are collectively referred to as biochemical sensors. Biochemical sensors provide useful data on many chemical and biological properties. However, they have several disadvantages, including fragility and, sometimes, high cost, as well as slow processing due to the required sample preparation. Biochemical sensors require direct contact with the sample or its liquid extract, whereas many physical sensors can take measurements from a distance. Thus, physical methods currently predominate in soil and crop sensing.

Sensor Calibration Basics

Calibration establishes the relationship between a sensor's output and the true corresponding values of the target quantities, as established by recognized measurement standards. It is necessary because (a) sensing systems can be distorted over time by physical stress and wear, contamination, corrosion, overheating, and other factors, preventing consistent measurements over time; and (b) sensing systems often measure the target property indirectly.

The calibration used to remove degradation effects of the sensing system is also called (instrument-)zeroing, blank-correction, or equipment calibration (Cammann, 1982; Cuadros-Rodríguez et al., 2001). This is usually done by measuring manufactured objects with known properties (etalons) under controlled conditions before deploying the sensor system in the field. Manual or numerical adjustments are applied to correct the system's output. For example, optical sensors are calibrated using Spectralon disks with 99% light reflectance, and pH electrodes are calibrated with pH 4.0 and pH 7.0 solutions. Indirect methods for assessing soil and plant properties are the most common approaches used in precision agriculture. Indirect measurement means that the sensor does not directly capture the target quantity (measurand) but instead measures one or more other quantities of the object linked to the measurand by a known relationship (Mari et al., 2023). The relationship between the sensor readings and the property of interest must be established

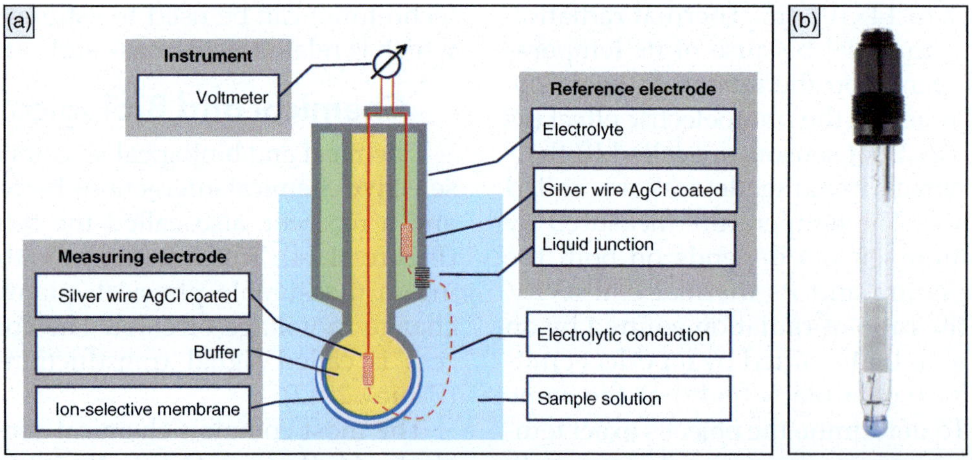

Figure 11.15. An ion-selective sensor system is an example of a chemical sensor. (a) Components of the system. The ion-selective membrane serves as the receptor. (b) pH electrode. Source: Courtesy of Robin Gebbers.

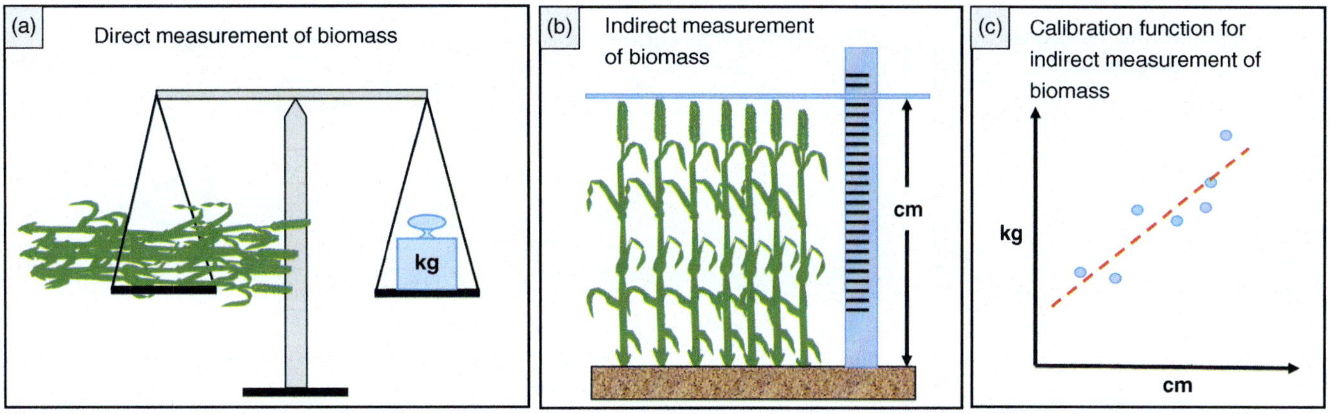

Figure 11.16. Examples of direct and indirect measurements. (a) Direct measurement of biomass through weighing. (b) Indirect measurement using height. (c) Calibration function, which links indirect measurements to direct measurements.

through a mathematical model called the calibration function. For example, aboveground plant biomass can be measured directly by cutting the plants and weighing them on a scale. Measuring plant height with a rising plate meter and relating it to biomass through a regression model is an indirect measurement approach (Figure 11.16).

The calibration function for indirect measurements relies on reference samples from plants or soils analyzed by reference (standard) methods, usually conducted in the laboratory. Reference methods are recognized as those that measure the target property more directly and precisely than alternative methods. The calibration function for indirect measurements can be simple, such as a linear equation, or complex, such as a neural network. Because proximal sensing is performed under uncontrolled field conditions, calibration is never free of error. Temperature, humidity, distance from the sample, vibrations, and even the operator may affect the measurements. Thus, sensor readings must be interpreted with care. In many cases, the correlation between the sensor output and the target property depends on the field conditions. Therefore, it is advisable to collect reference samples (ground truth) along with sensor measurements at the same time and analyze them using reference methods to calibrate the sensor readings for the specific field. This approach is called local calibration. However, if sensor readings are unaffected by ambient conditions and the sample matrix, it is possible to establish global (universal) calibrations. However, this is rarely the case.

Since reference sampling and consecutive analysis require extra effort, the number of samples should be kept as low as possible, and sample locations should be selected carefully. Samples should be selected to cover a wide range of sensor readings. They should not be outliers, spatially clustered, or affected by field operations. In addition, headlands, or field access points, should not be used as reference points (Adamchuk et al., 2011).

Data Fusion for Multi-sensor Calibration

To obtain more information, it is often beneficial to combine measurements from different sensors or from different times into a single database (Mitchel, 2012). This process, called data fusion, is designed to do the following:

- Improve accuracy or expand the footprint by using several sensors of the same type at the same time. The same properties are measured, for example, by two pH electrodes working in parallel.
- Obtain more information about an object by combining data from multiple sensors on that same object, for example, by combining soil EC and soil pH into a single dataset.

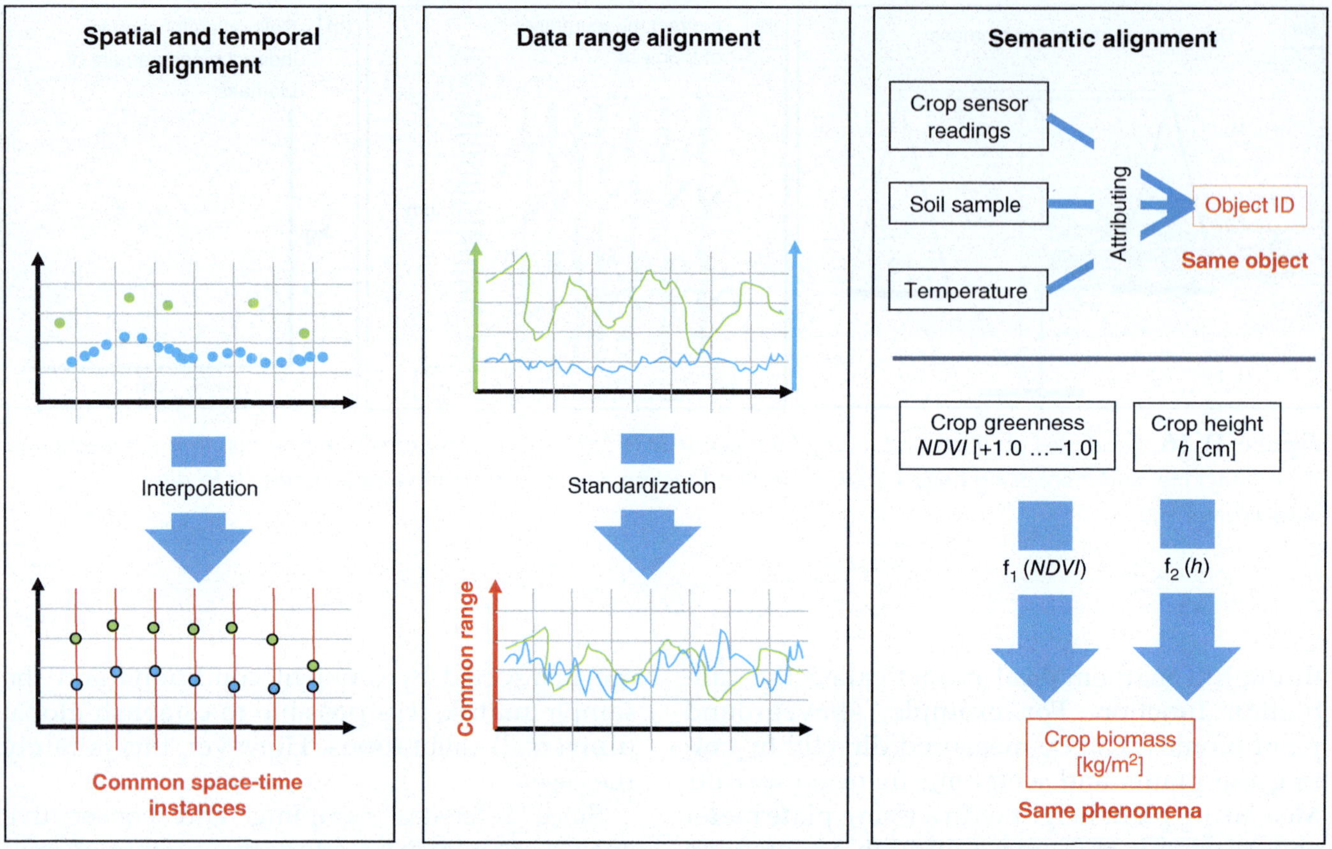

Figure 11.17. Objectives and approaches for aligning sensor data. Source: Courtesy of Robin Gebbers.

- Combine data collected at different spatial scales.
- Combine current and historical information.

Data fusion requires data alignment (Figure 11.17; Mitchel, 2012). Spatial alignment ensures that observations from multiple sources refer to the same locations, while temporal alignment ensures that observations are referenced to common points in time. Spatial and temporal alignment is typically accomplished through interpolation. For data visualization and certain data analysis methods, data range alignment is required. This transforms the data into a comparable range. References to objects and observed phenomena must be defined to establish databases that link tables from multiple data sources (Figure 11.17). For example, unique labels are used to address management zones as objects, and data from different sources are converted to a common unit, such as kilograms per square meter. This process, which requires a deeper understanding of the meaning of the data, is called semantic alignment.

Chapter Summary

Proximal crop and soil sensors detect physical, chemical, or biological phenomena—also called stimuli or signals—within a measurement distance of 2 m. These sensors are part of sensing systems that also include other components, such as power management, data processing, and communication. The sensing system facilitates the quantification of relevant crop and soil parameters and the integration of sensor readings into decision-making for fertilization, irrigation, plant protection, and other management measures. Sensing systems can be stationary for monitoring tasks or mobile for mapping. Proximal sensing is challenging because soils and plants are complex objects that are difficult to access due to their inherent mechanical resistance (in soils), fragility (in plants), and spatiotemporal variability, while the target properties are sometimes not well described even by laboratory-based reference methods. Because of these and other reasons, most sensors cannot measure the target properties directly. Usually, the primary sensor readings provide only an indication of the soil and plant properties relevant for decision-making. To rate the potential and limitations of sensors, it is necessary to understand the signal-generating processes.

Most proximal sensors used in agriculture are physical sensors that detect physical signals. These signals include EM radiation, mechanical stimuli, and acoustic stimuli. Electromagnetic radiation provides the richest source of information. It includes charge transfer, charge polarization, optical phenomena, and radioactivity. Chemical and biological sensors, unlike physical sensors, must be in direct contact with the target. They rely on direct, selective interaction between the object and a receptor. To facilitate this, sample processing is usually required. This reduces the flexibility of chemical and biological sensors, which, among other factors, limits their practical application. Because of the sensor's measurement principle and distortions from ambient conditions and other sources of noise, there is usually a trade-off between the quantity measured by the sensor and the quantity to be detected. Therefore, sensor calibration is essential. Calibration establishes a relationship between sensor readings and the known properties of a reference through a mathematical calibration model. In many cases, calibration must be performed "locally" for a given field and time. To improve understanding of soil and crops, it is often beneficial to combine data from multiple sources. Sensor data fusion requires data alignment, which may involve matching spatial and temporal instances through interpolation, transforming data to a common range of values, and attributing them to the same object or phenomenon.

Study Questions

1. What are the main components of a proximal sensing system?
2. List at least four types of proximal sensors.
3. What are some of the types of sensors your smartphone has?
4. Compare proximal and remote sensing in terms of spatial sampling elements, including their advantages and disadvantages.
5. Compare the support (footprint) of the antimony electrode with that of the electrical conductivity sensor shown in Figure 11.4.
6. True or false: Neutron radiation is commonly used in agricultural field sensors because of its precision and ease of handling.
7. Which region of the electromagnetic spectrum is primarily responsible for exciting electrons to higher energy states in atoms or molecules?
   A. Infrared
   B. Microwave
   C. Visible and ultraviolet
   D. Radio
8. True or false: X-ray fluorescence spectroscopy (XRF) is used to measure soil water content.
9. Match each sensor characteristic with its correct description:

Characteristic	Description
Selectivity	A. Ability to maintain consistent performance over time
Stability	B. Ability to detect a specific analyte without interference
Resolution	C. Smallest detectable change in the measured variable
Uncertainty	D. Degree of accuracy/correctness in sensor output

10. True or false: Smart sensors are capable of real-time data processing and decision-making without the need for post-processing.
11. Why is resonance important in the absorption of electromagnetic radiation by materials?
12. What is the principle behind the photoelectric effect used in optical sensors?

13. Discuss the advantages and risks of using radioactive radiation in agricultural sensing. How do safety concerns affect the choice of technology in the field?
14. How does reflectance differ from transmittance in terms of light interaction with a surface?
15. What is the difference between physical and chemical sensors?
16. What is the advantage of chemical and biological sensors as compared with physical sensors?
17. What is calibration, and why is it necessary?
18. What are some advantages to fusing data from multiple sensors?
19. How can data collected at different points within a field be fused?

ACKNOWLEDGMENT

This paper was funded in part by the USDA National Institute of Food and Agriculture (Multistate Project S1090) Program under grant number 7008110, National Science Foundation under grant numbers 2202706 and 2026431, Kentucky Small Grain Association, South Dakota State University, USDA Natural Resources Conservation Service Conservation Innovation Grants (NR233A750011G019), USDA National Institute of Food and Agriculture through the Multi–State Hatch project (accession no. 7008110), American Coalition for Ethanol, GEVO, South Dakota Corn Utilization Council, South Dakota Soybean Research and Promotion Council, and Federal Ministry of Education and Research (BMBF) of Germany under grant number 031B1069A. The use of trade names is for the convenience of the reader and does not imply endorsement by the authors.

REFERENCES

Abdul-Munaim, A. M., & Abdulmunem, O. M. (2025). A review of detection, monitoring, and analysis techniques in terahertz sensing for sustainable agriculture. *Sensing and Imaging, 26*(124). http://dx.doi.org/10.1007/s11220-025-00652-z.

Adamchuk, V. I., Viscarra Rossel, R. A., Marx, D. B., & Samal, A. K. (2011). Using targeted sampling to process multivariate soil sensing data. *Geoderma, 163*(1–2), 63–73. http://dx.doi.org/10.1016/j.geoderma.2011.04.004.

Baldridge, A. M., Hook, S. J., Grove, C. I., & Rivera, G. (2009). The ASTER spectral library version 2.0. *Remote Sensing of Environment, 113*(4), 711–715. https://doi.org/10.1016/j.rse.2008.11.007.

Bănică, F.-G. (2012). *Chemical sensors and biosensors: Fundamentals and applications*. John Wiley & Sons.

Cammann, K. (1982). Comments on instrument-zeroing and blank-correction methods in trace analysis. *Fresenius' Zeitschrift für analytische Chemie, 312*, 515–516. https://link.springer.com/article/10.1007/BF00636068.

Cuadros-Rodríguez, L., Gámiz-Gracia, L., Almansa-López, E., & Laso-Sánchez, J. (2001). Calibration in chemical measurement processes: I. A metrological approach. *Trends in Analytical Chemistry, 20*(4), 195–206. https://www.sciencedirect.com/science/article/abs/pii/S0165993600000935?via%3Dihub.

de Gruijter, J. J., McBratney, A. B., & Taylor, J. (2010). Sampling for high resolution soil mapping. In R. A. Viscarra Rossel, A. McBratney, & B. Minasny (Eds.), *Proximal soil sensing* (pp. 3–14). Springer, Dordrecht.

Everett, M. E. (2013). *Near-surface applied geophysics*. Cambridge University Press.

Fraden, J. (2016). *Handbook of modern sensors: Physics, designs, and applications*. Springer International Publishing.

International Organization for Standardization (ISO). (2007). *Optics and photonics—Spectral bands. ISO 20473:2007*. International Organization for Standardization.

Jócsák, I., Végvári, G., & Vozáry, E. (2019). Electrical impedance measurement on plants: A review with some insights to other fields. *Theoretical and Experimental Plant Physiology, 31*(3), 359–375. http://dx.doi.org/10.1007/s40626-019-00152-y.

Legendre, P., & Legendre, L. (2012). *Numerical ecology* (3rd ed.). Elsevier.

Loewer, M., Günther, T., Igel, J., Kruschwitz, S., Martin, T., & Wagner, N. (2017). Ultra-broad-band electrical spectroscopy of soils and sediments—A combined permittivity and conductivity model.

Geophysical Journal International, *210*(3), 1360–1373. https://doi.org/10.1093/gji/ggx242.

Mari, L., Wilson, M., & Maul, A. (2023). *Measurement across the sciences. Developing a shared concepts system for measurement* (2nd ed.). Springer Nature Switzerland AG.

McGrath, M. J., & Ní Scanaill, C. N. (2013). *Sensor technologies: Healthcare, wellness and environmental applications*. Apress.

McInnes, K. J. (2002). 5.1 Temperature. In *Methods of soil analysis: Part 4 physical methods, 5.4* (pp. 1183–1199). John Wiley & Sons, Ltd.

Meerdink, S. K., Hook, S. J., Roberts, D. A., & Abbott, E. A. (2019). The ECOSTRESS spectral library version 1.0. *Remote Sensing of Environment*, *230*(111196). https://doi.org/10.1016/j.rse.2019.05.015.

Mitchel, H. B. (2012). *Data fusion: Concepts and ideas* (2nd ed.). Springer.

Morris, A. S., & Langari, R. (2021). *Measurement and instrumentation. Theory and application* (3rd ed.). Elsevier.

Schirrmann, M., Gebbers, R., Kramer, E., & Seidel, J. (2011). Soil pH mapping with an on-the-go sensor. *Sensors*, *11*(1), 573–598. https://doi.org/10.3390/s110100573.

Van Haeverbeke, M., De Baets, B., & Stock, M. (2023). Plant impedance spectroscopy: A review of modeling approaches and applications. *Frontiers in Plant Science*, *14*, 1187573. https://doi.org/10.3389/fpls.2023.1187573.

Viscarra Rossel, R. A., Adamchuk, V. I., Sudduth, K. A., McKenzie, N. J., & Lobsey, C. (2011). Chapter Five—Proximal soil sensing: An effective approach for soil measurements in space and time. In *Advances in Agronomy* (Vol. 113, pp. 243–291). http://dx.doi.org/10.1016/B978-0-12-386473-4.00005-1.

Waak, M. (2016). *Das N-Düngekonzept von Agri Con*. https://www.wrrl-mv-landwirtschaft.de/export/sites/wrrl/de/downloads/Fachliche-Empfehlungen-Landwirtschaft/Duengung-Sensortechnik-und-Messverfahren/N-Duengungskonzept-AC_16_Sensortage-2016.pdf.

Yang, Z., Albrow-Owen, T., Cai, W., & Hasan, T. (2021). Miniaturization of optical spectrometers. *Science*, *371*(6528), eabe0722. http://doi.org/10.1126/science.abe0722.

Yu, L., Wang, W., Zhang, X., & Zheng, W. (2016). A review on leaf temperature sensor: Measurement methods and application. In D. Li & Z. Li (Eds.), *Computer and computing technologies in agriculture IX* (Vol. 478, pp. 216–230). Springer International Publishing. http://dx.doi.org/10.1007/978-3-319-48357-3_21.

Proximal Soil Sensing 12

Asim Biswas, Robin Gebbers, Viacheslav Adamchuk, Gaurav Jha, Katsutoshi Mizuta, Slawomir Kurpaska, Maciek Gliniak, Ce Yang, Naoya Kadota, and Hiteshkumar Bhogilal Vasava

Chapter Purpose

This chapter provides an overview of proximal soil sensing technologies and their agronomic applications. The goal of this chapter is to help students and practitioners understand the fundamental principles behind proximal soil sensing, evaluate the capabilities and limitations of different sensing systems, and learn how to integrate information collected by these technologies into the decision process.

Introduction to Soil Properties and Sensing

Proximal crops and soil sensors detect heat, light, changes in nutrient concentrations, proteins, organisms, enzymes, physical properties, and/or electrochemical potential generated by physical, chemical, or biological phenomena that are within 2 m of the target (Viscarra Rossel et al., 2011). These sensors collect data about specific parameters to help the grower make informed decisions. Understanding soil properties, many of which can be measured using soil sensors, is fundamental to sustainable agricultural production and environmental stewardship (Adamchuk et al., 2018, 2021). Table 12.1 provides information on measurements and the related sensors.

Physical properties, such as soil aggregation and structure, can be estimated with a proximal sensor. Good soil structure protects the soil's ability to hold nutrients and water, increases resilience to wind and water erosion, and ensures that biological life thrives in the soil (plants and microorganisms). Soil structure is the physical arrangement of the soil particles into aggregates, but it is difficult to measure directly. However, indirect measures of soil structure can be obtained by measuring soil texture (amount of sand, silt, and clay), soil strength, bulk density, soil organic carbon, water infiltration, and the water- and air-filled porosity. Based on these measurements, it is possible to make inferences about how the soil structure influences soil and environmental health. For example, measuring soil strength with a penetrometer determines the energy requirement for a root to penetrate a soil layer. If too much energy is required, the result may be lower yields. In this case, management strategies would be formulated to reduce the compaction problem.

Chemical properties regulate the availability and mobility of essential nutrients required for plant growth. Measuring soil chemical properties can include a direct estimation of the nutrients found in the soil and the associated fertilizer requirements. For example, soil pH (the concentration of H^+ in the soil solution) can have an impact on the availability of nutrients, while cation exchange capacity can affect how easily those nutrients are retained or lost. Proximal soil

Table 12.1. Soil Properties and Recommended Sensing Approaches

Soil property	Primary sensors	Secondary sensors	Typical accuracy	Critical factors	Management applications	Calibration requirements
Clay content	EMI, Vis-NIR	Gamma-ray, EC	$R^2 > 0.80$	Moisture correction, mineralogy	Management zones, drainage	Site-specific, seasonal
Soil pH	ISE, pH electrodes	Vis-NIR spectroscopy	±0.2 pH units	Temperature, electrode condition	Variable lime application	Daily calibration checks
Organic matter	Vis-NIR, MIR	EMI, Gamma-ray	$R^2 > 0.85$	Surface preparation, moisture	Carbon assessment, fertility	Local soil library needed
Soil moisture	TDR, capacitance	GPR, EMI	±2–3% volumetric	Soil type, temperature	Irrigation scheduling	Soil-specific curves
Bulk density	Penetrometer	Gamma-ray, GPR	$R^2 > 0.70$	Strong moisture dependency	Compaction management	Moisture correction essential
Salinity	EMI, EC sensors	ISE, TDR	$R^2 > 0.75$	Temperature, texture effects	Irrigation management	Multi-depth calibration
Available P	X-ray fluorescence, Vis-NIR	Spectroscopy	$R^2 > 0.60$	Extraction method dependent	Precision fertilization	Extraction method specific
Available K	Gamma-ray, X-ray fluorescence	ISE	$R^2 > 0.70$	Total vs. available forms	Nutrient management	Clay mineral considerations

Note. EC, electrical conductivity; EMI, electromagnetic induction; GPR, ground-penetrating radar; ISE, ion-selective electrode; MIR, mid-infrared spectroscopy; TDR, time domain reflectometry; Vis-NIR, visible/near-infrared spectroscopy.

sensing system can be used to evaluate the soil physical and chemical properties, such as soil pH, soil drainage, and soil moisture, to ensure they are most conducive to beneficial microorganisms. A high EC suggests that seed germination will slow for many plants, whereas a high pH indicates that the availability of Fe and P will be low. Based on these measurements, it is possible to make more informed management decisions.

Assessments of biological properties include identifying and differentiating the living organisms in the soil. There are millions of bacteria, fungi, algae, viruses, and protozoa contained in each teaspoon of soil. These microorganisms have a direct impact on plant and environmental health (Bhattarai et al., 2025). In addition, larger organisms such as earthworms, insects, and small vertebrates can assist in recycling nutrients. Because micro- and macro-organisms have many important functions, researchers and growers have begun to formulate management strategies that encourage the growth of the most beneficial organisms. Proximal soil sensing systems can be used to evaluate the soil physical and chemical properties, such as soil pH, soil drainage, and soil moisture, to ensure they are most conducive to beneficial microorganisms. These technologies are also being developed to predict microbial communities and enzymes with the use of optical sensor systems. However, proximal soil sensors can also directly quantify the abundance of microorganisms in the soil by measuring respiration with carbon dioxide (CO_2). An overview of the proximal sensing technologies are in Table 12.2

Sensor Resolution

Soil sensors that measure soil properties can help to reduce laboratory costs, accelerate the decision-making process, and improve the spatial and temporal resolution of collected information by making sampling quicker and less difficult (Viscarra Rossel et al., 2011). Selecting the appropriate proximal sensor requires an understanding of the different resolutions and deployment approaches. Sensors with higher-resolution data means there is more information and it is more detailed. For example, common spatial data collected from a field may include a composite soil sample of every half-acre by half-acre grid space in a field. To produce higher spatial resolution from these data, a grower may use a tractor-mounted EM38 sensor that can be driven over the field. This approach can collect more data, more often and increase the spatial and temporal resolution of the dataset (Figure 12.1).

Student Exercise

Use the data in case study 1 in Chapter 31 to explore the importance of sample resolution. This dataset contains soil test, apparent electrical conductivity (EC_a), elevation, and corn and soybean yield data from a 30.5 by 30.5 m grid spacing in a 35 ha field. There are over 600 sampling point in the field. Remove every other pH or EC_a value and compare the images with half of the values with all of the images.

Different spatial and temporal resolutions are needed for answering different questions. For example, capturing chemical and biological soil properties may need a higher spatial and temporal resolution dataset if being used to delineate management zones. However, the identification of soil physical properties such as texture and structure across a single field may not need as many samples taken as often. For more information of spatial and temporal variability or soil-sampling strategies please see Chapters 5 and 6.

Deployment Methods

Each sensor and sampling method offers distinct advantages and limitations that must be carefully considered. The three primary deployment approaches are (a) stop-and-go, (b) static, and (c) on-the-go. The stop-and-go method involves collecting detailed measurements at predetermined locations before moving to the next sampling point. This deployment method allows for multiple sensor readings at each location and requires stable positioning of the sensor at the sampling location for accurate measurements. Sensors use in the stop-and-go approach include instruments such as cone penetrometers that measure soil physical properties such as compaction or bulk density. This approach may include a portable

Table 12.2. Overview of Proximal Soil Sensing Technologies

Technology	Measurement principle	Physical basis	Primary applications	Deployment methods	Typical accuracy	Key advantages	Main limitations	Cost range
Electrical conductivity	Current flow through direct contact electrodes	Ohm's law—voltage/current relationship	Texture, salinity, moisture, and management zones	All three methods	$R^2 > 0.75$	Proven technology, real-time results	Multiple factors affect readings	$5,000–15,000
Electromagnetic induction	Induced eddy currents without soil contact	Faraday's law of electromagnetic induction	Profile characterization, compaction detection	On-the-go, static	$R^2 > 0.70$	No ground contact, multi-depth	Temperature sensitive, complex interpretation	$15,000–40,000
Visible/near-infrared spectroscopy	Light absorption by soil constituents (400–2,500 nm)	Molecular absorption and scattering	Organic matter, texture, mineralogy	All three methods	$R^2 > 0.85$	Multiple properties simultaneously	Moisture effects, calibration needs	$8,000–25,000
Ground penetrating radar	Electromagnetic wave reflection from interfaces	Dielectric property contrasts	Soil layering, moisture, compaction layers	Stop-and-go, on-the-go	$R^2 > 0.75$	Noninvasive, depth profiling	Limited clay penetration, expensive	$20,000–50,000
Ion-selective electrodes	Electrochemical potential measurement	Nernst equation—ion activity response	pH, specific nutrients, salinity	Stop-and-go, static	±0.2 pH units	Direct chemical measurement	Maintenance intensive, drift issues	$3,000–12,000
Penetrometers	Mechanical resistance measurement	Force-displacement relationships	Compaction, bulk density, hardpan	Stop-and-go	$R^2 > 0.80$	Direct physical measurement	Moisture dependent, slow	$2,000–8,000
Gamma-ray spectroscopy	Natural radioactivity detection (K-40, U-238, Th-232)	Nuclear decay processes	Mineralogy, parent material, potassium	On-the-go, static	$R^2 > 0.65$	Large sensing volume, penetrates residue	Limited specificity, safety concerns	$25,000–60,000
Gamma-ray bulk density sensor	A low-activity (Na-22) gamma-ray emission	Gamma-ray backscatter via Compton scattering	Bulk density	Stop-and-go	0.09~0.14 g/cm^3	Multi-depth, real-time results, rapid, portable	Moisture dependent, depth limited to ~80 cm, point measurement	$23,000–30,000
Portable x-ray fluorescence	Characteristic x-ray emission analysis	Atomic electron excitation	Elemental composition, heavy metals	Stop-and-go	Varies by element	Direct elemental analysis, rapid	Matrix effects, surface only	$15,000–45,000
Time domain reflectometry	Electromagnetic wave travel time	Dielectric constant effects on wave velocity	Moisture content, bulk density	Static, stop-and-go	±2% volumetric	High accuracy, minimal calibration	Point measurement, installation needs	$1,000–5,000

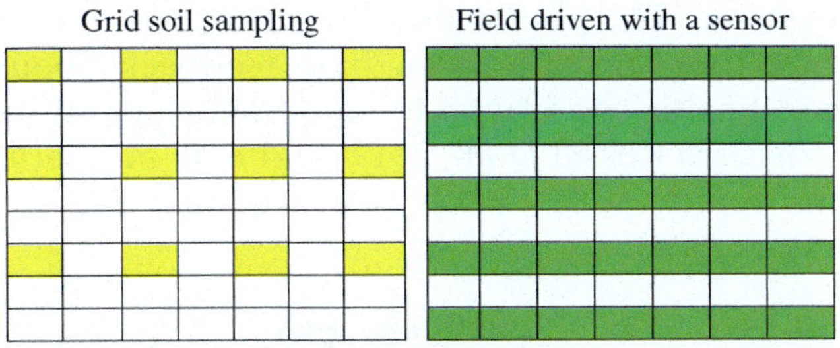

Figure 12.1. Comparison of grid-based soil sampling (left) with EM38 sensor mapping for apparent electrical conductivity (EC_a). The EC_a map was used to identify management zones and, through a conversion equation, estimate saturated paste electrical conductivity (EC_e). Sensor-based mapping provides greater spatial detail than grid sampling (He et al., 2018). A field where soil samples are collected, sent to a laboratory, and analyzed (left), compared to a field where an EM38 sensor is driven across in strips to determine EC_a (right). Based on the EC_a map, management zones were identified (Adapted from He et al., 2018). Based on a derived equation, the EC_a is converted to EC_e. The sensor map provides much more information in a grid sample.

gamma-ray sensor or an x-ray fluorescence analyzer that provides detailed elemental composition analysis.

In contrast, static sensors are installed at fixed locations to provide detailed temporal changes in soil and environmental conditions for extended periods. Examples of static sensors include a soil moisture sensor to monitor irrigation management, gas flux sensors for measuring microbial respiration, and comprehensive weather stations that monitor soil temperature and other environmental conditions. The primary advantages of a static sensor system are that they capture high-resolution, highly detailed temporal data that can be used to make extremely informed management decisions by producing large datasets suitable for statistical analysis and trend identification. However, because these systems are fixed at a single location, the data collected are site-specific and may not be relevant to other locations.

On-the-go sensors collect information continuously while moving across fields during normal agricultural operations. These systems are typically integrated with GPS receivers for accurate spatial referencing. The primary advantages of on-the-go systems include the cost-effective acquisition of data with high spatial resolution and the ability to combine data collected from the sensor with other data collected at the same GPS coordinates. These systems can collect thousands of measurements per hectare during normal field operations (e.g., planting, fertilizing, and harvesting), creating detailed maps of soil variability patterns that would be impossible to achieve through conventional sampling approaches. However, on-the-go systems face some challenges as data collection can be skewed by environmental conditions such as high or low soil moisture levels, surface roughness, and travel speed.

Measuring Soil Texture

Soil texture depends on the amounts of sand, silt, and clay in the soil. More clay content in the soil increases soil moisture and nutrient retention. By measuring the EC_a using an electromagnetic (EM) sensor (Chapter 11), the collected information provides an indirect estimate the clay content. This estimate is possible because the ability of a soil to transmit an electrical charge increases with the percentage of clay. When evaluating EC in a laboratory chemical analysis, it is important to consider the approach. For example, the EC determined by using a saturated paste (EC_e) will be double the value of an $EC_{1:1}$ that is made by mixing 10 g of soil with 10 mL of water because of the differing ratios of soil to water solution.

EC_a measures the bulk soil matrix outside of the laboratory. It can be measured with EM induction (EMI) and electrical resistivity sensors. EMI instruments are a noncontact method for measuring soil EC. An alternating current in the transmitter coil generates a primary EM field, causing an eddy current within the soil matrix. The eddy current, in turn, generates a secondary EM field within the receiving coil. The secondary currents are sensitive to multiple factors, including soil moisture, clay content, salinity, and temperature (Adamchuk et al., 2004).

EM systems incorporate multiple receiver coils at different orientations and separations from transmitters, enabling simultaneous measurement at different soil depths while traveling at speeds up to 15–20 km/hr. These sensors can therefore be mounted onto moving vehicles, such as tractors, to take measurements. EMI is widely used in on-the-go sensing technology to estimate soil textural changes, and it has been used in texture mapping across diverse soil types to account for local variations in clay mineralogy and organic matter content. Case study 1 in Chapter 31 contains EM data that can be used to explore the relationships between yields and EC_a.

Soil texture can also be evaluated using spectroscopy. Spectroscopy is the measurement of light that is reflected, transmitted, or emitted from an object. Spectroscopic approaches can be useful because different bonds (C-H, N-H, C-C, and O-H) have different molecular signatures. For example, in the near-infrared (NIR) region (700–2,500 nm) functional groups including C-H, N-H, and O-H vibrate the most, making them very specifically detected by NIR spectroscopy (Shin et al., 2025). Because clay minerals also exhibit distinctive molecular signatures in different wavelengths, a quantitative prediction of clay content that is independent of electrical properties can be made (Viscarra Rossel et al., 2016). Additionally, the proportion of sand, silt, and clay in the soil can be assessed by using gamma-ray spectroscopy, taking advantage of the natural radioactivity from soil minerals (e.g., K-40, Th-232, U-238).

Soil Organic Matter

Organic matter assessment also uses spectroscopy, leveraging the distinctive characteristics of soil organic compounds such as the C-H and C-C molecular bonds. When light reaches the soil or crop, radiation causes individual molecular bonds to vibrate, which means the soil or crop absorbs radiation. The amount of light absorbed by a soil sample depends on the properties present, resulting in characteristic peaks/valleys at certain wavelengths.

On-the-go optical sensing systems have emerged as powerful tools for measuring soil organic matter by assessing and interpreting these wavelengths. Currently, on-the-go systems that measure soil organic matter in real time are commercially available as tractor-mounted systems. The power of these measurements is that they provide immediate feedback about soil health conditions to equipment operators (Viscarra Rossel et al., 2019) and may be useful in rapid soil health assessments (Vavlas et al., 2024).

Soil Strength

Soil strength assessments require an integrated approach that can distinguish between natural soil consolidation/aggregation and management-induced compaction problems. If the soil is too compact, roots and water cannot penetrate, and the soil will lose productivity. Penetrometers are rod- or cone-shaped tools that can be pushed into the soil to provide a direct assessment of soil mechanical resistance to penetration/compaction. These soil sensors can be used to estimate bulk density when properly calibrated for moisture content (Figures 12.2 and 12.3). Current automated penetrometer systems can collect high-resolution measurements to a depth of 45 cm or more while recording GPS coordinates. This ability allows for the creation of detailed compaction maps that identify problem areas requiring management attention. In research studies or comparison between different sites,

Figure 12.2. Measuring the bulk density (top images) and soil resistance (bottom images) at a point within a field. Source: With permission of Asim Biswas.

Figure 12.3. Measuring soil resistance using an on-the-go sensor (from PFB Figure 9.5).

standard protocols (ASABE Standard S313.3) can be used to collect and compare the data (ASABE Standards, 2017), which allows growers to make management decisions across site locations. Emerging technologies for rapid density estimation at deeper depths, including the commercially available Medusa RhoC5 and Multi-Sensor Core Logger, utilize a gamma-ray backscatter to measure soil resistance instead of physical pressure (Pepers et al., 2024; Smith et al., 2024). While these systems offer promising advancements, their spatial coverage remains limited owing to the time-intensive processes of positioning, measurement, and relocation. As a result, they are best suited for detailed site characterization or validation of other measurement approaches.

Soil strength can also be assessed by determining the pulling force, also known as draft resistance, during tillage operations (Figure 12.5). Both approaches provide direct information about soil physical conditions affecting root growth and equipment operation. An important consideration when interpreting penetrometer or draft resistance data is the impact of soil moisture effects. In general, resistance decreases with increasing moisture and root penetration is restricted if the penetrometer resistance is greater than 2.0 MPa. To compare penetrometer values across dates and soils, it is important follow standardized measurement protocols.

Soil Moisture

Soil moisture is an important characteristic to measure because it has a direct impact on productivity. Soil moisture can be measured using a wide range of technologies including, time domain reflectometry (TDR), gypsum block, and tensiometers. TDR instruments measure a material's ability to store electrical energy when exposed to an electrical field. The instrument consists of a transmission line and physical wires and rods that guide the direction and path of the electrical energy. Measurement of the dielectric constant, the material's ability to store electrical energy, can predict the volume of water in the soil because solid particles such as minerals and rock have a much smaller dielectric constant (ranging from 3 to 6) than water, which has a value of 80. TDR sensors typically achieve accuracy within 2–3% of volumetric water content when properly calibrated.

A gypsum block also uses electrical resistance to estimate soil water. It consists of two electrodes mounted in a small block of gypsum. Wires are attached to electrodes that are connected to a handheld reader. The sensor is buried in the soil at the depth to be monitored. The system works because the moisture content of the block is in equilibrium with the moisture in the surrounding soil. Therefore, the electrical resistance in the block is indirectly related to the moisture content and can be estimated. This system, as well as the TDR system, allows for continuous monitoring to detect rapid changes in soil conditions that might be missed by periodic sampling. This ability makes these systems extremely beneficial for optimizing irrigation schedules and can reduce water consumption by 20–30%.

Other approaches to measure soil water include the use of a tensiometer, an airtight ceramic cup attached to a vacuum gauge at one end. When placed in the soil, water in the soil moves into the cup, and the extent to which the soil holds onto the moisture indicates how much soil water is available to the plant. For example, at field capacity the tension will be close to zero. Tensiometers can be useful for quick soil moisture assessments when field scouting, especially to determine how much water is available to plant roots and how easily the roots can access the water. The underlying principle of the tension measurement is its mimicry of how much energy the plant root must exert to take in water at the observed soil moisture level.

Soil pH

Soil pH, the measurement of the concentration of H^+ ions in the soil solution, can be mapped across a field to measure how the soil chemical properties change nutrient availability and plant growth in different areas of the field as well as help in management decisions such as variable-rate lime application. Soil pH mapping has emerged as one of the most successful applications of electrochemical sensing in agriculture, with significant economic benefits from making soil amendments more site-specific, reducing application cost and time. These technologies use ion-selective electrodes that respond to specific ion activities through selective membranes. These sensors create a reliable on-the-go pH measurement system through automated soil-sampling mechanisms that ensure consistent electrode-soil contact while minimizing contamination effects. Advanced systems incorporating automated cleaning and calibration procedures can achieve measurement frequencies of two to three samples per minute while maintaining accuracy comparable to laboratory methods.

The electrochemical principle underlying pH sensors is based on measuring the potential difference between two electrodes within a closed circuit in the probe. This difference creates a linear slope in which every change in the electric potential (the voltage difference) can be correlated to a change in pH (change in H^+ ion concentration in the soil). Modern pH sensing systems incorporate temperature compensation algorithms and drift correction procedures to maintain accuracy under variable field conditions. Proper electrode maintenance, including regular calibration with standard buffer solutions and replacement of reference electrodes, remains critical for reliable measurements.

Soil Nutrients

Student exercise. Estimate the amount of mineralized N if the soil is 3% soil organic matter (SOM), where the SOM is 58% C, the C:N ratio is 10, and 2% of the organic C mineralizes (min) annually.

$$\frac{2{,}000{,}000\,\text{lb soil}}{\text{acres}} \times \frac{2\,\text{lb SOM}}{100\,\text{lb soil}} \times \frac{0.58\,\text{lb C}}{1\,\text{lb SOM}}$$

$$\times \frac{1\,\text{lb N}}{10\,\text{lb C}} \times \frac{0.02\,\text{lb min}}{1\,\text{lb N}} = 46.4\,\text{lb N}$$

Nutrient status assessment requires sophisticated integration of multiple sensing technologies owing to the complexity of nutrient cycling processes and the various chemical forms in which nutrients exist in soils. As outlined in Table 12.1, different nutrients require different sensing approaches based on their chemical behavior and availability mechanisms. For N assessment the challenge lies in measuring the mineral N and estimating mineralized N.

Currently, real-time sensors using EM technologies that can estimate the amount of nutrients in the soil are being developed. Spectroscopic approaches can also be used to identify soil nutrients and chemical composition, including inorganic or N fertilizers in the soil. Owing to the complexity of molecular signatures in soil (e.g., the molecular signature for NO_3 may be intertwined with the molecular signature for H_2PO_4), machine learning is often used in the workflow to extract information. While complicated, these spectroscopic options do not require contact with the soil. Advancements of this technology may eventually make sampling of soil nutrients rapid and easy.

In addition, ion-selective electrodes mounted on a tractor can be used to identify specific soil nutrients. The method is based on the chemical composition of a membrane that allows the electrode to sense only one nutrient at a time. For example, an ion-specific membrane can be used to measure nitrate and K^+ (Kim et al., 2006). However, these sensors are not yet ready for deployment.

Soil Salinity Mapping

Soil salinity arises from an excess of soluble salts (Ca^{2+}, K^+, Mg^{2+}, and Na^+) in the soil. Therefore, it is common to measure these soil properties to determine salinity concerns (Figure 12.4). In laboratory settings, soil salinity is typically measured through the direct measurement of soil EC in a soil solution (also called electrolytic conductivity). However, mobile geophysical tools, such as EMI, can determine the soil conductivity in its natural state. Many EMI instruments have multifrequency or multi-depth measurement capabilities, which can provide detailed information about salt distribution throughout the soil profile (Figure 12.4). However, to convert EC_a measured using an EM meter to EC_e, the relationship between the field values (EC_a) and laboratory-derived values (EC_e) should be determined. For example, based on the EC_e of a saturated paste extract, He et al. (2018) converted EM38 EC_a values to EC_e. Based on these calculated values, salinity management zones were created.

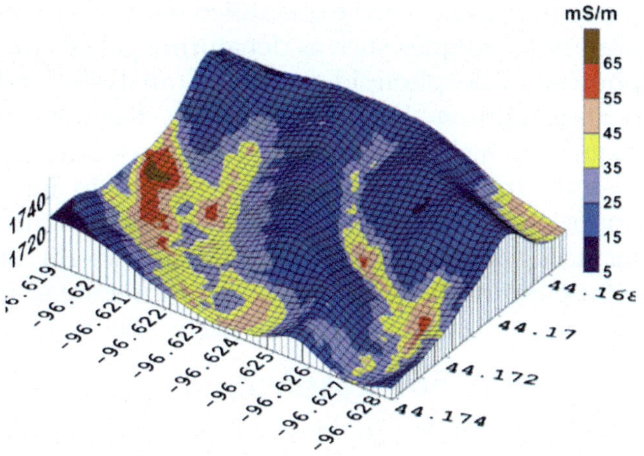

Figure 12.4. EC_a mapping overlaid on field elevation, demonstrating how soil electrical properties relate to topographic features and providing insights for management zone delineation and drainage planning (Courtesy of D. Clay). Note that the dataset used to create this image is in case study 1 in Chapter 31.

Identifying Enzymes and Microbial Communities

Soil enzymes play a crucial role in soil quality, function, and health by playing key roles in the biochemical processes involved in building soil health. To some extent, proximal sensors that use visible and NIR spectroscopy can provide clues about the soil enzymes. For example, β-1,4-glucosidase enzyme has been strongly correlated with total organic carbon because of its ability to degrade cellulose and complex organic compounds from dead plant tissue (Tiwari et al., 2019). This enzyme can be detected using a 1,538 nm wavelength that corresponds to the molecular signatures of organic matter in which the enzyme is primarily found (Qaswar et al., 2025). Additionally, acid phosphatase and alkaline phosphatase are important enzymes for phosphorus cycling. The key signature of these enzymes has been observed at a 718 nm wavelength, which correlates with concentrations of iron oxides; these oxides are chemical compounds that can "fix" phosphorus in the soil, making it unavailable (Qaswar et al., 2025).

Currently, laboratory techniques using specific assays can directly test for these enzymes. Additionally, entire microbial populations and their abundances can be quantified with other laboratory techniques such as denaturing gel electrophoresis, phospholipid fatty acid analysis, and next-generation DNA sequencing. Field techniques using portable spectroscopy devices to identify microbial communities known to have important and beneficial relationships with plants, including plant growth promoting bacteria or arbuscular mycorrhizal fungi, are future areas of research and exploration (Fontal et al., 2025).

Soil Respiration

Soil respiration provides clues about the health of the soil. In general, a stable rate of soil respiration suggests that the soil is healthy. Soil respiration can be measured using gas flux measuring. These monitoring systems measure CO_2 that is emitted from the soil and typically appear as static sensors that are deployed in the field. These systems provide detailed information about soil biological activity and environmental impacts of management practices, including important relationships between soil moisture conditions and N_2O emissions (Figure 12.5) (Bhattarai et al., 2025).

Emerging Technologies

Nanotechnology-based sensors offer exceptional sensitivity for specific ions through carbon nanotube and graphene sensors, enabling continuous monitoring with unprecedented precision. Biosensors incorporating enzyme systems and microbial fuel cells enable detecting real-time biological activity. DNA-based sensors provide direct microbial community measurements for comprehensive soil health. Artificial intelligence integration through deep learning approaches extracts patterns from complex sensor data, while edge computing enables real-time field processing for decision-making (Padarian et al., 2020).

Answering Management Questions with Proximal Soil Sensing

The goal of proximal soil sensing technologies is to reduce the need for classical soil analysis. However, this goal has not yet been achieved. The current status of the various measurements and sensors are outlined in Tables 12.1 and 12.2. Significant advances have been made for some measurements. For example, Figure 12.4 shows that the relationship between elevation and the EC_a has clear patterns that can be used for management purposes.

Multi-sensor Platforms and Data Fusion

The successful implementation of proximal soil sensing requires careful consideration of technology integration, economic factors, and practical deployment strategies. This section provides actionable guidance for practitioners seeking to

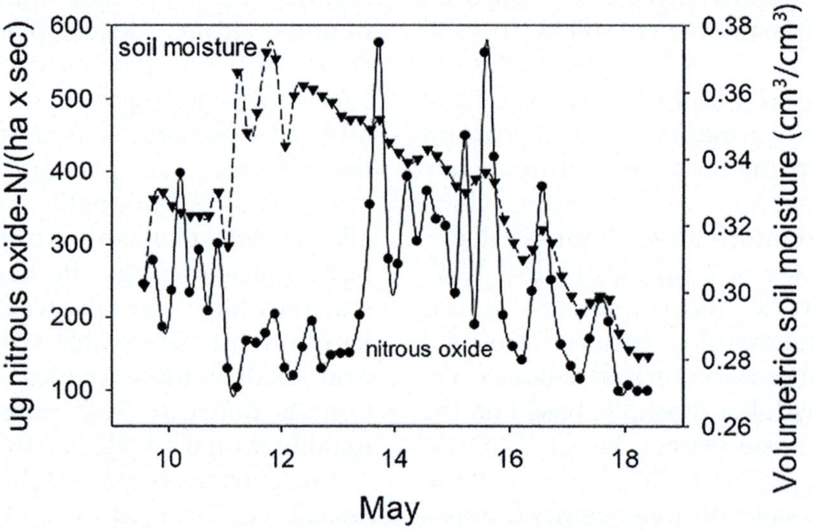

Figure 12.5. Soil moisture and nitrous oxide emissions over time, demonstrating how continuous monitoring at a static site reveals relationships between soil moisture and nitrous oxide emissions (Thies et al., 2019). Courtesy of D. Clay (South Dakota State University).

implement sensing technologies effectively, while maximizing return on investment (Table 12.3). Commercial systems combine multiple sensing technologies on a single platform for simultaneous measurement of multiple soil properties. The Veris MSP3 system integrates EC, pH measurement, and optical sensing, demonstrating practical sensor fusion applications. Field studies show that multi-sensor platforms can reduce total measurement costs by 30–40% relative to deploying individual sensors separately, while also providing enhanced data quality through cross-validation.

Data integration employs machine learning techniques combining sensor streams, while accounting for measurement scale, resolution, and uncertainty differences. Current approaches include random forest algorithms, support vector machines, and neural networks that can properly handle a combination of different sensor data. Successful integration requires standardized data formats, real-time processing capabilities, and quality control procedures for automated validation (Mulla, 2013).

The key to successful data fusion lies in understanding how to best use different sensors together to collect data. Specifically, the sensors need to be complementary to each other. In addition, different technologies may be better suited for different size farms (Table 12.3). For example, EMI provides broadscale variability patterns, while ion-selective electrodes offer precise chemical measurements at specific locations. Combining these approaches through targeted sampling strategies provides both comprehensive spatial coverage and accurate local calibration.

Implementation of Precision Agriculture Technology

When making the decision to implement precision agriculture, it is important to consider the direct and indirect costs and benefits. For example, on a 200 ha farm the direct costs might be approximately $4,000 annually versus $15,000 for the purchase of a sensor and $2,000 for processing the information. The indirect costs might include annual expenses of $1,000 for software and $1,500 for maintenance, repair, replacement, and calibration. However, these calculations do not consider the impact of more detailed measurement with a proximal sensor on the return on investment (Khosla et al., 2025). For example, pH-guided lime application systems demonstrate 15–30% reductions in lime usage, while achieving uniform correction. Nitrogen management using optical sensors combined with soil conductivity can improve efficiency by 10–20%, while reducing environmental losses.

Current Challenges

Key limitations requiring continued research include developing robust calibration approaches to accommodate soil variability, standardizing measurement protocols to enable data integration across platforms, establishing validation methodologies, and ensuring measurement reliability. Integration challenges with digital agriculture systems require advances in data management, processing capabilities, and communication protocols for seamless operation across different technology platforms.

Cloud computing platforms provide scalable resources for massive sensor datasets, while Internet of Things technologies transform connectivity capabilities. Digital twins incorporating real-time sensor data with mathematical models represent emerging opportunities for sophisticated agricultural system simulation and management optimization.

Summary and Conclusions

Proximal soil sensing provides essential information for site-specific soil management through comprehensive soil characterization across spatial and temporal scales. Integration of multiple technologies with advanced data processing enables unprecedented insights into soil dynamics previously impossible to achieve economically.

Key advantages include real-time field measurements without sample preparation, high spatial resolution mapping capabilities, cost-effective alternatives to laboratory analysis, and integration potential with automated management systems. Critical limitations include cost, the need for time-consuming site-specific calibration, and in some cases, the requirement for technical expertise to interpret data. Success factors for implementation include understanding both the potential and the constraints of current technologies, validating sensor-based recommendations with ground-truth data, maintaining realistic expectations about measurement accuracy, and ensuring proper calibration and maintenance.

Table 12.3. Practical Implementation Framework for Proximal Soil Sensing

Implementation factor	Small farms (<50 ha)	Medium farms (50–200 ha)	Large farms (>200 ha)	Key decision criteria	Success indicators
Technology selection	Simple sensors (EC, pH, penetrometer)	Integrated platforms (EMI + spectroscopy)	Full multi-sensor systems with GPS	Cost per hectare, operator skill, maintenance needs	>15% input efficiency improvement
Deployment strategy	Service provider or rental	Owned seasonal equipment	Dedicated integrated systems	Capital availability, usage hours, support needs	>80% field coverage efficiently
Calibration approach	Regional libraries	Site-specific 10–15 samples/field	Adaptive continuous validation	Soil variability, crop diversity, accuracy needs	Predictions within ±15% of lab values
Data management	Simple mapping software	Cloud platforms with mobile access	Integrated farm management systems	Technical expertise, data volume, integration	Maps generated within 24 hr
Economic threshold	>$150 · ha^{-1} · yr^{-1} crop value	>$75 · ha^{-1} · yr^{-1} crop value	>$50 · ha^{-1} · yr^{-1} crop value	Break-even analysis, risk tolerance	Positive return on investment in 3–5 yr
Training requirements	4–8 hr initial	16–24 hr comprehensive	40+ hr advanced interpretation	Available time, staff turnover, background	>90% successful deployments
Maintenance level	Minimal calibration checks	Monthly procedures	Dedicated technical support	Expertise, service availability, downtime	<5% maintenance failures

Study Questions

1. Compare deployment methods: Analyze advantages and limitations of stop-and-go, static, and on-the-go measurement approaches with specific examples.
2. Explain apparent conductivity: Why do electromagnetic sensors measure "apparent" rather than "true" electrical conductivity, and why is calibration essential?

Discussion Question

Evaluate future potential: Assess the statement "Proximal soil sensing will eventually eliminate laboratory analysis discussing strengths and limitations."

Multiple Choice Questions

1. Primary advantage of proximal sensing:
   a. Can be as accurate as lab analysis
   b. Real-time field measurements without sample preparation
   c. Costs nothing after purchase
   d. Works identically in all conditions
2. Electromagnetic sensors measure:
   a. Soil pH precisely
   b. Specific nutrients directly
   c. Apparent electrical conductivity
   d. Biological activity levels
3. "Apparent" conductivity means:
   a. Sensors lack precision
   b. It is related to the electrical conductivity
   c. True conductivity unmeasurable
   d. Intentionally imprecise design
4. Most chemically specific sensors:
   a. Electromagnetic induction
   b. Ground penetrating radar
   c. Ion-selective electrodes
   d. Mechanical penetrometers
5. Management zone delineation:
   a. Creates legal boundaries
   b. Divides fields by similar soil properties for management
   c. Determines equipment size
   d. Establishes crop rotations

ACKNOWLEDGMENTS

Support provided by the University of Guelph, South Dakota State University, USDA-NIFA (Multistate Project S1090) Program under (7008110), Kentucky Corn Grower Association, and the Precision Farming Systems community in the American Society of Agronomy. Trade names are for convenience and do not imply endorsement.

REFERENCES

Adamchuk, V., Ji, W., Rossel, R. V., Gebbers, R., & Tremblay, N. (2018). Proximal soil and plant sensing. In D. K. Shannon, D. E. Clay, & N. R. Kitchen (Eds.), *Precision agriculture basics* (pp. 119–140). ASA-CSSA-SSSA. https://doi.org/10.2134/precisionagbasics.2016.0093.

Adamchuk, V. I., Biswas, A., Huang, H. H., Holland, J. E., Taylor, J. A., Stenberg, B., Wetterlind, J., Singh, K., Minasny, B., Fidelis, C., Yinil, D., Sanderson, T., Snoeck, D., & Field, D. J. (2021). Soil sensing. In R. Kerry & A. Escolà (Eds.), *Sensing approaches for precision agriculture* (pp. 93–132). Springer. https://doi.org/10.1007/978-3-030-78431-7_4.

Adamchuk, V. I., Hummel, J. W., Morgan, M. T., & Upadhyaya, S. K. (2004). On-the-go soil sensors for precision agriculture. *Computers and Electronics in Agriculture, 44*(1), 71–91. https://doi.org/10.1016/j.compag.2004.03.002.

ASABE Standards. (2017). *ASABE S313.3 FEB1999 (R2017): Soil cone penetrometer*. American Society of Agricultural and Biological Engineers.

Bhattarai, D., Pandit, S., Kafle, R., Nleya, T., Clay, D. E., & Clay, S. A. (2025). Synergistic effects of biochar and plants can reduce greenhouse gas emissions from salt affected soil. *Scientific Reports, 15*, 8879. https://doi.org/10.1038/s41598-025-93198-9.

Fontal, A., Borràs, S., Cañas, L., Pozdniakova, S., & Rodó, X. (2025). Benchmarking laser-induced fluorescence and machine learning for real-time identification of bacteria in bioaerosols. *EGUsphere*. [preprint]. https://doi.org/10.5194/egusphere-2025-2484.

He, Y., DeSutter, T. M., Norland, J., Chatterjee, A., Casey, F., & Clay, D. E. (2018). Prediction of soil sodicity and the development of soil management zones in low-relief sodic soils. *Precision Agriculture, 19*, 858–875. https://doi.org/10.1007/s11119-018-9561-x.

Khosla, R., Fleming, K., Delgado, J. A., Shaver, T., & Westfall, D. G. (2025). Use of site-specific management zones to improve nitrogen management for precision agriculture. *Journal of Soil and Water Conservation, 57*, 513–518. https://doi.org/10.1080/00224561.2002.12457486.

Kim, H. J., Hummel, J. W., & Birrell, S. J. (2006). Evaluation of nitrate and potassium ion-selective membranes for soil macronutrient sensing. *Transactions of the ASABE, 49*(3), 597–606.

Mulla, D. J. (2013). Twenty five years of remote sensing in precision agriculture: Key advances and remaining knowledge gaps. *Biosystems Engineering, 114*(4), 358–371. https://doi.org/10.1016/j.biosystemseng.2012.08.009.

Padarian, J., Minasny, B., & McBratney, A. B. (2020). Machine learning and soil sciences: A review aided by machine learning tools. *Soil, 6*(1), 35–52. https://doi.org/10.5194/soil-6-35-2020.

Pepers, K. H., van Egmond, F., Koomans, R., Teuling, K., Staats, G., & van Os, G. (2024). Validation of a new gamma ray soil bulk density sensor. *European Journal of Soil Science, 75*(4), e13542. https://doi.org/10.1111/ejss.13542.

Qaswar, M., Kuramae, E. E., & Mouazen, A. M. (2025). Spatial prediction of soil microbial enzyme activities using on-the-go Vis–NIR spectroscopy. *European Journal of Soil Science, 76*(6), e70225. https://doi.org/10.1111/ejss.70225.

Shin, S. K., Lee, S. J., & Park, J. H. (2025). Predictions of soil properties using vis-NIR spectroscopy combined with machine learning: A review. *Sensors, 25*(16), 5045. https://doi.org/10.3390/s25165045.

Smith, S. A., Quinteros, S., Bilici, C., & Gjerde P. E. (2024). Linking multi-sensor core logger data with in-situ and laboratory testing: A North Sea case study. In *Proceedings of the 7th International Conference on Geotechnical and Geophysical Site Characterization* (Vol. 18, p. 21).

Thies, S., Bruggeman, S., Clay, S. A., Mishra, U., Hatfield, G., Kumar, S., & Clay, D. E. (2019). Midmorning point sampling may not accurately represent nitrous oxide emissions following fertilizer application. *Soil Science Society of America Journal, 83*, 339–347. https://doi.org/10.2136/sssaj2018.08.0313.

Tiwari, R., Dwivedi, B. S., Sharma, Y. M., Sharma, A., & Dwivedi, A. K. (2019). Activities of β-glucosidase, phosphatase and dehydrogenase as soil quality indicators: A review. *International Journal of Current Microbiology and Applied Sciences, 8*(6), 834–846. https://doi.org/10.20546/ijcmas.2019.806.101.

Vavlas, N.-C., Porre, R., Meng, L., Elhakeem, A., van Egmond, F., Kooistra, L., & De Deyn, G. (2024). Cover crop impacts on soil organic matter dynamics and its quantification using UAV and proximal sensing. *Smart Agricultural Technology, 9*, 100621. https://doi.org/10.1016/j.atech.2024.100621.

Viscarra Rossel, R. A., Adamchuk, V. I., Sudduth, K. A., McKenzie, N. J., & Lobsey, C. (2011). Proximal soil sensing: An effective approach for soil measurements in space and time. *Advances in Agronomy, 113*, 243–291. https://doi.org/10.1016/B978-0-12-386473-4.00005-1.

Viscarra Rossel, R. A., Behrens, T., Ben-Dor, E., Brown, D. J., Demattê, J. A. M., Shepherd, K. D., Shi, Z., Stenberg, B., Stevens, A., Adamchuk, V., Aïchi, H., Barthès, B. G., Bartholomeus, H. M., Bayer, A. D., Bernoux, M., Böttcher, K., Brodský, L., Du, C. W., Chappell, A., . . . Ji, W. (2016). A global spectral library to characterize the world's soil. *Earth-Science Reviews, 155*, 198–230. https://doi.org/10.1016/j.earscirev.2016.01.012.

Viscarra Rossel, R. A., Lee, J., Behrens, T., Luo, Z., Baldock, J., & Richards, A. (2019). Continental-scale soil carbon composition and vulnerability modulated by regional environmental controls. *Nature Geoscience, 12*(7), 547–552. https://doi.org/10.1038/s41561-019-0373-z.

Proximal Plant Sensing Systems

13

Ce Yang, Alireza Sanaeifar, Lang Qiao, Sravanthi Bachina, Asim Biswas, Robin Gebbers, Viacheslav Adamchuk, Gaurav Jha, Katsutoshi Mizuta, Slawomir Kurpaska, Maciek Gliniak, and Dharani Suresh Babu

Chapter Purpose

This chapter provides an overview of the developments and applications of sensors positioned within 2 m of measured plants. These sensors, referred to as proximal sensing technologies, collect high-resolution data for a variety of purposes, including determining fertilizer recommendations, detecting stress, identifying plant diseases, and phenotyping plants. Key advantages of these sensors are that the information they provide has high spatial resolution and can be collected from individual plants or even from portions of leaves. When combined with tissue sampling or analysis, the sensor data can help confirm underlying yield-limiting factors. Detailed information on various proximal sensors is provided.

Definition of Proximal Plant Sensing

Plant sensing can be conducted at distances ranging from direct contact with a plant to sensors placed in space. This chapter discusses proximal plant sensors that are placed in contact with plants or within a 2-m distance (Figure 13.1). The primary goal of proximal plant sensing is to convert a plant's physical or biochemical signals into data that can guide management decisions. Proximal plant sensors can be used for many purposes, such as agronomists using them to identify limiting factors and breeders using them to select plants with improved resistance to pests (Viscarra Rossel et al., 2011).

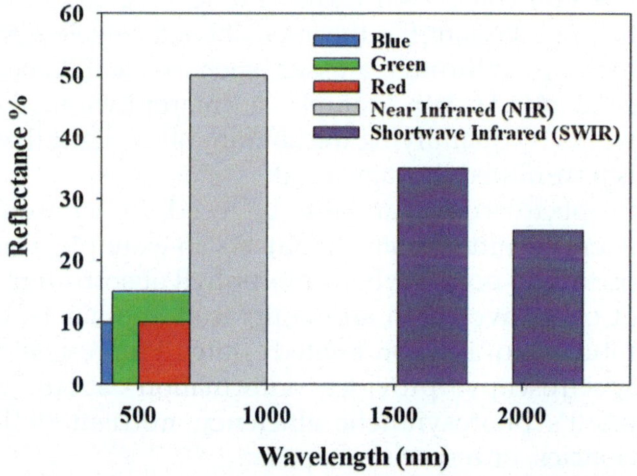

Figure 13.1. Leaf reflectance spectrum showing key absorption features of chlorophyll and plant biochemicals (Adapted from Sanaeifar et al., 2023).

The Electromagnetic Spectrum in Proximal Plant Sensing Systems

Proximal plant sensing systems rely on specific regions of the electromagnetic spectrum to provide information about plant health and physiological status (e.g., photosynthetic activity, metabolism, and stress). For example, plants that appear yellow (chlorotic) have reduced chlorophyll content due to various factors, including nutrient deficiencies (S, Fe, Zn, and N), pesticide damage, water stress, or diseases. Further, plants exposed to high temperature, as measured with an active infrared sensor, may be experiencing water stress (Clay et al., 2006).

The electromagnetic spectrum encompasses all possible frequencies of electromagnetic radiation, including gamma rays, X-rays, ultraviolet, visible light, infrared, microwaves, and radio waves (Chapters 11 and 14). Each category of electromagnetic radiation is characterized by distinct wavelengths. For example, visible light contains blue, green, and red bands with wavelengths of 450–485 nm, 500–565 nm, and 625–750 nm, respectively. When energy within the visible and other bands strikes a target, it can be absorbed, reflected, or transmitted. The specific pattern and amount of reflected radiation across multiple wavelengths form a spectral signal that can be used to identify physiological traits or stress responses in plants.

Spectroscopy

Spectroscopy involves the measurement of light that is reflected, transmitted, or emitted from the surface of an object (Chapters 14 and 15). When spectral radiation strikes a plant, a portion is absorbed by pigments such as chlorophyll, while the remainder is reflected or transmitted (light movement through the leaf). In spectral sensing, transmission is also a measure of relative radiation because it quantifies the intensity of a wavelength that moves through a tissue, such as a leaf (Gates et al., 1965). For example, a chlorophyll meter works by quantifying the amount of red light that is transmitted through a leaf.

Spectroscopy can also be used to measure energy emitted from a plant. As an example, fluorescence occurs when chlorophyll absorbs light at one wavelength and emits it at another as it returns to its non-excited state. Fluorescence spectroscopy provides information about a plant's photosynthetic efficiency, nutrient deficiencies, or responses to pests.

Spectral Analysis of Photosynthesis

Photosynthesis is the process by which most plants, algae, and cyanobacteria convert sunlight, water, and CO_2 into glucose and oxygen. In eukaryotic cells, this transformation occurs in chloroplasts. Absorption refers to the process of "intaking" light. The two primary pigments involved in photosynthesis are chlorophyll a and chlorophyll b. These pigments absorb light at slightly different wavelengths of the spectrum. Chlorophyll a absorbs at ≈429 nm and 659 nm and reflects some blue-green light, while chlorophyll b absorbs most efficiently at ≈455 and 642 nm and transfers that energy to chlorophyll a. Chlorophyll a is universal among photosynthesizing species; chlorophyll b is present in most but not all species. Together, these pigments are responsible for the strong absorption in the blue and red regions of the electromagnetic spectrum, while other pigments, such as carotenoids and xanthophylls, strongly absorb the blue region of the spectrum (~445 nm) (Figure 13.1). Most plants appear green because chlorophyll does not absorb energy at green wavelengths.

The absorption of blue and red energy as measured by a chlorophyll meter has been used to evaluate plant health (Table 13.1). This meter measures the transmittance of light through the leaf and enables an assessment of the chlorophyll potential (Parry et al., 2014). The chlorophyll meter can be used to assess factors that limit photosynthesis.

Multispectral cameras or sensors that provide true (blue, green, and red), false (green, red, and near-infrared [NIR]), and normalized difference vegetation index (NDVI) images of target plants are commonly used to assess their health (Table 13.1). Tractor-mounted sensors, such as GreenSeeker® (Trimble, Inc., Westminster, CO, USA), often use RGB cameras that are specific for red and NIR wavelengths. These sensors emit bursts of red and infrared light and then measure the amount of light at these wavelengths reflected from plant surfaces. Unlike the chlorophyll meter, GreenSeeker does not come into physical contact with the plant (Samborski et al., 2009). The NIR and red reflectance values are calculated with the equation $NDVI = \frac{(NIR - Red)}{(NIR + Red)}$. The NDVI value can indicate plant health by monitoring how much red is absorbed and how much NIR is reflected from the leaf. This ability relies on the fact that the leaf is made up of two layers, one composed of palisade mesophyll cells and the other of spongy mesophyll cells. The palisade mesophyll cells

Table 13.1. Proximal Sensors that are Used in Research and Application.

Sensor	Category	Data Collection	Example of Available Products	Common Management Uses	Cost	Limitations
Chlorophyll or Fluorometer	Handheld	Numerical	Fluor PenPocket PEA; N-sensor	Nitrogen or drought stress	Medium	Not suitable for large field applications
SPAD Meter	Handheld	Numerical	MinoltaSPAD 502	Nitrogen stress and general plant health	Low	Requires calibration
Thermal Sensor or Imager	Handheld, vehicle mounted	Numerical	FLIR; Tyledyne	Plant canopy temperature	Medium	Requires calibration
NDVI or other VI Sensor	Handheld, vehicle mounted	Numerical	GreenSeeker; CropcCircle; ISARIA	Vegetation detection, plant stress, plant health	Medium to High	Best return on investment in large farms
RGB and Multispectral Cameras	Handheld, vehicle mounted	Imagery of visible reflectance bands	DSLR, smartphone, UAV-mounted cameras	General plant health or phenotyping	Low	Better for later growth stages
Hyperspectral	Lab equipment or handheld	Reflectance bands within 350-2500 nm	ASD Field SpecSpectra Vista	Nutrient mapping or general plant health	High	Extensive data analysis
Lidar	Vehicle mounted	3D structures	VelodyneRIEGL	Plant height or to quantify lodging	High	Computationally intensive data analysis
Solar Induced Fluorescence	Vehicle mounted or tower mounted	Chlorophyll fluorescence	FLoX BoxOCO-3	Photosynthetic rate measurement or transpiration rate	High	High set up cost and labor intensive
Raman	Lab equipment or handheld	Numerical	Ocean Insight QEPro-Raman	Monitor plant metabolites and nutrient levels	High	Requires careful lab sample preparation
LIBS	Lab equipment or handheld	Wavelength intensity	SciAps Z-300 LIBS AnalyzerApplied Spectra J200 LIBS	Identify heavy metal contamination and assess soil nutrients	High	Interpretation varies by cell composition requiring expertise
Electrochemical	Probe	Numerical	Atlas Scientific Nitrate sensorSentek Drill & Drop Probe	Measure metabolites in plant tissue	High	Requires many samples to make conclusions
Terahertz spectroscopy	Lab equipment	Numerical	TeraView 4000Advantest TAS7500	Mapping internal features of the plant to track water loss	High	Requires advanced modeling techniques

contain the plant chlorophyll, and the spongy mesophyll cells create large air pockets. The energy of wavelengths such as NIR bounces around and can be scattered in such air pockets, which means healthy plants can reflect more NIR. If a plant is unhealthy, the mesophyll will collapse, reducing NIR reflectance. While there are nuances to the use of NDVI, it is generally understood that high values indicate good plant health and low values indicate poor health.

In addition to reflectance-based sensors such as GreenSeeker®, recent advances in low-cost digital imaging have expanded the role of RGB cameras as practical proximal sensing tools for continuous plant monitoring. Automated imaging systems such as phenocams and digital cameras have improved temporal resolution and reduced observer bias in phenological research (Crimmins & Crimmins, 2008; Richardson, 2019). Commercially available wildlife trail cameras represent a cost-effective and weather-resistant extension of this concept, enabling repeated, nondestructive canopy observation under field conditions (Newey et al., 2015). When mounted at canopy height and configured for time-lapse acquisition, these systems provide high-frequency observations of canopy structure, leaf coloration, and growth dynamics throughout the season. Unlike handheld meters that provide single-point measurements, time-lapse RGB systems capture both spatial and temporal variability, supporting whole-canopy assessment in heterogeneous cropping systems.

RGB imaging systems, while limited to the visible portion of the electromagnetic spectrum, still capture valuable information related to chlorophyll concentration, pigment composition, and canopy structure. Variations in red, green, and blue reflectance can be used to derive vegetation indices such as the green chromatic coordinate (GCC) and other color-based metrics that serve as proxies for canopy greenness and phenological stage (Sonnentag et al., 2012). These indices offer a practical and accessible means of tracking plant development and stress responses at high temporal resolution. In this way, RGB systems complement multispectral and hyperspectral platforms by emphasizing frequency of observation and operational scalability over spectral granularity.

The effectiveness of RGB-based proximal sensing depends on the objective evaluation of image quality. Field conditions introduce variability in illumination, moisture, and wind, which can affect image clarity and color fidelity. Quantitative metrics such as blur (image sharpness), color cast (deviation from true color balance), and noise (random signal variation) can be used to standardize performance across camera models.

Measuring Nitrogen and Water Stress

NDVI and chlorophyll meters are often used to assess the magnitude of N stress in plants. The magnitude of N stress is determined by dividing the NDVI or chlorophyll value in an N-limited zone by the value in a well-fertilizead zone using the equation $NSI = \frac{N(Limited\ zone)}{N(Well-fertilized\ zone)}$. This ratio is often called the N sufficiency index (NSI; Hawkins et al., 2007). If the NSI is <0.9, additional N may be required. However, factors other than N, such as disease or water stress, can affect the NSI ratio; therefore, users must always consider variability within the system to make the best management decisions (Blackmer & Schepers, 1994).

When using these sensors, it is important to follow appropriate protocols to convert the information into sound management decisions. These protocols are often provided by the manufacturer. For example, when a handheld chlorophyll meter is used to evaluate N response in corn, the recommendation is to sample 30–40 plants on the uppermost fully expanded leaf (halfway between the base and the tip). Additionally, corn rapidly takes up N between the V8 (eight collared leaves) and VT (tassel) vegetative stages. Consequently, the best time to measure N response in corn may be closer to V8, ensuring that a management decision can be made and implemented before the rapid N uptake period closes.

To evaluate water stress, plant temperature, gas exchange, or both are measured through proximal sensing. This method is used because high temperatures lead to greater water loss through the stomata (leaf openings used for gas exchange). Thermal sensors measure absolute temperature or relative temperature differences in a plant. The assumption underlying this measurement is that a well-watered plant is better able to regulate its temperature through transpiration, the movement of water through the plant. In contrast, a water-stressed plant is more likely to close its stomata, reducing the transpiration rate and causing canopy temperature to rise. Owing to the simplicity of this technique, thermal sensors are among the oldest sensor systems used in agriculture to measure water stress. Thermal sensors are now available as electronic thermometers called thermocouples, which are relatively inexpensive and can be attached to a leaf to monitor heat and water stress (Table 13.1). These sensors are often integrated into larger sensor systems to monitor temperature effects.

Heat pulse sensor systems can also be used to measure plant temperature, which serves as an indirect indicator of water stress. These sensors measure the direction of sap flow in the plant stem (Miner et al., 2017). Two thermometers are placed below and above a heat source, and the heated water in the xylem moves upward or downward past one of them. The time it takes for the heat to reach each thermometer indicates the rate and direction of sap flow, providing insight into the plant's water transport and stress status.

Water stress can also be measured using a porometer (Rusconi et al., 2025). Rusconi et al. (2025) showed that porometer measurements between 50 and 300 mmol $\cdot$ m^{-2} $\cdot$ s^{-1}) are highly correlated with stomatal conductance measured with an infrared gas analyzer. Specifically, water-stressed plants have lower values, whereas well-watered plants have higher values. However, Rusconi et al. (2025) also reported that porometer values can have high variability, a limitation that can be offset by increasing the number of replications.

While thermometers and porometers are sensors that must be in direct contact with an object, infrared thermometers do not require physical contact with a leaf. They contain a lens that focuses the radiation onto a detector sensitive to wavelengths of about 700–1,400 nm. Thermal cameras can cover a similar spectral range of about 800–1,400 nm with pixel-scale information. Owing to the high-resolution data they produce, these cameras are typically used in research to assess water stress in crops (Zia et al., 2011). However, thermal imaging in outdoor conditions can be challenging because of environmental factors such as wind and cloud cover.

Many trail cameras also include embedded infrared temperature sensors that record ambient temperature alongside each image. Although not designed as precision thermometric instruments, these sensors have been shown to reliably track diurnal temperature patterns and correlate strongly with canopy-level thermistor measurements (Swann et al., 2004; Welbourne et al., 2016). Temperature monitoring plays a critical role in evaluating plant water stress and canopy energy balance (Jones et al., 2009). Comparisons with canopy-level probes and nearby weather stations, as well as trail-camera infrared sensors, indicate that while absolute temperature values may differ due to housing effects and radiation exposure, temporal trends are consistently preserved. This dual functionality, simultaneous RGB image acquisition and temperature logging, positions trail cameras as integrated proximal sensing platforms capable of supporting both phenotypic assessment and microclimate monitoring.

Fluorescence Sensing

Electromagnetic radiation that strikes a plant can be used to drive photosynthesis, be emitted as heat, be transmitted through plant tissue, or be remitted as light. The fate of electromagnetic radiation provides clues about a plant's photosynthetic efficiency. When a chlorophyll molecule absorbs the energy of a photon, it becomes excited, and as it returns to its normal state, it releases some of that energy as light. This emitted energy is called fluorescence. Fluorescence sensing (Table 13.1) is based on the detection of this energy. The amount of fluorescence is influenced by the plant species and the stress the plant is undergoing. A study by Soufi et al. (2025) provides an example of how fluorescence sensing can be used. They reported that reduced fluorescence can indicate drought stress, which can be alleviated by applying N fertilizer.

Maximum fluorescence intensity occurs at a wavelength just beyond the red absorption peak of the chlorophyll. For shorter wavelengths (before the peak), the fluorescence intensity declines rapidly, whereas at longer wavelengths, the intensity decreases more slowly, creating a long tail that extends into the NIR region.

A smaller, secondary fluorescence peak also appears around 730 nm. Fluorescence sensors work by exciting plant molecules with light at a specific wavelength, causing them to emit light at a different, longer wavelength when they return to their ground state. The change in fluorescence intensity is a signal that is directly related to the concentration of certain molecules. For example, in agriculture, fluorescence sensing has emerged as a powerful tool for providing early indicators of drought, nutrient deficiency, and disease. Portable fluorometers, handheld devices that rapidly assess crop health across large fields, can produce spatial maps of photosynthetic efficiency to guide targeted agronomic interventions. In addition, as sensor miniaturization and data processing capabilities continue to advance, fluorescence-based sensing systems are expected to become integral components of smart farming, enabling automated crop monitoring and data-driven agricultural decision-making.

Raman Spectroscopy

Raman spectroscopy (Table 13.1) is a proximal sensing technique that provides a detailed analysis of a plant's chemical and physical properties. In this technique, a laser is used to scatter molecules (known as inelastic scattering). Each molecule has a characteristic vibrational energy state, which in turn provides a fingerprint for that molecule. This technology has the potential to allow nondestructive in-field analysis of fruit ripeness, grain quality, and nutritional content.

Raman analysis can also be used for evaluating plant metabolite activities and discriminating between proteins, lipids, carbohydrates, flavonoids, and alkaloids (Chandra et al., 2024). However, this analysis is only possible after careful sample preparation, since fresh plant biomass and soils contain substances that create a "fluorescence background" that often masks other spectral features. Work is being done to overcome these limitations.

Laser-induced Breakdown Spectroscopy

Laser-induced breakdown spectroscopy (LIBS) is a powerful analytical tool that enables fast on-site elemental analysis. In LIBS, a high-energy laser pulse is focused on a minimal area of plant tissue. The laser energy leads to rapid local heating and ablation (removal of the thin surface layer) of the sample, producing a transient micro-plasma. As this plasma cools, the excited atoms and ions relax back to their ground states and emit light at wavelengths that are characteristic of their elemental identities (Miziolek et al., 2006; Singh & Thakur, 2020). The plasma is rich in electrons, and its emission spectrum can be recorded with a spectrometer coupled to a sensitive detector. Because each element emits light at different wavelengths, the resulting spectrum acts as a fingerprint of the sample's elemental composition. This feature is especially useful in plant studies, with a single measurement being sufficient to reveal information about the primary nutrients and trace elements (Hahn & Omenetto, 2012).

An advantage of LIBS is that it requires minimal sample preparation, unlike conventional analytical methods that may involve prolonged digestion or extraction. LIBS can analyze solid, liquid, and gaseous samples with minimal delay after ablation. In the context of plant analysis, tissue samples may be evaluated immediately in the field, giving the near-real-time data needed for quick management decisions. LIBS has been applied to monitor heavy metal contamination, such as cadmium accumulation in rice tissues, and thus improve food safety and safety remediation measures (Xie et al., 2015).

Although speed and simplicity are strong points, there are technical challenges to LIBS. For example, the formation and development of laser-induced plasma can depend on the physical and chemical properties of the plant tissue. Differences in moisture, cell structure, or other factors that may form can make interpreting spectral data challenging. LIBS is a potent and versatile tool for proximal plant sensing, offering fast, accurate, and minimally destructive analysis. With ongoing improvements, LIBS is positioned to deepen our understanding of plant health.

Terahertz Spectroscopy and Imaging

Terahertz (THz) spectroscopy (Table 13.1) has emerged as a promising tool because it offers nondestructive on-site analysis of plant tissues. THz radiation has a frequency range of 0.1–10 THz, which lies between the microwave and infrared regions of the electromagnetic spectrum. The THz technique is highly effective for detecting small changes in plant hydration, and it allows researchers to monitor leaf water status in real time without damaging the plant (Tonouchi, 2007). Unlike conventional optical techniques, THz radiation is non-ionizing with low photon energy and is thus safe for biological samples (Ferguson & Zhang, 2002).

In a typical THz time-domain spectroscopy arrangement, a broadband THz pulse produced by short laser pulses is directed onto the plant sample. The interaction between the THz wave and the tissue results in an amplitude and phase shift in the THz pulse. These changes are recorded and examined to derive the refractive index and absorption coefficient, which are related to the tissue water content (Wang et al., 2017). For instance, THz imaging has proven to be an effective way to track water loss in drought-stressed leaves and provides a quick and noninvasive way to direct irrigation decisions (Li et al., 2020). Further, advancements in source and detector technologies have improved the spatial resolution and speed of the THz imaging systems. Together with data processing techniques, including machine learning and chemometrics, THz imaging is now powerful enough to identify small structures in plant tissues. This integration makes mapping vascular bundles, cell layers, and other internal features easier, which is essential for understanding plant physiology and optimizing crop management (Afsah-Hejri et al., 2020).

Overall, THz spectroscopy and imaging offer a unique combination of safety, sensitivity, and speed for proximal plant sensing. With continuous improvements in instrumentation and the evolution of appropriate data analysis techniques, THz methods are expected to gain increasing prominence in noninvasive monitoring of plant health.

Electrochemical Sensors

Electrochemical sensors convert chemical signals used by the plant, such as hydrogen peroxide, into electrical signals. These chemical signals might include oxidation-reduction reactions, ion migration, and charge transfer processes at the electrode surface (Liu et al., 2025). Electrochemical sensors can be fabricated as flexible electrode materials that adhere to plant surfaces, enabling continuous monitoring of key physiological indicators in plant tissues without affecting normal plant growth. Meanwhile, by integrating multiple electrodes and signal processing units, simultaneous detection of multiple targets is possible (Xu et al., 2025). In addition, real-time monitoring of the plant signaling molecules can help managers make decisions about plant diseases, pests, or environmental stress.

Three-dimensional Reconstruction of Plant Objects

Three-dimensional (3D) reconstruction is a high-resolution technique for estimating leaf area, plant height, canopy architecture, and biomass. One common strategy for 3D reconstruction involves the use of LiDAR (Table 13.1). The acronym stands for "light detection and ranging." A LiDAR system is a distance sensor that includes a laser and an optical sensor. The laser emits a beam of light with high energy density. When the light is reflected from a target it bounces back to the sensor. The distance between the sensor and the target is then related to the time required for the laser to travel to the target and back.

LiDAR technology can be used to determine the way biotic and abiotic stresses affect plant growth and development. For example, in an examination of fruit trees, a detailed 3D model can be helpful in deciding if a fungicide treatment is needed (Figure 13.2). LiDAR sensors have also been extensively tested in vineyards and tree crops to estimate leaf shape and area as an input for spraying and other measures. Additional details for this technique and how its use in agriculture are provided by a previous study (Farhan et al., 2024).

Conclusion

Overall, proximal plant sensors have changed the way growers and researchers collect and analyze plant data, allowing them to make rapid decisions about fertilizer recommendations, assess plant stress, identify plant diseases, and determine phenotype up close and in the field. Research on these sensors has yielded information that can be used to identify detailed biochemical and physical changes in plants, thereby

Figure 13.2. Three-dimensional structure of corn obtained by LiDAR mounted on drone (With permission of Asim Biswas).

improving understanding of stressors. In addition to advances in spectral, thermal, and structural sensing technologies, the emergence of affordable time-lapse RGB imaging systems has expanded the practical accessibility of continuous, non-destructive canopy monitoring. This information has increased, and will continue to increase, the ability of growers to manage agriculture in a site-specific, cost-effective, and sustainable way.

Chapter Questions

1. State some examples of active and passive sensors.
2. Explain how a transmittance-based chlorophyll meter differs in function from reflectance-based sensors such as GreenSeeker?
3. How can THz spectroscopy distinguish between healthy and drought-stressed leaves?
4. In what way does LiDAR outperform RGB imaging for 3D canopy reconstruction in high-density crop systems?
5. Name one of the plants' signaling molecules detected by electrochemical sensors.
6. Which of the following is a key advantage of proximal sensing over satellite sensing?
   a. Lower cost
   b. High spatial resolution
   c. Larger area coverage
   d. Wide spectral coverage
7. What is the principle behind Raman spectroscopy?
   a. Emission of fluorescence per millisecond
   b. Reflectance of light energy per millisecond
   c. Inelastic scattering of monochromatic light per millisecond
   d. Elastic scattering of light per millisecond
8. What is the ratio used to calculate nitrogen sufficiency index (NSI)?

ACKNOWLEDGMENTS

Support for this document was provided by the Precision Farming Systems community in the American Society of Agronomy, South Dakota State University, USDA-NRCS-CIG (NR233A750011G019), USDA-NIFA (Multistate Project S1090) Program under (7008110), American Coalition for Ethanol, the South Dakota Corn Utilization Council, and the South Dakota Soybean Research and Promotion Council. The use of trade names is for the convenience of the reader and does not imply endorsement by the authors.

REFERENCES

Afsah-Hejri, L., Akbari, E., Toudeshki, A., Homayouni, T., Alizadeh, A., & Ehsani, R. (2020). Terahertz spectroscopy and imaging: A review on agricultural applications. *Computers and Electronics in*

Agriculture, *177*, 105628. https://doi.og/10.1016/j.compag.2020.105628.

Blackmer, T. M., & Schepers, T. S. (1994). Techniques for monitoring crop nitrogen status in corn. *Communications in Soil Science and Plant Analysis*, *25*, 1791–1800.

Chandra, A., Kumar, V., Chandra, U., Dada, R., Qamar, I., Goel, V. K., & Agarwal, S. (2024). Unveiling the molecular secrets: A comprehensive review of Raman spectroscopy in biological research. *ACS Omega*, *9*, 50049–50063.

Clay, D. E., Kim, K.-I., Chang, J., Clay, S. A., & Dalsted, K. (2006). Characterizing water and N stress in corn using remote sensing. *Agronomy Journal*, *98*, 579–587.

Crimmins, T. M., & Crimmins, M. A. (2008). Monitoring plant phenology using digital repeat photography. *Environmental Management.*, *41*, 949–958. https://doi.org/10.1007/s00267-008-9086-6.

Farhan, S. M., Yin, J., Chen, Z., & Memon, M. S. (2024). A comprehensive review of LiDAR applications in crop management for precision agriculture. *Sensors (Basel)*, *24*(16), 5409. https:/doi.org/10.3390/s24165409.

Ferguson, B., & Zhang, X. C. (2002). Materials for terahertz science and technology. *Natural Materials*, *1*, 26–33. https://doi.org/10.1038/nmat708.

Gates, D. M., Keegan, H. J., Schleter, J. C., & Weidner, V. R. (1965). Spectral properties of plants. *Applied Optics*, *4*, 11–20. https://doi.org/10.1364/AO.4.000011.

Hahn, D. W., & Omenetto, N. (2012). Laser-induced breakdown spectroscopy (LIBS), part II: Review of instrumental and methodological approaches to material analysis and applications to different fields. *Applied Spectroscopy*, *66*, 347–419. https://doi.org/10.1366/11-06574.

Hawkins, J. A., Sawyer, J. E., Barker, D. W., & Lundvall, J. P. (2007). Using relative chlorophyll meter values to determine nitrogen application rates for corn. *Agronomy Journal*, *99*, 1034–1040.

Jones, H. G., Serraj, R., Loveys, B. R., Xiong, L., Wheaton, A., & Price, A. H. (2009). Thermal infrared imaging of crop canopies for the remote diagnosis and quantification of plant responses to water stress in the field. *Functional Plant Biology.*, *36*, 978–989. https://doi.org/10.1071/FP09123.

Li, B., Wang, R., Ma, J., & Xu, W. (2020). Research on crop water status monitoring and diagnosis by terahertz imaging. *Frontiers in Physics*, *8*, 571628. https://doi.org/10.3389/fphy.2020.571628.

Liu, W., Zhang, Z., Geng, X., Tan, R., Xu, S., & Sun, L. (2025). Electrochemical sensors for plant signaling molecules. *Biosensor and Bioelectronics*, *267*, 116757. https://doi.org/10.1016/j.bios.2024.116757.

Miner, G. L., Ham, J. M., & Kluitenberg, G. J. (2017). A heat-pulse method for measuring sap flow in corn and soybean using 3D-printed sensor bodies and low-cost electronics. *Agricultural and Forest Meteorology*, *246*, 86–97.

Miziolek, A. W., Palleschi, V., & Schechter, I. (2006). *Laser induced breakdown spectroscopy*. Cambridge University Press. https://doi.org/10.1017/CBO9780511541261.

Newey, S., Davidson, P., Nazir, S., Fairhurst, G., Verdicchio, F., Irvine, R. J., & Van Der Wal, R. (2015). Limitations of recreational camera traps for wildlife management and conservation research: A practitioner's perspective. *Ambio.*, *44*, 624–635. https://doi.org/10.1007/s13280-015-0713-1.

Parry, C., Blonquist, J. M., & Bugbee, B. (2014). In situ measurement of leaf chlorophyll concentration: Analysis of the optical/absolute relationship. *Plant Cell & Environment*, *37*, 2508–2520.

Richardson, A. D. (2019). Tracking seasonal rhythms of plants in diverse ecosystems with digital camera imagery. *New Phytologist.*, *222*, 1742–1750. https://doi.org/10.1111/nph.15591.

Rusconi, C., Confalonieri, R., Movedi, E., Gazzoli, A., Arrigoni, G., Brocco, G., Consentino, A. D., Foglia, T., Lombardo, F., Mandelli, B., Pavasini, M., Pigni, G., & Paleari, L. (2025). Porometer for estimating stomal conductance in maize: Determination of trueness and precision according the ISO 5725. *Biosystems Engineering*, *250*, 158–162.

Samborski, S. M., Trembley, T., & Fallon, E. (2009). Strategies to make use of plant sensor-based diagnostic information for nitrogen recommendations. *Agronomy Journal*, *101*, 800–816.

Sanaeifar, A., Yang, C., de la Guardia, M., Zhang, W., Li, X., & He, Y. (2023). Proximal hyperspectral sensing of abiotic stress in plants. *Science of the Total Environment*, *861*, 160652.

Singh, J. P., & Thakur, S. N. (2020). *Laser-induced breakdown spectroscopy*. Elsevier.

Sonnentag, O., Hufkens, K., Teshera-Sterne, C., Young, A. M., Friedl, M., Braswell, B. H., & Richardson, A. D. (2012). Digital repeat photography for phenological research in forest ecosystems. *Agricultural and Forest Meteorology.*, *152*, 159–177. https://doi.org/10.1016/j.agrformet.2011.09.009.

Soufi, M. H., Mobasser, H. R., Mehraban, A., Ganjali, H. R., & Miri, K. (2025). Corn (*Zea mays* L.) production and chlorophyll fluorescence affected by drought stress, N fertilization, and genotype. *Journal of Trace Elements and Minerals*, *12*(100237).

Swann, D. E., Hass, C. C., Dalton, D. C., & Wolf, S. A. (2004). Infrared-triggered cameras for detecting wildlife: An evaluation and review. *Wildlife Society Bulletin.*, *32*, 357–365.

Tonouchi, M. (2007). Cutting-edge terahertz technology. *Nature Photonics*, *1*, 97–105. https://doi.org/10.1038/nphoton.2007.3.

Viscarra Rossel, R. A., Adamchuk, V. I., Sudduth, K. A., McKenzie, N. J., & Lobsey, C. (2011). Proximal soil sensing: An effective approach for soil measurements in space and time. *Advances in Agronomy*, *113*, 237–282. https://doi.org/10.1016/B978-0-12-386473-4.00005-1.

Wang, K., Sun, D.-W., & Pu, H. (2017). Emerging non-destructive terahertz spectroscopic imaging technique: Principle and applications in the agri-food industry. *Trends in Food Science and Technology*, *67*, 93–105. https://doi.org/10.1016/j.tifs.2017.06.001.

Welbourne, D. J., Claridge, A. W., Paull, D. J., & Lambert, A. (2016). How do passive infrared triggered camera traps operate and why does it matter? Breaking down common misconceptions. *Remote Sensing in Ecology and Conservation.*, *2*, 77–83. https://doi.org/10.1002/rse2.20.

Xie, P., Deng, J., Zhang, H., Ma, Y., Cao, D., Ma, R., Liu, R., Liu, C., & Liang, Y. (2015). Effects of cadmium on bioaccumulation and biochemical stress response in rice (*Oryza sativa* L.). *Ecotoxicology and Enviromental Safety*, *122*, 392–398. https://10.1016/j.ecoenv.2015.09.007.

Xu, S., Huang, X., Liang, X., & Lu, H. (2025). Application of wearable sensors in crop phenotyping and microenvironment monitoring. *Chemical Engineering Journal*, *505*, 159059. https://doi.org/10.1016/j.cej.2024.159059.

Zia, S., Spohrer, K., Wenyong, D., Spreer, W., Romano, G., Xiongkui, H., & Muller, J. H. (2011). Monitoring physiological responses to water stress in two maize varieties by infrared thermography. *International Journal of Agricultural and Biological Engineering*, *4*, 7–15.

Remote Sensing for Site-specific Crop Management

14

Richard Ferguson, Brian Wardlow, Jackson Stansell, Gaurav Jha, Leonardo Bastos, Guillermo Balboa, and Sheetal Kumari

Chapter Purpose

This chapter is a revision of Ferguson and Rundquist (2018) and provides an overview of the development of the remote sensing tools used to assess crop and soil health. Today, a wide variety of sensing options that allow close observation of the crop are available to crop producers and researchers alike. Remote sensing information can be collected by unmanned aerial vehicles, manned aircraft, and satellite platforms.

Definition of Remote Sensing

In the fifth century B.C.E., Socrates is said to have stated that "man must rise above the Earth—to the top of the atmosphere and beyond—for only thus will he fully understand the world in which he lives." This amazing bit of acumen could not be implemented until man developed the means of rising above the terrestrial surface: first, by lighter-than-air balloon technology, then by airplanes and satellites. Indeed, once man developed the means of studying the Earth from aerial and satellite perspectives, "remote sensing" was born, and science advanced in a multitude of new directions. The term remote sensing was coined in a white paper prepared by the staff of the Geography Branch of the Office of Naval Research in the early 1960s. The chief architect of that paper was Evelyn L. Pruitt, who was searching for terminology to include regions of the electromagnetic spectrum beyond the visible range. In addition, she felt that the term "aerial" was too limited in view of the potential for viewing the Earth from space. Thus, the term remote sensing was introduced (Pruitt, 1979).

What exactly is remote sensing? Many individuals have attempted to describe the nature, scope, and meaning of the term (Fussell et al., 1986), but one representative definition suggests that it is "the measurement or acquisition of information of some property of an object or phenomenon, by a recording device that is not in physical contact with the object or phenomenon under study" (Colwell, 1983). This lengthy description can be simplified by recognizing that remote sensing consists of

- Technology and techniques (for data collection and sometimes mapping)
- Data collection done by means of instrumentation (sensors)

- Sensors carried on "platforms" (i.e., satellites, manned and unmanned aircraft, and in situ/ground-based applications)
- Platforms positioned at a distance (i.e., non-contact, nondestructive recording at distances ranging from a few meters to many thousands of kilometers)
- Sensors operating in various parts of the electromagnetic spectrum (e.g., using visible light, infrared energy).

"Proximal sensing" (Chapters 11, 12, and 13) is very similar to remote sensing, with one exception: The sensors are positioned very close to the target, ranging from those in physical contact with the target to a few meters away. Proximal sensing includes investigators carrying instruments into the field for data collection (and deploying them in hand-held fashion), and sensors mounted on farm implements or other mechanical devices, such as all-terrain vehicles.

Figure 14.1. A panchromatic (black & white) aerial image from a study site located in Nebraska. (Courtesy University of Nebraska-Lincoln Conservation and Survey Division).

A Brief History of Remote Sensing in Agriculture

Remote sensing has long been used to monitor and analyze agricultural activities (Goodman, 1959). In the United States, general crop inventories using aerial photographs in the 1920s and '30s by the U.S. Department of Agriculture (USDA), and the U.S. Soil Conservation Service (now the Natural Resources Conservation Service [NRCS]) used aerial imagery in soil survey mapping. Many of these images were like the panchromatic image shown in Figure 14.1.

Things changed dramatically with the launch of the first Earth-resources–oriented satellites in the early 1970s, when it became possible to monitor agricultural lands over broad geographic areas. For example, the Large Area Crop Inventory Experiment (LACIE) was the first U.S. government–sponsored program aimed at examining the feasibility of using satellite data, specifically from the Landsat-1 vehicle, to estimate wheat production over large regions of the world (Erickson, 1984). The idea was proposed by the National Research Council in 1960, and with the 1972 launch of the first Landsat sensor, their proposal became a reality. The LACIE program was operated jointly by NASA, NOAA, and USDA. During 1974–1975, the emphasis of the work was on developing multi-date spectral "signatures" for wheat and yield-estimation models for the U.S. Great Plains. Subsequently, activity was expanded to include Canada and the Soviet Union. Numerous other satellite-based initiatives followed the successes of LACIE, including the USDA's modern-day GEO Global Agricultural Monitoring (GEOGLAM) program that integrates various satellite observations for global monitoring of different crop type conditions (Whitcraft et al., 2015).

Platforms: Aircraft, Satellite, or Unmanned Aerial Vehicles

Today, sensors collect remote sensing data using piloted aircraft, satellites, and unmanned aerial vehicles (UAVs) operating at up to thousands of kilometers above the terrestrial surface. Information collected from these systems is useful to those in many disciplines, including agronomists, plant breeders, biologists, and farmers and their advisors (Clay et al., 2024). A partial listing of current remote sensing satellites is provided in Table 14.1.

Sensor information, when combined with advanced machine learning and artificial intelligence (AI) analytical techniques, improves prediction accuracy at a fraction of the cost of human-collected information. This information is used to assess crop health, phenology, pest or disease pressure, and fertilizer requirements at an operational scale that was not feasible in 2010 (Mogili & Deepak, 2018; Rejeb et al., 2022). A discussion about advanced machine learning and AI analytical techniques is beyond the scope of this chapter and is available in Chapter 27.

Imaging or Non-imaging Sensors

Sensors used in data collection can be imaging or non-imaging. Imaging sensors measure

Table 14.1. Selected and unclassified satellite information: currently available satellite remote sensing platforms.

Name	Launched	Bands	Resolution (meters)	Revisit frequency (days)[a]
ASTER	Dec. 1999	14 bands	VNIR (15), SWIR (30), TIR (90)	16
EO-1 Hyperion	2000	Hyperspectral (400–2500 nm)	30	16
EROS A	Dec. 2000	Panchromatic	2	3–6
FORMOSAT 2	May 2004	Panchromatic, 4-band MS	2 pan, 8 MS	1
EROS B	April 2006	Panchromatic	0.5	3–6
WorldView 1	Sept. 2007	Panchromatic	0.5	1–5
GeoEye 1	Sept. 2008	Panchromatic, 4-band MS	0.41 pan, 1.65 MS	2–8
Deimos 1	July 2009	3-band MS	22	2–3
WorldView 2	Oct. 2009	8 bands	0.46 pan, 1.85 MS	1–4
Pleiades 1	Dec. 2011	Panchromatic, 4-band MS	0.5 pan, 2 MS	1–13
Kompsat 3	May 2012	Panchromatic, 4-band MS	0.7 pan, 2.8 MS	1–4
Landsat 8	Feb. 2013	11 bands	15 pan, 30 MS	16
Landsat 9	Sept. 2021	9 bands	15 pan, 30 MS	16
SkySat 1,2	Nov. 2013, July 2014	4-band MS	2	1
SkySat 4–7	2016	8-band MS	2	1
SkySat 8–13	2017	8-band MS	2	1
SPOT 7	June 2014	Panchromatic, 4-band MS	2.2 pan, 8.8 MS	1
WorldView 3	Aug. 2014	29 bands	0.31 pan, 1.24 MS	1–5
TripleSat	July 2015	Panchromatic, 4-band MS	0.8 pan, 3.2 MS	1
Sentinel 2 A, B, C	2015	12-band MS	10–30	10
FORMOSAT 5	Aug. 2017	4-band MS	2 pan, 4 MS	
Pleiades Neo	2021, 2022	6-band MS	0.3 pan, 1.2 MS	1
PlanetScope	2018	8-band MS	3.0 MS	1

Note: The table provides information about the sensor (name), when launched, bands, resolution, and revisit time.
Abbreviations: ASTER, Advanced Spaceborne Thermal Emission and Reflection Radiometer; EO-1, Earth Observing-1; EROS A, Earth Resources Observation Satellite A; FORMOSAT 2, FORMOSA Satellite 2; MS, multispectral; Pan, panchromatic (black and white); SPOT 7, Satellite pour l'Observation de la Terre 7; SWIR, short-wave infrared; TIR, thermal infrared; VNIR, visible to near infrared.
[a] Revisit times can be dependent on the degree of deviation from the nadir view.

radiation intensity as a function of location, enabling the production of a two-dimensional (2D) image. Examples produced by imaging sensors are shown in Figures 14.1 and 14.2. Non-imaging sensors either do not measure radiation intensity or do not measure it as a function of location, and so they do not produce an image. Today, most remote sensing data is collected in a non-imaging digital format that may contain information from multiple sections of the electromagnetic spectrum (Rundquist & Samson, 1988). This information can be used to create single- or multiband images or processed using analytical techniques to provide information for crop breeding or to identify yield-limiting factors.

Active and Passive Sensors

Sensors are classified as either passive or active. Passive sensors use naturally available energy, such as sunlight. Ordinary digital cameras measure reflected sunlight, using detectors sensitive to the red, green, and blue (RGB) portions of the visible spectrum, to produce conventional color photographs. Thus, the digital camera is a passive sensor. Other types of energy, such as heat, exist naturally in the environment and can be sensed. Multispectral cameras, capable of capturing RGB, as well as non-visible wavelengths, such as red edge and near infrared (NIR), and hyperspectral cameras, capable of capturing wavelengths of between 250 and 15,000 nm, are other examples of passive sensors. However, they are limited by light availability and are restricted to daylight conditions for optimal results.

Figure 14.2. Color-infrared image, South Central Research Station, University of Nebraska, Clay County, Nebraska, circa 1983. (Courtesy the Center for Advanced Land Management Information Technologies [CALMIT], University of Nebraska-Lincoln).

Active sensors, on the other hand, produce their own energy (at specific wavelengths) to illuminate the target and then measure the reflected signal. An advantage of active sensors is that they can operate independently of sunlight, making them less affected by weather conditions. There are also handheld sensors in the market capable of measuring vegetation indices, such as the Normalized Difference Vegetation Index (NDVI; GreenSeeker®) or both NDVI and the Normalized Difference Red Edge (NDRE; Rapidscan®). Radar and Lidar (light detection and ranging) are other examples of active sensors. Radar operates in the microwave portion of the spectrum, generating thousands of microwave energy pulses per second, and part of those energy pulses are reflected by ground targets at various levels of intensity. The energy that is reflected and returned from the target to the sensor system is recorded and used to produce an image of radar backscatter. Lidar is another type of sensor that uses a laser to measure distance from the sensor to the target. Lidar can be used to determine the 3D structure of a plant.

Sensors: Making Use of Emitted and Reflected Energy

The digital camera and radar system are good examples of sensors that measure reflected energy. Reflected energy consists of the wavelengths from electromagnetic radiation that are not absorbed by an object but instead bounce off its surface. Reflected energy can only be measured when there are wavelengths, such as in sunlight, that bounce off a surface. Both camera and radar systems depend on reflectance for data collection, and reflectance can be analyzed in a non-imaging manner or used to produce various types of images. Other sensors make use of emitted rather than reflected energy. Emitted energy, which is different from reflected energy, can be measured at any time and is measured as the radiation that comes from an object's energy, for example, temperature. Many of these sensors detect energy at specific wavelengths that are naturally emitted by all terrestrial targets. The Earth, for example, gives off thermal energy (i.e., heat) at specific wavelengths, and thermal infrared sensors are tuned to capture those wavelengths that can produce images of surface temperature patterns across portions of the Earth's surface (e.g., across a field of corn as a means of examining the health of the canopy).

Optical and Radio Remote Sensing Sensors

Optical sensing generally collects reflected sunlight from crops and soil across the visible (400–700 nm), NIR (700–1400 nm), and shortwave infrared (SWIR; 1400–3000 nm) wavelengths. The sensors used to collect this information can be placed on ground-based, aerial, and space-based platforms. Research has shown that these wavelengths are particularly effective for analyzing photosynthetic activity and detecting plant stress.

Radio remote sensing or microwave sensing operates in the longer wavelength portion of the spectrum (typically 1 mm–1 m) and includes active and passive sensing techniques, such as synthetic aperture radar (commonly known as SAR; Reddy, 2018; Richards, 2009). Radio sensors can penetrate cloud cover and vegetation to some extent and can provide information on soil moisture and crop structural characteristics even under adverse weather conditions (Babaeian et al., 2019; Petropoulos et al., 2015). In summary, optical sensing is like a high-definition camera that captures vivid details of crop health under clear conditions, while radio sensing is like radar, reliably detecting soil moisture and field structure through clouds and darkness.

Basic Principles of Electromagnetic Radiation: Seeing What Cannot Be Seen

A virtual "sea" of electromagnetic radiation, some of it natural, and some human-made, is all around us. If we could see all the radiation moving through our personal spaces, it would likely surprise us. The basic unit of electromagnetic radiation is the "photon" or "quantum." Photons are generated as electrons shift their orbits up and down, depending on their excitation levels, as a result of atomic and molecular actions and reactions such as those occurring on the surface of the sun. When an electron orbit shifts downward, photons are given off (emitted) in a wave pattern at the speed of light from the source (the Sun or an active sensor) to a physical element that absorbs, reflects (scatters), or transmits the energy.

All of the energy types are contained within the "electromagnetic spectrum" (Figure 14.3). Notice that the visible spectrum, which is the portion to which our eyes are sensitive, occupies a very small portion of the total spectrum. In other words, we see only a small part of all the radiation across this spectrum that surrounds us. The ultraviolet region contains wavelengths shorter

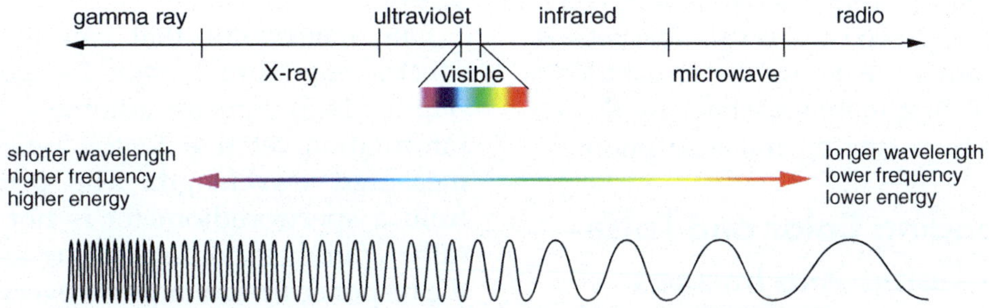

Figure 14.3. The electromagnetic spectrum. (*Source:* NASA Goddard Space Flight Center).

than those in the visible region, while the infrared region contains wavelengths longer than those in the visible region. One unit of measure for wavelength is the nanometer (nm), one 1-millionth of a meter (1×10^{-9}). The visible region spans wavelengths ranging from approximately 400 to 700 nm. The blue region ranges from 450 to 495 nm, green from 495 to 570 nm, yellow from 570 to 590 nm, orange from 590 to 620 nm, and red from 620 to 750 nm. The wavelength range for the NIR region is 700–1400 nm. The middle-infrared range lies between 1400 and 2500 nm, and the principal region for TIR data collection is between 8000 and 14,000 nm. Figure 14.3 shows that shorter wavelengths have higher frequencies and possess more energy per photon than longer wavelengths. Thus, a long wavelength will have a lower frequency and lower energy compared with a shorter wavelength. In agriculture, information collected from the visible, red edge, NIR, middle-infrared, and (to a lesser extent) TIR wavelengths can be very important.

By absorbing and reflecting energy, the Earth's atmosphere modifies reflected signals received from the Earth. Consider, for example, a satellite attempting to detect a signal in the visible region of the spectrum from a wheat field. The sunlight must pass through the atmosphere twice, once from the Sun to the Earth, and once from the wheat field to the satellite. During these two passes, atmospheric constituents (water, particulates, ozone, etc.) cause a portion of the sunlight to be scattered and absorbed. Thus, there are specific "atmospheric windows" within the electromagnetic spectrum where remote sensing is possible because scattering and absorption are minimal, and other portions of the spectrum where remote sensing is not possible because the atmosphere significantly modifies the signal. For example, information from the TIR regions (3000–5000 nm, and 8000–14,000 nm) is valuable because scattering is minimal, whereas the remainder of the TIR region is of minimal use because the signal is degraded by scattering and absorption.

Understanding Color and False-Color Composite Images

True-color composite (TCC) images are based on information from the RGB bands, which match the visible spectrum detected by human vision. The TCC image, shown at the bottom left of Figure 14.4, makes it easy to interpret features based on their natural colors. TCC images help farmers identify obvious issues, such as areas of sparse vegetation or standing water.

The FCC goes beyond visible wavelengths by incorporating the NIR band along with the red and green bands (Figure 14.4). This combination is helpful because healthy vegetation reflects a large amount of NIR light. In the FCC image shown in Figure 14.4, healthy vegetation appears in bright shades of red, while stressed or sparse vegetation appears in darker shades. This contrast helps distinguish between healthy growth areas and areas under potential stress, even if those differences are not visible to the naked eye.

The Spectral Signature

The spectral signature is the set of repeatable spectral characteristics for individual targets and materials. In the case of the visible spectrum, signatures can be the visual characteristics (i.e., color in the RGB spectrum) of certain objects. For example, we all know that most tree leaves are green in summer and most ripened apples are red (as seen by the human eye). The reason vegetation is green and apples are red is that they reflect relatively large amounts of green and red wavelengths, respectively. At the same time, if a target is seen as red, it absorbs the other two primary visible colors, blue and green, more than it absorbs red. Our experience shows that the color pattern described for tree leaves and apples is generally repeatable and true, so we can safely regard these as "signatures." However, information from beyond the visible bands can be used to develop more meaningful, detailed spectral signatures. This information can be collected using multispectral or even hyperspectral sensors.

One instrument that can measure detailed spectral signatures is called a spectroradiometer (Figure 14.5). Spectroradiometers collect digital information relative to the reflectance at each measured wavelength. The principal product from a spectroradiometer is not an image but, rather, numbers corresponding to the strength of the response from a target in every wavelength to which the instrument is sensitive. Those numbers can be used to create a chart often referred to as a spectral curve or spectral profile. As a result,

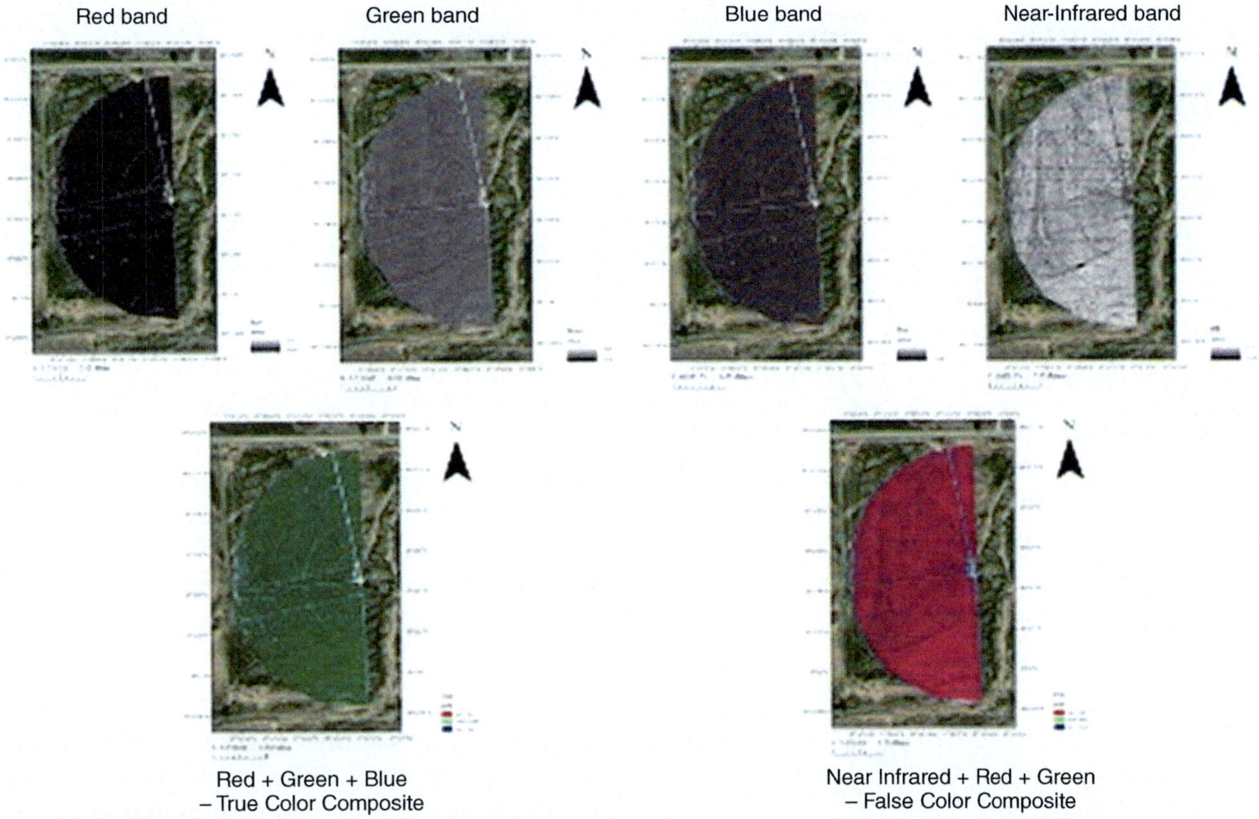

Figure 14.4. Multispectral images of a field showing individual bands (red, green, blue, NIR) and their composites. The TCC combines RGB bands for a natural color view, while the false-color composite (FCC) combines green (visualized as blue), red (visualized as green), and NIR (visualized as red). (Courtesy Udit Debangshi, Digital Agronomy Research Team, Kansas State University).

a spectroradiometer provides the user with both the intensity (or magnitude) of reflectance from the target and the spectral distribution of the energy being reflected (Figure 14.5).

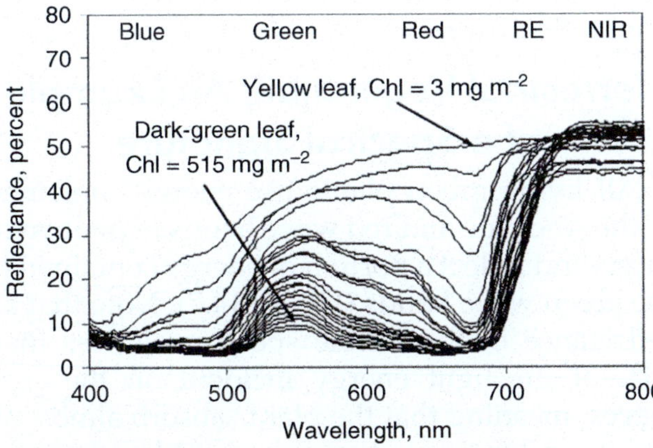

Figure 14.5. Effects of grape leaf chlorophyll content on canopy reflectance (Steele et al., 2008).

Understanding Multispectral and Hyperspectral Sensing

Multispectral sensing typically captures information from 3 to 10 bands of the electromagnetic spectrum. Generally, these bands are in portions of the spectrum where water absorption reduces reflectance (Figure 14.6). Many multispectral sensors collect reflectance information from the RGB and NIR bands. Information from these bands can be used to develop false-color (Figure 14.7) and NDVI images. Multispectral images may be used by a farmer or crop advisor to assess general plant health because these sensors work similarly to a digital camera, using only a few filters, such as sepia, grayscale, and color. Each filter gives a distinct but general view, capturing only the most important aspects of the scene (Figure 14.7). This makes these sensors useful for broad insights, but they may miss finer details.

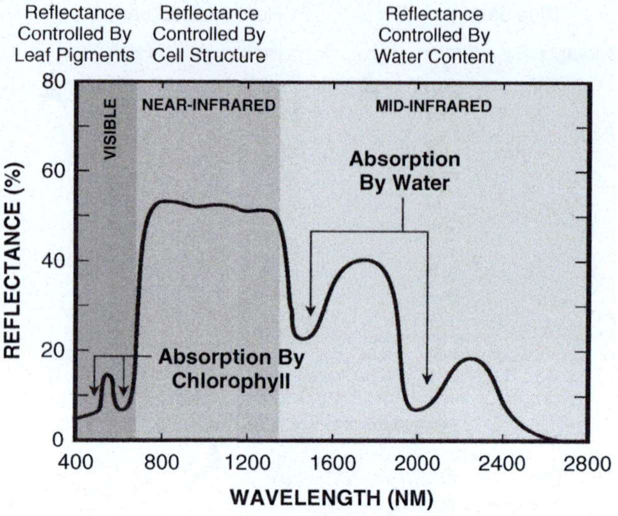

Figure 14.6. Primary factors influencing plant canopy reflectance.

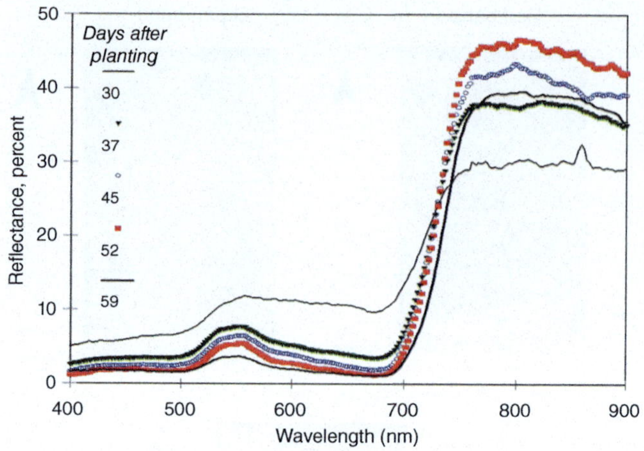

Figure 14.8. Corn canopy reflectance as influenced by days after planting. This is a spectral profile generated by a spectroradiometer. Such diagnostic curves are important in understanding the physical basis for spectral response. (Courtesy CALMIT, University of Nebraska-Lincoln).

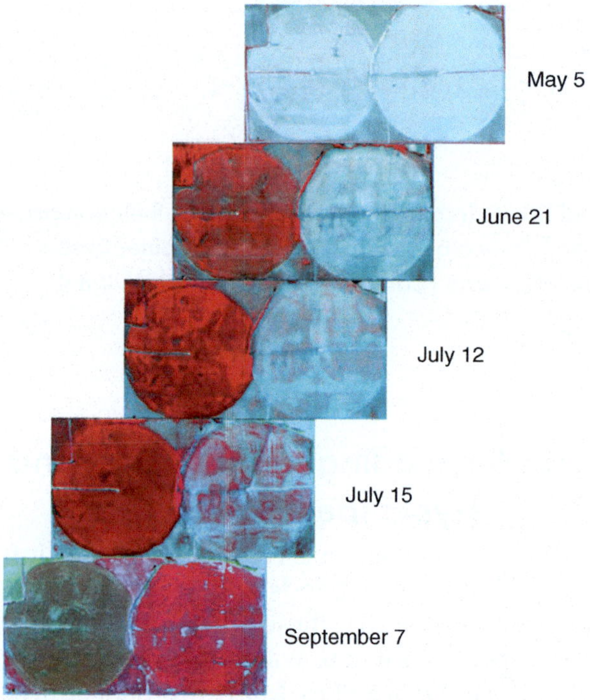

Figure 14.7. False-color near-infrared comparison of corn (left field) and soybean (right field) canopies throughout the growing season (CALMIT AISA sensor system). In these images, red indicates more biomass, blue and green less biomass.

Hyperspectral sensing can capture the finer details that multispectral cameras may miss. These sensors capture hundreds of narrow, closely spaced bands across the spectrum, allowing for the detection of subtle differences in material composition. Hyperspectral sensing aims to reveal small differences that can be specifically linked to a limiting factor. It is often used by researchers to gain greater insight into the fine details of a scene, revealing key information that can be used in the larger study. Just as a microscope shows details invisible to the naked eye, hyperspectral sensing provides a detailed "fingerprint" of each crop, potentially helping detect early signs of disease, nutrient deficiency, or even crop type. Data collected by a hyperspectral sensor can be displayed in a chart with wavelength on the x-axis and reflectance on the y-axis. Spectral charts, such as Figure 14.8, are often considered diagnostic because they display areas with low (high absorption) and high (low absorption) reflectance.

Terrestrial Vegetation: An Example of a Spectral Signature

All living green plants use energy contained within the blue and red wave bands in photosynthesis and reflect a portion of energy contained in the green wave bands (Figure 14.7). Note that all reflectance in the visible spectrum is less than 10% of the light energy incident on the corn leaves, meaning that the plants absorb almost all the incoming sunlight in the visible spectrum, including green light. Interestingly, the highest reflectance (i.e., lowest absorption) occurs in the (non-visible) NIR spectral region. The high reflectance in the NIR energy is attributed to the internal structure of the leaf (i.e., its spongy mesophyll layer). Thus, we can regard such a spectral profile as the unique signature in the visible and NIR regions of the spectrum.

False-color images also can be used to provide a signature (Figure 14.8). For example, Figure 14.8 compares the spectral profile of an irrigated corn field (left column of circular pivots) with that of an irrigated soybean field (right column of circular pivots). Notice how the color patterns in both fields change over time, with the bright red tones being indicative of actively photosynthesizing green vegetation, and a strong signal in the NIR region.

What Do Spectral Signatures Tell Us?

The principal factor controlling reflectance in the visible region of the spectrum is pigments, primarily chlorophyll (Figure 14.6). Thus, knowing where to look (spectrally) for information about green vegetation is critical. In short, if information about pigments is wanted, then one would examine the signal in the visible region, check the NIR region for information about cell structure and canopy architecture, and examine the middle-infrared region for information regarding the leaf water content.

The TIR region can provide information about the temperature of a plant canopy, which is related to the general plant condition. Stressed plants, such as those subjected to a moisture deficit, can be detected using thermal sensors (Swain et al., 2012). For example, Figure 14.5 depicts a series of spectral profiles that illustrate the effect of increasing chlorophyll on the signal from grape leaves. The lines in the graph summarize chlorophyll densities ranging from a minimum of 3 mg m^{-2} to a maximum pigment content of 515 mg m^{-2}. The main point here is that increasing chlorophyll causes a pronounced (and orderly) increase in absorption (i.e., a decrease in reflectance) in the visible region of the spectrum. The effect is especially dramatic in the red region.

Vegetation Indices

Vegetation indices (VI) that combine information from multiple bands can be used to investigate plant stress (Figure 14.9). More than 150 vegetation indices have been developed, each for a different purpose (Table 14.2). The most widely known vegetation index is the Normalized Difference Vegetation Index (NDVI) (Rouse et al., 1974). The NDVI uses the relationship of reflectance in the visible (normally red) and NIR regions, using the formula, NDVI = (NIR – Red)/(NIR + Red). Values for NDVI range between −1.0 and +1.0.

Vegetation indices can be designed for specific purposes. For example, the Soil Adjusted Vegetation Index (SAVI) was developed to minimize the way soil brightness influences canopy reflectance in the red and NIR regions (Huete, 1988). A more recent vegetation index that is being used in agriculture is the NDRE index, which replaces reflectance in the red band with reflectance in the red edge. The equation for

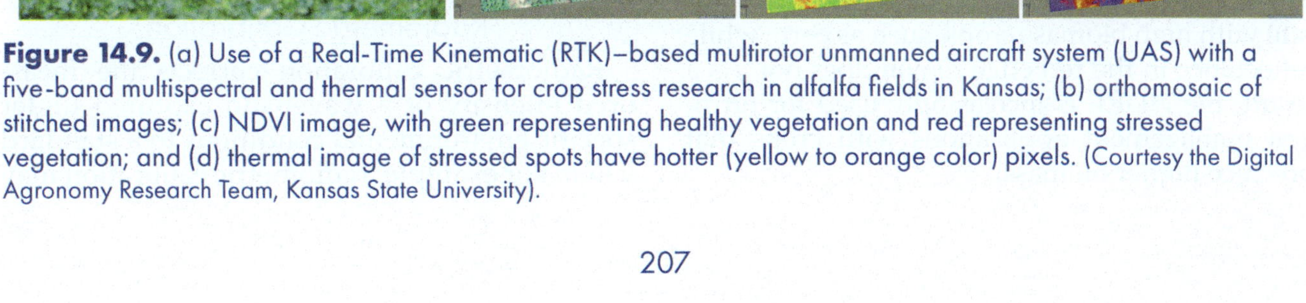

Figure 14.9. (a) Use of a Real-Time Kinematic (RTK)–based multirotor unmanned aircraft system (UAS) with a five-band multispectral and thermal sensor for crop stress research in alfalfa fields in Kansas; (b) orthomosaic of stitched images; (c) NDVI image, with green representing healthy vegetation and red representing stressed vegetation; and (d) thermal image of stressed spots have hotter (yellow to orange color) pixels. (Courtesy the Digital Agronomy Research Team, Kansas State University).

Table 14.2. Selected vegetation indices, abbreviations, equations, and references.

Index	Abbreviation	Equation	References
Normalized Difference Vegetation Index	NDVI	$(R_{760} - R_{670})/(R_{760} + R_{670})$	Rouse et al. (1974)
Normalized Difference Red Edge	NDRE	$(R_{760} - R_{720})/(R_{760} + R_{720})$	Barnes et al. (2000)
Modified Soil Adjusted Vegetation Index	MSAVI	$0.5[2R_{800} + 1 - SQRT((2R_{800} + 1)^2 - 8(R_{800} - R_{670}))]$	Qi et al. (1994)
Green Normalized Difference Vegetation Index	GNDVI	$(R_{800} - R_{550})/(R_{800} + R_{550})$	Gitelson et al. (1996)
Green Red Vegetation Index	GRVI	$(R_{550} - R_{670})/(R_{550} + R_{670})$	Motohka et al. (2010)
Modified Chlorophyll Absorption in Reflectance Index	MCARI	$[(R_{700} - R_{670}) - 0.2(R_{700} - R_{550})](R_{700}/R_{670})$	Daughtry et al. (2000)
Normalized Difference Index	NDI	$(R_{780} - R_{710})/(R_{780} - R_{680})$	Datt (1999)
Optimized Soil Adjusted Vegetation Index	OSAVI	$(1 + 0.16)(R_{800} - R_{670})/(R_{800} + R_{670} + 0.16)$	Rondeaux et al. (1996)
Transformed Chlorophyll Absorption in Reflectance Index	TCARI	$3 * [(R_{700} - R_{670}) - 0.2 * (R7_{00} - R_{550})(R_{700}/R_{670})]$	Haboudane et al. (2002)
Chlorophyll Index	CI	$(R_{880}/R_{590}) - 1$	Gitelson et al. (2005)
Chlorophyll Index Red Edge	CIRE	$(R_{760}/R_{720}) - 1$	Gitelson et al. (2005)
Datt	Datt	$(R_{760} - R_{720})/(R_{760} - R_{670})$	Datt (1999)
MERIS Terrestrial Chlorophyll Index	MTCI	$(R_{760} - R_{720})/(R_{720} - R_{670})$	Dash and Curran (2004)
Soil Adjusted Vegetation Index	SAVI	$[(R_{800} - R_{670})/(R_{800} + R_{670} + L)] * (1 + L)$	Huete (1988)

Abbreviation: MERIS, Medium Resolution Imaging Spectrometer; R, reflectance.

NDRE is (NIR – Red Edge / NIR + Red Edge). Values for NDRE range between –1.0 and +1.0. The amount of reflected radiation in the red edge spectral band changes in response to changes in the chlorophyll pigments of a plant's leaves, which is often used as a proxy for plant nitrogen status. Reflectance in the red band tends to saturate with high biomass crops, such as corn, while reflectance in the red edge region does not. As a result, the NDRE is increasingly used for nitrogen management applications with crops that produce higher biomass.

Radiometric, Spatial, and Crop-Response Calibration

To ensure that accurate information is collected, radiometric, spatial, and crop-response calibration should be conducted.

Radiometric Calibration

Radiometric calibration corrects the measured intensity of a waveband captured under specific environmental conditions to a standard value. For example, radiometric calibration may correct the measured intensity for solar irradiance (intensity and geometry), atmospheric absorption, scattering, and sensor drift. It is important to radiometrically calibrate remote sensing data regardless of the platform (e.g., satellite, manned aircraft, UAV) used to capture the data. The importance of radiometric calibration depends on the platform. Satellites are particularly vulnerable to atmospheric absorption, scattering, and sensor drift (change in sensor performance over time) because the altitude at which satellites orbit the Earth requires a long electromagnetic radiation path through the atmosphere and because sensor performance changes with time. Data from manned aircraft suffer from similar calibration challenges as satellites, though with fewer concerns about atmospheric conditions. Manned aircraft must also contend with sensor position and solar illumination geometry challenges resulting from platform movement during typical operations, especially under windy conditions. Radiometric calibration of data captured from UAVs is particularly challenging for a variety of reasons, including the following:

- Multiple images of the field must be captured and stitched together to form an orthomosaic of the area of interest (AOI).
- Wind-influenced changes in platform orientation can influence relative sensor position and illumination geometry between images.
- Intermittent cloud cover can influence the reflectance signature.

Radiometric calibration requires capturing the sensor response to a known standard target under the same or known environmental conditions at an ideal sensor position. Several techniques can be used to achieve radiometric calibration practically. One method of radiometric calibration is to place known reflectance panels on the ground within the image scene to be captured by a satellite, manned aircraft, or UAV. This enables post-processing corrections to reflectance intensity measurements by scaling measured values to the known reflectance panel value(s). Another technique is to adjust the measured values based on calibrations that were conducted before and/or after a mission. This approach can correct the measured values for sensor drift. Other approaches place an illumination detector and a gyroscope on the sensing platform to monitor illumination conditions and sensor position during data collection. A combination of pre- and post-mission calibration with in-mission illumination and sensor position measurement is an effective approach to radiometric calibration. Artificial intelligence techniques and sensor fusion via harmonization (Chapter 27) may also be used for radiometric calibration purposes at scale if there is a highly reliable standard with which to harmonize.

Spatial Calibration

Spatial calibration is generally easier to achieve than radiometric calibration since accurate Global Positioning System (GPS) systems are available. Two common methods used to conduct spatial calibration are to identify the location of ground (ground control points [GCPs]) and feature control points. GCPs are objects placed within the AOI that will be captured in data collected during the mission. The measured location of the GCPs can be compared with the location information within the image. Ground control points may be combined with surface reflectance control panels to facilitate simultaneous radiometric and spatial calibration. Similarly, features (ideally with known coordinates) already established within an AOI (e.g., an irrigation system) may be used as control points. If there are differences between the known and measured locations, spatial corrections may be required. How to conduct spatial corrections is beyond the scope of this chapter.

Practical concerns regarding calibration techniques for agriculture include the feasibility of executing a particular technique under typical field conditions and the reliability of that technique to produce timely information. Some factors to consider include the following:

- The size of the AOI from which data is being captured
- The variability of environmental conditions during the data capture mission
- The number of individual sensors being used to produce data
- The homogeneity of the AOI from which data is collected.

For example, to accelerate data collection, a crop consultant could use four UAVs to capture multispectral data over different portions of a field. However, due to small differences in sensor efficiency, each sensor will produce slightly different results, making it difficult to combine the four images into a single whole-field image. This highlights how important the calibration approach is and that the user ensures it effectively mitigates data quality concerns imposed by the equipment, time requirements, and field conditions.

Crop-Response Calibration

A final important aspect of calibration is to calibrate to a particular crop response. This is because many factors (e.g., nutrient deficiency, water stress, pests) may have similar effects on reflectance and it is often challenging to separate the effects of one factor from another. This generally results in the remote sensing image being used to identify problem areas and to scout, inspect, and diagnose the actual issue. This problem of separating different factors' effects may be solved by identifying the unique fingerprint of the factors impacting reflectance. However, the fingerprint approach has proven challenging because reflectance changes as plants age.

The calibration of remote sensing to crop response can be affected by sensor and platform types, the vegetation index used, and other factors. For example, a tractor-mounted active sensor may have a different assessment of crop nutrition differences than a satellite-mounted passive sensor. Similarly, different vegetation indices will differentiate crop nutrition levels differently. This is commonly observed when red band–based vegetation indices, such as NDVI, are compared with indices that do not use the red band, such as the Green Normalized Difference Vegetation Index (GNDVI) or NDRE. In the case of NDVI, high biomass and chlorophyll concentrations in a crop field can saturate the red band, making NDVI less sensitive in differentiating high crop vigor conditions than other indices that do not use the red band.

Sensor Fusion and AI

Sensor fusion is the practice of combining data from multiple sensors into a single derivative dataset. Generally, sensor fusion is performed to improve spatial, spectral, or temporal resolution relative to all collected information. This process can combine data from different sensors with different spatial and temporal resolutions. For example, Landsat missions have a spatial resolution of 30 m and a temporal resolution of 16 days, while the Moderate Resolution Imaging Spectroradiometer (MODIS) mission has a spatial resolution of 250–1000 m and a temporal resolution of near daily. Fusing data from both Landsat and MODIS missions allows the creation of satellite imagery with the spatial resolution of Landsat (30 m) and the temporal resolution of MODIS (near daily).

Sensor fusion may also be performed to standardize data, enhance data accuracy, provide a unified and comprehensive data source, or increase coverage area. Sensor fusion has been empowered by the use of AI techniques. Artificial intelligence may be applied to remote sensing data to create derivative data or to analyze raw or derivative data. Examples of AI applications for data analysis include classification, feature extraction, and change detection, among others. Potential applications of these techniques to agriculture include field boundary delineation, classification of irrigated and rainfed cropland, crop type classification, crop progress quantification, and irrigation system characterization.

Practical examples of sensor fusion in the public and private sectors abound. NASA has leveraged sensor fusion techniques to produce public derivative datasets through its Harmonized Landsat and Sentinel-2 (HLS) project that leverages Landsat 8, Landsat 9, Sentinel-2A, and Sentinel-2B data to provide data products that are harmonized and available for analysis at least twice weekly (NASA; https://www.earthdata.nasa.gov/esds/harmonized-landsat-sentinel-2). Houborg and McCabe (2018) developed the Cubesat-enabled Spatio-Temporal Enhancement Method (CESTEM) to perform radiometric calibration across a Planet Labs (a private remote sensing company) satellite constellation by leveraging data from that constellation alongside MODIS and Landsat 8 data, and Planet Labs now offers a Fusion Monitoring product using similar techniques to provide moderate-resolution, daily data of the Earth's surface (https://learn.planet.com/rs/997-CHH-265/images/Planet%20Fusion%20Monitoring%20Datasheet_Letter_Web.pdf). Planet Labs also provides data from their PlanetScope constellation harmonized to Sentinel-2 data to overcome the challenges of different relative spectral response between sensors on satellites in the constellation (Planet Labs; https://assets.planet.com/docs/scene_level_normalization_of_planet_dove_imagery.pdf). Another interesting application of sensor fusion is super-resolution. Super-resolution is the process by which the resolution of a highly reliable, scalable, and available data source (such as satellite data) is increased by using an algorithm trained on source data aligned with a higher-resolution target dataset. Super-resolution has been observed to have four times higher resolution when combined with Sentinel-2 RGB and SPOT-7 imagery using generative adversarial networks (Puri & Kotze, 2022). These advancements are important to consider for agricultural applications because of their potential to improve quality and spatial and temporal resolution.

Integrating Remotely Sensed Data with Other Geospatial Data

Combining remote sensing data with other geospatial data layers has the potential to improve precision prescriptions. Other geospatial data may include field operations (harvest, planting, tillage, application, irrigation), soil properties (chemical, physical, biological), terrain characteristics (elevation, slope, aspect), crop properties (genetics, planting date), machine operations (fuel consumption, hours per operation), and weather (precipitation, temperature), among others. This data may be captured and stored in standard or proprietary formats, generally characterizable as vector or raster data. For additional information about vector and raster data, see Chapters 8 and 9.

Yield or harvest data is one example of data that may be integrated with remote sensing data for crop performance evaluation (Chapter 4). Three significant challenges of integrating yield and remote sensing data are (a) aligning spatial resolution, (b) data format compatibility, and (c) handling capture timing misalignment. Ensuring that geospatial data layers are in the same projection is important for accurate integration and analysis. More information and details on this issue can be found in Chapter 7.

To align spatial resolution, all data is generally aligned to the lowest resolution because that is the lowest common denominator in the dataset. Using the lowest resolution should make it easy to select the nearest value from other datasets. If the data are in a point format, this is more challenging, and interpolation may be needed. Interpolation (Chapter 9) is also required to increase the resolution of lower-resolution datasets to match that of higher-resolution datasets, or to get all data into a raster format. Data integration is ultimately a process of ensuring compatibility, alignment, and correctness and ultimately relies on good decision-making about how the data will be integrated.

Integrating other geospatial data with remote sensing data has the potential to improve precision prescriptions. For example, when using remote sensing to determine crop variable rate nitrogen requirements, the best recommendations may come from integrating datasets from remote sensing, field history, soil/landscape, and weather. The integration of these data layers can generate nitrogen recommendations that are more profitable than using a single layer alone (Colaco et al., 2024). A geographic information system (GIS) is required for integrating these different geospatial information data together. Additional information on GIS is available in Chapters 8 and 9.

Nutrient Management

Various remote sensing techniques, such as canopy color analysis, visible light remote sensing, multispectral remote sensing, and hyperspectral remote sensing, allow for nondestructive and stable detection of soil and crop nutrient stresses in real time (Daughtry et al., 2000; Samreen et al., 2023). Many chapters in this book discuss the details of using remote and proximal sensing techniques to manage plant nutrient requirements. For more information, see Chapters 4 and 13.

One of the barriers to the adoption of remote sensing in nutrient management is the challenge of separating nutrient stress from other types of stress (Barnes et al., 2000). Shiratsuchi et al. (2011) showed that the Datt and MERIS Terrestrial Chlorophyll Index (MTCI; Table 14.2) reflectance indices could distinguish nitrogen deficiency relatively accurately, even in the presence of moderate water stress (Dash & Curran, 2004; Datt, 1999). Other indices that have had success are GNDVI and Chlorophyll Index Red Edge (CIRE; Clay et al., 2006; Martins, 2022). Stansell (2021) used multispectral image data to detect approaching crop nitrogen demand and recommend nitrogen applications to proactively address that demand. The problem with identifying nitrogen stress using remote sensing is confounded by different cultivars having different canopy architectures and nitrogen stress signatures. This problem can be partially solved by using an appropriate calibration approach.

Physical and virtual calibration have been used in nutrient management. Physical calibration methods require ensuring that part of the crop has more or less of a particular limiting factor than what it needs to perform at its best. One example of this is the nitrogen-rich strip concept to calibrate for crop response to additional nitrogen (Varvel et al., 1997). Another example is the nitrogen-poor, or zero-nitrogen, strip for the same purpose. A recent approach to physical calibration involved pairing nitrogen-poor and nitrogen-rich blocks adjacent in specific field zones expected to respond to nitrogen differently for site-specific calibration to crop nitrogen response (Stansell, 2021). However, physical calibration raises financial concerns about the cost of unfertilized and nitrogen-rich blocks.

Virtual reference concepts have instead been proposed to mitigate the concerns about physical references. Virtual reference is a statistical technique in which an AOI is established to produce a reference value for calibrating and determining the crop response without physically needing to create a reference strip in the field. One such virtual reference approach is the statistical technique developed by Holland and Schepers (2013). This technique uses the cumulative 95th percentile of vegetation index values in an AOI to establish a reference value that may be treated as a virtual nitrogen-rich strip for calibrating crop nitrogen response. While this approach is effective for determining how to spatially vary the rate of nitrogen given the specific known nitrogen demand values, the risk of the approach is that it may perceive yield-limitation that is not due to nitrogen and still indicate a nitrogen response. Currently, while virtual calibration is easier to implement at scale, physical calibration offers the highest certainty of isolating and quantifying specific yield-limiting factors. It is likely that further investigation of virtual reference concepts and yield-limiting factor isolation using big data will be pursued with the continued proliferation of AI.

Research is currently being conducted to determine whether hyperspectral reflectance or leaf fluorescence can be used for evaluating other nutrient deficiencies (Adams et al., 2000; Okyere et al., 2023; Osborne et al., 2002). One approach that looks promising is to use remote sensing of soybean and corn to identify areas within fields of elevated pH for chlorosis management (Rogovska & Blackmer, 2009). Often, these areas are relatively stable over time, with chlorosis severity and geographic extent varying with weather conditions. Such mapping can allow cost-effective use of iron (Fe) fertilizer application or cultivar selection within zones for chlorosis management (Kyaw et al., 2008).

Weed Management

Weed infestations often occur in patches within fields, related to underlying soil properties and/or the occurrence of previous weed infestations (Chen, Xia, et al., 2024; Clay et al., 1999). This chapter discusses the general remote sensing aspects of weed management. For greater detail, see Chapter 18. Consequently, mapping weed occurrence has the potential to lead to effective site-specific management. One approach to creating a weed map is to note weed occurrence while performing other field operations, such as cultivation or harvest. A second approach is to use remote sensing to identify areas with high weed infestations (Chang et al., 2004). The use of sensors has followed two approaches. The first approach is to generate a map for future control operations. The second approach is the use of sensors for real-time detection and control (Chapters 11, 12, 13, and 18). Several challenges arise in using remotely sensed data for weed management during a growing season. Early detection of weeds is critical for effective control, but very small plants are difficult to detect with satellite systems. In addition, the delay between sensing and map creation may reduce the efficiency of the control measures.

Remote sensing can readily detect the presence of non-crop plants between rows once they reach a critical size. Identification of weeds within rows is a greater challenge, as is the actual identification of weed species. Greater success in separating weeds from crops and identifying weed species has resulted from proximal sensing approaches (Chapter 13), which have used both spectral reflectance and leaf shape analysis for identification (Vrindts et al., 2002). Soil background reflectance confounds early-season detection of weeds and separation from crop reflectance and should also be taken into consideration (Chapter 9; Thorp & Tian, 2004).

Given the image resolution and the turnaround time required for satellite imagery, most recent research has focused on UAV-based remote sensing to create control maps that can be implemented during the current growing season. Various authors have explored approaches such as Object-Based Image Analysis (OBIA), pixel-based Artificial Neural Networks (ANN), or machine learning algorithms with RGB sensor imagery, allowing the use of low-cost imagery systems (Chen, Hein, et al., 2024; Chen, Xia, et al., 2024; Nikolic et al., 2021; Peteinatos et al., 2020; Sapkota et al., 2020). Gao et al. (2018) combined row detection based on pixel features with OBIA for mapping weeds in corn with an RGB sensor, resulting in high correlation with ground-truth data. Mesias-Ruiz et al. (2024) demonstrated the adoption of facial recognition software with high-resolution RGB imagery for early growth stage weed detection, finding 99.1% accuracy in identifying weed species.

Generally, authors report significant herbicide savings with site-specific weed management based on remote sensing data. de Castro et al. (2013) found great potential for reducing herbicide use by site-specific application informed by remote sensing. Mink et al. (2018) found up to 90% herbicide savings in corn and 43% in sugar beet when using UAV-based RGB and multispectral maps of creeping thistle and curled dock, while Nikolic et al. (2021) showed a significant theoretical reduction in herbicide application with maps developed from a drone-mounted RGB sensor. In general, weed management can be greatly improved through the use of remote and proximal sensing technologies.

Disease Management

Early detection of plant disease is needed to minimize yield losses. To provide this information, research is being conducted at scales ranging from individual leaves to regions (Clay et al., 2024). The present chapter discusses general aspects of disease management with remote sensing technology. For a more detailed discussion, see Chapter 19. At the leaf scale, spectral images analyzed using AI are being used to identify specific diseases, whereas at the regional scale, Yuan et al. (2016) has assessed powdery mildew infestation in winter wheat with an overall mapping accuracy of 78%. Chen et al. (2018) also assessed wheat rust, while Chen, Hein, et al. (2024) evaluated wheat streak mosaic virus, both at regional scales.

An area of active research is to identify which wavebands or groups of wavebands are most useful (Khanal et al., 2017; Mahlein et al., 2013). Research is being conducted using multispectral, hyperspectral, and thermal bands with AI image processing techniques in a variety of crops (Calderon et al., 2013; Clay et al., 2024; Franke & Menz, 2007; Khanal et al., 2017; Mahlein et al., 2013; Mirandilla et al., 2023). To date, the findings are plant and disease specific. The problem is confounded by where on the plant the disease occurs. For example, with some crops and diseases, initial infection may occur on lower leaves, while canopy reflectance is dominated by reflectance from upper leaves. This makes early detection a challenge. In more recent work, Feng et al. (2016) found reflectance in wavebands in the 580- to 710-nm range to be sensitive to wheat powdery mildew and could be used to effectively monitor for powdery mildew infestation in wheat during the middle and late growth stages. Other studies identified a potential for remote detection of basal stem rot in oil palm (Santoso et al., 2011), grapevine leafroll disease in grapes (Hou et al., 2016), fungal diseases in sugar beet (Mahlein et al., 2010), wheat curl mite–vectored viruses in wheat (Stilwell et al., 2013), citrus greening disease (Huanglongbing; Kumar et al., 2012), and sudden death syndrome in soybean (Raza et al., 2020). A few studies have used machine and deep learning approaches to identify a variety of diseases based on reflectance data collected by UAVs (Kouadio et al., 2023; Morellos et al., 2020; Zhao et al., 2024). In general, managing crop diseases using remote or proximal sensing can improve detection, allowing growers to locate and target areas of disease before yield loss.

Insect Management

As with diseases and weeds, insect infestation can occur in patches, or hot spots, that may or may not have a relationship to underlying soil variation (Ellsbury et al., 1998). This chapter provides a general discussion of insect management. For a more detailed discussion, see Chapter 17. Remote sensing can help identify these locations within fields for control before infestations spread to other areas of the field. Typically, remote sensing for insect management has occurred at regional or continental scales. Brewster et al. (1999) used Landsat imagery to map whitefly breeding habitat in California's Imperial Valley. Klein et al. (2023) used Landsat and Sentinel-2 imagery to identify land areas that are attractive for locust breeding and to monitor for treatment needs. Qi et al. (2024) used satellite imagery to assess fall armyworm habitat suitability across Africa. Willers et al. (2005) used satellite multispectral imagery to develop maps of likely habitat for cotton insect infestation. Abd El-Ghany et al. (2020) and Abd Kharim et al. (2022) used remote sensing to assess stem borer in rice; Thapa et al. (2022) used satellite remote sensing to assess defoliation in the forest; Fitzgerald et al. (2004) investigated remote sensing for the detection of spider mites in cotton; Mirik et al. (2012) investigated Russian wheat aphid in winter wheat, and Puig et al. (2015) investigated the use of UAV imagery to detect white grub in sorghum. Additionally, in greenhouse environments, Nguyen and Nansen (2020) used hyperspectral imagery to detect leaf miner infestation over time and in various management regimes, and Yang et al. (2009) used visible (VIS) and NIR vegetation indices to differentiate stress in wheat caused by either greenbugs or Russian wheat aphids. Overall, insect management can be improved through the use of remote and proximal sensing technologies, enabling growers to identify hotspot locations for targeted management.

Soil Assessment

Remote sensing of soil properties can occur indirectly using sensing tools for soil itself, or indirectly by assessing crop canopy properties as indicators of the underlying soil resource. Examples of using the crop canopy to infer soil properties are from Rogovska and Blackmer (2009) and Kyaw et al. (2008). They reported that canopy reflectance in corn and soybeans could be used to identify high-pH areas prone to iron chlorosis. Machine learning approaches can be used to process remote sensing data to estimate soil quality indicators (Diaz-Gonzalez et al., 2022), soil moisture (Ford & Quiring, 2019), and soil degradation (Wang et al., 2023). A review of 28 journal articles reported that the use of UAV platforms increased the accuracy of soil property prediction compared with satellite imagery (Angelopoulou et al., 2019). Using thermal, radar, multispectral, and hyperspectral technologies, remote sensing can estimate soil moisture by detecting temperature differences and radar backscatter, guiding irrigation planning in water-limited regions. Remote sensing further supports salinity assessment by detecting salt-stressed soils and guides nutrient management by evaluating vegetation health through indices, such as NDVI, indirectly reflecting soil nutrient levels. Soil with high salt content may be detected and mapped due to its high surface reflectance. Both remote and proximal sensing methods have been used to estimate and map soil organic matter content (Angelopoulou et al., 2019; Bhatti et al., 1991; Roberts et al., 2010).

Water Management

The use of remote sensing for crop water management can include sensing soil properties to evaluate the potential for soil water supply, but more often its use is related to in-season sensing of crop canopy properties (Chapters 15 and 16). This chapter provides a general overview of the use of remote sensing for water stress management. For a more detailed discussion, see Chapters 15 and 16. The Crop Water Stress Index (CWSI) has been used for this purpose (Idso et al., 1981). Determining the CWSI requires knowing the difference in canopy and ambient air temperatures, which may be remotely sensed, and the vapor pressure deficit, which normally requires meteorological information at the field or nearby (Jackson et al., 1988). Pinter et al. (2003) noted that many stresses to plants can result in elevated plant temperatures that are often correlated with plant water status and potential yield reduction. Plants under water stress also exhibit changes in canopy architecture, such as leaf rolling, which can influence the use of canopy reflectance information to manage other plant stresses, such as nitrogen deficit.

Pinter et al. (2003) noted that the remote estimation of evapotranspiration, combining ground-based meteorological information with remote measurement of reflected and emitted radiation, followed by estimation of latent energy exchange, had considerable promise for estimating crop water use over broad areas. However, this estimation is limited by a lack of sensors with adequate spatial and temporal resolution from satellite or aircraft platforms. Increasingly, there is interest in the use of UAVs to assess stress, particularly water stress, due to the potential for relatively inexpensive data collection with high temporal density with UAVs. Remote sensing in thermal bands is effective using unmanned aircraft system (UAS) platforms (Bellvert et al., 2014; Berni et al., 2009). More detailed information on the use of remote sensing for water management in cropping systems is available in Chapters 15 and 16.

Increased Use of UAVs in Agriculture

Significant growth in the use of UAVs for crop remote sensing has occurred since the Federal Aviation Administration (FAA) established regulatory requirements for the commercial use of UAVs in agriculture in 2016. For some remote sensing applications, UAVs provide significant advantages over manned aircraft or satellite platforms. Unmanned aerial sensors can be deployed quickly to take advantage of good weather for data collection or to quickly assess unanticipated issues. For simple crop scouting applications, UAVs can be promptly deployed over a field by a producer or crop consultant at a relatively low altitude to scan the field for issues during the growing season. For scouting purposes, there may be no need for nadir (downward) views; an oblique view across a field at low altitude with a standard camera may suffice to provide information on areas in the field that need closer examination on the ground.

Opportunities and Limitations

Over the past 20 years, remote sensing for agriculture has evolved significantly, especially with new developments in satellite platforms, the continued growth in the use of UAVs, and the expanded capabilities in terms of data analytics for site-specific management (Figure 14.10). New advances in telecommunications, electronics, and software engineering allow access to higher spatial and temporal imagery resolution, as well as expanded computational power for data analysis. These enable more precise and timely monitoring of crop conditions. Newer platforms feature multispectral and hyperspectral bands that allow better assessment of crop health and soil conditions. In the past, high-resolution satellite imagery was costly and less accessible. Today, public missions like Landsat and Sentinel provide free access to high-quality imagery, greatly expanding the user base and utility of precision agriculture. Small satellite constellations have enabled rapid, cost-effective coverage and more continuous data, filling previous temporal gaps (Khanal et al., 2020; Maes & Steppe, 2019).

The limits of aerial sensing include the need to schedule around cloud cover. However, the use of UAVs presents an opportunity to have greater flexibility in when images are taken, optimizing data collection. However, UAVs are not without their own limitations. For example, UAVs are currently unable to capture images of large swaths of land at once. This can still be done only by satellites or manned aircraft. On the other hand, UAV costs are decreasing; they can assist in crop scouting, systems can be developed for individuals or commercial providers, and they can provide high-resolution, real-time information quickly (Figures 14.9 and 14.10; Weiss et al., 2020). Additionally, when combined with advances in computational capabilities and new analytical techniques, UAVs can help reduce adoption barriers to precision techniques by creating useful, site-specific management strategies that growers can use annually.

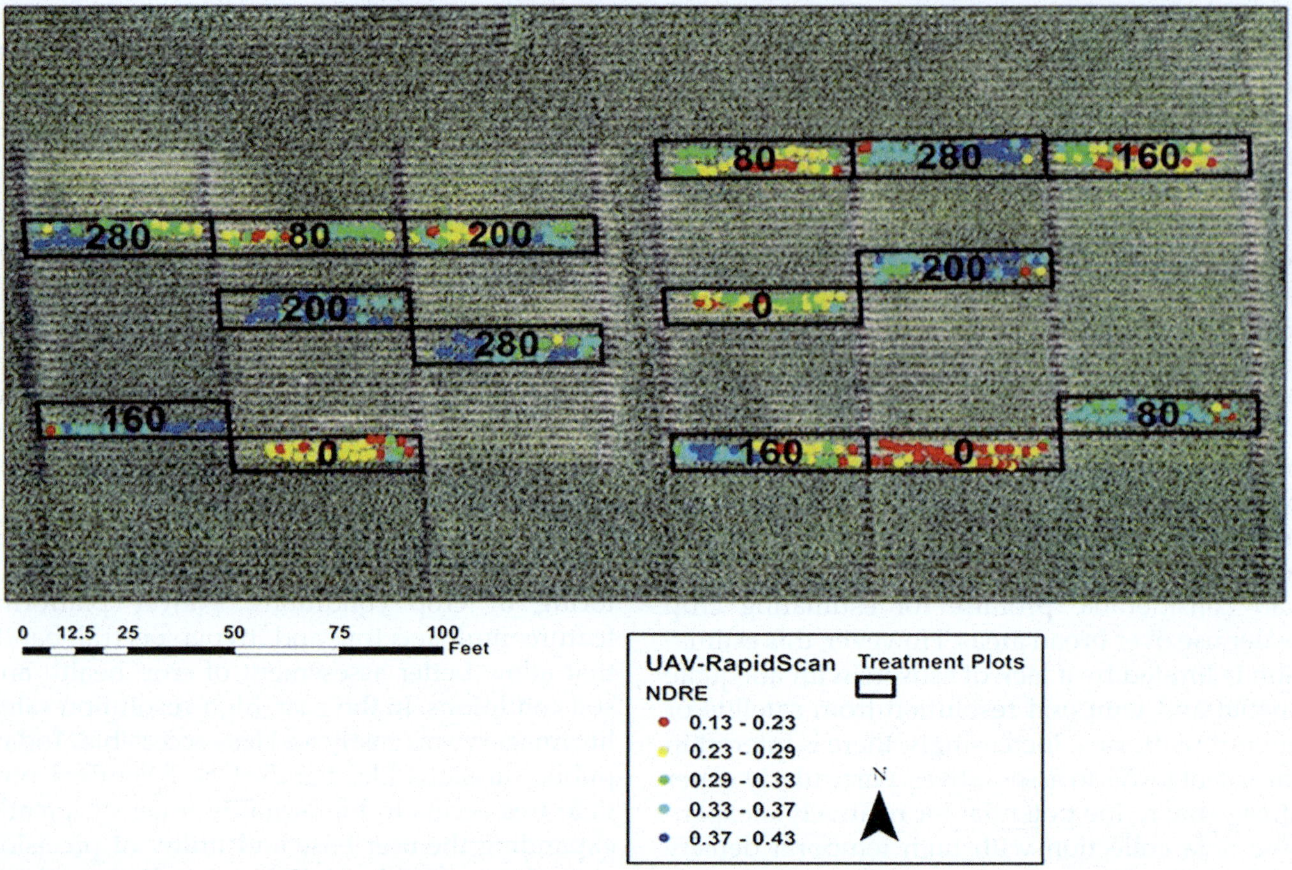

Figure 14.10. Use of a UAV-mounted multispectral active canopy sensor to assess nitrogen status of corn. Background image is from a UAS-mounted high-resolution standard RGB camera. Normalized Difference Vegetation Index Red Edge (NDRE) values superimposed over the background image were collected with a UAS-mounted active sensor operated at a distance approximately 1.5 m over the canopy. Numbers within each treatment plot are fertilizer nitrogen rate in pounds per acre. (Courtesy the University of Nebraska-Lincoln).

Despite improvements in all platforms for remote data acquisition, there are still limitations, including atmospheric interference, the high cost for high-resolution (e.g., 50-cm) data, and tailored visiting time in the case of satellite platforms. In addition, UAV limitations are associated with flight time, area covered, weather-sensitivity, and regulatory restrictions. It is likely there will be continued integration between remote sensing, proximal sensing, and vehicle-mounted and in situ sensing platforms. Management systems will more fully integrate sensor information from these various platforms to provide timely, and increasingly more automated, control of inputs such as fertilizer, irrigation, and pesticides (Mulla, 2013; Sishodia et al., 2020). As technologies continue to advance, there will be limitations (see Chapters 2 and 3 for a more detailed discussion); however, as research progresses, these limitations will be reduced, presenting progressively greater opportunities to manage agriculture systems globally.

ACKNOWLEDGMENTS

This paper was partially funded by National Science Foundation grants 2202706 and 2026431, South Dakota State University, the University of Nebraska Experiment Station, the Kansas State Experiment Station, USDA Natural Resources Conservation Service Conservation Innovation Grants (NR233A750011G019), the USDA Agriculture and Food Research Initiative, the American Coalition for Ethanol, GEVO, the

South Dakota Corn Utilization Council, and the South Dakota Soybean Research and Promotion Council. The use of trade names is for the convenience of the reader and does not imply endorsement by the authors.

Study Questions

1. What portion of the electromagnetic spectrum constitutes the visible range (visible to the human eye)?

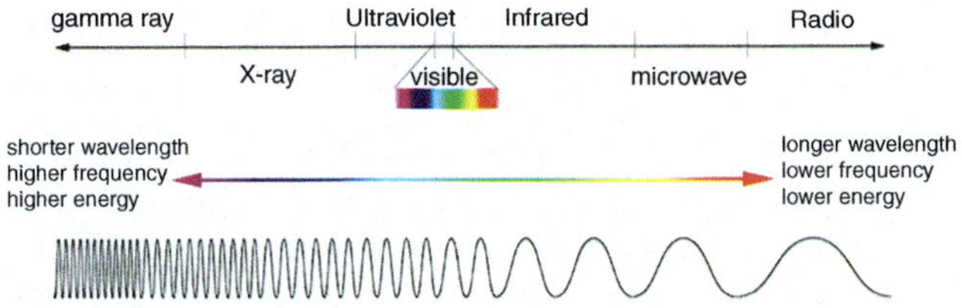

Source: NASA GSFC.

2. Provide a definition of "vegetation index."
3. Differentiate "remote" sensing from "proximal" sensing.
4. What is the term for a sensor that produces its own energy for sensing?
5. What is the principal wavelength range for thermal infrared sensing?
6. Which remote sensing platform is best suited to the coverage of many square miles on a weekly basis?
7. Which remote sensing platform can be readily deployed on short notice for crop scouting?

REFERENCES

Abd El-Ghany, N. M., Abd El-Aziz, S. E., & Marei, S. S. (2020). A review: Application of remote sensing as a promising strategy for insect pests and disease management. *Environmental Science and Pollution Research*, *27*(27), 33503–33515. https://doi.org/10.1007/s11356-020-09517-2.

Abd Kharim, M. N., Wayayok, A., Abdullah, A. F., Shariff, A. R. M., Husin, E. M., & Mahadi, M. R. (2022). Predictive zoning of pest and disease infestations in rice fields based on UAV aerial imagery. *Egyptian Journal of Remote Sensing and Space Science*, *25*(3), 831–840. http://doi.org/10.1016/j.ejrs.2022.08.001.

Adams, M. L., Norvell, W. A., Philpot, W. D., & Peverly, J. H. (2000). Spectral detection of micronutrient deficiency in 'Bragg' soybean. *Agronomy Journal*, *92*(2), 261–268. https://doi.org/10.2134/agronj2000.922261x.

Angelopoulou, T., Tziolas, N., Balafoutis, A., Zalidis, G., & Bochtis, D. (2019). Remote sensing techniques for soil organic carbon estimation: A review. *Remote Sensing*, *11*(6), 676. https://doi.org/10.3390/rs11060676.

Babaeian, E., Sadeghi, M., Jones, S. B., Montzka, C., Vereecken, H., & Tuller, M. (2019). Ground, proximal, and satellite remote sensing of soil moisture. *Reviews of Geophysics*, *57*(2), 530–616. https://doi.org/10.1029/2018RG000618.

Barnes, E. M., Clarke, T. R., Richards, S. E., Colaizzi, P. D., Haberland, J., Kostrzewski, M., Waller, P., Choi, C., Riley, E., Thompson, T., Lascano, R. J., Li, H., & Moran, M. S. (2000). Coincident detection of crop water stress, nitrogen status and canopy density using ground-based multispectral data [CD]. *Proceedings of the Fifth International Conference on Precision Agriculture*, No. 1356. International Society of Precision Agriculture.

Bellvert, J., Zarco-Tejada, P. J., Girona, J., & Fereres, E. (2014). Mapping crop water stress index in a 'Pinot-noir' vineyard: Comparing ground measurements with thermal remote sensing from an unmanned aerial vehicle. *Precision Agriculture*, *15*, 361–376. https://doi.org/10.1007/s11119-013-9334-5.

Berni, J. A. J., Zarco-Tejada, P. J., Sepulcre-Canto, G., Fereres, E., & Villalobos, F. (2009). Mapping canopy conductance and CWSI in olive orchards using high-resolution thermal remote sensing imagery. *Remote Sensing of Environment*, *113*(11), 2380–2388. https://doi.org/10.1016/j.rse.2009.06.018.

Bhatti, A. U., Mulla, D. J., & Frazier, B. E. (1991). Estimation of soil properties and wheat yields on complex eroded hills using geostatistics and thematic mapper images. *Remote Sensing of Environment*, *37*(3), 181–191. https://doi.org/10.1016/0034-4257(91)90080-P.

Brewster, C., Allen, J., & Kopp, D. (1999). IPM from space: Using satellite imagery to construct regional crop maps for studying crop–insect interaction. *American Entomologist*, *45*(2), 105–117. https://doi.org/10.1093/ae/45.2.105.

Calderón, R., Navas-Cortés, J. A., Lucena, C., & Zarco-Tejada, P. J. (2013). High-resolution airborne hyperspectral and thermal imagery for early detection of *Verticillium* wilt of olive using fluorescence, temperature and narrow-band spectral indices. *Remote Sensing of Environment*, *139*, 231–245. https://doi.org/10.1016/j.rse.2013.07.031.

Chang, J., Clay, S. A., Clay, D. E., & Dalsted, K. (2004). Detecting weed-free and weed-infested areas of a soybean field using near-infrared spectral data. *Weed Science*, *52*(4), 642–648. https://doi.org/10.1614/WS-03-074R1.

Chen, D., Hein, G. L., Adams-Selin, R., Wang, L., Zhang, J., Zhou, X., Ma, H., McMechan, J., & Shi, Y. (2024). Spatial relationship between pre-harvest hail and the impact from the wheat streak mosaic disease complex by using remote sensing data. *Crop Protection*, *179*, 106627. https://doi.org/10.1016/j.cropro.2024.106627.

Chen, D., Shi, Y., Huang, W., Zhang, J., & Wu, K. (2018). Mapping wheat rust based on high spatial resolution satellite imagery. *Computers and Electronics in Agriculture*, *152*, 109–116. https://doi.org/10.1016/j.compag.2018.07.002.

Chen, P., Xia, T., & Yang, G. (2024). A new strategy for weed detection in maize fields. *European Journal of Agronomy*, *159*, 127289. https://doi.org/10.1016/j.eja.2024.127289.

Clay, D. E., Brugler, S., & Joshi, B. (2024). Will artificial intelligence and machine learning change agriculture: A special issue. *Agronomy Journal*, *166*(3), 791–794. https://doi.org/10.1002/agj2.21555.

Clay, D. E., Kim, K.-I., Chang, J., Clay, S. A., & Dalsted, K. (2006). Characterizing water and nitrogen stress in corn using remote sensing. *Agronomy Journal*, *98*(3), 579–587. https://doi.org/10.2134/agronj2005.0204.

Clay, S. A., Lems, G. J., Clay, D. E., Forcella, F., Ellsbury, M. M., & Carlson, C. G. (1999). Sampling weed spatial variability on a fieldwide scale. *Weed Science*, *47*(6), 674–681. https://doi.org/10.1017/S0043174500091323.

Colaço, A. F., Whelan, B. M., Bramley, R. G. V., Richetti, J., Fajardo, M., McCarthy, A. C., Perry, E. M., Bender, A., Leo, S., Fitzgerald, G. J., & Lawes, R. A. (2024). Digital strategies for nitrogen management in grain production systems: Lessons from multi-method assessment using on-farm experimentation. *Precision Agriculture*, *25*, 983–1013. https://doi.org/10.1007/s11119-023-10102-z.

Colwell, R. N. (1983). *Manual of remote sensing* (2nd ed.). American Society for Photogrammetry and Remote Sensing.

Dash, J., & Curran, P. J. (2004). The MERIS terrestrial chlorophyll index. *International Journal of Remote Sensing*, *25*(23), 5403–5413. https://doi.org/10.1080/0143116042000274015.

Datt, B. (1999). Visible/near infrared reflectance and chlorophyll content in Eucalyptus leaves. *International Journal of Remote Sensing*, *20*(14), 2741–2759. https://doi.org/10.1080/014311699211778.

Daughtry, C. S. T., Walthall, C. L., Kim, M. S., de Colstoun, E. B., & McMurtrey, J. E., III. (2000). Estimating corn leaf chlorophyll concentration from leaf and canopy reflectance. *Remote Sensing of Environment*, *74*(2), 229–239. https://doi.org/10.1016/S0034-4257(00)00113-9.

de Castro, A. I., López-Granados, F., & Jurado-Expósito, M. (2013). Broad-scale cruciferous weed patch classification in winter wheat using QuickBird imagery for in-season site-specific control. *Precision Agriculture*, *14*, 392–413. https://doi.org/10.1007/s11119-013-9304-y.

Diaz-Gonzalez, F. A., Vuelvas, J., Correa, C. A., Vallejo, V. E., & Patino, D. (2022). Machine learning and remote sensing techniques applied to estimate soil indicators – Review. *Ecological Indicators*, *135*, 108517. https://doi.org/10.1016/j.ecolind.2021.108517.

Ellsbury, M. M., Woodson, W. D., Clay, S. A., Malo, D., Schumacher, J., Clay, D. E., & Carlson, C. G. (1998). Geostatistical characterization of the spatial distribution of adult corn rootworms (Coleoptera: Chrysomelidea) emergence. *Population Ecology*, *27*(4), 910–917. https://doi.org/10.1093/ee/27.4.910.

Erickson, J. D. (1984). The LACIE experiment in satellite aided monitoring of global crop production. Chapter 8. In G. M. Woodwell (Ed.), *The role of terrestrial vegetation in the global carbon cycle: Measurement by remote sensing* (pp. 191–217). John Wiley & Sons Ltd.

Feng, W., Shen, W., He, L., Duan, J., Guo, B., Li, Y., Wang, C., & Guo, T. (2016). Improved remote sensing detection of wheat powdery mildew using dual-green vegetation indices. *Precision Agriculture*, *17*, 608–627. https://doi.org/10.1007/s11119-016-9440-2.

Ferguson, R., & Rundquist, D. (2018). Remote sensing site-specific crop management. In D. K. Shannon, D. E. Clay, & N. R. Kitchen (Eds.), *Precision agriculture basics* (1st ed., pp. 103–117). ASA, CSA, and SSSA.

Fitzgerald, G. J., Maas, S. J., & Detar, W. R. (2004). Spider mite detection and canopy component mapping in cotton using hyperspectral imagery and spectral mixture analysis. *Precision Agriculture*, *5*, 275–289. https://doi.org/10.1023/B:PRAG.0000032766.88787.5f.

Ford, T. W., & Quiring, S. M. (2019). Comparison of contemporary in situ, model, and satellite remote sensing soil moisture with a focus on drought monitoring. *Water Resources Research*, *55*(2), 1565–1582. https://doi.org/10.1029/2018WR024039.

Franke, J., & Menz, G. (2007). Multi-temporal wheat disease detection by multi-spectral remote sensing. *Precision Agriculture*, *8*, 161–172. https://doi.org/10.1007/s11119-007-9036-y.

Fussell, J., Rundquist, D. C., & Harrington, J. A., Jr. (1986). On defining remote sensing. *Photogrammetric Engineering and Remote Sensing*, *52*(9), 1507–1511.

Gao, J., Liao, W., Nuyttens, D., Lootens, P., Vangeyte, J., Pižurica, A., He, Y., & Pieters, J. G. (2018). Fusion of pixel and object-based features for weed mapping using unmanned aerial vehicle imagery. *International Journal of Applied Earth Observation and Geoinformation*, *67*, 43–53. https://doi.org/10.1016/j.jag.2017.12.012.

Gitelson, A. A., Kaufmann, Y. J., & Merzlyak, M. N. (1996). Use of a green channel in remote sensing of global vegetation from EOS-MODIS. *Remote Sensing of Environment*, *58*(3), 289–298. https://doi.org/10.1016/S0034-4257(96)00072-7.

Gitelson, A. A., Viña, A., Ciganda, V., Rundquist, D. C., & Arkebauer, T. J. (2005). Remote estimation of canopy chlorophyll content in crops. *Geophysical Research Letters*, *32*(8), L08403. https://doi.org/10.1029/2005GL022688.

Goodman, M. S. (1959). A technique for the identification of farm crops on aerial photographs. *Photogrammetry Engineering*, *25*, 131–137.

Holland, K. H., & Schepers, J. S. (2013). Use of a virtual-reference concept to interpret active crop canopy sensor data. *Precision Agriculture*, *14*, 71–85. https://doi.org/10.1007/s11119-012-9301-6.

Hou, J., Li, L., & He, J. (2016). Detection of grapevine leafroll disease based on 11-index imagery and ant colony clustering algorithm. *Precision Agriculture*, *17*, 488–505. https://doi.org/10.1007/s11119-016-9432-2.

Haboudane, D., Miller, J. R., Tremblay, N., Zarco-Tejada, P. J., & Dextraze, L. (2002). Integrated narrow-band vegetation indices for prediction of crop chlorophyll content for application to precision agriculture. *Remote Sensing of Environment*, *81*, 416–426.

Houborg, R., & McCabe, M. F. (2018). A Cubesat enabled spatio-temporal enhancement method (CESTEM) utilizing Planet, Landsat and MODIS data. *Remote Sensing of Environment*, *209*, 211–226. https://doi.org/10.1016/j.rse.2018.02.067.

Huete, A. R. (1988). A soil-adjusted vegetation index (SAVI). *Remote Sensing of Environment*, *25*(3), 295–309. https://doi.org/10.1016/0034-4257(88)90106-X.

Idso, S. B., Jackson, R. D., Pinter, P. J., Jr., Reginato, R. J., & Hatfield, J. L. (1981). Normalizing the stress-degree-day parameter for environmental variability. *Agricultural Meteorology*, *24*, 45–55. https://doi.org/10.1016/0002-1571(81)90032-7.

Jackson, R. D., Kustas, W. P., & Choudhury, B. J. (1988). A re-examination of the crop water stress index. *Irrigation Science*, *9*, 309–317. https://doi.org/10.1007/BF00296705.

Khanal, S., Fulton, J., & Shearer, S. (2017). An overview of current and potential applications of thermal remote sensing in precision agriculture. *Computers and Electronics in Agriculture*, *139*, 22–32. https://doi.org/10.1016/j.compag.2017.05.001.

Khanal, S., Kc, K., Fulton, J. P., Shearer, S., & Ozkan, E. (2020). Remote sensing in agriculture—accomplishments, limitations, and opportunities. *Remote Sensing*, *12*(22), 3783. https://doi.org/10.3390/rs12223783.

Klein, I., Uereyen, S., Eisfelder, C., Pankov, V., Oppelt, N., & Kuenzer, C. (2023). Application of geospatial and remote sensing data to support locust management. *International Journal of Applied Earth Observation and Geoinformation*, *117*, 103212. https://doi.org/10.1016/j.jag.2023.103212.

Kouadio, L., El Jarroudi, M., Belabess, Z., Laasli, S.-E., Kadir Roni, M. Z., Amine, I. D. I., Mokhtari, N., Mokrini, F., Junk, J., & Lahlali, R. (2023). A review on UAV-based applications for plant disease detection and monitoring. *Remote Sensing*, *15*(17), 4273. https://doi.org/10.3390/rs15174273.

Kumar, A., Suk Lee, W., Ehsani, R. J., Albrigo, L. G., Yang, C., & Mangan, R. L. (2012). Citrus greening disease detection using aerial hyperspectral and multispectral imaging techniques. *Journal of Applied Remote Sensing*, *6*(1), 063542. https://doi.org/10.1117/1.JRS.6.063542.

Kyaw, T., Ferguson, R. B., Adamchuk, V. I., Marx, D. B., Tarkalson, D. D., & McCallister, D. L. (2008). Delineating site-specific management zones for pH-induced iron chlorosis. *Precision Agriculture*, *9*, 71–84. https://doi.org/10.1007/s11119-008-9059-z.

Maes, W. H., & Steppe, K. (2019). Perspectives for remote sensing with unmanned aerial vehicles in precision agriculture. *Trends in Plant Science*, *24*(2), 152–164. https://doi.org/10.1016/j.tplants.2018.11.007.

Mahlein, A.-K., Rumpf, T., Welke, P., Dehne, H.-W., Plümer, L., Steiner, U., & Oerke, E.-C. (2013). Development of spectral indices for detecting and identifying plant diseases. *Remote Sensing of Environment*, *128*, 21–30. https://doi.org/10.1016/j.rse.2012.09.019.

Mahlein, A.-K., Steiner, U., Dehne, H.-W., & Oerke, E. C. (2010). Spectral signatures of sugar beet leaves for the detection and differentiation of diseases. *Precision Agriculture*, *11*, 413–431. https://doi.org/10.1007/s11119-010-9180-7.

Martins, M. (2022). *Integration of remote sensing approaches for in-season nitrogen management* [Doctoral dissertation, Louisiana State University and Agricultural and Mechanical College]. LSU Doctoral Dissertations. https://doi.org/10.31390/gradschool_dissertations.5939.

Mesías-Ruiz, G. A., Peña, J. M., de Castro, A. I., Borra-Serrano, I., & Dorado, J. (2024). Cognitive computing advancements: Improving precision crop protection through UAV imagery for targeted weed monitoring. *Remote Sensing*, *16*(16), 3026. https://doi.org/10.3390/rs16163026.

Mink, R., Dutta, A., Peteinatos, G. G., Sökefeld, M., Engels, J. J., Hahn, M., & Gerhards, R. (2018). Multi-temporal site-specific weed control of *Cirsium arvense* (L.) Scop. and *Rumex crispus* L. in maize and sugar beet using unmanned aerial vehicle based mapping. *Agriculture*, *8*(5), 65. https://doi.org/10.3390/agriculture8050065.

Mirandilla, J. R., Yamashita, M., Yoshimura, M., & Paringit, E. C. (2023). Leaf spectral analysis for detection and differentiation of three major rice diseases in the Philippines. *Remote Sensing*, *15*(12), 3058. https://doi.org/10.3390/rs15123058.

Mirik, M., Ansley, R. J., Michels, G. J., Jr., & Elliott, N. C. (2012). Spectral vegetation indices selected for quantifying Russian wheat aphid (*Diuraphis noxia*) feeding damage in wheat (*Triticum aestivum* L.). *Precision Agriculture*, *13*, 501–516. https://doi.org/10.1007/s11119-012-9264-7.

Mogili, U. R., & Deepak, B. B. V. L. (2018). Review on application of drone systems in precision agriculture. *Procedia Computer Science*, *133*, 502–509. https://doi.org/10.1016/j.procs.2018.07.063.

Morellos, A., Tziotzios, G., Orafanidou, C., Pantazi, X. E., Sarantaris, C., Maliogka, V., Alexandridis, T. K., & Moshou, D. (2020). Non-destructive early detection and quantitative severity stage classification of tomato chlorosis virus (ToCV) infection in young tomato plants using VIS–NIR spectroscopy. *Remote Sensing*, *12*(12), 1920. https://doi.org/10.3390/rs12121920.

Motohka, T., Nasahara, K. N., Oguma, H., & Tsuchida, S. (2010). Applicability of green-red vegetation index for remote sensing of vegetation phenology. *Remote Sensing*, *2*(10), 2369–2387. https://doi.org/10.3390/rs2102369.

Mulla, D. J. (2013). Twenty five years of remote sensing in precision agriculture: Key advances and remaining knowledge gaps. *Biosystems Engineering*, *114*(4), 358–371. https://doi.org/10.1016/j.biosystemseng.2012.08.009.

Nguyen, H. D. D., & Nansen, C. (2020). Hyperspectral remote sensing to detect leafminer-induced stress in bok choy and spinach according to fertilizer regime and timing. *Pest Management Science*, *76*(6), 2208–2216. https://doi.org/10.1002/ps.5758.

Nikolić, N., Rizzo, D., Marraccini, E., Gotor, A. A., Mattivi, P., Saulet, P., Persichetti, A., & Masin, R. (2021). Site- and time-specific early weed control is able to reduce herbicide use in maze—a case study. *Italian Journal of Agronomy*, *16*(4), 1780. https://doi.org/10.4081/ija.2021.1780.

Okyere, F. G., Cudjoe, D., Sadeghi-Tehran, P., Virlet, N., Riche, A. B., Castle, M., Greche, L., Simms, D., Mhada, M., Mohareb, F., & Hawkesford, M. J. (2023). Modeling the spatial-spectral characteristics of plants for nutrient status identification using hyperspectral data and deep learning methods. *Frontiers in Plant Science*, *14*, 1209500. https://doi.org/10.3389/fpls.2023.1209500.

Osborne, S. L., Schepers, J. S., Francis, D. D., & Schlemmer, M. R. (2002). Detection of phosphorus and nitrogen deficiencies in corn using spectral radiance measurements. *Agronomy Journal*, *94*(6), 1215–1221. https://doi.org/10.2134/agronj2002.1215.

Peteinatos, G. G., Reichel, P., Karouta, J., Andújar, D., & Gerhards, R. (2020). Weed identification in maize, sunflower and potatoes with the aid of convolutional neural networks. *Remote Sensing*, *12*(24), 4185. https://doi.org/10.3390/rs12244185.

Petropoulos, G. P., Ireland, G., & Barrett, B. (2015). Surface soil moisture retrievals from remote sensing: Current status, products & future trends. *Physics and Chemistry of the Earth, Parts A/B/C, 83–84*, 36–56. https://doi.org/10.1016/j.pce.2015.02.009.

Pinter, P. J., Jr., Hatfield, J. L., Schepers, J. S., Barnes, E. M., Moran, M. S., Daughtry, C. S. T., & Upchurch, D. R. (2003). Remote sensing for crop management. *Photogrammetric Engineering and Remote Sensing*, *69*(6), 647–664. https://doi.org/10.14358/PERS.69.6.647.

Pruitt, E. (1979). The Office of Naval Research and Geography. *Annals of the Association of American Geographers*, *69*(1), 103–108. https://doi.org/10.1111/j.1467-8306.1979.tb01235.x.

Puig, E., Gonzalez, F., Hamilton, G., & Grundy, P. (2015). Assessment of crop insect damage using unmanned aerial systems: A machine learning approach. *21st International Congress on Modelling and Simulation*. The Modeling and Simulation Society of Australia and New Zealand. https://www.mssanz.org.au/modsim2015/F12/puig.pdf.

Puri, J. S., & Kotze, A. (2022). Evaluation of SRGAN algorithm for superresolution of satellite imagery on different sensors. *AGILE GIScience Series*, *3*, 57. https://doi.org/10.5194/agile-giss-3-57-2022.

Qi, J., Chehbouni, A., Huete, A. R., Kerr, Y. H., & Sorooshian, S. (1994). A modified soil adjusted vegetation index. *Remote Sensing of Environment*, *48*(2), 119–126. https://doi.org/10.1016/0034-4257(94)90134-1.

Qi, T., Dong, Y., Li, X., Zhao, M., & Huang, W. (2024). Fall armyworm habitat analysis in Africa with multi-source earth observation data. *Computers and Electronics in Agriculture*, *225*, 109283. https://doi.org/10.1016/j.compag.2024.109283.

Raza, M. M., Harding, C., Liebman, M., & Leandro, L. F. (2020). Exploring the potential of high-resolution satellite imagery for the detection of soybean sudden death syndrome. *Remote Sensing*, *12*(7), 1213. https://doi.org/10.3390/rs12071213.

Reddy, G. P. O. (2018). Satellite remote sensing sensors: Principles and applications. In G. Reddy & S. Singh (Eds.), *Geospatial technologies in land resources mapping, monitoring and management. Geotechnologies and the environment, Vol. 21*. (pp. 21–43). Springer.

Rejeb, A., Abdollahi, A., Rejeb, K., & Treiblmaier, H. (2022). Drones in agriculture: A review and bibliometric analysis. *Computers and Electronics in Agriculture, 198,* 107017. https://doi.org/10.1016/j.compag.2022.107017.

Richards, J. A. (2009). *Remote sensing with imaging radar* (pp. 172–173). Springer-Verlag.

Roberts, D. F., Adamchuk, V. I., Shanahan, J. F., Ferguson, R. B., & Schepers, J. S. (2010). Estimation of surface soil organic matter using a ground-based active sensor and aerial imagery. *Precision Agriculture*, *12*, 82–102. https://doi.org/10.1007/s11119-010-9158-5.

Rogovska, N., & Blackmer, A. M. (2009). Remote sensing of soybean canopy as a tool to map high pH, calcareous soils at field scale. *Precision Agriculture*, *10*, 175–187. https://doi.org/10.1007/s11119-008-9087-8.

Rondeaux, G., Steven, M., & Baret, F. (1996). Optimization of soil-adjusted vegetation indices. *Remote Sensing of Environment*, *55*, 95–107.

Rouse, J. W., Jr., Haas, R. H., Scheel, J. A., & Deering, D. W. (1974). Monitoring vegetation systems in the Great Plains with ERTS. *Proceedings, 3rd Earth Resource Technology Satellite (ERTS) Symposium*, *1*, 48–62. NASA.

Rundquist, D. C., & Samson, S. A. (1988). *A guide to the practical use of aerial color-infrared photography in agriculture*. Educational Circular #7,. In *Conservation and survey division, Institute of Agriculture and Natural Resources*. University of Nebraska-Lincoln.

Samreen, T., Tahir, S., Arshad, S., Kanwal, S., Anjum, F., Nazir, M. Z., & Sidra-Tul-Muntaha (2023). Remote sensing for precise nutrient management in agriculture. *Environmental Sciences Proceedings*, *23*(1), 32. https://doi.org/10.3390/environsciproc2022023032.

Santoso, H., Gunawan, T., Jatmiko, R. H., Darmosarkoro, W., & Minasny, B. (2011). Mapping and identifying basal stem rot disease in oil palms in North Sumatra with QuickBird imagery. *Precision Agriculture*, *12*, 233–248. https://doi.org/10.1007/s11119-010-9172-7.

Sapkota, B., Singh, V., Cope, D., Valasek, J., & Bagavathiannan, M. (2020). Mapping and estimating weeds in cotton using unmanned aerial systems-borne imagery. *AgriEngineering*, *2*(2), 350–366. https://doi.org/10.3390/agriengineering2020024.

Shiratsuchi, L., Ferguson, R., Shanahan, J., Adamchuk, V., Rundquist, D., Marx, D., & Slater, G. (2011). Water and nitrogen effects on active canopy sensor vegetation indices. *Agronomy Journal*, *103*(6),

1815–1826. https://doi.org/10.2134/agronj2011.0199.

Sishodia, R. P., Ray, R. L., & Singh, S. K. (2020). Applications of remote sensing in precision agriculture: A review. *Remote Sensing*, *12*(19), 3136. https://doi.org/10.3390/rs12193136.

Stansell, J. (2021). *Development and automation of a sensor-based fertigation management framework for improved nitrogen use efficiency and profitability in irrigated row crop production systems* [Master's thesis, University of Nebraska-Lincoln]. Digital Commons, University of Nebraska-Lincoln. https://digitalcommons.unl.edu/embargotheses/207/.

Steele, M., Gitelson, A. A., & Rundquist, D. (2008). Nondestructive estimation of leaf chlorophyll content in grapes. *American Journal of Enology and Viticulture*, *59*, 299–305. https://doi.org/10.5344/ajev.2008.59.3.299.

Stilwell, A. R., Hein, G. L., Zygielbaum, A. I., & Rundquist, D. C. (2013). Proximal sensing to detect symptoms associated with wheat curl mite-vectored viruses. *International Journal of Remote Sensing*, *34*(14), 4951–4966. https://doi.org/10.1080/01431161.2013.783945.

Swain, S., Rundquist, D., Arkebauer, T. J., Narumalani, S., & Wardlow, B. (2012). Non-invasive estimation of relative water content in soybean leaves using infrared thermography. *Israeli Journal of Plant Sciences*, *60*(1–2), 25–36. https://www.tandfonline.com/doi/abs/10.1560/IJPS.60.1-2.25.

Thapa, B., Wolter, P. T., Sturtevant, B. R., & Townsend, P. A. (2022). Linking remote sensing and insect defoliation biology—A cross-system comparison. *Remote Sensing of Environment*, *281*, 11236. https://doi.org/10.1016/j.rse.2022.113236.

Thorp, K. R., & Tian, L. F. (2004). A review on remote sensing of weeds in agriculture. *Precision Agriculture*, *5*, 477–508. https://doi.org/10.1007/s11119-004-5321-1.

Varvel, G. E., Schepers, J. S., & Francis, D. D. (1997). Ability for in-season correction of nitrogen deficiency in corn using chlorophyll meters. *Soil Science Society of America Journal*, *61*(4), 1233–1239. https://doi.org/10.2136/sssaj1997.03615995006100040032x.

Vrindts, E., De Baerdemaeker, J., & Ramon, H. (2002). Weed detection using canopy reflection. *Precision Agriculture*, *3*, 63–80. https://doi.org/10.1023/A:1013326304427.

Wang, J., Zhen, J., Hu, W., Chen, S., Lizaga, I., Zeraatpisheh, M., & Yang, X. (2023). Remote sensing of soil degradation: Progress and perspective. *International Soil and Water Conservation Research*, *11*(3), 429–454. https://doi.org/10.1016/j.iswcr.2023.03.002.

Weiss, M., Jacob, F., & Duveiller, G. (2020). Remote sensing for agricultural applications: A meta-review. *Remote Sensing of Environment*, *236*, 111402. https://doi.org/10.1016/j.rse.2019.111402.

Whitcraft, A. K., Becker-Reshef, I., & Justice, C. O. (2015). A framework for defining spatially explicit earth observation requirements for a global agricultural monitoring initiative (GEOCLAM). *Remote Sensing*, *7*(2), 1461–1481. https://doi.org/10.3390/rs70201461.

Willers, J. l., Jenkins, J. N., Ladner, W. L., Gerard, P. D., Boykin, D. L., Hood, K. B., McKibben, P. L., Samson, S. A., & Bethel, M. M. (2005). Site-specific approaches to cotton insect control. Sampling and remote sensing analysis techniques. *Precision Agriculture*, *6*, 431–452. https://doi.org/10.1007/s11119-005-3680-x.

Yang, Z., Rao, M. N., Elliott, N. C., Kindler, S. D., & Popham, T. W. (2009). Differentiating stress induced by greenbugs and Russian wheat aphids in wheat using remote sensing. *Computers and Electronics in Agriculture*, *67*(1–2), 64–70. https://doi.org/10.1016/j.compag.2009.03.003.

Yuan, L., Pu, R., Zhang, J., Wang, J., & Yang, H. (2016). Using high spatial resolution satellite imagery for mapping powdery mildew at a regional scale. *Precision Agriculture*, *17*, 332–348. https://doi.org/10.1007/s11119-015-9421-x.

Zhao, D., Cao, Y., Li, J., Cao, Q., Li, J., Guo, F., Feng, S., & Xu, T. (2024). Early detection of rice leaf blast disease using unmanned aerial vehicle remote sensing: A novel approach integrating a new spectral vegetation index and machine learning. *Agronomy*, *14*(3), 602. https://doi.org/10.3390/agronomy14030602.

Remote Sensing for Water Stress Management

15

Lorena N. Lacerda and Meetpal Kukal

Chapter Purpose

This chapter is new to this edition of the Precision Agriculture Basics textbook. It provides an overview of how aerial and satellite remote sensing can be used to monitor and manage crop water stress. This chapter also discusses how the principles of crop physiology allow for the direct and indirect estimation of plant water status. Owing to the variety of sensors and the challenges of using remote sensors in water stress monitoring, this chapter should be read in combination with Chapter 16, Precision Irrigation.

Introduction to Crop Water Stress Monitoring Using Remote Sensing

Insufficient water can rapidly lead to wilting; reductions in shoot length, leaf size, and plant height; and reduced plant health yields (Hussain et al., 2019; Kuwayama et al., 2018; Seleiman et al., 2021). The effect of water on yield makes accurate monitoring and efficient management of crop water status essential. Classically, water stress monitoring has relied on soil moisture and plant-based sensors (Dong et al., 2024); for more information on these sensors, see Chapters 12 and 13. However, soil- and plant-based measurement systems are expensive and are not designed to track water stress over large areas. An alternative to soil-based sensing is to use remote sensing to measure changes in plant responses to water stress. Owing to remote sensing technology advancements, monitoring water stress over large areas is feasible.

Water and Stomata

Effective water stress monitoring requires a clear understanding of water's role in plants. Water affects nearly every physiological process within a plant, and therefore, it is highly valued and conserved within the plant (Swaef et al., 2022). Water-stressed plants close their stomata, which reduces their photosynthetic capacity (Wu et al., 2022). Stomata are small pores on the leaf surface through which gas exchange happens. Specifically, carbon dioxide (CO_2) enters a plant via its stomata, while oxygen (a waste product of photosynthesis) and water (lost through transpiration) exit by the same path. The stomata include guard cells that surround the pore opening. These cells close in times of water and heat stress to prevent the excessive loss of water through transpiration (Iqbal et al., 2022; Wahab et al., 2022; Wu et al., 2022). The opening and closing of the guard cells rely on the adjustment of turgor pressure in the cells. This pressure is controlled by the influx and efflux of ions and water across the plasma membrane, facilitated by protein channels called aquaporins. These ion fluxes are triggered by environmental stimuli such as water or heat stress. For example, water stress triggers hormones in the plant to transport ions such as potassium or sodium out of the guard cells, which moves water out of these cells and reduces their turgor pressure. The reduced pressure ultimately results in the deflation of the guard cells, blocking the opening of the stomata and preventing excess water from leaving the plant (Ding & Chaumont, 2020).

Wilting is a sign that the plant is undergoing water or heat stress. Plants wilt because they are losing water faster than they are absorbing it. Different plants have different sensitives to water stress. For example, C3 plants such as wheat, soybeans, and sunflowers are more sensitive to water and heat stress than C4 plants such as corn, sugarcane, and sorghum. The difference is due to the physiological mechanisms of photosynthesis whereby C4 plants physically separate the light and dark reactions maintaining the efficiency of Rubisco, the key enzyme for carbon fixation. Rubisco is not efficient at fixing CO_2 into sugars when the stomata are closed for long periods; however, a C4 plant is able to maintain efficiency and conserve water by separating the two photosynthesis reactions (Khamis et al., 2025). This ability gives C4 crops an advantage over C3 plants, which do not have a separation of these reactions. C4 plants are extremely useful in certain regions such as in hot and dry locations in the southern United States.

Student problem: Precipitation use efficiency can be estimated using the equation $\frac{\text{Bushels}}{\text{inches}} = \frac{\text{Yield}}{\text{ET}}$, where ET (evapotranspiration) is estimated using the equation, ET = Precipitation−runoff−deep water drainage −Δ in soil water (Clay et al., 2014; Kim et al., 2008). If runoff, deep drainage, and Δ in soil water approach zero, then precipitation use efficiency is yield/precipitation.

Compare the precipitation use efficiencies of a C4 plant (corn) and a C3 plant (soybean) grown in South Dakota. If the corn yield is 150 bu/ac and the soybean yield is 65 bu/ac, what is the precipitation use efficiency if the growing season precipitation is 20 inches? Note a corn bushel is 56 lb at 15.5% moisture, whereas a soybean bushel is 60 lb at 13.5% moisture.

$$\text{Corn PUE} = \frac{150\,\text{bu}}{\text{acre} \times 20\,\text{inches}} = \frac{7.5\,\text{bu corn}}{\text{acre} \times \text{inch}}$$

$$\text{Soybean PUE} = \frac{65\,\text{bu}}{\text{acre} \times 20\,\text{inches}} = \frac{3.25\,\text{bu soybean}}{\text{acre} \times \text{inch}}$$

Water Stress Assessment Using Multispectral and Hyperspectral Imaging

Multispectral sensing typically captures from 3 to 10 bands, including reflectance information from the blue, green, red, and near infrared (NIR) bands from the electromagnetic spectrum. These bands can be used to calculate the normalized difference vegetation index (NDVI) or crop water stress index (CWSI). By measuring wavelengths between 250 and 15,000 nm, hyperspectral sensors can capture the finer details that multispectral cameras may miss. For more detailed information on multispectral and hyperspectral sensors, please see Chapter 14.

Drivers of Canopy Reflectance and Emissivity

The type and amount of energy reflected or absorbed by a plant can be used to assess the abiotic and biotic stresses that it is undergoing (Figure 15.1). This information can be collected using remote sensing sensors that collect multispectral information from the green, red, and NIR bands. For example, NIR reflectance (700–1,400 nm) increases with plant health based on each leaf having two layers: the palisade mesophyll cell layer on top and the spongy mesophyll cell layer below. In a healthy plant, the spongy mesophyll has large air pockets in which

Figure 15.1. Side-by-side comparison of thermal infrared (left) and RGB (right) images of a crop field collected with a WIRIS Pro SC camera (Workswell Infrared Cameras and Systems, Prague, CZ). Courtesy of Lorena Nunes Lacerda.

wavelengths such as NIR bounce around and scatter, leading to the reflection of higher levels of NIR. In an unhealthy plant, the spongy mesophyll cells collapse, which reduces scatter and thus reflectance. NDVI can be calculated with NIR and red wavelengths using the equation, $\mathrm{NDVI} = \frac{(\mathrm{NIR} - \mathrm{Red})}{(\mathrm{NIR} + \mathrm{Red})}$. This vegetative index is commonly used to indicate plant health by monitoring the amount of red that is absorbed and the amount of NIR that is reflected from the leaf. This absorption and scattering, along with the presence of leaf pigments such as chlorophyll (a and b), anthocyanins, and carotenoids, are important factors influencing how much light in blue, green, and red wavelengths is reflected. Carotenoids affect absorbance/reflectance in the blue wavelength/band, anthocyanins absorb light in the green and blue regions, and chlorophyll strongly absorbs light in the blue and red regions. An important characteristic of chlorophyll is the strong reflectance of green light. The higher the chlorophyll content in the plant, the greener and healthier it will appear. For more detailed information on vegetative indices and the use of proximal and remote sensors to assess plant health, please see Chapters 13 and 14.

Water Stress on Reflectance and Emissivity

Water stress can be monitored using different wavelengths in the electromagnetic spectrum. For example, within the visible spectrum (400–700 nm), leaf pigment content is the main driver of leaf reflectance (Gitelson & Merzlyak, 1998). Healthy leaves appear green because there is a spectral gap between 500 nm and 600 nm in which chlorophyll a and b can absorb efficiently. This absorption results in green light being reflected and a generally positive association between the color green and the health of a plant (Virtanen et al., 2020).

The infrared region is divided into near, mid, and longwave infrared, each providing different levels of useful information. NIR is "near" and can indicate general changes in leaf structure (i.e., collapse of the spongy mesophyll) in a water-stressed plant. The mid-infrared region, also known as short-wave infrared, can give crucial insights into the crop water status because water inside the leaf absorbs light in this part of the spectrum. Therefore, when using short-wave infrared, the less light reflected, the higher the water content because the water is absorbing the light. The long-wave infrared range is also known as the thermal infrared radiation, with plant temperature being primarily measured as heat radiation. Water-stressed plants often have higher temperatures than non-water-stressed plants because leaves are cooled by transpiring water through the stomata. Therefore, when long-wave infrared is measured, a greater emission of heat radiation indicates higher plant temperature levels.

Thermal Cameras to Detect Water Stress

Canopy temperature is a well-known indicator of crop water stress (Cohen et al., 2005). The correlation between canopy temperature and water stress is based on the observation that water transpiration cools the leaf surface. However, when plants experience water stress, transpiration decreases and the temperature increases (González-Dugo et al., 2006; Ihuoma & Madramootoo, 2017). Because water-stressed plants often have higher temperatures than non-water-stressed plants, temperature is often measured with infrared radiation using remote sensing (Figure 15.1). All objects on earth emit infrared radiation that can be captured and quantified by thermal sensors (Gade & Moeslund, 2014). Thermal sensors can detect infrared energy emitted in the range of 700–1,400 nm. These sensors were initially developed for military use owing to their capabilities for nighttime surveillance and later became available for commercial use in the 1980s. Currently, thermal cameras provide valuable information that can inform management decisions, using a variety of camera types (Table 15.1).

Thermal cameras are categorized based on the type of detector they use. For agricultural applications, most thermal imaging cameras use thermal detectors. These detectors, called microbolometers, depend on electrical resistance to produce an electrical output from the absorption of electromagnetic radiation (Gade & Moeslund, 2014; FLIR, 2008). They absorb electromagnetic radiation, particularly in the long-wave infrared range, and convert it into thermal energy. Microbolometers are a type of resistor, often made from vanadium oxide (VOx) or amorphous silicon (a-Si). These detectors work in a three-step process. First, when the infrared radiation strikes the VOx or the a-Si, a change in the electrical resistance is produced. Second, this change in resistance is converted to electrical signals. Third, the signal is processed into an image by a complementary metal-oxide semiconductor input cell.

Thermal cameras can be mounted on unmanned aerial vehicles (UAVs) and come in two categories: cooled and uncooled. Cooled thermal sensors use a cryocooler to lower the temperature of their camera to a cryogenic level (below −150 °C). Uncooled sensors do not have a cooling system. As a result, the camera can heat up because of the ambient temperature, leading to thermal-induced noise in the images (FLIR, 2015). Uncooled cameras are more affordable than cooled cameras because they have fewer components, are simpler to manufacture, and are lighter and smaller in size. Therefore, uncooled thermal cameras are most often used in UAV systems. However, because these cameras allow for thermal-induced noise, the temperature measured from the target object tends to be less accurate than that measured using a camera with a cooling system. A partial list of the popular thermal infrared cameras used in UAV platforms and their main specifications are listed in Table 15.1.

Cryogenically cooled sensors are usually seen in satellite platforms, such as the thermal infrared sensor (TIRS) used in the Landsat satellites (Irons et al., 2012). A partial list of currently available satellite platforms that have a thermal band along with their main specifications are listed in Table 15.2. A disadvantage of satellite systems is that they often have relatively coarse resolutions, making it difficult to detect small differences.

Canopy Temperature and Crop Water Stress Index

The crop water stress index (CWSI) was created to simplify the identification of water-stressed plants when using remote sensing. The CWSI is calculated based on canopy temperature, taking into consideration the air temperature and the difference between the actual water content in the air and the vapor pressure deficit. The CWSI outputs values from 0 to 1, with 0 representing a well-watered crop and 1 representing a crop with severe stress. The simplified version of the CWSI equation is CWSI = $(T_{canopy} - T_{wet})/(T_{dry} - T_{wet})$, where T_{canopy} represents the canopy temperature measured with any infrared instrument, including thermal cameras; T_{wet} is a baseline that represents a fully transpiring well-watered crop; and T_{dry} is a baseline that represents a stressed crop.

Table 15.1 Thermal Cameras for Use with UAVs.

Model	Manufacturer	Thermal accuracy (°C)	Wavelength (μm)	Thermal sensitivity (mK)
Boson+	Teledyne FLIR	±5	8–14	≤30
Zenmuse H20T	DJI	±2	8–14	≤50
Zenmuse XT2	DJI	±5	7.5–13.5	≤50
WIRIS Pro SC	Workswell	±2	7.5–13.5	≤30
CD319SPX	Seek Thermal	±5	7.8–14	65
ICI 8640 P	Infrared Cameras Inc.	±1	7–14	≤20

Table 15.2 Satellite Platforms with a Thermal Infrared Band.

Launch date	Name	Band(s)	Wavelength (μm)	Spatial resolution (m)	Revisit time
1999 (active)	MODIS	20–25, 27–36	3.66–4.549 6.535–14.385	1,000	1 day
1999 (active)	ASTER	10–14	8.125–11.65	90	16 days
2013 (active)	Landsat 8 (TIRS)	10 and 11	10.6–11.19 and 11.5–12.51	100	16 days
2021 (active)	Landsat 8 (TIRS)	10 and 11	10.6–11.19 and 11.5–12.51	100	16 days
2016 (active)	Sentinel 3	7–9	3.74–3.96, 10.85–11.65, and 11.95–12.95	1,000	27 days
Scheduled for 2026	TRISHNA	TIR1–TIR4	8.65–11.6	57	3 days

Multiple methods exist for calculating the dry and wet baseline temperatures. A common method is to use canopy temperatures of a known nonstressed field area and then use the difference between the canopy temperature and the vapor pressure deficit to calculate the wet baseline (Irmak et al., 2000). The dry baseline can be calculated by adding a crop-dependent number of degrees to the air temperature, such as 5 °C for cotton (Cohen et al., 2005), 4.65 °C for corn (Irmak et al., 2000), and 4 °C for citrus (González-Dugo et al., 2013).

Synthetic Aperture Radar

Remote sensing uses active and passive sensors. The multispectral, hyperspectral, and thermal sensors discussed in preceding sections are passive sensors. Passive sensors measure emitted or reflected energy from a target. In contrast, synthetic aperture radar (SAR) is an active sensor that emits an energy pulse and then measures and records the amount of energy that is reflected. Such sensors can be mounted on UAVs, or the information can be downloaded from satellite images such as Sentinel-1 (Meyer, 2019). The active energy emitted by SAR allows image collection at all times of the day regardless of cloud cover. These features give SAR an advantage over passive sensors, which are highly affected by cloud cover and dependent on sunlight (Cruz et al., 2022).

In SAR, microwaves are directed toward the target, and the backscatter of the light signal is measured by a radar antenna. The radar antenna contains an aperture through which backscatter signals are recorded. The antenna aperture records the backscatter from the emitted signals as the aircraft or satellite moves along their path, and the recorded strength and time delay of the backscatter signal is processed and converted into synthetic images of the specific target in the swath area (Cruz et al., 2022; Meyer, 2019).

SAR data have been widely used for monitoring and mapping different land cover types (Tsokas et al., 2022) and for assessing soil water because the signals are sensitive to plant biomass and soil dielectric properties (Gopi & Periasamy, 2023, Chapters 11 and 12). The dielectric property of a material governs how materials respond to an electric field. For example, in vegetation, both the canopy structure and the dielectric constant of the tissue affect how the emitted pulse will be reflected and/or transmitted as backscatter (Konings et al., 2019). In general, radar backscatter increases with water, and this characteristic can be used to evaluate the need for irrigation (Parag et al., 2024).

Opportunities and Challenges of OpenET

The application of remote sensing technology by using satellites to observe crop stress and water use from space has been widely explored for several decades (Estes et al., 1978; Idso et al., 1977). Even though there has been significant progress, practical challenges such as uncertainty in data accuracy, labor-intensive data processing, and scale differences between satellite data and actual plant water requirements have limited its widespread use.

OpenET was introduced in 2021 to meet some of these challenges, and it has revolutionized irrigation and water management by integrating remote sensing ET models with cloud computing (Melton et al., 2022). Measuring ET, which encompasses evaporation from the soil and transpiration from the plant, allows researchers and growers to track how water is moving through the whole system. Additionally, by using these models with cloud computing, data processing can be performed faster and easier. OpenET employs six established and scientifically validated models, each designed to estimate ET precisely at a high resolution down to 30 m and covering much of the United States. These models include ALEXI/DisALEXI (Anderson et al., 2007; Anderson et al., 2018), eeMETRIC (Allen et al., 2005, 2007, 2011), geeSEBAL, PT-JPL (Fisher et al., 2008), SIMS (Melton et al., 2012; Pereira et al., 2020), and SSEBop (Senay et al., 2013; Senay, 2018). Most of these models were designed using vegetation indices and land surface temperatures.

The individual ET models within OpenET are not new, but their unified integration into one platform, along with consistent satellite input data and an easy-to-use interface, represents a novel and practical development. The six models leverage data from the Landsat satellites, particularly surface reflectance data and thermal infrared measurements, that are crucial for accurately understanding surface energy dynamics. Because these remote sensing ET models measure reflectance at a detailed, field-level scale, they can detect variations in crop health, growth, and stress.

Consequently, OpenET can provide real-time, site-specific insights for individual fields and even specific areas within fields. This information has the potential to help farmers and water resource managers make more informed decisions throughout the growing season. For example, the OpenET and the Farm and Ranch Management Support (FARMS) system can generate field-specific ET reports, and the OpenET datasets are also available via an application programming interface.

To ensure the reliability of these ET estimates, Volk et al. (2024) compared OpenET's results with ground-measured values and found strong agreement between the two measurements. However, OpenET using SAR faces important challenges related to its accuracy and reliability. These issues include the following:

1. ET may be underestimated during the winter and spring.
2. Lower accuracy occurs in semi-arid climates with higher cloud cover compared with Mediterranean climates with less cloud cover.
3. Combining multiple models generally improves accuracy, but the individual models can produce vastly different results than the combined model.
4. The complex canopy structure in orchards and vineyards causes uncertainty in the ET estimates.

Challenges of Satellite and UAV Remote Sensing for Water Monitoring

The use of remote sensing for in-season irrigation management has several challenges. The first challenge is that uncooled thermal sensors placed on UAVs result in temperature drift (Aragon et al., 2020). The implication of temperature drift is that large temperature differences may exist between two consecutive images. This phenomenon limits the application of UAV-thermal data in precision agriculture applications. Despite efforts to offset this limitation (Virtue et al., 2021; Yuan & Hua, 2022), solutions have yet to be identified.

Another challenge for remote sensing in water management is that the cooled thermal sensors placed in satellites have low spatial and temporal resolution, which hinders making optimal management decisions from the data (Lacerda et al., 2021). Cloud cover can further decrease the availability of usable images, which limits the application of these images for in-season crop monitoring. For example, MODIS has a revisit time of 1 day and a spatial resolution of 1 km, which does not offer the level of detail needed for many decisions.

To address these challenges, techniques to utilize information collected by satellites with high revisit times have been explored (Cohen et al., 2017). These techniques are commonly known as

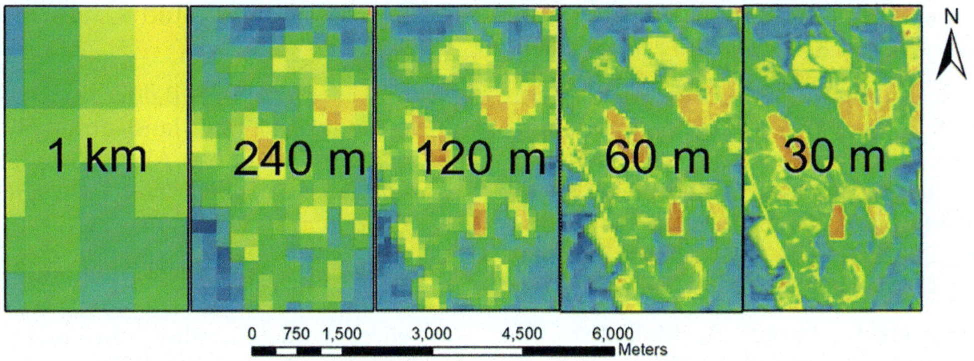

Figure 15.2 Results of a coarse satellite thermal image sharpened to multiple higher spatial resolutions, after an image fusion technique. Courtesy of Lorena Nunes Lacerda.

thermal sharpening and are based on the premise that land surface temperature is inversely related to vegetation cover. In other words, land surface temperature decreases with higher vegetation cover and increases when vegetation cover is low. The basic premise of this method is that high spatial resolution multispectral images can fill gaps in low-resolution thermal images (Huryna et al., 2019; Figure 15.2). Although great effort is being made in exploring and developing strategies to improve both UAV and satellite thermal imaging, further advancements are still needed to overcome these challenges.

Conclusion

This chapter discusses the impact of water stress in the plant and describes remote sensing techniques to measure water stress. Many advances in technology have made it feasible to use remote sensing and associated models to assess water stress in production fields. However, further advancements are still needed to overcome challenges such as the coarse resolution in the thermal imagery.

Study Questions

1. Why are stomata important for plant water management?
2. What are the limitations of thermal scanners placed on a UAV?
3. What are the limitations of thermal scanners placed on space-based platforms?
4. What are the benefits of SAR?
5. What are the disadvantages of SAR?

ACKNOWLEDGMENTS

Support for this document was provided by the Precision Farming Systems community in the American Society of Agronomy, South Dakota State University, USDA-NRCS-CIG (NR233A750011G019), USDA National Institute of Food and Agriculture (NIFA) under the Multi-State Hatch project [Project Number: KY006151-7008110-S1090], American Coalition for Ethanol, the South Dakota Corn Utilization Council, and the South Dakota Soybean Research and Promotion Council. The use of trade names is for the convenience of the reader and does not imply endorsement by the authors.

REFERENCES

Allen, R. G., Pereira, L. S., Howell, T. A., & Jensen, M. E. (2011). Evapotranspiration information reporting: I. Factors governing measurement accuracy. *Agricultural Water Management*, *98*(6), 899–920.

Allen, R. G., Tasumi, M., Morse, A., & Trezza, R. (2005). A Landsat-based energy balance and evapotranspiration model in Western US water rights regulation and planning. *Irrigation and Drainage Systems*, *19*, 251–268.

Allen, R. G., Tasumi, M., & Trezza, R. (2007). Satellite-based energy balance for mapping evapotranspiration with internalized calibration (METRIC)—Model. *Journal of Irrigation and Drainage Engineering*, *133*(4), 380–394.

Anderson, M., Gao, F., Knipper, K., Hain, C., Dulaney, W., Baldocchi, D., Eichelmann, E., Hemes, K., Yang, Y., Medellin-Azuara, J., & Kustas, W. (2018). Field-scale assessment of land

and water use change over the California Delta using remote sensing. *Remote Sensing*, *10*(6), 889.

Anderson, M. C., Norman, J. M., Mecikalski, J. R., Otkin, J. A., & Kustas, W. P. (2007). A climatological study of evapotranspiration and moisture stress across the continental United States based on thermal remote sensing: 2. Surface moisture climatology. *Journal of Geophysical Research: Atmospheres*, *112*(D11), D10117.

Aragon, B., Johansen, K., Parkes, S., Malbeteau, Y., Al-Mashharawi, S., Al-Amoudi, T., Andrade, C. F., Turner, D., Lucieer, A., & McCabe, M. F. (2020). A calibration procedure for field and UAV-based uncooled thermal infrared instruments. *Sensors*, *20*, 3316. https://doi.org/10.3390/s20113316.

Clay, D. E., Clay, S. A., Reitsma, K. D., Dunn, B. H., Smart, A. J., Carlson, C. G., Horvath, D., & Stone, J. L. (2014). Does the conversion of grasslands to row crop production in semi-arid areas threaten global food security? *Global Food Security*, *3*, 22–30. https://doi.org/10.1016/j.gfs.2013.12.002.

Cohen, Y., Agam, M., Klapp, I., Karnieli, A., Beeri, O., Alchanatis, V., & Sochen, N. (2017). Future approaches to facilitate large-scale adoption of thermal based images as key input in the production of dynamic irrigation management zones. *Advances in Animal Biosciences*, *8*, 546–550.

Cohen, Y., Alchanatis, V., Meron, M., Saranga, Y., & Tsipris, T. (2005). Estimation of leaf water potential by thermal imagery and spatial analysis. *Journal of Experimental Botany*, *56*, 1843–1852. https://doi.org/10.1093/jxb/eri174.

Cruz, H., Vestias, M., Monteiro, J., Neto, H., & Duarte, R. P. (2022). A review of synthetic-aperture radar image algorithms and implementation: A computational perspective. *Remote Sensing*, *14*, 1258.

Ding, L., & Chaumont, F. (2020). Are aquaporins expressed in stomatal complexes promising targets to enhance stomatal dynamics? *Frontiers in Plant Science*, *11*, 458. https://doi.org/10.3389/fpls.2020.00458.

Dong, H., Dong, J., Sun, S., Bai, T., Zhao, D., Yin, Y., Shen, X., Wang, Y., Zhang, Y., & Wang, Y. (2024). Crop water stress detection based on UAV remote sensing systems. *Agriculture Water Management*, *303*, 109059. https://doi.org/10.1016/j.agwat.2024.109059.

Estes, J. E., Jensen, J. R., & Tinney, L. R. (1978). Remote sensing of agricultural water demand information: A California study. *Water Resources Research*, *14*(2), 170–176.

Fisher, J. B., Tu, K. P., & Baldocchi, D. D. (2008). Global estimates of the land–atmosphere water flux based on monthly AVHRR and ISLSCP-II data, validated at 16 FLUXNET sites. *Remote Sensing of Environment*, *112*(3), 901–919.

FLIR Systems, Technical note. (2015). Cooled vs Uncooled thermal imaging: discover the difference. Available online: chrome-extension://efaidnbmnnnibpcajpcglclefindmkaj/http://www.flirmedia.com/MMC/THG/Brochures/RND_038/RND_038_EN.pdf (accessed on 5 March, 2025)

FLIR Systems, Technical Note. (2008). Uncooled detectors for thermal imaging cameras. *Infrared Imaging News*, *14.*, Available online: http://www.flirmedia.com/MMC/CVS/Appl_Stories/AS_0015_EN.pdf. (accessed on 6 March, 2025)

Gade, R., & Moeslund, T. B. (2014). Thermal cameras and applications: A survey. *Machine Vision and Applications*, *25*, 245–262. https://doi.org/10.1007/s00138-013-0570-5.

Gitelson, A. A., & Merzlyak, M. N. (1998). Remote sensing of chlorophyll concentration in higher plant leaves. *Advances in Space Research*, *22*, 686–692.

González-Dugo, M. P., Moran, M. S., Mateos, L., & Bryant, R. (2006). Canopy temperature variability as an indicator of crop water stress severity. *Irrigation Science*, *24*, 233–240. https://doi.org/10.1007/s00271-005-0022-8.

González-Dugo, V., Zarco-Tejada, P., Nicolás, E., Nortes, P. A., Alarcón, J. J., Intrigliolo, D. S., & Fereres, E. (2013). Using high resolution UAV thermal imagery to assess the variability in the water status of five fruit tree species within a commercial orchard. *Precision Agriculture*, *14*, 660–678. https://doi.org/10.1007/s11119-013-9322-9.

Gopi, N. M., & Periasamy, S. (2023). Appraising the crop health response to water stress from enhanced crop and soil estimates using SAR data and machine learning approaches. *International Journal of Remote Sensing*, *44*, 4190–4216.

Huryna, H., Cohen, Y., Karnieli, A., Panov, N., Kustas, W. P., & Agam, N. (2019). Evaluation of TsHARP utility for thermal sharpening of sentinel-3 satellite images using Sentinel-2 visual imagery. *Remote Sensing*, *11*(19), 2304. https://doi.org/10.3390/rs11192304.

Hussain, S., Hussain, S., Qadir, T., Khaliq, A., Shraf, U., Parveen, A., Saqib, M., & Rafiq, M. (2019). Drought stress in plants: An overview on implications, tolerance mechanisms and agronomic mitigation strategies. *Plant Science Today*, *6*, 389–402.

Idso, S. B., Jackson, R. D., & Reginato, R. J. (1977). Remote sensing for agricultural water management and crop yield prediction. *Agricultural Water Management*, *1*(4), 299–310.

Ihuoma, S. O., & Madramootoo, C. A. (2017). Recent advances in crop water stress detection. *Computers and Electronics in Agriculture*, *141*, 267–275. https://doi.org/10.1016/j.compag.2017.07.026.

Iqbal, N., Fatma, M., Gautam, H., Sejar, Z., Rasheed, F., Khan, I. R., Sofo, A., & Khan, N. A. (2022). Salicylic acid increases photosynthesis of drought grown mustard plants effectively with sufficient-N via regulation of ethylene, abscisic acid, and nitrogen-use efficiency. *Journal of Plant Growth Regulation*, *41*, 1966–1977. https://doi.org/10.1007/s00344-021-10565-2.

Irmak, S., Haman, D. Z., & Bastug, R. (2000). Determination of crop water stress index for irrigation timing and yield estimation in corn. *Agronomy Journal*, *92*, 1221–1227.

Irons, J. R., Dwyer, J. L., & Barsi, J. A. (2012). The next Landsat satellite: The Landsat data continuity mission. *Remote Sensing of Environment*, *122*, 11–21.

Khamis, G., Alsherif, E. A., Korany, S. M., Aldailami, D. A., Aloufi, A. S., Khalaf, M. H., Hamed, S. M., Mariduena-Zavala, M. G., Al Jaouni, S. K., & Selim, S. (2025). Drought stress differentially influences growth, physiology, and metabolite accumulation in *Triticum aestivum* (C3) and *Amaranthus caudatus* (C4) plants. *BMC Plant Biology*, *25*(1), 1199. https://doi.org/10.1186/s12870-025-07022-7.

Kim, K.-I., Clay, D. E., Carlson, C. G., Clay, S. A., & Trooien, T. (2008). Do synergistic relationships between nitrogen and water influence the ability of corn to use nitrogen derived from fertilizer and soil? *Agronomy Journal*, *100*, 551–556. https://doi.org/10.2134/agronj2007.0064.

Konings, A. G., Rao, K., & Steele-Dunne, S. (2019). Macro to micro: Microwave remote sensing of plant water content for physiology and ecology. *New Phytologist*, *223*(3), 1166–1172.

Kuwayama, Y., Thompson, A., Bernknopf, R., Zaitchik, B., & Vail, B. (2018). Estimating the impact of drought on agriculture using the U.S. drought monitor. *American Journal of Agricultural Economics*, *101*, 193–210. https://doi.org/10.1093/ajae/aay037.

Lacerda, L. N., Cohen, Y., Snider, J., Huryna, H., Liakos, V., & Vellidis, G. (2021). Field scale assessment of the TsHARP technique for thermal sharpening of MODIS satellite images using VENµS and Sentinel-2-derived NDVI. *Remote Sensing*, *13*(6), 1155. https://doi.org/10.3390/rs13061155.

Melton, F. S., Huntington, J., Grimm, R., Herring, J., Hall, M., Rollison, D., Erickson, T., Allen, R., Anderson, M., Fisher, J. B., & Kilic, A. (2022). OpenET: Filling a critical data gap in water management for the western United States. *JAWRA Journal of the American Water Resources Association*, *58*, 971–994.

Melton, F. S., Johnson, L. F., Lund, C. P., Pierce, L. L., Michaelis, A. R., Hiatt, S. H., Guzman, A., Adhikari, D. D., Purdy, A. J., Rosevelt, C., & Votava, P. (2012). Satellite irrigation management support with the terrestrial observation and prediction system: A framework for integration of satellite and surface observations to support improvements in agricultural water resource management. *IEEE Journal of Selected Topics in Applied Earth Observations and Remote Sensing*, *5*, 1709–1721.

Meyer, F. (2019). Spaceborne synthetic aperture radar—Principles, data access, and basic processing techniques. In A. Flores, K. Herndon, R. Thapa, & E. Cherrington (Eds.), *SAR handbook: Comprehensive methodologies for forest monitoring and biomass estimation*. NASA.

Parag, M., Lottering, R., Peerbhay, K., Agjee, N., & Poona, N. (2024). The use of synthetic aperture radar technology for crop biomass monitoring: A systematic review. *Remote Sensing Applications: Society and Environment*, *33*, 101107. https://doi.org/10.1016/j.rsase.2023.101107.

Pereira, L. S., Paredes, P., Melton, F., Johnson, L., Wang, T., López-Urrea, R., Cancela, J. J., & Allen, R. G. (2020). Prediction of crop coefficients from fraction of ground cover and height. Background and validation using ground and remote sensing data. *Agricultural Water Management*, *241*, 106197.

Seleiman, M. F., Al-Suhaibani, N., Ali, N., Akmal, M., Alotaibi, M., Refay, Y., Dindaroglu, T., Abdul-Wajid, H. H., & Battaglia, M. L. (2021). Drought stress impacts on plants and different approaches to alleviate its adverse effects. *Plants*, *28*, 259. https://doi.org/10.3390/plants10020259.

Senay, G. B. (2018). Satellite psychrometric formulation of the Operational Simplified Surface Energy Balance (SSEBop) model for quantifying and mapping evapotranspiration. *Applied Engineering in Agriculture*, *34*(3), 555–566.

Senay, G. B., Bohms, S., Singh, R. K., Gowda, P. H., Velpuri, N. M., Alemu, H., & Verdin, J. P. (2013). Operational evapotranspiration mapping using remote sensing and weather datasets: A new parameterization for the SSEB approach. *JAWRA Journal of the American Water Resources Association*, *49*(3), 577–591.

Swaef, T., Pieters, O., Appeltans, S., Borra-Serrano, I., Coudron, W., Couvreur, V., Garre, S., Lootens, P., Nicolai, B., Pols, L., Cast, C. S., Salogovic, J., Haeverbeke, M. V., Stock, M., & Weffels, F. (2022). On the pivotal role of water potential to model plant physiological processes. *Silico Plants*, *4*, diabo38.

Tsokas, A., Rysz, M., Pardalos, P. M., & Dipple, K. (2022). SAR data applications in earth observation: An overview. *Expert Systems with Applications*, *205*, 117342. http://dx.doi.org/10.1016/j.eswa.2022.117342.

Virtanen, O., Constantinidou, E., & Tyystjärvi, E. (2020). Chlorophyll does not reflect green light—How to correct a misconception. *Journal of Biological Education*, *56*(5), 552–559.

Virtue, J., Turner, D., Williams, G., Zeliadt, S., McCabe, M., & Lucieer, A. (2021). Thermal sensor calibration for unmanned aerial systems using an external heated shutter. *Drones*, *5*(4), 119. https://doi.org/10.3390/drones5040119.

Volk, J. M., Huntington, J. L., Melton, F. S., Allen, R., Anderson, M., Fisher, J. B., Kilic, A., Ruhoff, A., Senay, G. B., Minor, B., & Morton, C. (2024). Assessing the accuracy of OpenET satellite-based evapotranspiration data to support water resource and land management applications. *Nature Water*, *2*, 193–205.

Wahab, A., Abdi, G., Saleem, M. H., Ali, B., Ullah, S., Shah, W., Mumtaz, S., Yasin, G., Muresan, C. C., & Marc, R. A. (2022). Plants' physio-biochemical and phyto-hormonal responses to alleviate the adverse effects of drought stress: A comprehensive review. *Plants*, *11*, 1620. https://doi.org/10.3390/plants11131620.

Wu, J., Wang, J., Hui, W., Zhao, F., Wang, P., Su, C., & Gong, W. (2022). Physiology of plant responses to water stress and related genes: A review. *Forests*, *13*(2), 324. https://doi.org/10.3390/f13020324.

Yuan, W., & Hua, W. (2022). A case study of vignetting nonuniformity in UAV-based uncooled thermal cameras. *Drones*, *6*(12), 394. https://doi.org/10.3390/drones612039.

Precision Irrigation Technology and Management

16

Meetpal S. Kukal, Emily Bedwell, Jasmine Neupane, and Unius Arinaitwe

Purpose

This chapter explores advances in precision technologies for irrigation management. Precision irrigation includes the use of soil, canopy, and atmospheric sensors to assess plant health, as well as advanced water application infrastructure to improve farm-level profitability. The chapter also addresses the challenges of soil and weather variability and regulatory constraints, while highlighting opportunities for integrating real-time decision support systems and automated irrigation in agronomic and specialty crop management.

Need for Precision Irrigation Technologies in Irrigation

Precision irrigation includes the use of soil, canopy, and atmospheric sensors to assess plant health, in addition to the use of improved irrigation infrastructure to optimize water efficiency. Water is essential for global food production; approximately 70% of the world's fresh water is used to produce 40% of the world's food supply on 24% of the croplands (Mehta et al., 2024). In the United States, approximately 42% of the fresh water is used for irrigation (Hrozencik & Aillery, 2021). Irrigation plays a critical role in areas with unreliable rainfall, especially in semi-arid and arid regions where irrigated crops can yield two to three times more than rainfed crops (Kukal & Irmak, 2019). In these lands, precision irrigation can potentially increase water use efficiency, stabilize crop yields by providing a reliable water supply, and protect agricultural production from drought, heat stress, and extreme weather events (Kukal & Irmak, 2020). If successful, precision irrigation can produce more food with less water (McDonald & Girvetz, 2013; Zeng & Ren, 2022). Precision irrigation is one approach to mitigating the decline in water and land availability while also meeting the increasing global food demand.

Improving Water Use Efficiency

Water use efficiency is the ratio of the amount of food produced to the amount of water used to produce that food. Water use efficiency can be increased through a number of strategies that researchers continue to develop, including the following:

- Improving the crop's genetic capacity to extract water from the soil
- Reducing the amount of water lost to leaching, runoff, and evaporation
- Improving technologies that can identify and match the plants' water requirement

Water use efficiency, and therefore, irrigation, is complex. This is due to the numerous interactions between the soil system and the plant. Soil can be thought of as a leaky sponge, causing water to move below the plant rooting zone when more water is applied than the soil can retain. Soil sensors can be used to determine how much water can be retained in the soil, whereas plant sensors are used to assess the plants' requirements. The amount of water available to plants in the soil is referred to as plant-available water (PAW). Plant-available water depends on the soil texture and is defined as the difference between the water content at field capacity (−1/3 bar) and at the plants' permanent wilting point (−15 bar). In fine sand, PAW typically ranges from 0.7 to 1 inch/ft (5.8 to 8.3 cm/m), whereas in silt loams or loamy soils PAW ranges from 1.8 to 2.6 inches/ft (15 to 21.7 cm/m). Although the water may be in the soil, it is not equally available to plants. As the soil dries, the remaining water becomes increasingly difficult for the plant roots to extract. At this point, water stress can occur. Water stress generally begins to occur once 30%–70% of the available water in the soil has been depleted. This depends on the crop type and environmental conditions. The percentage of PAW that the soil is allowed to reach before requiring irrigation is the management allowable depletion (MAD). The water held in the soil above the MAD level is called readily available water (RAW). The RAW can be calculated by multiplying the PAW by the MAD. The MAD value depends on the drought tolerance of the plant. For example, the MAD value commonly used in corn production is 50%. Through irrigation scheduling, the grower can maintain soil water at the optimal MAD level for the crop being grown, optimizing growth by avoiding water stress during critical periods. The critical period of water stress can result in a significant reduction in yield. For example, for soybean, the critical period is between the full bloom and seed filling stages. If the plant is water stressed between these key growth stages, it may result in significantly reduced bean yield. For corn, two critical periods of water stress are known: during vegetative growth (V8) in mid to late July, and during the reproductive stage known as the milk stage. If water stress occurs at these points, grain fill can be slowed, resulting in much lower yields. Therefore, given the impact of water stress on yield, especially during these critical periods, efficient irrigation that is optimized for the crop's needs is key.

Soil Sensors

In situ soil sensors are among the most popular advances offered by precision technologies. These sensors can measure a variety of soil parameters, including volumetric water content, soil matric potential, soil temperature, and electrical conductivity. Among the various types of sensors available, soil water content is a critical parameter for data-driven irrigation management. This chapter covers the general need for and use of sensors to detect and mitigate water stress. For a more detailed discussion on proximal soil sensors, see Chapter 12. Soil water content can be measured at specific points with probes, such as soil water potential sensors (tensiometers), resistivity-based sensors (e.g., gypsum block or granular matrix sensors), and capacitance sensors, as well as sensors based on time-domain reflectometry (TDR) and frequency domain reflectometry (FDR; Figure 16.1).

Precision irrigation scheduling is based on monitoring the soil water content in the root zone. When soil water is depleted to a preestablished irrigation trigger or refill point (depending on crop-specific sensitivity to stress), irrigation water is applied to replenish the soil water deficit. For effective soil moisture-based irrigation scheduling, it is critical to identify this trigger point (Figure 16.2), which is defined by the soil's field capacity and the crop's permanent wilting point (Jensen & Allen, 2016).

While the fundamentals of soil water sensing have not changed recently, there have been significant improvements in the electronics and data communication protocols. These improvements in soil sensors have reduced labor costs required for successful precision irrigation techniques (Kukal et al., 2019). For example, instead of walking out to check a sensor, information

Figure 16.1. Installation of a soil moisture sensor at 6 inches (15.2 cm) below the soil surface. (Image credits: Dr. Meetpal Kukal).

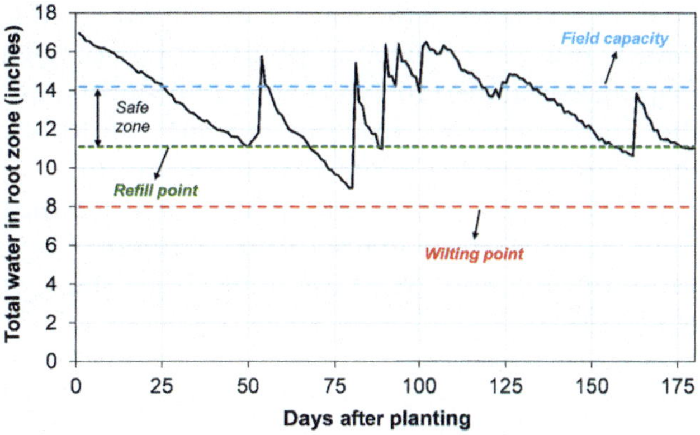

Figure 16.2. Time series of total root zone soil water (in inches) from 0 to 175 days after planting, aggregated from multiple soil depths. Wilting point: −15 bars (red line); field capacity: −1/3 bars (field capacity). The irrigation refill point lies between them (green line) and is typically considered synonymous with MAD. (Image credits: Dr. Meetpal Kukal).

Figure 16.3. A thermal infrared thermometer (marked with a yellow box) mounted on a center pivot system monitoring a barley canopy. (Image credits: Dr. Meetpal Kukal).

can be delivered wirelessly to a home computer or a phone application. While the convenience of this is unprecedented, it may also be important to consider having more than one sensor located at a single site. This is because many soil moisture sensors are exposed to a relatively small area of soil, making them vulnerable to spatial variability.

Crop Canopy Sensors

There are various crop canopy sensors that can assist with irrigation scheduling. In agronomic crops, a major class of canopy sensors consists of thermal infrared sensors. These sensors are typically positioned at an oblique angle to the canopy (Figure 16.3) to minimize the inclusion of warmer ground temperatures in their readings (Clawson & Blad, 1982). These sensors simply report the temperature of a crop canopy, which is influenced by transpiration. Plants cool themselves through transpiration, the process by which water evaporates through stomata during carbon dioxide uptake. Any reduction in water availability may decrease transpiration and increase canopy temperature (Idso et al., 1981; Jackson et al., 1981). This temperature deviation is influenced not only by the plant's physiological responses but also by wind speed, humidity, and the Sun's position. In calm, humid conditions, the temperature difference may be minimal. In contrast, windy conditions, as well as dry conditions, can make the temperature difference more pronounced and readily detectable (Jackson et al., 1981).

Because canopy temperature increases in response to water stress, this measurement can serve as an early indicator of the need for irrigation. Several field-tested approaches have been developed that rely on these temperature measurements. For example, one method involves triggering an irrigation application when the range of canopy temperatures, determined from multiple measurements, exceeds a specific threshold (Clawson & Blad, 1982). Another approach compares the actual canopy temperature with that of a well-watered reference crop. When the difference exceeds a preset threshold, it signals that irrigation should be initiated.

The Crop Water Stress Index (CWSI) is a widely used temperature-based assessment approach that is calculated by normalizing the difference between the actual canopy temperature and air temperature against a baseline established by well-watered and no-water conditions (Alderfasi & Nielsen, 2001; Irmak et al., 2000; Jackson et al., 1988). When the CWSI exceeds a critical value, it indicates that the plant is experiencing significant water stress (Nielsen, 1990; Stegman & Soderlund, 1992). A key drawback of using indexes such as the CWSI, which rely on canopy temperature as an irrigation trigger, is that water stress is already present. If the goal is to avoid water stress entirely, growers need to be able to predict water stress events before they occur. This is problematic in systems where irrigation applications take several days to complete, as the portion of the field irrigated last may experience prolonged water stress.

Measuring Evapotranspiration

Crop evapotranspiration (ET)—the sum of water evaporation (E) and plant transpiration (T)—provides important information for precision irrigation. Typically, crop ET is estimated using mathematical formulations that integrate the atmosphere's demand for water with soil and crop conditions. For irrigation management applications, a widespread method used by practitioners is the two-step method reported in FAO-56 (Allen et al., 1998) and more recently outlined by Jensen and Allen (2016). Evapotranspiration estimates are often used to replace detailed soil moisture measurements.

The first step is to measure the atmospheric evaporative demand, also called reference ET (ET_{os} or ET_{rs}; El-Naggar et al., 2020). Within the United States, reference ET is generally measured by automated agricultural weather stations. These

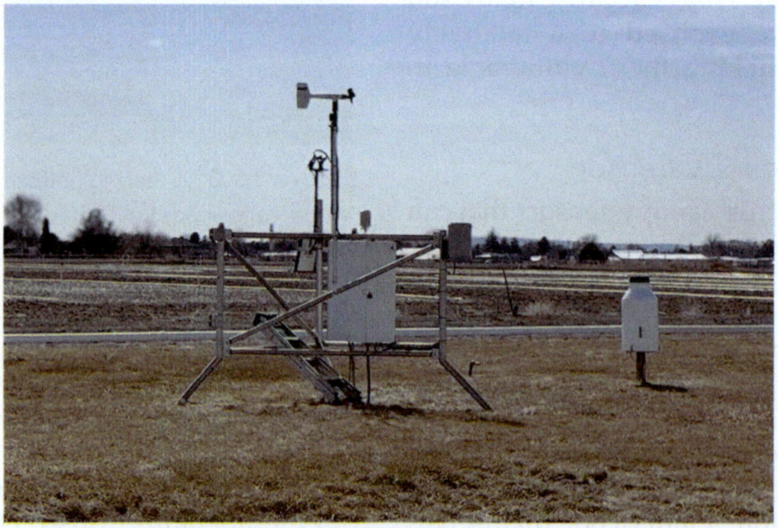

Figure 16.4. An automated weather station monitoring air temperature, relative humidity, wind speed and direction, solar radiation, and precipitation. (Image credits: Dr. Meetpal Kukal).

stations (Figure 16.4) need to follow research and industry standards. Some major ET networks in the United States are the California Irrigation Management Information System, Arizona Meteorological Network, High Plains Regional Climate Center, Bureau of Reclamation's AgriMet Network, New Mexico State University Climate Center, Colorado Agricultural Meteorological Network, North Dakota Agricultural Weather Network, Kansas Mesonet, Oklahoma Mesonet, and Missouri Agricultural Weather Station Network. More recently, major commercial manufacturers of environmental sensors have introduced relatively cheaper alternatives into research-grade standard weather sensors, improving accessibility and affordability.

The second step in measuring plant ET is to estimate the water removed and the amount of water remaining in the soil. In these calculations, the crop coefficient (K_c) is used to adjust the reference ET values to estimate ET. The equation commonly used for this calculation is $ET_c = ET_{ref} \times K_c$, where ET_c is the actual crop ET rate (crop water use), ET_{ref} is the reference ET, and K_c is the crop coefficient. Crop coefficients depend on the plants' growth stage and generally increase as the season progresses from germination to midseason. Crop coefficient values for selected crops are available in Pereira et al. (2025). The removal values are then subtracted from the amount of PAW in the soil. Sample calculations are provided by Ostrem et al. (2016).

Remote Sensing Observations

Remote sensing technologies can provide detailed, high-resolution images that help assess crop health, irrigation uniformity, crop water status, and ET. Remote sensing aims to incorporate a spatial element into crop monitoring, making it possible to monitor large acreages efficiently. For example, limitations posed by the small field of view of ground-based canopy sensors can be overcome using information collected from satellites (Bhatti et al., 2020; Melton et al., 2012; Senay, 2018; Xue et al., 2020) and unmanned aerial vehicles (Chávez et al., 2020; Mokari et al., 2022; Nieto et al., 2019). For more detailed discussion on the use of remote sensing in crop growth or water stress management, see Chapters 14 and 15, respectively.

Modernization of Irrigation Systems Infrastructure

Irrigation system infrastructure improvements have improved irrigators' ability to fine-tune water management and meet productivity and profitability goals. Such improvements have occurred in many systems, as discussed below.

Improvements in Gravity Irrigation Systems

Gravity-fed irrigation systems have long relied on manual operations and flood irrigation methods. These outdated practices have historically achieved a water use efficiency of only about 50% due to significant losses from runoff and deep percolation. Recent technological advancements have transformed these systems into modern, automated networks that integrate precise metering, high-flow discharge techniques, sensor technology, and real-time data analytics to deliver water on demand, closely aligning with the crop's actual consumptive needs. A primary example includes sensors that monitor the movement of water as it advances across a field. By pinpointing the exact moment the water front reaches a specific location, the grower can track the infiltration rate and the uniformity of water application, providing critical "advance time" data to the grower. This information allows irrigation managers to adjust cut-off times accurately, ensuring the water is applied evenly and efficiently across the entire field. These sensors employ a variety of technologies, from simple float switches and optical sensors to more sophisticated pressure transducers and remote telemetry devices (e.g. WETStake Flood Irrigation Water Notifier). Typically installed at predetermined points along furrows or borders, they trigger a signal when water arrives and relay real-time data to a central control unit. In some setups, this data is used to automatically halt irrigation once the desired advance time is reached, thereby minimizing runoff and reducing water waste. This precision not only conserves water but also reduces crop stress from uneven or excessive watering.

Furrow irrigation is one of the main methods of supplying water to crops through shallow, evenly spaced furrows in the soil. However, in conventional furrow systems, areas close to the water inlet receive an excessive amount of water, which then percolates deeply beyond the root zone, leading to significant water losses, also known as deep percolation loss (Bryant et al., 2017; Musick et al., 1987). Additionally, tail-water runoff is a common issue in which water is lost as it drains away from the surface and away from the field itself. A practical solution to these concerns is to use a cyclical on-and-off water delivery method called surge irrigation. Surge irrigation means applying water in intervals, turning it on and off instead of applying it continuously, by alternating water between different furrows or sections using automated valves. This intermittent application helps to mitigate excessive water application at the source end, reducing deep percolation and runoff while improving overall irrigation efficiency and uniformity (Bryant et al., 2017; Gudissa & Edossa, 2014; Horst et al., 2007).

Another advancement in precision irrigation to optimize efficiency is a computer program called Computerized Hole Selection (CHS). Computerized Hole Selection represents a significant technological advancement aimed at improving the conventional furrow irrigation system, particularly those used in the Mid-Southern United States. In conventional furrow irrigation systems, a single, uniform hole is traditionally used along the length of the poly-tubing in the furrow. The issue with that irrigation method is that it does not account for the natural decline in pressure as the distance from the water source increases. This results in uneven water distribution and irregular furrow advances. By using CHS, the program calculates optimal hole sizes by factoring in the field's shape, poly-tubing length, and elevation changes, thereby ensuring consistent water discharge across all furrows. This tailored approach not only creates a more uniform irrigation pattern but also minimizes tail-water runoff, ultimately enhancing application efficiency and reducing overall water use (Bryant et al., 2017; Henry et al., 2020).

Improvements in Pressurized Irrigation Systems

In sprinkler irrigation systems, a primary outcome has been improved application efficiency. One approach to doing this is called Low Energy Precision Application (LEPA), which delivers water directly onto the soil surface at low pressures. This approach minimizes evaporation and wind drift because only a limited portion of the soil is wetted, which in turn reduces water loss from the soil surface (Lyle & Bordovsky, 1983). Operating at reduced pressures also leads to substantial energy savings by lowering pumping requirements. However, rapid water application can increase the risk of ponding and runoff. If ponding and runoff are problems, consider using smaller sprinkler nozzles to lower the water application rate (Jones & Baumhardt, 2003).

Another approach is Low Elevation Spray Application (LESA), which uses a suspended sprinkler or spray head to deliver water 3–5 ft (0.961–1.5 m) above the soil surface (Figure 16.5).

Figure 16.5. A LESA system sprays water 3–5 ft (0.961 – 1.5 m) above the soil surface. This system has a double gooseneck and is located in Southern Idaho. (Image credits: Dr. Meetpal Kukal).

A LESA system operates at low pressures to reduce evaporation and minimize wind-drift losses. Generally, LESA results in a relatively uniform distribution of water over the field, allowing for more consistent infiltration. This uniformity helps mitigate issues related to uneven watering and can enhance overall crop performance.

Surface drip irrigation is another approach in which small, controlled quantities of water are applied directly to the soil surface near the root zone. This technique reduces water loss caused by evaporation, runoff, and deep percolation. As a result, crops experience reduced moisture stress, promoting healthier plant growth and potentially higher yields. This approach also reduces weed pressure, as the soil surface between rows remains relatively dry, discouraging weed germination and growth. A significant benefit is that it can be used in fertigation, where nutrients dissolved in irrigation water are applied directly to the crop's root zone. This ensures precise nutrient delivery, minimizing fertilizer waste and maximizing nutrient uptake by plants.

Subsurface drip irrigation (SDI) can deliver water directly into the crop root zone through buried polyethylene tubing, commonly called drip tape or drip line (Figure 16.6). These tubes contain small openings, or emitters, placed at regular intervals. The efficiency of the system relies on correct design and operation, and it can achieve irrigation efficiencies exceeding 95% (Lamm & Rogers, 2017). Switching from furrow irrigation to SDI often improves water conservation. For example, where a furrow system might require 23 inches (0.58 m) of water to meet a crop's needs, an SDI system would require just under 16 inches (0.40 m). This reduction in water usage also lowers pumping costs and energy demand. However, actual water savings depend on field conditions, soil types, and existing irrigation methods. One important advantage of SDI is reduced labor. Once installed, SDI requires a management effort similar to that of a center pivot system but significantly less than that of traditional surface irrigation. It can be automated, further reducing labor demands. The subsurface placement of drip lines also allows field operations to continue uninterrupted, even during active irrigation cycles. The consistently dry soil surface limits weed emergence and reduces soil crusting.

Figure 16.6. A subsurface drip line exposed to monitor emitter discharge in a subsurface drip irrigation system. (Image credits: Dr. Suat Irmak).

Mobile drip irrigation (MDI) is an emerging technology in which drip lines attached beneath pivots dispense water directly onto the soil surface. Drip tapes typically vary in length, becoming longer as they extend from the central pivot. The spacing and emitter flow rates are designed to match crop water demand. One benefit of MDI is reduced water loss due to evaporation, wind drift, and excessive deep-soil infiltration. This targeted watering approach enhances overall irrigation efficiency, supports consistent crop growth, and produces yields similar to those using conventional watering but with less water.

Spatial Variability and Variable Rate Irrigation

Soil moisture variability (Figure 16.7) can result from many factors, including spatial changes in elevation, soil texture, plant growth, drainage, water infiltration, irrigation and precipitation, tillage, and nutrient management (Irmak et al., 2022; Wilson et al., 2004). Each of these factors impacts the rate at which plants use water, which in turn affects the amount of water remaining in the soil. Traditional irrigation methods often apply water uniformly across fields, which can result in some areas receiving too little and others too much. Precision irrigation seeks to correct these problems by varying the rate and

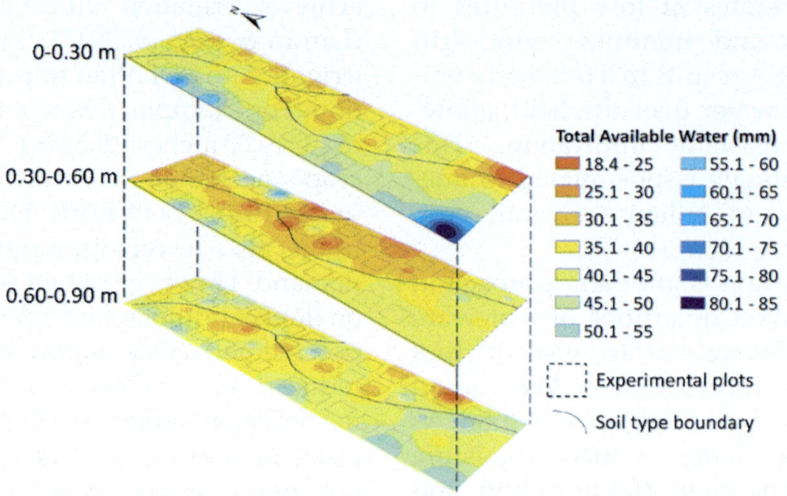

Figure 16.7. Spatial variability in total available water (in millimeters) in the top 0.90 m (35.4 inches) of the soil profile, measured in 0.30-m (11.8-inch) increments, determined by extensive soil sampling and laboratory analysis. (Image credits: Dr. Meetpal Kukal, Dr. Suat Irmak, and Trevor Hinn).

quantity of water applied to specific field areas according to their precise needs.

One of the goals of precision irrigation technology is to address spatial variability. One approach is to vary the rate at which the equipment travels across the field or to vary the rate at which water is applied from a nozzle. Modifying these values reduces runoff and increases water infiltration. For example, areas with slopes have a natural moisture gradient in which the summit area is drier and prone to runoff, while the toe- and footslope areas retain more moisture. To address this variability, the application rate and pivot speed can be adjusted. Increasing the pivot speed and reducing the application rate over the toe- and footslope areas that need less moisture can save water, whereas reducing speed and increasing the rate across summit areas can provide more water where it is needed. This method is cost-effective because it typically requires only minor adjustments to existing equipment. In contrast, zone-controlled variable rate irrigation (VRI) systems divide fields into distinct management zones, each independently irrigated through a series of valves placed along the pivot. This method permits a higher degree of control than by adjusting speed.

Individual zones can receive highly specific amounts of water according to their unique soil conditions, crop needs, or topographical features. While offering greater precision, zone control systems are more complex and expensive (O'Shaughnessy et al., 2019). All VRI systems require detailed irrigation prescription maps. These maps guide irrigation equipment by indicating how much water each section or zone should receive. Creating zone maps involves combining field-specific data with soil property measurements, historical yield data, topographical information, and remote sensing imagery (Haghverdi et al., 2015). Each information layer provides details about the soil. For example, soil maps provide information on moisture retention capabilities, nutrient availability, and overall soil health—all of which influence irrigation scheduling. Historical yield maps provide prior information about areas where the crop under- or overperforms. Topographical maps reveal elevation changes and slopes that affect water infiltration and runoff, while remote sensing provides information on plant health.

Adopting VRI technologies can improve water use efficiency by ensuring that precise amounts of water are applied to meet specific crop needs while reducing waste and runoff. The net result can be healthier crops and higher yields. Additionally, precision irrigation techniques can contribute to energy savings by reducing the overall volume of water pumped and distributed, thereby lowering operational costs. Environmentally, precision irrigation reduces the risk of excess nutrient runoff and leaching, protecting nearby ecosystems and water sources from contamination.

Advances in Communication

Technological advances allow real-time communication between the equipment and the manager. This communication can enable farmers to respond quickly to issues such as system malfunctions or interruptions in irrigation (Steele, 2009). These notifications help prevent crop stress or loss caused by unexpected equipment failures. Detailed data on water usage, irrigation pressures, and rainfall records facilitate better-informed decisions. Moreover, remote management provides growers with greater flexibility, allowing them to control irrigation from anywhere using smartphones or computers (Figure 16.8). This convenience reduces the need for frequent on-site checks, freeing farmers to focus on other tasks. Although precision technology requires an initial investment and specialized knowledge, the long-term benefits of reduced operating costs and increased crop productivity can justify it.

Data-Driven Irrigation Water Management

Software applications, or "apps," for smartphones are available to support data-driven precision irrigation management. These apps often require location, crop types, planting and harvesting dates, and irrigation system details. They can often provide real-time irrigation scheduling recommendations by accessing publicly available weather and soil data (Figure 16.9). Some specialized apps may improve these recommendations by using actual soil moisture measurements (Bordovsky et al., 2017). These include Irrigation Scheduler Mobile, SI CropFit, Water Irrigation Scheduler for Efficiency (WISE) application, KanSched, Wisconsin Irrigation Scheduling Program, and Irrigation Management Assistant (IMA) Tool, to name a few; however, many more are available. A number of studies demonstrated that apps can be used in precision irrigation management (Andales et al., 2014; Bordovsky et al., 2017; Brad & Phil, 2017; Cahn et al., 2015; Carlson, 2019; Han, 2016; Irmak et al., 2010; Peters et al., 2013; Rogers, 2012; Sanford & Panuska, 2015; Scherer & Morlock, 2008; Stevens, 2014; Vellidis et al., 2016; Wright, 2018). These apps have the potential to reduce costs compared with manual operations.

Precision Irrigation of Specialty Crops

For high-value crops such as perennial woody plants, including almonds, pecans, citrus, and grapes, it may be desirable to monitor water status using dendrometers and pressure chambers. Dendrometers measure changes in the diameter

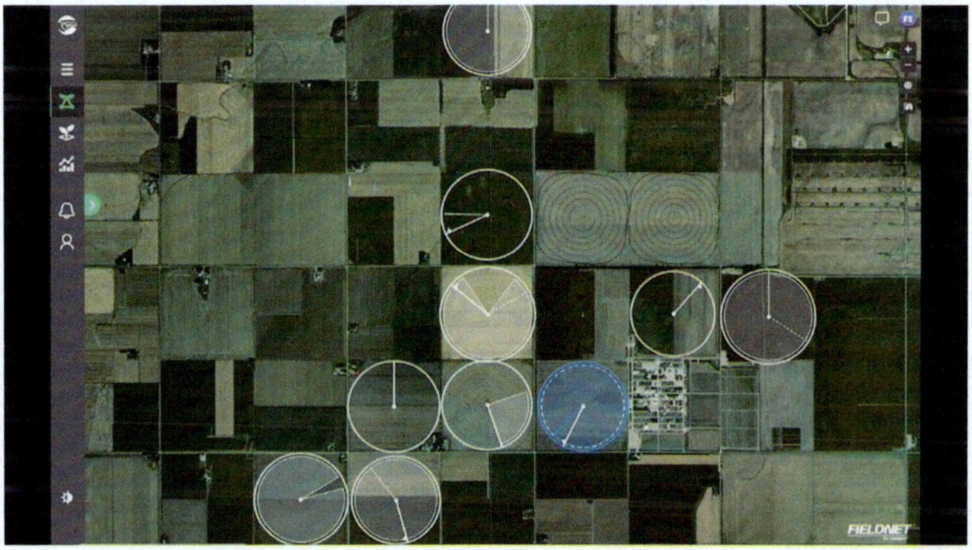

Figure 16.8. A screenshot of a commercial dashboard allowing a manager to track multiple center pivot systems simultaneously.

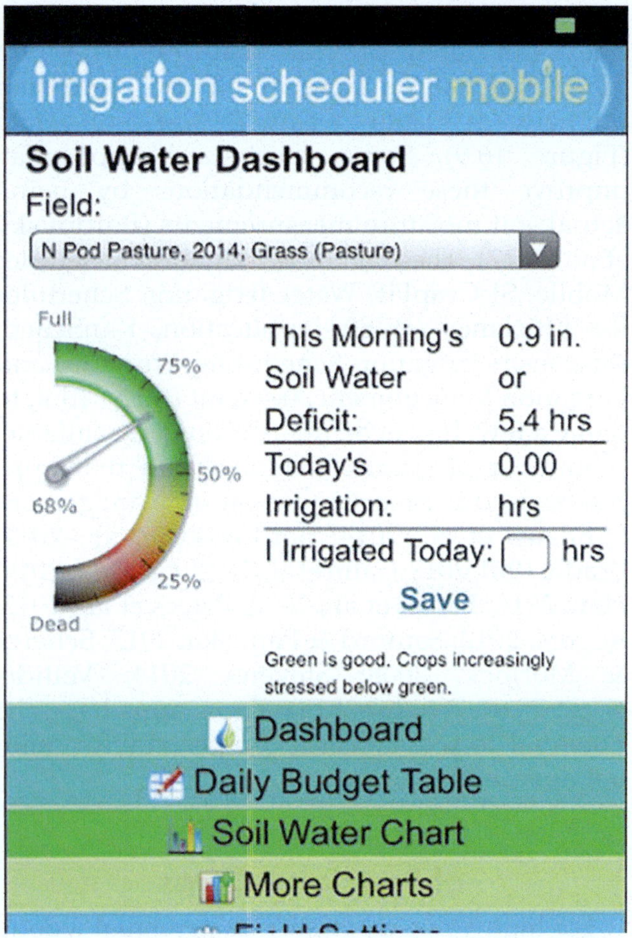

Figure 16.9. A screenshot of the Irrigation Scheduler Mobile smartphone app showing real-time soil water deficit for a grass pasture field.

of a tree trunk. These sensors can be used to provide real-time measurements of water stress. However, these systems are awkward, and the data are difficult to interpret.

Like the dendrometer, a pressure chamber measures the water stress in the plant, not the soil. In the pressure chamber approach, a leaf is cut and placed inside the pressure chamber. Pressure is then used to force sap from the leaf, and when sap is forced out, the leaf water potential (LWP) is reached. When using pressure chambers, it is important to measure plant water stress at the appropriate time of day and on the correct plant part to ensure accurate and meaningful results. For grapes, this is generally around solar noon. The measurement represents the LWP. This measurement is useful because it incorporates soil and environmental factors affecting the entire root zone (Fulton et al., 2014).

The LWP fluctuates throughout the day and generally decreases as the soil water also decreases (Kume et al., 2007). Despite its accuracy, the labor-intensive nature of midday LWP measurements has limited their broader adoption in precision irrigation. Recent technological advances have led to the development of continuous LWP monitoring sensors to address this limitation. These sensors primarily include osmometers and microtensiometers (Figure 16.10). Osmometer-based sensors detect pressure fluctuations resulting from osmotic changes in the sensor's fluid chamber, mediated by a semipermeable membrane, which allows water exchange with the plant xylem (Meron et al., 2015). These pressure variations correlate with stem water potential. Conversely, microtensiometers use tensiometric principles to measure the chemical potential of water, leveraging the thermodynamic equilibrium between liquid water under tension and its vapor phase (Pagay, 2014).

Dendrometers offer another viable approach for monitoring plant water status to support precision irrigation scheduling. These instruments measure daily trunk shrinkage, defined as the daily variation between the maximum and minimum trunk diameters. Increased soil water depletion or greater ET demand results in more pronounced daily trunk shrinkage.

Figure 16.10. A commercially available microtensiometer embedded into a tree's woody tissue, where it directly measures stem water potential. Courtesy of FloraPulse.

Current Barriers for Precision Irrigation Adoption

While precision irrigation technologies have strong potential, there are barriers to their widespread adoption (Figure 16.11). This chapter discusses the barriers to irrigation technology adoption in general. For a more detailed evaluation of barriers to the adoption of precision agriculture, see Chapter 2. In irrigation technologies, the main barriers include economic, technical, educational, infrastructural, and behavioral factors (Drewry et al., 2019; El Bilali & Allahyari, 2018; Wolfert et al., 2017). Economic barriers are prominent due to the significant upfront investment required for precision irrigation technologies. High initial costs, coupled with uncertainty about economic returns, discourage many producers, particularly those managing small- to medium-scale farms. Financial incentives and cost-sharing programs provided by federal and state agencies alleviate this barrier to some extent. Technical challenges include the complexity of integrating diverse technologies, the lack of compatibility across manufacturers' equipment, and inadequate interoperability, which cause frustration and inefficiencies. Moreover, operating complex systems, such as VRI systems, requires specialized knowledge and ongoing training for farmers, raising potential hurdles in adoption and operational effectiveness.

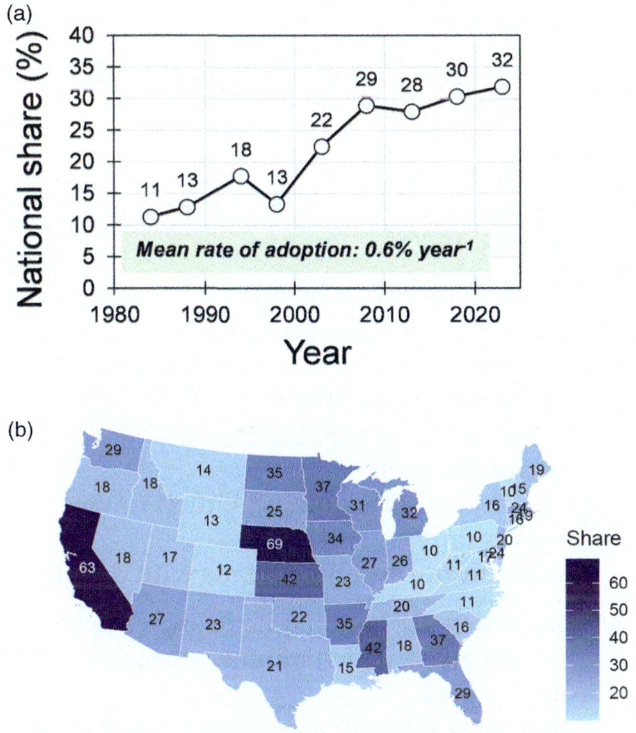

Figure 16.11. (a) Growth in adoption (percentage) of scientifically based data-driven irrigation scheduling methods in the United States during 1984–2023, and (b) geographical distribution of the share (percentage) of irrigators that use one or more data-driven irrigation scheduling methods across the United States as of 2023. Data used for these visualizations have been adopted from the USDA Irrigation and Water Management Surveys (1984–2023). For the definition of data-driven methods used here, we include soil moisture sensing devices, plant moisture sensing devices, scheduling services, daily evapotranspiration reports, and computer simulation models. (Image credits: Dr. Meetpal Kukal).

Due to equipment malfunctions, calibration issues, or sensor placement errors, obtaining accurate and reliable sensor data is often difficult. Precision systems also require regular maintenance, adding further operational burden. Educational barriers compound these challenges because many farmers lack adequate access to training, practical demonstrations, and the technical support needed to understand and operate advanced precision irrigation systems efficiently. Producers often express uncertainty about interpreting sensor-generated data and decision support outputs, leading to limited trust and utilization. Moreover, infrastructural limitations, particularly the lack of reliable internet connectivity, persist in many rural regions. Unfortunately, this connectivity is often crucial for real-time monitoring and data analysis, limiting the efficacy of advanced tools (Strover et al., 2021). Behavioral factors also play a crucial role (Daberkow & McBride, 2003). Traditional attitudes, risk aversion, and resistance to change among farmers contribute significantly to the slow adoption of new technologies. Concerns about cybersecurity, data privacy, ownership, and potential misuse contribute to hesitancy in adopting precision irrigation technologies, which rely heavily on extensive data collection, storage, and sharing across platforms. Last, not all agricultural fields exhibit sufficient variability to justify implementing such technologies, highlighting the importance of thorough field evaluation and site-specific assessments prior to adoption.

Conclusion

Water stress can be extremely damaging to yield; however, using conventional irrigation strategies may not completely prevent this yield loss and may result in lost profits due to inefficient water use. This chapter discussed the main ways researchers and growers are working together to address this variability with new technologies. These technologies are improving the efficiency of water use across the field and enhancing profitability by avoiding excess costs that occur when water is lost from the field after application. New data-driven techniques in irrigation are advancing rapidly, and the barriers to adopting many of these technologies are being rapidly lowered, producing new opportunities across arid, semi-arid, and humid regions to improve crop production.

Study Questions

1. In fine sandy soils, typical plant-available water (PAW) per foot of depth is closest to
   a. 0.2–0.4 inch/ft
   b. 0.7–1.0 inch/ft
   c. 1.8–2.6 inches/ft
   d. 3.0–3.5 inches/ft
2. Management allowable depletion (MAD) represents
   a. total soil water at wilting point
   b. fraction of plant-available water (PAW) that can be depleted before irrigation
   c. water lost during deep percolation
   d. difference between RAW and field capacity
3. Which soil moisture sensing technology measures the travel time of an electromagnetic pulse along a probe?
   a. Gypsum block
   b. Tensiometer
   c. Capacitance sensor
   d. Time-domain reflectometry (TDR)
4. Thermal infrared canopy sensors can detect water stress primarily because stressed plants
   a. emit more long-wave radiation due to higher leaf temperature
   b. display lower chlorophyll content
   C. increase reflectance in the near-infrared band
   d. exhibit reduced stomatal conductance causing canopy cooling
5. Low Energy Precision Application (LEPA) sprinkler packages are characterized by
   a. high-pressure operation with fine-mist nozzles
   b. water applied 10 ft above the canopy
   c. very low-pressure heads wetting a narrow soil band
   d. buried emitters delivering subsurface water
6. Computerized Hole Selection (CHS) in polytube irrigation adjusts hole diameters to compensate for
   a. variable soil texture
   b. declining pressure along tubing length
   c. changes in water salinity
   d. slope-induced runoff at the tail-end of furrows
7. Subsurface drip irrigation typically achieves field-level application efficiencies
   a. below 60%
   b. around 70%
   c. above 95%
   d. exactly 100%
8. Variable rate irrigation (VRI) by speed control modifies application depth by
   a. changing nozzle pressure while pivot speed stays constant
   b. altering travel speed of the machine across the field
   c. opening and closing individual tower valves only
   d. substituting sprinkler heads with drip tapes
9. Smartphone irrigation scheduling "apps" generally need which minimum inputs from the user?
   a. soil sand–silt–clay percentages only
   b. local weather station ownership details
   c. crop type, location, planting date, and system characteristics
   d. satellite image mosaics at 10-cm resolution
10. In the two-step FAO-56 approach widely used for irrigation scheduling, which prerequisite variable must be estimated before calculating crop evapotranspiration (ET_c)?
    a. Soil matric potential
    b. Reference evapotranspiration (ET_o or ET_r) from weather data
    c. Canopy temperature
    d. Soil electrical conductivity

ACKNOWLEDGMENTS

Support for this document was provided by the Precision Farming Systems community of the American Society of Agronomy, South Dakota State University, USDA Natural Resources Conservation Service Conservation Innovation Grants (NR233A750011G019), American Coalition for Ethanol, the South Dakota Corn Utilization Council, and the South Dakota Soybean Research and Promotion Council. The use of trade names is for the convenience of the reader and does not imply endorsement by the authors. The primary author also acknowledges the University of Idaho Agricultural Experiment Station and the USDA National Institute of Food and Agriculture Research Capacity Fund (Hatch) Accession Number 7010874 for their support of this activity.

REFERENCES

Alderfasi, A. A., & Nielsen, D. C. (2001). Use of crop water stress index for monitoring water status and scheduling irrigation in wheat. *Agricultural Water Management*, *47*(1), 69–75. https://doi.org/10.1016/S0378-3774(00)00096-2.

Allen, R. G., Pereira, L. S., Raes, D., & Smith, M. (1998). *Crop evapotranspiration. Guidelines for computing crop water requirements. FAO irrigation and drainage paper 56. Food and Agriculture Organization of the United Nations.*

Andales, A. A., Bauder, T. A., & Arabi, M. (2014). A mobile irrigation water management system using a collaborative GIS and weather station networks. In L. R. Ahuja, L. Ma, & R. J. Lascano (Eds.), *Practical applications of agricultural system models to optimize the use of limited water* (Vol. 5, pp. 53–84). ASA,CSA,SSSA.

Bhatti, S., Heeren, D. M., Barker, J. B., Neale, C. M. U., Woldt, W. E., Maguire, M. S., & Rudnick, D. R. (2020). Site-specific irrigation management in a sub-humid climate using a spatial evapotranspiration model with satellite and airborne imagery. *Agricultural Water Management*, *230*, 105950. https://doi.org/10.1016/j.agwat.2019.105950.

Bordovsky J., Stoker, B., Bilnoski, P., Garcia, C., Porter, D., Ale, S., Biggers, K., & Wall, J. (2017). *Dashboard for Irrigation Efficiency Management (DIEM).* https://www.irrigation.org/IA/FileUploads/IA/Resources/TechnicalPapers/2017/DashboardForIrrigationEfficiencyManagement.pdf.

Brad, J. & Phil, H. (2017) *Irris scheduler home page.* Purdue Research Foundation. http://www.purdue.edu/agsoftware/irrigation.

Bryant, C. J., Krutz, L. J., Falconer, L., Irby, J. T., Henry, C. G., Pringle, H. C., III, Henry, M. E., Roach, D. P., Pickelmann, D. M., Atwill, R. L., & Wood, C. W. (2017). Irrigation water management practices that reduce water requirements for Mid-South furrow-irrigated soybean. *Crop, Forage & Turfgrass Management*, *3*(1), 1–7. https://doi.org/10.2134/cftm2017.04.0025.

Cahn, M., Hartz, T., Smith, R., Noel, B., Johnson, L., & Melton, F. (2015). *CropManage: An online decision support tool for irrigation and nutrient management* (pp. 5–6). *Proceedings of the Western Nutrient Management Conference.*

Carlson, L. (2019). *Automatic soil moisture monitors and AgriMet: Dynamic tools irrigation scheduling.* https://waterquality.montana.edu/farm-ranch/irrigation/irrigation-tools/monitors-agrimet.html.

Chávez, J. L., Torres-Rua, A. F., Woldt, W. E., Zhang, H., Robertson, C. C., Marek, G. W., Wang, D., Heeren, D. M., Taghvaeian, S., & Neale, C. M. U. (2020). A decade of unmanned aerial systems in irrigated agriculture in the Western US. *Applied Engineering in Agriculture*, *36*(4), 423–436. https://elibrary.asabe.org/abstract.asp?aid=51678.

Clawson, K. L., & Blad, B. L. (1982). Infrared thermometry for scheduling irrigation of corn. *Agronomy Journal*, *74*(2), 311–316. https://doi.org/10.2134/agronj1982.00021962007400020013x.

Daberkow, S. G., & McBride, W. D. (2003). Farm and operator characteristics affecting the awareness and adoption of precision agriculture technologies in the US. *Precision Agriculture*, *4*(2), 163–177. https://doi.org/10.1023/A:1024557205871.

Drewry, J. L., Shutske, J. M., Trechter, D., Luck, B. D., & Pitman, L. (2019). Assessment of digital technology adoption and access barriers among crop, dairy and livestock producers in Wisconsin. *Computers and Electronics in Agriculture*, *165*, 104960. https://doi.org/10.1016/j.compag.2019.104960.

El Bilali, H., & Allahyari, M. S. (2018). Transition towards sustainability in agriculture and food systems: Role of information and communication technologies. *Information Processing in Agriculture*, *5*(4), 456–464. https://doi.org/10.1016/j.inpa.2018.06.006.

El-Naggar, A. G., Hedley, C. B., Horne, D., Roudier, P., & Clothier, B. E. (2020). Soil sensing technology improves the application of irrigation water.

Agricultural Water Management, *228*, 105901. https://doi.org/10.1016/j.agwat.2019.105901.

Fulton, A., Grant, J., Buchner, R., & Connell, J. (2014). *Using the pressure chamber for irrigation management in walnut, almond and prune* (pp. 8503). University of California Division of Agriculture and Natural Resources Publication #. https://ucanr.edu/sites/default/files/2018-05/283498.pdf.

Gudissa, H. D., & Edossa, D. C. (2014). Evaluation of surge and cutback flow furrow irrigation systems for pepper (*Capsicum annuum*) production. *Irrigation and Drainage*, *63*(4), 463–473. https://doi.org/10.1002/ird.1828.

Haghverdi, A., Leib, B. G., Washington-Allen, R. A., Ayers, P. D., & Buschermohle, M. J. (2015). Perspectives on delineating management zones for variable rate irrigation. *Computers and Electronics in Agriculture*, *117*, 154–167. https://doi.org/10.1016/j.compag.2015.06.019.

Han, J. C. (2016). *Development of CornSoyWater, a web-based irrigation app for corn and soybean*. [Doctoral dissertation, The University of Nebraska-Lincoln]. ETD Collection for University of Nebraska-Lincoln. AAI10247096. https://digitalcommons.unl.edu/dissertations/AAI10247096.

Henry, C. G., Krutz, L. J., Mane, R., & Simpson, G. D. (2020). The Arkansas "most crop per drop" contest: An innovative extension method to improve irrigation water management adoption. *Transactions of the ASABE*, *63*(6), 2083–2088. https://doi.org/10.13031/trans.13964.

Horst, M. G., Shamutalov, S. S., Goncalves, J. M., & Pereira, L. S. (2007). Assessing impacts of surge-flow irrigation on water saving and productivity of cotton. *Agricultural Water Management*, *87*(2), 115–127. https://doi.org/10.1016/j.agwat.2006.06.014.

Hrozencik, R. A., & Aillery, M. (2021). *Trends in US irrigated agriculture: Increasing resilience under water supply scarcity*. USDA-Economic Research Service. https://www.ers.usda.gov/publications/pub-details?pubid=102927.

Idso, S. B., Jackson, R. D., Pinter, P. J., Jr., Reginato, R. J., & Hatfield, J. L. (1981). Normalizing the stress-degree-day parameter for environmental variability. *Agricultural Meteorology*, *24*, 45–55. https://doi.org/10.1016/0002-1571(81)90032-7.

Irmak, S., Haman, D. Z., & Bastug, R. (2000). Determination of crop water stress index for irrigation timing and yield estimation of corn. *Agronomy Journal*, *92*(6), 1221–1227. https://doi.org/10.2134/agronj2000.9261221x.

Irmak, S., Kukal, M. S., & Sharma, K. (2022). Soil moisture heterogeneity and sensor deployment in uniformly managed field with unitextural soil. *Agronomy Journal*, *114*(3), 1800–1816. https://doi.org/10.1002/agj2.21064.

Irmak, S., Rees, J. M., Zoubek, G. L., van DeWalle, B. S., Rathje, W. R., DeBuhr, R., Leininger, D., Siekman, D. D., Schneider, J. W., & Christiansen, A. P. (2010). Nebraska Agricultural Water Management Demonstration Network (NAWMDN): Integrating research and extension/outreach. *Applied Engineering in Agriculture*, *26*(4), 599–613. https://doi.org/10.13031/2013.32066.

Jackson, R. D., Idso, S. B., Reginato, R. J., & Pinter, P. J., Jr. (1981). Canopy temperature as a crop water stress indicator. *Water Resources Research*, *17*(4), 1133–1138. https://doi.org/10.1029/WR017i004p01133.

Jackson, R. D., Kustas, W. P., & Choudhury, B. J. (1988). A reexamination of the crop water stress index. *Irrigation Science*, *9*, 309–317. https://doi.org/10.1007/BF00296705.

Jensen, M. E., & Allen, R. G. (2016). *Evaporation, evapotranspiration, and irrigation water requirements*. ASCE Manuals and Reports on Engineering Practice No. 70. (2nd ed.). American Society of Civil Engineers. https://ascelibrary.org/doi/pdf/10.1061/9780784414057.fm.

Jones, O. R., & Baumhardt, R. L. (2003). Furrow dikes. In B. A. Stewart & T. A. Howell (Eds.), *Encyclopedia of water science* (pp. 317–320). Marcel Dekker.

Kukal, M. S., & Irmak, S. (2019). Irrigation-limited yield gaps: Trends and variability in the United States post-1950. *Environmental Research Communications*, *1*(6), 061005. https://doi.org/10.1088/2515-7620/ab2aee.

Kukal, M. S., & Irmak, S. (2020). Impact of irrigation on interannual variability in United States agricultural productivity. *Agricultural Water Management*, *234*, 106141. https://doi.org/10.1016/j.agwat.2020.106141.

Kukal, M. S., Irmak, S., & Sharma, K. (2019). Development and application of a performance and operational feasibility guide to facilitate adoption of soil moisture sensors. *Sustainability*, *12*(1), 321. https://doi.org/10.3390/su12010321.

Kume, T., Takizawa, H., Yoshifuji, N., Tanaka, K., Tantasirin, C., Tanaka, N., & Suzuki, M. (2007). Impact of soil drought on sap flow and water status of evergreen trees in a tropical monsoon forest in northern Thailand. *Forest Ecology and Management*, *238*(1–3), 220–230. https://doi.org/10.1016/j.foreco.2006.10.019.

Lamm, F. R., & Rogers, D. H. (2017). Longevity and performance of a subsurface drip irrigation system. *Transactions of the ASABE*, *60*(3), 931–939. https://doi.org/10.13031/trans.12237.

Lyle, W. M., & Bordovsky, J. P. (1983). lepa irrigation system evaluation. *Transactions of the ASAE*, *26*(3), 776–781. https://doi.org/10.13031/2013.34022.

McDonald, R. I., & Girvetz, E. H. (2013). Two challenges for U.S. irrigation due to climate change: Increasing irrigated area in wet states and increasing irrigation rates in dry states. *PLoS One*, *8*(6), e65589. https://doi.org/10.1371/journal.pone.0065589.

Mehta, P., Siebert, S., Kummui, M., Deng, Q., Ali, T., Marston, L., Xie, W., & Davis, K. F. (2024). Half of the twenty-first century global irrigation expansion has been in water stressed regions. *Nature Water*, *2*, 254–261. https://doi.org/10.1038/s44221-024-00206-9.

Melton, F. S., Johnson, L. F., Lund, C. P., Pierce, L. L., Michaelis, A. R., Hiatt, S. H., Guzman, A., Adhikari, D. D., Purdy, A. J., Rosevelt, C., Votava, P., Trout, T. J., Temesgen, B., Frame, K., Sheffner, E. J., & Nemani, R. R. (2012). Satellite irrigation management support with the terrestrial observation and prediction system: A framework for integration of satellite and surface observations to support improvements in agricultural water resource management. *IEEE Journal of Selected Topics in Applied Earth Observations and Remote Sensing*, *5*(6), 1709–1721. https://doi.org/10.1109/JSTARS.2012.2214474.

Meron, M., Goldberg, S. Y., Solomon-Halgoa, A., & Ramon, G. (2015). Embedded stem water potential sensor. In J. V. Stafford (Ed.), *Precision Agriculture '15* (pp. 527–532). Wageningen Academic. https://doi.org/10.3920/978-90-8686-814-8_65.

Mokari, E., Samani, Z., Heerema, R., Dehghan-Niri, E., DuBois, D., Ward, F., & Pierce, C. (2022). Development of a new UAV-thermal imaging-based model for estimating pecan evapotranspiration. *Computers and Electronics in Agriculture*, *194*, 106752. https://doi.org/10.1016/j.compag.2022.106752.

Musick, J. T., Walker, J. D., Schneider, A. D., & Pringle, F. B. (1987). Seasonal evaluation of surge flow irrigation for corn. *Applied Engineering in Agriculture*, *3*(2), 247–251. https://doi.org/10.13031/2013.26683.

Nielsen, D. C. (1990). Scheduling irrigations for soybeans with the Crop Water Stress Index (CWSI). *Field Crops Research*, *23*(2), 103–116. https://doi.org/10.1016/0378-4290(90)90106-L.

Nieto, H., Kustas, W. P., Torres-Rúa, A., Alfieri, J. G., Gao, F., Anderson, M. C., White, W. A., Song, L., del Mar Alsina, M., Prueger, J. H., McKee, M., Elarab, M., & McKee, L. G. (2019). Evaluation of TSEB turbulent fluxes using different methods for the retrieval of soil and canopy component temperatures from UAV thermal and multispectral imagery. *Irrigation Science*, *37*(3), 389–406. https://doi.org/10.1007/s00271-018-0585-9.

O'Shaughnessy, S. A., Evett, S. R., Colaizzi, P. D., Andrade, M. A., Marek, T. H., Heeren, D. M., Lamm, F. R., & LaRue, J. L. (2019). Identifying advantages and disadvantages of variable rate irrigation: An updated review. *Applied Engineering in Agriculture*, *35*(6), 837–852. https://doi.org/10.13031/aea.13128.

Ostrem, D., Trooien, T., & Hay, C. (2016). Chapter 33: Irrigating corn in South Dakota. In D. E. Clay, C. G. Carlson, S. A. Clay, & E. Byamukama (Eds.), *iGrow corn: Best management practices*. South Dakota State University. https://extension.sdstate.edu/sites/default/files/2019-09/S-0003-33-Corn.pdf.

Pagay, V. (2014). *Physiological responses of grapevine shoots to water stress and the development of a microtensiometer to continuously measure water potential.* [Doctoral dissertation, Cornell University]. Cornell Theses and Dissertations. https://hdl.handle.net/1813/37138.

Pereira, L. S., Allen, R. G., Paredes, P., López-Urrea, R., Raes, D., Smith, M., Kilic, A., & Salman, M. (2025). *Crop evapotranspiration. Guidelines for computing crop water requirements. FAO Irrigation and Drainage Pager, No. 56 Rev. 1* (2nd ed.). Food and Agriculture Organization of the United Nations. https://doi.org/10.4060/cd6621en.

Peters, R. T., Hoogenboom, G., & Hill, S. (2013). *Simplified irrigation scheduling on a smart phone or web browser*. Irrigation Association. https://www.irrigation.org/IA/FileUploads/IA/Resources/TechnicalPapers/2013/SimplifiedIrrigation Scheduling OnASmartPhoneOrWebBrowser.pdf.

Rogers, D. H. (2012). Introducing the web-based version of KanSched: An ET-based irrigation scheduling tool. *Proceedings of the 2012 Central Plains Irrigation Conference* (pp. 230–236). CPIC.

Sanford, S., & Panuska, J. (2015). Irrigation management in Wisconsin. The Wisconsin Irrigation Scheduling Program (WISP). University of Wisconsin Cooperative Extension Publication A3600–01 (revised). University of Wisconsin. https://fyi.extension.wisc.edu/cropirrigation/files/2019/11/A3600-01-19.pdf.

Scherer, T. F., & Morlock, D. J. (2008). A site-specific web-based irrigation scheduling program. *2008 ASABE Meeting, 2008 Providence* (paper #083589). American Society of Agricultural and Biological Engineers.

Senay, G. B. (2018). Satellite psychrometric formulation of the operational simplified surface energy balance (SSEBop) model for quantifying and mapping evapotranspiration. *Applied Engineering in Agriculture, 34*(3), 555–566. https://doi.org/10.13031/aea.12614.

Steele, D. (2009). Pivot irrigation: Wireless networking and remote sensor monitoring. In *USCID Fifth International Conference* (pp. 143–153). U.S. Committee on Irrigation and Drainage.

Stegman, E. C., & Soderlund, M. (1992). Irrigation scheduling of spring wheat using infrared thermometry. *Transactions of the ASAE, 35*(1), 143–152. https://doi.org/10.13031/2013.28581.

Stevens G. (2014). *Crop Water Use App*. University of Missouri Extension. http://ag3.agebb.missouri.edu/horizonpoint/cropwater/.

Strover, S., Riedl, M. J., & Dickey, S. (2021). Scoping new policy frameworks for local and community broadband networks. *Telecommunications Policy, 45*(10), 102171. https://doi.org/10.1016/j.telpol.2021.102171.

Vellidis, G., Liakos, V., Andreis, J. H., Perry, C. D., Porter, W. M., Barnes, E. M., Morgan, K. T., Fraisse, C., & Migliaccio, K. W. (2016). Development and assessment of a smartphone application for irrigation scheduling in cotton. *Computers and Electronics in Agriculture, 127*, 249–259. https://doi.org/10.1016/j.compag.2016.06.021.

Wilson, D. J., Western, A. W., & Grayson, R. B. (2004). Identifying and quantifying sources of variability in temporal and spatial soil moisture observations. *Water Resources Research, 40*(2). https://doi.org/10.1029/2003WR002306.

Wolfert, S., Ge, L., Verdouw, C., & Bogaardt, M.-J. (2017). Big data in smart farming–A review. *Agricultural Systems, 153*, 69–80. https://doi.org/10.1016/j.agsy.2017.01.023.

Wright, J. (2018). *Irrigation scheduling checkbook method*. University of Minnesota Extension. https://extension.umn.edu/irrigation/irrigation-scheduling-checkbook-method.

Xue, J., Bali, K. M., Light, S., Hessels, T., & Kisekka, I. (2020). Evaluation of remote sensing-based evapotranspiration models against surface renewal in almonds, tomatoes and maize. *Agricultural Water Management, 238*, 106228. https://doi.org/10.1016/j.agwat.2020.106228.

Zeng, R., & Ren, W. (2022). The spatiotemporal trajectory of US agricultural irrigation withdrawal during 1981–2015. *Environmental Research Letters, 17*(10), 104027. https://doi.org/10.1088/1748-9326/ac91e2.

Precision Insect Measurement and Management

17

Ivan Grijalva, Brian McCornack, and Sharon A. Clay

Chapter Purpose

This chapter discusses many topics, including what precision Integrated Pest Management (IPM) is, as well as conventional and advanced sampling techniques for accurate identification of insect pests using IPM and machine learning, population dynamics, and economic thresholds for populations. Accurate insect identification and effective sampling techniques are needed to assess potential yield losses and aid in making appropriate management decisions. Precision agriculture tools, such as unmanned vehicles, imagery, and machine learning, can be integrated with traditional information to improve management.

Evolution of Integrated Pest Management

Integrated Pest Management (IPM) is a sustainable decision-making approach often used to manage a variety of pests, including insects, weeds, or diseases, by integrating cultural practices with advanced technologies to mitigate damage. Precision IPM includes a suite of techniques that also use this integrated approach of technology and cultural practices, but expands the concept to consider site-specific differences in creating pest management plans using precision agriculture technologies and sensor-based measurements. The precision IPM approach can be optimized by

1. Understanding the biology and life cycle of insect pests
2. Collecting accurate information on the pest through scouting
3. Recording and analyzing the collected information about the pest
4. Using the collected information to make an appropriate management decision.

This chapter will focus on using IPM for managing herbivorous insects (Kogan, 1998; Oerke, 2006). For discussions on disease or weed management, see Chapters 18 and 19.

Integrated Pest Management

Plant damage caused by insects can be mitigated by implementing insect-specific management practices. These practices must account for the expected return on investment in managing the insect pests. This includes taking into consideration the expected short- and long-term success of management, the scale and scope of the problem, and an understanding of pest biology and life cycle. Mediation techniques that consider these factors are often highlighted as IPM. IPM strategies use knowledge of proper insect scouting and identification, the biology of insects, the susceptibility of the crop, the effectiveness of applied treatments, and site-specific environmental conditions to assess optimal solutions. IPM first identifies the action threshold, the point at which measures must be taken to avoid crop loss. This threshold is determined by assessing crop damage through systematic field scouting, collecting and organizing the information, and analyzing the information to decide on the mitigation strategy. Today, data analytics for IPM have been improved by machine learning (ML) and artificial intelligence (AI) technologies (see the section "Automated Insect Identification Using ML") to more quickly identify the insect, increasing the speed at which a mitigation strategy can be identified.

General Crop Damage Mechanisms Caused by Insects

Herbivorous insects reduce crop yield by consuming the leaves, flowers, and roots of the plant. For example, caterpillars of the fall armyworm (*Spodoptera frugiperda*) consume corn leaves, while root-feeding larvae of the corn rootworm (*Diabrotica* spp.) damage the plant's roots. This feeding activity not only directly diminishes the plant's ability to photosynthesize and develop properly but also causes wounds and structural damage that weaken the plant's overall health. These injuries can also serve as entry points for secondary pests or pathogens.

In addition to direct feeding damage, many insects can act as disease vectors, or transmitters of disease, increasing the ability of disease to spread in a field. For instance, aphids, such as the green peach aphid (*Myzus persicae*), transmit viruses like Potato leafroll virus, while whiteflies (*Bemisia tabaci*) are notorious for transmitting tomato yellow leaf curl virus and other viral infections.

Insect Scouting

Insect scouting and identification are necessary to determine the action threshold. Insects invade fields in distinct ways and have different timelines for their impact on growing crops. For example, grasshoppers may invade from a neighbor's field, whereas other insect pests, such as aphids, may be transported by the wind. These disparate transport processes require different scouting techniques. For example, if the pest is transported from a neighbor's field, then scouting the field perimeter may be useful. However, scouting for cereal-sorghum aphids, which are transported by the wind, requires that the interior, not just the perimeter, of the field be investigated (Figure 17.1). The activity of some insect pests is based on air and/or soil temperatures, and therefore their activity and life cycles can be predicted using Growing Degree Days (GDDs). GDDs are the combination of temperature and time, expressed as heat units, used to understand the growth and development of plants and other organisms, such as insects. The accumulation of these heat units over a period of time is referred to as GDDs. For example, 50% of the second generation of the European corn borer hatch is expected when the GDDs (base 54°F [12°C]) reach 1550. This pest overwinters as full-grown larvae in corn stalks, and in central North America, they emerge in late May or early June. Corn seed maggots feed on the germinating seeds and have a lower base temperature for activity: 40°F (4°C). Wire worms become active at 50°F (10°C), and as the season progresses, the single-generation/year population starts to emerge when 1422 GDDs (base 50°F [10°C]) are reached. For the stalk borer, gypsy moth, and alfalfa weevil, the commonly used base values for GDD calculations are 41°F (5°C), 50°F (10°C), and 48°F (9°C), respectively. For more details and information on how to calculate GDDs, see Chapter 32.

In addition to predicting activity after the growth and development of adult insects, other considerations include that some insect pests cause more crop damage in the larval stage while others have a second (or third) generation in a year that may cause the most severe damage. Thus, identifying the pest and determining the current stage in its life cycle is important for selecting appropriate management practices to minimize damage.

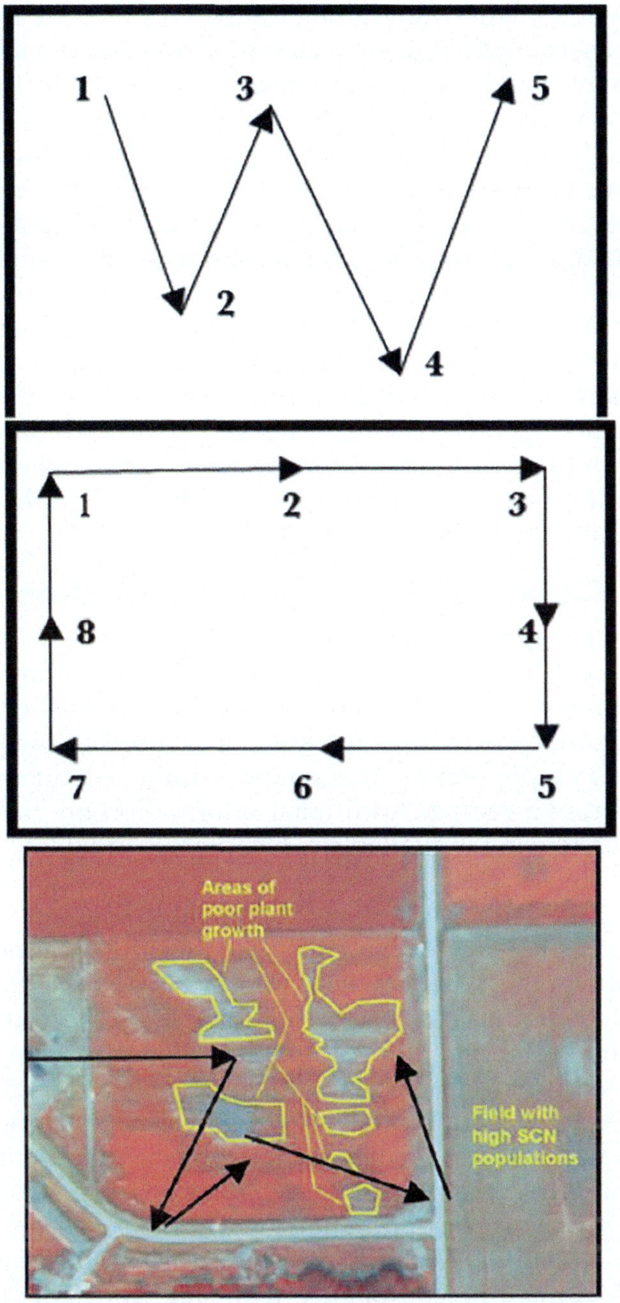

Figure 17.1. Scouting approaches that can be used for different insect pests. A W-pattern may be useful for assessing the interior of a field for insects that fly or are carried on wind currents. Field border sampling may be useful to track the movement of ground-dwelling insects that invade from neighboring fields. When remote sensing identifies plants in poor health, targeted sampling might be useful. (Courtesy South Dakota State University).

Technologies and Cultural Practices Used in IPM

Once damage has been observed during scouting and it has been determined that action is needed, various technologies and cultural practices can be used to mitigate the damage. Often, these technologies and practices use knowledge of insect pest biology. Some of these practices include technologies crafted specifically for this purpose, while others are cultural practices used for a variety of reasons and have been observed to help in pest control.

For instance, many technologies target a specific stage in a pest's life cycle, killing the insect at a specific vulnerable point that either prevents reproduction or prevents the population from spreading. However, it is important to note that management strategies that attack the insect at biologically vulnerable points in its life cycle must be used carefully, and the application instructions must be fully understood. This is because some technologies that use pest biology to mitigate a pest's impact also carry the opportunity for the development of pest resistance. For example, a prevention strategy to reduce rootworm populations in corn and cotton is to plant a transgenic *Bt* variety that produces cry proteins. Cry proteins are highly specific to the target pest, but continued use over 20 years has resulted in some resistant populations that were once controlled. This is a cautionary example of the overuse of a successful biological technology.

Other prevention techniques use biological approaches and are not as likely to lead to pest resistance. These practices include cultural methods, such as crop rotation, managing planting dates, and/or intercropping. Such methods are designed to target the pest's life cycle by removing or limiting the resources it needs to survive and reproduce at critical times. For example, to prevent corn rootworm infestations, a grower might rotate corn with soybeans for a season, disrupting the pest's food source and life cycle. Overall, cultural practices take into account the pests' biology and requirements, actively reducing opportunities for them to thrive. However, when these practices fail to prevent insect populations from exceeding the economic threshold, the point at which profits are reduced, the use of insecticides is warranted. Insecticides also target the biology of the pest, and a complete understanding of the application instructions is essential to keep the applicator safe and to avoid future opportunities for the insect pest to develop resistance. Clearly, a good understanding of the species and its biology is needed to choose the best mitigation strategy and avoid excessive yield loss.

Measuring Populations

Insects are very diverse, and determining the population density and distribution of each insect species may require unique sampling protocols tailored to its biology. Measuring the insect population in a field is challenging. The same species of pest is generally monitored or evaluated differently depending on its life stage. For example, corn rootworm adults, which can fly, are often monitored using sticky or pheromone traps arranged in grids across a corn field. After capturing a threshold number within a set period, appropriate management can begin. Conversely, rootworm larvae and root ratings are conducted by washing roots to perform a "float test" (i.e., larvae float to the surface when submerged in water), and roots are rated using a standardized scale once they are washed and cleaned.

Aphids, thrips, stink bugs, and other small, soft-bodied insects can be sampled using sticky traps to assess densities across environments and host plants. Large suction-traps that are part of a larger geographic region network (https://suctiontrapnetwork.org/about) are used to monitor flying pests. These traps can be used to determine establishment dates and trigger sampling or scouting events. In some cases, aphid species are carried by winds from southern regions (e.g., cereal-sorghum aphids), while others overwinter in nearby habitats (e.g., soybean aphids on buckthorn) (Parry, 2013; Severtson et al., 2015). These coordinated trapping efforts can provide short-term (within-season) and longitudinal (between-season) information on temporal and spatial changes in population densities.

Based on the crop, pests and beneficial arthropods can be sampled with sweep nets or whole-plant counts along transects or by visually observing the foliage. Belowground pests, which may spend part or all of their lives in the soil, are typically sampled manually through soil collection (Park & Tollefson, 2006). Soil samples are typically collected at least once a growing season; however, for some species, periodic sampling throughout the season may be needed. Whenever feasible, the data collected should include a description of how, where, and when the sample was collected. Combining the location (latitude and longitude) with the measurement is called georeferencing. The georeferenced data can be used to identify zones or hotspots where additional resources are needed (Fleischer et al., 1999; Hafeez et al., 2023).

Georeferencing Sample Collection

To apply targeted treatments to areas with insect populations that exceed the economic threshold, sampling points should be established. Georeferenced information can be used to track insect population movement within a field or across a region. Additional information on georeferencing is available in Chapters 8 and 9.

Record Keeping

In IPM, record keeping provides information about the where, when, and what of the IPM remediation that was used (Table 17.1). Records should also include the calculations, scouting records, insect species, climate information (temperature and rainfall), field location, scouting dates, rotations, crop plant and associated genetic capacity, plant growth stage, pest and crop growth stages, population levels, and any recommendations. Accurate records are also critical for documenting compliance with state and federal laws, and the information they contain can be used to conduct an economic analysis of the practices. An important component of record keeping is the correct identification of the insect. Insect species can be identified using appropriate books or extension bulletins. Insect taxonomic keys help identify the insect order and species by answering key questions. Currently, there are around 29 orders, and there are likely between 10 and 30 million insect species. While some common species can be identified after training, many species are unidentified. Manual identification of insect species can be a slow and

Table 17.1. An example of record keeping during scouting to efficiently collect information to make management decisions

Location in field and climate information	Notes on crop damage and date	Identification of insect	Life stage	Visual counts
Northeast corner, recent precipitation event (~5 cm), 25 °C	Damage to leaves at V6 stage in 30 plants found June 18, 2025	European corn borer	Adults	2

expensive process. An alternative approach is to use ML approaches. Machine learning approaches, discussed below, focus on current pest identification and data collection for insect management. For a more general overview of ML and AI in agriculture, see Chapter 27.

Automated Insect Identification Using ML

Currently, most insect identification relies on visual observations (Høye et al., 2021). This approach is relatively slow and relies on experts to identify insects one at a time based on experience and knowledge. Machine learning techniques provide an alternative approach. Entomologists and computer scientists have been conducting insect identification-based ML and AI research for decades (Martineau et al., 2017). Machine learning can assist in data analysis by creating mathematical equations that find complex relationships in large datasets (Jordan & Mitchell, 2015). A special feature of these algorithms is that they can learn from their own experience. In entomology, these techniques have been successfully used to automate the identification of pests primarily visually, but they can also include identification based on sound bites, for example, bee identification from the images and buzz sound of the wings (Arbuckle et al., 2001). In that study, Arbuckle et al. (2001) differentiated bee species using images of wings with 95% accuracy. Almost 20 years later, Ribeiro et al. (2021) had similar results but was able to include the auditory wing buzz as a feature to enhance identification. Based on the buzzing sound of the bee wings, they were able to identify the bee species associated with tomato with an accuracy of 73%. Deep learning, a subfield of ML that can use neural networks called convolutional neural networks (CNNs) is also being used to identify insects (Amarathunga et al., 2021; Høye et al., 2021). This approach has been successfully used in extracting features from images and then using the features to perform an insect classification or detection (Albawi et al., 2017; Hansen et al., 2020; Motta et al., 2020; O'Shea & Nash, 2015).

Object classification requires that the object in an image (i.e., an insect) has been previously characterized and assigned an identification value (ID). During classification, an object is predicted to belong to a group of images that fit an assigned ID (Li et al., 2021; Spiesman et al., 2021; Arroyo, 2021. Kusrini et al., 2020). On the other hand, object detection goes a step further by not only classifying the object but also determining its location within a sensor image, commonly represented using a bounding box. More details of these steps in insect identification using object detection can be found in the literature (Grijalva et al., 2023a, 2025b; Li et al., 2021; Padilla et al., 2020; Vega et al., 2021; Wang et al., 2022). Using object detection for insect identification is currently an active area of research (Ahmad et al., 2022; Du et al., 2022; Wang et al., 2022), and additional information on ML is available in Chapter 27.

Pest Monitoring with Unmanned Vehicles and Large-scale Data Collection

Recent advances in agriculture and technology have enabled precision IPM to implement targeted treatments, thereby improving the speed and accuracy of grower decision-making (Figure 17.2). By leveraging the scalability of ML, data collected from unmanned vehicles can be analyzed with CNN models, allowing researchers and farmers to monitor and make decisions across large areas more efficiently. For example, Grijalva et al. (2023a, 2023b) proposed the use of CNN classification models to automatically

Figure 17.2. A ground robot designed for targeted insecticide application to control aphids in sorghum fields. (Photo courtesy Ajay Sharda).

classify sorghum aphid densities on leaves into six classes based on established standard threshold levels for spraying. Notably, the models correctly classified aphids as above or below threshold spray density over 97% of the time. Therefore, the final output model in their study can be used in unmanned vehicles to perform real-time aphid identification and to locate and treat infestations found on individual plants, resulting in targeted treatments, which change the way we manage pests (Figure 17.3).

Implementing targeted treatments has the potential to reduce the total amount of pesticides applied to land and water systems, which in turn can reduce their impact on surrounding lands that contain non-target organisms. The combination of traditional IPM approaches with new technologies may improve the efficiency and sustainability of current pest management strategies (Dara, 2019). We envision a future where automated systems identify and control insect pests when their populations exceed the economic threshold. As these related technologies continue to advance, we can expect more sophisticated applications in pest management in the future.

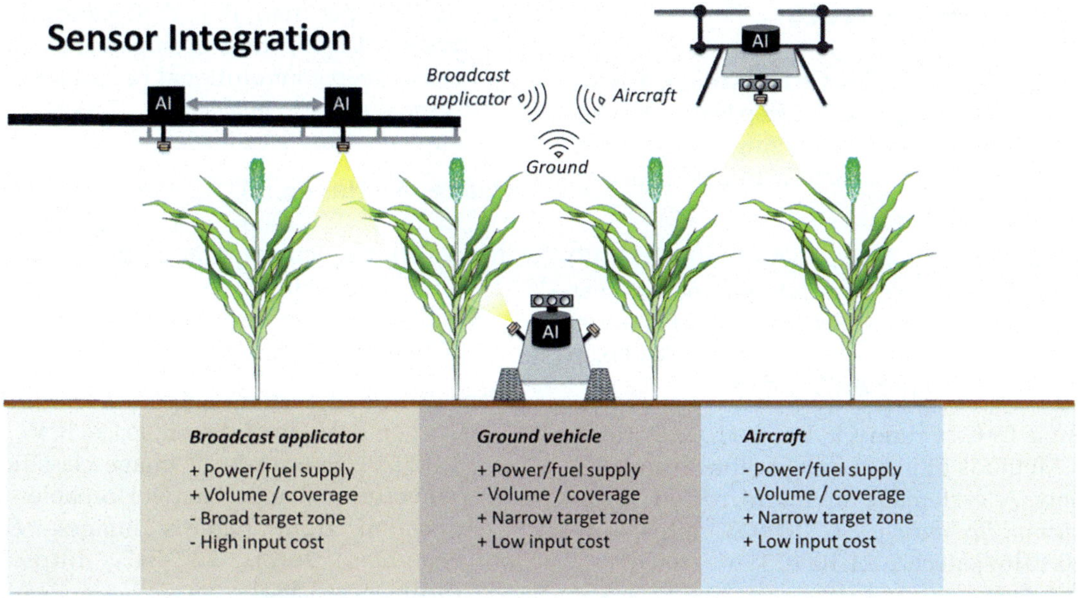

Figure 17.3. The future of IPM is merging AI with unmanned vehicles for pest management. (Photo courtesy Brian McCormack and Ivan Grijalva).

Summary

IPM has continued to advance through technological advancements over the years. Most recently, ML and computer vision models integrated into robotics and autonomous systems have the potential to detect and manage pests using less labor and time (Talaviya et al., 2020). Currently, ML detection models are being developed and tested (Li et al., 2019; Li et al., 2022; Thenmozhi & Srinivasulu Reddy, 2019; Tibbetts, 2018; Zhang et al., 2023). When implemented, these systems have the potential to reduce the risk of insect pesticide resistance, decrease the use of insecticides, and improve food and environmental security (Talaviya et al., 2020). These technologies are being used to extend the core principles of IPM toward a more efficient, optimal future of agriculture.

Study Questions

1. What is IPM?
2. Why is insect biology important to understand?
3. What are some examples of scouting techniques used to determine insect populations (See Figure 17.1)?
4. How could new technologies be used to inform insect management decisions?
5. Why is machine learning important in IPM?
6. What is a CNN?
7. What is the difference between object classification and object detection?

ACKNOWLEDGMENTS

This paper was partially funded by the National Science Foundation grants 2202706 and 2026431, South Dakota State University, USDA Natural Resources Conservation Service Conservation Innovation Grants (NR233A750011G019), USDA Agriculture and Food Research Initiative, the American Coalition for Ethanol, GEVO, the South Dakota Corn Utilization Council, and the South Dakota Soybean Research and Promotion Council. The use of trade names is for the convenience of the reader and does not imply endorsement by the authors. This chapter has been approved by the director of the Louisiana Agricultural Experiment Station as manuscript no. 2025-234-40422.

REFERENCES

Ahmad, I., Yang, Y., Yue, Y., Ye, C., Hassan, M., Cheng, X., Wu, Y., & Zhang, Y. (2022). Deep learning based detector YOLOv5 for identifying insect pests. *Applied Sciences*, *12*(7), 10167. https://doi.org/10.3390/app121910167.

Albawi, S. Mohammed, T. A., & Al-Zawi, S. (2017). Understanding of a convolutional neural network. *2017 International Conference on Engineering and Technology (ICET)*, pp. 1–6. https://doi.org/10.1109/ICEngTechnol.2017.8308186.

Amarathunga, D. C., Grundy, J., Parry, H., & Dorin, A. (2021). Methods of insect image capture and classification: A systematic literature review. *Smart Agricultural Technology*, *1*, 100023. https://doi.org/10.1016/j.atech.2021.100023.

Arbuckle, T., Schröder, S., Steinhage, V., & Wittmann, D. (2001). Biodiversity informatics in action: identification and monitoring of bee species using ABIS. In *Proceedings of the 15th International Symposium on Informatics for Environmental Protection* (pp. 425–430). EMPA.

Arroyo, J. C. T. (2021). Coleoptera classification using convolutional neural network and transfer learning. *International Journal of Engineering Trends and Technology*, *69*(5), 1–5. https://doi.org/10.14445/22315381/IJETT-V69I5P201.

Dara, S. K. (2019). The new integrated pest management paradigm for the modern age. *Journal of Integrated Pest Management*, *10*(1), 12. https://doi.org/10.1093/jipm/pmz010.

Du, J., Liu, L., Li, R., Jiao, L., Xie, C., & Wang, R. (2022). Towards densely clustered tiny pest detection in the wild environment. *Neurocomputing*, *490*, 400–412. https://doi.org/10.1016/j.neucom.2021.12.012.

Fleischer, S. J., Blom, P. E., & Weisz, R. (1999). Sampling in precision IPM: When the objective is a map. *Phytopathology*, *89*(11), 1112–1118. https://doi.org/10.1094/PHYTO.1999.89.11.1112.

Grijalva, I., Adams, H. B., Clark, N., & McCornack, B. (2025b). A fine-tuned deep learning model for detecting Japanese beetles in soybeans using unmanned aircraft systems UAS and mobile imaging. *Machine Learning with Applications*, *21*, 100711. https://doi.org/10.1016/j.mlwa.2025.100711.

Grijalva, I., Spiesman, B. J., & McCornack, B. (2023a). Computer vision model for sorghum aphid detection using deep learning. *Journal of Agriculture and Food Research*, *13*, 100652. https://doi.org/10.1016/j.jafr.2023.100652.

Grijalva, I., Spiesman, B. J., & McCornack, B. (2023b). Image classification of sugarcane aphid density using deep convolutional neural networks. *Smart Agricultural Technology*, *3*, 100089. https://doi.org/10.1016/j.atech.2022.100089.

Hafeez, A., Husain, M. A., Singh, S. P., Chauhan, A., Khan, M. T., Kumar, N., Chauhan, A., & Soni, S. K. (2023). Implementation of drone technology for farm monitoring & pesticide spraying: A review. *Information Processing in Agriculture*, *10*(2), 192–203. https://doi.org/10.1016/j.inpa.2022.02.002.

Hansen, O. L. P., Svenning, J.-C., Olsen, K., Dupont, S., Garner, B. H., Iosifidis, A., Price, B. W., & Høye, T. T. (2020). Species-level image classification with convolutional neural network enables insect identification from habitus images. *Ecology and Evolution*, *10*(2), 737–747. https://doi.org/10.1002/ece3.5921.

Høye, T. T., Ärje, J., Bjerge, K., Hansen, O. L. P., Iosifidis, A., Leese, F., Mann, H. M. R., Meissner, K., Melvad, C., & Raitoharju, J. (2021). Deep learning and computer vision will transform entomology. *Proceedings of the National Academy of Sciences of the United States of America,*, *118*(2), Article e2002545117. https://doi.org/10.1073/pnas.2002545117.

Jordan, M. I., & Mitchell, T. M. (2015). Machine learning: Trends, perspectives, and prospects. *Science*, *349*(6245), 255–260. https://doi.org/10.1126/science.aaa8415.

Kogan, M. (1998). Integrated pest management: Historical perspectives and contemporary developments. *Annual Review of Entomology*, *43*, 243–270. https://www.annualreviews.org/content/journals/10.1146/annurev.ento.43.1.243.

Kusrini, K., Suputa, S., Setyanto, A., Agastya, I. M. A., Priantoro, H., Chandramouli, K., & Izquierdo, E. (2020). Data augmentation for automated pest classification in Mango farms. *Computers and Electronics in Agriculture*, *179*, 105842. https://doi.org/10.1016/j.compag.2020.105842.

Li, R., Wang, R., Xie, C., Chen, H., Long, Q., Liu, L., Zhang, J., Chen, T., Hu, H., Jiao, L., Du, J., & Liu, H. (2022). A multi-branch convolutional neural network with density map for aphid counting. *Biosystems Engineering*, *213*, 148–161. https://doi.org/10.1016/j.biosystemseng.2021.11.020.

Li, R., Wang, R., Xie, C., Liu, L., Zhang, J., Wang, F., & Liu, W. (2019). A coarse-to-fine network for aphid recognition and detection in the field. *Biosystems Engineering*, *187*, 39–52. https://doi.org/10.1016/j.biosystemseng.2019.08.013.

Li, W., Zheng, T., Yang, Z., Li, M., Sun, C., & Yang, X. (2021). Classification and detection of insects from field images using deep learning for smart pest management: A systematic review. *Ecological Informatics*, *66*, 101460. https://doi.org/10.1016/j.ecoinf.2021.101460.

Martineau, M., Conte, D., Raveaux, R., Arnault, I., Munier, D., & Venturini, G. (2017). A survey on image-based insect classification. *Pattern Recognition*, *65*, 273–284. https://doi.org/10.1016/j.patcog.2016.12.020.

Motta, D., Santos, A. Á. B., Machado, B. A. S., Ribeiro-Filho, O. G. V., Camargo, L. O. A., Valdenegro-Toro, M. A., Kirchner, F., & Badaró, R. (2020). Optimization of convolutional neural network hyperparameters for automatic classification of adult mosquitoes. *PLoS One*, *15*(7), Article e0234959. https://doi.org/10.1371/journal.pone.0234959.

Oerke, E. C. (2006). Crop losses to pests. *Journal of Agricultural Science*, *144*(1), 31–43. https://doi.org/10.1017/S0021859605005708.

O'Shea, K., & Nash, R. (2015). *An introduction to convolutional neural networks*. arXiv. https://doi.org/10.48550/arXiv.1511.08458.

Padilla, R., Netto, S. L., & da Silva, E. A. B. (2020). A Survey on performance metrics for object-detection algorithms. In *2020 International Conference on Systems, Signals and Image Processing (IWSSIP)* (pp. 237–242). IEEE. https://doi.org/10.1109/IWSSIP48289.2020.9145130.

Park, Y.-L., & Tollefson, J. J. (2006). Spatial distributions of corn rootworm (Coleoptera: Chrysomelidae) eggs and larvae: Implications for sampling. *Journal of the Kansas Entomological Society*, *79*(2), 129–135. https://www.jstor.org/stable/25086312.

Parry, H. R. (2013). Cereal aphid movement: General principles and simulation modelling. *Movement Ecology*, *1*, 14. https://doi.org/10.1186/2051-3933-1-14.

Ribeiro, A. P., da Silva, N. F. F., Mesquita, F. N., de Araújo, P., C. S., Rosa, T. C., & Mesquita-Neto, J. N. (2021). Machine learning approach for automatic recognition of tomato-pollinating bees based on their buzzing-sounds. *PLoS Computational Biology*, *17*(9), Article e1009426. https://doi.org/10.1371/journal.pcbi.1009426.

Severtson, D., Flower, K., & Nansen, C. (2015). Nonrandom distribution of cabbage aphids (Hemiptera: Aphididae) in dryland canola (Brassicales: Brassicaceae). *Environmental Entomology*, *44*(3), 767–779. https://doi.org/10.1093/ee/nvv021.

Spiesman, B. J., Gratton, C., Hatfield, R. G., Hsu, W. H., Jepsen, S., McCornack, B., Patel, K., & Wang, G. (2021). Assessing the potential for deep learning and computer vision to identify bumble bee species from images. *Scientific Reports*, *11*, 7580. https://doi.org/10.1038/s41598-021-87210-1.

Talaviya, T., Shah, D., Patel, N., Yagnik, H., & Shah, M. (2020). Implementation of artificial intelligence in agriculture for optimisation of irrigation and application of pesticides and herbicides. *Artificial Intelligence in Agriculture*, *4*, 58–73. https://doi.org/10.1016/j.aiia.2020.04.002.

Thenmozhi, K., & Srinivasulu Reddy, U. (2019). Crop pest classification based on deep convolutional neural network and transfer learning. *Computers and Electronics in Agriculture*, *164*, 104906. https://doi.org/10.1016/j.compag.2019.104906.

Tibbetts, J. H. (2018). The frontiers of artificial intelligence. *BioScience*, *68*(1), 5–10. https://www.jstor.org/stable/90017988.

Vega, M, Benítez, D. S., Pérez, N., Riofrío, D., Ramón, G., & Cisneros-Heredia, D. (2021). Coccinellidae beetle specimen detection using convolutional neural networks, In *2021 IEEE Colombian Conference on Applications of Computational Intelligence (ColCACI)* (pp. 1–5). IEEE. https://doi.org/10.1109/ColCACI52978.2021.9469588.

Wang, R., Li, R., Chen, T., Zhang, J., Xie, C., Qiu, K., Chen, P., Du, J., Chen, H., Shao, F., Hu, H., & Liu, H. (2022). An automatic system for pest recognition and forecasting. *Pest Management Science*, *78*(2), 711–721. https://doi.org/10.1002/ps.6684.

Zhang, T., Li, K., Chen, X., Zhong, C., Luo, B., Grijalva, I., McCornack, B., Flippo, D., Sharda, A., & Wang, G. (2023). Aphid cluster recognition and detection in the wild using deep learning models. *Scientific Reports*, *13*, 13410. https://doi.org/10.1038/s41598-023-38633-5.

18 Targeted Weed Management Using Precision Agriculture

J. Anita Dille, Rodrigo Werle, Thomas R. Butts, Lynn Sosnoskie, and Thierry Besançon

Purpose

This chapter is a new chapter that was not included in *Precision Agriculture Basics,* first edition. It discusses how site-specific weed management builds on prior investments in weed science and does not eliminate the need to follow labeled rates and rotate active ingredients. In this chapter, students will learn about creating weed management maps and the benefits and barriers to adopting site-specific weed management.

Introduction

A weed is a plant that is growing where it is not desired. Weeds often have rapid growth rates, have high seed production, and can survive in adverse environments. Weeds reduce crop yields through many mechanisms including competition for basic resources (sunlight, nutrients, and water), downregulation of many genes in the crop plant, and providing habitat for insects and disease (Dhadly et al., 2025; Horvath et al., 2023; Srinivasan et al., 2014). In addition, weed seeds, leaf tissue, and stem fragments can contaminate harvested grain, and the presence of dense weed stands can physically interfere with harvest. Site-specific weed management (SSWM) builds on classical weed management approaches and requires weed identification and use of labeled herbicide rates and encourages rotation of herbicide active ingredients within and across years.

Effective herbicide programs have been the cornerstone of weed management across cropping systems (Duke, 2012). However, the repeated use of a single active ingredient or herbicides with the same mode of action (i.e., the overall manner by which the herbicide kills the weed) has contributed to the global rise of herbicide-resistant weeds, now documented at over 534 unique cases (species × herbicide mode of action) globally (Heap, 2025). Moreover, off-site movement of herbicides carries a significant risk of injury or damage to nontarget organisms or habitat. Conversely, use of mechanical tillage practices is often undesirable because it can degrade soil structure and health and increase wind and water erosion (Boyd et al., 2022; Fennimore et al., 2016; Fennimore & Cutulle, 2019). For some high-value crops, hand-weeding is an effective method of weed removal, although it is costly and labor intensive. An alternative approach is to use site-specific or targeted weed control technologies that enable growers to collect and utilize spatial information about crops, weeds, and field conditions.

Spatial and Temporal Variability of Weed Populations and Communities

Weed populations and communities are not uniformly or equally distributed in fields; instead, they have unique spatial and temporal variability and patterns. Even in a well-managed field, the long lifespan of weed seeds results in a continuing weed problem. Emerged seedlings often form patches with distinct high-density centers that gradually transition to lower-density edges, varying in location across the field (Dieleman & Mortensen, 1999). Weed seedling germination and emergence vary temporally over the growing season in response to environmental conditions (Werle, Bernards, et al., 2014, Werle, Sandell, et al., 2014), which can also differ across a field. Conditions such as seasonal temperature, soil moisture, field topography (including slope and aspect), soil texture, surface crop residue, and the presence of existing crop plants all influence these patterns (Marschner et al., 2024). Weed communities within a field are usually composed of diverse species and have varying life cycles (annual, biennial, perennial), distinct environmental requirements for germination and emergence, and unique dormancy and persistence traits. This diversity further highlights the unevenness in spatial and temporal distribution that must be addressed when developing any effective weed control program.

Opportunities and Advances in Technologies for Targeted Weed Management

Site-specific weed management systems have the potential to complement traditional practices by providing season-long control with greater efficiency. These technologies enable targeted interventions rather than traditional broadcast herbicide applications or field-scale tillage passes, which reduces operational costs and improves management precision. By minimizing inputs, targeted approaches can address regulatory demands, enhance application accuracy, lower carbon footprints, and respond to public concerns about sustainability. A recent survey of 475 weed scientists across the United States and Canada highlighted strong interest in advancing SSWM technologies (Brainard et al., 2023). Over half (53%) of respondents identified SSWM as a high priority for research and outreach, with 19% ranking it as their top priority.

Despite the enormous potential of precision agriculture tools for weed management, there are practical and financial roadblocks to the adoption of SSWM by growers and their service providers. While these technologies are quickly evolving, research to ground-truth their agronomic feasibility for integration into cropping systems, recommendations for effective application strategies, and potential environmental and economic benefits are still needed.

Site-Specific Weed Management: Scouting and Sampling

For some users, SSWM adoption requires measurement, understanding, and assessment of the weed community in their fields. This work generally shows that economic feasibility increases with spatial variability. For example, less herbicide will be applied or less soil disturbed by tillage with less dense or more patchy weed populations across a field. In the past, determining spatial variability was accomplished by physically scouting fields and documenting what weed species were present and where they were in the field. Emerged weed plants were easily observed, while gathering information on weed seedbanks and soil properties would take more effort to sample, process, and analyze before knowing the extent of variability. With advances in vision-guided on-the-go systems (see Section Vision-Guided On-the-go Sensing), gathering information on weed species occurrence or other field conditions has become much more automated.

Sensors Used in SSWM

There are multiple ways to generate maps of spatial information collected from fields to support SSWM activities. See Chapters 5, 6, 8, and 9 for more details on identifying, sampling, and collecting data on spatial and temporal variability. Data collected from on-the-ground sampling at discrete points, such as weed species observations, soil cores for weed seedbank, or physical and chemical soil property assessments, can be interpolated to generate geospatial maps. Proximal sampling will generate data from vision-guided on-the-go imagery, while remote sensing will collect data from unmanned aerial vehicles (UAVs), airplanes, and satellites. When combined with locations, these measurements can be used to create maps of weed distributions, soil properties, or crop performance. The maps can then be used to develop prescription maps and to conduct financial assessments. See Section Example of Prescription Map Development Using Pix4DFields for demonstration of a step-by-step approach to using remote sensing imagery to generate a prescription map.

A variety of sensors (red, green, blue [RGB], hyperspectral, multispectral, thermal) mounted on proximal or remote sensing tools generate images from which data can be extracted and used to calculate a variety of vegetative indices (Chapters 13, 14; Xue & Su, 2017). Depending on spatial and spectral resolution, it may be possible to assess the magnitude and scope of the weed population or community of concern within and across fields.

Spatial resolution depends on the platform and can range from ~10 mm^2 on a UAV to >100 m^2 on a satellite (Spaeth et al., 2024). Different resolutions are useful for different purposes. For example, high-resolution data collected from UAVs can detect individual plants or assess the seed emergence rate of crop plants (Rasmussen et al., 2021), while low-resolution information is useful for conducting a whole- or multiple-field assessment of concerns such as off-target herbicide movement or overall levels of weed infestation. Generally, cost, processing requirements, and expertise needed will increase with resolution, and the data files can quickly become incredibly large (encompassing terabytes of data), resulting in significant requirements for computer processing power and data storage. Conversely, satellite imagery can be inexpensive (often free for certain resolutions), cover substantial acreages, and can have computer storage and processing cost less than UAV imagery. Limitations associated with satellite imagery include lower-resolution capabilities, time gaps between image collection dates (typically 3–5 days), and loss of images owing to cloud cover that negatively affected the collection of timely and useful information (Matese et al., 2015). Overall, the choice between proximal and remote sensing tools for weed management comes down to the user's needs.

Vision-Guided On-the-go Sensing

Rather than waiting for a map to be generated, advancements in computer vision and artificial intelligence (AI) can be used to detect weed plants in green-on-brown (GOB) and green-on-green scenarios (Anne et al., 2024; Avent et al., 2024; Barnhart et al., 2022, 2025; Spaeth et al., 2024). Green-on-brown is a straightforward scenario in which green plants (usually weeds) are detected against a fallow or soil background, usually in a burndown or preemergence weed control application situation. Green-on-green is a more advanced scenario in which weed plants are actually detected and separated from green crop plants based on algorithms, typically in a postemergence weed control application situation. Training an algorithm to separate weed plants from crop plants has an accuracy of >80% (Figure 18.1; Barnhart et al., 2022).

Another approach for separating crops and weeds is the use of computer-vision techniques. The patterns in crops and weeds within and between rows can be used to guide weeding implements (Jiao et al., 2024). Some of these implements allow for greater field operation speeds without causing crop damage. Another approach is to modify crop plants, that is, alter their appearance through physical, chemical, or biological methods, to optimize this process. For example, crop-specific markers such as

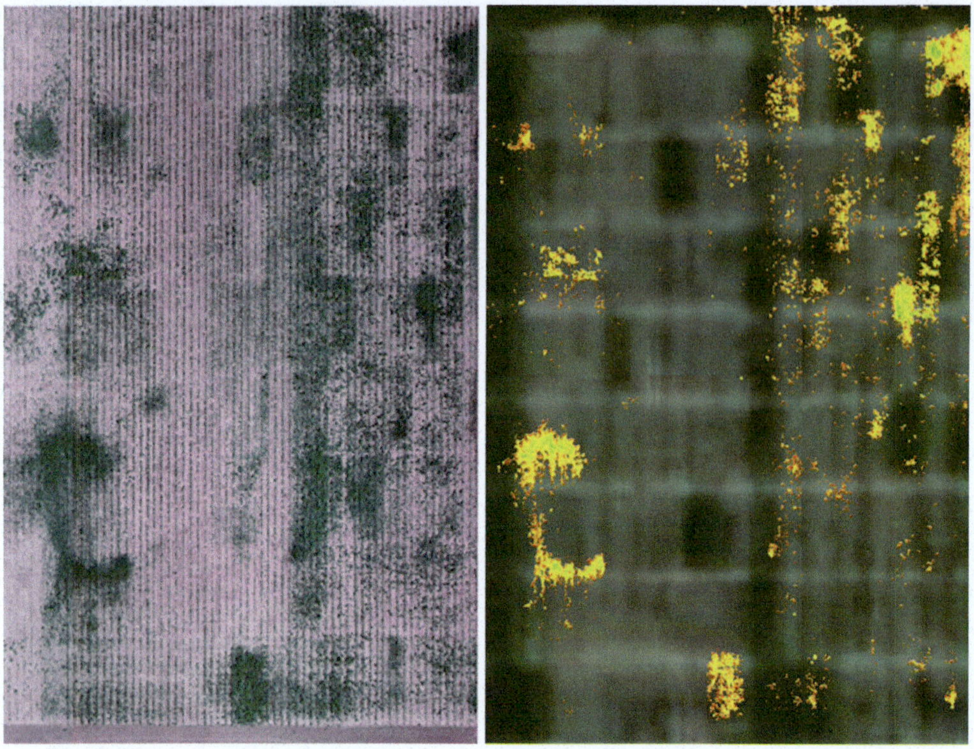

Figure 18.1. Detection and discrimination of weeds within a corn crop using analysis of multispectral imagery collected from UAV. Courtesy of Dr. Thomas R. Butts and Ms. Jada Davis.

fluorescent tracers can be used as a tool to enable fast, accurate crop recognition (Jiao et al., 2024; Su et al., 2019) and to treat the non-marked plants as weeds.

Site-specific Weed Management: Tools and Practices

SSWM uses multiple approaches, including cultural, mechanical, and chemical weed control tactics (Table 18.1), to avoid crop losses and reduce the long-term impact of weeds (Norsworthy et al., 2012; Swanton & Weise, 1991).

Cultural practices for weed management focus on ways to enhance crop competitiveness against weeds. This goal could be accomplished by changing crop seeding rates with corresponding levels of fertilizer in parts of the field with greater weed occurrence, which may include variable rate application (VRA) seeding or the use of soil fertility algorithms. Tailoring cultural practices to manage weeds presents a unique challenge and opportunity because many models do not account for weed distribution in the field.

Table 18.1. Proposed site-specific weed management options that can be implemented at various times during the annual cropping cycle.

Weed management options	Timing of application	
	Pre-emergent to the crop	Post-emergent to the crop
Cultural	Variable rate seeding of crops to be competitive in weedy areas	Variate rate application of fertilizers according to crop needs
Mechanical	Localized tillage practices (disc, cultivate)	Intra-row weeding implements (hoeing), robotic mowers, vision-guided lasers or weeders
Chemical	Variable rate application of soil residual herbicides based on soil texture, organic matter, pH; GOB applications of foliar herbicides	Targeted sprayer or drone applications of foliar-applied herbicides (green-on-green) based on weed detection

Mechanical practices for weed management include any tools that cause damage to weed plants, ranging from hand-weeding and removal to large-scale tillage tools for a field. Plowing, discing, or hoeing are effective for uprooting, burying, or damaging weed tissues, but these tools disrupt soil structure, incur high energy costs, stimulate weed seed germination, and disperse vegetative and reproductive propagules within and between fields. Mowing and flaming remove or destroy aboveground plant biomass but may not fully eliminate weeds, particularly grasses or perennials, that have belowground meristems and perennial propagules that could regrow.

Pairing physical or thermal weed control implements with vision-guided, AI-driven technology provides growers with a precise and efficient alternative to hand-weeding or loss of weed control within crop rows. This technological approach can lower costs, reduce physical strain on workers, and allow the labor force to focus on higher-value tasks. There are multiple mechanical tools (i.e., actuators) that have been explored or are being used commercially for nonchemical SSWM, particularly in vegetable production systems (Fennimore et al., 2016; Jiao et al., 2024). These tools include reciprocating knives, rotating discs or shovels, hoeing brushes, propelled grit, and electric, flame, steam, hot oil, and laser weeders on field-scale equipment. Robotic weeders are uncrewed vehicles designed to autonomously or semi-autonomously navigate through crops, to identify weeds, and execute targeted plant removal from fields with minimal human intervention. Many robotic weeders are smaller, more portable systems that have been developed, and they employ advanced technologies such as computer vision, AI, and several mechanical tools.

Chemical practices for SSWM include soil- and foliar-applied herbicides. VRAs for soil residual herbicides are possible as products have a range of rates identified on the product label. These rates might be based on soil texture (coarse, medium, and fine soil classifications) or soil organic matter (OM) content. For example, the herbicide TriCor 4F needs to be applied more heavily to a site with over 4% OM content relative to a site between 2 and 4%; coarse-textured soils with less than 2% OM content are recommended to have no product applied at all, which shows the role texture plays in prescriptions (Table 18.2). Despite some fields having immense soil variability (Figure 18.2), the standard practice is to broadcast a fixed rate of residual herbicide across the whole field. This practice can lead to poor herbicide performance in some areas and excess rates in other areas; therefore, it is recommended to identify in-field variation before herbicide application.

The efficacy of soil-applied residual herbicides can be optimized at the commercial level by

Table 18.2. Soil-residual herbicide label example (adapted from Tricor 4F, UPL Limited, King of Prussia, PA) illustrating a variable rate structure dependent on soil texture and organic matter content.

Pints of TriCor 4F per acre			
Soil texture	Organic matter		
	Less than 2%	2–4%	Over 4%
Coarse soils (sandy loam, loamy sand)	Do not use[c]	0.75	1
Medium soils[a] (loam, silt loam, silt, sandy clay, sandy clay loam)	0.75–1	1–1.25	1.25–1.5
Fine soils[a] (silty clay, silty clay loam[b], clay, clay loam)	1–1.25	1.25–1.5	1.5–1.75
Mississippi Delta only	1.5	1.75	2

[a] For control of lambsquarters, redroot pigweed, and wild mustard, and for suppression of green, yellow and giant foxtails on alkaline (calcareous) soils in Nebraska, Minnesota, South Dakota, and North Dakota only, apply TriCor 4F at rates of 0.5 pt/ac on medium soils and 0.5 to 0.75 pt/ac on fine soils regardless of soil organic matter percentage (use 0.75 pt only where soil pH is less than 7.5 and weed pressure is heavy). The 0.5 pt/ac rate of TriCor 4F alone can be applied regardless of soil pH. For control of other weeds listed on this label used TriCor 4F at full rates specified in the table above, but note that crop injury may occur on soils have a calcareous surface area or a pH of 7.5 or higher.

[b] Silty clay loam soils are transitional soils and may be classified as medium-textured soils in some regions of the United States.

[c] Refer to the appropriate section of this label for use of TriCor 4F on soybeans in coarse soils with 0.5% or more organic matter in certain states.

Figure 18.2. Unmanned aerial vehicle images display within-field soil variability at two locations in Rush County, Indiana. Courtesy of Dr. Thomas R. Butts and Mr. Alexander Meuth.

selecting rates of the herbicide according to soil parameters (Gundy & Dille, 2022). However, to effectively accomplish this optimization, a method to accurately quantify and delineate the soil variability must be developed (see Chapters 5 and 6). Soil sampling for soil texture and OM has been shown to be more effective than relying on the Soil Survey Geographic (SSURGO) database (Vagedes et al., 2024). Additionally, electrical conductivity measurements paired with grid-based soil test results (pH, texture, OM) could be used to develop herbicide rate management zones (Figure 18.3; Gundy & Dille, 2022; Vagedes et al., 2024). Future research should explore the use of remote sensing to replace grid-based soil sampling in the development of herbicide rate management zones. An additional layer of information would be to enumerate weed seedbanks and add a weed density factor when determining herbicide rate management zones.

Following the creation of herbicide management zones, field prescription maps need to be created and/or uploaded into the operator's software of choice. There may be several equipment options for applying the treatments, but ideally a pulse-width modulation (PWM) sprayer would be used (Chapter 26). Sprayers with PWM are able to more rapidly adjust the flow rate compared with other sprayer types (Butts et al., 2019). Applicators should mix for the lowest herbicide

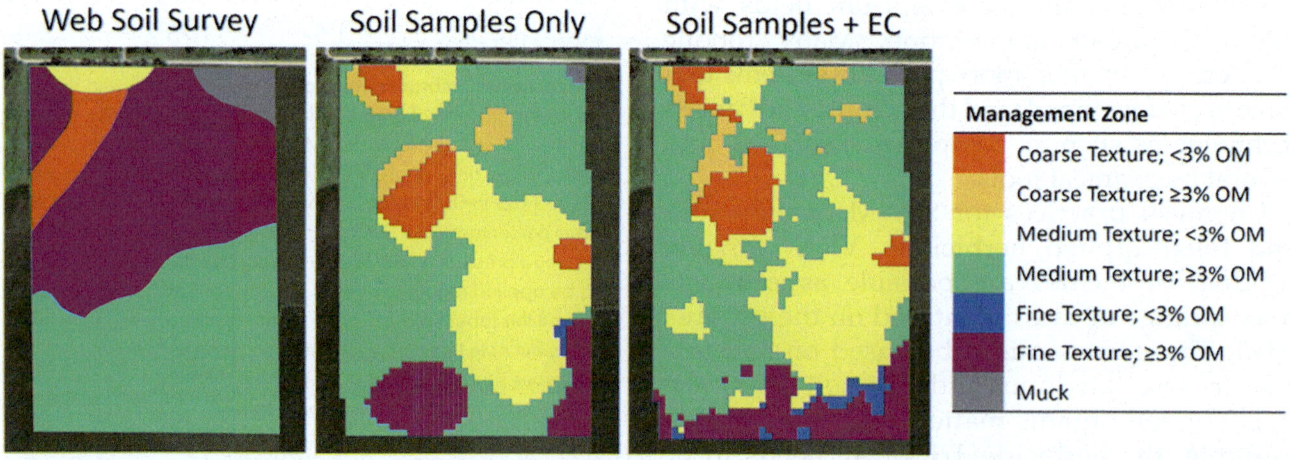

Figure 18.3. Illustration of herbicide rate management zones created from a field in Warren County, Indiana, using (a) Soil Survey Geographic database, (b) soil samples (texture and organic matter [OM]) only, and (c) soil samples + electrical conductivity (EC) measurements. Courtesy of Ms. Rose Vagedes and Dr. Bryan G. Young.

rate present in the field (which will correspond to the greatest travel speed and lowest pulsing duty cycle). This approach will allow for a decrease in sprayer speed or an increase in pulsing duty cycle to increase the herbicide rate as needed based on the prescription map. In the future, sprayers equipped with dual spray booms may be an improved system for variable rate soil-applied residual herbicides (ONE SMART SPRAY sprayer; e.g., Spaeth et al., 2024). In the dual boom system, a base (minimum) rate can be applied from one boom for a broadcast rate, while the second boom, fed from the same solution tank, could use PWM to apply additional herbicide as needed, based on changes in soil properties or occurrence of weed patches. Another enhancement for using a dual tank system would allow herbicide active ingredients to be applied in specific areas within a field, while other active ingredients could be placed in alternative field areas. This specificity may be largely beneficial where regulatory concerns require buffer zones to mitigate off-target impacts for individual herbicide active ingredients. Overall, VRA of soil-applied residual herbicides can be an effective tool to reduce potential crop injury, maximize herbicide effectiveness, and enhance environmental stewardship when high soil variability, variable spatial distribution of weed seed, or both are present within a field.

Targeted spraying systems use vision-based on-the-go sensing for foliar-applied herbicides and first became commercially available in the 1990s and allowed for GOB applications (i.e., WeedSeeker®, Trimble Ag). The WeedSeeker is a light-activated spraying system that functions by emitting a combination of red and near-infrared light through light-emitting diodes, detecting the reflected light from green vegetation in the field, and activating the solenoid connected to each nozzle to actively spray an area that includes the detected green vegetation. Such technology can be implemented for several GOB applications, such as preplant burndown applications of foliar herbicides, for nonselective weed control during fallow periods (before or after crop harvest), and before crop emergence. These technologies are particularly useful in no-till agriculture. The benefits were demonstrated in Montana, a semi-arid region that uses no-till agriculture to conserve soil moisture; because tillage is not an option, foliar herbicides often need to be applied. However, challenges include widespread occurrence of herbicide-resistant weeds. By using a targeted spray system such as WeedSeeker, there was a reduction of 23–55% in total foliar-applied herbicide in the region; targeted spraying systems were able to more efficiently control established weeds in postharvest wheat (Sharma et al., 2023).

Similarly, the WEED-IT (Rometron B.V., Steenderen, The Netherlands) system can be retrofitted to any sprayer and is also a GOB targeted spraying system. It uses sensors that detect chlorophyll fluorescence, that is a near-infrared signal emitted in response to blue light (450–500 nm; Figure 18.4). The WEED-IT system has been used for weed detection in fallow fields for burndown application of nonselective herbicides, similar to the WeedSeeker (Fischer et al., 2020; Genna et al., 2021; Zanin et al., 2022). Both WEED-IT and WeedSeeker spray systems used 53% less herbicide for dryland weed control in the Pacific Northwest compared with a broadcast spray application (Genna et al., 2021). The modularity of the WEED-IT allows for implementation on smaller spraying equipment typically used in perennial cropping systems such as orchards and vineyards (IR-4, 2021). In perennial cropping systems, spray nozzles can be directed at the ground and towards the brown trunks of trees or vines, avoiding interference from non-photosynthetic

Figure 18.4. WEED-IT™ targeted spraying system mounted on an ATV for postemergence herbicide application in highbush blueberry. Courtesy of Dr. Thierry E. Besançon.

surfaces. Sosnoskie et al. (2025) reported that the WEED-IT system reduced herbicide application volumes up to 50% relative to banded applications in highbush blueberry, while achieving comparable weed control efficacy for weed seedling. It was important to target weed seedlings because the WEED-IT sprayer did not deliver enough herbicide to achieve complete control of larger plants. However, researchers are working to address these challenges (Sosnoskie et al., 2023).

Green-on-green applications have become a reality with more recent advancements in computer processing, machine learning, AI, and image recognition (e.g., Blue River John Deere See & Spray™ Ultimate, Figure 18.5), expanding the opportunities for technology use beyond GOB applications. These novel systems are capable of differentiating crops from weeds in real time. This technology relies on RGB sensors mounted on the sprayer that collect information processed by the spraying system, thus detecting weeds and activating nozzles all within milliseconds. Another benefit of these technologies is the creation of "as-applied" maps (based on when the nozzles are turned on or off), which can be used to describe weed distributions within fields for subsequent planning (Figure 18.6). Research conducted in Germany and Hungary with the ONE SMART SPRAY sprayer technology reported that 10–55% less foliar-applied herbicide product was used for postemergence targeted weed control in sugar beet, corn, and sunflower fields (Spaeth et al., 2024). Research conducted in Arkansas and Mississippi with the John Deere See & Spray technology was able to reduce the foliar herbicides applied by 28–62% compared with traditional broadcast applications (Avent et al., 2024).

In addition to the ground-based targeted spraying systems, spray UAVs (Figures 18.7 and 18.8) can apply some herbicides aerially (Hunter et al., 2020b). Spray UAVs have been shown to provide effective coverage even with reduced spray volumes (Butts et al., 2024; Hunter et al., 2020a), which has resulted in studies demonstrating equivalent weed control or crop desiccation as ground-based applications (Lopes et al., 2024; Liao et al., 2019; Meng et al., 2019). Spray UAVs can provide flexibility for small irregularly shaped fields or apply treatments when wet weather prevents commercial ground-based spray equipment from entering the field. However, the commercial implementation of spray UAVs is limited by battery life and payload carrying capacities.

Although spray UAVs provide a unique approach to overcoming some problems with ground-based targeted spray applications, they do have their own set of challenges. Spray UAV applicators require significant licensure through the Federal Aviation Administration as well as state pesticide regulatory committees, which can take time to acquire. Many herbicide labels do not permit aerial applications, nor do they allow

Figure 18.5. Blue River John Deere See & Spray Ultimate with real-time camera-based weed detection systems for site-specific weed management. Courtesy of Dr. Rodrigo Werle.

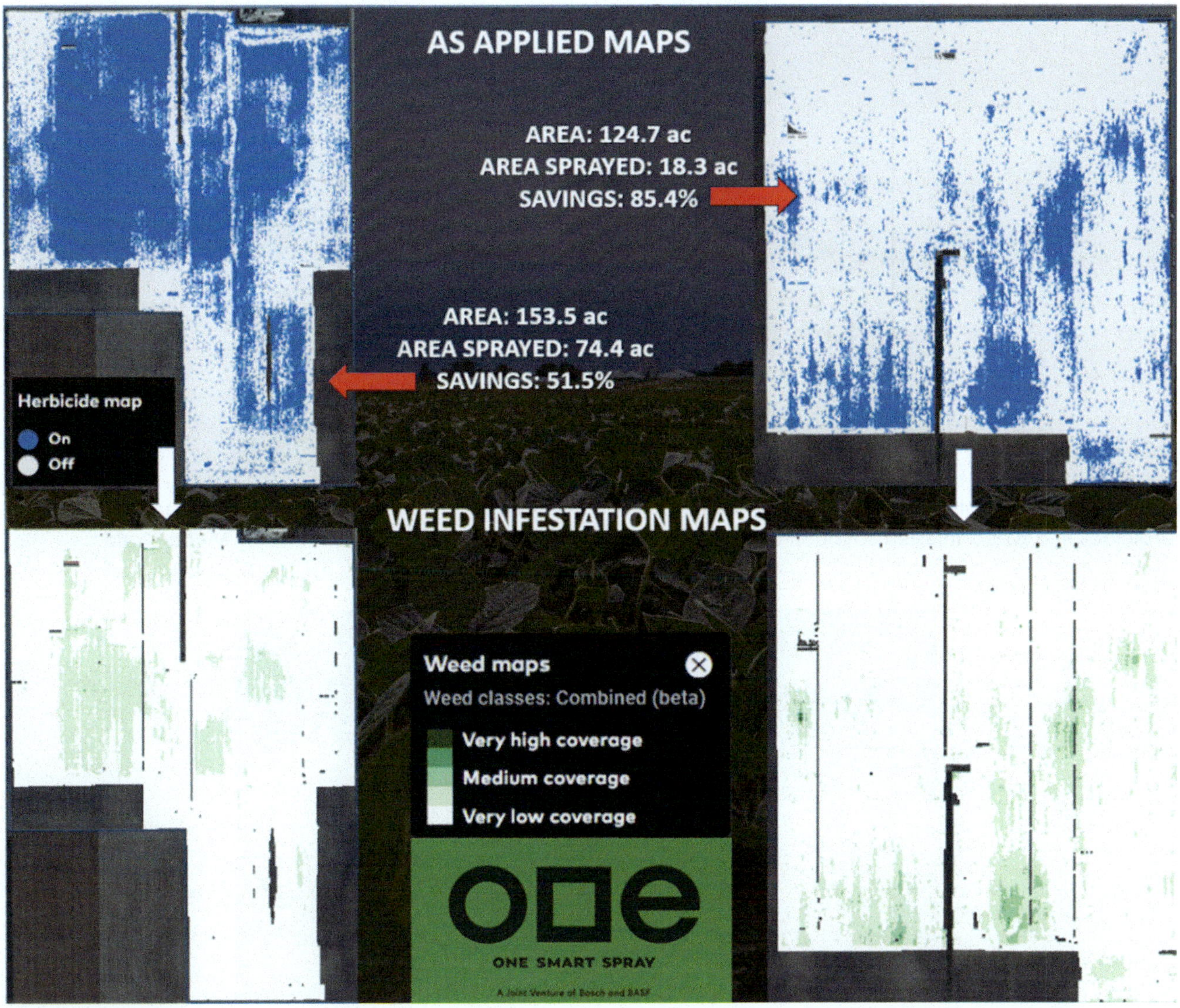

Figure 18.6. As-applied and weed infestation maps generated by the ONE SMART SPRAY (Bosch-BASF joint venture) through BASF-Xarvio Digital Farming Solutions Field Manager following applications in two commercial fields in Illinois, USA. Courtesy of Mr. Zaim Ugljic and Mr. Kalvin Miller.

for the low spray volumes typically utilized by spray drones (<3 gallons per acre [GPA], 27.9 L/ha). Additionally, the swath width has been highly variable, although research has shown that effective coverage can be obtained (Figure 18.9). Therefore, care must be used when developing application protocols, and each operator must understand how different application parameters (nozzle selection, flight speed, flight height, wind speed, wind direction, etc.) will influence the deposition pattern.

Targeted spraying systems can improve weed management efficiency and cut operational costs by reducing herbicide inputs, application time (fewer refills), wear and machinery cost, and off-target spray movement. However, before purchasing and adopting any new technology, it is essential to consider equipment and subscription costs, the agronomic efficacy and efficiency of the technology, and the maintenance requirements and availability of technical support. Most importantly, the herbicide savings are very dependent on the weed infestation level, which can vary widely within and across fields (Anne et al., 2024; Avent et al., 2024; Barnhart et al., 2025; Genna et al., 2021; Sharma et al., 2023; Spaeth et al., 2024).

Figure 18.7. Example of a spray UAV (drone) equipped with hydraulic nozzles (DJI Agras T30) applying herbicides in a weed-infested soybean field. Courtesy of Dr. Thomas R. Butts and Mr. Hunter Medenwald.

Figure 18.8. Example of a spray UAV (drone) equipped with rotary atomizer/spinning disk nozzles (DJI Agras T40) applying preemergence soil residual herbicides. Courtesy of Dr. Rodrigo Werle.

Example of Prescription Map Development Using PIX4Dfields

It is possible to utilize remote sensing imagery (UAV or satellite) and machine learning algorithms to create prescription and application maps. In this example, we use the software program PIX4Dfields (https://www.pix4d.com/product/pix4dfields) to describe both the concept and the completion of this task.

First, imagery should be imported and a field boundary created around the intended application area (Figure 18.10). Imagery can come from a variety of sources including UAVs and satellites and must be georeferenced. Depending on the imagery source, the software can also handle RGB, thermal infrared, and multispectral data. As an important note, image collection should occur at a high enough resolution to allow for detection of weeds at appropriate sizes for effective control (4 to 6 in. [9–13 cm] in height).

Next, the "Magic Tool," which is a built-in feature for PIX4Dfields, is equipped with an AI algorithm and is used to make "weed" versus "no weed" determinations. The user selects a minimum of three "wanted" cells (weeds) and three "unwanted" cells (crop, bare ground, etc.) to train the model for the image. Additional cells can then be classified to fine-tune the selection process (Figures 18.11 and 18.12). Additional

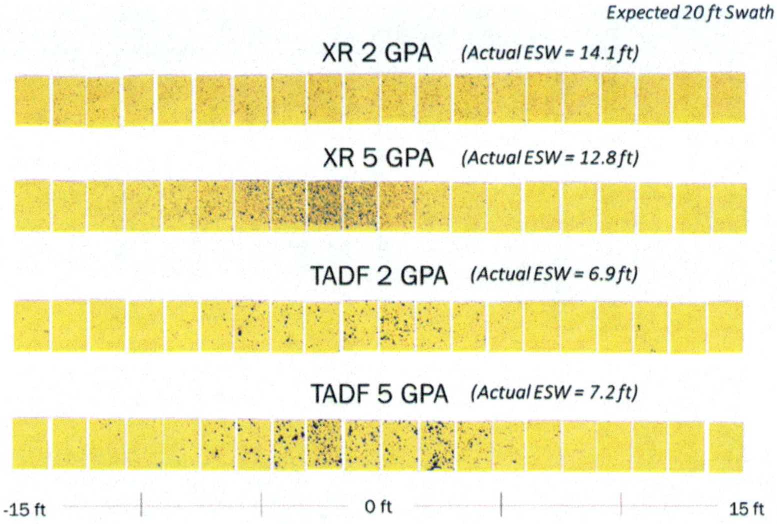

Figure 18.9. Spray card visualization of the variability in swath width distribution from a spray UAV (drone) at different spray volumes (2 and 5 GPA) and different equipped nozzles (XR, TADF). Abbreviation: ESW, effective swath width. Courtesy of Dr. Thomas R. Butts.

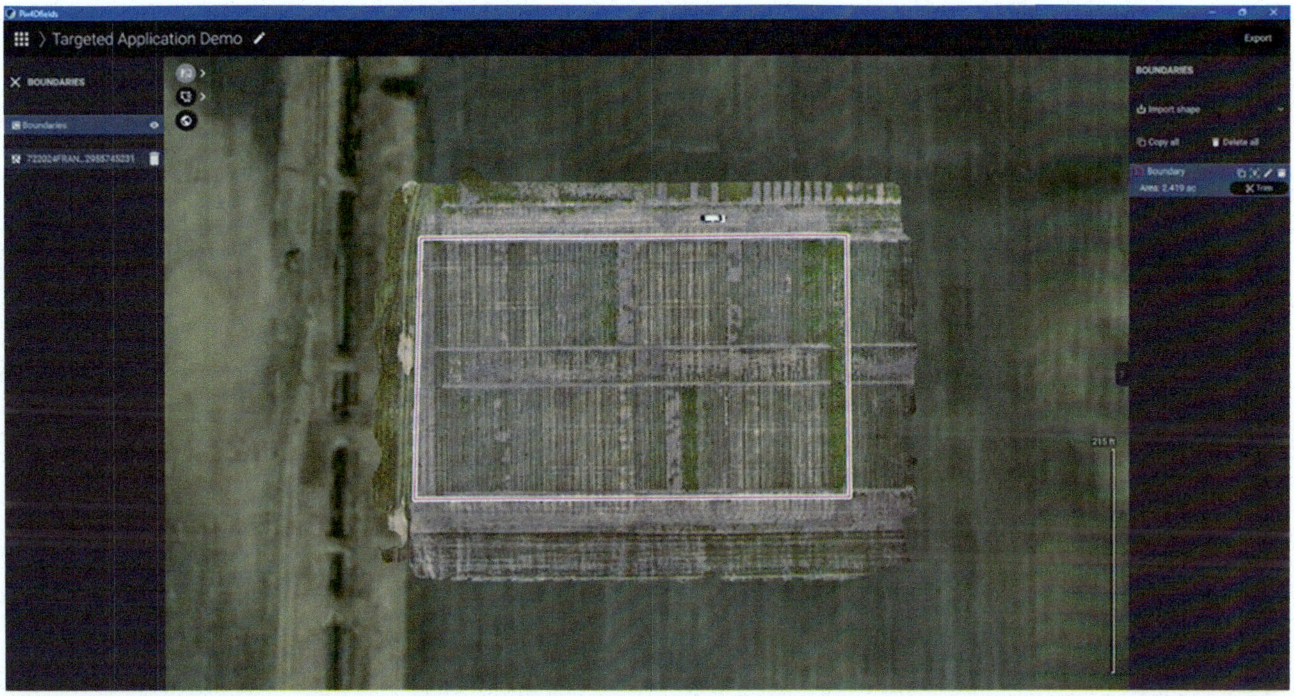

Figure 18.10. Importation of aerial imagery into the PIX4Dfields software package and creation of field boundary (pink border) in which subsequent processes will occur. Courtesy of Dr. Thomas R. Butts.

manual selections can be included as needed, and buffers around the marked areas can be increased (more coverage area, more likely to encompass all weeds) or decreased (less coverage area, more herbicide spray savings; Figure 18.13).

Following weed detection and buffer selection, the annotation layer containing the selected weeds is created (Figure 18.14). From there, a grid size can be selected, and an operation layer created for the targeted operation. The selected grid size should be approximately based on the

Figure 18.11. Using artificial intelligence, the machine learning "Magic Tool" detects and marks the weeds within a manually selected grid size area (1.5 × 1.5 ft [0.46 × 0.46 m]). A total of 22 individual grids were selected using the tool and labeled as either "wanted" (weeds) or "not wanted" (crop, bare ground, etc.). Courtesy of Dr. Thomas R. Butts.

Figure 18.12. An entire field overview of the initial annotation created, marking weeds within the field boundary following the use of the artificial intelligence algorithm. Courtesy of Dr. Thomas R. Butts.

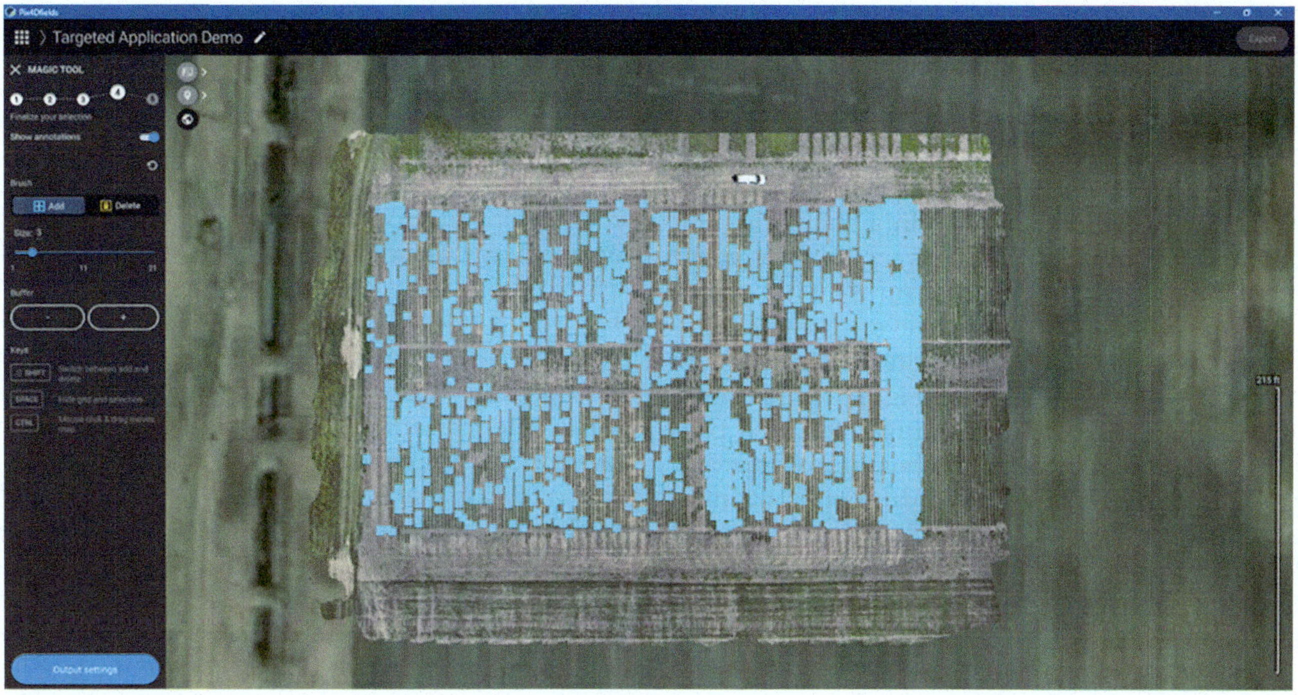

Figure 18.13. Following the creation of the initial annotation layer, an option to enlarge the targeted spray area with a user-defined buffer (e.g., be conservative in detection estimations) is provided. In this example, an additional 3 ft (0.91 m) spray buffer was added around all initial identified spray areas. Courtesy of Dr. Thomas R. Butts.

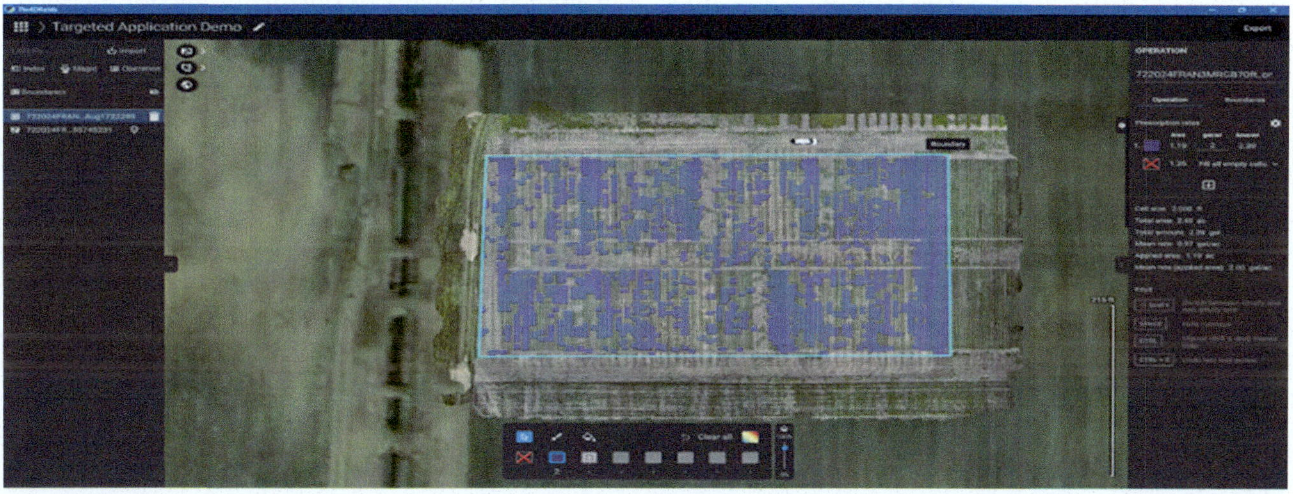

Figure 18.14. A targeted operation is created following the finalization of the annotation layer. In this example, the initial field size was 2.45 ac (1 ha) and the target spray volume was 2 gal/ac (18.8 L/ha); following the use of the targeted tool, this field has an estimated spray area of 1.19 ac (0.48 ha, approximately 51% spray output savings) with a total estimated spray output of 2.39 gal (9.05 L). Courtesy of Dr. Thomas R. Butts.

resolution capabilities of the spray equipment (individual nozzle and/or boom section control). The software will also allow the user to input a preferred spray volume (GPA), and based on the map, it will provide estimates of the applied area, total amount of spray solution needed, and the area with "empty cells" from which estimates of savings can be calculated (Figure 18.14). As a final step, the application map can then be exported as a file readable for the user's specific spray equipment. Files can be exported directly to a spray drone or for any GPS-driven implement (such as a PWM sprayer).

The PIX4Dfields program is user-friendly, exports the application map in a shapefile, and opens opportunities for growers to implement targeted weed control measures without an immense start-up cost. However, the AI algorithm is not foolproof, and more refinements are needed. As can be seen in Figure 18.15, some areas without weeds were sprayed (shaded purple) and some areas with weeds were not sprayed. Efficiency may be improved by better training with the Magic Tool. Further research is needed to evaluate the accuracy and effectiveness of these prescription maps on resulting weed control in the field.

Finally, users should be aware of the computer processing power requirements, connectivity capabilities, and the time delay between collecting imagery data, post-processing in the software, and getting the prescription map to the sprayer in the field. Any time delays can result in larger, more difficult-to-control weeds, as well as newly emerged weeds not previously detected.

Assessment of Weed Control and Crop Injury After Applications

Proximal- and remote-sensing imagery can be used after application to evaluate effectiveness of implemented weed control tactics and to document areas of crop damage following herbicide applications (Henry et al., 2004; Huang et al., 2015, 2016; Robles et al., 2010; Zhang et al., 2021; Figure 18.16). A variety of sensors, vegetation indices, and commercially available software tools enable this detection. Generally, herbicide injury identification using imagery tools follows two primary approaches: (1) large-scale evaluation and (2) small-scale evaluation. A large-scale evaluation uses remote sensing to quantify the general level of injury severity across a field and pinpoints the location and severity of injury (Denton et al., 2024). A small-scale evaluation includes the assessment of individual plants or leaves, using proximal sensing for precise diagnosis of specific herbicide injuries (Silva et al., 2019). Combining these approaches offers a strategy for detecting and diagnosing herbicide injury. Once potential injury hotspots are identified, a grower may want to perform a visual inspection to understand the nature of the injury zones.

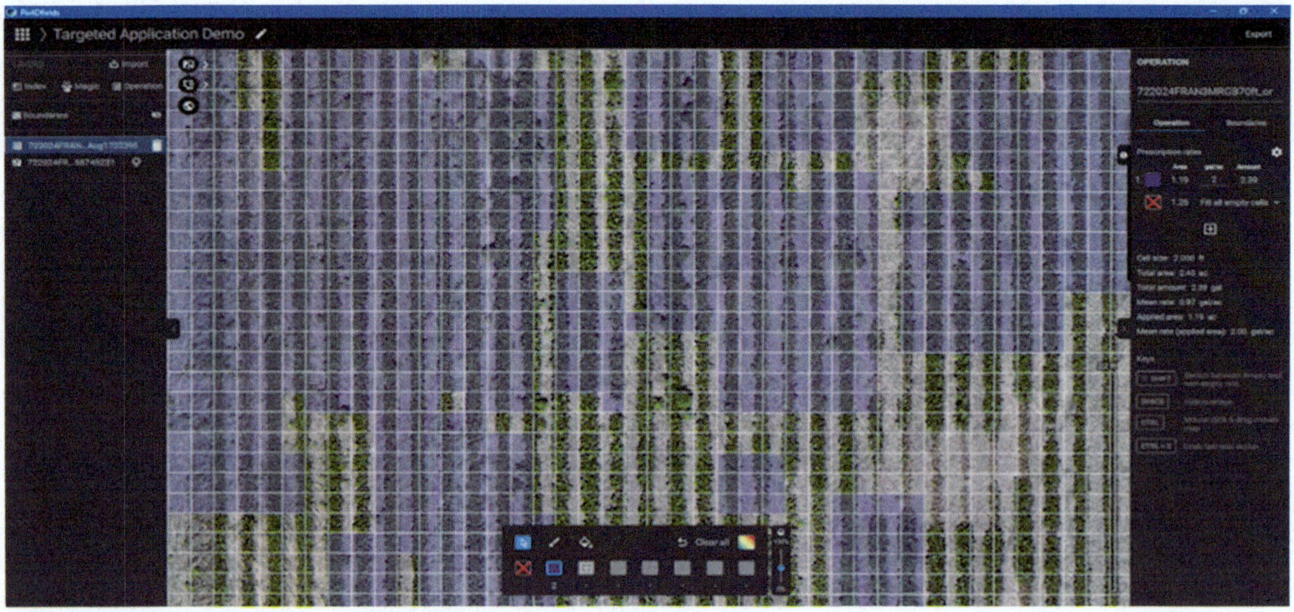

Figure 18.15. Despite the power and simplicity of this tool, it is not perfect. Demonstrated here are areas designated to be sprayed (purple) despite many of the grids having no weeds, while other grids with large, identifiable weeds present are not within a designated spray area. Courtesy of Dr. Thomas R. Butts.

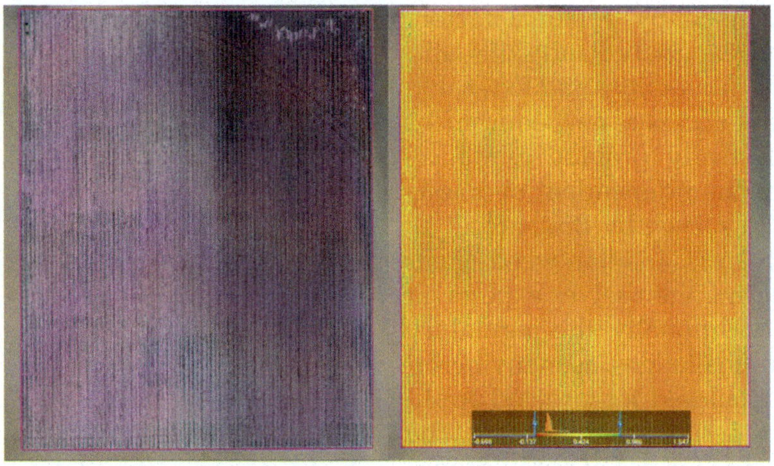

Figure 18.16. Spatial herbicide injury detected within a research trial on V1 soybean from multispectral imagery collected using UAVs. Darker orange indicates less vegetation (areas with a greater impact from a soil residual herbicide), while brighter yellow and green areas indicate more healthy vegetation (areas with less of an impact from the soil residual herbicide). Courtesy of Dr. Thomas R. Butts and Ms. Abigail Norsworthy.

Additionally, a researcher may seek out a more detailed analysis at this stage using high-resolution hyperspectral imaging, which can help in assessing the agent responsible for the damage. To confirm injury, the observed responses should be compared with known standards (Shumway et al., 2025).

Detecting Herbicide Resistance

Hyperspectral and multispectral imaging are emerging as powerful tools for detecting herbicide resistance in weeds (Huang et al., 2022; Jones et al., 2023a, 2023b; Reddy et al., 2014). The goal is to identify unique spectral signatures that differentiate resistant and susceptible weed biotypes. These nondestructive imaging methods not only enhance the early detection of herbicide resistance but also provide actionable insights for developing SSWM strategies that reduce reliance on broadcast herbicide applications and mitigate resistance spread.

Considerations for Adoption

The level of adoption of SSWM practices is influenced by a combination of barriers and drivers (Anastasiou et al., 2023; Bastiaans et al., 2008; Dentzman, 2023; Feisthauer et al., 2024; Fennimore et al., 2016; Hanson et al., 2022; Hurley & Frisvold, 2016; Kendall et al., 2022; Masi et al., 2023; Townsend & Noble, 2022). These barriers include:

1. High initial costs that include training, subscription fees, software, and the purchase of equipment.
2. The risk of switching from a proven and well-tested weed control approach that has been used for many years to an untested approach that may have numerous complications.
3. The financial stability of the provider as well as their ability to provide service, parts, and technical advice into the future.
4. Communication difficulties across time zones in rural areas where inconsistent cellular coverage and unreliable internet infrastructure are always problems.

Commercial providers or organizations interested in purchasing novel weed control tools should also consider factors such as weather and soil conditions, field size and shape, and crop type and value that could reduce their usefulness.

The integration of new SSWM tools may require shifts in general crop production practices, such as changes to arrangement of crop plants, to facilitate effective use (e.g., row spacing or individual plant arrangement). The adoption of precision technology can also involve a significant learning curve for growers. Consequently, the integration of this equipment into farming operations will be shaped by personal factors such as age, education, experience and skills, and interest in innovation. Universities and Extension programs may not have access to new equipment early enough to effectively provide guidance, leaving growers to make independent decisions ("buyer beware"); in a survey by Brainard et al. (2023), only 8% of US and Canadian weed scientists reported considering themselves experts on SSWM technology. Additionally, the regulatory environment presents challenges, as many herbicide labels are not yet tailored for targeted applications by ground- or aerial-based systems. The regulations and licensure for use of spray drones need to be considered. The adoption of SSWM technology may not eliminate the need for broadcast herbicides, tillage and cultivation, or human labor. Many technologies will be effective and perform well as part of an integrated weed management program.

Conclusions

Opportunities to implement precision agricultural tools for targeted weed management are expanding rapidly with advances in computer-based vision, machine learning, and AI. However, it remains necessary to still use fundamentals of weed science such as confirming weed identification, following herbicide label rates, rotating herbicide active ingredients, selecting diverse and integrated tools for weed control, and following up with evaluation of success. Through the novel precision agricultural tools, we can maintain excellent levels of weed control, limit exposure of chemicals to the environment, and be economical in our approaches.

Study Questions

1. What is a weed?
2. Why are weed populations spatially and temporally variable?
3. Why does SSWM reduce herbicide application rates?
4. What are the risks of not applying a preemergence herbicide to a portion of the field?
5. What is green-on-brown?
6. What is green-on-green?
7. Explain the process of creating a weed spray map.
8. Explain the benefits and barriers to adopting SSWM.

ACKNOWLEDGMENTS

Support for this document was provided by Kansas State University, USDA-IR4, the Precision Farming Systems community in the American Society of Agronomy, South Dakota State University, USDA-NRCS-CIG (NR233A750011G019), and USDA National Institute of Food and Agriculture (NIFA) under the Multi-State Hatch project (Project Number: KY006151-7008110-S1090). The use of trade names is for the convenience of the reader and does not imply endorsement by the authors.

REFERENCES

Anastasiou, E., Fountas, S., Koutsiaras, M., Voulgaraki, M., Vatsanidou, A., Barreiro-Hurle, J., Di Bartolo, F., & Gómez-Barbero, M. (2023). Precision farming technologies on crop protection: A stakeholders survey. *Smart Agricultural Technology*, *5*, 100293.

Anne, P., Gasser, S., Gottl, M., & Tanner, S. (2024). The reduction of chemical inputs by ultra-precise smart spot sprayer technology maximizes crop potential by lowering phytotoxicity. *Frontiers in Environmental Economics*, *3*, 1394315. https://doi.org/10.3389/frevc.2024.1394315.

Avent, T. H., Norsworthy, J. K., Patzoldt, W. L., Schwartz-Lazaro, L. M., Houston, M. M., Butts, T. R., & Vazquez, A. R. (2024). Comparing herbicide application methods with See & Spray™ technology in soybean. *Weed Technology*, *38*(e74). https://doi.org/10.1017/wet.2024.70.

Barnhart, I. H., Lancaster, S., Goodin, D., Spotanski, J., & Dille, J. A. (2022). Use of open-source object detection algorithms to detect Palmer amaranth (*Amaranthus palmeri*) in soybean. *Weed Science*, *70*, 648–662.

Barnhart, I. H., Proctor, C. A., Werle, R., Miller, K. A., Lancaster, S. R., Roozeboom, K. L., & Dille, J. A. (2025). Evaluating the ONE SMART SPRAY for weed control in midwestern corn and soybean crops. *Weed Science*, *73*(1), e85. https://doi.org/10.1017/wsc.2025.10058.

Bastiaans, L., Paolini, R., & Baumann, D. T. (2008). Focus on ecological weed management: What is hindering adoption? *Weed Research*, *48*, 481–491.

Boyd, N. S., Moretti, M. L., Sosnoskie, L. M., Singh, V., Kanissery, R., Sharpe, S., Besançon, T., Culpepper, S., Nurse, R., Hatterman-Valenti, H., & Mosqueda, E. (2022). Occurrence and management of herbicide resistance in annual vegetable production systems in North America. *Weed Science*, *70*, 515–528.

Brainard, D. C., Haramoto, E. R., Leon, R. G., Kells, J. J., Van Wychen, L. R., Devkota, P., Jugulam, M., & Barney, J. N. (2023). A survey of weed research priorities: Key findings and future directions. *Weed Science*, *71*, 330–343.

Butts, T. R., Butts, L. E., Luck, J. D., Fritz, B. K., Hoffmann, W. C., & Kruger, G. R. (2019). Droplet size and nozzle tip pressure from a pulse-width modulation sprayer. *Biosystems Engineering*, *178*, 52–69. https://doi.org/10.1016/j.biosystemseng.2018.11.004.

Butts, T. R., Fritz, B. K., Davis, J. A., & Spurlock, T. N. (2024). Spray coverage and deposits from a remotely piloted aerial application system using various nozzle types. *Frontiers in Agronomy*, *6*, 1493799. https://doi.org/10.3389/fagro.2024.1493799.

Denton, S., Raper, T., Dodds, D., Main, C., Duncan, L., & Mueller, T. (2024). Auxin injury on cotton, I: Aerial reflectance data, crop injury, and yield. *Agronomy Journal*, *116*, 2952–2966. https://doi.org/10.1002/agj2.21698.

Dentzman, K. (2023). Barriers to the effective management of herbicide-resistant weeds: Perspectives from crop advisors. *Outlooks on Pest Management*, *34*, 194–198.

Dhadly, D. K., Kavalappara, S. R., McAvoy, T., Severns, P. M., Simmons, A. M., Srinivasan, R., & Bag, S. (2025). Weed hosts: Hidden reservoirs for whitefly-transmitted viruses threatening Georgia cucurbits. *Plant Disease*, *109*, 2352–2361. https://doi.org/10.1094/PDIS-12-24-2730-RE.

Dieleman, J. A., & Mortensen, D. A. (1999). Characterizing the spatial pattern of *Abutilon theophrasti* seedling patches. *Weed Research*, *39*, 455–467.

Duke, S. O. (2012). Why have no new herbicide modes of action appeared in recent years? *Pest Management Science*, *68*, 505–512.

Feisthauer, P., Hartmann, M., & Börner, J. (2024). Behavioral forces driving farmers' intentions to adopt spot spraying for sustainable weed control. *Journal of Environmental Management*, *353*, 120218.

Fennimore, S. A., & Cutulle, M. (2019). Robotic weeders can improve weed control options for specialty crops. *Pest Management Science*, *75*, 1767–1774.

Fennimore, S. A., Slaughter, D. C., Siemens, M. C., Leon, R. G., & Saber, M. N. (2016). Technology for automation of weed control in specialty crops. *Weed Technology*, *30*, 823–837.

Fischer, J. W., Thorne, M. E., & Lyon, D. J. (2020). Weed-sensing technology modifies fallow control of rush skeletonweed (*Chondrilla juncea*). *Weed Technology*, *34*, 857–862. https://doi.org/10.1017/wet.2020.76.

Genna, N. G., Gourlie, J. A., & Barroso, J. (2021). Herbicide efficacy of spot spraying systems in fallow and postharvest in the Pacific Northwest dryland wheat production region. *Plants*, *10*, 2725. https://doi.org/10.3390/plants10122725.

Gundy, G. J., & Dille, J. A. (2022). Implementing variable-rate herbicide applications based on soil physical properties in grain sorghum. *Precision Agriculture*, *23*, 768–790. https://doi.org/10.1007/s11119-021-09860-5.

Hanson, E. D., Cosette, M. K., & Roberts, D. C. (2022). The adoption and usage of precision agriculture technologies in North Dakota. *Technology in Society*, *71*, 102087.

Heap, I. (2025). *The international herbicide-resistant weed database*. Retrieved May, 5, 2025, from https://www.weedscience.org.

Henry, W. B., Shaw, D. R., Reddy, K. R., Bruce, L. M., & Tamhankar, H. D. (2004). Remote sensing to detect herbicide drift on crops. *Weed Technology*, *18*, 358–368.

Horvath, D. P., Clay, S. A., Swanton, C. J., Anderson, J. V., & Chao, W. S. (2023). Weed-induced crop yield loss: A new paradigm and new challenges. *Trends in Plant Science*, *28*, 567–582. https://doi.org/10.1016/j.tplants.2022.12.014.

Huang, Y., Reddy, K. N., Thomson, S. J., & Yao, H. (2015). Assessment of soybean injury from glyphosate using airborne multispectral remote sensing. *Pest Management Science*, *71*, 545–552.

Huang, Y., Yuan, L., Reddy, K. N., & Zhang, J. (2016). In-situ plant hyperspectral sensing for early detection of soybean injury from dicamba. *Biosystems Engineering*, *149*, 51–59.

Huang, Y., Zhao, X., Pan, Z., Reddy, K. N., & Zhang, J. (2022). Hyperspectral plant sensing for differentiating glyphosate-resistant and glyphosate-susceptible johnsongrass through machine learning algorithms. *Pest Management Science*, *78*, 2370–2377.

Hunter, J. E., Gannon, T. W., Richardson, R. J., Yelverton, F. H., & Leon, R. G. (2020a). Coverage and drift potential associated with nozzle and speed selection for herbicide applications using an unmanned aerial sprayer. *Weed Technology*, *34*, 235–240. https://doi.org/10.1017/wet.2019.101.

Hunter, J. E., Gannon, T. W., Richardson, R. J., Yelverton, F. H., & Leon, R. G. (2020b). Integration of remote-weed mapping and an autonomous spraying unmanned aerial vehicle for site-specific weed management. *Pest Management Science*, *76*, 1386–1392. https://doi.org/10.1002/ps.5651.

Hurley, T. M., & Frisvold, G. (2016). Economic barriers to herbicide-resistance management. *Weed Science, Special Issue*, 585–594.

IR-4 Project. (2021). *Food crops database*. Retrieved October 2, 2024, from https://ir4app.cals.ncsu.edu/Ir4FoodPub/IS_Number?ISnum=IS00002.

Jiao, J., Zang, Y., & Chen, C. (2024). Key technologies of intelligent weeding for vegetables: A review. *Agriculture*, *14*(8), 1378.

Jones, E. A., Austin, R., Dunne, J. C., Cahoon, C. W., Jennings, K. M., Leon, R. G., & Everman, W. J. (2023a). Utilization of image-based spectral reflectance to detect herbicide resistance in glufosinate-resistant and glufosinate-susceptible plants: A proof of concept. *Weed Science*, *71*, 11–21.

Jones, E. A., Austin, R., Dunne, J. C., Leon, R. G., & Everman, W. J. (2023b). Discrimination between protoporphyrinogen oxidase-inhibiting herbicide-resistant and herbicide-susceptible redroot pigweed (*Amaranthus retroflexus*) with spectral reflectance. *Weed Science*, *71*, 198–205.

Kendall, H., Clark, B., Li, W., Jin, S., Jones, G. D., Chen, J., Taylor, J., Li, Z., & Frewer, L. I. (2022). Precision agriculture technology adoption: A qualitative study of small-scale commercial "family farms" located in the North China Plain. *Precision Agriculture*, *23*, 319–351.

Liao, J., Zang, Y., Luo, X., Zhou, Z., Lan, Y., Zang, Y., Gu, X., Xu, W., & Hewitt, A. J. (2019). Optimization of variables for maximizing efficacy and efficiency in aerial spray application to cotton using unmanned aerial systems. *International Journal of Agricultural and Biological Engineering*, *12*, 10–17. https://doi.org/10.25165/j.ijabe.20191202.4288.

Lopes, L. L., Cunha, J. P. A. R. D., Nomelini, Q. S. S., & de Alvarenga, C. B. (2024). Control of *Urochloa decumbens* using glyphosate applied by remotely piloted aircraft and ground sprayer with different spray nozzles. *Plants*, *13*(6), 757. https://www.mdpi.com/2223-7747/13/6/757.

Marschner, C. A., Colucci, I., Stup, R. S., Westbrook, A. S., Brunharo, C. A. C. G., DiTommaso, A., & Mesgaran, M. B. (2024). Modeling weed seedling emergence for time-specific weed management: A systematic review. *Weed Science*, *72*, 313–329. https://doi.org/10.1017/wsc.2024.25.

Masi, M., Di Pasquale, J., Vecchio, Y., & Capitano, F. (2023). Precision farming: Barriers of variable rate technology adoption in Italy. *Land*, *12*(5), 1084.

Matese, A., Toscano, P., Di Gennaro, S. F., Genesio, L., Vaccari, F. P., Primicerio, J., Belli, C., Zaldei, A., Bianconi, R., & Gioli, B. (2015). Intercomparison of UAV, aircraft and satellite remote sensing platforms for precision viticulture. *Remote Sensing*, *7*, 2971–2990. https://doi.org/10.3390/rs70302971.

Meng, Y., Song, J., Lan, Y., Mei, G., Liang, Z., & Han, Y. (2019). Harvest aids efficacy applied by unmanned aerial vehicles on cotton crop. *Industrial Crops and Products*, *140*, 111645. https://doi.org/10.1016/j.indcrop.2019.111645.

Norsworthy, J. K., Ward, S. M., Shaw, D. R., Llewellyn, R. S., Nichols, R. L., Webster, T. M., Bradley, K. W., Frisvold, G., Powles, S. B., Burgos, N. R., Witt, W. W., & Barrett, M. (2012). Reducing the risks of herbicide resistance: Best management practices and recommendations. *Weed Science, Special Issue*, 31–62.

Rasmussen, J., Azim, S., & Nielsen, J. (2021). Pre-harvest weed mapping of *Cirsium arvense* L. based on free satellite imagery—The importance of weed aggregation and image resolution. *European Journal of Agronomy*, *130*, 126373. https://doi.org/10.1016/j.eja.2021.126373.

Reddy, K. N., Huang, Y., Lee, M. A., Nandula, V. K., Fletcher, R. S., Thomson, S. J., & Zhao, F. (2014). Glyphosate-resistant and glyphosate-susceptible Palmer amaranth (*Amaranthus palmeri* S. Wats.): Hyperspectral reflectance properties of plants and potential for classification. *Pest Management Science*, *70*, 1910–1917.

Robles, W., Madsen, J. D., & Wersal, R. M. (2010). Potential for remote sensing to detect and predict herbicide injury on water hyacinth (*Eichhornia crassipes*). *Invasive Plant Science and Management, 3*, 440–450.

Sharma, P., Kumar, V., & Jha, P. (2023). Efficacy and economics of light-activated sensor-controlled sprayer for site-specific weed control. *Agronomy Journal, 115*, 2590–2599. https://doi.org/10.1002/agj2.21420.

Shumway, C., Scott, B., & Boyd, J. (2025). *Plant database: Herbicide injury image database*, University of Arkansas System Research and Extension. Retrieved June 3, 2025, from https://plants.uada.edu/herbicide/select/.

Silva, A. R. d., Freitas, M. A. M. d., Galvani, M. E., de Souza Costa, D., da Silva Araújo, L., de Almeida Rocha, R., dos Santos, P. V., & Galvani Filho, M. E. (2019). Proximal sensing estimation of glyphosate injury on weeds in central Brazil. *Journal of Applied Remote Sensing, 13*, 044524. https://doi.org/10.1117/1.JRS.13.044524.

Sosnoskie, L. M., Batts, R. G., & Besancon, T. E. (2025). An evaluation of targeted spraying for reducing herbicide use in highbush blueberry. *HortTechnology, 35*, 214–222.

Sosnoskie, L. M., Maloney, E. C., & Besancon, T. E. (2023). Targeted herbicide applications for weed control in grapes. *Fruit Quarterly, 31*(4), 25–31.

Spaeth, M., Sökefeld, M., Schwaderer, P., Gauer, M. E., Sturm, D. J., Delatrée, C. C., & Gerhards, R. (2024). Smart sprayer a technology for site-specific herbicide application. *Crop Protection, 177*, 106564.

Srinivasan, R., Riley, D., Diffie, S., Shrestha, A., & Culbreath, A. (2014). Winter weeds as inoculum sources of tomato spotted wilt virus and as reservoirs for its vector, *Frankliniella fusca* (Thysanoptera: Thripidae) in farmscapes of Georgia. *Environmental Entomology, 43*, 410–420.

Su, W. H., Fennimore, S. A., & Slaughter, D. C. (2019). Fluorescence imaging for rapid monitoring of translocation behaviour of systemic markers in snap beans for automated crop/weed discrimination. *Biosystems Engineering, 186*, 156–167.

Swanton, C. J., & Weise, S. F. (1991). Integrated weed management: The rationale and approach. *Weed Technology, 5*, 657–663.

Townsend, L. C., & Noble, C. (2022). Variable rate precision farming and advisory services in Scotland: Supporting responsible digital innovation? *Sociologia Ruralis, 62*, 212–230.

Vagedes, R. V., Ackerson, J. P., Johnson, W. G., & Young, B. G. (2024). Management zone classification for variable-rate soil residual herbicide applications. *Precision Agriculture, 25*, 1688–1713. https://doi.org/10.1007/s11119-024-10130-3.

Werle, R., Bernards, M. L., Arkebauer, T. J., & Lindquist, J. L. (2014). Environmental triggers of winter annual weed emergence in the Midwestern United States. *Weed Science, 62*, 83–96. https://doi.org/10.1614/WS-D-13-00091.1.

Werle, R., Sandell, L. D., Buhler, D. D., Hartzler, R. G., & Lindquist, J. L. (2014). Predicting emergence of 23 summer annual weed species. *Weed Science, 62*, 267–279. http://www.jstor.org/stable/43700656.

Xue, J., & Su, B. (2017). Significant remote sensing vegetation indices: A review of developments and applications. *Journal of Sensors, 2017*, 1353691. https://doi.org/10.1155/2017/1353691.

Zanin, A. R. A., Neves, D. C., Teodoro, L. P. R., da Silva, C. A., da Silva, S. P., & Teodoro, P. E. (2022). Reduction of pesticide application via real-time precision spraying. *Scientific Reports, 12*, 5638. https://doi.org/10.1038/s41598-022-09607-w.

Zhang, T., Huang, Y., Reddy, K. N., Yang, P., Zhao, X., & Zhang, J. (2021). Using machine learning and hyperspectral images to assess damages to corn plant caused by glyphosate and to evaluate recoverability. *Agronomy, 11*(3), 583.

Plant Disease Identification and Management Using Remote Sensing

19

Febina Mathew, Harikat Kaur, Milsha George, Karthika Mohan, Taofeek Mukaila, Nitha Rafi, and Sharon Clay

Chapter Purpose

This chapter explores the growing importance of precision agriculture, imaging technologies, and artificial intelligence–driven tools in plant disease diagnosis. It also examines how these technologies support disease management through site-specific strategies that integrate cultural practices, genetic resistance, precision chemical application, and biological control. Case studies and validated applications in both greenhouse and field conditions demonstrate how these innovations enhance crop health, reduce input use, and promote more efficient and sustainable agricultural systems.

Plant Disease

Plant disease refers to any condition that disrupts the normal functioning of plants. Plant diseases are typically caused by organisms (pathogens) such as fungi, bacteria, nematodes, or viruses. These organisms can significantly reduce crop yields and quality, posing a major threat to agricultural productivity. For a plant disease to develop, three critical components, known as the disease triangle, must be present: a susceptible host plant, a virulent pathogen, and an environment conducive to disease development.

A prominent example of disease in soybeans is sudden death syndrome (SOS), caused by the soilborne fungus *Fusarium virguliforme*. Sudden death syndrome occurs throughout all soybean-growing states in the United States and manifests as leaf chlorosis (yellowing), necrosis (browning), and premature defoliation (Westphal et al., 2008). In severely affected fields, SDS can lead to yield losses of up to 80% (Baetsen-Young et al., 2021), illustrating the substantial impact plant diseases can have on crop production (Figure 19.1). Sudden death syndrome is most severe in environments with cool, wet, poorly drained soils early in the season, when the causal fungus and susceptible soybean varieties are both present. If any one of these factors is missing, such as wet soil, the disease is unlikely to develop even if both the host and fungus are present.

Effectively managing plant diseases requires not only understanding how they develop, but also how they spread within and between crop fields. Pathogens are transmitted from infected plants to healthy ones, potentially causing rapid and widespread outbreaks. This transmission

Figure 19.1. Soybean leaves infected with the *F. virguliforme* cause sudden death syndrome in soybeans. Courtesy of North Dakota State University, Nitha Rafi.

can occur through various means, including air, water, soil, or insect vectors. For example, fungal spores like those of *Puccinia* can travel long distances on the wind, while bacteria may be dispersed by rain splashes or irrigation. Insects, such as aphids and whiteflies, spread viruses as they move between infected and healthy plants. Human activities, including the use of contaminated tools and the movement of infected plant materials, also play a major role in spreading diseases across fields and into new areas.

Once introduced, the spread and severity of plant diseases are largely shaped by environmental conditions that either promote or suppress pathogen activity. Key factors such as temperature, humidity, rainfall, wind, and soil conditions significantly influence disease development (Agrios, 2005).

Many fungal diseases, including powdery mildew, thrive in humid environments where surface moisture on plants facilitates the germination and dispersal of ascospores (sexual spores) and the subsequent infection of host tissues, while free moisture can inhibit the growth of conidia (asexual spores; Mieslerová et al., 2022). In contrast, charcoal rot, a significant disease caused by the soilborne fungus *Macrophomina phaseolina* and which affects a wide range of crops (including soybean, corn, and sorghum), is unlike diseases that favor moist environments in that charcoal rot tends to develop under hot and dry conditions (Marquez et al., 2021).

Prolonged soil saturation creates ideal conditions for soilborne fungal-like organisms, particularly oomycetes, which can cause diseases, such as damping-off in seedlings. Among the most common is *Pythium* spp., which can cause both pre-emergence damping-off, where seeds rot before germination, and postemergence damping-off, which is marked by water-soaked lesions around the collar region of seedlings. Postemergence damping-off results in the decay of collar tissues, causing plants to droop and collapse. Such symptoms are commonly observed in crops like cucumber, soybean, and tomato (Matthiesen & Robertson, 2023).

Beyond broader environmental conditions, specific soil characteristics also play a crucial role in disease development. Soil properties, including pH and nutrient content, can influence a plant's susceptibility to infection. Improper pH levels can limit nutrient uptake, weakening the plant's defenses against pathogens (Barrow & Hartemink, 2023). For example, Diaporthe (Phomopsis) seed decay of soybean, caused by *Diaporthe longicolla*, becomes more severe in potassium-deficient soils as plants struggle to absorb nutrients vital for proper seed development (Jeffers et al., 1982).

Plant Pathogens as Biotic Stressors

Plant pathogens function as biotic (i.e., living) stressors that compromise plant health and productivity (Agrios, 2005). Early identification and diagnosis of these pathogens during field scouting can inform effective management strategies and help prevent yield loss. Pathogens typically infect plants by entering through natural openings (e.g., stomata) or wounds, initiating the infection process. Mechanical injuries can serve as points of entry, and insects often act as vectors, especially for diseases caused by plant viruses (see Chapter 17).

Among pathogens, fungi are the most prevalent and are responsible for widespread plant diseases such as rusts, blights, and molds. Examples of fungal genera include *Fusarium*, *Sclerotinia*, *Botrytis*, and *Puccinia*, all of which are associated with significant agricultural losses worldwide.

Bacterial pathogens, similarly enter through stomata or wounds and cause diseases like bacterial blight, wilt, and cankers (Ellis et al., 2008). Notable genera such as *Xanthomonas* and *Pseudomonas* are major concerns in crops like soybean, wheat, and corn, where they cause leaf streak and blight.

Viruses are obligate parasites, meaning they require living plant cells to replicate, and this typically results in systemic (i.e., able to move throughout the plant) infections. Common viral symptoms include mosaic patterns, leaf distortion, and severe stunting (Parthasarathy 2024). For instance, tobacco mosaic virus (TMV) causes mottling of leaves, resulting in light- and dark-green or yellow patches, and affects crops such as wheat, soybean, and tomato. Similarly, tomato yellow leaf curl virus (TYLCV) is a major constraint in tomato production (Agrios, 2005).

Nematodes are microscopic roundworms that attack plant roots, interfering with water and nutrient uptake, leading to symptoms such as root rot, wilting, and nutrient deficiencies. The soybean cyst nematode (*Heterodera glycines*) is one of the most damaging nematodes in soybean, causing yield losses by disrupting root function and reducing the plant's overall ability to absorb nutrients and water (Poromarto et al., 2010).

Complicating the diagnosis of plant diseases is the fact that many symptoms caused by pathogens closely resemble those triggered by abiotic (e.g., non-living) stressors (e.g., nutrient imbalances, water stress, chemical phytotoxicity, physical or environmental damage). For instance, nutrient deficiencies may lead to yellowing leaves or stunted growth when essential elements are lacking (see Chapters 5, 6, and 14). Water-related stress, whether from drought or waterlogging, can also result in wilting, root damage, and chlorosis—symptoms commonly mistaken for disease (see Chapters 15 and 16).

Chemical exposure, such as herbicide injury, can cause leaf burns, root stunting, and growth inhibition, while mechanical damage from wind, hail, or extreme temperature fluctuations can further stress plants. These non-infectious stressors not only mimic disease symptoms but also predispose plants to secondary infections, complicating diagnosis and management. Therefore, accurately distinguishing between biotic and abiotic causes of plant stress is essential. Thorough assessments of environmental and soil conditions along with proper pathogen identification, are necessary to correctly diagnose plant disease symptoms and implement effective disease management strategies.

Plant Disease Diagnosis and Pathogen Identification

While traditional methods of identifying plant diseases have relied heavily on visual inspection and manual sampling, advances in technology are transforming how diseases are identified and monitored. Tools such as molecular techniques, ground-based sensors, unmanned aerial vehicles (UAVs), and satellite imagery, as well as geographic information systems (GIS) and the Global Positioning System (GPS), now enable more efficient, large-scale monitoring that can complement and enhance traditional ground scouting efforts (Agrios, 2005). The emergence of artificial intelligence (AI) has also further improved plant disease diagnostics. Automated systems leveraging machine learning (ML) and deep learning (DL) algorithms can now analyze large datasets and high-resolution images to detect disease symptoms with high accuracy, often before visible signs become severe (Jafar et al., 2024). Chapter 27 provides a broader discussion with examples. Nonetheless, because these remote sensing tools may not precisely identify the underlying cause of observed disease symptoms, they need to be used alongside conventional diagnostic techniques. For example, in the early stages, SDS can appear as localized disease hotspots in low-lying or compacted areas of the soybean field, where environmental conditions like poor drainage and high soil moisture favor pathogen development (Figure 19.2). Ground scouting in these areas allows farmers, often working alongside plant pathologists, crop consultants, or agronomists, to differentiate SDS from other potential stressors that may cause similar aboveground symptoms. However, molecular diagnostic tools and expert analysis are necessary to confirm the presence of SDS.

Visual Examination

Visual signs of pathogens and symptoms on plants are often the first clues indicating a potential disease. Common symptoms include chlorosis, necrosis, lesions, wilting, stunted

Figure 19.2. A commercial soybean field illustrating early symptoms and how within-field, environmental, and pathogen variability can create localized conditions conducive to the development and increased risk of sudden death syndrome. Courtesy of North Dakota Stale University, Nitha Rafi.

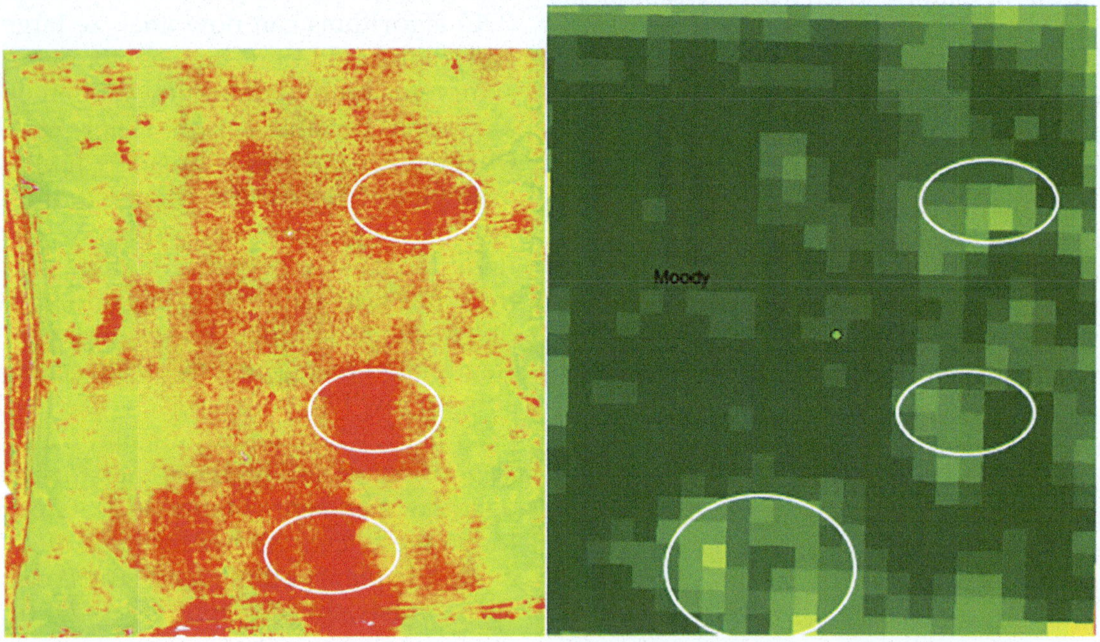

Figure 19.3. demonstrates the use of UAVs and satellite imagery to detect and monitor white mold in a soybean field in Moody County, South Dakota. The left panel shows areas of reduced crop health highlighted in red, indicating zones likely affected by the disease. The right panel complements this by pinpointing specific locations of white mold detection through spectral reflectance data shown in light green. Together, these contrasting images illustrate how remote sensing technologies, including vegetation indices and thermal imaging, can be used to detect the presence of disease before visible symptoms appear, enabling timely, targeted management interventions to reduce yield loss. Courtesy of South Dakota State University, Emmanuel Byamukama.

growth, cankers, root rot, blights, powdery growth, plant decline, and, in severe cases, plant death. The specific symptoms can vary widely depending on the type of pathogen involved, whether fungus, bacteria, virus, or nematodes, as discussed earlier. In contrast, signs refer to the actual visible presence of the pathogen itself. Examples include fungal mycelium or spores on infected tissues (such as powdery mildew), bacterial ooze from plant wounds, or nematode cysts on roots. Recognizing both signs and symptoms is crucial for accurate disease diagnosis.

However, visual examination has its limitations. The accuracy of symptom identification often depends on environmental conditions, the timing of infection, and the plant's growth stage and can result in inconclusive or inconsistent diagnoses. This subjectivity complicates disease identification (Nutter et al., 2002). Although visual symptoms and signs are valuable for early detection, they are frequently insufficient for accurate diagnosis, especially when symptoms overlap among different pathogens or resemble those caused by abiotic factors in the initial stages of infection.

Molecular Diagnostics

Molecular diagnosis involves detecting specific DNA, RNA, or protein markers associated with pathogens, enabling highly sensitive and specific disease identification. While a detailed discussion of molecular diagnostics is beyond the scope of this chapter, it is important to note that molecular diagnostic tools are widely used to identify plant pathogens accurately. Among the most commonly used molecular techniques are end point/standard polymerase chain reaction (PCR), quantitative PCR (qPCR), nested PCR, multiplex PCR, and digital droplet PCR. Other advanced methods include loop-mediated isothermal amplification (LAMP), isothermal nucleic acid amplification, nucleic acid sequence-based amplification (NASBA), and DNA macro- and microarrays.

Furthermore, next-generation sequencing (NGS) is a powerful technology that helps scientists detect and identify pathogens. These methods support early disease diagnosis and can identify infections before visible symptoms appear (Hariharan and Prasannath, 2021).

However, molecular diagnostics also have limitations. They often require specialized laboratory equipment and skilled personnel, and they can be expensive for routine use. Additionally, many molecular tests are designed to detect specific known pathogens, meaning they may miss emerging or unknown pathogens unless methods like NGS are used. There is also the potential for false positives or negatives if protocols are not carefully optimized.

Imaging Technologies and AI-driven Diagnostic Tools

Imaging technologies are nondestructive, sensor-based methods that use optical techniques to detect and identify plant diseases (Mahlein, 2016). These technologies enable disease assessment by capturing reflectance data across various spectral bands (Figure 19.3). Infected plants often show yellowing or reddening in the visible spectrum due to chlorophyll degradation, resulting in changes to the reflectance in the red, green, and blue (RGB) bands (Krezhova et al., 2017). Moreover, diseased plants can also show differences in reflectance in the near-infrared (NIR) and shortwave infrared (SWIR) bands (García-Vera et al., 2024).

Remote sensing technologies, such as hyperspectral imaging, multispectral sensors, and thermography, offer powerful tools for diagnosing plant diseases. This section focuses on identifying plant disease using these tools. See Chapters 13, 14, and 15 for broader discussions on sensors, remote sensing for crops, and remote sensing for water stress, respectively. Figure 19.3 illustrates the use of UAVs and satellite-based remote sensing to identify white mold in a soybean field. These technologies capture high-resolution images and reflectance data that can reveal subtle changes in plant health before symptoms become apparent to the naked eye.

To effectively use spectral reflectance measurements for disease identification, it is crucial to determine the specific spectral wavelengths or bands—across both the visible and non-visible regions of the electromagnetic spectrum—that are most closely associated with the disease. These regions include the ultraviolet (UV: ~100–400 nm), visible (VIS: ~400–700 nm), near-infrared (NIR: ~700–1100 nm), SWIR (~1100–2500 nm), and thermal infrared (TIR: ~8000–12,000 nm) ranges. For instance, significant differences in spectral reflectance between healthy and SDS-infected soybean plants have been observed in

Figure 19.4. A sunflower trial demonstrating the effects of foliar fungicide treatment. The control plot (right) exhibited symptoms of Phomopsis stem canker, evident as brown discoloration on the stems. In contrast, the fungicide-treated plot (left) showed reduced disease development and healthier plants, characterized by greenish-yellow stems compared with the browned stems in the control plots. This comparison highlights the effectiveness of fungicide application in managing Phomopsis stem canker. Courtesy of North Dakota State University, Karthika Mohan.

the 500–750 nm (visible and red edge) and 1400–2350 nm (SWIR) ranges when hyperspectral imaging was applied under greenhouse conditions (Brown et al., 2024). However, validation under field conditions is necessary to ensure the practical application of these findings.

Several studies have demonstrated the field validation of spectral approaches for disease identification. Spectral disease indices (SDIs) for sugar beet diseases, such as the Cercospora Leaf Spot Index (CLSI), Sugar Beet Rust Index (SBRI), and Powdery Mildew Index (PMI), were developed under greenhouse conditions and validated in the field using the most relevant wavelengths, between 450 and 950 nm, offering an effective method for early disease identification (Mahlein et al., 2013). Similarly, handheld sensors have proven helpful in the field for assessing the severity of Phomopsis stem canker and the use of foliar fungicides in sunflowers (Figure 19.4). Guidini et al. (2021) demonstrated that the Normalized Difference Vegetation Index (NDVI), a common indicator of crop health (see Chapters 13 and 14), measured using a GreenSeeker sensor across all fungicide treatments, strongly and negatively correlated with the Disease Severity Index ($r = -0.68$; $P < 0.0001$), indicating that higher Phomopsis stem canker severity was associated with lower NDVI values and, therefore, lower crop health.

In addition, AI systems use convolutional neural networks (CNNs) and other ML techniques, such as support vector machine (SVM), *k*-nearest neighbor (KNN), and partial least squares regression (PLSR), which can be used to classify healthy versus diseased plants and assess disease severity (García-Vera et al., 2024). Table 19.1 provides examples of sensors commonly used for plant disease identification under greenhouse or field conditions.

How Precision Agriculture Helps in Monitoring and Predicting Disease Spread

Building on the importance of recognizing signs and symptoms, remote sensing technologies offer a powerful means to detect disease-

Figure 19.5. Conceptual schematic diagram illustrating how plant reflectance across different wavelengths can reveal disease signatures, enabling early detection and precise monitoring of plant health. AI-generated illustration; Courtesy of North Dakota State University, Taofeek Mukaila.

related changes that may not be visible to the naked eye. Since diseased plants often produce distinct spectral signatures, choosing the appropriate sensor is critical for accurate detection. Sensors mounted on satellites, UAVs, and agricultural machinery measure reflected light, which varies due to changes in chlorophyll content, moisture, and cellular structure in infected plants. Key spectral bands, including red, green, and NIR, are used to identify these alterations (Figure 19.5). By processing remote sensing data, disease signatures can be developed to enable early detection and monitor disease progression over time. The choice of sensor depends mainly on the size of the study area: High-resolution sensors are ideal for smaller plots, whereas lower-resolution sensors are more suitable for extensive fields. Striking the right balance between sensor capabilities and detection needs is essential. For further details on remote sensing applications, see Chapters 14 and 15.

With advances in imaging technology, cameras mounted on satellites, UAVs, and airborne platforms now capture high-resolution images that reveal detailed information about crop health. The reflected-light data collected by these sensors provides real-time insights into how temperature, humidity, and moisture influence disease development. Vegetative indices like the NDVI, Green Normalized Vegetation Index (GNDVI), and Soil Adjusted Vegetation Index (SAVI) are processed with ML tools to detect changes in crop health and predict high-risk disease areas (Xue et al., 2019). For example, Zheng et al. (2018) used Sentinel-2 multispectral imagery along with the Red Edge Disease Stress Index (REDSI) to detect and monitor yellow rust in winter wheat at both the canopy and regional scales. Additionally, GIS can be used to generate whole-field disease risk maps, as demonstrated by Plata et al. (2022), who developed a UAV-based GIS system to create heat maps of Cassava Phytoplasma Disease (CPD) severity.

How Precision Agriculture Helps in Decision Support Systems for Plant Disease

Decision Support Systems (DSS) are tools that help farmers make informed crop management decisions to protect plant health and boost yields. These systems combine data from various sources, such as weather forecasts, soil conditions, and drone or satellite imagery, to predict disease outbreaks. Based on this information, experts can provide targeted disease management recommendations.

When embedded within precision agriculture frameworks, a DSS facilitates the implementation of disease management strategies. A notable example is the Soybean Rust–Pest Information Platform for Extension and Education (SBR-PIPE), which combines weather forecasts and spore monitoring data to alert growers about soybean rust outbreaks and suggests optimal fungicide application timing (Sikora et al., 2014). Similarly, nematode-specific tools like NemaDecide and NemDST help manage plant-parasitic nematodes

Table 19.1. Examples of Sensors Used for Plant Disease Identification.

Sensors	Examples	Sensor Validation (Model, Greenhouse, Field)	References
RGB imaging	Soybean rust (*Phakopsora pachyrhizi*), Calonectria leaf blight (*Calonectria pteridis*) of eucalyptus, tobacco leaf scorch disease (*Xylella fastidiosa*), wheat leaf blast (*Pyricularia oryzae*), potato late blight (*Phytophthora infestans*)	Diseased leaves obtained from artificial inoculation or field collection. Images taken under controlled lighting and a homogeneous background for model validation.	Alves et al. (2022)
Multispectral sensors	Banana Fusarium wilt (*Fusarium oxysporum* f. sp. *cubense* race 4)	Field validation.	Zhang et al. (2022)
Hyperspectral sensors	Grapevine vein-clearing virus	Field validation.	Nguyen et al. (2021)
	Bacterial leaf spot (*Xanthomonas perforans*) of tomato	Greenhouse validation.	Zhang et al. (2024)
Thermography	Septoria tritici blotch in wheat (*Zymoseptoria tritici*)	Field validation.	Wang et al. (2019)
Combination of AI-driven diagnostic tools, image processing algorithms, and machine learning	Septoria leaf spot of wheat (*Zymoseptoria tritici*), wheat rusts (yellow rust and brown rust: *Puccinia striiformis* f. sp. *tritici* and *Puccinia recondita* f. sp. *tritici*), and tan spot of wheat (*Pyrenophora tritici-repentis*)	Field validation.	Johannes et al. (2017)
Deep CNN transfer learning model	Bacterial leaf blight of rice (*Xanthomonas oryzae* pv. *oryzae*), leaf scald (*Microdochium oryzae*), brown spot (*Bipolaris oryzae*), narrow brown spot (*Cercospora oryzae*), rice leaf blast (*Magnaporthe oryzae*)	Computational processing of images from the rice leaf disease dataset was assessed online; no field validation yet.	Latif et al. (2022)

by providing detailed data on pest population dynamics and the use of specific disease management tools (Abd-Elgawad, 2024; Been et al., 2005; Pun et al., 2023).

ML and AI are used for disease forecasting by integrating satellite imagery and Internet of Things (IoT)–based field sensors. These models can now predict outbreaks with over 85% accuracy (Mahlein, 2016). When combined with historical disease data and weather variables, their predictive power supports proactive interventions (Juroszek & von Tiedemann, 2013). Artificial intelligence-driven DSS also enables real-time diagnostics and recommends targeted treatments, lowering input costs and improving profits, in addition to increasing the environmental and economic sustainability of an operation. For instance, Yuan et al. (2025) achieved greater environmental and economic efficiency for multiple growers through a mobile-based platform that applied DL to forecast wheat yellow rust in the United Kingdom using historical weather and disease data. The Fully Connected Neural Networks (FCNNs) models achieved 80% accuracy, while Long Short-Term Memory (LSTM) models exceeded 85% accuracy by capturing seasonal patterns. These models are now embedded in cloud platforms that sync with real-time weather data and send mobile alerts, allowing growers to act swiftly with precise management strategies, saving time and labor on scouting and reducing input costs.

Disease Management

There is no one-size-fits-all strategy for all diseases. Effective disease management typically involves an integrated approach that combines cultural practices, chemical treatments, biological control, and the use of disease-resistant plant varieties. Given the variability within agricultural fields, ranging from differences in soil type and microclimate to crop growth stages, farmers can use site-specific interventions focused on areas with the greatest disease pressure.

Combining Remote Sensing with Cultural Practices

Remote sensing and GIS are widely used to identify high-risk zones for disease, enabling more targeted interventions such as adjusted planting dates, strategic variety selection, and localized soil amendments (Gebbers & Adamchuk, 2010). By integrating field-based sensor data with high-resolution digital elevation models (DEMs), detailed disease risk maps can be generated to support informed management decisions. At the field scale, subtle elevation differences captured by DEMs influence microtopography and soil moisture distribution. These microenvironmental variations can significantly affect disease prevalence, particularly for water-related soilborne pathogens. Combining DEM data with sensor information allows for precise identification of high-risk zones, facilitating site-specific disease management.

A major advantage of remote sensing technologies is their capacity to optimize crop rotation through geospatial analysis of historical yield data. This enables farmers to create risk maps from remote sensing data and tailor cultural practices to risk-associated areas. For example, rotation schedules can be adjusted based on site-specific pathogen inoculum or plant density, or seeding rates can be reduced in certain field areas, row spacing increased, or tillage and residue management selectively applied. Data-driven crop rotation can reduce soilborne pathogen inoculum by up to 40% over multiple growing seasons (Mulla, 2013).

Targeted tillage, guided by spatial data, is another cultural practice enhanced by precision agriculture. While conventional tillage can help manage soilborne pathogens like *Rhizoctonia solani*, it often disrupts soil structure and beneficial microbial communities. To mitigate soil degradation, researchers in North Dakota used DEMs, soil electrical conductivity (EC) mapping, and georeferenced disease surveys to identify zones prone to pathogen infection due to poor drainage and soil compaction. In these high-risk areas, deep tillage was selectively applied to break pathogen life cycles, while low-risk zones were managed with minimal or no tillage to preserve soil health (Franzen et al., 2011).

In no-till systems, where pathogens often survive on crop residues, spectral imaging combined with ML has been successfully used to identify pathogen hotspots, enabling targeted residue management or site-specific deep plowing to reduce disease pressure (Gozdowski et al., 2021). Similarly, precision irrigation systems that regulate soil moisture based on real-time data help mitigate conditions favorable to soilborne pathogens such as *Phytophthora* spp. and *R. solani* (Mahlein et al., 2019). Advanced systems integrating evapotranspiration models and soil moisture sensors enable efficient water use while reducing pathogen activity.

Genetic Resistance and Precision Breeding

Genetic resistance remains a fundamental pillar of plant disease management, and recent advances in precision breeding have accelerated the development of resistant crop varieties. High-throughput phenotyping, using UAVs and spectral imaging, has enhanced the efficiency of selecting for disease-resistant traits.

The integration of DNA marker technologies with big data analytics has transformed plant breeding by enabling precise identification and tracking of genes responsible for disease resistance. DNA markers are specific genomic sequences linked to desirable traits, and analyzing large datasets of genetic and phenotypic information allows breeders to efficiently pinpoint resistance genes and accelerate the development of improved varieties. Molecular tools like CRISPR-Cas9 enhance this process by enabling targeted, precise editing of plant DNA; guided by RNA sequences, the Cas9 protein makes exact cuts or modifications, allowing researchers to understand gene functions and directly introduce resistance traits without the lengthy process of traditional breeding. These precision breeding technologies facilitate sustainable crop improvement by enabling heritable modifications without introducing foreign DNA, thereby reducing dependence on chemical pesticides. Successful applications have included disrupting susceptibility genes in rice to confer resistance to bacterial blight and rice blast, reducing Fusarium head blight in wheat via marker-assisted selection, engineering tomato varieties resistant to powdery mildew and bacterial speck through gene knockouts, and editing citrus promoters to block disease activation: All demonstrating the power of integrating molecular genetics and big data to accelerate disease-resistant crop development.

Map-based cloning remains a valuable tool for dissecting resistance mechanisms by identifying mutant genotypes that confer resistance to pathogens. For example, Hurni et al. (2015) discovered the *Htn1* gene cluster on chromosome 8 in corn, linked to resistance against *Setosphaeria turcica*, the fungus responsible for northern corn leaf blight (NCLB). Within this cluster, the gene *ZmWAK-RLK1* was shown to be critical for defense signaling, as its disruption increased susceptibility to NCLB. Gene editing technologies like CRISPR have shown great promise, such as reducing bacterial blight by over 60% in rice through edits targeting *Xanthomonas oryzae* (Zaidi et al., 2018), although challenges remain, including regulatory hurdles, off-target effects, public acceptance, and the need for extensive field validation. Nevertheless, precision gene editing, combined with advances in phenotyping and synthetic biology, continues to revolutionize disease resistance strategies and holds great potential for future crop improvement. In this evolving landscape, ML models increasingly support genomic selection by helping prioritize gene targets and predict trait performance, further accelerating breeding efforts.

Chemical Control with Precision Application

Fungicides, bactericides, and nematicides remain essential components of plant disease management, but the adoption of precision agriculture technologies has revolutionized how these chemicals are applied. Advanced variable-rate application systems now allow for site-specific treatments, reducing chemical inputs while maintaining efficacy. For instance, precision fungicide applications guided by hyperspectral imaging and sensor systems have been shown to reduce chemical use by up to 40% without compromising disease control (West et al., 2022). Global Positioning System–guided sprayers with real-time sensor feedback can further enhance uniformity and prevent overapplication, cutting both costs and environmental impact.

Weather-based DSS also optimize fungicide timing. In potato crops, predictive weather models help to time fungicide applications for late blight more effectively, reducing unnecessary treatments by 35% (Shtienberg, 2013). Similarly, in soybeans, integrating predictive modeling with UAV-guided spraying has reduced fungicide use by 35% and minimized yield losses by 25% (Reis et al., 2021).

Precision spraying technologies have resulted in a broad reduction in pesticide use, ranging from 12% to 96%, while improving disease control and operational efficiency (Zanin et al., 2022). In Brazil's Cerrado region, Zanin et al. (2022) evaluated real-time, site-specific spraying in soybean and corn. They demonstrated that precision sprayers, equipped with real-time sensors, applied pesticides only where plants were detected, resulting in a significant reduction in pesticide use during the early stages of soybean and corn growth. The cost savings were 2.3 times greater than those achieved by applying pesticides across the whole area with a traditional sprayer. The addition of UAV surveillance and AI further enhanced accuracy, allowing dynamic adjustments based on disease severity (Zanin et al., 2022). Another notable example is the management of cotton root rot, caused by *Phymatotrichopsis omnivora*, in Texas. Using high-resolution satellite imagery, researchers developed prescription maps that identified infested zones, allowing fungicides to be applied only where needed. This site-specific approach led to substantial reductions in fungicide use, lower production costs, and promoted sustainability at scale (Yang et al., 2017).

Biological Control

The integration of precision agriculture with biological control strategies is transforming plant disease management. Unmanned aerial vehicle–based applications of biocontrol agents, such as *Trichoderma* spp. and *Bacillus subtilis*, have enhanced coverage and delivery accuracy, improving their overall effectiveness. Studies have shown that *Trichoderma harzianum* and *Bacillus subtilis* significantly inhibit *Fusarium verticillioides* growth, reducing disease incidence by over 70% in *Vicia faba* cultivars and enhancing plant defense through increased antioxidant enzyme activities (Attia et al., 2025). Liu et al. (2022) further demonstrated that co-cultivation of *Trichoderma atroviride* SG3403 and *Bacillus subtilis* 22 inhibited *Fusarium graminearum* growth by 61.5% *in vitro* and reduced wheat infection by 66.2% when applied as a seed coating. Additionally, the development of smart bioreactor systems (controlled devices for growing helpful microbes) allows beneficial microbes to be grown and applied when needed, giving crops in high-risk areas a reliable source of natural protection (Palladino et al., 2024). These bioreactors use advanced sensors, real-time data monitoring, and automation to continuously adjust parameters, such as pH, temperature, and nutrient levels, maintaining optimal conditions for cell growth. They reduce the need for manual intervention and enhance process efficiency. Traditional bioreactors, by contrast, require manual adjustments and lack automated feedback systems, making them less precise and more labor intensive.

Summary

Understanding plant diseases, including their causal pathogens, development, spread, and the influence of environmental factors, is essential for effective disease management. Precision agriculture provides farmers with the tools to manage diseases more sustainably and efficiently, protecting crop health, preserving the environment, and ultimately improving yields.

Integrated disease management leverages multiple data layers, including field-based scouting, field history information, remote sensing, and AI-driven disease identification tools, to provide accurate field assessments. The integration of these data layers enables more precise and effective disease management, combining targeted chemical treatments with crop rotation, tillage management, disease-resistant cultivars, and remote sensing technologies—empowering farmers to make informed decisions that enhance disease management while promoting sustainability and efficiency in agricultural production.

Case Study: Early Identification of Sudden Death Syndrome in Soybean Using Hyperspectral Imaging

Reference: Brown, M. T., Oh, S., Rainey, K. M., & Telenko, D. E. P. (2024). Presymptomatic leaf reflectance of *Fusarium virguliforme*-infected soybean plants in greenhouse conditions. *PhytoFrontiers*, 4(4), 544–552.

Background

On an early summer morning, Jason, an intern at Dr. Jacob's plant disease diagnostics laboratory, accompanied the research team to a commercial soybean field near the facility. While the current soybean crop showed visually healthy green foliage, the farmer, Mr. Johnson, expressed concern based on prior experience. The field had been planted with soybeans last season as well, during which time it suffered a yield loss of nearly 80% due to sudden death syndrome (SDS). Given this history, Mr. Johnson was worried about the potential early onset of the pathogen in this season's crop.

Disease Overview: What Is SDS?

Dr. Jacob explained that, in the United States, SDS is primarily caused by the soilborne fungus *Fusarium virguliforme*. The fungus infects soybeans, especially under cool, moist conditions. However, foliar symptoms—such as interveinal chlorosis and necrosis—only become visible much later, during the flowering and pod development stages. By this time, substantial internal damage has already occurred.

The Problem: How Can SDS Be Detected Before Symptoms Appear?

Jason asked, "If the plant appears healthy, how can farmers detect SDS?" Dr. Jacob explained that traditional detection methods require destructive sampling of the roots followed by laboratory tests, such as quantitative polymerase chain reaction (qPCR) assays. However, these methods are time-consuming and impractical for large-scale field monitoring. "What if," Dr. Jacob suggested, "the plants were already reflecting subtle signs of stress—not visible to the human eye but detectable with advanced imaging technology?"

Hypothesis: Can Hyperspectral Imaging Detect SDS Before Symptoms Appear?

Dr. Jacob suggested using hyperspectral imaging, a technology that can see more than the human eye is capable of detecting. Diseased plants often undergo subtle changes in reflectance due to physiological stress, even before visible symptoms manifest. The question then arose: Could hyperspectral data provide an early diagnosis of SDS, enabling detection before the typical foliar signs of the disease appear?

Experiment

A greenhouse experiment was conducted:

- Soybean plants were grown and inoculated with *F. virguliforme* (treatment) or left uninfected (control).
- Starting 21 days after transplanting, weekly hyperspectral images were taken using a gantry-mounted system through Day 42. A gantry-mounted system is a setup in which a sensor is placed on an overhead frame that moves along a track to take images.
- At the same time the images were taken, infection of the root by *F. virgulirome* was confirmed by taking plant root samples and using a lab test (quantitative polymerase chain reaction assay) to analyze them.
- The collected images were labeled and analyzed using machine learning models to distinguish diseased plants from healthy plants.
- Three classification algorithms were tested: (a) Partial Least Squares Discriminant Analysis (PLS-DA)—a method used to find patterns in the images and separate different groups; (b) Support Vector Machine (SVM)—a technique to classify data by finding the best boundary between different categories; and (c) Random Forest (RF)—a method that uses multiple decision trees to make predictions or classifications based on image patterns.
- The ReliefF algorithm was used to identify the wavelengths corresponding to SDS.

Key Findings

- Spectral bands related to *F. virguliforme* root infection were identified, which included the visible and red-edge regions (500–750 nm) and the shortwave infrared (SWIR) region (1400–2350 nm).
- The PLS-DA, SVM, and RF classification models successfully distinguished healthy plants from *F. virguliforme*-infected ones, achieving mean F1 scores, a score of accuracy for the model, of over 79%, indicating strong performance in accurately identifying both healthy and infected plants.
- PLS-DA performed the best, with accuracy (F1) scores ranging from 83.1% to 85.3% and consistency (kappa statistics) between 0.43 and 0.54. This suggests that the model is reliable in detecting infection of soybean roots caused by *F. virguliforme* based on the images it was trained on.

Implications for Disease Management

This study highlights the potential of leaf spectral data in detecting root infections caused by *F. virguliforme* in soybean. Such an approach could support the development of effective management tools by enabling breeders and plant pathologists to screen for root rot resistance more efficiently. While the findings are promising, further field validation is necessary—particularly using hyperspectral imaging to identify resistant varieties. Importantly, the study also identifies key spectral features that could be targeted for practical field applications in disease monitoring and resistance screening.

Study Questions

1 Define plant disease.

2 Explain how precision agriculture technologies contribute to the early identification of plant diseases.

3 Provide a successful example where precision spraying technologies have led to a significant reduction in pesticide use.

ACKNOWLEDGMENTS

We thank the North Dakota Soybean Council; the National Sunflower Association; the South Dakota Oilseed Council; the United Soybean Board, the USDA National Institute of Food and Agriculture (NIFA), Hatch Multistate Project under accession no. 7009296; and the USDA-NIFA, Hatch project no. ND02250 for support of this document. Additionally, this material was supported in part for Dr. Milsha George by the U.S. National Science Foundation (NSF; NSF EPSCoR cooperative agreement no. 2316502) and Karthika Mohan by the USDA NIFA–Critical Agricultural Research and Extension (award no. 2024-69008-42655). Any opinions, findings, and conclusions or recommendations expressed in this material are those of the author(s) and do not necessarily reflect those of the NSF. In addition, support for this document was provided by the Precision Farming Systems community in the American Society of Agronomy, South Dakota State University, USDA Natural Resources Conservation Service Conservation Innovation Grants (NR233A750011G019), the American Coalition for Ethanol, the South Dakota Corn Utilization Council, the South Dakota Soybean Research and Promotion Council, and USDA-NIFA, Multistate Project S1090 under no. 7008110. The use of trade names is for the convenience of the reader and does not imply endorsement by the authors.

REFERENCES

Abd-Elgawad, M. M. M. (2024). Upgrading strategies for managing nematode pests on profitable crops. *Plants (Basel).*, *13*(11), 1558. https://doi.org/10.3390/plants13111558.

Agrios, G. N. (2005). *Plant pathology* (5th ed., pp. 4–882). Elsevier Academic Press.

Alves, K. S., Guimarães, M., Ascari, J. P., Queiroz, M. F., Alfenas, R. F., Mizubuti, E. S. G., & Del Ponte, E. M. (2022). RGB-based phenotyping of foliar disease severity under controlled conditions. *Tropical Plant Pathology*, *47*, 105–117. https://doi.org/10.1007/s40858-021-00448-y.

Attia, M. S., El-Wakil, D. A., Hashem, A. H., Al-Askar, A. A., AbdElgawad, H., Alotaibi, R. S., Abdel-Kader, S. A., & Abdelaziz, A. M. (2025). Investigating the activity of *Bacillus subtilis* and *Trichoderma harzianum* to mitigate Fusarium wilt disease of diverse cultivars of *Vicia faba*. *Scientific Reports*, *15*, 16093. https://doi.org/10.1038/s41598-025-99381-2.

Baetsen-Young, A. M., Swinton, S. M., & Chilvers, M. I. (2021). Economic impact of fluopyram-amended seed treatments to reduce soybean yield loss associated with sudden death syndrome. *Plant Disease*, *105*(1), 78–86. https://doi.org/10.1094/PDIS-04-20-0792-RE.

Barrow, N. J., & Hartemink, A. E. (2023). The effects of pH on nutrient availability depend on both soils and plants. *Plant and Soil*, *487*, 21–37. https://doi.org/10.1007/s11104-023-05960-5.

Been, T. H., Schomaker, C. H., & Molendijk, L. P. G. (2005). NemaDecide: A decision support system for the management of potato cyst nematodes. In A. J. Haverkort & P. C. Struik (Eds.), *Potato in progress: Science meets practice* (pp. 154–167). Wageningen Academic Publishers.

Brown, M. T., Oh, S., Rainey, K. M., & Telenko, D. E. P. (2024). Presymptomatic leaf reflectance of *Fusarium virguliforme*-infected soybean plants in greenhouse conditions. *PhytoFrontiers*, *4*(4), 544–552. https://doi.org/10.1094/PHYTOFR-09-23-0121-R.

Ellis, S. D., Boehm, M. J., & Coplin, D. (2008). Bacterial diseases of plants. *Agriculture and Natural Resources* (PP401.06), 1–4. https://www.ncpedia.org/sites/default/files/documents/bacterial-diseases-fact-sheet.pdf.

Franzen, D. W., Kitchen, N. R., Holland, K., Schepers, J. S., & Shapiro, C. A. (2011). Precision agriculture: NRCS support for emerging technologies. *Computers and Electronics in Agriculture, 75*, 1–10. https://doi.org/10.1016/j.compag.2010.10.009.

García Vera, Y. E., Polochè Arango, M. A., Mendivelso-Fajardo, C. A., & Gutiérrez-Bernal, F. J. (2024). Hyperspectral image analysis and machine learning techniques for crop disease detection and identification: A review. *Sustainability, 16*(14), 6064. https://doi.org/10.3390/su16146064.

Gebbers, R., & Adamchuk, V. I. (2010). Precision agriculture and food security. *Science, 327*(5967), 828–831. https://doi.org/10.1126/science.1183899.

Gozdowski, D., Jablonski, B., & Samborski, S. (2021). The use of machine learning in predicting pathogen spread in no-till agriculture. *Precision Agriculture, 22*, 534–550. https://doi.org/10.1007/s11119-021-09806-9.

Guidini, R., Braun, N., Markell, S., Harveson, B., and Mathew, F. (2021). Evaluation of fungicides for their efficacy against Phomopsis stem canker of sunflower. Proceedings of the 42nd Annual Sunflower Research Forum, Fargo, ND. January 8–9, 2020.

Hariharan, G., & Prasannath, K. (2021). Recent advances in molecular diagnostics of fungal plant pathogens: A mini review. *Frontiers in Cellular and Infection Microbiology, 10*, 600234. https://doi.org/10.3389/fcimb.2020.600234.

Hurni, S., Scheuermann, D., Krattinger, S. G., Kessel, B., Wicker, T., Herren, G., Fitze, M. N., Breen, J., Presterl, T., Ouzunova, M., & Keller, B. (2015). The maize disease resistance gene *Htn1* against northern corn leaf blight encodes a wall-associated receptor-like kinase. *Proceedings of the National Academy of Sciences of the United States of America, 112*(28), 8780–8785. https://doi.org/10.1073/pnas.1502522112.

Jafar, A., Bibi, N., Naqvi, R. A., Sadeghi-Niaraki, A., & Jeong, D. (2024). Revolutionizing agriculture with artificial intelligence: Plant disease detection methods, applications, and their limitations. *Frontiers in Plant Science, 15*, 1356260. https://doi.org/10.3389/fpls.2024.1356260.

Jeffers, D. L., Schmitthenner, A. F., & Kroetz, M. E. (1982). Potassium fertilization effects on Phomopsis seed infection, seed quality, and yield of soybeans. *Agronomy Journal, 74*(5), 886–890. https://doi.org/10.2134/agronj1982.00021962007400050028x.

Johannes, A., Picon, A., Alvarez-Gila, A., Echazarra, J., Rodriguez-Vaamonde, S., Navajas, A. D., & Ortiz-Barredo, A. (2017). Automatic plant disease diagnosis using mobile capture devices, applied on a wheat use case. *Computers and Electronics in Agriculture, 138*, 200–209. https://doi.org/10.1016/j.compag.2017.04.013.

Juroszek, P., & von Tiedemann, A. (2013). Climate change and potential future risks through wheat diseases: A review. *European. Journal of Plant Pathology, 136*, 21–33. https://doi.org/10.1007/s10658-012-0144-9.

Krezhova, D., Velichkova, K., Petrov, N., & Maneva, S. (2017). *The effect of plant diseases on hyperspectral leaf reflectance and biophysical parameters. RAD Conference Proceedings* (Vol. 2, pp. 269–275). https://doi.org/10.21175/RadProc.2017.55.

Latif, G., Abdelhamid, S. E., Mallouhy, R. E., Alghazo, J., & Kazimi, Z. A. (2022). Deep learning utilization in agriculture: Detection of rice plant diseases using an improved CNN model. *Plants, 11*(17), 2230. https://doi.org/10.3390/plants11172230.

Liu, H., Li, T., Li, Y., Wang, X., & Chen, J. (2022). Effects of *Trichoderma atroviride* SG3403 and *Bacillus subtilis* 22 on the biocontrol of wheat head blight. *Journal of Fungi, 8*(12), 1250. https://doi.org/10.3390/jof8121250.

Mahlein, A.-K. (2016). Plant disease detection by imaging sensors—Parallels and specific demands for precision agriculture and plant phenotyping. *Plant Disease, 100*(2), 241–251. https://doi.org/10.1094/PDIS-03-15-0340-FE.

Mahlein, A.-K., Kuska, M. T., Thomas, S., Wahabzada, M., Behmann, J., Rascher, U., & Kersting, K. (2019). Quantitative and qualitative phenotyping of disease resistance of crops by hyperspectral sensors: Seamless interlocking of phytopathology, sensors, and machine learning is needed! *Current Opinion in Plant Biology, 50*, 156–162. https://doi.org/10.1016/j.pbi.2019.06.007.

Mahlein, A.-K., Rumpf, T., Welke, P., Dehne, H.-W., Plümer, L., Steiner, U., & Oerke, E.-C. (2013). Development of spectral indices for detecting and identifying plant diseases. *Remote Sensing of Environment, 128*, 21–30. https://doi.org/10.1016/j.rse.2012.09.019.

Marquez, N., Giachero, M. L., Declerck, S., & Ducasse, D. A. (2021). *Macrophomina phaseolina*: General characteristics of pathogenicity and methods of control. *Frontiers in Plant Science*, *12*, 634397. https://doi.org/10.3389/fpls.2021.634397.

Matthiesen, R. L., & Robertson, A. E. (2023). Effect of infection timing by four *Pythium* spp. on soybean damping-off symptoms with and without cold stress. *Plant Disease*, *107*(12), 3975–3983. https://doi.org/10.1094/PDIS-01-23-0082-RE.

Mieslerová, B., Cook, R. T. A., Wheater, C. P., & Lebeda, A. (2022). Ecology of powdery mildews—Influence of abiotic factors on their development and epidemiology. *Critical Reviews in Plant Sciences*, *41*(6), 365–390. https://doi.org/10.1080/07352689.2022.2138044.

Mulla, D. J. (2013). Twenty five years of remote sensing in precision agriculture: Key advances and remaining knowledge gaps. *Biosystems Engineering*, *114*(4), 358–371. https://doi.org/10.1016/j.biosystemseng.2012.08.009.

Nguyen, C., Sagan, V., Maimaitiyiming, M., Maimaitijiang, M., Bhadra, S., & Kwasniewski, M. T. (2021). Early detection of plant viral disease using hyperspectral imaging and deep learning. *Sensors*, *21*(3), –742. https://doi.org/10.3390/s21030742.

Nutter, F. W., Jr., Guan, J., Gotlieb, A. R., Rhodes, L. H., Grau, C. R., & Sulc, R. M. (2002). Quantifying alfalfa yield losses caused by foliar diseases in Iowa, Ohio, Wisconsin, and Vermont. *Plant Disease*, *86*(3), 269–277. https://doi.org/10.1094/PDIS.2002.86.3.269.

Palladino, F., Marcelino, P. R. F., Schlogl, A. E., José, Á. H. M., Rodrigues, R. D. C. L. B., Fabrino, D. L., Santos, I. J. B., & Rosa, C. A. (2024). Bioreactors: Applications and innovations for a sustainable and healthy future—A critical review. *Applied Sciences*, *14*(20), 9346. https://doi.org/10.3390/app14209346.

Parthasarathy, S. (2024). *Fundamentals of plant pathology*. CRC Press.

Plata, I. T., Panganiban, E. B., Alado, D. B., Taracatac, A. C., Bartolome, B. B., & Labuanan, F. R. E. (2022). Drone-based geographical information system (GIS) mapping of cassava pythoplasma disease (CPD) for precision agriculture. *International Journal of Emerging Technology and Advanced Engineering*, *12*(2), 1–9. https://doi.org/10.46338/ijetae0222_01.

Poromarto, S. H., Nelson, B. D., & Goswami, R. S. (2010). Effect of soybean cyst nematode on growth of dry bean in the field. *Plant Disease*, *94*(11), 1299–1304. https://doi.org/10.1094/PDIS-05-10-0326.

Pun, T. B., Neupane, A., & Koech, R. (2023). A deep learning-based decision support tool for plant-parasitic nematode management. *Journal of Imaging*, *9*(11), 240. https://doi.org/10.3390/jimaging9110240.

Reis, E. M., Forcelini, C. A., & Balardin, R. S. (2021). Integrated approaches for controlling Asian soybean rust: A review. *Field Crops Research*, *267*, 108163. https://doi.org/10.1016/j.fcr.2021.108163.

Shtienberg, D. (2013). Will decision-support systems be widely used for the management of plant diseases? *Annual Review of Phytopathology*, *51*, 1–16. https://doi.org/10.1146/annurev-phyto-082712-102324.

Sikora, E. J., Allen, T. W., Wise, K. A., Bergstrom, G., Bradley, C. A., Bond, J., Brown-Rytlewski, D., Chilvers, M., Damicone, J., DeWolf, E., Dorrance, A., Dufault, N., Esker, P., Faske, T. R., Giesler, L., Goldberg, N., Golod, J., Gómez, I. R. G., Grau, C., & Zidek, J. (2014). A coordinated effort to manage soybean rust in North America: A success story in soybean disease monitoring. *Plant Disease*, *98*(7), 864–875. https://doi.org/10.1094/PDIS-02-14-0121-FE.

Wang, Y., Zia-Khan, S., Owusu-Adu, S., Miedaner, T., & Müller, J. (2019). Early detection of *Zymoseptoria tritici* in winter wheat by infrared thermography. *Agriculture*, *9*(7), 139. https://doi.org/10.3390/agriculture9070139.

West, J. S., Bravo, C., Oberti, R., Lemaire, D., Moshou, D., & McCartney, H. A. (2022). The prospects for optical detection and quantification of disease symptoms in precision crop protection. *Annual Review of Phytopathology*, *60*, 191–213. https://doi.org/10.1146/annurev-phyto-080521-090103.

Westphal, A., Abney, T. S., Xing, L. J., & Shaner, G. (2008). Sudden death syndrome of soybean. *Plant Health Instructor*, *8*. http://dx.doi.org/10.1094/PHI-I-2008-0102-01.

Xue, J., Leung, Y., & Fung, T. (2019). An unmixing-based Bayesian model for spatio-temporal satellite image fusion in heterogeneous landscapes. *Remote Sensing*, *11*(3), 324. https://doi.org/10.3390/rs11030324.

Yang, C., Odvody, G. N., Thomasson, J. A., Isakeit, T., Minzenmayer, R. R., Drake, D. R., Norton, R., Barnes, E. M., & Nichols, R. L. (2017). *Site-specific management of cotton root rot using airborne and satellite imagery and variable rate technology*, Site-specific

Management of Cotton Root Rot—Cotton Incorporated. https://www.cottoninc.com/cotton-production/ag-research/plant-pathology/site-specific-management-cotton-root-rot.

Yuan, Z., Zhang, Y., Bi, G., & Yang, P. (2025). Exploring the feasibility of deep learning models for long-term disease prediction: A case study for wheat yellow rust in England. *arXiv*. https://doi.org/10.48550/arXiv.2501.15677.

Zaidi, S. S., Mukhtar, M. S., & Mansoor, S. (2018). Genome editing: Targeting susceptibility genes for plant disease resistance. *Trends in Biotechnology*, *36*(9), 898–906. https://doi.org/10.1016/j.tibtech.2018.04.005.

Zanin, A. R. A., Neves, D. C., Teodoro, L. P. R., da Silva Júnior, C. A., da Silva, S. P., Teodoro, P. E., & Baio, F. H. R. (2022). Reduction of pesticide application via real-time precision spraying. *Scientific Reports*, *12*, 5638. https://doi.org/10.1038/s41598-022-09607-w.

Zhang, J., Huang, Y., Pu, R., Gonzalez-Moreno, P., Yuan, L., Wu, K., & Huang, W. (2019). Monitoring plant diseases and pests through remote sensing technology: A review. *Computers and Electronics in Agriculture*, *165*, 104943. https://doi.org/10.1016/j.compag.2019.104943.

Zhang, S., Li, X., Ba, Y., Lyu, X., Zhang, M., & Li, M. (2022). Banana Fusarium wilt disease detection by supervised and unsupervised methods from UAV-based multispectral imagery. *Remote Sensing*, *14*(5), 1231. https://doi.org/10.3390/rs14051231.

Zheng, Q., Huang, W., Cui, X., Shi, Y., & Liu, L. (2018). New spectral index for detecting wheat yellow rust using Sentinel-2 multispectral imagery. *Sensors*, *18*(3), 868. https://doi.org/10.3390/s18030868.

Yield Monitoring and Mapping

20

John P. Fulton, Elizabeth Hawkins, Kent Shannon, Sunoj Shajahan, and Simerjeet Virk

Chapter Purpose

Yield monitors enable farmers to not only estimate crop yield, but also to identify the specific location in fields where it is produced. The georeferenced data can be used for many purposes, including estimating the nutrient removal by the harvested crop, assessing profitability, developing management zones, identifying yield-limiting factors, and analyzing data from on-farm studies. This chapter focuses on the operational principles and advancements of yield monitoring systems, offers guidance on preparing for harvest, and accurately record yield data. The chapter also covers information on common errors in yield monitor data, techniques for error removal, how to use yield data to improve your understanding of field variability, and the use of cloud-based computing tools for data analysis.

Overview of the Yield Monitor Harvesting Process

Traditionally, crop yields have been monitored very roughly, relying on volume or weight measurements made when the commodity was delivered to the buyer. For example, the weight and moisture content of a wagon or truckload of grain were determined on delivery to the buyer. Based on that weight, the yield was determined using the following equation:

$$Yield\left(\frac{\text{bu}}{\text{ac}}\right) = \frac{Grain\ weight(\text{lb}) \times Density\ of\ grain\left(\frac{\text{bu}}{\text{lb}}\right)}{Harvested\ area(\text{acres})}$$

Based on the yield and measured moisture content, the harvested yield was then adjusted to standard moisture using the following equation:

$$Yield\ at\ standard\ moisture\ percentage = \frac{(Harvested\ yield) \times (1 - moisture\ as\ a\ decimal)}{(1 - the\ standard\ moisture\ as\ a\ decimal)}$$

In this equation, the percentage moisture values need to be converted to decimals, and the appropriate standard moisture values for the grain should be used. Different grains have different standard moisture percentages. For example, the standard moisture percentages for corn, soybean, and wheat are 15.5, 13.0, and 13.5%, and the standard bushel weights for corn, soybean, and wheat are 56, 60, and 60 lb (25.4, 27.2, and 27.2 kg), respectively (Table 20.1).

In the early 1990s, measurements shifted from an area-based average yield to instantaneous estimates of yields and moisture contents throughout a field. This transformation involved merging information from multiple sensors on a combine to produce a yield estimate. To understand the value of each yield estimate, a basic understanding of the harvesting process is needed. The harvesting process varies slightly depending on the type of crop, but the general stages for grain crops remain consistent. The following outlines the six-step process of a combine while harvesting grain (Figure 20.1).

1. Crops are harvested by **cutting** them from their stalks or stems. This is done mechanically via a sickle for crops like soybean or small grains. Corn harvesting involves two distinct processes: The ear is removed from the stalk using a snapping roll, while the stalk is cut by the deck plate (Figure 20.1).
2. The harvested crop is gathered and directed toward the **feeder house** of the combine. The gathering mechanism, usually an auger or a draper belt, collects the harvested crop and moves it to the center of the combine. From there, the feeder house lifts the crop and feeds it into the threshing mechanism.
3. **During threshing**, the grain is separated from the chaff, stalks, stems, and husks. This process uses either a conventional cylinder or a rotor threshing mechanism.
4. The threshed grain is then **cleaned** to remove debris from the harvested material.
5. The cleaned grain is then transported to the **grain tank** using a cross auger and a clean grain conveyor (also called a clean grain elevator). The yield monitoring system is positioned and utilized at this stage of the combine harvesting process.
6. The material that is not grain is then chopped and spread over the soil surface. The goal of this process, called **residue management**, is to evenly distribute the residue across the harvest swath.

As technology evolves, new yield monitor systems are continually being developed. In 2024, commercially available yield monitor systems were available for corn, wheat, oat, soybean, rice,

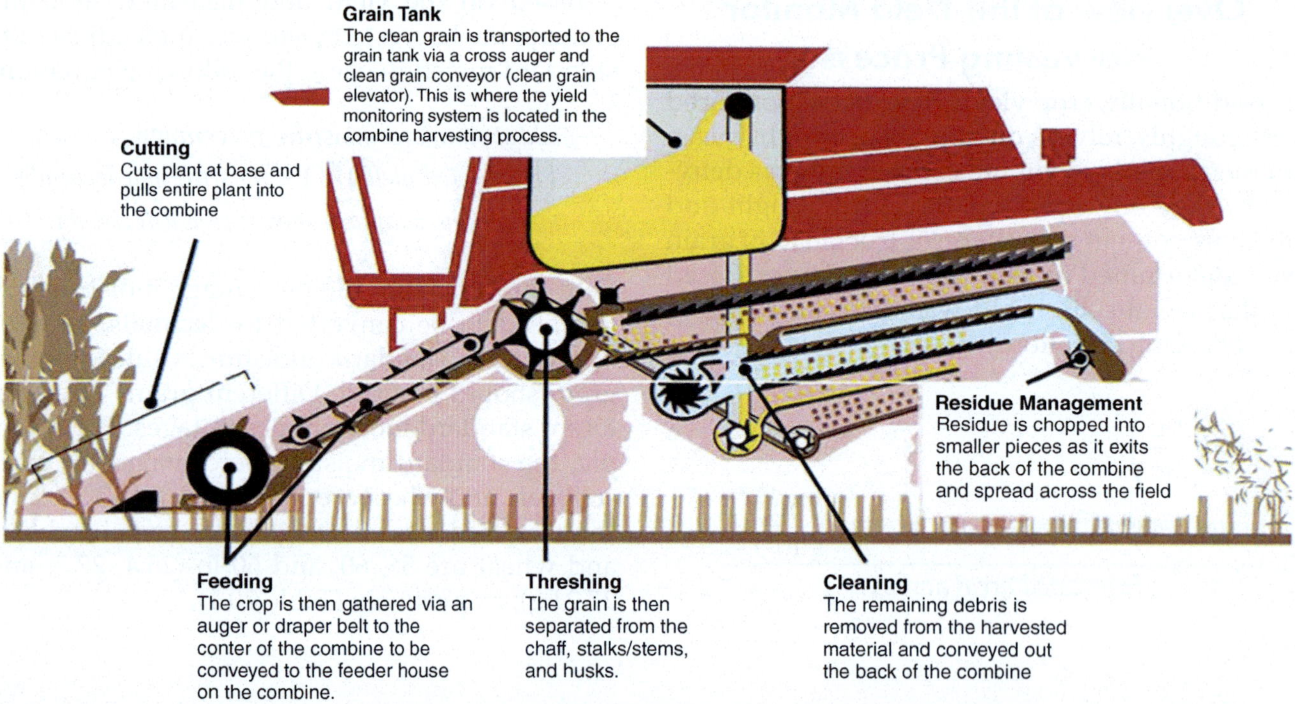

Figure 20.1. This depiction of how a combine works is based on the illustration from https://ilfbpartners.com/farm/how-do-combines-work/attachment/how-a-combine-works-flyer_7-27_red_p.

cotton, sugarcane, forages, potato, onion, alfalfa, and grass (Andrade-Sanchez & Heun, 2013; Franzen & Humburg, 2016; "John Deere HarvestLab™ 3000," 2024; Luck & Fulton, 2014; Magalhães & Cerri, 2007; "RiteYield," 2024). These systems estimate crop yield and produce information that can be used to generate yield maps.

Grain Yield Monitors

The grain yield monitor system typically contains a user interface (UI), Global Navigation Satellite System (GNSS) or Global Positioning System (GPS) receiver, mass flow sensor, and moisture sensor (Figure 20.2). Using the information collected from these components, we can calculate a yield estimate for specific locations. However, the area is not a fixed value; it is dependent on the header width and the combine speed. The accuracy of the calculated yield depends on the accuracy of the components discussed below.

Mass Flow Sensors

Mass flow sensors can generally be categorized into two types: (a) those using impact force measurement (based on Newton's Second Law), and (b) those based on volumetric flow measurements. Impact force sensors are generally used for dried grain products, as they are less sensitive to mechanical handling than fleshy crops. For grains, the impact plate is positioned near the top of the clean grain elevator (Figure 20.3), where it measures the force or change in momentum generated as the grain impacts the plate after being accelerated off the elevator conveyor paddle.

The measured force is a function of the mass and velocity of the bulk material as it contacts the impact plate (Figure 20.3). Since the geometry of the elevator is known or can be measured, the grain velocity can be estimated using the elevator speed (measured by a sensor on the elevator) and the elevator geometry (based on the distance between the top of the elevator and the impact plate). Because yield monitors convert the impact sensor readings into mass estimates, mass flow sensors need to be calibrated properly on a regular basis.

Sensor calibration should be conducted during harvest preparation, which is crop-specific, as different crops require distinct calibration settings. Calibration involves harvesting and weighing multiple loads of grain, then recording the load weights in the yield monitor in-cab display. When calibrating this sensor, it is best to select field areas that are relatively uniform and level. Each load should contain at least 3,000 lb (1,360 kg) of grain. In areas producing a 150-bu/ac corn crop (9.39 Mg/ha), this requires

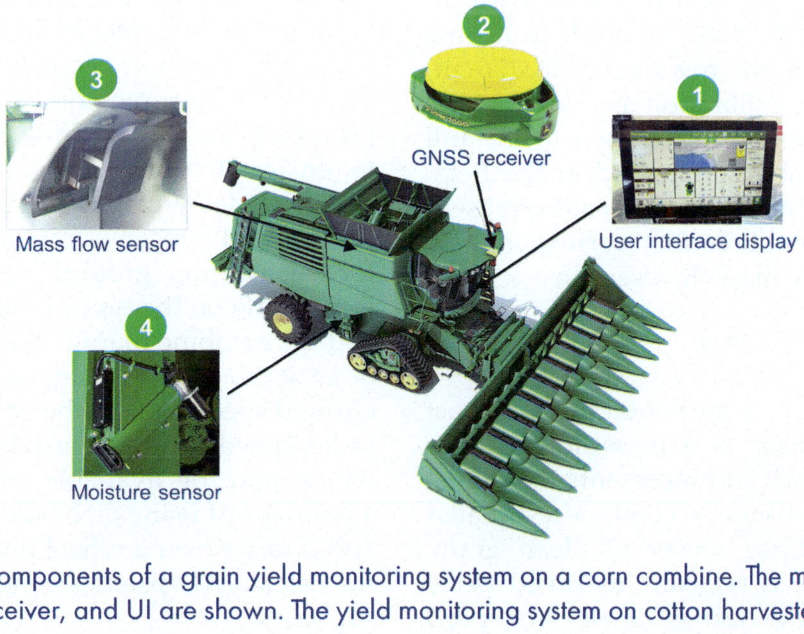

Figure 20.2. Basic components of a grain yield monitoring system on a corn combine. The moisture sensor, mass flow sensor, GNSS receiver, and UI are shown. The yield monitoring system on cotton harvesters has a similar configuration.

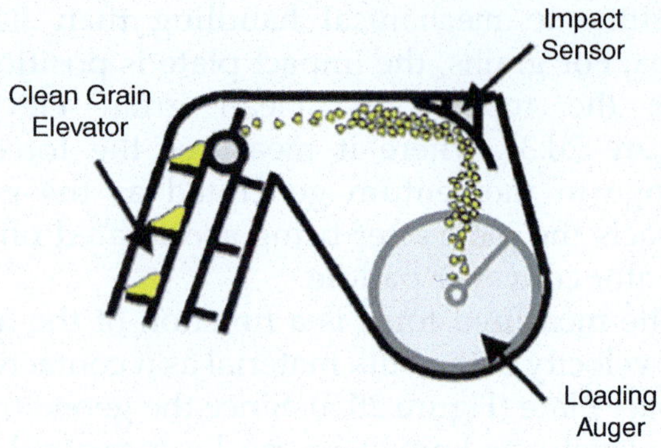

Figure 20.3. Example of an impact style plate and its mounting location at the top of the clean grain elevator. (Courtesy Ag Leader).

approximately 800 ft (274 m) of travel with an 8-row head, or 520 ft (158.5 m) with a 12-row head. In this system, the rows are 30 in. (76 cm) apart. An effective calibration process should expose the yield monitoring system to the full range of yield variability expected in the field. Therefore, each calibration load should represent a distinct yield condition that corresponds to data captured by the mass flow sensor.

$$F = \dot{m} * v \quad (20.1)$$

Where, the force, F, is measured by an impact style grain flow sensor; $\dot{m}$ is mass flow rate (in kilograms per second) and represents the rate of clean grain flow through a combine estimated using an impact force sensor of clean grain; and the velocity, v, is typically assumed to be constant (based on the clean grain elevator speed [in revolutions per minute] or measured separately in meters per second).

There are various methods for collecting calibration loads that simulate high-, medium-, and low-yield conditions. These techniques include harvesting loads at varying speeds or with full and reduced header widths in constant-yield areas. Calibrating the mass load sensor by collecting harvest loads from both high- and low-yield areas is less desirable and should be avoided. The recommended number of calibration loads varies by manufacturer, but it is generally advisable to use more calibration loads to effectively capture the range of conditions. The loads should be collected on the same day and under consistent weather conditions whenever possible. Details for several different calibration methods are provided by Franzen and Humburg (2016).

Differentially Corrected GNSS or GPS Receiver

Information collected by differentially corrected GNSS or differential GPS (DGPS) receivers is used to determine the harvester's spatial locations as it moves through the field. See Chapter 7 for a broader discussion on GNSS or GPS. Differential correction is a technique used to improve the accuracy of estimated locations; it uses a signal from a base station with a known location to correct the locations of the reported values. Typically, the GNSS or GPS receiver is mounted in the center of the vehicle on top of the cab (Figure 20.2). It is critical to enter the correct machine measurements into the UI when prompted for the machine configuration offsets to ensure that the proper material handling and flow delay can be correctly calculated. The GNSS or GPS receiver can also serve as a sensor for measuring ground speed and distance. Depending on the type of yield monitor installed and the combine manufacturer, it can sometimes be easier to obtain ground speed from the GNSS. Ground speed can also be obtained using either a radar speed sensor or a wheel counter, both of which may be available on the combine. The advantage of using GNSS information or ground speed radar over a wheel rotation counter is that the calculated speeds are not affected by wheel slippage.

Grain Moisture Sensor

Since most grain crops are sold on a wet basis, accurate measurement of the moisture is required for accurate yield estimation. Equations and example calculations are available in Reese and Carlson (2017), and step-by-step directions for measuring grain and hay moisture contents using sensors, ovens, and microwaves are available in Anderson and Grant (1993). It is important to note that when measuring grain moisture, it is essential to follow product-specific protocols because

1. Different protocols must be followed for different grains.
2. Oven drying may require one to three days.
3. Calibrated grain moisture sensors should provide acceptable levels of accuracy if the proper protocols are followed (Anderson & Grant, 1993).

Grain moisture sensors generally calculate the moisture percentage based on the dielectric properties of the harvested grain, which are related to the grain's ability to store energy in an electrical field, which in turn is directly related to grain moisture content. These sensors are often referred to as capacitance-based moisture sensors since capacitors also store energy in an electric field. The dielectric constant is measured with a calibrated capacitance sensor, and the grain moisture content is calculated using established relationships. It should be noted that temperature has a direct effect on moisture sensing, thereby temperature is measured in addition to the dielectric constant. In most cases, a thermocouple is integrated into the capacitance moisture measuring device to measure grain temperature and adjust for grain temperature changes during harvest. The dielectric constant is the ratio of the capacitance of a material between two plates to the capacitance of air between the same two plates. Capacitance is determined by measuring the phase shift between an applied AC voltage and the current flowing into the material between the two plates.

Most moisture-sensing devices are placed either in the clean grain auger (fountain augers) or on the side of the clean grain elevator (Figures 20.4 and 20.5). The most desirable location for the moisture sensor is on the side of the clean grain elevator. This position offers easier mounting and access for maintenance, as well as limited grain flow through the device. This configuration samples only from the elevator

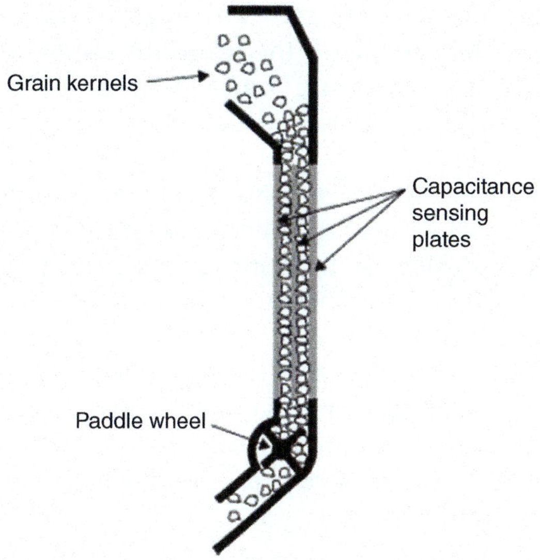

Figure 20.4. A side-mounted moisture sensor.

Figure 20.5. A side-mounted moisture sensor attached on the side of the clean grain elevator. Source: Midstate Machinery Brokers.

instead of measuring the entire grain stream. To mount the sensor on the fountain auger, a portion of the flighting must be removed, and capacitance is measured between a fin and the auger tube.

To calibrate the moisture sensor, four to six samples of grain are collected from the hopper. These samples can be stored in a 5-quart pail or a large coffee can. The moisture content of these samples should be measured using a calibrated moisture meter, such as the John Deere Moisture Tester SW08120. The average moisture content value should be entered into the yield monitor display before entering the load weight for the calibration load.

User in-Cab Display Interface

The UI serves as a link between the farmer and the sensors and is commonly referred to as the in-cab display. In many situations, the peripheral components are grouped into a working system (Figure 20.6). The UI is typically located in the combine cab and is used to do the following:

1. Convert the output from various sensors into data for storage, display, and later use.
2. Provide the equipment location from the GNSS receiver (also called GPS).
3. Provide access to external storage devices.
4. Provide a display and keyboard that can be used to monitor and control the devices associated with the yield monitor system.
5. Provide real-time harvest information. The UI stores the collected information and uses this information to calculate the yield estimate.

Yield is a function of mass collected over a measured area. Thus, the header width is used in combination with the speed of the harvester (or distance traveled based on speed from one yield point to the next). The mass flow rate of the material flowing through the machine during this same period is combined with the area and moisture content to estimate yield for each point in the field (Reese & Carlson, 2017). Accurate estimations require the proper calibration of all sensors.

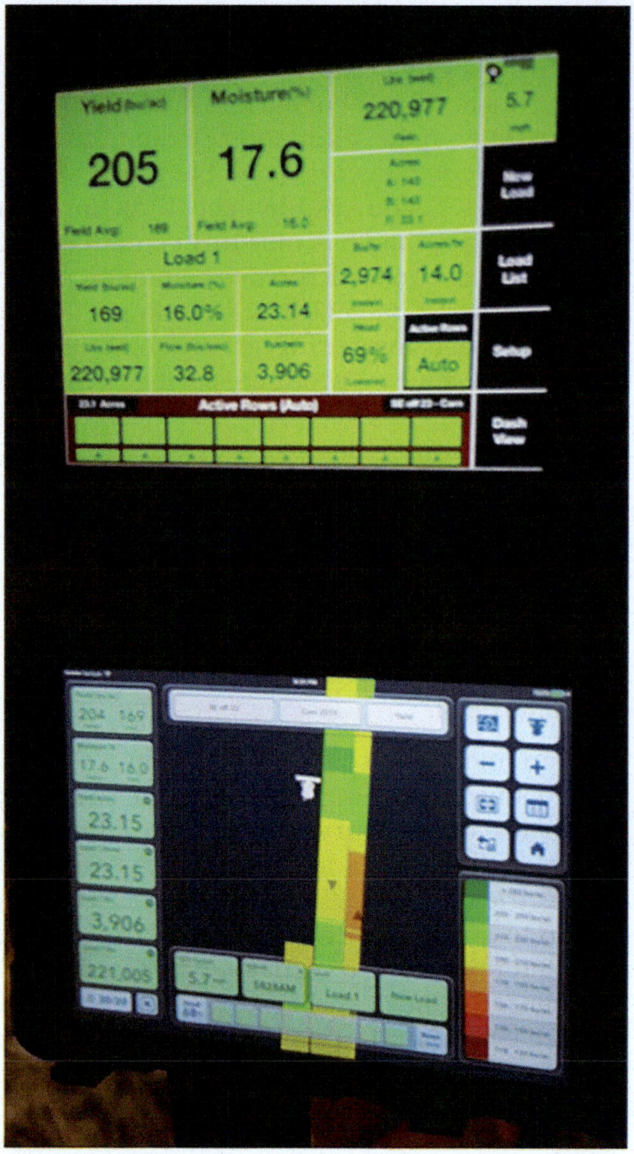

Figure 20.6. In-cab display showing yield and moisture percentage.

Measuring Material Volume Flow with Sensors

Mass flow sensors, as previously discussed, can be categorized into sensors using impact force measurement (based on Newton's Second Law) and sensors based on volumetric changes. The principles of impact for measurement can be applied in a variety of different types of sensors to determine the volumetric measurement of the grain.

Measuring Volume with a Photoelectric Sensor

A photoelectric sensor estimates the volume of product that passes through a light beam. These sensors have a light emitter and a light sensor, which are mounted on opposite sides of the duct or elevator that directs the product flow. The ratio of time that the light beam is broken versus unobstructed is related to the volume of the product. The light beam is interrupted when the harvested product passes through the path between the emitter and the sensor (Figure 20.7). The timing of these interruptions provides a method for determining the volumetric flow of the product. On a grain combine, the photoelectric sensor measures the volume of grain contained on each paddle of the clean grain elevator. On a grain harvesting combine, one disadvantage to optical volume sensors is that operating the combine on a sloped portion of the field causes the grain to shift to one side or the other on the elevator paddle.

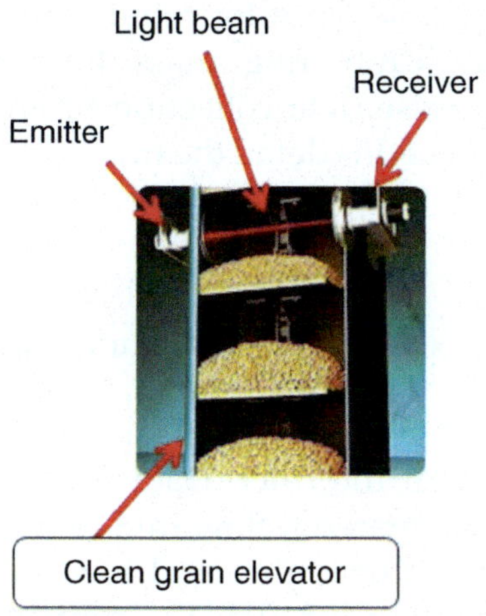

Figure 20.7. Illustration of a photoelectric / optical sensor for estimating the volume of clean grain being conveyed in the clean grain elevator. On one side of the clean grain elevator is mounted a light emitter and on the opposite side a receiver.

Using Microwave, Radiation, and Gamma Rays to Estimate Volume

Microwave and radiation sensors work similarly to photoelectric sensors; however, instead of using a light source, they rely on microwaves and gamma radiation, respectively. Microwave sensors are used in John Deere cotton harvesters. The sensor is located on the cotton conveyance duct just behind the cab. Microwave mass flow sensors emit beams of microwave energy that illuminate the stream of flowing cotton, which is then detected by the same sensors. These sensors require a poly or plastic type of duct because metal would disrupt the microwaves. The number of microwaves reflected on the sensor increases with the amount of cotton passing through the duct.

Radiation sensors require a detector positioned opposite the radiation source (typically gamma-ray radiation), similar to how optical sensors operate. The detector measures the reduction in the radiation as the grain flows between the source and the detector. The presence of grain between the source and the detector decreases the amplitude of the radiation that is measured at the detector. The change in the signal is correlated to the yield, and it is independent of moisture content. The United States currently prohibits the use of these sensors due to their use of a radioactive source.

Volumetric Flow Sensing Alternatives

Volumetric flow can also be measured using a paddle wheel grain flow sensor. Although this method has not yet been incorporated into any commercially available grain combines in the United States, it may provide viable options for measuring grain flow in the future. This type of sensor is not used commercially in grain combines because it restricts the volume of grain flow (cubic meters per second), which slows down the ground speed of the harvest.

Paddle wheel grain flow sensors are typically mounted at the base of the loading auger tank. They work by rotating at a controlled speed. For example, when the area between two adjacent paddles becomes filled, they rotate to expose the region between the next two adjacent paddles. To estimate the yield, the volume between adjacent paddles and the number of paddle revolutions must be known.

Additional Onboard Sensors

In addition to the core components of yield monitoring systems, several auxiliary sensors play a crucial role in enhancing data accuracy and operational efficiency. These sensors, measuring fan speed, elevator speed, and header position, provide valuable input for refining material flow calculations, calibrating mass flow rates, and ensuring precise data collection timing. The following subsections detail the function and importance of each of these supplemental sensors.

Fan Speed Sensor

A fan speed sensor is included in the yield monitoring system on some harvesters, such as cotton pickers. This type of sensor measures air speed through the duct. Data from this sensor improves the material flow rate calculations and helps account for higher and lower volumes of material moving through the conveyance duct. Simply put, this sensor is essential for calculating the velocity of the material as it moves past the yield sensors.

Elevator Speed Sensor

The elevator speed is determined using a simple magnetic sensor located on one of the separator shafts of the clean grain elevator. This sensor generates a square wave with a frequency that is proportional to the speed at which a ferrous (iron-containing) block passes the magnetic pickup. The clean grain elevator speed is used in some yield monitoring systems as part of the calibration for mass flow rates. By monitoring the speed of a shaft that is directly coupled to the elevator drive, a simple ratio can be used to determine the frequency at which the paddles pass the impact plate. This information can also be used for signal conditioning of the mass flow rate.

Header Position Sensing

Most yield monitoring systems use the position of the header (i.e., "down" during harvesting operations and "up" when turning) as the basis for starting and stopping data collection. This is accomplished by adding a potentiometer or electronic switch to the feeder housing or header of the harvester, or by detecting the activation of the header height control switch at the operator station. The header position switch is a critical component, as it signals the UI whether to start or stop data collection. By working in conjunction with properly set start and stop delays, this sensor can reduce the collection of zero "yield data" points from being incorporated into the yield map for areas already harvested or for no-crop areas. If the zero values were not removed during data cleaning, their inclusion would lower the average yield for the field. The additional area traveled while the header is erroneously in the "down" position increases the estimated harvest area, which in turn reduces the yield estimate.

Overview of Yield Monitor Data

The collection of yield monitor data involves proper equipment calibration to ensure accurate measurements, as well as the effective collection and storage of yield monitor data. Yield monitor data is valuable because it helps farmers better understand how crop yields vary across a field and supports more informed decisions for the next growing season.

Yield Monitor Data Formats

When planning to use a yield monitor, it is important to consider how the data will be stored and processed. Different yield monitor systems use different formats for storing data. For example, Ag Leader may store data in a .yld file format, whereas John Deere may use a .gsd format. Different formats may have different software requirements. Information on the various formats can be found in Hawkins et al. (2017).

Data Collected and Archived by Yield Monitors

Once a yield monitor has been properly installed and calibrated, it allows producers to observe many parameters "on the go" during harvest. Instantaneous yield, average yield, mass flow rate through the combine, area harvested, and moisture content can be displayed on the monitor with the touch of a button. Yield monitors also store a summary of this data for each load and field. These data summaries can be downloaded using a USB flash drive and transferred to an office computer. The exported summary files can be opened in a spreadsheet for further data analysis. When the yield monitor is linked to a differential GNSS (DGNSS) receiver, geographically referenced data is available for download to the office computer. This data can be expanded and exported to several software packages. The nature of the data recorded by the yield monitor and how these data are related to the sensors should be taken into account. The following "advanced" export data format is nearly universal:

ddd.dddddd, dd.dddddd, mm.mm, tttttttt, n, lll, www, cc.c, kk, ppppp, ssssss, Fnn:bbbbbbbb, Ln:bbbbbbbb, gggggggggg, sss, ppp, aaaa

where ddd.dddddd represents the longitude (in decimal degrees, +East and –West); dd.dddddd represents the latitude (in decimal degrees, +North and –South); mm.mm represents grain mass flow (in pounds per second); tttttttt represents GPS time (in seconds; in Unix timestamp format); n represents the cycle period (in seconds); lll represents the distance traveled in cycle period (in inches); www represents the effective swath width (in inches); cc.c represents the moisture content (in percentage wet basis); kk represents status (in bits 0 through 4- header down); ppppp represents the pass number, ssssss represents the yield monitor serial number, Fnn:bbbbbbbb represents the field identifier (ID; number and name); Ln:bbbbbbbb represents the load ID (number and name); gggggggggg represents grain type, sss represents GPS status, ppp represents the positional dilution of precision (PDOP); and aaaa represents the altitude (in feet).

A typical exported data string might look like the following:

–95.478086,40.270372,13.60,1,445,876,045,1,77, 420, 9.3,33,3,212,959,"F0:901","L0:CLARK","SOYBEANS",7,0,872.5

It is obvious from this string that Clark harvested soybeans from field 901. Not as obvious is the yield in bushels per acre. As one looks at the character string, it can be determined that 13.60 lb (6.17 kg) of soybean at a moisture content of 9.3% were harvested as the combine traveled 77 in. in a 1-s cycle (7.05 km/hr).

Character string: –95.478086,40.270372,**13.60**,1 445876045,**1**,**77**,**420**, **9.3**,33,3,212959,"F0:901"," L0:CLARK",

The effective combine header width was set by the operator at 420 in. (35 ft; 10.7 m). To arrive at the marketable yield in bushels per acre for this 1-s cycle, the mass flow rate of grain at a particular moisture content must be corrected to reflect marketable bushels. To accomplish this, the wet mass flow value must be multiplied by a ratio of moisture content differences.

$$m_{\text{market}} = \frac{(100\% - MC_{\text{harvest}}) \times m_{\text{harvest}}}{(100\% - MC_{\text{market}})}$$

where m_{market} is the corrected mass flow rate of grain in pounds per second; m_{harvest} is the mass flow of moist grain at a moisture content of MC_{harvest} (%); and MC_{market} represents the moisture content used as a basis for marketing (Table 20.1). For soybean, the marketing moisture percentage is 13.0%. Sample calculations are available in Reese and Carlson (2017).

Next, the yield is determined on a per-acre basis. To accomplish this, the corrected mass flow rate is multiplied by the data logging interval and then divided by the area harvested during that logging interval. The form of this calculation is as follows:

$$Y = \frac{(m_{\text{market}} \times t_{\text{sample}})}{d \times w \times p_{\text{grain}}}$$

where Y is the corrected yield (in bushels per acre); t_{sample} is the data logging interval (in seconds); d is distance traveled in the data logging period (in inches); w is the effective width of the combine header (in inches); and p_{grain} is the mass density of a particular grain (in pounds per bushel). Typical mass densities used for conversion from mass or weight to the volumetric measure of bushels and marketable moisture are shown in Table 20.1. Substituting in the appropriate numbers and conversion factors results in the following:

$$\frac{45.7\ \text{bu}}{\text{ac}} = \frac{\frac{14.13\ \text{lb}}{\text{s}} \times 1\ \text{s}}{77\ \text{in.} \times 420\ \text{in.} \times \frac{60\ \text{lb}}{\text{bu}}} \times \frac{144\ \text{in.}^2}{1\ \text{ft}^2} \times \frac{43{,}560\ \text{ft}^2}{1\ \text{ac}}$$

Table 20.1. Grain weights and moisture contents for different grain crops

Crop	Weight lb/bu	Specified moisture content %	Weight at 0% moisture content lb
Corn	56	15.5	47.32
Soybean	60	13.0	52.20
Wheat	60	13.5	51.90
Barley	48	14.5	41.04
Oat	32	14.0	27.52
Rye	56	14.0	48.16

Source: PAB Table 5.1.

Thus, for the previously exported yield data string, the average marketable yield for this logging interval is 45.7 bu/ac (3199 kg/ha). The reported yield has been corrected to a marketable volume of grain per acre. The standard weight and moisture contents are provided in Table 20.1.

Cotton Yield Monitors

The two primary types of sensors used in cotton yield monitors are optical and microwave sensors. Each of these systems works similarly, with the main difference being the type of energy (e.g., optical vs. low-energy microwave) used to measure the amount of cotton flowing through the ductwork on a cotton picker. In 2024, the microwave-type yield monitoring system was the only commercially available system for cotton harvesters. The optical sensor that was used for several years for cotton yield monitoring has been discontinued by the manufacturer. At a minimum, components of the yield monitor system include the following:

1. GNSS or GPS receiver using differential correction
2. In-cab display
3. Header height sensor
4. Ground speed sensor if not using the GNSS or GPS for speed measurements
5. Flow sensor mounted on the duct.

Based on information from these components, yield estimates are computed and reported as pounds per acre (weight per area) of seed cotton (e.g., cotton bolls without seeds removed). Additional calculations of cotton yield include lint yield in pounds per acre, which is calculated by determining the turnout percentage (the percentage of lint remaining after the cotton seeds have been removed during the ginning process). The turnout percentage is a constant value inputted into the in-cab display setup, typically ranging from 38 to 42%, based on the cotton variety. The operator can manually adjust the turnout percentage using the in-cab display, depending on their understanding of the cotton variety and ginning process. Some farmers and their consultants prefer the yield to be calculated as bales per acre, representing the number of ginned bales of lint cotton per acre. Nominally, a bale of lint cotton weighs 480 lb. (218 kg), but as previously mentioned, the operator can adjust this accordingly within the display to reflect the local bale weight for their operation. Like other harvesters, cotton pickers manufactured for the North American market have yield monitoring technology embedded into them (Figures 20.8–20.10). For example, all John Deere Cotton Pickers and Baler models are equipped

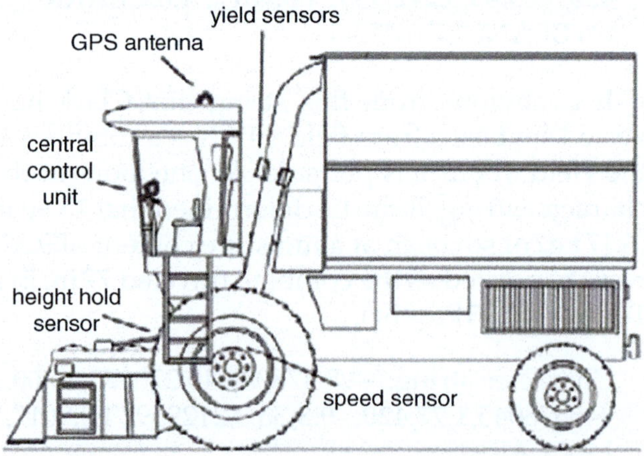

Figure 20.8. Sensors used to measure yield in a cotton picker. Source: Markinos et al., 2005 / Springer Nature.

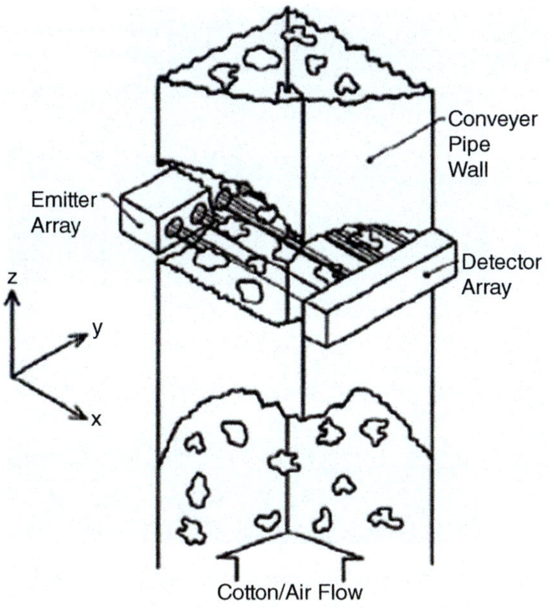

Figure 20.9. Illustration of an optical sensor array measuring cotton from spindles to the basket. The output or detection of cotton is converted to a yield estimate Source: Markinos et al., 2005 / Springer Nature.

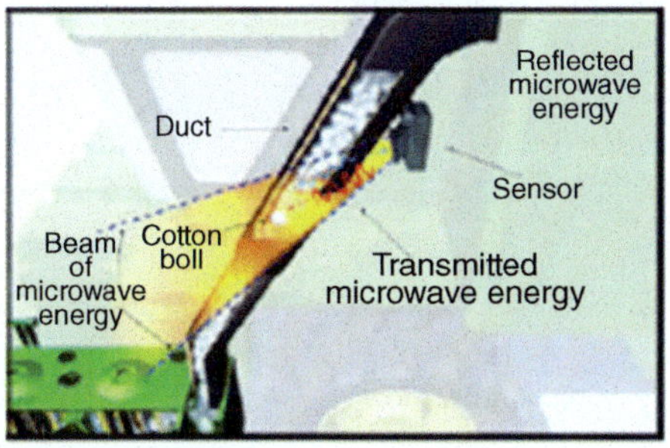

Figure 20.10. Illustration of microwave technology measuring the amount of cotton flowing through the ductwork on a cotton picker.

with yield monitoring capabilities. Cotton yield monitoring systems require calibration that involves harvesting a basket or a module of cotton, weighing it, and then inputting this data into the in-cab display. Modern cotton harvesters are also equipped with an onboard module weighing system that measures and records the weight of each module before offloading it from the harvester. Systems with real-time calibration capabilities are also being developed. Flow rate and cotton variety must be taken into account during calibration.

Other Yield Monitor Systems

Commercial systems for monitoring and mapping the yields of forage and specialty crops are also currently available. For yield monitoring for forage crops, these systems use near-infrared sensors installed on the discharge spout to estimate whole-plant dry matter during harvest. These sensors also provide information on other useful properties, including moisture, protein content, fiber, and cut length. This real-time yield information allows the operator to fine-tune harvester settings to adjust chop length and make other processor adjustments to ensure optimal compaction and forage quality.

For yield monitoring for specialty crops, kits containing load cells and an onboard controller can be outfitted to the harvest machinery that uses belt- or conveyor-type systems to transport the harvested material. Like other yield monitoring systems, these systems use equipment speed, belt or conveyor speed, and GPS position to record and display real-time yield estimates during harvest. These retrofit yield kits are currently available from various technology or third-party companies, including GK Technologies, Vantage, Topcon, and Greentronics (Figure 20.11). These systems can be used to monitor pea, bean, grape, sugar beet, potato, and onion yields.

Preparing to Harvest a Field

To minimize errors with a grain yield monitor collection system, the following steps are recommended:

1. Inspect the clean grain elevator chain for wear, making sure that the mass flow sensor and the associated impact plate are in good condition and clean.
2. Inspect all wiring harnesses for signs of wear or damage. If necessary, securely attach any loose wiring to avoid damage.
3. Inspect and clean the moisture sensor.
4. Check the yield monitor display for any errors and inspect the wires for rodent damage. Check the data card and preloaded field names and hybrids.

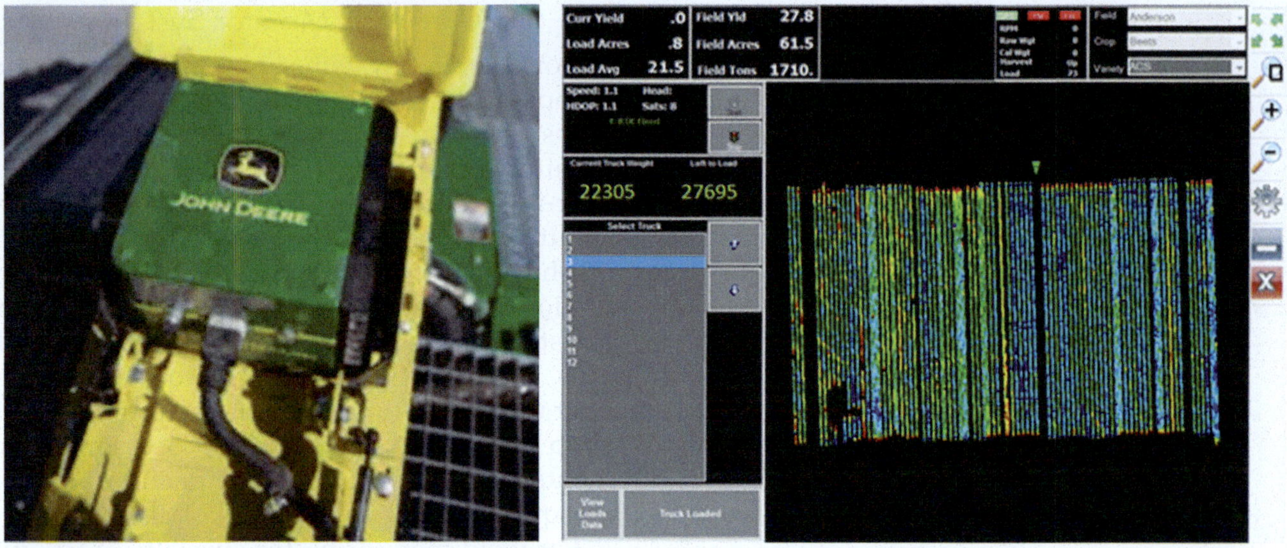

Figure 20.11. (a) John Deere HarvestLab 3000 sensors mounted on the discharge spout of a self-propelled harvester. (Courtesy John Deere). (b) An example yield map for sugar beet using one of the retrofit yield kits (GK Technologies) on a sugar beet harvester. (Courtesy GK Technologies).

5. Check for any firmware updates for various components, including the in-cab display.
6. Review the owner's manual as a refresh on operation and calibration.
7. Calibrate the sensors.

Variables that Influence Yield Estimates

Yield estimates are affected by several factors, including crop type (e.g., corn, soybean, wheat), grain moisture, grain density or test weight, roll and pitch of the harvester (which may involve gyro or other inertial sensors), and the quantity of data used in the calibration (Figure 20.12). Generally, increasing the amount of data improves the accuracy of yield predictions. For example, in two-point calibration, the mass flow may be overestimated at the points between the calibration values. This problem can be prevented by incorporating multiple points into the calibration curve. To achieve multipoint calibration, it is necessary to include additional points with different flow rates. The advantage of multipoint calibration is that it improves accuracy, while the disadvantage is that it takes more time to conduct the calibration.

Properly calibrating the yield monitor is critical to the accuracy and precision of harvest data. Improperly calibrated yield moni-

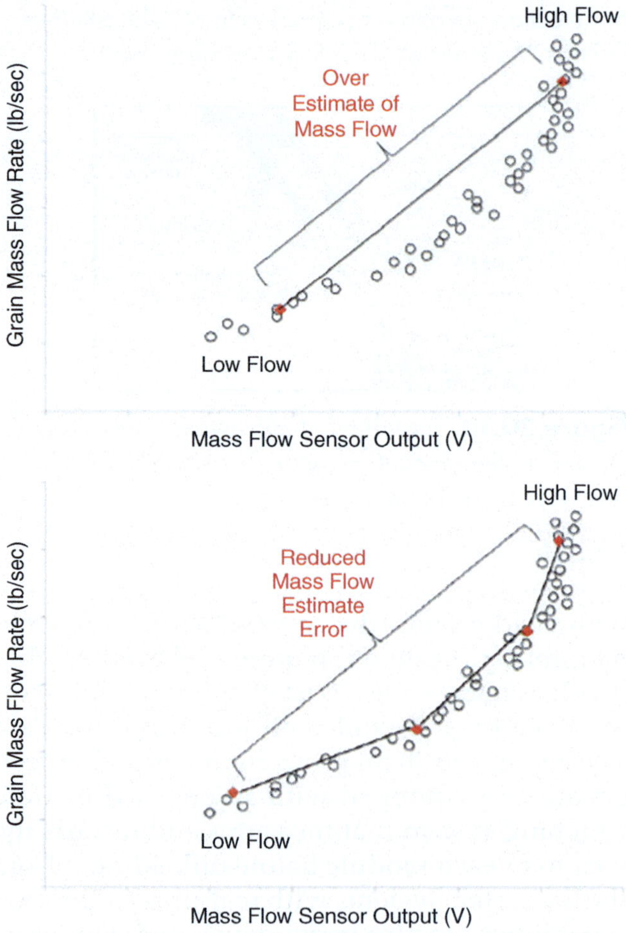

Figure 20.12. The mass flow sensor response is nonlinear, with the illustration showing the difference between a two-point and multipoint calibration procedure. (Courtesy Luck & Fulton, 2014).

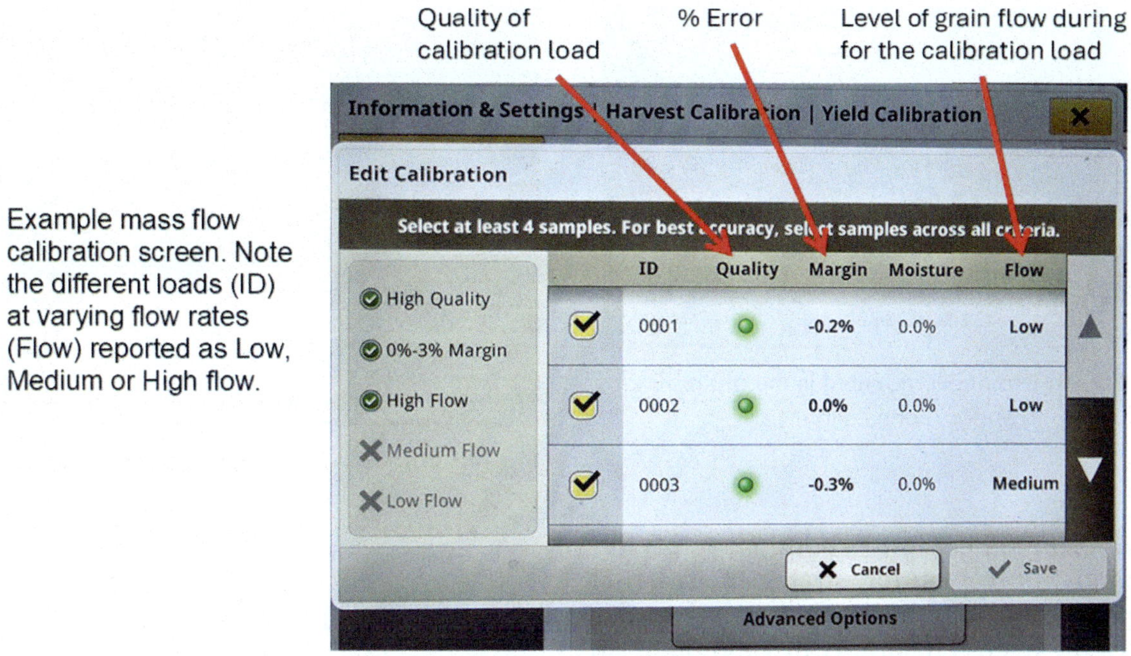

Figure 20.13. The mass flow calibration screen. (Courtesy John Fulton).

tors can result in erroneous data that may lead to incorrect conclusions and decisions. A yield monitor calculates a yield estimate from data generated by multiple sensors, including mass flow, moisture, and temperature sensors. Calibrating all sensors helps ensure the accuracy of the data.

The calibration of the mass flow sensor requires harvesting individual loads of grain and verifying the weight using accurate scales. The yield monitor software uses these verified weights to generate a calibration curve that aligns with the display's weights (Figure 20.13). The recommended process varies by manufacturer, but many use a multipoint method that requires four to six loads to complete the calibration procedure. The harvest loads used for calibration should represent the full range of expected yield levels. Varying the speed and swath width can be used to simulate areas with high or low yields. Moisture and temperature sensors should be calibrated by taking readings using an accurate moisture meter and thermometer, then entering the offsets into the monitor.

The accuracy of a yield monitor is affected by several factors, including crop and weather conditions, as well as the harvester and yield monitor components. To maintain accurate calibration throughout the harvest season, yield estimates should be regularly checked against an accurate scale, and the monitor should be recalibrated if any discrepancies are detected.

Weigh Wagons and Scaled Grain Carts

Weigh wagons and scaled grain carts offer a convenient way to measure the amount of harvested grain. The scales on these wagons or carts should be checked for accuracy against a known weight or by using scale tickets from the elevator. These weights can be compared with the yield monitor–predicted weights.

Auto-calibration and Auto-Area Technology

Auto-calibration for grain yield monitors is an important advancement that allows monitors to automatically adjust and fine-tune their settings based on real-time feedback. Auto-calibration reduces the need for "traditional" calibration. Auto-calibration ensures precise yield measurements by monitoring and updating the mass flow sensor calibration curve in response to physical changes in the harvested grain. This process is important for ensuring quality yield map data, enabling farmers and consultants to

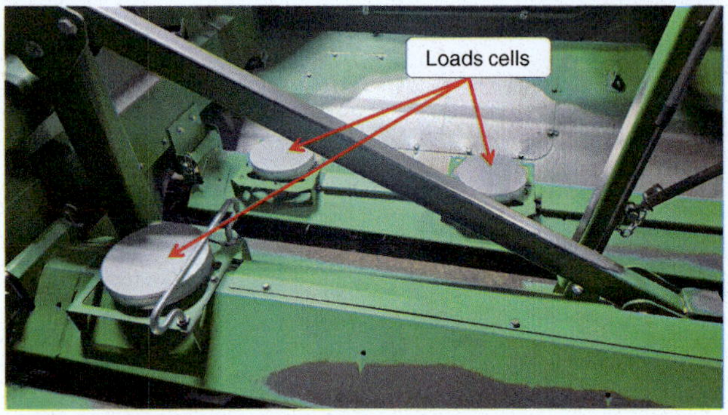

Figure 20.14. Load cells mounted in the grain bin of a grain combine. (Courtesy John Fulton).

make informed decisions about crop performance and future planting strategies. This not only saves time and reduces labor costs but also improves the overall reliability of yield data. An example of auto-calibration technology is John Deere's ActiveYield. It continuously calibrates the mass flow sensor using load cells installed in the grain tank (Figure 20.14). The load cells in the grain tank estimate the change in the weight of the grain as the grain tank fills. This information, when combined with data from the moisture and mass flow sensors, is used to adjust the calibration of the mass flow sensor. As a result, errors in mass flow estimates are minimized, particularly as grain test weight and physical properties vary across a field. This technology differs from that used in the past in that the mass flow sensor calibration curve is updated in the field, rather than relying on a single calibration curve.

In the past, the size of the harvested area was based on information the operator entered into the in-cab display. Today, most yield monitor manufacturers use the coverage map to define the area per yield point. Machines from companies like Ag Leader, John Deere, and Precision Planting can adjust the cut width based on the previous pass or coverage map, eliminating the need for the harvest operator to manually adjust the cut width during harvest. Cut width adjustments are needed when using a partial header width or operating in point rows (Figure 20.15). An example of this technology is Ag Leader's AutoSwath, which automatically adjusts the cut width and, consequently, the area estimate for each yield map point.

Figure 20.15. Illustration of a combine operating in point rows. (Courtesy Ag Leader).

Yield Monitor Data

Cleaning Yield Monitor Data

Yield monitors create files containing large amounts of spatial information, including the type of monitor, longitude, latitude, field name, object ID, track, swath width, distance, elevation, header switch status, GPS differential status, time, *Y* and *X* offset, material flow rate, harvester speed, productivity (in acres per hour), grain moisture content, type of crop harvested, and date. As mentioned earlier, this data is compiled to create a yield map. However, before using the data, any erroneous data must be removed. Erroneous data is collected when the combine slows down or speeds up, as well as in the end rows. For example, if 10 bu (670 kg) of soybeans move from the grain auger through the clean grain elevator, and the combine covers 0.1 ac (0.04 ha), then the yield is 100 bu/ac (6,705 kg/ha). However, if the same amount of grain is transferred but the combine slows down and covers only 0.005 ac (0.002 ha), then the calculated yield is 2,000 bu/acre. The 2,000 bu/ac (134,118 kg/ha) is not correct and needs to be removed from the dataset. This process is commonly referred to as cleaning.

Raw yield monitor data can also be affected by other problems, such as flow delays (Fig. 20.16), partial swaths, abrupt velocity changes, positional GPS inaccuracies, among others. Yield data points that originated under error-prone conditions must be assessed, and potentially removed, before using yield data to

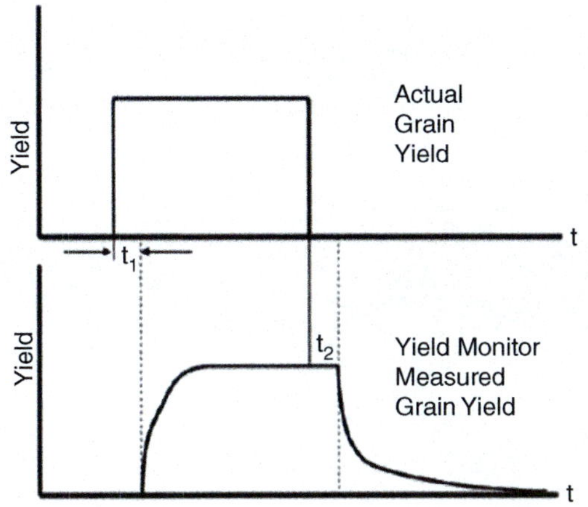

Figure 20.16. From PAB Figures 5.13 and 5.14.Actual grain yield in a field versus the measured yield through a combine by a mass or volumetric sensor. It takes time for the combine to "fill up" and grain flow stabilizes through the combine, and then it takes time for the combine to "empty out" at the end of a pass.

assess, modify, or implement crop management practices. Yield Editor (Sudduth et al., 2019) is a widely used software for evaluating and cleaning yield monitor data. It includes 12 different filters to identify and potentially remove erroneous data points. The software is free and was developed for Windows operating systems. Step-by-step examples for cleaning yield monitor data are available in Khosla and Flynn (2008). Cleaning of yield data to remove or correctly adjust yield estimates is critical, particularly when using these data in field analytical tools and prescriptive agriculture processes, as described by Luck and Fulton (2014).

Improving the Accuracy of Yield Monitor Data

The following list outlines protocols for improving yield data estimates. This list is not fully comprehensive, but it does highlight common causes of errors in yield estimates.

1. **Using old calibration data.** Systematic errors are produced by using old calibration data. Over time, calibrations can shift due to machine wear or sensor output drift. To minimize these errors, routine calibration is recommended.
2. **Calibration loads do not span the flow rates.** When calibrating the yield monitor system, it is important to calibrate the system for the full range of expected values.
3. **Poor calibration of the moisture, temperature, and speed sensors.** Poor calibration of these sensors results in incorrect estimates. Accurate estimates require a properly calibrated temperature sensor. If the ground speed estimates are not correct, the reported values are also likely to be inaccurate.
4. **Inaccurate setting of the number of rows.** The operator must correctly identify the swath width of the combine. Not correctly identifying the number swath width results in yield estimates that are either too low or too high.
5. **Sudden ground speed changes.** Rapid changes in the combine speed can result in erroneous yield estimates. Yields will be underestimated in areas where speeds are increasing and overestimated in areas where speeds are decreasing.
6. **Wear of the flighting in the clean grain elevator.** As the flighting wears, the ejection speed of the grain leaving the elevator declines, which reduces the estimated yields. Recalibrating the mass flow sensor reduces this error. If the auger flighting is replaced, the mass flow sensor must be recalibrated.
7. **Changes in the speed setting on the clean grain elevator.** If the speed settings on the clean grain elevator are changed, it is important to recalibrate the mass flow sensor.
8. **Buildup of plant residue or other debris on the sensor plate.** With operation, residue or debris may build up on the sensor plate. This can result in inflated yield estimates. Recalibration reduces this issue, which is often a greater problem in soybean than in corn. For additional information, see Luck and Fulton (2014).

Using and Understanding Yield Monitor Data

To understand and use yield monitor data, the data are often combined with other layers (Fig. 20.17). Geographic information systems (GIS) can assist in the process (Chapters 8 and 9). For example, overlaying yield data on an eleva-

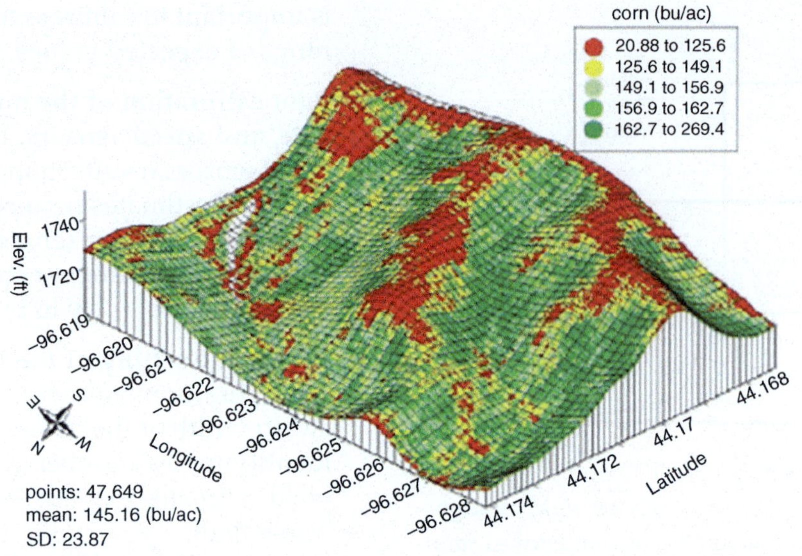

Figure 20.17. From PAB Figures 5.13 and 5.14. A yield map superimposed on an elevation map. The low maize yields in the summit/shoulder locations cannot be eliminated, but the magnitude of the different can be reduced (Courtesy South Dakota State University).

tion map makes it easier to identify yield-limiting factors when inspecting these maps (Hansen et al., 2013). Based on the interpretation or on-farm testing, the farm manager can implement site-specific treatments that will enhance yields and profitability in these areas (Chapter 21).

Yield monitor data can be used for many different applications, including diagnosing crop production problems, assessing the effectiveness of a wide range of inputs, selecting varieties or hybrids, conducting on-farm studies, conducting profitability assessments, and identifying management zones (Clay et al., 2017a,b; Kyveryga et al., 2018). However, it is important to note that a yield monitor map provides information only about yield, and the map itself cannot identify the factors impacting yield. The power of the yield map is enhanced by combining it with a scouting report, elevation maps, or on-farm testing. For the field shown in Figure 20.16, subsequent research showed that yields were reduced by water stress in the upland location (Mishra et al., 2008). In a different study, Hansen et al. (2013) combined gene expression data with yield and elevation maps. That study demonstrated that the corn downregulated nutrient uptake and its ability to respond to pests in response to increasing water stress. When using yield monitor data and associated spatial layers, the objective is to make forward-looking management decisions based on retrospective observations. This requires the data to be archived in a safe location. See Chapter 22 for information on data management. It is also important to recognize the value of older data. For this reason, it is best to retain all data for future analysis, even if erroneous or anomalous points are removed during cleaning.

Displaying Data

Printing yield maps can be the first step in postharvest analysis. The maps can provide clues into the effectiveness of the previous year's inputs (pesticides, fertilizers, and genetics). A positive aspect of diagnosing crop production problems with yield maps is that it can be used to quickly document trends and identify ways to improve future management practices. Some companies offer previous yield and profitability maps as tools for improving prescriptions for the upcoming growing season. These digital tools help integrate production and business information. These maps are generated using GIS and farm management information software (FMIS). However, it is important to remember that maps are visual representations of numbers. Therefore, we need to take into account the limitations of data visualization. This includes using appropriate range values, number of ranges, and color

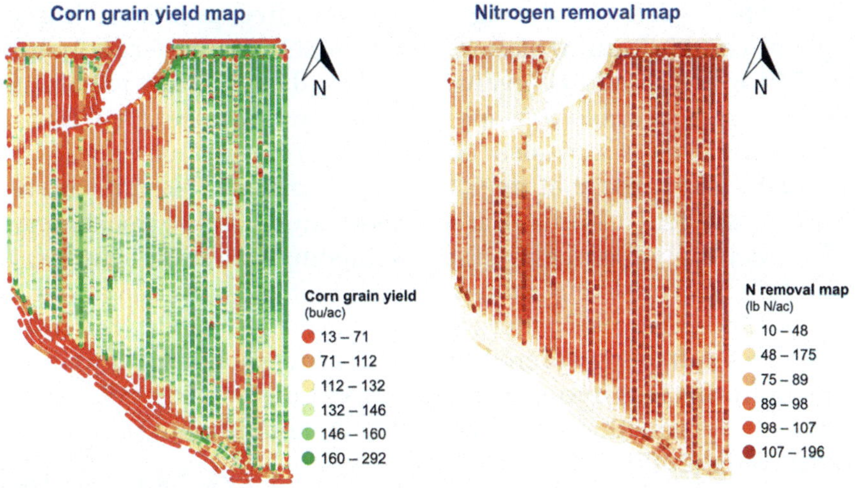

Figure 20.18. Illustration of the yield map and the corresponding nitrogen (N) removal map. The N removal is calculated by multiplying each yield data point by a factor of 0.67. These maps are being used as information in creating fertilizer prescriptions for the following year.

schemes. The range of values should be based on agronomic principles and management style. The number of ranges and their values will determine how variability is represented on a map. Generally, only five to seven ranges can be visually distinguished. Fewer ranges make it difficult to visualize the existing variability. Range values can be set using multiple methods, including Equal Points, Natural Breaks, Even Intervals, Predefined Crop, Standard Deviation, and Percent of Average.

Each display method has its advantages, but the practitioner must decide the best way to display their data. For example, if a manager would not make a change for less than 10 bu of soybeans per acre (0.626 Mg/ha), then it makes no sense to create yield ranges less than that. The last aspect to consider in creating maps is the color scheme. While individual preferences may vary, a common scheme is green–yellow–red, where green indicates high yield and red indicates low yield. The color selection will determine how the yield map is visualized and interpreted. Chapters 8 and 9 offer further information about creating maps with GIS.

Creating Nutrient Removal Maps

Although it may be somewhat challenging, yield data can be used to estimate nutrient removal (Clay et al., 2016, 2017a,b). Every bushel of grain removed contains nitrogen, phosphorus, potassium, and micronutrients. While the exact levels of these nutrients in the grain may be unknown, they can be estimated (Figure 20.18). By using yield monitor data along with estimates of nutrient levels in the grain, one can create nutrient removal maps. For example, corn removes approximately 0.67 lb (0.3 kg) of nitrogen, 0.35 lb (0.16 kg) of phosphorus, and 0.25 lb (0.11 kg) of potassium per bushel. These figures are used as multiplying factors for spatial nutrient removal calculations. Similarly, multiplying factors are available for every macro- and micro-nutrient (https://agphd.com/resources/nutrient-removal-charts/corn-grain-and-stover-nutrient-removal-charts). However, these factors may vary by region and crop type. These maps can be used to improve fertilizer recommendations.

Creating Management Zones

In precision farming, a management zone is a subfield area characterized by a relatively homogeneous combination of yield-limiting factors. Management zones can be based on data collected by a yield monitor (Clay et al., 2017a,b). The management zone concept can be applied to a wide variety of issues, such as selecting hybrids or varieties, determining seeding rates, and building treatment maps for nutrient and pest problems. Different problems (e.g., fertility, weeds, diseases, or insect infestations) may require different management zones. An example of converting yield monitor data into management zones is provided by Clay et al. (2017a,b). However, since management zone development can be local in nature, data layers such as yield maps should be evaluated to create zones that are applicable to a specific production area. Profit maps can also be created based on annual yield maps that depict areas of profit or loss across a field. Conducting a multiyear analysis can identify which parts of a field, or the field as a whole, that are losing money under current practices. Once identified, alternative management strategies can be considered in these areas to turn them from a loss into a profit. Information for calculating production costs can be found in Fausti and Wang (2017).

Sensors that Measure Grain Quality

There are currently commercially available sensors that measure grain quality properties, including protein, scratch, and oil. These sensors are based on near-infrared (NIR) technology, which operates in a part of the electromagnetic spectrum that falls just beyond the visible light range. Examples of grain quality sensors are the John Deere HarvestLab 3000, the New Holland NutriSense system, and the CropScan 3300H sensor. The New Holland NutriSense system measures various grain properties, including moisture content, crude protein, crude fat, starch, neutral detergent fiber (NDF), acid detergent fiber (ADF), and ash. Grain quality maps offer value to farmers by providing detailed insights into the variability of grain quality across their fields. Additionally, these sensors may provide important information for marketing.

Additional technologies include sensors that "look ahead" at the standing crop and make measurements. Look-ahead sensors mounted on combine headers are also being developed to assess additional grain parameters, as well as to count corn ears and measure their size, assess stalk diameter, and determine the harvested population. In the future, sensors mounted to look ahead of the combine could provide row-by-row yield estimates to greatly improve the resolution of yield maps. Currently, yield monitor estimates are limited to the wide-cut widths of modern combines, preventing farmers from evaluating row-by-row variability. New sensors are being developed to assess crop and field conditions during grain harvest. These sensors will have the potential to provide invaluable insights into how to optimize practices, improve efficiency, and ultimately increase productivity.

Finally, advances in computer vision and artificial intelligence (AI) have enabled real-time monitoring of harvest losses, particularly for grain crops. This information is valuable for assessing harvest machinery performance and adjusting combine settings, either manually or automatically, to maximize harvest equipment performance. This harvest–loss data is also recorded spatially, allowing for an understanding of how varying field or crop conditions affect harvest equipment performance.

Yield Data Storage

The transfer and storage of yield data used to be a highly manual process that involved data cards and USB sticks. Today, more options are available for the wireless and telematic management of data. Connected machines facilitate the transfer of yield data to cloud-based farm management information systems or apps for processing, visualization, and decision-making. Most agriculture or technology companies that offer yield monitoring solutions provide capabilities for accessing and viewing real-time harvest data via mobile applications or online.

Advanced Technologies in Modern Yield Monitoring

The integration of AI, advanced sensors, and automation technologies is transforming the agricultural harvesting landscape. Modern combines are no longer just mechanical workhorses; they are intelligent systems capable of real-time decision-making, precision adjustments, and autonomous operation. Additionally, imagery from unmanned aerial vehicles (UAVs) and drones enhances our understanding of yield estimates by providing high-resolution aerial perspectives that complement ground-based sensor data. This type of data is used to generate predictive estimates of yield and crop biomass before harvest, enabling combines to make predictive adjustments to settings and ground speed during operation to maximize field capacity (in hectares per hour) and improve grain quality. From AI-driven yield estimation and speed automation to robotic autonomy and sub-header resolution, these innovations enhance efficiency, reduce human error, and offer farmers actionable insights. This section explores the latest advancements driving the evolution of harvesting technology and transforming the capture, interpretation, and application of data through analytics to inform farm decisions. Additionally, imagery from UAVs and drones is improving our understanding of yield estimates.

Artificial Intelligence Adjusting and Improving Yield Estimates

Advancements in sensor technologies and the integration of AI algorithms have greatly transformed how crop yields are measured during harvest. Today's combines are equipped with a suite of mechanical and vision sensors, along with automation systems that streamline operations and enhance both yield data accuracy and harvesting efficiency. Artificial intelligence is crucial to this transformation, facilitating real-time assessment of grain quality and harvest conditions. These intelligent systems continuously adjust onboard parameters, such as threshing settings, to maintain optimal performance, reduce grain loss, and maximize material quality.

John Deere's Predictive Ground Speed Automation is an example of this innovation. Using forward-facing stereo cameras and preharvest satellite imagery, the system generates predictive field maps that estimate crop biomass and yield. These maps allow the combine to automatically adjust its ground speed before entering areas with varying crop conditions, ensuring consistent feed rates and reducing operator fatigue. Similarly, Harvest Settings Automation leverages AI to fine-tune rotor speed, fan speed, and sieve settings according to crop volume and loss sensor data. These automated adjustments maintain grain quality and minimize losses—all without requiring manual intervention. Artificial intelligence will continue to be integrated into harvesters to help automate features and settings and to inform the autonomy platform for autonomous harvesting.

Advances in Combine Speed Automation

Modern combines feature two forward-looking cab-mounted stereo cameras that capture images while AI models measure the crop height and volume ahead of the harvester (Figure 20.19). This data, combined with preharvest satellite imagery, enables the combine to generate a predictive yield map, which can inform recommendations for adjusting speed.

Advances in Harvest Setting Automation

Automation streamlines the adjustment of processing settings on the go. Based on the crop volume measured ahead of time, settings like rotor speed, fan speed, concave clearance, chaf-

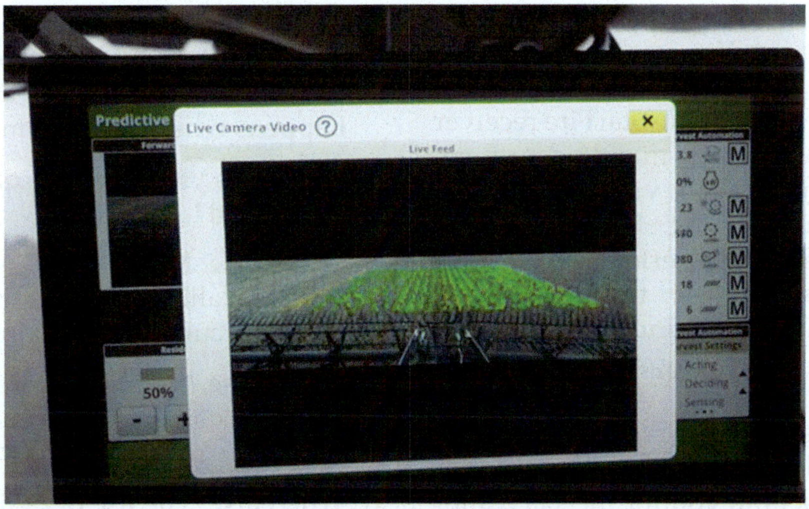

Figure 20.19. Live camera feed of crop volume measured with computer vision and AI models. Two stereo cameras are mounted in front of the harvester combine, streaming image data into the AI workflow. ("New Combine Technology," 2024).

fer, and sieve components are automatically adjusted. All these automated data-driven adjustments ensure consistent performance, reduce grain loss, and overcome human fatigue during the harvesting process.

Improvements in Yield Mapping

Advances in technology are expected to improve the accuracy and usefulness of yield maps. Modern combines equipped with real-time adjustments and continuous monitoring offer detailed and reliable data that farmers can use to make informed decisions for the upcoming year. This level of precision in yield mapping also allows for the optimization of production practices.

Advances in Sub-Header Resolution

Wider corn headers, such as the 16-row (40 ft) and 32-row (80 ft) models, record yield as a single point across the entire swath, resulting in reduced data granularity. Meanwhile, imaging satellites provide finer details with resolutions as sharp as 3 m per pixel. Future research could explore the integration of satellite data and AI to enhance sub-header yield mapping for improved field insights. This includes real-time error detection during harvest and cleaning yield data, as well as providing accurate yield estimates for individual rows instead of relying on data averaged from the combine.

Robotic Autonomy in Agriculture

Crop harvesting has the potential to be transformed by automation. The core of the autonomous system is the high-accuracy DGNSS devices, such as the John Deere StarFire receiver. The tractors equipped with mounted cameras offer a comprehensive 360° view around the tractor, accurately identifying obstacles and calculating distances. High-performance edge-compute modules and AI neural network models process image data and deliver results in under a second, facilitating safety and quick decision-making.

This autonomous harvesting technology offers both traditional and autonomous modes of operation, allowing farmers to control the harvesting equipment. Automation may enable farm managers to switch between hands-on control and fully automated operation during peak harvest seasons or when there are labor shortages. Autonomous yield harvesting may also boost efficiency and integrate smoothly with the farmer's need for control and timely decision-making.

Summary

Yield monitors are a critical component of precision management. Accurate yield monitor data requires proper maintenance and calibration of the yield monitoring components. Data collected from yield monitors can be used for a wide variety of purposes, including creating management zones and nutrient removal maps. The advancements in computer vision and artificial intelligence are becoming important for improving yield monitor data quality, increasing harvest efficiency, and enabling automated speed and machine setting adjustments. The future of yield monitors could focus on achieving subheader resolution of yield.

Study Questions

1. What sensors need to be calibrated when preparing to harvest a field?
2. What factors influence calibration constants?
3. You harvest 50,000 lb of corn from a field at 22% moisture. How many pounds and bushels of corn do you have at 15.5% moisture?
4. What will happen to the yield monitor data when the combine slows down? What happens when the combine speeds up?
5. Why does yield monitor data need to be cleaned?
6. **True/False:** Mass flow sensors on combines must be calibrated for each crop type to ensure accurate yield estimates.
7. **True/False:** Grain moisture sensors estimate moisture content by measuring the dielectric properties of the grain.
8. **True/False:** The use of radiation sensors for yield monitoring is permitted in the United States.

9. **True/False:** Yield monitor data does not need to be cleaned before being used for management decisions.
10. **True/False:** AI algorithms in modern combines can automatically adjust ground speed and harvest settings in real time.

Fill in the Blank

11. The process of adjusting sensor accuracy by harvesting and weighing multiple loads of grain is called __________.
12. The __________ sensor is typically mounted at the top of the clean grain elevator and measures the force of grain impacting a plate.
13. __________ correction is a technique used to improve the accuracy of GNSS or GPS measurements by comparing data from a mobile receiver with a stationary base station.
14. The __________ provides a link between the farmer and the sensors and is commonly referred to as the in-cab display.
15. Imagery from __________ or drones is improving our understanding of yield estimates by providing high-resolution aerial perspectives.

Multiple Choice

16. Which of the following is NOT a component of a typical grain yield monitoring system?
    a. Mass flow sensor
    b. Moisture sensor
    c. GNSS or GPS receiver
    d. Soil pH sensor
17. What is the main reason for cleaning yield monitor data before analysis?
    a. To reduce file size
    b. To remove erroneous data points caused by speed changes, partial swaths, or GPS errors
    c. To improve combine speed
    d. To calibrate the moisture sensor
18. What technology is used in John Deere's Predictive Ground Speed Automation to generate predictive field maps?
    a. Infrared sensors
    b. Forward-facing stereo cameras and preharvest satellite imagery
    c. Soil moisture probes
    d. Weigh wagons
19. Which sensor is used to estimate the volumetric flow of clean grain by detecting interruptions in a light beam?
    a. Microwave sensor
    b. Photoelectric sensor
    c. Impact plate sensor
    d. Load cell
20. What happens to yield monitor data when the combine slows down?
    a. Yield estimates decrease
    b. Yield estimates increase
    c. Yield estimates remain unchanged
    d. Data is deleted

ACKNOWLEDGMENTS

This paper was funded in part by the National Science Foundation grant numbers 2202706 and 2026431, South Dakota State University, USDA Natural Resources Conservation Service Conservation Innovation Grants (NR233A750011 G019), USDA Agriculture and Food Research Initiative, American Coalition for Ethanol, GEVO, the South Dakota Corn Utilization Council, and the South Dakota Soybean Research and Promotion Council. The use of trade names is for the convenience of the reader and does not imply endorsement by the authors.

REFERENCES

Anderson, B., & Grant, R (1993). *G93–1168 moisture testing of grain, hay, and silage*. Extension paper 1312. University of Nebraska-Lincoln. https://digitalcommons.unl.edu/extensionhist/1312.

Fausti, S., & Wang, T. (2017). Chapter 12. Cost of crop production. In D. E. Clay, S. A. Clay, & S. A. Bruggeman (Eds.), *Practical mathematics for precision farming* (pp. 191–200). ASA, CSSA, SSSA. https://doi.org/10.2134/practicalmath2017.0032.

Franzen, A., & Humburg, D. (2016). Chapter 50: Calibrating yield monitors. In D. E. Clay, C. G. Carlson, S. A. Clay, & E. Byamukama (Eds.), *iGrow corn: Best management practices. South Dakota.* South Dakota State University. https://extension.sdstate.edu/sites/default/files/2019-09/S-0003-50-Corn.pdf.

Hansen, S., Clay, S. A., Clay, D. E., Carlson, C. G., Reicks, G., Jarachi, Y., & Horvath, D. (2013). Landscape features impact on soil available water, corn biomass, and gene expression during the late vegetative stage. *The Plant Genome*, *6*(2), plantgenome2012.11.0029. https://doi.org/10.3835/plantgenome2012.11.0029.

Hawkins, E., Fulton, J., & Port, K. (2017). *Tips for calibrating grain yield monitors—Maximizing value of your yield data*. ANR-8. The Ohio State University. https://ohioline.osu.edu/factsheet/anr-8.

John Deere HarvestLab™ 3000. (2024). John Deere. https://www.deere.com/en/technology-products/precision-ag-technology/data-management/harvest-lab-constituent-sensing.

Khosla, R., & Flynn, B. (2008). Understanding and cleaning yield monitor data. In S. Logsdon, D. E. Clay, D. Moore, & T. Tsegaye (Eds.), *Soil science: Step-by-step field analysis* (pp. 113–130). ASA, CSSA, SSSA. https://doi.org/10.2136/2008.soilsciencestepbystep.c9.

Kyveryga, P. M., Mueller, T. A., & Mueller, D. S. (2018). On-farm replicated strip trials. In D. K. Shannon, D. E. Clay, & N. R. Kitchen (Eds.), *Precision agriculture basics* (1st ed., pp. 189–207). ASA, CSSA, SSSA.

Luck, J. D., & Fulton, J. P. (2014). *Best management practices for collecting accurate yield data and avoiding errors during harvest*. EC2004. University of Nebraska-Lincoln. http://extensionpublications.unl.edu/assets/pdf/ec2004.pdf.

Magalhães, P. S. G., & Cerri, D. G. P. (2007). Yield monitoring of sugar cane. *Biosystems Engineering*, *96*(1), 1–6. https://doi.org/10.1016/j.biosystemseng.2006.10.002.

Markinos, A. T., Gemtos, T. A., Pateras, D., Toulios, L., Zerva, G., & Papaeconomou, M. (2005). The influence of cotton variety in the calibration factor of a cotton yield monitor. *Operations Research*, *5*(1), 165–176. https://doi.org/10.1007/BF02944169.

Mishra, U., Clay, D., Trooien, T., Dalsted, K., Malo, D., & Carlson, C. G. (2008). Assessing the value of using a remote sensing-based evapotranspiration map in site-specific management. *Journal of Plant Nutrition*, *31*(7), 1188–1202. https://doi.org/10.1080/01904160802134491.

New combine technology – Predictive ground speed automation [Video]. (2024). John Deere. YouTube. https://www.youtube.com/watch?v=t8GTVVRtB6Q.

Reese, C. L., & Carlson, C. G. (2017). Chapter 17. Understanding grain moisture percentages and nutrient contents for precision grain management. In D. E. Clay, S. A. Clay, & S. A. Bruggeman (Eds.), *Practical mathematics for precision farming* (pp. 241–246). ASA, CSSA, SSSA. https://doi.org/10.2134/practicalmath2017.0030.

RiteYield: Yield monitor for vegetable harvesters. (2024). Greentronics. https://greentronics.com/products/riteyield.

Sudduth, K. A., Drummond, S. T., & Myers, D. B. (2019). *Yield Editor 2.0.7. United States Department of Agriculture. Model.* https://hdl.handle.net/10113/AA22668.

On-farm Replicated Strip Trials

21

Peter M. Kyveryga, Matthew E. Caroll, and
Takashi S. T. Tanaka

Chapter Purpose

For many, precision farming starts with conducting on-farm research to improve site-specific management decisions. This chapter provides an overview of how to plan, design, conduct, and analyze on-farm replicated strip trials. While encouraging farmers to follow basic principles of field experimentation—such as replication, randomization and blocking—when executing on-farm trials, it is not feasible to implement standard protocols across all participating farms because of the varying requirements of different pieces of equipment and software. Consequently, practical considerations are needed for different projects. In this chapter, specific examples are presented on how to summarize data from individual locations and on combining data from multiple locations. Digital agriculture technologies are rapidly advancing, yet the fundamental principles of on-farm research will remain the same in the future.

Precision Agriculture Technology and On-farm Research

In the past, small-plot field (5–10 ft wide and 20–50 ft long), greenhouse, and laboratory experiments were the primary methods for conducting agronomic research (Figure 21.1). These studies provided excellent information, and in many situations, the equipment scientists used was like that used on the farm. However, over time, field equipment and farms have expanded in size and capacity. Many farmers now own and operate combines equipped with yield monitors and global navigation satellite systems, as well as fertilizer applicators and sprayers capable of applying variable-rate treatments. In addition, unmanned aerial vehicles (UAVs) can capture high-resolution imagery at critical points throughout the year. These technologies allow farmers to test new products, technologies, and management practices using on-farm testing techniques (Figure 21.2). In many on-farm research fields, treatments are applied in strips across the entire field in an experimental design referred to as a strip trial (Figure 21.2). As with all experiments, they are most successful when they are replicated and based on carefully constructed questions or hypotheses. While precision agriculture technologies enable farmers to conduct on-farm studies, not all farmers are comfortable with the on-farm research process. However, university researchers, extension specialists, and industry precision-farming agronomists can assist with planning and data analysis. The current trend is to organize farmers into local groups or networks that serve as platforms for on-farm participatory research and learning. On-farm research networks offer new ways to integrate science and new technologies with farmers' personal knowledge and know-how (Chapman et al., 2016; Kyveryga, 2019; Lacoste et al., 2022). Additional information on integrating science and new tech-

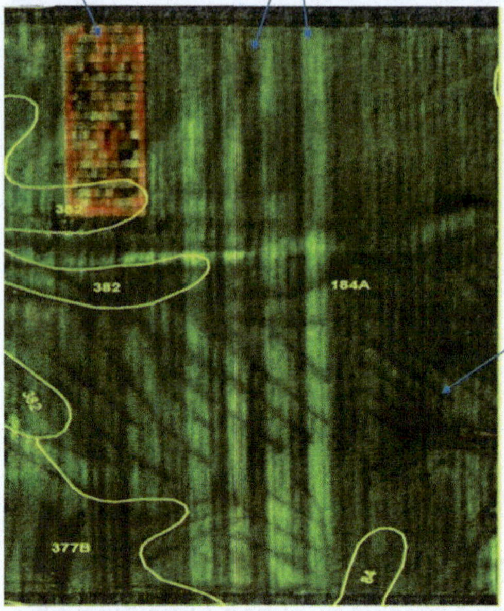

Figure 21.1. Small-plot and replicated strip-trial experiments. A small-plot experiment is shown in the upper left corner, while an on-farm replicated strip trial with two nitrogen (N) fertilizer rate treatments is shown in the center. The dark angled lines represent the location of drainage tiles, and the yellow lines denote soil map units. The dark vertical strips indicate side-dressed applications, fertilizer applied in a band onto the surface of the soil during the growing season. Note that because of plot size, the drainage tiles may affect yield variability more in the small-plot experiment than in the strip-trial experiment.

nologies into on-farm research is available in Anderson (1996), Exner & Thompson (1998), Kyveryga et al. (2018), Ramsay & Schafer (2013), and Rempel (2002).

Planning Experiments

Ask the Right Research Question

Some of the best experiments start by defining the problem and research questions. However, asking the right question usually requires doing some homework. In other words, one should begin by learning what kind of research has already been conducted. Past information can be obtained by interviewing experts and reading scientific papers and reports. Keeping the research question simple is key to success. Large complex studies are not well suited for on-farm experiments, as their experimental complexity increases the likelihood of errors and the application of multiple treatments is limited by the availability of equipment on a specific farm. When designing an experiment, it is important to know the answers to the following questions:

1. What is the question or hypothesis?
2. How many treatments and replications are required?
3. What are the specific requirements for the product to be tested?
4. What are the costs and expected results?
5. What is the available land area for the treatments?
6. What is the size of the equipment available for the study?
7. What is the time commitment for the researcher and the farmer?
8. Is there a need for additional resources to analyze data and summarize results?
9. What is the risk of yield losses in the experimental area?
10. How will inconveniences, such as the requirement to leave field headlands unplanted until the last planting date for the planting date study, affect other farm operations?

Formulate a Research Hypothesis

A research *hypothesis* is a simple statement that outlines what researchers and farmers plan to discover from their research. Two complementary hypotheses exist for research questions. The first is the *null hypothesis*, which usually states that no differences exist among treatments. The second is the *alternative hypothesis*, which contradicts the null hypothesis, stating that if the null hypothesis is rejected, then the differences could be attributed to the treatment effect (Table 21.1).

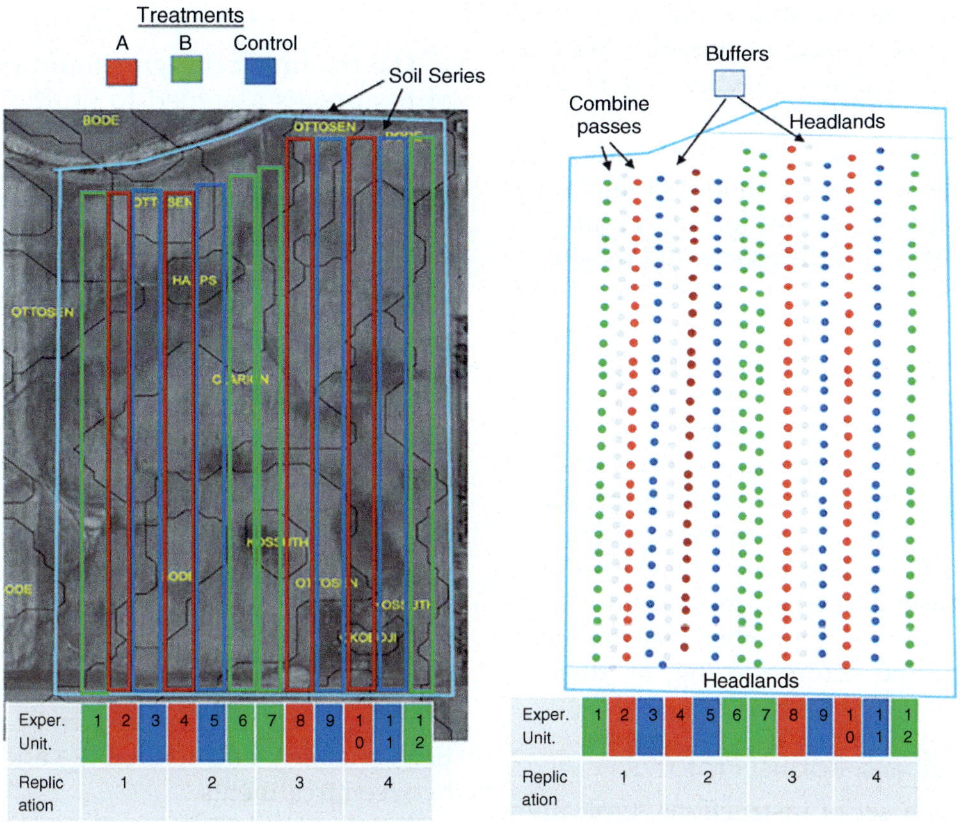

Figure 21.2. The chart on the left shows an on-farm trial with three treatments (A, B, and C) arranged in a randomized complete block design (RCBD) with four replications. The soil series map is overlaid with the aerial imagery of the soil surface. The chart on the right displays individual yield monitor observations for the three treatments.

Table 21.1. Examples of research questions and the corresponding null and alternative hypotheses for two on-farm trials. The experiments were associated with the impact of seeding and planting rates on yield.

	Seed treatment example	Planting rate example
Research question	Is there a yield difference between the two seed treatments?	Does a higher seeding rate result in greater yields than the farmer's current rate?
Null hypothesis	Both seed treatments have the same yields.	The mean yields from the high seeding rates and the farmer's current seeding rate are the same.
Alternative hypothesis	One of the seed treatments has a statistically different yield.	The average yield associated with the higher seeding rate is statistically discernible from the farmer's current seeding rate.

Selecting Treatments for a Field Experiment

Treatment is a variable of interest that is manipulated by the experimenter. Treatment selection usually follows logically from the research hypothesis, prior knowledge, or personal beliefs. For example, if you want to test whether increasing the planting rate results in higher yields, the experiment may involve two planting rates: a control rate and an adjusted planting rate. The control provides a reference for your standard practice of 31,500 corn seeds/ac (77,805 seeds/ha) or a test seeding rate of 34,500 seeds/ac (85,250 seeds/ha). Additional treatments with planting rates higher or lower than the control can also be considered. When choosing the treatments, take into account the equipment, land requirements, and cost of conducting the experiment.

Identifying the Variables to be Measured

Once the research question has been identified and the treatments selected, the most appropriate response variables to measure must be chosen. A response variable is the factor that changes in response to the treatment. For example, the response variable may include changes in soil health, soil nutrients, disease level, soil erosion, crop reflectance, vegetation indices, or yields. A response variable should be measured or collected if it is important for interpreting the study's findings or if it improves the sensitivity of the analysis. For example, collecting disease severity observations from a fungicide trial could help explain the observed yield response to the fungicide, or gathering soil nutrient information from a fertilizer experiment could also be beneficial. A common error is failing to collect data that is essential to the stated hypothesis. Some practical considerations for determining which variables to measure include ease of measurement (considering time and cost), the required accuracy and precision, and whether measurements should be repeated before, during, or after the growing season.

Develop a Robust Protocol

A protocol is a set of instructions, typically a one- to two-page document for conducting the on-farm experiment. Protocols should clearly state the objectives, specify the treatments, outline the data to be collected, provide a field map or treatment plan, and state the expectations. When developing the protocol, it is essential to highlight equipment restrictions and other considerations.

Designing the On-farm Experiment

Well-designed on-farm replicated experiments should enable testing of the research hypothesis. The treatments, often referred to as *factors*, may include different rates of fertilizers, fungicides, insecticides, herbicides, cover crops, or types of tillage. Categories of each factor are commonly referred to as *levels*. For example, fertilizer or seeding-rate experiments may contain three or more rates or levels. The treatments can be applied to plots ranging from those containing a single plant to those containing thousands of plants (Figure 21.2). For strip-trial research, treatments should be applied perpendicular to the slope, and plot sizes should correspond to the available field equipment.

Treatment Design

The treatment design should provide a map of treatments assigned to multiple experimental units (i.e., plots) within a field. There are many ways to design on-farm trial experiments. In most experimental designs, replication and randomization are critical. For statistical reasons, the experimental units or plots should be as similar as possible. Where feasible, all paired treatments should have the same size and dimensions.

A widely used experimental design for on-farm trials is the *randomized complete block design* (RCBD; Figure 21.2). Each block, often referred to as a replication, contains all the treatments. The treatments are randomly assigned to one plot within a block. Blocking is the process of grouping experimental units within a field, typically based on a gradient of soil properties, prior history, or other characteristics. The purpose of blocking is to minimize known or unknown spatial variability and to enable valid comparisons between treatments.

A *Split-plot* design is an experimental design that includes at least two treatment types, one that is difficult to change or apply, and another that is easy to modify during the experiment (Figure 21.3). For example, a trial may include substantial tillage treatments (which are difficult to change) that are split by nitrogen (N) rates (which are easy to change). Some on-farm strip trials focus on evaluating different site-specific or variable rate recommendations. Typically, these trials involve multiple rates or treatments. One example would be to compare a variable rate with a farmer's normal practice. The key point is that the variable rate treatments differ according to soil, crop canopy, topography, or management.

Often, a *buffer* is placed between the plots. Buffers are areas where treatments are not applied to prevent cross-contamination between treatments (see Figure 21.3). Cross-contamination can occur when one treatment affects the response of an adjacent treatment. An example would be pesticide applied to one treatment drifting onto an adjacent treatment.

In addition to the *RCBD* and *split-plot design*, other experimental designs can also be used for

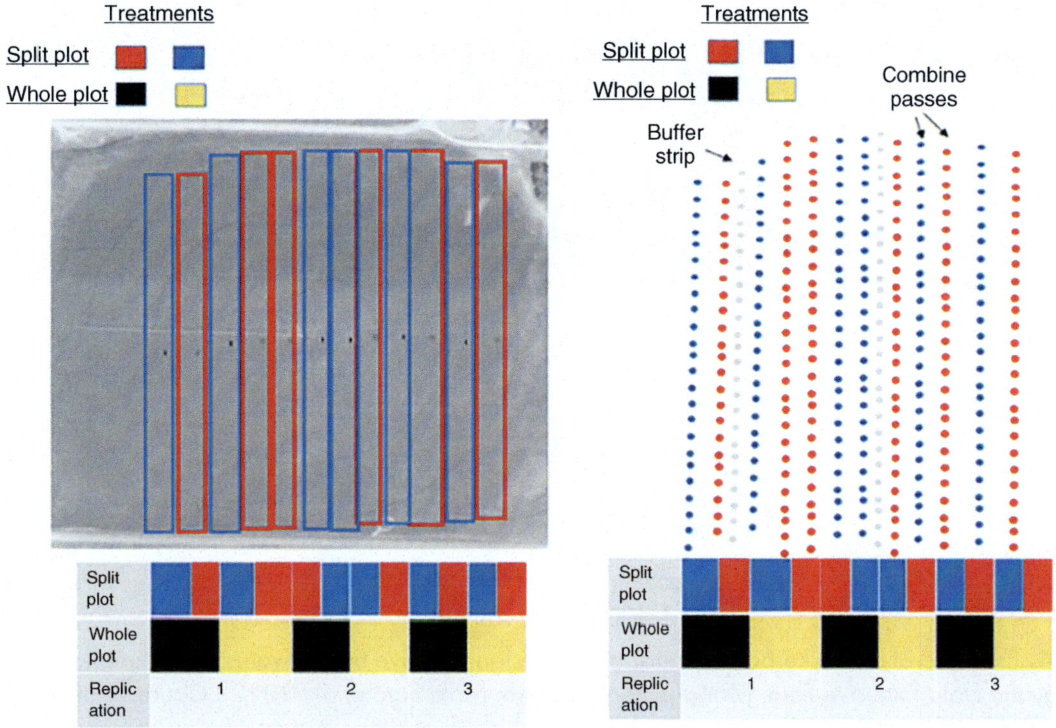

Figure 21.3. The chart on the left shows an on-farm trial with two factors and two treatment levels arranged in a split-plot design. The experiment includes three replications in a center pivot irrigated field. The whole plot treatments may involve no-till versus conventional tillage, while the split-plot treatments could focus on seeding or fertilizer rates. The split-plot treatments are nested within the whole plots as subplots. Treatments at the whole and subplot levels can be arranged in a RCBD. The chart on the right illustrates the individual yield monitor observations for each treatment.

on-farm experimentation. Examples of these other experimental designs include strip trials featuring long field strips of different treatments (Figure 21.3), a chessboard or checkerboard design using small experimental units or plots that are randomized across entire fields (Figure 21.4), and blocks of small plots located within specific areas within the fields (not shown). It is crucial to use experimental and treatment designs that best fit the research objectives or hypotheses and are practical for farmers to implement.

Replication

Replication refers to the number of times each treatment is repeated. If the experimental design is a randomized block design, then the number of blocks equals the number of replications. Replications are needed to capture variation and to perform statistical analyses. Variability in field experiments is typically attributed to some combination of bias (systematic error), noise, spatial variability, measurement errors, varying environmental conditions, equipment issues, human error, and poorly designed experiments.

Farmers often compare treatments by splitting a field into two parts, known as the "split-field," half-field design, or side-by-side method, in which one part of a field receives one treatment, and the other part receives a different treatment. Because these comparisons are made without replications, they cannot be statistically compared. For example, one side of the field receiving "Fungicide Treatment A" may differ from the other side receiving "Fungicide Treatment B" due to spatial soil variability. The resulting differences in yield may not be due to the fungicide treatments but rather, to different soil types. Replications minimize the effect of

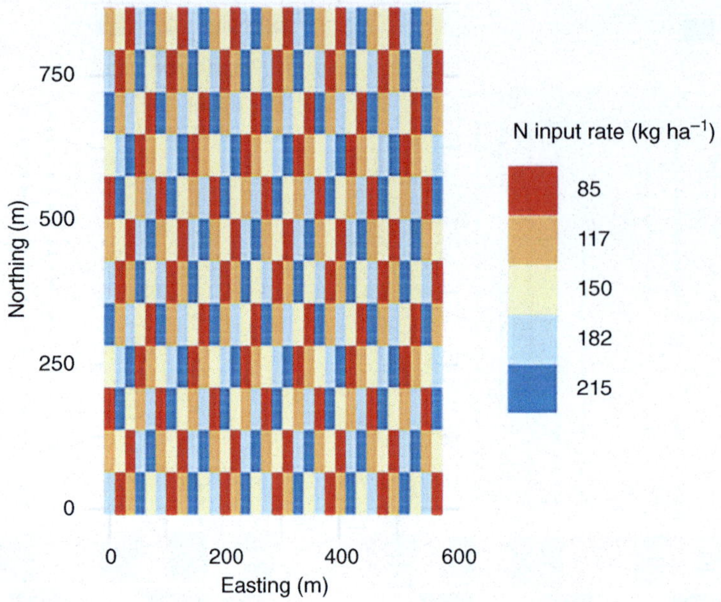

Figure 21.4. Example of a checkerboard on-farm trial design with five urea N rates. This experiment was conducted by the Data Intensive Farm Management Program (Kakimoto et al., 2022). Courtesy of University of Illinois Urbana-Champaign.

external factors that are not relevant to the study. It is highly recommended to replicate all treatments at least four times in each trial. Some types of trials require more than four replications to fully capture the entire field for the spatial analysis of yield responses.

Randomization

Randomization means that treatments are applied to each plot in a random manner, without any specific order or organization. For example, if the experimental design is a randomized complete block, then each treatment is randomly assigned to a plot within each block. In the past, randomization was assigned based on a flip of a coin or by drawing treatment labels from a hat. Today randomization software can be used to randomly assign treatments. Randomization is used to minimize bias from unknown factors that could affect the response variables being measured and to help avoid bias (Figure 21.5) from factors other than the treatments. These factors may include previous manure applications, historical field boundaries, extremely large within-field variability, nonuniform irrigation, or uneven crop residue distribution.

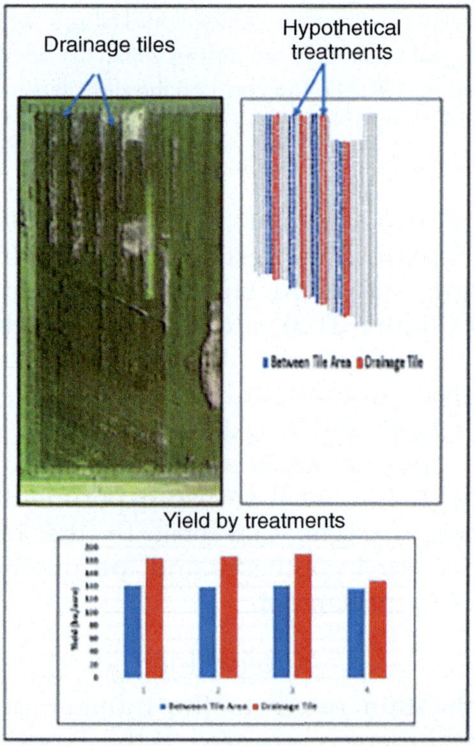

Figure 21.5. Potential bias arising from tile drainage that is aligned with the treatments. In this situation, the tile drainage lines could increase yield by removing gravitational water from the soil.

Figure 21.6. This field had high spatial variability in soil texture and soil organic matter (SOM). The center portion of the field had a sandy soil texture and low SOM. The treatment applied to this field was with or without $CaSO_4 \cdot H_2O$-gypsum. The early July aerial imagery of the corn canopy indicated a potential sulfur deficit in the center and upper right areas of the field.

Methods of Selecting Field Sites

Field selection for on-farm trials depends on the product or practice being tested. Some trials are targeted to specific geographical areas that have certain characteristics. For example, a lime experiment might be targeted to soils with low (i.e., acidic) soil pH (Figure 21.6), whereas fungicide treatments may be targeted to soils with poor drainage. Other treatments may be based on scouting reports, soils maps, or satellite imagery (Table 21.2).

A soil map unit is the basis for the soil map (Chapter 10). Each map unit is represented by a unique symbol or letter. These letters have different meanings. For example, BaA indicates that the dominant soil is a Beltsville silt loam (Ba) with a slope of between 0 and 2% (A), whereas BaB may mean that the dominant soil is a Beltsville silt loam (Ba) with a slope of between 2 and 5% (B; Brewer, 2011).

Table 21.2. Publicly Available Resources to Aid in Site Selection for On-farm Research.

Measurement	Webpage	Web address
County soil surveys	Geospatial Data Gateway—NRCS	https://gdg.sc.egov.usda.gov
	Web Soil Survey	http://websoilsurvey.sc.egov.usda.gov/App/WebSoilSurvey.aspx
Aerial or satellite imagery	Google Earth	https://earth.google.com
	National Agriculture Imagery Program	https://naip-usdaonline.hub.arcgis.com/
Topography	Digital elevation model (DEM)	https://lta.cr.usgs.gov/LIDAR

Self-generated or purchased information can also aid in site selection. This information may include historical yield monitor data, in-season aerial or satellite images, crop canopy reflectance maps, soil testing and soil fertility maps, and apparent soil electrical conductivity (EC_a) readings, as well as scouting reports on weeds, insects, or diseases. Farmers can also provide information about areas to avoid, such as old manure piles, animal confinement areas, and headlands. In addition, areas where previous on-farm trials have been conducted should be avoided.

Tools Used to Design and Implement On-farm Trials

Agronomists, researchers, and farmers can use different commercial and publicly available tools to create treatment plans that take into account field dimensions, equipment size, crop type, row spacing, field characteristics, historical field management, and many other factors (Tables 21.3 and 21.4). These tools are cloud or Geographic Information System (GIS) based and can be linked to soil maps, historical aerial imagery, and crop yield or history data. For more information and a general discussion of

Table 21.3. Common Variables and Tools or Techniques Required to Collect Data for On-farm Trials.

Variable	Tools required
Yield	Yield monitor or weigh wagon, grain moisture analyzer
Crop characteristics (e.g., stand counts, growth stages, biomass, plant height)	Field guide, tape measure, shovel for digging plant roots, visual observations, photos
Crop canopy spectral properties	Canopy sensors, chlorophyll meters, aerial imagery
Grain quality	Visual assessment, grain moisture analyzer, lab analysis
Soil test values	Hand and hydraulic soil probes and lab analysis
Disease levels	Field guide, visual assessment, lab analysis for nematodes and difficult-to-identify diseases
Insect levels	Field guide, visual assessment, sweep net, sticky cards or sheets, pheromone or pit traps
Weed types and counts	Field guide, visual assessment

these tools, see Chapters 7–9. Guidance or AB lines are used to create treatment maps, ensuring that planting, fertilization, spraying, and harvest patterns are spatially aligned within the fields.

Data Collection and Interpretation

Once the on-farm trials have been established, data needs to be collected. The most commonly collected information layers include crop yields, measures of crop quality, plant stand counts, disease ratings, crop canopy reflectance, seeding rates, emergence rates, and soil nutrient levels (Tables 21.3 and 21.4). Some factors can be controlled, while others cannot. For those that can be controlled, strategies need to be developed to ensure they do not limit yield or quality. For example, if the study focuses on planting populations and an outbreak of aphids infests the field, the entire field should be sprayed with insecticides to reduce any negative effect of the aphids on yield. For more information on analysis and interpretation, see Chapters 8 and 9.

Table 21.4. Tools to Collect Information on External Factors that May Affect an Experiment.

Factor	Tool
Climate (e.g., temperature, growing degree days, relative humidity, rainfall, soil moisture, wind speed and direction, leaf wetness, hail) Weather extremes (e.g., flood, drought, frost, excessive heat, hail)	Field-specific weather stations, rain gauges Local weather reports State-wide reports (e.g., Mesonet or Climate Corp) Weather companies (e.g., SkyBit)
Soil and topography (e.g., pH, ponding, nitrate and phosphorus concentration, erosion)	Hand probe (e.g., soil samples for nutrient analysis and compaction) Sensors (e.g., pH, EC, specific ion, temperature, moisture) Lidar data, topographical maps Visual assessment
Current and historical field management (e.g., planting and harvest dates, variety or hybrid, fertilization rates, manure history, disease and pest history, crop rotation)	Paper forms Online data collection tools and forms Historical records Personal communication
Edge of field water (e.g., tile drain flow, sediment)	Water sample from tile drainage outlets Visual assessment of sediments

Equipment for Implementing Experiments

Planters equipped with variable-rate drives, hydraulic downforce, insecticide boxes, in-furrow liquid applicators, and other technologies are well suited for seed treatments, row spacing adjustments, variety or hybrid comparisons, and experiments involving in-furrow fertilizer, chemical, or biological products. Planters equipped with individual row seed boxes, two bulk tanks, or section shutoffs for in-furrow applications are used to establish *split-planter trials*. Split-planter trials are when a farmer sets up treatment comparisons with different sections of the planter. The most common split-planter trials consist of only two treatments, but it is also possible to apply three treatments in a single planter pass. A split-planter trial is one of the easiest types of trials to conduct because once the planter is loaded, the treatments can be applied across the entire field.

Prescription planting trials can be conducted if the planter is equipped with electric or hydraulic drive capabilities. A treatment prescription can be loaded manually or accessed via the cloud on the tractor monitor to control planting rates or activate the application of products for various treatments. The prescription is sent directly from the monitor to the planter, eliminating the need for farmers to adjust seeding rates or other treatments manually.

Farmers can also conduct trials by planting alternate passes with one treatment and returning to plant the skipped passes with a second treatment. They can do this by manually turning the in-furrow applicator on or off every other pass or by manually adjusting the seeding rates. Each crop presents different challenges when it comes to setting up trials with planters. The most common challenge is that the harvesting equipment does not match the width of the treatments.

Farmers use different types of agricultural machinery and equipment to conduct on-farm trials. These include ground and areal sprayers, nutrient and manure applicators, planters, tillage equipment, and grain and forage combines, as well as irrigation and remote sensing equipment. It is critical to take into account specific characteristics of the machinery and equipment available on the farm, such as size, clearance, calibration, productivity, transportation, and payload capacity, when planning and conducting on-farm trials.

Treatment Applications with Sprayers

Applying Treatments with Ground Sprayers

Ground sprayers are used to apply foliar fungicides, insecticides, herbicides, biologicals, and both micro- and macronutrients. With some modifications, sprayers can be used to seed cover crops. Like planter trials, sprayer trials are generally easy to conduct. A common application method is to spray with or along the rows. For example, a standard recommendation is to apply an insecticide treatment to a 12-row strip alongside an untreated 12-row strip. Another recommendation is to apply the treatment in wider swaths to ensure that sufficient data can be collected from the center of the wider swath (Figure 21.7).

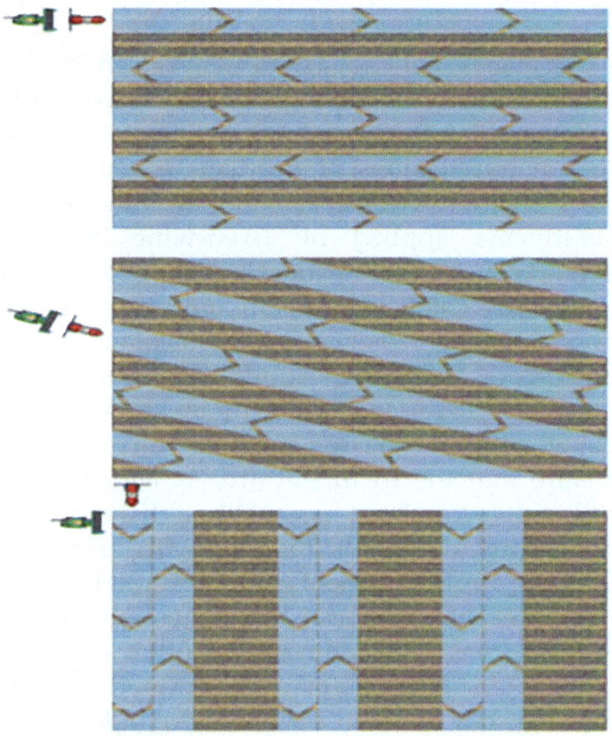

Figure 21.7. Treatment application pattern and harvesting the treatment strips using a grain combine. The upper figure shows that applications and harvesting are done along the crop rows. In the middle figure, applications and harvesting are done at an angle, while in the lower figure, applications are applied across the rows and harvesting is done along the rows.

There are several challenges associated with applying chemicals or products using sprayers. It is important to obtain and follow the labeled rates and protocol directions. For example, pesticides must be applied according to the labeled rates. Failure to adhere to the label requirements may be unlawful. Combining multiple products can save both time and money; however, it is important to ensure that the products being mixed are compatible and do not adversely affect the crops. Be sure to include the appropriate control comparison when evaluating combined products. Sometimes it may be appropriate to compare Products A+B to Product A alone instead of, or in addition to, having a true untreated control.

Using Aerial Applicators

Airplanes, helicopters, and UAVs can be used to apply many of the same products as ground sprayers. Aerial applications enable mid- to late-season applications of products when ground sprayers may not be able to get through the field. Treatments applied by fixed-wing airplanes should cover an area wider than a single combine pass across the field to ensure uniform treatment application. This is because the airplane's spray delivery systems are calibrated to ensure that each pass overlaps with the next. Moreover, using multiple passes with an airplane will ensure that the treatment area is wider than a full, single combine pass. Although helicopters offer similar benefits to fixed-wing airplanes, they can apply treatments more accurately within fields because they can maneuver more easily around trees, power lines, and other obstacles. Helicopters have a relatively limited payload capacity, but they can easily reload and refuel at the edge of the field.

Treatments applied with an aerial applicator rather than a ground sprayer or high-clearance applicator may be more difficult to apply accurately. In addition, variable winds and applicator speeds can pose challenges when applying aerial treatments. For example, if a product is applied on a windy day, the treatments may not be accurately placed (most pesticide labels specify that spraying be performed in wind conditions of 10 or 15 mph or less). To account for potential drift, wider strip swaths should be created. To avoid problems, it is important to communicate clearly with the applicator about the plot plan.

Depending on the field layout and nearby obstacles (e.g., wind turbines, trees, cell phone towers), aerial applicators may be unable to apply a product at a constant rate to the entire field. When using a fixed-wing aerial applicator, select fields with fewer obstructions and long, straight rows to reduce variability. Many aerial applicators use light-bar technology and can record their application data. To ensure that the application map data is collected and accessible, consult with the applicator before spraying.

Nutrient Applicators

Different types of nutrient applicators are used in trials designed to identify optimal fertilizer rates, the best nutrient forms, appropriate application timing, or the ideal placement of nutrients. These trials are implemented using fertilizer carts, floaters, manure applicators, toolbars, or in-season high-clearance sprayers. Using Global Positioning System (GPS) and flow meter–capable equipment with nutrient applicators is preferable.

All equipment requires calibration. For trials involving dry and liquid fertilizers, wind speed can pose challenges during the application of replicated strips. Lime or dry fertilizers should be applied on a less windy day, or wider strip swaths should be created to account for potential drift.

Spinner spreaders often have overlapping swaths. If a single pass is used for a treatment, the edges of the swath may receive lower fertilizer rates than the center of the swath. Proper calibration minimizes inconsistent product distribution.

Manure Applicators

Like nutrient applicators, different types of manure applicators can be used for on-farm research. To minimize ammonia loss due to volatilization, it is essential to inject manure into the soil and to calibrate manure application rates for accuracy. Some modern manure applicators for liquid swine or beef slurry can be equipped with near-infrared sensors to monitor the amount of applied N, phosphorus (P), potassium (K), and sulfur (S).

Tillage Implements

It is important to recognize that a tillage experiment may require several years to assess improvements in soil health and yield. Therefore, the planning should be done accordingly. Tillage treatments are typically conducted in the fall or spring, making the time window for establishing tillage trials generally wider than that for other types of trials. Tillage passes should go with the rows, if possible, to ensure one or more combine swaths. If the field is tilled at an angle, experimental units for tillage treatments should be wide enough to collect yield data.

Comparing two different tillage systems, such as vertical tillage and deep tillage, requires using different equipment. Implement width, tillage depth, machine compaction, and other factors must be taken into account to reduce or eliminate potential errors. In these experiments, it is important to ensure that implement widths are wide enough to allow full planting and harvest passes.

In addition, different tillage methods may require adjustments to the planter to properly manage residue, soil penetration, seed-to-soil contact, and closure of the seed trench. Suitable coulters or row cleaners should be used for different soil conditions within different tillage treatments. New tillage equipment should be tested, and coulters or cleaners should be adjusted before use. Sufficient weight and ground contact must be maintained on the gauge wheels to ensure good seed-to-soil contact for an even plant stand.

Remote Sensing Equipment Using Crop Canopy Sensors

Testing crop canopy sensors that measure changes in crop reflectance usually requires establishing a reference, that is, calibration or N-rich strips with slightly above-optimal N status within a field. Reference strips are generally applied before planting to allow the crop to develop canopy reflectance patterns specific to each variety or hybrid. The calibration strips are sensed before variable-rate applications. When conducting trials with crop sensors or testing other variable rate prescriptions, accurately recording nutrient application rates is essential.

Irrigation System Equipment

Common irrigation systems include center pivot (sprinkler systems), furrow, and subsurface irrigation. This chapter briefly discusses on-farm trials using this equipment; however, for a broader discussion of irrigation systems, see Chapter 16. Water regime treatments in on-farm trials using furrow and subsurface irrigation are easier to establish than those with center pivot irrigation. If on-farm trials include irrigation treatments, the irrigation schedule should be developed collaboratively with the farmer.

If a trial does not include irrigation treatments, the optimal amount and uniform distribution of water are crucial, as extreme deviations below or above the optimal water amount can affect treatment results. Excessive irrigation can lead to loss of N or other nutrient deficiencies, while applying too little water can reduce yield potential. Additionally, uneven water applications can increase variability. The amount of rainfall, nitrate-N, and salinity (EC_w) of the irrigation water should be measured in on-farm irrigation studies.

Yield Monitor Harvesting Equipment

Yield data should be collected with a properly calibrated yield monitor. Weigh wagons can also be used to measure the yields of a plot or strip. Additional information on yield monitors is available in Chapters 4 and 20.

A harvest plan should be created during the planning stage. It is important that the entire trial is harvested on the same day using the same combine to avoid yield monitor calibration differences. The combine header width should be aligned with each treatment to ensure full or complete, harvest passes. However, if this is not possible, planning and establishing wider treatment areas, along with minimizing the harvest angle, can help negate these problems. To prevent confusion, reviewing the spatial as-applied map and harvest plan before harvest is essential.

Equipment for Collecting Accurate Information

When conducting on-farm research, a key goal is to collect accurate information. Collecting accurate yield monitor data begins with cleaning the yield monitor data (Chapter 20) and paying close attention to the details. For example, aerial imagery can be used to identify anomalies such as water ponding areas, changes in varieties or hybrids, applications that do not conform to the protocol, and N skips. Further, it can pinpoint other management or equipment issues that may have affected some treatment areas but not others. If aerial photos indicate that the yield data for one treatment was affected by factors other than the treatments, consider removing the confounded data, as a third variable may cause an overlap in responses and make it difficult to determine the true effect of the treatment. For example, when testing for a response to N fertilizer, a previous manure application could be a confounding variable.

Visual observations made while scouting throughout the season can also be used for quality control. Many on-farm protocols require additional trips through the field to assess plant stand counts, disease or insect levels, weed pressure, or plant root development. Scouting trips should be used to identify potential problems.

As-applied or as-planted data for each trial recorded with a GPS-enabled monitor can be overlaid with yield data and aerial imagery using GIS software (Chapters 8 and 9). A clean-yield harvest pass refers to a pass within a single treatment using the same variety or hybrid, harvested on the same day, while holding all other factors constant except the treatment being studied. Yield observations for the headlands and the first approximately 50 ft (15.2 m) of each pass are removed to adjust for the delay in combine grain flow.

Data Analyses and Interpretations

Although there are many different tools and methods for conducting statistical analyses and summarizing information, data analysis can be a daunting task. Farmers, consultants, and even scientists alike are often frustrated by this process. The goal of data analysis is to distinguish the signal from the noise. The *signal* is the treatment effect on the measured factor, while the *noise* primarily consists of random variation or other unidentified sources of error. For example, the reflectance image of a crop plant contains reflectance information from both the plant and the soil. The reflectance image from the plant may represent the signal, while the reflectance from the soil may represent noise.

During analysis and interpretation, one expectation is to establish relationships that can be extended beyond the research plots to the entire population. The *target population* may include all possible fields or conditions within a particular geographic area or specific management practices. Key descriptive statistics for summarizing observations from data samples include the mean (average), standard deviation (spread of the data), median (midpoint of the data distribution), minimum and maximum values, and the range.

After removing outliers (data points that substantially differ from the mean or the majority of the other data points) and errors, the first step is to determine whether the response variable is continuous or categorical. Continuous variables are numeric and have an infinite number of values, while categorical variables consist of a finite number of distinct groups, such as treated versus untreated areas. The type of response variable will assist in selecting the appropriate data analysis technique.

The second step is to check data distribution. This can be done by calculating the skewness or symmetry of the distribution and kurtosis values to assess whether the data have heavy or light tails relative to a normal distribution. Plotting histograms (using a spreadsheet) or box plots (using statistical software) of the variables of interest will reveal whether the data are normally distributed and present a symmetrical bell-shaped curve (Figure 21.8). To compare treatments, it is often assumed that the data follow a normal distribution, that there is a homogeneity of variance, and that the measurements are independent.

Many yield monitor datasets are not normally distributed. Data transformation or the selection of other distributions should be considered if the data are not normally distributed, which is a common occurrence in fields with large spatial variability. Other distributions should be used for count data (non-negative integers), categorical or binary data, maximum or minimum values, and bounded data like ratios and percentages.

When analyzing data, it is best to use an appropriate statistical package such as SAS, R, or JMP. Please consult professionals about writing the correct code to obtain relevant output statistics.

Two-treatment Comparisons

When analyzing experiments with two treatments, the observed differences should undergo statistical analysis. This analysis is used to determine whether the collected data supports the null hypothesis. In the statistical analysis, the software typically reports the p values. The p value, a number between 0 and 1, indicates the probability of a statistical difference between the treatments, assuming that the null hypothesis of no significant difference is true. If the p value is extremely small, then the difference was likely caused by the treatments. A p value of 0.05 means that 95% of the time, the data support the null hypothesis. A p value of 0.5 means that 50% of the time the data supports the null hypothesis. p Values greater than 0.05 or 0.1 indicate that the measured data do not support the null hypothesis. A general classification of p values in terms of the evidence of statistically

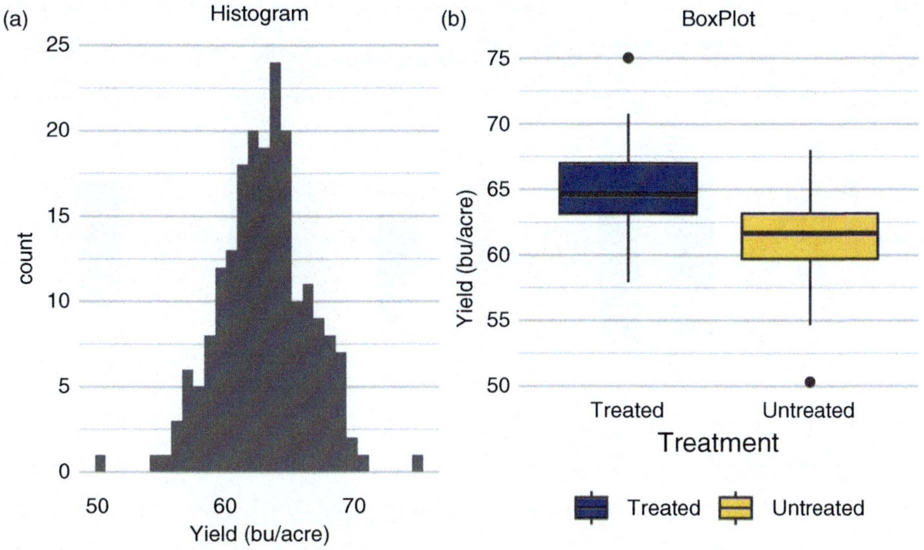

Figure 21.8. Example of a data histogram and box plot, with a box showing data between 25th and 75th percentiles.

discernible yield difference is as follows: greater than 0.10—no evidence of significant yield difference, 0.05–0.10—some evidence of significant yield difference, less than 0.05—strong evidence of significant yield difference.

If the treatments are not statistically different when, in fact, they are actually different, this is called a false negative. Conversely, if the statistical analysis indicates that the treatments are different when they are not actually different, this is called a false positive. The risk of false positives and false negatives is reduced by increasing the number of replications and by repeating the experiment.

An Example of a Nonsignificant Difference

A trial was conducted on corn with a foliar fungicide, using six treatment replications and two fungicide treatments. The two treatments were (a) fungicide-treated (red), and (b) untreated (blue; Figure 21.9). The null hypothesis stated that the yields for the two treatments were statistically similar, meaning that the difference between the two treatments was zero. The average yield response from the fungicide was 3.1 bu/ac (190 kg soybean/ha). The percentage of samples with less than and more than 3.1 bu/ac yield difference in the distribution of random noise indicates that the calculated p value was 0.36. This means that the results from two treatments were probably not different. A p value greater than 0.10 means that the yield response was not statistically significant, so the null hypothesis would not be rejected, suggesting that the yield differences may have occurred by chance.

An Example of Significant Difference

In this on-farm trial, a foliar fungicide was compared with the untreated control in a soybean field (Figure 21.10). The null hypothesis was that the yields for both treatments are similar. In this example, the observed yield response between the treated and untreated areas was 2.3 bu/ac (154 kg/ha). The percentage of samples with less than and more than 2.3 bu/ac (154 kg/ha) in the distribution of random noise is 10%, so the calculated p value is 0.10. Although the yield difference was less than the 3.1 bu/ac (207 kg/ha) shown in Figure 21.9, in this case, the difference between the two treatments was statistically significant, so the null hypothesis was rejected.

Paired t-Test for Two-Treatment Trials

An alternative to the randomization test is the t-test, which can be calculated in Microsoft Excel and is based on the same statistical logic as the randomization test mentioned above; however, the distribution of random noise is theoretically represented by a bell-shaped t-distribution. Figure 21.11 illustrates a protocol for executing a

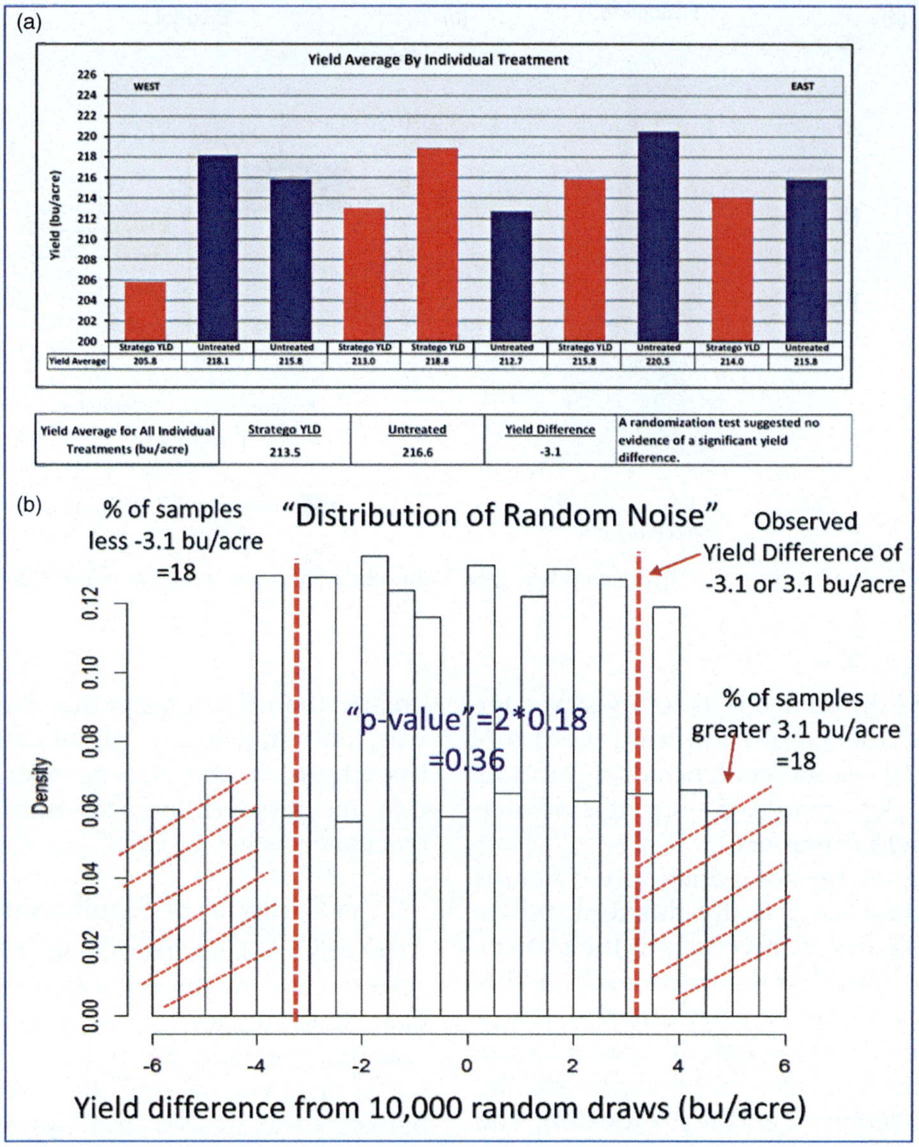

Figure 21.9. Example of detecting a significant yield difference ($\alpha = 0.10$) between two treatments using a randomization test. The *p* value is the probability of observing a yield difference as extreme as the observed yield difference (i.e., plus or minus 3.1 bu/ac) under the assumption that the null hypothesis is true. The randomization or distribution of random noise indicates potential yield differences when the null hypothesis is true, indicating that such yield differences are likely due to random chance.

paired *t*-test in a two-treatment trial using a RCBD. A paired option for the test indicates that treatments are compared in pairs within each block, thereby increasing the power of the test to detect significant yield differences.

Multiple Treatment Comparisons

The randomization test and the paired *t*-test are more difficult to use when on-farm trials have more than two treatments. In this case, a multiple treatment comparison, such as the least significant difference (LSD) test, can be used. The main idea is to generate a statistic that will indicate whether the treatment difference meets a threshold for a significant difference. When displaying these differences, a value known as LSD can be provided or two means may have different letters next to the data value, such as the letters "a,"

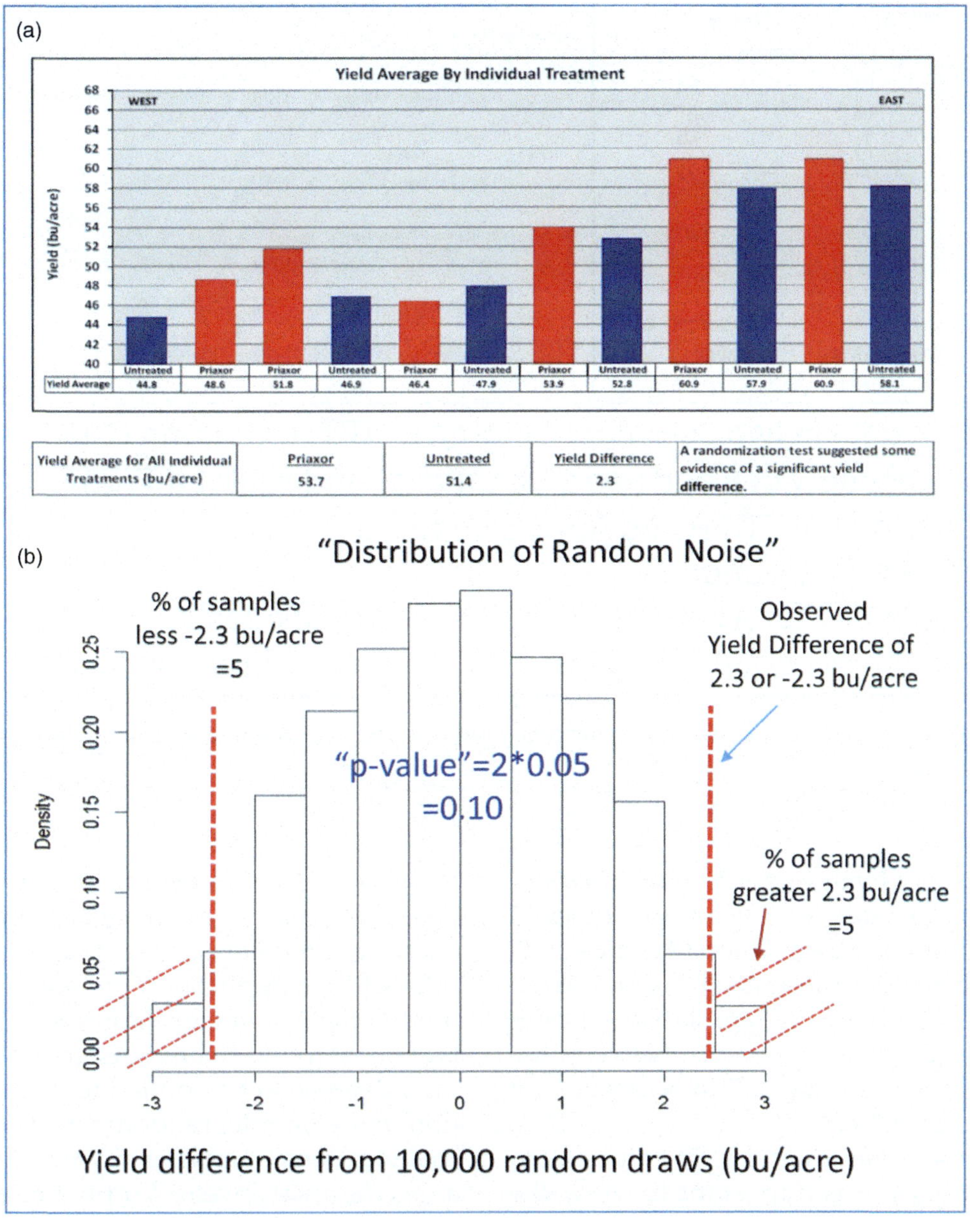

Figure 21.10. Example of detecting a nonsignificant yield difference ($\alpha = 0.10$) between two treatments using a randomization test. The *p* value is the probability of a yield difference as extreme as the observed yield difference (i.e., plus or minus 2.3 bu/ac), assuming that the null hypothesis is true. The randomization or distribution of random noise indicates potential yield differences when the null hypothesis is true; such yield differences are likely due to random chance.

"b," and "c" to indicate the differences. If the difference between the two means is greater than the LSD value, then the two means are considered statistically different.

A caution with the LSD test is that it should only be used when there is at least one pair of treatments that is significantly different (Figure 21.11). Otherwise, the test may indicate significant yield differences when none are present. Many tests, such as the LSD test and Tukey's Honest Significant Difference (HSD) test, can potentially produce different statistical inferences.

Multiple Treatment Comparison Example

Four N application rates—100, 150, 200, and 250 lb N/ac (111, 167, 223, 279 kg N/ha)—are compared in this on-farm trial (Figure 21.11). By

Replication	TRT A		TRT B
	bu/acre		
1	65		65
2	68		67
3	71		69
4	74		71
Average	69.5		68.0
	Paired t-test		Two sample equal variance
p-value	*0.103*	p-value	*0.543*

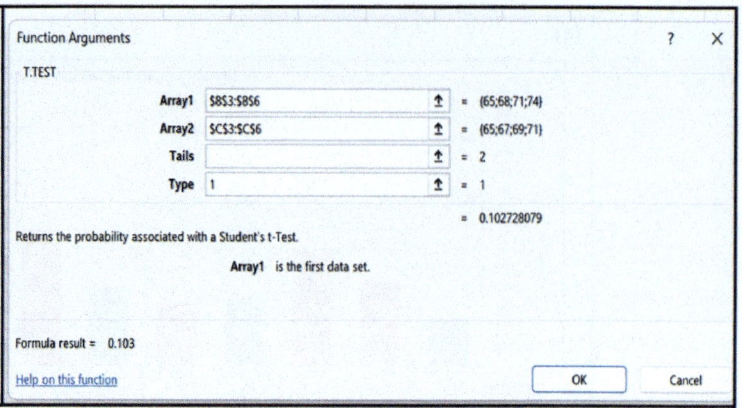

1. Put yield data in two columns: Column B -TRT A, Column C-TRT B
2. Select an empty cell and use a function f_x and select a TTEST
3. Select Array 1 as TRT A and Array 2 as TRT B
4. Tails: 2-tail distribution
5. Type: paired=1 for a complete randomized block design.

Type: Two sample equal variance=2 for other experimental designs.

Figure 21.11. A protocol to use a paired *t*-test function in Microsoft Excel when analyzing yield data from on-farm trials following RCBDs.

calculating the LSD value of 5.2 bu/ac (348 kg/ha) at a 10% significance level ($\alpha = 0.10$), the yield differences among the three lowest N rates (100, 150, and 200 lb N/ac, represented by different letters such as "a," "b," and "c") are statistically significant, while the yield difference between the two highest N rates (200 and 250 lb N/ac) are not statistically discernible.

In addition to *p* values and LSD values from the statistical tests, it is important to show the effect size (average yield differences, treatment means), confidence intervals for the means, and within- and across-treatment variability values.

Analysis of Multilocation Trials

Although analyses often focus on individual on-farm trials, a greater benefit is gained from analyzing data from multiple locations. Similar to individual trials, observations from multilocation trials can be analyzed using either a RCBD or a split-plot design (additional information is available earlier in this chapter).

During the analysis of multilocation trials, it is necessary to define the population or area of inference and determine whether the tested factors should be treated as *fixed* or *random* effects. Fixed effects are those where the treatment inferences are predetermined (or "fixed"), such as the rates of lime application or chemical application or the type of tillage. Random effects refer to those where the treatments are not fixed, allowing inference to be applied to the entire distribution. For example, random effects might include trial locations (implying any trial location within a specific area) or years (implying any year within a specific period). Whenever a field experiment involves multiple observations from experimental units, subsamples from individual plants (e.g., plant leaves, stems, roots), or individual soil cores, it is important to treat those as random effects. In statistical analyses, random effects are represented as a normal distribution with a mean of zero and a positive variance.

Multilocation Analysis Example

Multilocation on-farm trials comparing 15- and 30-in. row spacing for soybeans in Iowa were conducted (Table 21.5). The trials were analyzed separately, and paired treatment comparisons within each block were performed using a randomization test. The last column of Table 21.5

Table 21.5. Summary of On-farm Replicated Strip Trials Comparing 15- and 30-in. Soybean Row Spacing in Iowa in 2014. To Convert Bushels per Acre to Megagrams per Hectare (bu/ac to Mg/ha), Multiply Bushels per Acre by 0.06706.

Trial designation	Number of replications	Mean yield response	SD	Randomization test	
		bu/ac		*p* Value[a]	Evidence of significant difference
A	26	9.3	4.0	0.0001	Strong
B	3	1.9	1.2	0.24	No evidence
C	3	4.8	3.0	0.25	No evidence
D	3	−1.0	1.5	0.50	No evidence
E	4	1.6	1.3	0.25	No evidence
F	4	−1.1	2.1	1.0	No evidence
Pooled	—	**2.6**	**2.0**	—	—

[a] *p* Values from a randomization test; no evidence of statistically significant yield difference if $p > 0.10$, some evidence, $p = 0.01–0.10$; strong evidence, $p < 0.01$.

presents statistical inferences for testing the null hypothesis that the yield difference is equal to zero in each trial.

A more useful analysis is to express field- and across-field–level mean yield responses as random effects, using normal distributions, each with its own mean and standard deviation or variance. The 5th, 10th, 50th (median), 80th, and 90th percentiles of these yield response distributions will then be calculated. Confidence intervals that include zero or negative values indicate little evidence for a significant yield difference. In general, an 80% confidence interval is narrower than a 90% confidence interval.

An Example of Pooling Data Across Several Fields

The same trials comparing 15-in. (38 cm) and 30-in. (76 cm) soybean row spacing are analyzed by partially pooling or sharing information across trials (Table 21.6). For Trial A, there is an 80% chance that the true field-level mean yield response for 15-in. row (38 cm) spacing versus 30-in. (76 cm) row spacing will fall between 0.6 (40 kg/ha and 10th quantile) and 9.6 bu/ac (643 kg/ha and 90th quantile) and a 90% chance that it will fall between 0 and 11.1 bu/ac (744 kg/ha). When pooling information across trials, two of the six trials have 90% confidence intervals

Table 21.6. Percentiles of Distributions of Yield Responses from Six On-farm Replicated Strip Trials with Soybean Row Spacing Treatments of 15 versus 30 in. Conducted in Iowa in 2014. To Convert Bushels per Acres (bu/ac) to Kilograms per Hectare (kg/ha) Multiply Bushels per Acres (bu/ac) by 67.1. This Calculation Assumes that a Bushel of Soybeans Weighs 60 lbs.

Trial designation	Adjusted SD[a]	Adjusted yield response for different quantiles				
		5th	10th	50th	90th	95th
	bu/ac					
		Within-field level				
A[b]	3.0	0.0	0.6	3.5	9.6	11.1
B	1.0	0.1	0.4	1.7	3.1	3.4
C	2.4	−0.1	0.2	3.0	5.7	7.6
D	1.4	−2.3	−1.9	0.1	1.6	2.2
E	1.1	−0.2	0.1	1.5	3.0	3.3
F	1.7	−2.7	−2.0	0.2	2.1	2.7
		Across-field level				
	2.9	−0.6	0	1.6	4.0	5.4

[a] These summaries were estimated using hierarchical analyses, where the means and standard deviations for the two different levels were modeled as common random distributions.

[b] Trial A had hail damage during the summer, which may have affected yield response.

that do not include zero, suggesting some evidence of a discernible yield difference between treatments with 15-in. (38.1 cm) and 30-in. (76.2 cm) row spacing. However, the mean yield response of 1.6 bu/ac (107 kg/ha and 50th quantile) for the "across-field level" has considerable uncertainty, as the 90% interval includes a negative value.

Example of Spatial Analyses of Wheat Trials

This section presents an example of spatial data analysis of two-treatment on-farm trials examining the effect of fungicide applications on wheat grain yield during the 2019 and 2023 harvest seasons in the United Kingdom. Each trial included three replicated strips of fungicide application and an untreated control. The width of the individual strips ranged from 90 ft (27 m) to 150 ft (45.7 m). The yield data were cleaned by removing headlands, harvest row ends, overlapping harvest widths, and locally extreme data points (Roques et al. 2022).

Appropriate statistical analyses should be applied to evaluate the treatment effect on grain yield. Multiple yield data points from a yield monitor are aggregated as mean values within each strip or experimental unit. These aggregated values are then analyzed using conventional statistical methods, such as ordinary least squares (OLS) regression and *F* or *t*-tests. However, this approach may violate the assumption that the response variable (grain yield) is spatially independent if multiple observations within experimental units are used in the analyses. Grain yield can vary significantly across different areas, especially in large fields, and some areas within strips may be related or correlated. Consequently, spatial statistics or linear mixed-effects model (LMM) analyses are often recommended to account for potential correlation within experimental units or strips. Unlike OLS regressions, which account only for the average or fixed effect of treatments, linear mixed-effects regressions consider not only fixed treatment effects but also account for correlation within blocks, strips, or fields.

Table 21.7 compares the conventional OLS model with spatial LMM. In the 2019 trial, the yield response to fungicide application was greater than in the 2023 trial. Notably, in 2019, the conventional OLS model did not indicate a significant fungicide effect, whereas the spatial LMM estimated a mean yield value that was twice as high as that of the OLS model. This discrepancy is likely due to spatially higher yields in certain areas of the control strips, which were not accounted for by the conventional OLS. In the 2023 trial, the spatial LMM estimated a mean value that was nearly half of that estimated by the OLS model. The OLS model indicated a significant effect, whereas the spatial LMM did not.

Table 21.7. Using Conventional OLS Analysis and Spatial LMM Analysis for On-farm Trials with Wheat in the United Kingdom. To Convert Metric Tons per Hectare (t/ha) to Bushels per Acre (bu/ac), Multiply Metric Tons per Hectare (t/ha) by 19.62. This Calculation is Based on The Assumption that Each Bushel Weighs 60 lbs.

	The estimated effect of fungicide on yield (t/ha)[a]	
	Conventional OLS	Spatial LMM
2019 Trial	0.76 ± 0.42	1.50 ± 0.14*
2023 Trial	0.24 ± 0.15*	0.11 ± 0.13

[a] Estimated mean ± SD.
*Statistical significance of the treatment effect at one-sided $p < .10$.

Overall, each analytical approach produced markedly different outcomes, potentially resulting in different decisions. Implementing spatial statistics can be challenging for practitioners, as it requires specialized knowledge and substantial computational resources. Additionally, user-friendly software for such analyses is limited in availability. Therefore, consulting with statistical experts is recommended when analyzing strip-trial data using spatial analyses to ensure accurate and meaningful results.

Summarizing Data

Use Metadata and Research Databases

To interpret field data, information about the data collected is needed. *Metadata* refers to information that describes how the data or experiments were conducted, including details about calibration, the extent and severity of the problem, the timing and specifics of the applied treatments, and who performed the soil and plant analyses.

Metadata helps in interpreting the data. For example, foliar disease levels, weather conditions, plant growth stages, seeding and germination rates, estimated yields, and leaf area index can help explain yield differences between fungicide treatments. Connecting the dots in this example could lead to the development of a decision-making tool (e.g., excessive rainfall in July may contribute to more disease, which could increase the yield difference between fungicide treatments) rather than merely presenting unexplained yield observations.

Combining metadata with other layers of information and variables related to each on-farm trial can facilitate the development of research databases. Using research databases in statistical and economic analyses offers numerous potential benefits, such as identifying new research hypotheses or refining treatment effects as more data become available in the future.

Economic Analysis

In addition to statistical analysis, it is important to consider the economic and practical significance of research findings (Table 21.8). For a more detailed discussion of economics in precision agriculture, see Chapter 23. Economic considerations are important because statistically significant yield increases do not necessarily mean higher economic returns. Thus, if the yield, grain price, and input costs are known, the *break-even yield response, economic optimum rate, economic return,* or *return on investment (ROI)* values can be calculated. For an in-depth discussion on calculating the optimal economic rate of N application and seed, see Chapter 32.

In general, many inputs in crop production, including application, seed treatments, fungicides, insecticides, and micronutrients, among others, typically range between \$10 and \$30/ac (\$24.7–74/ha; Table 21.8). Keep in mind that all costs need to be factored into the analysis. For example, with a cover crop seeded by an airplane, the cost of a fixed-wing aircraft application might be \$10/ac (\$24.7/ha) and the cover crop seeds might cost \$30/ac (\$74/ha). Thus, the total cost of seeding the cover crop would be \$40/ac (\$98.8/ha). The *break-even yield response* is

Table 21.8. Estimated Costs of Inputs and Their Application in Calculating the Break-even Yield Response for On-farm Research. To Convert to Cost per Hectare (\$/ha), Multiply Cost per Acre (\$/ac) by 2.471.

Input	Cost per acre (\$)[a]
Ground application	15
Aerial application	20
Insecticide	10
Fungicide	10–15
Micronutrient	10–15
Seed treatment (insecticide plus fungicide)	10–20
Herbicide	10–20
Plant growth stimulators	10–20
Variable rate fertilizer or lime prescriptions	10–20
Cover crop seeds	15–30

[a] These are estimates. Actual values will vary from year to year and by location. Use the most accurate values possible when conducting economic analyses.

the yield value, in bushels per acre, required to cover the costs of the treatment.

Equation 1

Break-even yield response = cost of the input (in cost per acre [\$/ac])/price of unit of yield (cost per bushel [\$/bu]) [1]

- For example, the break-even yield response to cover the additional cost of \$20/ac (\$47.4/ha) in corn production is 20 (\$/ac)/4 (\$/bu) = 5 bu/ac (312 kg/ha)
    - To convert English bushels of corn per acre to metric (kilograms per hectare) assume each bushel weighs 56 lbs.
- The farmer needs to grow at least 5 bu/ac higher corn yield to justify the cost of the treatments. If corn is selling for \$46/bu, then this is \$3.3bu/ac or 4207 kg/ha.

Equation 2

The *economic return* (i.e., profit) can be estimated using Equation 2.

- Economic Return (\$/ac) = [Yield (bu/ac) × Grain Price (\$/bu)] – Input Cost (\$/ac) [2]

Return on Investment is the percentage of monetary gain in yield relative to the input cost per acre.

- ROI = 100 × Economic Return ($/ac) /Input Cost Per Acre ($/ac) [3]

ROI Example (Table 21.8):

- Two soybean row spacing (15- vs. 30-in. [38.1 vs. 76.2 cm])
- The adjusted across-field median yield response of 1.6 bu/ac (107.6 soybean kg/ha)
- The average ROI to the 15-in. row spacing was 44%
- Economic return = (1.6 bu/ac × $9/bu) – $10/ac = $4.40/ac ($10.9/ha)
- The ROI is calculated as 100 × 4.40/10, based on a $9/bu soybean price (296 kg soybean/ha yield response) and $10/ac ($24.7/ha) additional cost to use a 30-in. planter (72 cm) to drive across a field twice to plant 15-in. rows. In this scenario, the ROI will be much lower if the farmer needs to invest in a new 15-in. (38.1 cm) row planter.

It may be useful to estimate how to maximize the economic return per unit of area. To do so, the relationship between yield and input is often expressed as a production function or yield response curve. Calculating the maximum economic return per acre is then done by estimating the slope of the production function and making the slope equal to the ratio of the unit of input cost to the unit of crop price. The steeper the slope, the greater the yield response per unit of input. The *optimal rate* represents a point on the response curve at which the marginal return is equal to the marginal cost.

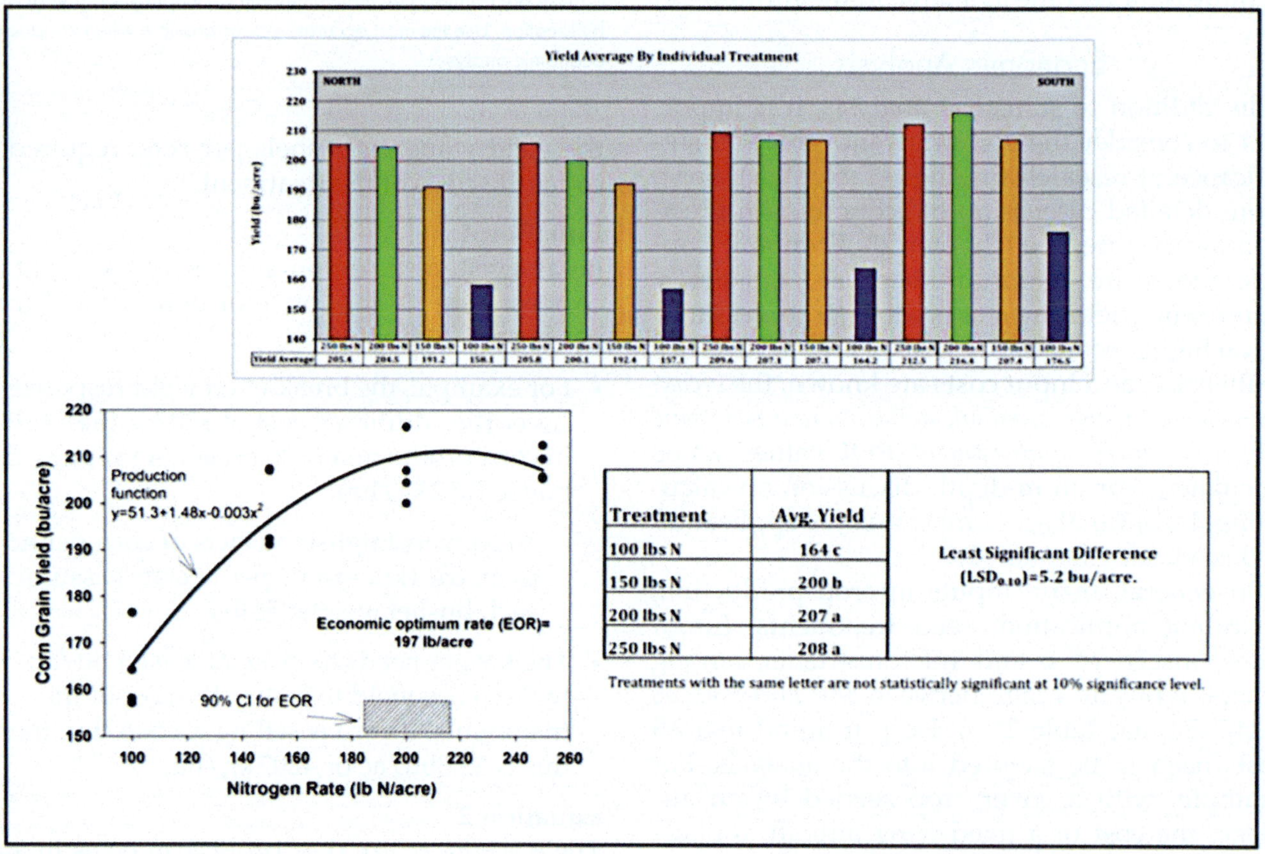

Figure 21.12. Example using the LSD test to identify significant yield differences between treatments with four N rates and estimating the economic optimal rate (EOR) of N fertilization and its confidence interval for the quadratic response function (data provided by Indiana InField Advantage, USA).

An Example of Determining the EOR from an On-farm Trial

For the on-farm trial with four N rates from Indiana, the EOR was estimated to be 197 lb. N/ac (220 kg N/ha), considering that 1 lb. (0.45 kg) of N costs \$0.50 and 1 bu of corn (25.3 kg) has a value of \$4 (Figure 21.12). A 90% confidence interval indicates that the true EOR would fall between 188 and 216 lb. N/ac (210–241 kg N/ha) at least 90% of the time.

To help farmers make better, more informed management decisions, data from multilocation on-farm trials can be used to extrapolate the observed yield responses for a broader area of interest. This type of analysis involves estimating risk, and it considers the whole distribution of potential yield responses under different soil conditions and weather scenarios (Figure 21.13).

Distribution of Yield Response Example

The predicted yield response for increasing the soybean seeding rates from 130,000 to 160,000 seeds/ac (from 321,000 to 395,000 seeds/ha) was based on 27 Iowa on-farm trials that were conducted in 2009 (Figure 21.12). The findings indicated that increasing the population could either decrease (a loss of 2 bu/ac [134 kg soybean/kg]) or increase (a gain of 4 bu/ac [268 kg soybean/ha]) yields in fields planted before and after May 20, respectively. The cumulative probability curves indicated a 60% chance of a positive economic return for soybean planted after May 20 and a 30% chance for plants planted before May 20.

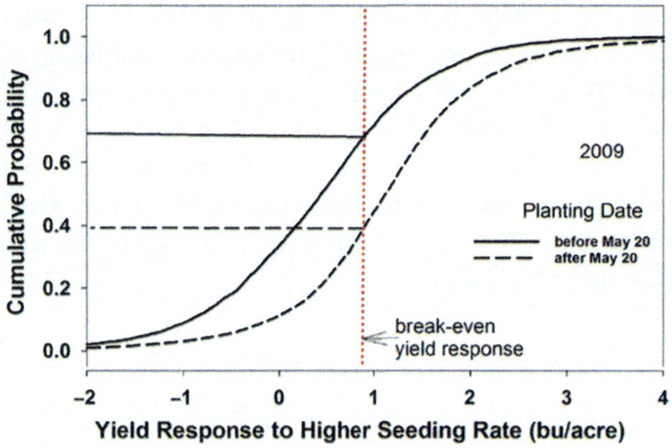

Figure 21.13. Field-level predictions of yield responses for unobserved or new situations. The cumulative probability functions were derived using yield differences estimated at 100-ft (30.5 m) grid patterns within each of 27 on-farm trial testing two soybean seeding rates, about 130,000 versus 160,000 seeds/ac (321,000 vs. 395,000 seeds/ha) with planting dates before and after May 20 across Iowa in 2009. The break-even yield response (0.9 bu/ac [60.6 kg/ha]) is represented as the red dashed line, which may shift to the right if seed costs increase or soybean prices decrease. The line can be shifted left if seed costs decrease or soybean prices increase. The two probability curves do not intersect, indicating that the curve for later planting dominates the one with early planting, suggesting strong evidence of a potential yield response with later soybean planting compared with earlier planting.

Join an On-farm Research Network

Although data collected in individual on-farm replicated trials can be valuable, joining or organizing an on-farm research network has many advantages. The most prominent advantage is increasing the ability to better summarize data. As part of a research network, data collected from individual trials will be combined with data from other similar trials, and the results will be made available. This may broaden the types of questions that can be asked, thereby increasing the statistical power to detect differences between treatments. If data are stored properly, they may be used later for even more complicated analyses.

For farmers and their advisors, localized information is critical for making better decisions. As a research network, there will be more input on the appropriate products or practices to test for increasing production.

There are other, less obvious benefits to joining or forming a research network. First, a research network can attract potential sponsors to cover treatment costs. Second, participating in a research network may increase access to existing research already collected in small-plot or greenhouse trials. While these data were not collected using the farmers' equipment, they may complement on-farm research data (Bissonnette et al., 2020, Kandel et al., 2018).

Finally, a research network provides a community for farmers and agronomists looking to improve the agronomic, environmental, and economic performance on their farms. Being part of a network allows farmers and agriculture professionals to learn not only from their own trials but also from the trial successes and failures of others.

Communication of Results in On-farm Research Networks

Table 21.9 lists several on-farm research networks or programs, along with their affiliations, websites, and the number of study locations conducted each year. These on-farm programs are associated with public universities or private grower organizations. Each program has developed unique methodologies for study protocols, data collection, and data summary. A critical part is communicating results to farmers and agronomists to aid in decision-making. For example, the on-farm program affiliated with the Iowa Soybean Association has developed a series of online, interactive decision-support tools (https://www.iasoybeans.com/what-we-do/productivity/tools). One of the interactive tools, Economic Viewer of On-Farm Studies (EVOS), enables farmers and agronomists to quickly compare agronomic and conservation practices side by side, assessing their impact on yield response and returns. The statistical summaries for this tool are derived from another online resource, Interactive Summaries of On-Farm Trials (https://analytics.iasoybeans.com/cool-apps/ISOFAST; Laurent et al., 2019), which provides summaries of individual trials as well as pooled data across each study category.

Table 21.9. Communication Methods for Selected Public On-farm Research Networks.

Group/network name and affiliation	Number of study locations per year	Official website and study results	Annual conferences/grower meetings
Data Intensive Farm Management Program, University of Illinois Urbana-Champaign	20–30	https://publish.illinois.edu/data-intensive-farm-managment	Yes
eFields On-Farm Research, The Ohio State University	25–30	https://digitalag.osu.edu/efields https://kx.osu.edu/ebarns/reports	Yes
Global On-Farm Experimentation Network	NA	https://www.gofen.org	NA
Farmers Datalab, Cornell University	NA	https://www.farmersdatalab.org	Yes
Nebraska On-Farm Research Network, University of Nebraska-Lincoln	50–100	https://resultsfinder.unl.edu	Yes
Nutrient Management Spear Program, Cornell University	20–30	http://nmsp.cals.cornell.edu/index.html	Yes
Research Center for Farming Innovation, Iowa Soybean Association	150–200	https://www.iasoybeans.com/research/results	Yes
South Dakota Soybean On-Farm Research Program	40–60	https://onfarmresearch.sdsoybean.org https://onfarmresearch.sdsoybean.org/trials	Yes
FarmPEP, ADAS, UK	200	https://farmpep.net/resource/guide-farmers-crop-trials	Yes
System-based Precision Agriculture for Sustainable Crop Production, Aarhus University, Denmark	<10	https://projects.au.dk/spears	No

Figure 21.14 presents a comparison of historical on-farm studies with four foliar fungicide products on soybeans. Fungicide application cost and grain price are provided by default in the tool, but these can be adjusted by users via a sliding rule tool or a dropdown menu. An interactive summary table and cumulative probability curves of yield responses offer basic statistical and economic summaries that are essential for decision-making for each fungicide product. More detailed information, including study objectives, experimental design, weather data, scouting data, and yield by each replication, is linked to the tool.

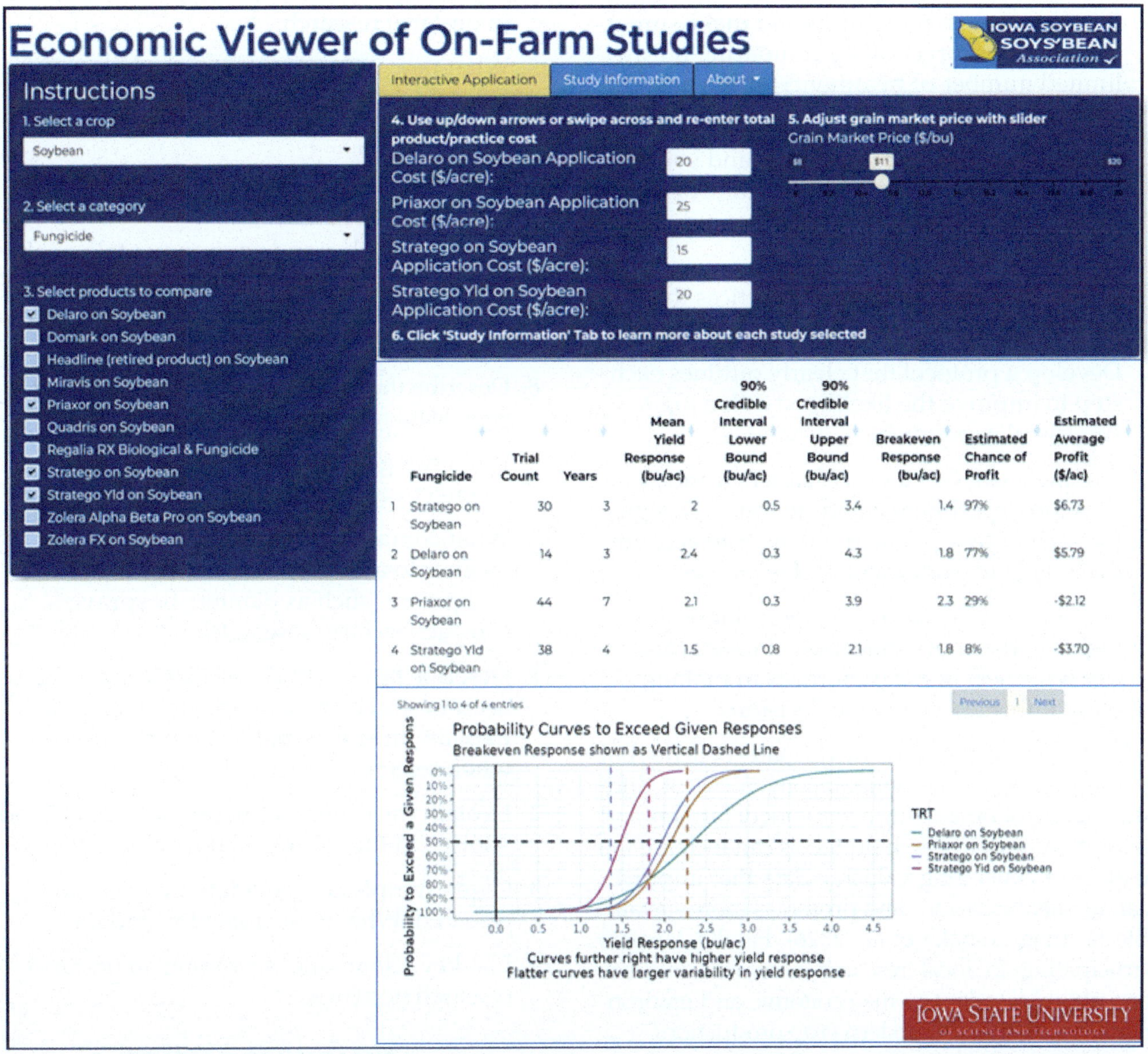

Figure 21.14. Economic Viewer of On-Farm Trial interactive tool (https://analytics.iasoybeans.com/cool-apps/EVOS) that compares different products and practices from on-farm studies in Iowa. The tool allows the user to select a crop, choose different practices, and adjust input costs and grain prices. The table and cumulative probability curves of yield response provide basic statistical and economic summaries. The input data for this tool comes from the online Interactive Strip Trial On-Farm Studies Tool (https://analytics.iasoybeans.com/cool-apps/ISOFAST; Laurent et al., 2019).

Conclusions

This chapter discussed on-farm experiments conducted by farmers using modern precision agriculture equipment and tools with assistance from researchers and local agronomists. The success of these on-farm trials largely depends on effective communication and a strong collaborative relationship among farmers, researchers, and technical providers. The keys to the success of on-farm trials are as follows:

1. Form a research hypothesis and make sure it is simple and practical by comparing only a limited number of treatments within a field.
2. Follow the principles of designed experiments by replicating treatments and using randomization, personal knowledge of within-field variability, or within-field management history.
3. Keep all other management practices the same, except for those used in treatments.
4. Develop a protocol that clearly outlines each step to improve the likelihood of having a successful experiment.
5. Use the latest GIS tools (Chapters 8 and 9) to develop experimental and treatment designs for each individual field, taking into account the available equipment and resources.
6. Use modern interactive analytical tools to improve the communication of results and assist farmers and agronomists in making more informed economic decisions.

On-farm research can provide many benefits, but at times, it can be daunting, inconvenient, or difficult. Joining or organizing a network of farmers and sharing on-farm research protocols can increase the chances of generating valuable data that improves management decisions and promotes more sustainable farming (Tanaka et al., 2026). Finally, farmers participating in local research networks increase their ability to adapt to the economic and environmental challenges of modern crop production.

Study Questions

1. List two benefits of on-farm replicated strip trials as a research tool in agronomic studies, specifically for farmers, agronomists, and researchers.
2. List key differences between on-farm trials conducted by farmers and small-plot controlled field experiments carried out by university researchers and graduate students.
3. What aspects of on-farm experiments require the most attention?
4. On-farm strip trials fall into the category of "learning by doing." List the roles of modern technologies, the internet, and social media in on-farm research.
5. What new technologies may be helpful for conducting on-farm research in the near future?
6. Why is a research hypothesis needed, and what are its key elements?
7. List key weather and environmental variables needed to interpret results from on-farm trials studying foliar fungicide applications on soybean.
8. Describe the role of aerial imagery in cleaning yield data from on-farm trials.
9. What can go wrong during the planning of an effective on-farm trial?
10. What common considerations should be taken into account when using farmers' equipment, such as planters or sprayers, to conduct on-farm research?
11. Describe how publicly available tools can be used to select on-farm research locations. Explain three external factors to consider when selecting a research site.
12. Explain the difference between "signal" and "noise" in the context of data analysis.
13. Give examples of metadata for on-farm research. What is the role of metadata?
14. List key advantages of joining an on-farm research network.

ACKNOWLEDGMENTS

This paper was funded in part by the Iowa Soybean Association, National Science Foundation grant numbers 2202706 and 2026431, South Dakota State University, USDA Natural Resources Conservation Service Conservation

Innovation Grants (NR233A750011G019), USDA-AFRI, the American Coalition for Ethanol, GEVO, the South Dakota Corn Utilization Council, and the South Dakota Soybean Research and Promotion Council. We appreciate Susie Roques (ADAS, UK) for providing the on-farm trial data used to test spatial statistics. The use of trade names is for the convenience of the reader and does not imply endorsement by the authors.

REFERENCES

Anderson, D. (1996). *On-farm research guidebook*. University of Illinois Extension.

Bissonnette, K. M., Marett, C. C., Mullaney, M. P., Gebhart, G. D., Kyveryga, P. M., Mueller, T. A., & Tylka, G. L. (2020). Effects of ILeVo seed treatment on *Heterodera glycines* reproduction and soybean yield in small-plot and strip-trial experiments in Iowa. *Plant Disease*, *104*(11), 2914–2920. https://doi.org/10.1094/PDIS-06-19-1132-RE.

Brewer, J. (2011, December 9). *What are soil map units, soil data mart, and web soil survey*. Webinar for UMD Extension.

Chapman, K., Kyveryga, P., Morris, T., & Menke, T. (2016). *Farmer network design manual: A guide for practitioners, advisors and research partners. 2015*. Environmental Defense Fund.

Exner, R., & Thompson, D. (1998). *The paired comparison: A good design for farmer-managed trials*. Practical Farmers of Iowa.

Kakimoto, S., Mieno, T., Tanaka, T. S. T., & Bullock, D. S. (2022). Causal forest approach for site-specific input management via on-farm precision experimentation. *Computers and Electronics in Agriculture*, *199*, 107164. https://doi.org/10.1016/j.compag.2022.107164.

Kandel, Y. R., Hunt, C. L., Kyveryga, P. M., Mueller, T. A., & Mueller, D. S. (2018). Differences in small plot and on-farm trials for yield response to foliar fungicide on soybean. *Plant Disease*, *102*(1), 140–145. https://doi.org/10.1094/PDIS-05-17-0697-RE.

Kyveryga, P. M. (2019). On-farm research: Experimental approaches, analytical frameworks, case studies and impact. *Agronomy Journal*, *111*(6), 2633–2635. https://doi.org/10.2134/agronj2019.11.0001.

Kyveryga, P. M., Mueller, T. S., & Mueller, D. S. (2018). On-farm replicated strip trials. In D. K. Shannon, D. E. Clay, & N. R. Kitchen (Eds.), *Precision agriculture basics* (pp. 189–207). ASA, CSSA, SSSA.

Lacoste, M., Cook, S., McNee, M., Gale, D., Ingram, J., Bellon-Maurel, V., MacMillan, T., Sylvester-Bradley, R., Kindred, D., Bramley, R., Tremblay, N., Longchamps, L., Thompson, L., Ruiz, J., García, F. O., Maxwell, B., Griffin, T., Oberthür, T., Huyghe, C., . . . Hall, A. (2022). On-farm experimentation to transform global agriculture. *Nature Food*, *3*(1), 11–18. https://doi.org/10.1038/s43016-021-00424-4.

Laurent, A., Kyveryga, P., Makowski, D., & Miguez, F. (2019). A framework for visualization and analysis of agronomic field trials from on-farm research networks. *Agronomy Journal*, *111*(6), 2712–2723. https://doi.org/10.2134/agronj2019.02.0135.

Ramsay, F., & Schafer, D. (2013). *The statistical sleuth: A course in methods of data analysis* (3th ed.). Duxbury Press.

Rempel, S. (2002). *On farm research guide*. The Garden Institute of Alberta.

Roques, S. E., Kindred, D. R., Berry, P., & Helliwell, J. (2022). Successful approaches for on-farm experimentation. *Field Crops Research*, *287*, 108651. https://doi.org/10.1016/j.fcr.2022.108651.

Tanaka, T. S. T., Colaço, A. F., Mieno, T., & Riley, S. S. (2026). Advanced data analytics for on-farm experimentation: methods, challenges, and opportunities for modern agronomic decision making. *Plant Production Science*, *29*(2), 271–292. https://doi.org/10.1080/1343943X.2026.2664854.

Precision Agriculture Data Management

22

Ali Mirzakhani Nafchi and John Fulton

Chapter Purpose

This chapter is a revision of Fulton and Port (2018) from the first edition. Precision Agriculture technologies have enabled growers, agronomists, and crop advisors to collect and analyze data from fields, machines, and other farm assets they manage. This information can be used to help make decisions about variety selection, fertilizer rates, pest management, drainage, tillage practices, planting rates, improve financial returns, and other production practices (Griffin et al., 2018). However, making these decisions depends on the ability to access data that has been stored for multiple years or may be stored on multiple company platforms. The purpose of this chapter is to provide an overview of precision agriculture data management.

Historical Perspective of Precision Agriculture Data

Since the early 1990s, with the introduction of computers, Geographic Information System (GIS), and the Global Positioning System (GPS), farmers have been creating detailed soil nutrient, crop yield, and pest management prescriptions. During this period, soil survey data were digitized, GPS accuracy improved, and remote sensing technology became widely accessible (Mulla, 2013). This allowed growers to move beyond general field observations to collect georeferenced, site-specific data while precisely controlling the application of inputs such as fertilizers, pesticides, and irrigation water. With significant advancements in computing power, cloud storage, and other technologies, digital agriculture has transformed how data is collected, stored, analyzed, and used. For example, sensors, drones, and satellite imagery can now provide real-time information on crop health, soil quality, and environmental conditions. Future projections suggest that the increasing use of smart farming will help growers make effective and profitable decisions. In addition, algorithms created using machine learning and artificial intelligence (AI) can help process this information into better decisions (Figure 22.1).

Big Data

Big Data refers to vast amounts of structured or unstructured data collected over time. This data presents many opportunities for complex analytics to extract valuable, relevant information for decision-making. Big Data is influencing health care, finance, and agriculture. Agricultural data is often referred to as Big Data because of the vast quantities of environmental data—such as yield, soil, and pest information—that can be used in decision-making (Fulton et al., 2018). For example, a 1,800-ac corn/soy farm that once used the previous year's field-average yield to set flat seeding or fertilizing rates (e.g., 34,000 seeds/ac, 200 lb. N/ac) now can stacks multiyear yield maps, grids soil electrical conductivity (EC), topography, weather, and remote sensing to build productivity zones and generate variable

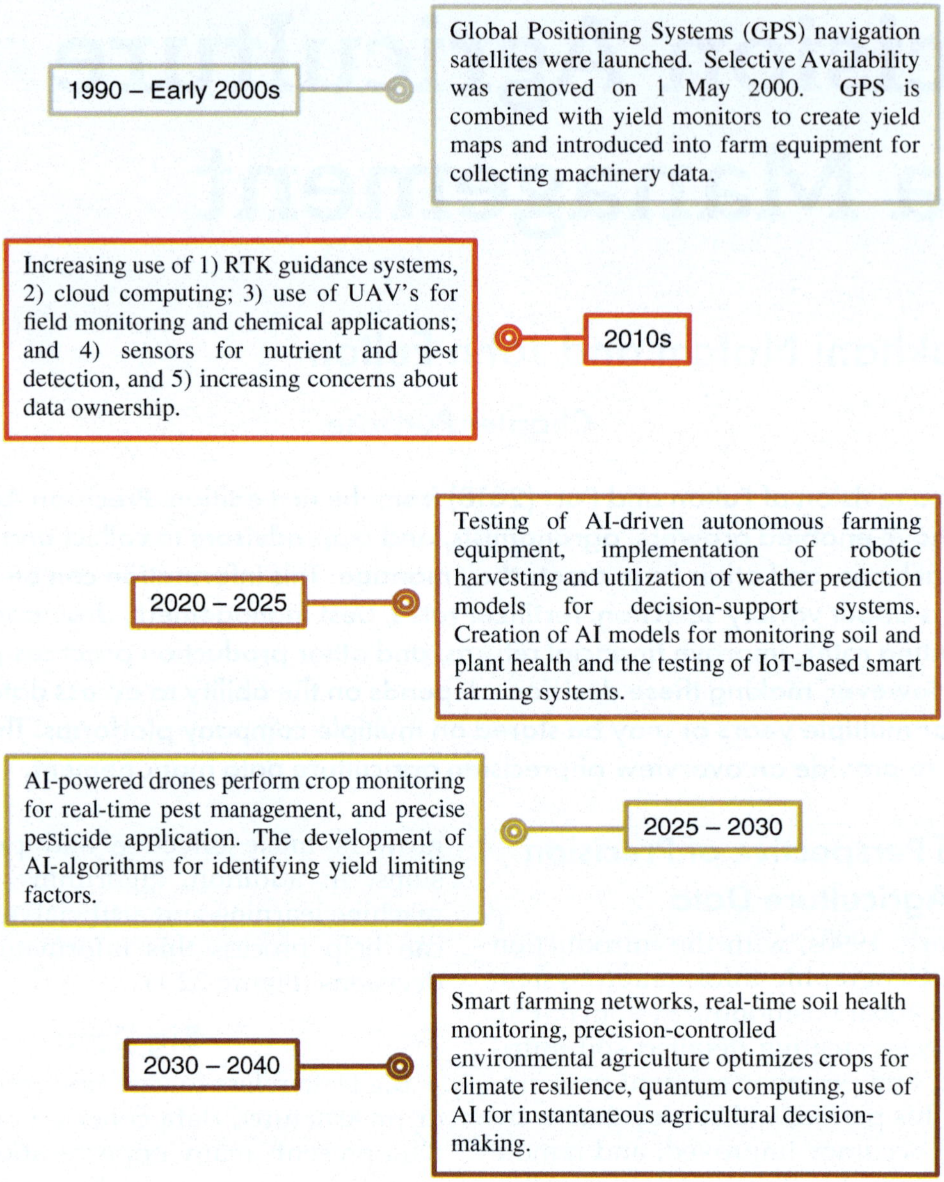

Figure 22.1. Timeline of the evolution of data management in agriculture from precision agriculture to AI-driven future agriculture. Courtesy of Ali Mirzakhani Nafchi.

rate prescriptions maps. Big Data has the potential to transform agriculture from using whole field recommendations to variable rate maps that optimize inputs and profitability while minimizing environmental impacts (Mayer-Schönberger & Cukier, 2013). Because of the importance of Big Data, data ownership and privacy concerns are widely discussed within the farming community (see the section "Type of Data Collected"). For example, the data can be used to estimate profitability and assess the effectiveness management using tools discussed in chapters 8, 9, 12, 13, 14, and 20. Unfortunately, with the vast amount of data most agricultural managers have access to, managing this information can be extremely difficult. The first step in effective management is to organize and store the data for future use. The organized and stored data can be analyzed using machine learning (ML) and AI algorithms developed for your farm (Chapter 27). While ML is an exciting development, this chapter focuses on the management of the data itself.

Types of Data Collected

Various data formats are sometimes interchangeable; however, different formats can pose issues when working across various platforms, interfaces, and software programs.

The digital data format (Table 22.1) impacts how it is stored, shared, and used (Chapters 8 and 9). Vector data is stored as a shapefile that uses the shp extension. Vector data contains points, lines, and polygons (e.g., field boundaries, roads), along with attribute tables, making them suitable for discrete objects and mapping boundaries. Raster data is stored with the TIFF [.tif/.tiff], IMG [.img], MrSID [.sid], extensions. Raster data stores contains information from grid cells or pixels (e.g., satellite imagery, yield maps, elevation), suitable for continuous surfaces and imagery. LiDAR data is stored with the las or .laz extensions. LiDAR contains dense three-dimensional (3D) point measurements (x, y, z, and often intensity) that can be used to create surface elevation maps, assess canopy structure, and conduct elevation modeling. Geodatabases (e.g., GDB [.Gdb], Personal GDB [.mdb]) are database containers that store multiple layers, attributes, and relationships in a single, structured package, helping organize complex geographic information systems (GIS) projects. Other formats (e.g., CSV [.csv], XML [.xml], NetCDF [.nc], KML/KMZ [.kml/.kmz]) are designed for sharing and interoperability. Document formats (e.g., ArcMap [.mxd], layer file [.lyr], PDF [.pdf], Word, Excel, PowerPoint) store layouts, symbology, reports, and presentations for communication.

Digital precision farming information combined with personal knowledge and on-farm testing can be used to explore and assess the implementation of precision agriculture treatments (Adamchuk et al., 2018). However, meeting this goal requires that the data be collected and archived for future reference (Figure 22.2). These data types include, but are not limited to the following:

- Non-numeric (qualitative) and numeric (quantitative) information.
- Soil survey data, providing detailed soil classifications and characteristics.

Table 22.1. Types of digital data, their formats, and extensions.

Category	Format	Extensions
Vector	Shapefile	*.shp
Raster	TIFF	*.tif, *.tiff
Raster	Arc/Info Coverage	*.e00
Raster	IMG	*.img
Raster	Layer File	*.lyr
Raster	Lidar	*.las, *.laz
Raster	MrSID	*.sid
Other	ArcMap Document	*.mxd
Other	Comma Separated Values	*.csv
Other	Extension Markup Language	*.xml
Other	File Geodatabase	*.gdb
Other	GIF	*.gif
Other	JPEG	*.jpg, *.jpeg
Other	KML	*.kml, *.kmz
Other	Microsoft Excel	*.xls
Other	Microsoft PowerPoint	*.ppt, *.pptx
Other	Microsoft Word	*.doc, *.docx
Other	Netcdf	*.nc
Other	PDF Document	*.pdf
Other	Personal Geodatabase	*.mdb
Other	PNG	*.png
Other	Rich Text Format	*.rtf
Other	Text	*.txt
Other	Windows Bitmap	*.bmp

- Crop yield maps, offering insights into variations in field productivity.
- Elevation, terrain features, and topography, aiding in water flow and erosion management.
- Soil properties, including EC, organic matter content, and moisture levels.
- Soil fertility metrics, such as nitrate, pH, phosphorus (P), potassium (K), sulfur (S), magnesium (Mg), and other essential micronutrients.
- High-resolution imagery captured via satellites, aircraft, or UAV enabling real-time field monitoring.

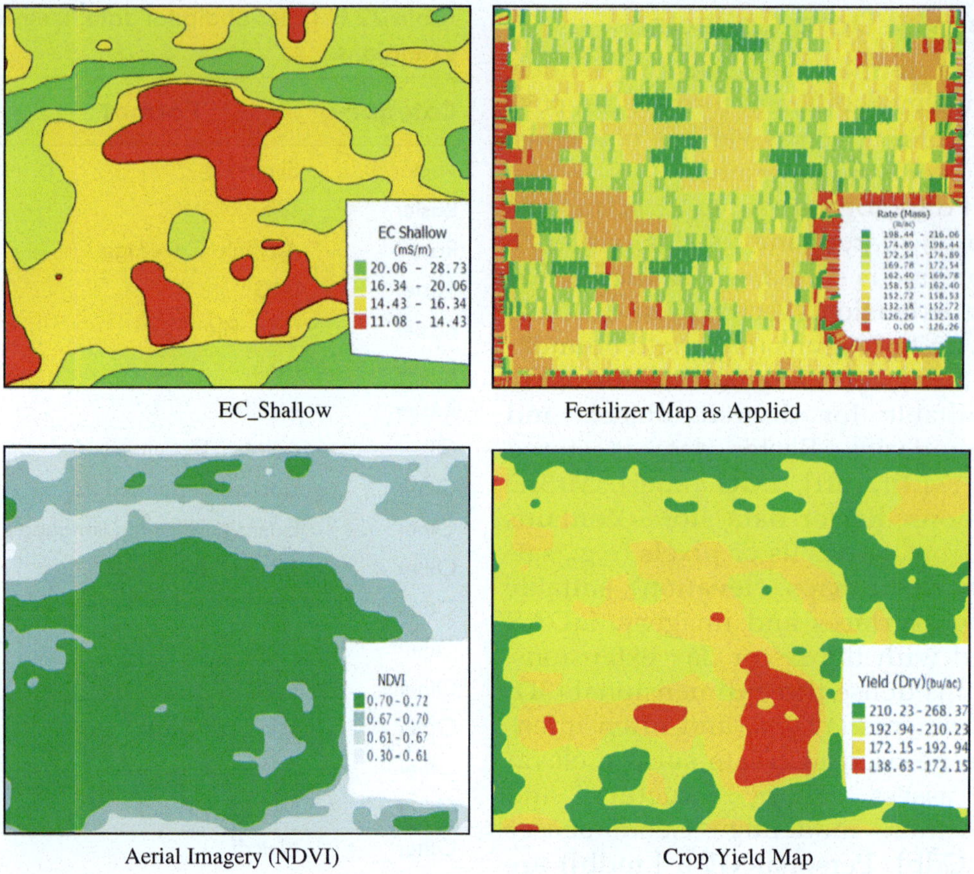

Figure 22.2. Examples of field maps collected and used by producers to manage at either the whole-field level or the site-specific level. Courtesy of Ali Mirzakhani Nafchi.

- As-applied and as-planted maps, documenting seed and input application patterns.
- Machinery performance data, tracking fuel consumption, engine load, rpm, and operational efficiency.
- Field scouting for detailing pest and disease observations for proactive management.

Managing and analyzing this information requires that it be accessed, inspected, and stored for future analysis.

As data is collected, measured, and analyzed, they are transformed into valuable information that can guide informed decision-making and strategic execution. Data management tools play a crucial role in streamlining this process. Organizing data and management methods for future use, that include using stable identifiers (e.g., F014), with standard folders organized by Year, Farm, and Field, and categorized by RAW/CLEAN/FarmData/Date/Boundaries/Soil/Planting/Fertility/Spraying Harvest/Imagery/Weather/Yield/Notes/. For these assessments, it is important that the data be accurate. Collecting accurate information requires that the field equipment be prepared for in-field operation, gcalibrated, and confirmed.

Protecting and Storing Data

Protecting and storing data for future use is a critical component of precision agriculture. The data you collect is valuable to you, but it could also be valuable to others. Growers are increasingly being contacted by third parties seeking to purchase or use their data. Especially as machine learning and AI algorithms advance, the acquisition of huge quantities of data will become even more important to third-party companies seeking to profit. It is important for growers to understand the legal issues related to data to ensure they store and protect their data effectively.

Farm data is legally unrivaled and non-excludable, meaning that a third party's access to the data does not prevent the original owner from also having access. This makes it legally challenging when the original owner of the data wants to deny another person access to their data. For example, a bushel of corn is tangible, and only one person can own a specific bushel. Therefore, if it was stolen, or given willingly, but the original owner changed their mind, there is a clear legal issue that can be quickly resolved by returning the bushel of corn to the owner. However, all forms of data, including farm data, are intangible; thus, data can be copied, sent, and used by others for profit without affecting the original owner's ability to use the data. It is helpful to equate farm data ownership to a Fair Isaac Corporation (FICO) credit score. Fair Isaac Corporation collects data on individuals and calculates their scores. Although FICO does not own the original data, it does own the newly generated credit score because it created it from the data collected (Goeringer, 2016). This results in legal issues. If the farmer no longer wants their data used by a third party, it is difficult, if not impossible, to definitively prevent the third party from using the data (Goeringer, 2016; Jouanjean et al., 2020; Kaur et al., 2022). This is further complicated by exploitative contracts and security breaches in which data are taken and used without the knowledge of the original owner.

Ensuring privacy and data ownership has become an ongoing challenge. One approach farmers can use when seeking to make third-party deals for their data while also protecting themselves is to work only with companies that follow the American Farm Bureau Federation's recommendations. The Ag Data Transparency (ADP) Evaluator (https://www.agdatatransparent.com) was created to help farmers more easily understand company data agreements. It enables them to determine whether the companies they work with uphold their data ownership and privacy rights. A company that meets the criteria set by the American Farm Bureau for data ownership and privacy rights will receive an ADP seal. In addition, farmers should become familiar with reading the terms and conditions of contracts and data storage agreements. If not stated in the contract, ensure that answers to important questions are documented in writing. This includes questions such as the following:

1. Will the data be shared with or sold to others outside this company?
2. How will the data be used?
3. Who owns the data once it has been shared?
4. Can I access the data for personal use once it has been shared?
5. How are data breaches handled?
6. Can I delete the data and void my contract?
7. Who in the company can I contact if I have additional questions about my data?

When contracting services, similar questions should be asked. For example, when submitting soil samples for analysis or hiring a crop advisor, ask who owns the data and how the data are protected. In addition, when selecting an appropriate storage mechanism (cloud, hardware, or company server), it is important to understand the security requirements. This includes using multi-factor authentication to access data stored in the cloud, using Application Program Interface (API) and software to enter or retrieve data, and considering backing up the data to a secondary source.

Some safe and reputable places for large data storage include imagery, yield maps, sensor or Internet of Things (IoT) data, and services such as Amazon S3 (AWS S3) as well as Google Cloud Storage and Archive or Coldline (for geospatial/remote sensing workflows; durable object storage). Tips include keeping a primary bucket and a secondary copy with a different provider or region, enforcing multi-factor authentication, using least-privilege access, and encrypting data with customer-managed keys where possible.

Data Sharing

Even though data sharing raises privacy and security challenges, it remains essential. Data sharing can be used to:

- Reduce duplication of data collection efforts,
- Enhance the speed and accuracy of decision-making,
- Provide new insights by combining large datasets for analysis,
- Facilitate better benchmarking for farm performance, and
- Enable collaborative problem-solving for issues such as water management and soil health.

By adopting best practices in data privacy, security, and ownership, farmers can harness the power of Big Data while safeguarding their interests. These best practices include providing only the data required for receiving services and defining who can access the data, for what purpose, for how long, and whether third-party sharing is allowed. Prohibiting secondary uses (e.g., marketing, resale, insurance or pricing models) and clarifying ownership and licensing.

Additional issues include converting information to different file formats that can be used across different software platforms. Establishing standardized protocols for data exchange between management and monitoring systems—such as farm management information systems, precision agriculture platforms, telematics or machine controllers, and sensor or IoT networks—ensures interoperability, reduces data silos, and improves decision-making across the operation. This will ensure that proper safeguards are in place to maintain data quality throughout the entire data-sharing process. Continued efforts toward transparency, clear data governance policies, and farmer-controlled data-sharing agreements are necessary to build trust and ensure the fair use of agricultural data.

Data Types

On-farm data collected by UAVs, satellites, and yield monitors, and manually by agronomists, can be stored on a local hard drive or cloud-based platforms. Cloud storage allows you to access data through an on-line network. This section will cover the different types of data an agriculture professional may encounter, including soil and plant data from soil testing laboratories, crop genetics, pesticides, seed treatments, and seeding rates. Additionally, machinery data may include equipment performance and efficiency metrics, yield monitoring and harvest analytics, and remote sensing imagery (captured by UAVs or satellites). Analyzing these data helps farmers enhance productivity, optimize resource use, improve economic returns, predict nutrient and irrigation requirements, and promote sustainable farming practices.

Agronomic Data

Agronomic data refers to digital information gathered from various agricultural operations and from monitoring soil and crop conditions at the field or sub-field level. In precision agriculture, yield maps, as-applied input maps, soil nutrient analyses, agronomists' ground-scouted information, and UAV- or space-based satellite images can be used to assess changes in soil and plant health (Figure 22.3).

These datasets typically include geospatial coordinates (latitude and longitude) linked to specific measured values. Current agronomic

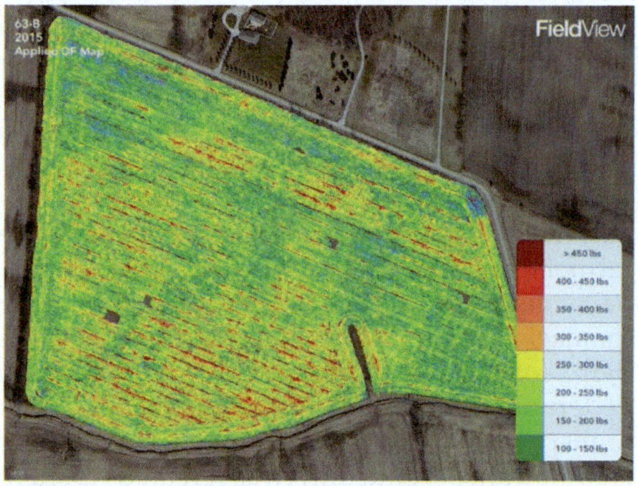

Figure 22.3. Illustration of planting downforce levels across a field. Note how soil characteristics during planting are represented in this type of as-planted data. Courtesy of John Fulton.

data integrates precision-defined management zones derived from advanced analytics, satellite imagery, and sensor technologies (Fulton & Port, 2018). Information within these zones commonly includes precise soil mapping attributes, variety selections, optimized planting populations, targeted pesticide and fertilizer application records, and detailed scouting insights. Each dataset provides unique information that can directly or indirectly help optimize variable rate applications, nutrient management strategies, and predictive analytics to enhance overall farm productivity and sustainability.

The field boundary is an important information layer that is used to identify areas within and outside the field. Farmers' and agronomists' data predominantly form operational boundaries representing the cultivated area within a field and are frequently collected using satellite imagery, UAVs, or farm machinery equipped with Real-Time Kinematic–GPS systems, providing high spatial accuracy (Chapter 7 and 9). Operational boundaries are commonly used to define precise acreage or hectares for farm management decisions and are typically created by integrating soil maps, satellite imagery, UAV imagery, and sensor-generated data layers. Other boundaries may be associated with legal land descriptions (legal parcels), USDA Common Land Units (CLUs), crop zones, and management zones (Clay, Kitchen, et al., 2017). The legal parcel represents the recorded area and is primarily used for tax assessments, real estate transactions, and legal ownership identification, and it often differs considerably from the operational area due to the inclusion of non-farmed land, woods, fence lines, or other landscape features. The USDA maintains CLUs, which represent the smallest units of land ownership or operational control eligible for enrollment in government programs, such as those administered by the Natural Resources Conservation Service (NRCS) or the Farm Service Agency (FSA).

Machine Data

Machine data is any data generated or collected by a machine. This includes, but is not limited to, autonomous tractors, harvesters, UAVs, robotic sprayers, fertilizer spreaders, and precision irrigation systems. Systems that collect digital information combine location data (measured with GPS) with actual measurements, including variable rate seeding and application maps, fuel consumption metrics, engine performance (rpm, torque, and engine load), ground speed, and equipment operational status. Data extraction and real-time analytics are often achieved through cloud-connected platforms, wireless telemetry, and the standardization of Controller Area Network (CAN) protocols.

Production Data

Production data from previous years may include field notes; cultivar and their success at different locations; seed treatments; historical yield records; crop scouting data; localized weather patterns, rainfall, and growing degree days; pest and disease records; input application timelines; and planting and harvest dates. This information is valuable and should be used when making decisions for the upcoming year.

Store and organize this production data for future use. This should include information about the cultivar and variety, as well as its associated growth requirements and traits. In addition, seed treatments, planting dates, soil conditions at planting, and soil test results should be recorded. Seed traits, for example, can be recorded and stored to provide many useful characteristics, including pest protection, herbicide tolerance, improved nutrient uptake, protection against iron deficiency chlorosis, and drought resistance. Before storing the data, it is recommended to organize the data by rating the success of the management strategy and indicating whether your expectations were met.

Real-Time Soil and Plant Sensor Data

Soil and plant sensors (discussed in Chapters 11, 12, and 13) can collect information at fixed or mobile locations and at various temporal scales. This information can be used for a variety of purposes, including determining when to irrigate, apply crop protection products, or add fertilizer. Depending on what data are collected and how, they can be used to develop prescriptive agriculture that targets

Table 22.2. Sample strategy for organizing seed trait data.

Management strategy	Data stored	Expectation	Did the strategy meet expectations?
Corn: Heat-tolerance	Variety: Hybrid name Specifications: tolerance and stability Seed treatment: fungicide + insecticide Planting date: 4/15/2025 Seeding rate: 34,000/ac Row spacing: 30 inch Previous crop: Soybean Tillage: Strip-till Soil at planting: 50 °F Soil type: Silt loam Soil test: pH 6.4; OM 3.1%; P 25 ppm; K 150 ppm Fertility program: 180 lb N/ac total; P and K per soil test recommendation Yield: 198 bu/ac @15.5% moisture	Maintain yield and reduce stress symptoms (leaf rolling, premature senescence) during sustained heat (>95 °F), with no yield penalty vs. adapted checks	Yes (6 bu/ac above non-trait check; reduced visible stress during heat window)
Soybean: Herbicide tolerance	Variety name Specifications: Herbicide tolerance system; label-compliant over-the-top applications Seed treatment: fungicide, inoculant Planting date: 5/10/2025 Seeding rate: 140,000/ac Row spacing: 15 inch Previous crop: Corn Soil at planting: 60 °F Soil test: pH 6.7; OM 3.8%; P 30 ppm; K 170 ppm Herbicide program: PRE, POST (products, rates, and timings logged) Weed spectrum: Waterhemp, foxtail pressure high In-season notes: Excellent crop safety; minor drift symptom at field edge from neighbor application Yield: 62 bu/ac 13% MC Dockage/quality: 0.8% FM	Achieve season-long control of resistant waterhemp with acceptable crop safety and fewer rescue passes	Yes (low escapes; no rescue pass needed)

Abbreviations: FM = foreign material; MC = moisture content; OM = organic matter; POST = postemergence; PRE = preemergence.

fertilizers, pesticides, and water to site-specific locations. Common technologies in prescriptive agriculture include soil and plant sensors, Global Navigation Satellite Systems (GNSS), IoT-based decision algorithms, and variable rate technologies. Information collected by the soil and plant sensors should be stored in the field database.

Remote Sensing Data

Remote sensing data from various sources, including satellites, UAVs, and aircraft, are available for precision agriculture (Ferguson & Rundquist, 2018). Details about remote sensing can be found in Chapters 14 and 15. For example, agricultural data mining extracts actionable information from large farm datasets—such as sensor, satellite, equipment, soil, weather, and market data—to reveal patterns and make predictions that improve decision-making, including boosting yield and profitability, optimizing input applications, and supporting sustainability goals.

Highlights of Data Use in Precision Agriculture

From the 1990s through 2010, the U.S. agriculture industry primarily relied on soil maps, yield maps, soil EC maps, and occasionally elevation maps to guide site-specific management. Grid or zone-precision soil sampling provided a method to evaluate variability in fertility and pH, which in turn informed the creation of prescription maps for variable rate technology applications. Many growers benefited from grid and zone sampling because it identified areas within fields that required varying intensities of management, regardless of soil type. Using data gathered through grid sampling, management zones were developed to optimize placement and improve management practices (Clay, Brase, et al., 2017). Management zone maps were then used to generate prescription, or "Rx," maps (Figure 22.4).

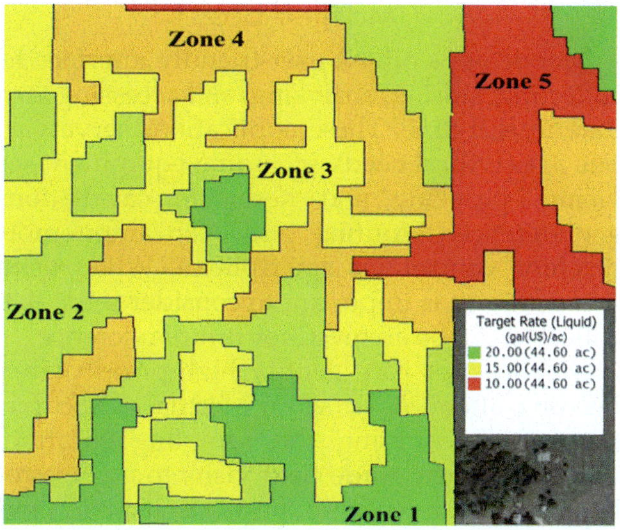

Figure 22.4. A Liquid N prescription (Rx) map (2022) outlining two management zones, each with three different amounts applied to each zone. Courtesy of Ali Mirzakhani Nafchi.

By 2010, these developments led operators to rely on intuitive digital interfaces and analytics platforms to interpret site-specific data streams, enabling real-time adjustments that maximize crop yield and resource efficiency. The driver's primary responsibility shifted toward monitoring equipment and making data-driven management decisions to maximize the productivity and profitability of the operation. This shift required skilled labor proficient in precision agriculture technologies, including those capable of maintaining and calibrating equipment, as well as those able to analyze the data collected to make decisions (Erickson et al., 2018). Technical colleges and 4-year degree institutions continue to focus on equipping students with the tools needed for success in precision agriculture, both on their own farms and in industry. This shift relied on data fluency, systems understanding, the ability to integrate agronomy with digital tools and demand for skills in precision equipment setup and troubleshooting, and the capacity to interpret analytics, quantify trade-offs, and make timely decisions under uncertainty.

Because precision agriculture has reduced required inputs and maximized resource use, there has been significant improvement in—and documentation of—sustainability. As of 2025, public awareness of how agriculture can benefit the environment through various management and soil health practices is mixed. Through the collection and use of data, agricultural professionals and researchers have a unique opportunity to show the public and policymakers how precision technologies can reduce the impact of agriculture on the environment (Chapter 25). For example, years of data tracking the adoption of reduced tillage in the U.S. Northern Great Plains Region have shown an increasing average rate of soil organic carbon (SOC) accumulation in the surface soils, a rate of 460 kg SOC / (ha per year), due to reduced soil disturbance, gradual crop increases, and widespread adoption of reduced or no-till practices (Clay et al., 2012). Policymakers must base their decisions on evidence. With the large quantities of data collected every day, the agriculture community can use this information to encourage policymakers to make decisions that will enhance growers' ability to improve their land, resulting in greater sustainability and productivity for all.

Mobile Apps

Mobile apps offer a user-friendly interface for collecting, storing, analyzing, and accessing farm data in real-time. These applications serve various agricultural needs, including equipment servicing, pesticide and herbicide calculations, georeferenced scouting, and even autonomous machine control via smartphone. When using these apps, it is important to consider if the data collected can be archived for future reference.

Mobile apps can identify high-priority areas needing attention by setting virtual GNSS pins (markers identifying the sampling location), which can then guide field visits to assess crop health, disease presence, pest activity, and environmental damage. Specialized farm apps include spray tank calculators that assist with precise chemical applications and budgeting tools that help estimate operational costs for specific machinery and farm operations. Additionally, marketplace apps offer access to used equipment sales and auctions, enabling more strategic business decisions. As technology advances, mobile applications will continue to be a cornerstone of modern farming.

Conclusion

Modern precision agriculture now generates large volumes of diverse data, from soil and crop measurements to machine performance and remote sensing, which must be effectively managed, organized, stored, and protected. Proper data management transforms this information into actionable insights that enhance decision-making, efficiency, profitability, and sustainability. The key elements of data stewardship include understanding data types and formats, storage systems, data accuracy, and safeguarding ownership, privacy, and security. In contrast, challenges can include data sharing, system interoperability, and the legal complexities of digital data ownership. As agriculture embraces advanced analytics, machine learning, AI, and cloud platforms, practitioners will increasingly need strong data literacy skills. By applying best practices in data management, growers and advisors can leverage data to advance precision agriculture and promote a sustainable, data-driven future.

Study Questions

1. Identify at least three types of georeferenced data commonly collected at the field level in precision agriculture today.
2. Data records can be categorized as either __________ or __________.
3. Provide four ways in which precision agriculture tools and data analytics can directly benefit farmers or agricultural consultants in current practice.
4. In which decade were computer-generated maps and the Global Positioning System (GPS) first widely integrated into agricultural practices?
5. Which two key technological advances have driven the transition toward modern "digital agriculture"?
6. Define the type of field boundary most commonly used in precision agriculture data management and explain what it represents.
7. List the four primary classifications of agricultural data and provide a contemporary example for each.
8. Identify four practical applications of remote sensing imagery in modern precision agriculture.
9. Explain the primary purpose of conducting multiyear data analyses in the context of precision agriculture.
10. Describe the purpose of data mining.
11. Discuss the importance of data compatibility and explain why data compatibility is slow.
12. List and briefly define each component of digital agriculture as it is currently understood.
13. Identify two advantages and two disadvantages associated with the practice of data sharing in agriculture today.

ACKNOWLEDGMENTS

Support for this document was provided by the Precision Farming Systems community in the American Society of Agronomy, South Dakota State University, USDA Natural Resources Conservation Service Conservation Innovation Grants (NR233A750011G019), American Coalition for Ethanol, the South Dakota Corn Utilization Council, the South Dakota Soybean Research and Promotion Council, and USDA National Institute of Food and Agriculture (Multistate Project S1090) Program under (7008110). The use of trade names is for the convenience of the reader and does not imply endorsement by the authors.

REFERENCES

Adamchuk, V., Ji, W., Rossel, R. V., Gebbers, R., & Trembley, N. (2018). Chapter 9: Proximal soil and plant sensing. In D. K. Shannon, D. E. Clay, & N. E. Kitchen (Eds.), *Precision agriculture basics* (pp. 119–140). ASA, CSSA, SSSA.

Clay, D. E., Chang, J., Clay, S. A., Stone, J., Gelderman, R. H., Carlson, G. C., Reitsma, K., Jones, M., Janssen, L., & Schumacher, T. (2012). Corn yields and no-tillage affects carbon sequestration and carbon footprints. *Agronomy Journal*, *104*(3), 763–770. https://doi.org/10.2134/agronj2011.0353.

Clay, D. E., Brase, T. A., & Reicks, G. (2017). Chapter 4: Mathematics of latitude and longitude. In D. E. Clay, S. A. Clay, & S. A. Bruggeman (Eds.), *Practical mathematics for precision farming* (pp. 65–73). ASA, CSSA, SSSA.

Clay, D. E., Kitchen, N. R., Byamukama, E., & Bruggeman, S. A. (2017). Chapter 7: Calculations supporting management zones. In D. E. Clay, S. A. Clay, & S. A. Bruggeman (Eds.), *Practical mathematics for precision farming* (pp. 123–135). ASA, CSSA, SSSA.

Erickson, B., Fausti, S., Clay, D. E., & Clay, S. A. (2018). Knowledge skills, and ability of precision agriculture: An industry survey. *Natural Sciences Education*, *47*, 1–11.

Ferguson, R., & Rundquist, D. (2018). Chapter 8: Remote sensing for site-specific crop management. In D. K. Shannon, D. E. Clay, & N. R. Kitchen (Eds.), *Precision agriculture basics* (pp. 103–117). ASA, CSSA, SSSA.

Fulton, J., Hawkins, E., Taylor, R., & Franzen, A. (2018). Chapter 5: Yield monitoring and mapping. In D. K. Shannon, D. E. Clay, & N. R. Kitchen (Eds.), *Precision agriculture basics* (pp. 63–77). ASA, CSSA, SSSA.

Fulton, J. P., & Port, J. (2018). Chapter 12: Precision agriculture data management. In D. K. Shannon, D. E. Clay, & N. R. Kitchen (Eds.), *Precision agriculture basics* (pp. 169–187). ASA, CSSA, SSSA.

Goeringer, P. (2016). *Farm data: Ownership and protections*. https://aglawpaul.medium.com/farm-data-ownership-andprotection-b872441153ba.

Griffin, T. W., Shockley, J. M., & Mark, T. B. (2018). Chapter 15: Economics of precision farming. In D. K. Shannon, D. E. Clay, & N. R. Kitchen (Eds.), *Precision agriculture basics* (pp. 221–230). ASA, CSSA, SSSA.

Jouanjean, M.-A., Casalini, F., Wiseman, L., & Gray, E. (2020). Issues around data governance in the digital transformation of agriculture: The farmers' perspective. *OECD Food, Agriculture and Fisheries Papers*, *146*. https://doi.org/10.1787/53ecf2ab-en.

Kaur, J., Hazrati Fard, S. M., Amiri-Zarandi, M., & Dara, R. (2022). Protecting farmers' data privacy and confidentiality: Recommendations and considerations. *Frontiers in Sustainable Food Systems*, *6*, 903230. https://doi.org/https://doi.org/10.3389/fsufs.2022.903230.

Mayer-Schönberger, V., & Cukier, K. (2013). *Big Data: A revolution that will transform how we live, work, and think*. Mariner Books, Houghton Mifflin Harcourt.

Mulla, D. J. (2013). Twenty five years of remote sensing in precision agriculture: Key advances and remaining knowledge gaps. *Biosystems Engineering*, *114*(4), 358–371. https://doi.org/10.1016/j.biosystemseng.2012.08.009.

Economics of Precision Agriculture

23

Tong Wang, Akinsola Oyebanji, and David Clay

Chapter Purpose

Driven by the promise of improved efficiency, reduced input waste, and greater control over field-level decision-making, the use of Global Positioning System–guided machinery, variable-rate input application, remote sensing, and real-time field monitoring is being integrated into commercial farm operations worldwide. Adopting these technologies is not inexpensive, and therefore, economic evaluation needs to be conducted prior to making a change. The financial returns from precision agriculture are often highly context dependent, varying across farm size, crop type, management intensity, local infrastructure, etc. This chapter will look at how to evaluate the economic returns of precision agriculture technologies using tools such as partial budgeting, cost–benefit analysis, net present value, internal rate of return, and the payback period. This chapter is an update and complement of the original chapter by Griffin et al. (2018).

Benefits and Efficiency of Precision Agriculture

Precision agriculture has helped farmers shift from using uniform field treatment to area-specific management, which has the potential to improve economic and environmental health. For example, by using variable-rate fertilizer application, growers can apply nitrogen (N) only to the zones that need it. The use of precision agriculture not only reduces unnecessary input costs but also increases the likelihood of achieving consistent, optimized yields. For farmers, tools like Global Positioning System (GPS)–guided tractors, variable-rate applicators, remote sensors, and yield monitors enable them to make precise decisions at the subfield level.

Agricultural production efficiency can be classified into technical, allocative, and economic efficiency (Watkins et al., 2014). Precision agriculture technologies enhance all three aspects by enabling farmers to use inputs more efficiently (technical efficiency), direct inputs toward areas with the highest returns (allocative efficiency), and increase profit per unit of investment (economic efficiency). For example, auto-steering systems reduce the number of overlapping passes required during operations such as planting, spraying, and harvesting. This leads to significant savings on fuel, seeds, and agrochemicals, while also minimizing fatigue among farmers and operators and enhancing labor efficiency.

The adoption of GPS-based systems can improve technical and allocative efficiency, resulting in a 6.3% reduction in fuel consumption and a reported savings of $733.85 per farm on an average farm size of 1,241 ac (502 ha; Bora et al., 2012). Yield maps and soil maps are valuable for improving technical efficiency, as they provide detailed information about field performance and spatial variability. Yield maps have been shown to increase technical efficiency by 1.1–7.2%, while soil maps contribute an additional 0.4–2.3% (Mosheim et al., 2021). Variable Rate Technology enables farmers to adjust input levels according to the spatial variability of the field, which can reduce fertilizer use by 15–25% while increasing yields by 10% (Farmonaut, 2025).

Economic Evaluation of Precision Agriculture

Despite the growing availability and economic potential of precision agriculture technologies, many producers remain uncertain about their economic value. Most farmers perceive the high investment cost for precision agriculture technologies to be a barrier to adoption (McFadden et al., 2023; Wang, Jin, & Sieverding, 2023). A survey conducted among producers in the Midwestern United States revealed that a considerable proportion lack a clear understanding of how various precision agriculture technologies can influence farm profitability (Wang & Cheye, 2022). This uncertainty is especially pronounced among non-adopters (Figure 23.1), and the perceived profit change is not enough to encourage them to invest in the technology (Barnes et al., 2019). It is reported that nearly 58.3% of non-adopters and 10.4% of adopters of variable-rate fertilizer application lacked a clear understanding of how this technology influences their profitability (Wang, Jin, Sieverding, Kumar, et al., 2023). Therefore, it is important for farmers to understand how to evaluate the economic feasibility of investing in precision agriculture technologies.

To help farmers, especially non-adopters, understand the economic benefits associated with precision agriculture technologies, this chapter introduces how to carry out economic analyses to identify where these technologies may significantly boost production efficiency. There are many approaches to conducting an economic analysis. This chapter discusses partial budgeting and discounted cash flow analysis. Partial budgeting analysis is a decision-making tool used to assess the impact of a practice change to an enterprise. It is a straightforward investment analysis tool that shows the expected change by isolating the financial impact made from the change. In this context, it accounts for the changes that come with adopting a specific

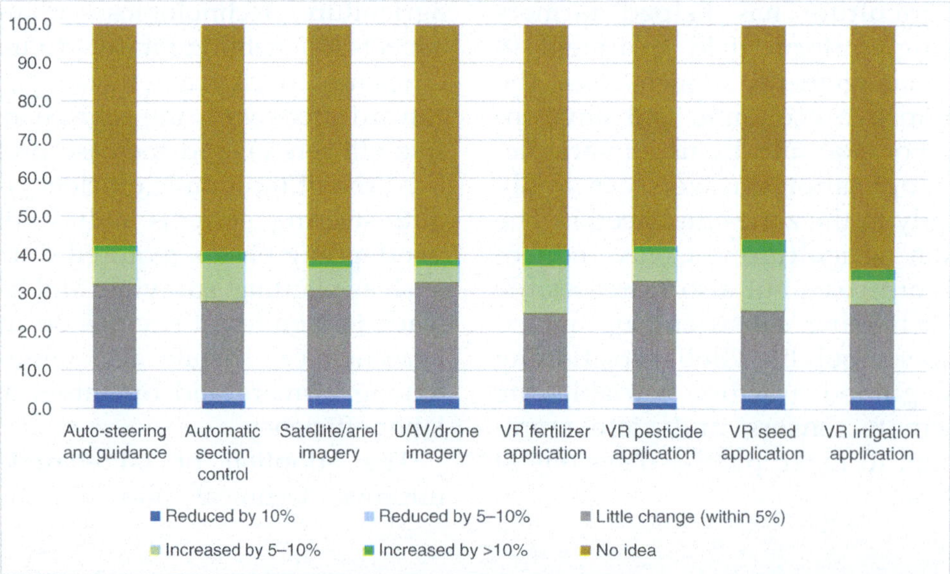

Figure 23.1. Perceived profit change for precision agriculture technologies by non-adopters Source: Wang & Cheye, 2021 / South Dakota State University.

precision agriculture technology. It accounts for both the positive and negative economic returns and the costs associated with adopting a given technology.

The basic steps of calculating the partial budget include the following:

1. List all *added returns* (e.g., increased yield or improved product quality) and *reduced costs* (e.g., savings on inputs, labor, fuel) that result from adopting the new technology. Then calculate the total positive effects from the listed factors.

2. List all *added costs* (e.g., cost of purchasing or maintaining the equipment, training costs) and any *reduced returns* (e.g., loss of income from discontinuing an existing activity). Then sum up all these factors to calculate the total negative effects.

3. Calculate the net financial impact (NFI) by subtracting the negative effect (total cost) from the positive effect (total benefit), using the equation below:

$$\begin{aligned} NFI &= Total\ benefit - Total\ cost \\ &= (Added\ income + Reduced\ costs) \\ &\quad - (Reduced\ income + Added\ costs) \end{aligned}$$

A positive NFI indicates that adopting a technology is expected to improve a farmer's profitability. However, a negative NFI value implies the change may reduce profitability. In cases of long-term investment, you should also consider alternative uses of the money and interest rates.

EXAMPLE 23.1 PARTIAL BUDGET ANALYSIS

A corn farmer managing 50 ac (20.3 ha) is considering replacing tractor-based spraying with a rented unmanned aerial vehicle (UAV) to apply treatment on a field. Initial analysis indicates that the UAV can cover the entire field for a cost of $750 per application. Since spraying is done three times per season, the total annual rental cost amounts to $2,250. Previously, the farmer needed $140/ac ($346/ha) on chemical inputs ($7,000 total), $9.19/ac ($22.7/ha) on labor and fuel ($459.50 total). The UAV should reduce the chemical costs by 28% and fuel and labor cost by 22%. In addition, due to improved timing and precision, corn yield should increase by 8%. Given an average yield of 177.3 bu/ac (11.1 Mg/ha) and a market price of $3.86/bu ($113/Mg), the farmer can determine whether the added income justifies the rental cost (Table 23.1).

Table 23.1. Partial budget analysis of renting a UAV to spray a 50-ac (20.3 ha) field

Positive economic changes		Negative economic changes	
Added income		Added cost	
Yield gain	$2,738	Drone rental	$2,250
Reduced cost		**Reduced income**	
Chemical use savings	$1,960	None	
Labor and fuel savings	$101		
Increase in net income	$4,799	**Decrease in net income**	$2,250
Estimated NFI (Net Income) = $4,799 – $2,250 = $2,549			

EXAMPLE 23.1 CALCULATIONS: BREAKDOWN

Step 1: Added income from yield gain
Extra yield: 8% × 177.3 bu = 14.184 bu/ac
Total extra for 50 ac: 14.184 bu × 50 ac = 709.2 bu
Added income: 709.2 bu × $3.86/bu = $2,738

Step 2: Reduced cost
Chemicals: 28% of $7,000 = $1,960
Labor & fuel: 22% × $459.50 = $101
Total reduced costs = $196 + $101 = $297

Step 3: Added cost
Drone rental: 3 × $750 = $2,250

Table 23.1 illustrates the partial budgeting analysis approach by calculating the economic changes, with positive changes and negative changes listed in separate columns.

Based on this analysis, there is a $785 return on the change in management. The farmer found that this added income would justify the rental cost of the UAV, and therefore, the economically sound decision is to use the UAV to spray his field.

EXAMPLE 23.2 COST–BENEFIT RATIO ANALYSIS

The cost–benefit ratio (CBRA) is an economic indicator used to assess the profitability and efficiency of the investment. While partial budgeting emphasizes net financial change, the CBRA demonstrates how effectively capital is used by providing a unit-based measure of return per dollar spent. The CBRA values are calculated using the equation,

$$CBRA = \frac{Total\ benefit}{Total\ cost} \quad (23.1)$$

The CBRA score is directly related to the attractiveness of the investment. In general, if the CBRA score is greater than 1, then the investment is considered economically sound; if the score is less than 1, then the investment is likely to fall short of recovering the investment. For instance, a CBRA score of 2 means an investment return of $2 for every $1 invested. To illustrate the use of the CBRA analysis, we can apply it to the financial outcomes presented in this example. The CBRA is calculated by dividing the total benefit (i.e., the increase in net income resulting from yield gains and cost savings) by the total cost associated with the investment (i.e., UAV rental fees), as in $4,799 / $2,250 = 2.13. This calculation resulted in a CBRA score of 2.13, meaning that the grower will have an investment return of $2.13 for every $1 invested in the UAV technology, making it a profitable option.

The CBRA can also be used as a decision tool in comparing multiple technologies and can help identify the best option for an operation. This can be used in situations where financial resources are constrained or dependent on external funding, such as loans or grants. This allows farmers to identify and prioritize investments and to determine whether they should rent or purchase the equipment by assessing which option is more cost-effective over a given period. Note that for long-term investments, CBRA should consider the time value of money, as explained in the next section.

Capital Budgeting

Capital budgeting is used to evaluate investment decisions with a useful life of more than 1 year. Through capital budgeting, the net present value, internal rate of return, and payback period associated with an investment decision can be determined. Capital budgeting includes the present value, the net present value, and the internal rate of return.

Net Present Value

When assessing investments that extend over several years, such as the purchase of precision agriculture technologies with high initial costs and long-term benefits, it is essential to account for the time value of the money spent, a fundamental concept in economics and finance. It reflects the idea that a dollar received today is worth more than a dollar received in the future due to its potential earning capacity. The value of today's money will not remain the same in the future due to inflation, interest rates, and opportunity costs. As a result, evaluating long-term investments solely on raw totals, without adjusting for the timing of those cash flows, can lead to misleading conclusions.

Evaluating long-term investments starts with introducing the discount rate. The discount rate serves as a tool to adjust future cash flows to their present-day equivalent value. This process, known as discounting, enables us to calculate the present value (PV) of benefits and costs that are expected to occur at later dates. Present value reflects how much a future amount of money is worth today, given a specific discount rate. The PV for T periods (PV_T) can be calculated using the formula

$$PV_T = \sum_{t=1}^{T} \frac{C_t}{(1+r)^t} \tag{23.2}$$

where

C_t is the net cash flow (revenue minus cost) at time period t.

t is any time period between 1 and T.

T is the total number of time periods being analyzed.

r *is* the discount rate.

Net present value (NPV) is a financial metric used to assess whether the present value of expected benefits from an investment outweighs the present value of its associated costs. The NPV for T periods (NPV_T) expands the PV calculation and compares the total present value of all future inflows. It is defined as the difference between the present value of all expected cash inflows (benefits) and the present value of all expected cash outflows (costs) over the life of an investment. It brings all future income and expenses into present-day terms using a discount rate, which typically reflects the opportunity cost of capital, prevailing interest rates, or inflation. The NPV_T is defined by the equation

$$NPV_T = \sum_{t=1}^{T} \frac{C_t}{(1+r)^t} - C_0 \tag{23.3}$$

where

C_t is the net cash flow (revenue minus cost) at time period t.

t is any time period between 1 and T.

T is the total number of time periods being analyzed.

r *is* the discount rate.

C_0 is the initial capital investment (e.g., cost of machinery).

When $NPV_T < 0$, the investment is not economically feasible, and when $NPV_T > 0$, the investment is considered profitable. However, it should be noted that a positive NPV_T does not always mean that it is the best use of funds.

Internal Rate of Return

Internal rate of return (IRR) is a metric for evaluating the long-term profitability of an investment. It is the discount rate at which the NPV_T of an investment becomes zero. In other words, the IRR is the discount rate of return at which the future stream of net benefits exactly offsets the initial investment cost:

$$0 = NPV_T = \sum_{t=1}^{T} \frac{C_t}{(1+IRR)^t} - C_0 \tag{23.4}$$

where

C_t is the net cash flow (revenue minus cost) at time period t.

t is any time period between 1 and T.

T is the total number of time periods being analyzed.

r *is* the discount rate.

C_0 is the initial capital investment (e.g., cost of machinery).

IRR is the internal rate of return to be calculated.

IRR is the internal rate of return that can be calculated using software such as Microsoft Excel.

The IRR can be compared directly with prevailing loan interest rates, inflation, or expected returns from alternative farm investments. For example, if a farmer is considering a \$15,000 investment in variable-rate fertilizer applicators, and the IRR is 16%, the loan rate is 12%, and the alternative investment return is 8%, then since the IRR exceeds both the loan rate and the alternative investment rate, the proposed precision agriculture investment will likely generate higher financial returns than other investments. If the IRR is lower than these benchmarks, it is often advisable for the farmer to reconsider the investment.

EXAMPLE 23.3 CAPITAL BUDGETING ANALYSIS

A corn farmer is considering whether to switch from conventional tractor-based spraying to a UAV-based application system for a farm of 200 ac (80.9 ha). The purchase price of a T50 UAV is $28,082, with an additional $6,028 for extra batteries and charging stations, and $1,613 on a Real-Time Kinematic (RTK) module. The initial investment cost also includes $3,000 for training and certification and $136 for applicator insurance. The software subscription costs $2,000 annually. Flight planning, spraying control, and battery replacements cost $4,949.85 per year.

The UAV operates at a capacity of 52 ac/hr (21 ha/hr), making it possible to complete spraying in about 4 hr. The farmer expects to use the UAV two to three times during the growing season. Before purchasing the UAV, the farmer would typically spend $140/ac ($346/ha) on chemical inputs and $9.19/ac ($22.7/ha) on labor and fuel for spraying. The average yield is 177.3 bu/ac (11.1 Mg/ha), and the market price of corn is $3.69/bu ($135/Mg). The use of the UAV is anticipated to reduce chemical use by 28% and labor and fuel costs by 22.03%, and to increase corn yield by 8%.

We can assess whether the investment is financially sound by calculating NPV_T, payback period, and IRR, but first, we need to do a few calculations.

EXAMPLE 23.3 CALCULATIONS: BREAKDOWN PER ACRE (HECTARE)

Chemical savings = $140/ac × 28% = $39.20/ac ($7,840 for 200 ac [81 ha])

Labor/fuel savings = $9.19/ac × 22% = $2.0218/ac ($404.36 for 200 ac [81 ha])

Yield gain = 177.3 bu/ac × 8% = 14.18 bu/ac (0.926 Mg/ha)

Revenue increase = 14.184 bu/ac × $3.69/bu = $52.339/ac ($129/ha, or $10,467.79 for 200 ac)

The initial investment costs, annual operating costs, and annual benefits are listed in Table 23.2.

The annual net benefit, which can also be referred to as the cash flow, will be calculated as Annual Benefits − Annual Operating Cost.

Annual benefits include Chemical Savings ($7,840) + Labor & Fuel Savings ($404.36) + Yield Improvement ($10,467.79).

Annual operating cost includes software subscription ($2,000) + Maintenance/battery repairs ($4,949.85).

Therefore, the annual net benefit in this example can be calculated as follows:

$$(\$7{,}840+\$404.36+\$10{,}467.79) - (\$2{,}000+\$4{,}949.85)=\$11{,}762.30.$$

Table 23.2. Financial Breakdown for Drone Spraying Investment on a 200-ac (81 ha) Corn Farm

	Amount
Initial investment costs	
UAV cost	$28,082
Battery	$6,028
Charging stations and RTK module	$1,613
Training and certification	$3,000
Applicator insurance	$136
Annual operating costs	
Software subscription	$2,000
Maintenance and battery repairs	$4,949.85
Annual benefits	
Chemical savings	$7,840
Labor and fuel savings	$404.36
Yield improvement	$10,467.79

This result suggests that the decision is profitable for the farmer. Let's continue the analysis by calculating the NPV_T to assess whether the present value of expected benefits from an investment outweighs the present value of its associated costs.

EXAMPLE 23.4 CALCULATING NET PRESENT VALUE

In this example, we will assume that the drone has an expected lifespan of 5 years with no salvage value and a discount rate of 10%.[1] The NPV_T can be calculated with Eq. (23.3), $NPV_T = \sum_{t=1}^{T} \frac{C_t}{(1+r)^t} - C_0$. In this equation, C_0 is $38,859, calculated as UAV cost + Battery + Charging stations and RTK module + Training and certification + Applicator insurance = $28,082 + $6,028 + $1,613 + $3,000 + $136; the net cash flow ($C_t$) is $11,762.3; r is the discount rate (10%), and the present value for the first 5 years is calculated using Eq. (23.2). Sample calculations are below and summarized in Table 23.3.

Table 23.3. Net cash flows and present values for the first 5 years

Year (t)	Net cash flow (C_t)	Discount $1/(1+r)^t$	Present value (PV_t)	Cumulative PV (PV_T)
1	$11,762.3	0.9091	$10,692.26	$10,692.26
2	$11,762.3	0.8264	$9,720.69	$20,412.95
3	$11,762.3	0.7513	$8,836.09	$29,249.04
4	$11,762.3	0.6830	$8,033.63	$37,282.67
5	$11,762.3	0.6209	$7,304.86	$44,587.53

For Years 1 and 2, cumulative PV can be calculated as

$$PV_1 = (\$11,762)\left[\frac{1}{(1+0.1)^1}\right] = \$10,692.26$$

$$PV_2 = \$10,692.26 + \$9,720.69 = \$20,412.95$$

Following Eq. (23.3), NPV_T (T = 5) can be calculated as

$$NPV_5 = \frac{C_1}{(1+r)^1} + \frac{C_2}{(1+r)^2} + \frac{C_3}{(1+r)^3} + \frac{C_4}{(1+r)^4} + \frac{C_5}{(1+r)^5} - C_0$$
$$= (\$10,692.26 + \$9,720.69 + \$8,836.09 + \$8,033.63 + \$7,304.86) - \$38,859$$
$$= \$44,587.53 - \$38,859 = \$5,729$$

The NPV_5 is $5,729 on 200 ac (81 ha). This shows that considering the time value of money, the UAV investment is expected to add more than $5,000 in value over the next 5 years. Therefore, the project is considered financially viable. This further confirms that this is a good financial investment for the farmer. Let's continue by determining how long it will take for the farmer to recoup his investment by calculating the payback period. We will additionally evaluate the farmer's long-term profitability on his investment using IRR.

Payback Period

To determine the payback period, we refer to the cumulative present values in Table 23.3. The initial investment of $38,859 is recovered between Years 4 and 5. By the end of Year 4, the farmer recovers $37,283 of the $38,859 invested, leaving a shortfall of $1,576. Comparing this shortfall to the Year 5 benefit shows that this remaining amount will be paid off in Year 5, as $1, 576/$7, 304 = 0.22, which means that about 22% of the fifth year's benefit is needed to fully recover the investment.

Therefore, we can infer that the payback period is roughly 4 years and 3 months. Beyond that point, all savings and extra revenue generated by the UAV will go toward profit.

Internal Rate of Return

According to Eq. (23.4), The IRR is calculated as,

$$0 = \frac{\$11,762.3}{(1+IRR)^1} + \frac{\$11,762.3}{(1+IRR)^2} + \frac{\$11,762.3}{(1+IRR)^3} + \frac{\$11,762.3}{(1+IRR)^4} + \frac{\$11,762.3}{(1+IRR)^5} - \$38,859.$$

[1] Note that the discount rate is affected by factors such as interest rate and could change over time. A higher discount rate lowers the present value and NPV_T.

	A	B	C
1	Period	Description	Cashflow
2	0	Initial investment	$(38,859.00)
3	1	Year 1	$ 11,762.30
4	2	Year 2	$ 11,762.30
5	3	Year 3	$ 11,762.30
6	4	Year 4	$ 11,762.30
7	5	Year 5	$ 11,762.30
8			
9			
10		Internal Rate of Return (IRR)	16%

Figure 23.2. Calculation of IRR using Excel.

This equation cannot be solved algebraically. It is typically solved in Microsoft Excel using the = IRR () function, as shown in Figure 23.2.

Using Excel, the IRR for the drone investment was calculated to be 16%. This means the investment is expected to generate an average annual return of 16% over its 5-year lifespan. Since this IRR is well above the chosen discount rate of 10%, the investment is considered financially attractive.

Overall, this indicates that the farmer would make a sound financial decision by investing in the UAV sprayer system. It will save him both time and money to incorporate this new technology into his operation, and he will have ensured the profitability of this investment through the calculations used in the example.

Role of Farm Size and Discount Rate

For small farm operators, alternative arrangements, such as renting a UAV, hiring custom spraying services, or forming cooperatives to share equipment and costs, may offer more financially viable pathways to adopt precision agriculture technologies (GeoPard Agriculture, 2025; Lowenberg-DeBoer, 2022). Other than farm size, the discount rate also impacts the present value of future benefits. For precision agriculture, the discount rate is determined by the loan interest rate. When the discount rate surpasses the IRR, the NPV turns negative, making the investment financially and economically unjustifiable.

Risk Management and Environmental Benefits

Precision agriculture is increasingly valued not just for its economic benefits but also for its ability to help farmers manage various risks while advancing environmental benefits. Amid increasing uncertainties from unpredictable weather to volatile markets, precision tools offer a proactive approach to making agricultural operations more economically resilient and sustainable. The adoption of precision technology has the potential to mitigate production-related risks. Technologies such as field sensors, GPS-guided machinery, and data-driven decision support systems enable farmers to respond quickly to crop stress, changing weather conditions, or pest outbreaks (Schimmelpfennig, 2016). These timely responses can improve productivity and reduce losses during critical stages of crop development. In addition, detailed yield maps and input records allow for better planning, giving farmers the confidence to invest in practices that suit each field's unique conditions.

Furthermore, price and market risks, often driven by shifting commodity prices or supply chain disruptions, can also be addressed through precision agriculture. By providing more accurate yield forecasts and cost assessments, precision agriculture helps farmers plan harvest timing, input purchases, and marketing strategies more effectively. Access to reliable production data also strengthens a farm's position in negotiating contracts or participating in insurance and risk-sharing programs. Legal and compliance-related risks can also be mitigated through precision agriculture technologies. The tools, such as georeferenced spray records or variable-rate application maps, serve as verifiable records of input use. This can be valuable when meeting regulatory requirements, participating in government programs, or addressing environmental concerns. This technology also plays a role in minimizing personal and operational risk. For example, automation reduces the physical demands placed on farm workers and can limit their exposure to hazardous chemicals or unsafe operating conditions. Monitoring systems enhance safety by providing real-time alerts on equipment performance or environmental hazards, thereby contributing to a safer overall working environment.

From a financial perspective, precision agriculture supports smarter investment decisions. Although the initial cost of adopting precision agriculture systems can be significant, the ability to target inputs more precisely often results in substantial savings over time. Reduced waste, better input timing, and improved yields contribute to stronger financial outcomes. In some cases, the detailed records generated by precision agriculture tools are now being used by insurance providers to offer customized coverage. This enables the policies to be more reflective of actual risks rather than general estimates.

Conclusion and Summary

This chapter examined the economic dimensions of adopting precision agriculture technologies, focusing on how farmers can determine whether such investments are financially justified. These technologies are designed to enhance farm operation by contributing to technical efficiency by improving input accuracy and reducing waste, supporting allocative efficiency by optimizing input distribution where returns are highest, and enhancing economic efficiency by increasing net profitability.

To evaluate these investments, we discussed key financial tools, such as partial budgeting. This was demonstrated to help assess the profitability of incremental or short-term changes by comparing added revenues and reduced costs to added expenses. CBRA provides a quick gauge of investment performance, indicating how much benefit is generated for every dollar invested. For larger or multiyear investments, NPV and IRR incorporate the time value of money and facilitate comparisons with alternative capital uses. Calculating the payback period further assists by estimating how long it will take to recover the initial investment through future cash flows. Note that the potential benefits of precision agriculture do not guarantee financial success. Economic returns could vary widely depending on factors such as farm size, crop type, market prices, and the extent to which the technologies are adopted and integrated into existing operations.

Study Questions

1. Distinguish between technical, allocative, and economic efficiency in precision agriculture, and explain how each contributes to farm performance.
2. Why do information-intensive technologies, such as yield monitors and soil sampling, have relatively lower adoption rates than automated technologies?
3. What are the main risks that precision agriculture technologies can help farmers manage?
4. A farmer is considering adopting an integrated Real-Time Kinematic (RTK) auto-steering system to reduce chemical input use and enhance crop yield over a period of 5 years. The initial investment cost is $18,000, with no salvage value at the end of its useful life. The farmer incurs an annual operating cost of $1,500 for software installation. Each year, this technology generates two types of economic benefits: (a) input cost savings of $3,000 from reduced pesticide and labor use and (b) additional revenue from yield improvement that amounts to $4,500.

   Assuming an annual discount rate of 8%,

   A. Calculate total benefit, total cost, and net benefit.

Table 23.A1. A list of cost and benefit items

Cost/benefit item	Amount
Initial investment cost	$18,000
Software installation cost	$1,500
Annual chemical and labor savings	$3,000
Annual yield improvement	$4,500

B. Calculate the net present value (NPV). The NPV is calculated using the formula, $NPV_T = \sum_{t=1}^{T} \frac{C_t}{(1+r)^t} - C_0$
Initial investment (C_0) = \$18,000
Net cash flow = \$6,000
Discount rate = 8%.

Table A23.2. Net cash flows and present values for the first 5 years

Year	Net cash flow (C_t)	Discount factor (8%) $1/(1+r)^t$	Present value (PV_t)	Cumulative PV (PV_T)
1	\$6,000	0.926	\$5,555	\$5,555
2	\$6,000	0.857	\$5,144	\$10,699
3	\$6,000	0.794	\$4,763	\$15,462
4	\$6,000	0.735	\$4,410	\$19,872
5	\$6,000	0.681	\$4,084	\$23,956

C. Based on your result in Part B, should the farmer invest in this technology? Provide an economic justification.

D. Calculate the cost–benefit ratio (CBR).

$$\text{CBR} = \text{Total Present Value of Benefits} / \text{Initial Investment} = PV_T / C_0 = \$23,956 / \$18,000 = 1.33.$$

E. Find out the payback period of this investment.

F. Suppose input savings decline to \$2,000 annually due to pesticide price change. Recalculate the NPV and reassess the investment decision.

ACKNOWLEDGMENTS

Support for this document was provided by the Precision Farming Systems community in the American Society of Agronomy, South Dakota State University, USDA Natural Resources Conservation Service Conservation Innovation Grants (NR233A750011G019), USDA National Institute of Food and Agriculture (Multistate Project S1090 Program under accession no. 7008110), American Coalition for Ethanol, the South Dakota Corn Utilization Council, and the South Dakota soybean Research and Promotion Council. The use of trade names is for the convenience of the reader and does not imply endorsement by the authors.

REFERENCES

Barnes, A. P., Soto, I., Eory, V., Beck, B., Balafoutis, A., Sánchez, B., Vangeyte, J., Fountas, S., van der Wal, T., & Gómez-Barbero, M. (2019). Exploring the adoption of precision agricultural technologies: A cross regional study of EU farmers. *Land Use Policy*, *80*, 163–174. https://doi.org/10.1016/j.landusepol.2018.10.004.

Bora, G. C., Nowatzki, J. F., & Roberts, D. C. (2012). Energy savings by adopting precision agriculture in rural USA. *Energy, Sustainability and Society*, *2*, 22. https://doi.org/10.1186/2192-0567-2-22.

Farmonaut. (2025, June 16). *How variable rate fertilizer boosts yield & profits: Precision agriculture's high-tech pathway to sustainable success*. https://farmonaut.com/precision-farming/how-variable-rate-fertilizer-boosts-yield-profits.

GeoPard Agriculture. (2025). *Factors affecting precision agriculture adoption rate*. https://geopard.tech/blog/factors-affecting-precision-agriculture-adoption-rates.

Griffin, T. W., Shockley, J. M., & Mark, T. B. (2018). Chapter 15. Economics of precision farming. In D. K. Shannon, D. E. Clay, & N. R. Kitchen (Eds.), *Precision agriculture basics* (pp. 221–230). ASA, CSSA, SSSA.

Lowenberg-DeBoer, J. (2022). *Economics of adoption for digital automated technologies in agriculture. Background paper for The State of Food and Agriculture 2022*. FAO Agricultural Development Economics Working Paper 22-10. FAO. https://doi.org/10.4060/cc2624en.

McFadden, J., Njuki, E., & Griffin, T. (2023). *Precision agriculture in the digital era: Recent adoption on U.S. farms*. USDA, Economic Research Service, EIB 248. https://ers.usda.gov/sites/default/files/_laserfiche/publications/105894/EIB-248.pdf?v=61234.

Mosheim, R., McFadden, J., & Schimmelpfennig, D. (2021). Precision agriculture: Profitability and resource stewardship. In *Agricultural Outlook Forum 2021* (No. 320996). USDA, Agricultural Outlook Forum.

Schimmelpfennig, D. (2016). *Farm profits and adoption of precision agriculture*. USDA, Economic Research Service, ERR-217. https://ers.usda.gov/sites/default/files/_laserfiche/publications/80326/ERR-217.pdf?v=11423.

Wang, T., & Cheye, S. (2022). *2021 Midwestern producer survey descriptive results*. Midwestern Farm and Producer Surveys and Policy Briefs. https://openprairie.sdstate.edu/midwestfarmsurvey/1.

Wang, T., Jin, H., Sieverding, H., Kumar, S., Miao, Y., Rao, X., Obembe, O., Nafchi, A. M., Redfearn, D., & Cheye, S. (2023). Understanding farmer views of precision agriculture profitability in the U.S. Midwest. *Ecological Economics*, *213*, 107950. https://doi.org/10.1016/j.ecolecon.2023.107950.

Wang, T., Jin, H., & Sieverding, H. L. (2023). Factors affecting farmer perceived challenges towards precision agriculture. *Precision Agriculture*, *24*, 2456–2478. https://doi.org/10.1007/s11119-023-10048-2.

Watkins, K. B., Hristovska, T., Mazzanti, R., Wilson, C. E., Jr., & Schmidt, L. (2014). Measurement of technical, allocative, economic, and scale efficiency of rice production in Arkansas using data envelopment analysis. *Journal of Agricultural and Applied Economics*, *46*(1), 89–106. https://doi.org/10.22004/AG.ECON.169037.

Soil Health in Precision Agriculture

24

Xi Zhang, Kristen Veum, Yuchuan Fan, Vander Luis Novaes Nunes, Wei Dai, Skye Brugler, and David E. Clay

Chapter Purpose

Farmers face the challenge of producing food for a rapidly growing population while simultaneously reducing their impact on the environment. Meeting food security and sustainability goals simultaneously requires focused training, scientific advances in new technologies, and the adoption of soil health practices that will improve the soil's ability to sustain plants, animals, and humans. The soil health practices most adopted by farmers include (a) reducing tillage intensity to minimize soil disturbance and erosion, (b) increasing crop rotation diversity to aid in weed management and pest control, and (c) planting cover crops to protect the soil from erosion and maximize living roots. These practices, when adopted in combination with genetic crop improvements, have been linked to increased soil organic matter, improved water dynamics, higher crop yields, and other benefits. This chapter provides an overview of (a) soil health and its importance; (b) the physical, chemical, and biological properties associated with soil health; (c) the spatial variability of soil health indicators; and (d) the link between soil health and precision farming.

Introduction to Soil Health

The concept of soil health first appeared in a peer-reviewed soil science journal in the early 1990s (Haberern, 1992). It has since been discussed in terms of its definition and practical applications as they relate to ecosystem services, which are the benefits provided by nature, directly or indirectly, such as clean air and water. Soil health is also often discussed in terms of how it relates to agricultural productivity (Mizuta et al., 2021). For example, the Food and Agriculture Organization of the United Nations defines soil health as the ability of the soil to sustain the productivity and environmental services of terrestrial ecosystems (FAO, 2020).

Soil health and its maintenance ensure the soil's continued capacity to function as a dynamic living system, supporting biological productivity, promoting environmental quality, and enhancing plant and animal health. This functionality arises from complex interactions among physical, chemical, and biological soil properties, all of which are influenced by broader ecological processes that vary by region.

Therefore, maintaining or improving soil health requires adopting farming practices that preserve the balance between the atmosphere, biosphere, hydrosphere, and lithosphere. When this balance is not maintained, soil health deteriorates, impacting the soil's ability to provide important ecosystem services, such as habitat for plants and animals, removal of pollutants, recycling of nutrients, and its capacity to serve as a medium to support plant growth and productivity (Figure 24.1).

Several factors influence the balance and maintenance of soil health, including the management of the soil through agricultural practices. The impact of farming practices on soil health varies regionally because different regions have different soil systems, making soil health relatively site specific. To maintain soil health, it is important to understand the soil system that is being farmed. For example, the original ecosystem of the U.S. Northern Great Plains was a tallgrass prairie. These prairies had perennial grasses with deep fibrous root systems. Agriculture in this region converted the tallgrass prairies into annual row crop production systems, which have much shallower roots than the original grasses. The shallower root systems contributed to some soil in the region developing an issue with excessive sodium (Na) and other salts moving to the soil surface as a result of a rising water table, bringing salts up from the original geologic material. The addition of Na and salts to the soil surface significantly decreased soil health and productivity in the impacted soils (Figure 24.2). To regain soil health, a remediation approach reintroduced perennial grasses into the areas of the soil system most impacted. Their deep fibrous root systems helped to naturally lower the water table, reducing movement of Na and other salts to the soil surface, and restoring the soil's health and productivity (Clay et al., 2022).

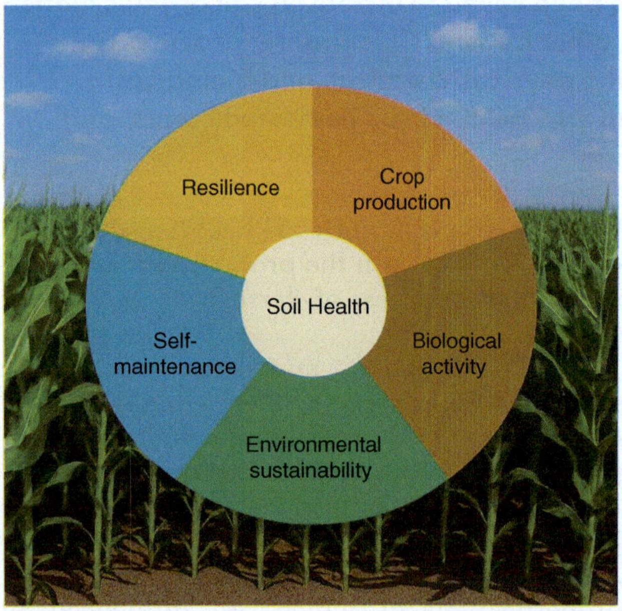

Figure 24.1. The importance of soil health to agroecosystems. Adapted from *Washington Soil Health Initiative*. (Courtesy Xi Zhang).

Similarly, in the tallgrass prairie region of central North America, intensive agricultural production relying on tillage and monoculture crops contributed to widespread erosion, which degraded soil health. In this region, implementing stacked conservation practices—including reduced tillage, diversified crop rotations, cover crops, and/or perennialization (e.g., growing grasses for biofuel)—in impacted soils, has shown potential for restoring soil health, although recovery may take several decades (Lord et al., 2024; Veum et al., 2015). A variety of factors determine soil productivity, and the maintenance of soil health is often site specific, requiring an understanding of the system. Protecting the soil requires understanding the relationships between the soil system and farm management.

The soil system is influenced by soil characteristics. These soil characteristics are driven by five primary soil-forming factors that operate at multiple scales: climate, biota, topography, time, and the original geologic "parent material" (Jenny, 1941). These factors influence soil classification systems across the globe (e.g., Soil Survey Staff, 1999) and can be used to identify management strategies to improve or maintain soil health at a given field location (Veum et al., 2025). Soil-forming factors interact with farming practices to drive differences in soil health at the field and sub-field scales (Karlen et al., 2017). For example, slope and landscape position (topography), combined with tillage, can impact aggregate stability, which is the ability of the soil to maintain its structure, and can

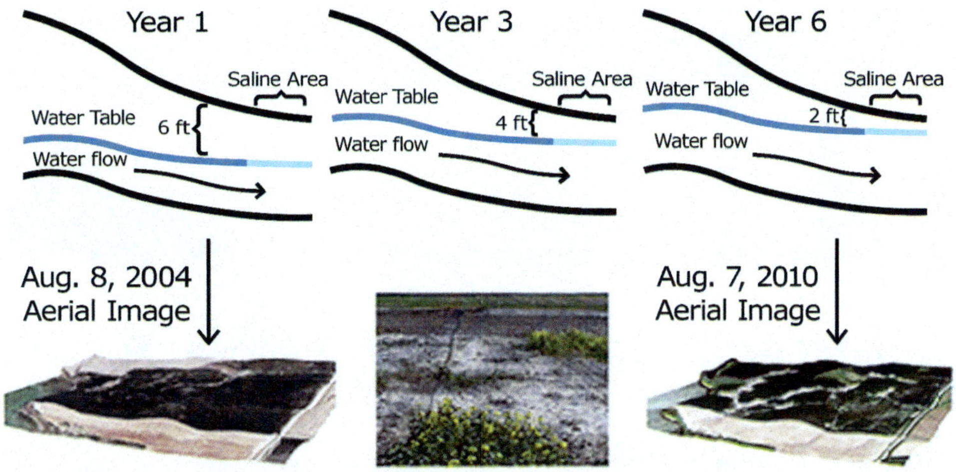

Figure 24.2. Diagram illustrating the formation and expansion of saline–sodic soils in cropped fields of the Northern Great Plains Region. (Courtesy of Blake Toennies).

also cause soil, nutrients, and organic matter to move from upland to bottomland positions by erosion (Figure 24.3).

Shifts in rainfall patterns or rising sea levels can also impact soil health. These changes can result in more intense and erratic rainfall, which increases soil erosion, disrupts nutrient cycling, weakens aggregate stability, degrades soil structure, and reduces water infiltration. In coastal areas, sea-level rise leads to saltwater intrusion, which increases soil salinity, impairing plant growth and microbial function. Waterlogging in low-lying regions creates anaerobic conditions, promoting greenhouse gas emissions (e.g., methane) and limiting root and microbial respiration. Together, these changes alter the physical, chemical, and biological properties of the soil, diminishing its ability to support plant productivity, regulate water and nutrients, and store carbon (C).

Components of Soil Health

As discussed above, soil health is heavily influenced by the unique soil characteristics of a particular system, which develop over time under the influence of soil-forming factors. Because of this uniqueness, managing soil health can be challenging. Thus, soil health indicators (Table 24.1) were developed to serve as proxy measurements. They are physical, chemical, and biological measurements of the soil that can be used to assess soil health relative to that of another field. These indicators provide greater spatial and/or temporal resolution to help understand and quantify soil health across regions with different soil health systems. This information can then be used to help optimize soil health decisions and increase agricultural productivity. A wide range of in-field and laboratory soil health indicators is available to farmers from service providers, such as crop consultants and fee-based laboratories. These soil health indicators, as discussed below, can be grouped into three general categories: physical, chemical, and biological properties.

Physical Properties

The physical properties of soil control the movement of mass and energy through the soil, root penetration, and resistance to wind and water erosion (Figure 24.4; Karlen & Stott, 1994). Aggregate stability represents the soil's ability to resist disintegration when exposed to external forces, including water, wind, and management practices that disturb the soil, such as tillage. Stable aggregates minimize erosion, improve air and water movement, and support root growth. The size and strength of soil aggregates are influenced by soil texture (i.e., the sand, silt, and clay composition) and farming practices (Veum et al., 2012).

Porosity, which refers to the volume of void spaces within the soil, also facilitates the transport of air and water. Soil compaction is the loss

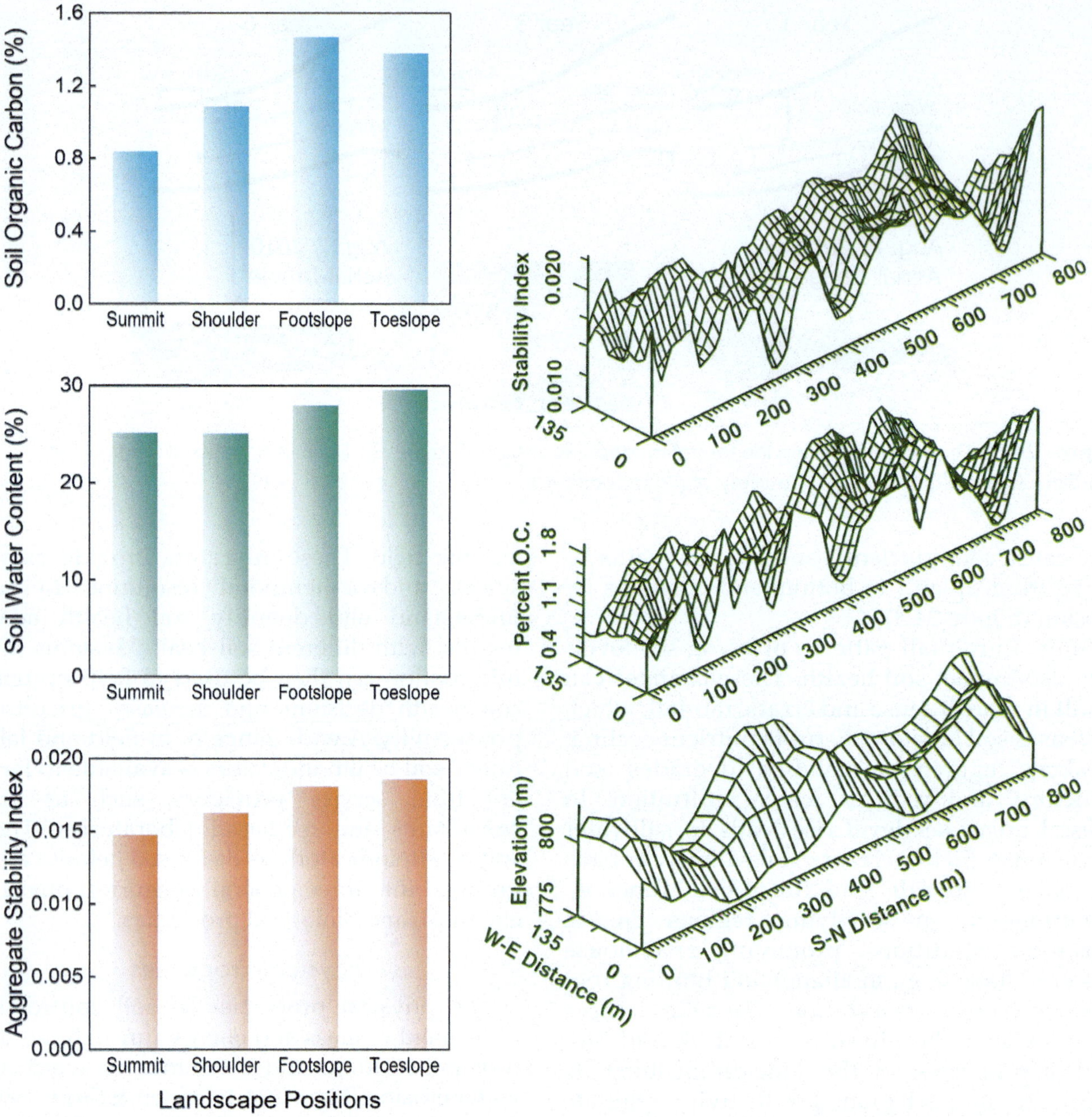

Figure 24.3. Topography effects on soil properties in croplands near Pullman, WA (Adapted from Pierson & Mulla 1990). *Left:* Soil organic carbon, water content, and aggregate stability are generally lowest on upper slope positions and highest on lower positions. *Right:* Aggregate stability and organic carbon are highest on foot and toe slopes (low elevations) and lowest on summits (high elevations).

of porosity, which can be monitored by measuring soil bulk density—the dry weight of the soil per unit volume. As porosity decreases, bulk density increases, leading to reduced gas exchange, decreased water infiltration and retention, and increased runoff. Bulk density generally decreases if soil health practices such as reduced tillage and cover cropping are implemented (Veum et al., 2015).

Available water capacity represents the soil's ability to retain water for plant use. It is the difference between the amount of soil water retained after free drainage (−10 to −33 kPa) and the permanent wilting point (−1,500 kPa). The amount of water that can be retained is a function of many factors and is generally greater in fine-textured soils than in coarse-textured soils. The soil's available water capacity can increase with the

Table 24.1. Selected soil health indicators and measurement methods.

Soil process	Soil health parameters	Measurement methods
Physical process	Aggregation (structural stability)	Wet aggregate stability
	Bulk density (porosity)	Cylindrical core
Chemical process	pH (acidity)	Active pH, reserve acidity
	Cation exchange capacity	Measured as the sum of exchangeable bases
	Electrical conductivity	The concentration of water-soluble salts in the soil
	Soil organic matter (nutrients cycling and C sequestration)	Dry combustion
	Readily available C pool	Particulate organic matter, permanganate oxidizable C (POXC), microbial biomass C
Biological process	Enzyme activity	β-Glucosidase, N-acetyl-β-D-glucosaminidase, phosphomonoesterases, arylsulfatase
	Microbial community diversity	Phospholipid fatty acid (PLFA)

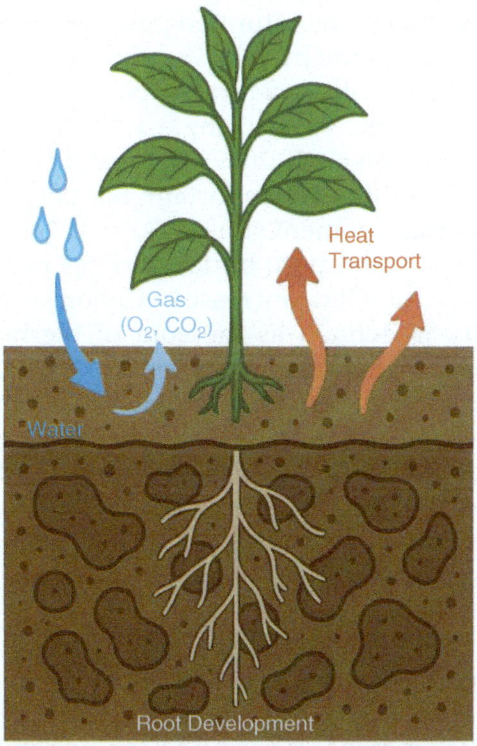

Figure 24.4. Well-structured soil supports the transport of air, water, and energy, as well as plant growth. Soil structure regulates water infiltration, gas exchange, heat transport, and root growth, creating the physical framework that sustains plant development and ecosystem function. (Courtesy Xi Zhang).

adoption of practices that reduce compaction and improve water-holding capacity, such as those that increase soil organic matter (SOM; Clay et al., 2014; Olness & Archer, 2005).

Chemical Properties

The chemical properties of soil influence its ability to supply essential nutrients for plant growth. These properties are typically measured in the spring or fall, when farmers seek fertilizer recommendations for the upcoming cropping season. Soils may be analyzed for pH, electrical conductivity (EC), soil macro- and micronutrients, SOM, and the sum of individually measured cations, referred to as the apparent cation exchange capacity (CEC_a; Blanco-Canqui et al., 2023; Chang et al., 2023). Each of these chemical factors is important and influences both short- and long-term productivity and sustainability in the soil. Soil pH and EC have direct impacts on plant growth and development. Soil pH determines the availability of essential plant nutrients, such as phosphorus (P; as phosphate [PO_4^{3-}]), by affecting the solubility of PO_4^{3-} compounds. At low pH (e.g., pH <6), P tends to bind with iron (Fe) ions (Fe^{3+}) and aluminum (Al) ions (Al^{3+}); in contrast, at high pH (e.g., pH >8), it forms insoluble complexes with calcium (Ca) ions (Ca^{2+}), making it most available to plants in the mildly acidic to neutral range (pH 6–7; Figure 24.5). Electrical conductivity, a measure

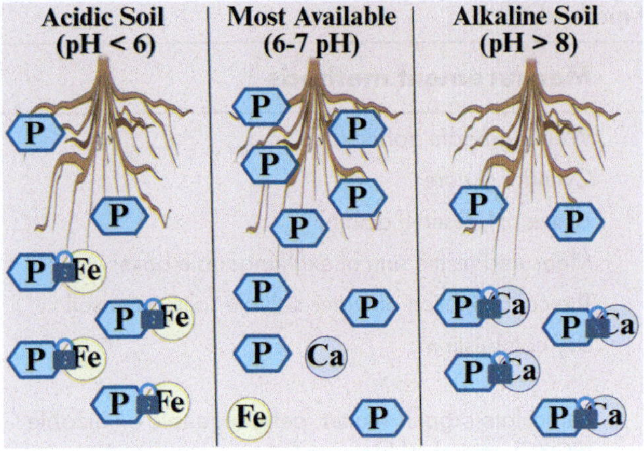

Figure 24.5. Soil pH can influence soil nutrient availability. Created in BioRender (Fischer, B. (2026)). (Courtesy Skye Brugler).

of how well a material conducts electricity, reflects the ability of the ions in water within the soil to carry an electrical current. Electrical conductivity can provide insight into the availability of dissolved nutrients and serve as an indicator of salinity. Both pH and EC have optimum levels that vary depending on site conditions and crops. For instance, while low soil pH may be viewed as detrimental in most agricultural contexts, it creates favorable conditions for acid-loving crops, such as blueberries, which would otherwise struggle in more neutral or alkaline soils. Similarly, while high soil EC may be viewed as harmful in most agricultural contexts due to salt stress and reduced water uptake in many plants, it increases the usefulness of salt-tolerant crops, such as barley, which can thrive in saline soils that would otherwise hinder the growth of less tolerant species.

Interpreting pH and EC values requires that the manager understand how these measurements are taken. This is because there are multiple methods for determining pH and EC. For example, there are two different forms of soil pH, active pH and buffer pH, which are measured and interpreted differently. Active pH is determined by mixing a 1:1 ratio of soil sample to deionized water. This ultimately determines the amount of hydrogen ions (H^+) that are free in the soil solution and provides a general idea of nutrient availability and soil management. However, there are many H^+ fixed to the surface of soil particles that are not measured in active acidity. Active acidity does not account for the H^+ that are not free in the soil water but are, instead, found on the soil exchange sites. Plant roots release H^+ into the soil to exchange cations with the surfaces of soil particles. A measure of this buffer pH is known as reserve acidity, which can be determined by mixing a soil sample with a solution of 0.01 M $CaCl_2$. The exchange with soil particles will release more H^+, resulting in a lower pH compared with measuring active acidity. Reserve acidity provides a much better indication of nutrient availability and whether greater management of soil pH is needed, such as the use of lime. The EC can be measured in two different ways: (a) using a 1:1 mixture of soil and deionized water, and (b) using a saturated paste of 100 g soil plus 65 mL of water. The use of a saturated paste is more representative of field conditions than using a 1:1 mixture, but it is less convenient and more expensive to produce. The 1:1 value (in deciSiemens per meter [dS/m]) can be converted to the saturated paste value by multiplying the value by 2.14. A value is considered saline–sodic with a saturated paste value of >4 dS/m or an EC value using a 1:1 solution of >2 dS/m (Carlson et al., 2019). Accurately interpreting these values is important for determining optimal management strategies.

The soil's ability to buffer changes in soil pH increases as CEC increases. Cation exchange capacity is defined as the sum of the negative charges on the surface of a soil particle, often approximated as the apparent CEC by summing the positive cations to estimate the negative charges the soil can hold. Both SOM and clays have negative charges and contribute to the CEC of soil. Soils with higher CEC can hold and supply more cations (e.g., Na^+, magnesium [Mg^{2+}], potassium [K^+], and Ca^{2+}) to plants and are generally considered more fertile. Producers can adopt soil health practices that increase soil CEC by adding more organic matter through manure, using cover crops, reducing tillage, and minimizing residue removal. Methods to measure CEC vary across laboratories, and soil acidity can affect CEC values. The CEC can be measured directly by determining the total negative charge of the soil, or, alternatively, the results can be reported as the CEC_a. Either way, CEC is often reported in milliequivalents per 100 grams of soil (mEq/100 g soil [$cmol_c$/kg soil]).

In dry climates or coastal areas, Na can represent a substantial proportion of the soil CEC cations. The ratio of milliequivalents of Na per 100 grams of soil to CEC_a multiplied by 100 is the exchangeable Na percentage (ESP). Soil health generally decreases with increasing ESP. Specifically, soil with high ESP tends to exhibit poor aggregation, increased surface crusting, and diminished crop emergence and growth. Similar to EC, crops vary in their tolerance to ESP, and different thresholds have been determined.

Soil organic matter is composed of plant and animal materials at various stages of decomposition. There are different pools of organic matter in the soil. These pools include (a) labile pools, which make the C and N from organic matter more actively available to plants and organisms, and (b) stable organic matter, which remains in the soil for a long period of time, contributing to long-term soil health. Soil organic matter is a key soil health indicator that serves as a reservoir for energy, water, and nutrients (Franzluebbers, 2002), impacts several other soil properties, and responds to changes in farming practices (Karlen & Stott, 1994). The adoption of reduced tillage and cover crops can increase SOM (Joshi et al., 2023, 2025), which in turn can improve soil water storage, nutrient cycling, and crop yield potential. Soil organic matter can be quantified directly or by measuring its primary component, soil organic carbon (SOC), through combustion, and a standard factor can be applied to convert between SOM and SOC: SOM × 0.58 = SOC (Sikora & Stott, 1997). By increasing SOM, a land manager can also enhance the retention of soil macronutrients (nitrogen [N], P, K, Ca, Mg, and sulfur [S]) and micronutrients (boron, copper, Fe, manganese, molybdenum, zinc, chlorine, and nickel). Several soil health farming practices are known to increase the SOM content, including reduced tillage, diverse crop rotations, and cover cropping (Nunes et al., 2020; Veum et al., 2022), although it may take several decades for measurable changes in SOM to be detected. For more sensitive indicators of changes in SOM quantity or quality, fractions of the SOM pool can be measured, such as permanganate oxidizable organic carbon (POXC) or total protein (Moebius-Clune et al., 2016).

Biological Properties

Biological properties define the living component of the soil and include the macro- and microorganisms living in the soil. These organisms ultimately contribute to organic matter decomposition, nutrient cycling, nutrient availability to plants, and overall ecosystem resilience (Figure 24.6; Bagnall et al., 2023; Blanco-Canqui et al., 2023). Key biological indicators provide insight into the size and activity of soil microorganisms and the soil microbial community, and they are closely connected to soil chemical and physical properties. As SOM is decomposed by soil microorganisms, nutrients and energy are released (Clark et al., 2019). The rate of SOM decomposition to carbon dioxide (CO_2) by soil microorganisms can be measured with simple laboratory tests, such as soil respiration, sometimes called C mineralization potential, CO_2 flush, or CO_2 burst (Dorsey, 2019; Franzluebbers & Veum, 2020; Franzluebbers et al., 2000). Different laboratories use various respiration protocols and report results using different units. Thus, comparisons of values from different sources should be approached with caution. In general, high microbial activity through soil respiration indicates a healthy, functioning ecosystem, and increased respiration has been linked to higher crop yields

Figure 24.6. Soil organisms influence soil biogeochemical processes. (Courtesy Xi Zhang).

in multiple cross-state studies (Crookston et al., 2023; Yost et al., 2018).

An alternative approach to quantifying soil respiration to assess biological activity is to measure soil enzyme activity. Popular enzyme tests are often related to the cycling of C (glucosidase), N (glucosaminidase), P (acid or alkaline phosphatase), and S (arylsulfatase). This method helps explain how valuable nutrients, such as C and N, move through the soil. Methods to measure these enzymes are described by Dick (1994), but these protocols are currently too expensive and time-consuming for most farmer-facing laboratories to adopt.

Measurement of microbial community diversity offers another window into the biological functions of soils. Multiple techniques are available to determine the types of microorganisms present in the soil; however, many of these methods are costly, rapidly evolving, and difficult to interpret. Phospholipid fatty acid (PLFA) and ester-linked fatty acid methyl ester (EL-FAME) analyses provide a snapshot of the microbial community at the functional group level. The PLFA approach can differentiate actinobacteria, gram-negative and gram-positive bacteria, arbuscular mycorrhizal fungi, and saprophytic fungi (Li et al., 2020). Although PLFA and EL-FAME biomarkers are affected by sample handling and storage (Veum et al., 2019), they are highly sensitive to management and provide an early indication of shifts in soil health status (Roper et al., 2025). Other molecular approaches using genomics or transcriptomics provide data-rich information related to plant–microbial interactions that could be useful for informing fertilizer management and other farming decisions. However, these methods are currently expensive and time-consuming for a farmer to use to assess soil health. This new frontier in soil health research has the potential to harness innovative technology to capture multi-scale insights into the impact of modern agricultural practices on complex plant–soil–microbial interactions (Zhang et al., 2025).

In contrast to expensive, sophisticated methods, low-tech in-field methods can also be effective for assessing soil health. For example, inspecting the soil for earthworms or macroinvertebrates can provide a real-time indication of soil health status (Weyers & Spokas, 2011). This is often done by cutting into the soil to identify earthworms or the empty spaces of their burrows. Earthworms break down plant residues and mix the soil and are generally considered a sign of healthy soil, although comparative observations should be made under similar soil moisture conditions.

Selecting Soil Health Measurements

To optimize and evaluate the impact of adopting soil health practices, it is important to select appropriate soil health indicators. Although a wide variety of methods exist for identifying variables (indicators), the process can be complex and difficult to navigate (Mizuta et al., 2018). One example involves using data-driven algorithms to develop a minimal dataset of highly relevant variables to maximize evaluation accuracy (e.g., yield or profit) while minimizing inputs (e.g., fertilizer application rates) and time and labor costs (Mizuta & Grunwald, 2022; Mizuta, Grunwald, Bacon, et al., 2022). The choice of indicators should take into account the objective or purpose of measuring soil health. Importantly, soil health assessment goes beyond simply comparing observed and estimated values of soil properties. For a farmer or crop advisor, tracking changes in EC or ESP may be appropriate to guide irrigation or fertilization practices. Additionally, a researcher may wish to evaluate soil health indices and assess the broader agricultural, ecological, and environmental benefits of implementing soil health practices. Resources such as the Web Soil Survey (Chapter 10) offer useful indicators and indices for soil health, but relying solely on publicly available databases may be insufficient to establish baseline levels due to their coarse resolution and infrequent updates (Chapter 10; Figure 24.7).

Soil health measurements can also be used to diagnose low agricultural productivity. This requires a holistic approach that incorporates the physical, chemical, and biological indicators (Table 24.2) discussed in this chapter as guided by the USDA Natural Resources Conservation Service (NRCS; Stott, 2019). Core indicators include SOM, aggregate stability, SOC (calculated by converting SOM into C using the equation SOM × 0.58), and soil respiration (the quantification of CO_2 from the soil surface) have been emphasized as foundational to robust soil

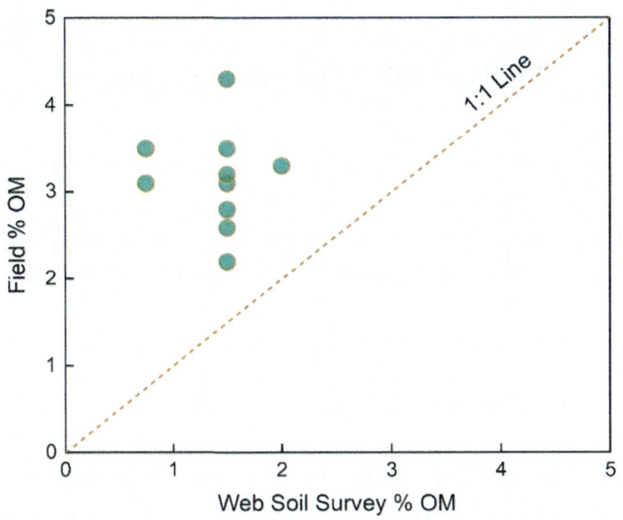

Figure 24.7. Comparison of Web Soil Survey percentage organic matter (%OM) and mean field %OM. Remade from Gergeni & Scasta (2019). (Courtesy Skye Brugler).

health evaluation (Bagnall et al., 2020, 2023; Derner et al., 2018). Collaboration among local stakeholders (e.g., producers, university scientists) to conduct repeated field sampling and monitor changes in key soil health indicators remains crucial for site-specific soil health management.

Soil Health Spatial Variability

Spatial variability refers to the natural and human-induced differences in soil conditions or characteristics across different locations within a single field. Soil health is also subject to spatial variability, making it more challenging to quantify soil health indicators across the field. For example, SOC and soil water contents are often higher in footslope positions (bottomlands) than in summit or shoulder positions (upland areas), as illustrated in Figures 24.3 and 24.8 (Cambardella & Karlen, 1999; Chang et al., 2003;

Table 24.2. Recommended soil health indicators and their interpretations.

Indicator	Function/what it measures	Interpretation/what it indicates
Aggregate stability	Soil structural stability and integrity.	Reflects the soil's capacity to maintain pore space for water infiltration and resist physical breakdown from disruptive forces such as heavy rainfall, wind, or machinery traffic.
Bulk density	The volume of the "voids" or pore spaces between soil particles.	Reflects a soil's level of compaction, where lower values indicate a healthy, porous structure that supports water infiltration and root growth, while high values signal restricted aeration and drainage.
Saturated hydraulic conductivity	The rate at which water moves through a completely saturated soil under a given hydraulic gradient.	Quantifies the "permeability" of the soil's plumbing system, representing the maximum speed at which water can drain through the profile once all pore spaces are filled with water.
Plant-available water	The amount of water held in the soil between field capacity (after excess water has drained) and the permanent wilting point that is accessible for plant root uptake.	Serves as a vital moisture reservoir that sustains plant growth during dry periods and depends on soil texture.
Soil organic matter (SOM)	The non-mineral fraction of the soil, which includes plant and animal residues at various stages of decomposition.	Acts as a biological "engine" that provides a slow-release source of nutrients (e.g., N and P) and stabilizes soil aggregates to improve water infiltration and resistance to erosion.
Total C	The combined amount of organic and inorganic carbon present in the soil.	Represents the sum of soil organic carbon (SOC) from decomposed life and soil inorganic carbon from mineral carbonates. It indicates organic matter cycling and C sequestration.

(Continued)

Table 24.2. (Continued)

Indicator	Function/what it measures	Interpretation/what it indicates
Water extractable organic C	The most mobile and bioavailable fraction of the SOM, specifically, the carbon (C) compounds that easily dissolve in water.	Represents the "active" energy supply that is immediately accessible to the soil microbial community.
Potential C mineralization	The amount of carbon dioxide (CO_2) released by soil microbes immediately after dry soil is rewetted.	Reflects the size and metabolic activity of the microbial community and the amount of "active" organic matter available for decomposition.
Permanganate oxidizable carbon (POXC)	The small portion of SOM that is readily available as a food and energy source for the soil microbial community.	Provides an early indicator of how management is enhancing or degrading SOM and supports informed decisions on soil management.
Total N	The entire pool of nitrogen (N) in the soil, including both the inorganic forms (NO_3^- and NH_4^+) that plants can use immediately and the much larger organic fraction locked in organic matter and microbial cells.	Serves as a measure of the soil's long-term "N bank," representing the total capital available for future fertility.
Potential N mineralization	The amount of plant-available N (e.g., NH_4^+) that a soil microbial community can release from the organic matter pool under ideal temperature and moisture conditions.	Serves as an estimate of the soil's "active" N bank, representing the biological capacity to supply N during the growing season beyond what is immediately present as mineral nitrate.
Water extractable organic N	The soluble and highly mobile fraction of organic N that dissolves easily in water.	Represents the N pool most readily available for microbial consumption and plant uptake.
Extractable P	The portion of P that is potentially available for plant uptake during the growing season.	Drives plant energy transfer, root development, and seed production.
Cation exchange capacity (CEC)	The soil's total capacity to hold and exchange positively charged ions (cations), effectively acting as the soil's "nutrient bank."	Serves as a fundamental indicator of soil fertility and buffering capacity. Soils with high CEC can store more nutrients and resist rapid changes in pH.
Electrical conductivity (EC)	The concentration of dissolved salts (ions) in the soil solution, which determines how easily the soil can conduct an electrical current.	Serves as a proxy for soil salinity and correlates with soil moisture and soil clay content.
pH	The concentration of hydrogen ions (H^+) in the soil solution, indicating whether a soil is acidic, neutral, or alkaline.	Serves as the "master variable" that dictates nutrient solubility and availability and microbial activity.
Exchangeable sodium (Na) percentage (ESP)	Quantifies the proportion of Na relative to the soil's CEC.	Indicates potential risks of sodicity, including poor infiltration, surface crusting, and erosion susceptibility.
Microbial biomass	The total mass of living microorganisms (mostly bacteria and fungi) in the soil.	Represents the "living" fraction of SOM and is typically reported as microbial biomass C or microbial biomass N, acting as a sensitive indicator of the soil's biological health and serves as an "early warning" of soil degradation.
Soil respiration	The CO_2 released from the soil surface.	Indicates how active the living biological community is at any given moment. Represents nutrient cycling and organic matter decomposition, reflecting the soil's ability to convert organic residues into plant-available nutrients.
Phospholipid fatty acid (PLFA)	The living microbial biomass and the community composition of the soil.	Acts as a "census" of the viable microorganisms, including bacteria and fungi, present in the soil at the time of sampling.

Table 24.2. (Continued)

Indicator	Function/what it measures	Interpretation/what it indicates
β-Glucosidase (BG)	The activity of an enzyme primarily responsible for the final step of cellulose degradation.	Breaks down cellobiose into glucose, which provides a critical, readily available energy source for the soil microbial community and serves as an indicator of the C cycle.
N-acetyl-β-D glucosaminidase (NAG)	The activity of an enzyme involved in the breakdown of chitin.	Serves as a primary indicator of N mineralization potential and fungal biomass.

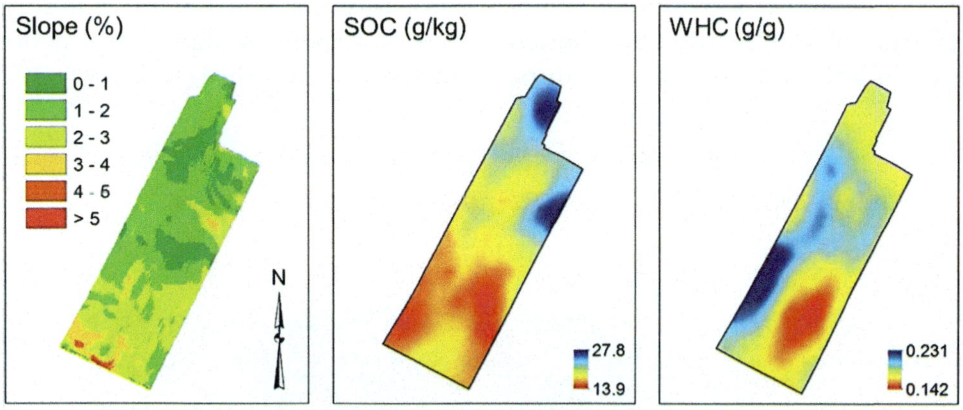

Figure 24.8. An example of sub-field scale spatial variation of SOC and water-holding capacity (WHC) in a field near New Brunswick, Canada (Zebarth et al., 2019).

Clay et al., 2001, 2005). Higher SOC and soil water content in footslope areas can influence plant growth and development and impact the rate at which SOC is mineralized, resulting in a change in biological activity in the soil (Khan et al., 2023). Clay et al. (2005) reported that 2.08% of the original SOC in the footslope position was mineralized annually, whereas in upland soil, only 0.95% of the original SOC was mineralized annually. In the central claypan region of the United States, a soil health assessment of varying management practices on claypan soils with less than a 1% slope found that multiple soil health indicators were highest in the toeslope, followed by those in the summit, while those in the backslope were lowest due to the erosion and redeposition of soil over decades (Veum et al., 2015). Therefore, to accurately characterize soil health in a field, it is important to identify spatial variability. Recognizing this spatial heterogeneity is essential for farmers to apply targeted interventions that enhance soil functions exactly where needed, prevent degradation, and optimize long-term productivity. For more information on the importance of spatial variability, see Chapter 5.

Given that the characterization of all parts of the field is important to accurately estimate soil health, techniques such as remote sensing, visual inspection, and sample collection followed by analysis can be used to characterize soil spatial variability, simultaneously supporting crop production and sustainability (Chang et al., 2022; Getahun et al., 2024; Khan et al., 2023; Niedbała et al., 2023). Advances in technology now provide more precise and cost-effective tools for evaluating soil health (Getahun et al., 2024; Niedbała et al., 2023). These technological improvements include portable sensors that enable real-time measurements of SOC, soil moisture, pH, and EC (see Chapters 11 and 12 for more information). For example, soil sensors that use near-infrared (NIR) and mid-infrared (MIR) spectroscopy, which enable rapid analysis of SOM and nutrient content without extensive sample preparation, are being developed to improve the evaluation of soil and crop health in the field. High-throughput microbial DNA sequencing has also opened new opportunities to understand the biological processes that govern soil functions, revealing detailed insights into microbial diversity and its functional potential and helping researchers create more site-specific fertilizer recommendations. Automated soil sampling systems, often integrated with the Global Positioning System (GPS), streamline the collection process, ensuring consistency and reducing labor requirements. Remote sensing technologies, including the use of satellites and unmanned aerial vehicles to collect data on key indicators, such as organic matter, soil moisture, and vegetation health, offer indirect yet valuable information about soil conditions (Zhang & Kovacs, 2012). Reflectance indices, such as the Normalized Difference Vegetation Index (NDVI), are widely used to estimate nutrient levels and soil productivity across large areas (Figure 24.9; Tucker, 1979). Additionally, advanced data analysis platforms and machine learning tools process large datasets from these technologies, identifying patterns and trends that inform precision management practices. Together, these innovations are transforming soil health monitoring, making it more efficient, scalable, and actionable for sustainable agriculture (Ackerson, 2022).

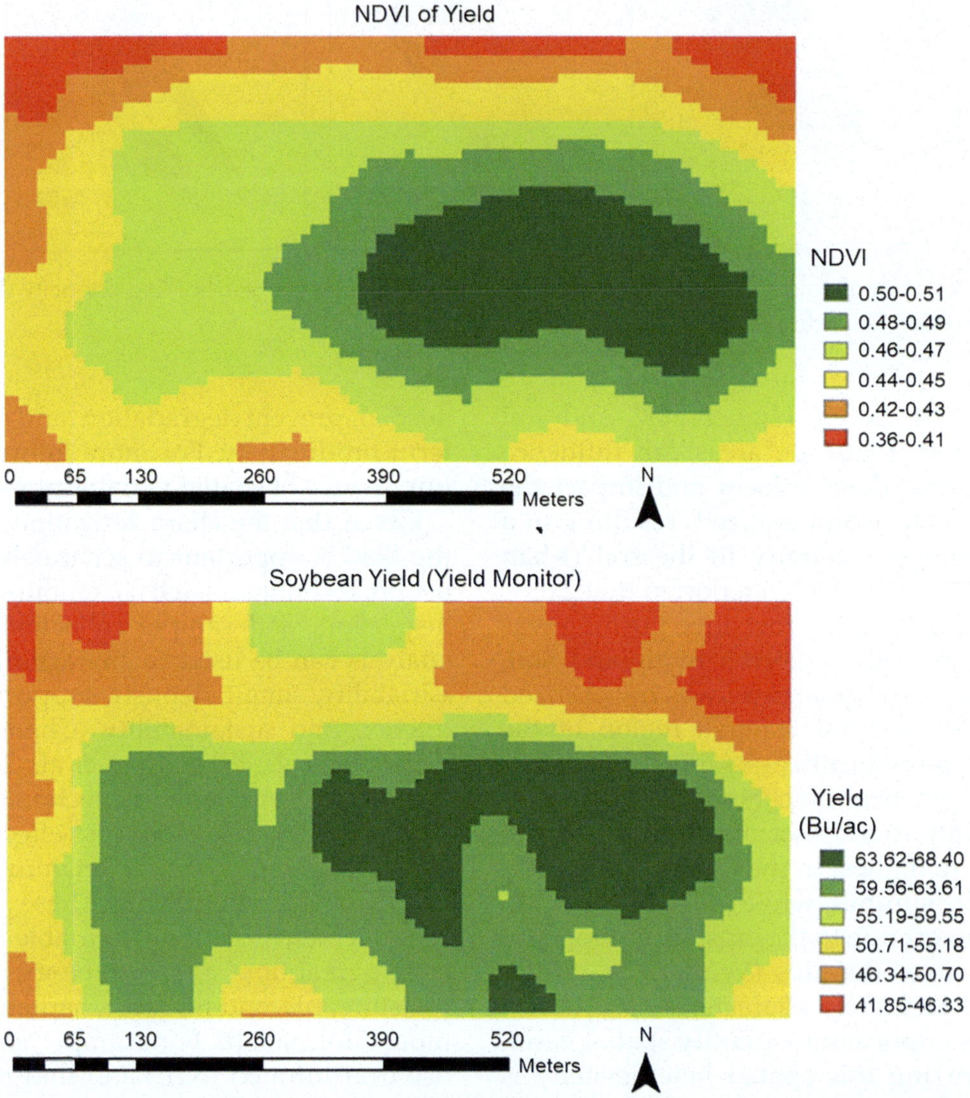

Figure 24.9. The spatial similarity between NDVI and crop yield in a South Dakota soybean field. Remade from Mfuka et al. (2020). (Courtesy Sravanthi Bachina).

Because soil health indicators are spatially heterogeneous, a well-designed sampling plan can provide a comprehensive understanding of soil health across the field (see Chapter 6 for more details on soil sampling techniques). Methods such as grid sampling or stratified random sampling, when combined with the GPS and a geographic information system (GIS), can be used to identify problem areas that limit crop performance (Brus & Heuvelink, 2007; Mulla & Schepers, 1997). This information can help develop a thoughtful sampling plan that supports targeted interventions, reduces waste, and enhances the overall efficiency and sustainability of agricultural operations.

Precision Agriculture Techniques to Enhance Soil Health

Precision agriculture leverages advanced technologies and data-driven approaches to optimize agricultural practices that enhance crop productivity while protecting the environment and soil health (Gamage et al., 2024; Getahun et al., 2024; Niedbała et al., 2023). These techniques improve the efficiency of resource use and address spatial variability within agricultural fields, contributing to better soil health and higher yield outcomes. Precision soil health practices include site-specific cover crops and tillage, variable rate technology (VRT), decision support systems (DSSs) integrated with artificial intelligence (AI), and precision irrigation and drainage systems (Guo, 2021; Lakhiar et al., 2024; Mulla & Schepers, 1997; Zhang & Kovacs, 2012).

Site-Specific Soil Management

Site-specific soil management involves tailoring agricultural practices to the unique characteristics and needs of different areas within a field (Figure 24.10; Mulla & Schepers, 1997). This approach acknowledges the inherent spatial variability of SOC, soil moisture, soil nutrients, and soil drainage across an agricultural landscape. By using tools such as GPS, GIS, and remote sensing, producers can map soil variability and design targeted practices. For instance, adjusting fertilizer applications or selecting specific crops based on localized soil conditions can enhance nutrient use efficiency and reduce environmental impacts. Site-specific management ensures that each area of a field receives the care it needs, promoting overall soil health and agricultural

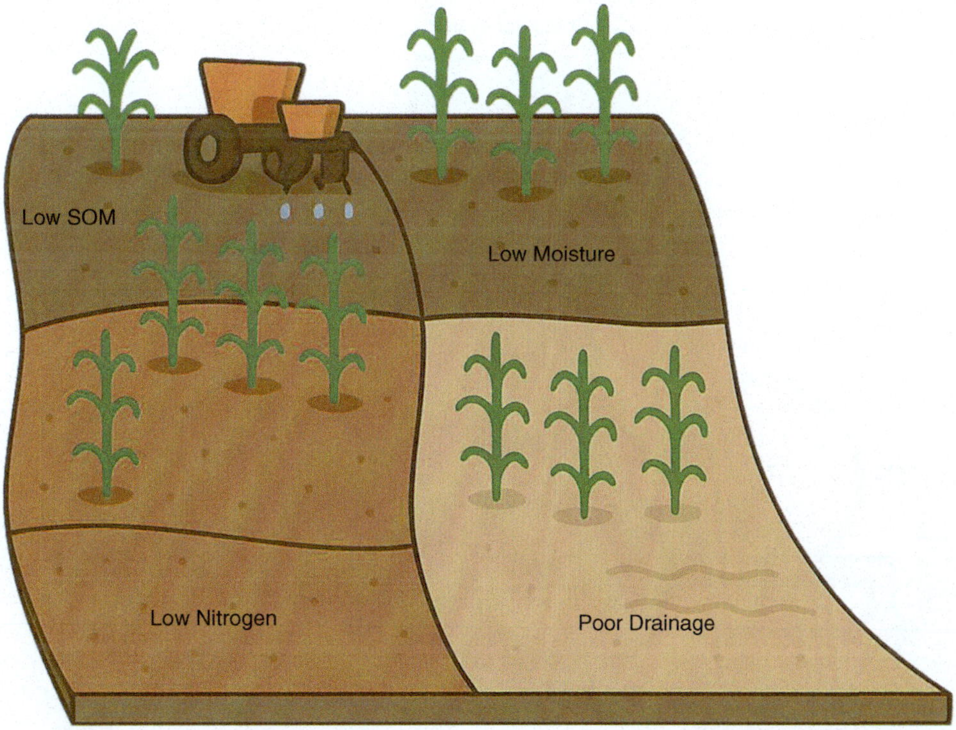

Figure 24.10. The delineation of a field into different management zones. (Courtesy Xi Zhang).

sustainability (Mulla & Schepers, 1997; Stott, 2019; Zhang & Kovacs, 2012).

Variable Rate Technology for Input Optimization

Variable rate technology enables the precise application of inputs, such as fertilizers, pesticides, and water, based on the specific requirements of different zones within a field. By integrating data from soil maps, sensors, and yield monitors, VRT systems (Figure 24.11) adjust application rates in real time, ensuring that resources are neither overused nor underapplied. This reduces nutrient runoff and improves efficiency. For example, applying N only where it is deficient enhances soil fertility without contributing to leaching or greenhouse gas emissions. Variable rate technology not only optimizes soil health but also lowers input costs, making it economically and environmentally beneficial (Campos et al., 2020; Grisso et al., 2011).

Use of DSS and AI

New AI-based data analysis tools are being developed to help producers understand the linkages between soil health and yields. In the future, these tools may combine information from soil sensors, weather stations, and crop models to recommend precise farming practices that optimize both soil and plant health. For instance, machine learning algorithms can forecast soil moisture depletion or nutrient deficiencies, enabling timely intervention. These technologies facilitate data-driven decision-making, reducing uncertainty and ensuring that management practices are aligned with the specific needs of the soil and crops (Pace et al., 2025). To learn more about machine learning and AI in agriculture, see Chapter 27.

Implementation of Precision Irrigation and Drainage Systems

This section discusses precision irrigation and drainage systems within the broader context of soil health. For more detailed information on precision water management, including precision irrigation, see Chapter 16. An important component of soil health is soil moisture at a level that minimizes nutrient losses and optimizes microbial activity (Al-Shammary et al., 2024; CAST, 2024). Soil moisture at a level between the field capacity and the permanent wilting point, known

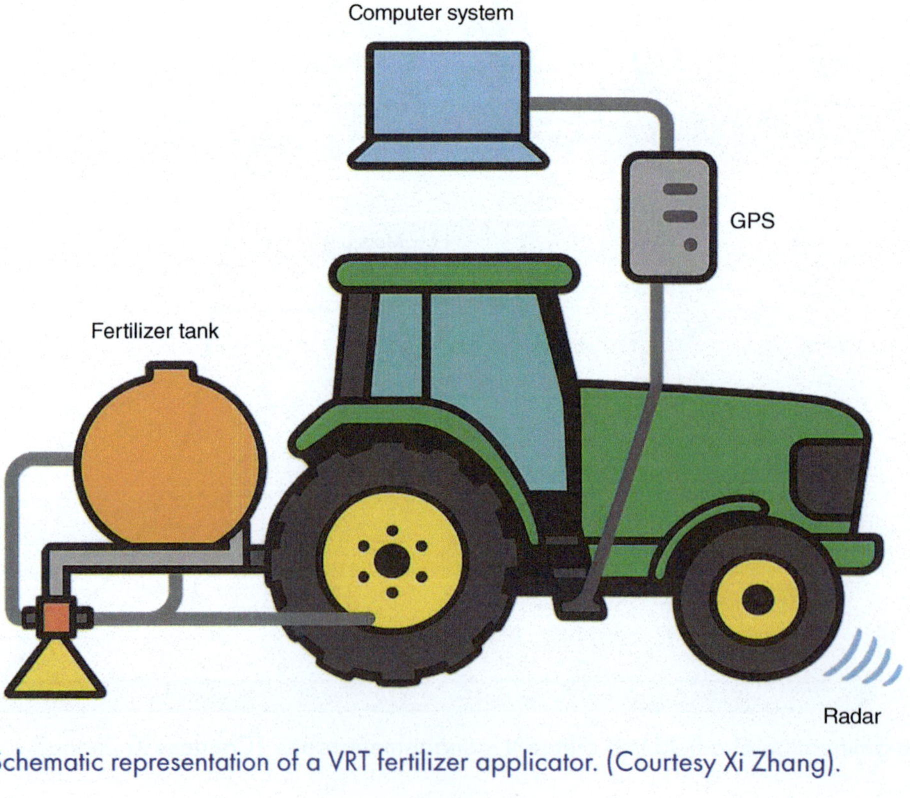

Figure 24.11. Schematic representation of a VRT fertilizer applicator. (Courtesy Xi Zhang).

as plant-available water, can improve crop productivity by delivering N to the roots through the soil solution. This can also reduce the amount of N that needs to be applied. Additionally, an appropriate combination of irrigation and drainage systems can mitigate issues such as salinity and soil erosion by efficiently channeling excess water away from crops and aerating the soil.

Ensuring adequate soil moisture can be accomplished using precision irrigation and drainage systems that manage water resources efficiently (Lakhiar et al., 2024). These systems use sensors and automated controls to deliver the right amount of water at the right time and locations in a field, preventing both under-irrigation and waterlogging (Figure 24.12). There are also natural ways to improve soil moisture retention and drainage. For example, greater SOM improves water retention, while the use of cover crops can improve soil drainage. These techniques and systems must be used on a site-specific basis to address the unique challenges of the specific field, increasing resource use efficiency by crops and improving soil health.

Figure 24.12. Sensors measure soil moisture to provide real-time data for irrigation decisions. (Courtesy Xi Zhang).

Management Techniques to Improve Soil Health

Agricultural management practices can enhance soil health or contribute to its deterioration (Bagnall et al., 2023; Chang et al., 2022; Karlen et al., 2019). The U.S. Dust Bowl is a well-known example of severe soil degradation that occurred between 1910 and 1930, when intensive tillage and lack of soil cover in the Great Plains Region contributed to the destruction of soil aggregates as a result of the rapid loss of SOM. A series of droughts in the 1930s resulted in crop failures and severe wind erosion. The loss of productive soil was not unique to the Great Plains or to the 1930s. Similar losses have occurred globally over time due to wind and water erosion, salinity, soil acidity, and toxic chemical contamination. A few primary agricultural management practices and principles for improving soil health have been successfully demonstrated across the climatic zones and crop types in the United States (Ankomah et al., 2024; Bagnall et al., 2021, 2023; Crespo et al., 2024; Kelly et al., 2021; Mitchell et al., 2015; Nunes et al., 2020; Ruis et al., 2025). These principles for managing soil health include minimizing disturbance and maximizing the presence of living roots, soil cover, and biodiversity. For example, tillage reduction and diversification through crop rotation and livestock integration are key practices for improving soil health (Figure 24.13). Diverse crop rotations break pest and disease cycles, enhance nutrient availability and cycling, and promote a more resilient microbial community. Pasture management, particularly rotational grazing, minimizes soil compaction, promotes nutrient cycling, and improves root systems that stabilize the soil (Teague & Kreuter, 2020). Integrating livestock into cropping systems further contributes to soil health by adding organic matter to the soil through manure. This increase in SOM stimulates root growth through grazing and enhances nutrient distribution and cycling. Pesticides have both beneficial and adverse impacts on soil health. Beneficial impacts include the reduction of pest populations that adversely affect yields, whereas adverse impacts include effects on non-target organisms. In addition, the slow decomposition of herbicides

Figure 24.13. An organic farm with pasture–corn–soybean rotation and wheat and rye as cover crops in Paxton, IL. (Courtesy Vander Luis Novaes Nunes).

can prevent crop establishment. Additional information on precision weed, insect, and disease management is available in Chapters 17, 18, and 19. These practices address the unique challenges and opportunities of each region, promoting sustainable farming and resilient agroecosystems.

Tillage Reduction

Implementing reduced tillage and no-till practices is a cornerstone of soil health management, primarily because it minimizes mechanical disturbance. Conventional tillage physically fractures the soil, accelerating the decomposition of organic matter. In contrast, no-till systems allow for the development of stable soil aggregates, which improve air and water movement and increase biological activity. This undisturbed structure, combined with the protective layer of crop residues left on the surface, significantly reduces erosion, enhances water infiltration, and fosters a "C sink" by sequestering organic carbon that would otherwise be released into the atmosphere, ultimately promoting a more sustainable and resilient agroecosystem.

Cover Crops

Cover crops provide soil cover and a living root system, serving as a mulching layer that reduces erosion, enhances organic matter, and supports microbial biodiversity during off-seasons (Joshi et al., 2023; Table 24.3). Cover crop residues help reduce evaporation, conserving soil moisture during drought conditions. They also enhance water infiltration by boosting soil living matter, which improves the soil's ability to absorb and retain water in the root zone (Figure 24.14). Deep-rooted species, such as radish, can penetrate compacted subsoil layers or hardpans. These taproots act as "biological tillage" tools, creating vertical macropores that persist long after the plant has died and decayed. In the central United States, growing cover crops between corn–soybean rotations has shown remarkable benefits. Cover crops reduce soil erosion during fallow periods, enhance organic matter content, and improve N cycling (Fioratti Junod et al., 2024; Kelly et al., 2021). Integrating a legume cover crop, such as crimson clover, enhances N cycling by forming a symbiotic relationship with *Rhizobium* bacteria, which capture atmospheric N and convert it into a plant-available form. Once the clover is terminated and its biomass decomposes, this biologically fixed N is mineralized and released into the soil, providing a natural, slow-release fertilizer source for the subsequent cash crop, such as corn. Nitrogen leaching occurs when excess nitrates, which are highly mobile in the soil, are washed downward by water, beyond the reach of plant roots and into the groundwater. Nonlegume cover crops, such as cereal rye, are exceptionally effective "scavenger" crops for preventing N leaching. Cereal rye absorbs excess nutrients during its growth, effectively locking N into the plant biomass and helping keep nutrients on the farm and out of the groundwater. Cover crop studies in the Midwest Region have reported increased microbial activity and reduced N leaching (Blanco-Canqui et al., 2025; Ruis et al., 2025).

Soil Amendments

Soil amendments, such as compost, biochar, or lime, further enrich soil fertility, improve structure, and enhance biological activity (Andrade et al., 2024). Together, these practices create a synergistic effect that boosts soil health, making it more productive and resilient to environmental challenges. Organic and inorganic fertilizers can have both beneficial and adverse effects on soil

Table 24.3. Selected cover crop species and their roles.

Type	Cover crop species	Primary functions/benefits	Root traits	Nitrogen source/ scavenger
Nonlegumes	Cereal rye (*Secale cereale*)	Excellent for erosion control and suppresses weeds.	Fibrous, extensive root system that improves soil structure and prevents compaction.	Scavenger
	Oats (*Avena sativa*)	Provides quick ground cover, improves soil tilth, and adds organic matter when terminated.	Fibrous roots that enhance topsoil aggregation.	Scavenger
	Wheat (*Triticum aestivum*)	Prevent erosion and add organic matter.	Fine root system that improves topsoil tilth.	Scavenger
	Sorghum–sudangrass (*Sorghum bicolor* × *S. bicolor* var. *sudanese*)	Suppresses weeds, adds large amounts of biomass, and improves soil organic matter.	Deep and aggressive root system that loosens subsoil.	Scavenger
Legumes	Crimson clover (*Trifolium incarnatum*)	Adds N to the soil, supports pollinators, and provides spring biomass.	The root system is simple, dominated by a taproot, and often well nodulated.	Source
	Hairy vetch (*Vicia villosa*)	Builds soil N, improves topsoil tilth, and creates a loose and friable soil structure.	Vining, fibrous roots with N-fixing nodules.	Source
	Field peas (*Pisum sativum* subsp. *arvense*)	Contributes N and provides spring forage.	Moderately deep taproot with lateral branching.	Source
Brassicas	Radish (*Raphanus sativus*)	Breaks up compacted soil layers ("bio-drilling").	Deep taproot that penetrates compacted layers.	Scavenger

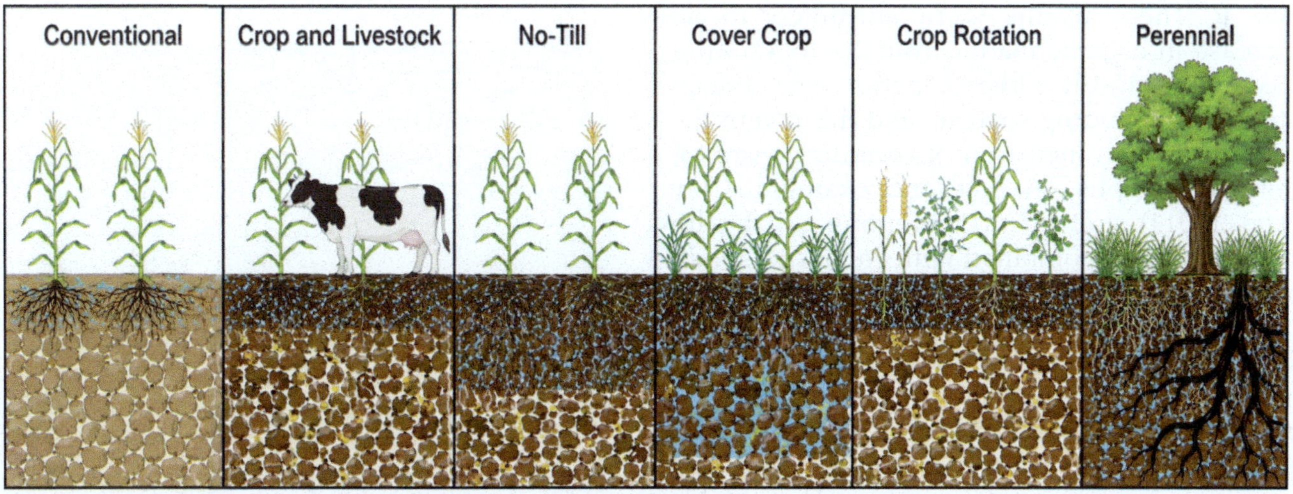

Figure 24.14. Soil water storage across various management practices. Remade from Basche & DeLonge (2019). Created in BioRender (Fischer, B. (2026)). (Courtesy Skye Brugler).

health. When applied appropriately, they replenish soil nutrients, promote plant growth, and improve crop yields. However, excessive or imbalanced fertilizer application can lead to nutrient leaching, soil acidification, and off-site movement of soil nutrients.

Soil Health and Enhanced Resilience to Drought and Flooding

Soil health practices that build soil structure, maintain surface cover, and sustain biological activity can improve resilience and adaptability to droughts and flooding (Ankomah et al., 2024; Clay et al., 2014; Kelly et al., 2021). One of the primary ways soil health practices impact drought and flood resilience is by increasing SOC, which can enhance soil structure and porosity. Improved soil structure facilitates rapid water infiltration during rainfall events while also increasing water storage in the root zone.

During drought periods, the higher soil water retention capacity ensures that more water remains available to crops, reducing the intensity and duration of water stress. Healthy soils also enhance root development and crop performance. A biologically active and physically stable soil environment allows roots to penetrate more deeply and uniformly, accessing water from greater soil depths. The synergy between plant roots and soil biota, such as mycorrhizal fungi, can further optimize water uptake efficiency.

There are several examples of how reducing tillage and implementing crop rotations can improve resilience. Anderson (2015) showed that the adoption of no-till farming and diversified crop rotation increased corn yields while reducing the need for supplemental water, N, and P. These results were attributed to an increase in organic matter from 2% to 4% after 23 years of no-till. Other benefits were associated with including soybean and dry pea in the rotation, which increased N fixation by symbiotic diazotrophs. In a second example, Clay et al. (2014) showed that adopting soil health practices when combined with enhanced genetics produced an $89.60/ac ($221.30/ha) impact on the South Dakota farm economy during the 2012 drought. However, the benefits of adopting soil health practices vary with location. For example, in the semi-arid regions of the North American Great Plains, cover crop-induced increases in transpiration can reduce yields (Moriles-Miller et al., 2025), whereas in humid environments, increased transpiration can reduce soil water ponding.

Adoption of Soil Health Practices

Appropriate soil health practices depend on soil, management, and weather conditions. For example, in the North American Northern Great Plains, an appropriate soil health practice may be to convert salt-affected degraded soils to perennial grasses (Clay et al., 2022), as well as using no-tillage and cover crops (Clay et al., 2014; Hansen et al., 2012). For many farmers in the North American Great Plains, no-till farming has reduced erosion, conserved moisture, promoted C sequestration, and maintained or improved productivity in water-limited environments (Hansen et al., 2012; Simão et al., 2023). In South Dakota, Nebraska, and Kansas, no-tillage was adopted on more than 45% of row crop land in 2017 (LaRose & Myers, 2019).

In the southeastern portion of North America, crop growth is limited by erosion, low concentrations of SOC, and degraded soil structure resulting from weathering and intensive agricultural management practices (e.g., residue removal, tillage; Baumhardt et al., 2015; Franzluebbers, 2005; Weidhuner et al., 2021). In this region, conservation tillage, when combined with cover cropping and diversified crop rotations, can improve soil health in a diverse cropping system (Figure 24.15; Adeli et al., 2024). Since the 1980s, no-till farming has been widely adopted by producers in Tennessee. With an adoption rate of approxi-

Figure 24.15. A field in Louisiana with winter rye as a cover crop ahead of soybean planting in the crop rotation. (Courtesy Xi Zhang).

mately 80%, the state now ranks among the national leaders in no-till agriculture (Jagadamma et al., 2021). These practices improve infiltration, soil water-holding capacity, and microbial biodiversity, all of which are critical for sustaining crops in regions prone to heavy rainfall and nutrient leaching. In Mississippi, no-tillage, when combined with cover cropping, can protect the soil surface, enhance N availability, and reduce compaction (Chang et al., 2023).

Alternate wetting and drying (AWD) is a water-management practice in irrigated rice production in which fields are allowed to dry until the water table drops to the soil surface through infiltration and evapotranspiration before being reflooded. Unlike traditional continuous flooding, AWD improves soil functions by transitioning the soil from an anaerobic (oxygen-poor) state to a periodic aerobic (oxygen-rich) state, which stimulates the activity of aerobic microbes and accelerates the decomposition of organic matter. This shift helps mitigate the accumulation of toxic organic acids, reduce the production of methane (CH_4), and minimize edge-of-field runoff (Henry et al., 2017). Recently, great interest has focused on rice produced under AWD management in the southern United States (e.g., Arkansas, Louisiana).

In the western United States, soil health is closely linked to water use efficiency. In this arid system, drip irrigation can improve water efficiency and help prevent salinity buildup. Additionally, the application of organic amendments, such as compost, has improved soil structure and nutrient content. Studies have reported increased SOC and better water retention, reducing the reliance on chemical inputs and enhancing overall soil health (Mitchell et al., 2015).

Agroforestry is a land management system that intentionally integrates trees or shrubs with crops and/or livestock on the same unit of land (Dollinger & Jose, 2018; Hillbrand et al., 2017). This practice enhances soil health by creating a diverse "closed-loop" system where tree roots stabilize soil structure and pull deep-stored nutrients to the surface, while falling leaf litter provides a continuous supply of organic matter. Agroforestry is recognized for its ability to improve water infiltration, reduce thermal stress on crops through shading, and significantly increase C sequestration compared with monoculture systems. By mimicking natural forest structures, it fosters a robust microbial community and provides a biological buffer against climate extremes, making it a cornerstone of regenerative agriculture in both tropical and temperate regions (e.g., Brazil, Cameroon, China, Ethiopia, India, Mexico, Nigeria, Tanzania).

Challenges with Soil Health Interpretation

The adoption of soil health management practices offers immense benefits for agricultural productivity and sustainability (Karlen et al., 2019; Lehmann et al., 2020). However, adoption is limited in many regions. One of the challenges limiting the widespread adoption of soil health practices is the lack of harmonized, standardized methods and protocols to facilitate soil health monitoring and goal setting for farmers. Historically, public and private laboratories have implemented their own adaptations of methods, and results often cannot be compared across laboratories. In addition, many methods are costly, labor-intensive, or require too much time or space to be profitable. Ongoing efforts are focused on developing high-throughput, streamlined, and lower-cost protocols that are reproducible and interpretable for farmers (e.g., Moebius-Clune et al., 2016).

For many decades, confidence in soil health interpretation tools was low due to models based on small, spatially limited datasets and reliance on "expert opinion." Research at both local and continental scales has shown that soil health indicators are strongly shaped by a region's climate and specific soil type. Because these natural conditions interact in complex ways, we must first account for the local environment before we can accurately measure how much a farmer's management practices are actually improving soil health. A robust, regionally relevant tool that accounts for spatial variability in soils at multiple scales is needed for meaningful, site-specific interpretation. To facilitate the interpretation of soil health indicators across a pedoclimatic gradient, an interpretive framework must provide a wide range of regionally relevant indicator options that can (a) account for inherent site-specific factors (e.g., edaphic and hydroclimatic factors), (b) be sensitive to anthropogenic activities, and (c) facilitate broad-scale monitoring to ensure sustainable agricultural management (Nunes et al., 2024). To this end, the Soil Health Assessment Protocol and Evaluation (SHAPE) tool was developed for scoring soil health values from any location across the United States on a scale from 0% to 100% and to set benchmark values that define a soil's potential. These benchmarks can be used to calculate soil health gaps and set goals for conservation management (Veum et al., 2025). The newest version of the tool (SHAPEv1.0S) incorporates a spatially explicit model to define peer-group scoring curves and site-specific benchmark values based on inherent soil characteristics (soil texture and soil taxonomy) and local climate conditions (mean annual precipitation and temperature), with proximity-based refinements. In other words, this approach captures regional variability in soil characteristics to improve site-specific SHAPE scores and benchmark values. Example SHAPEv1.0S peer-group scoring curves for SOC under variable climate conditions (including scores, 95th percentile benchmark values, and uncertainty bands) are illustrated in Figure 24.16a–l. In these examples, soil texture group, soil suborder group, and SOC content were held constant to highlight the effect of spatial proximity. Currently, SHAPE scoring

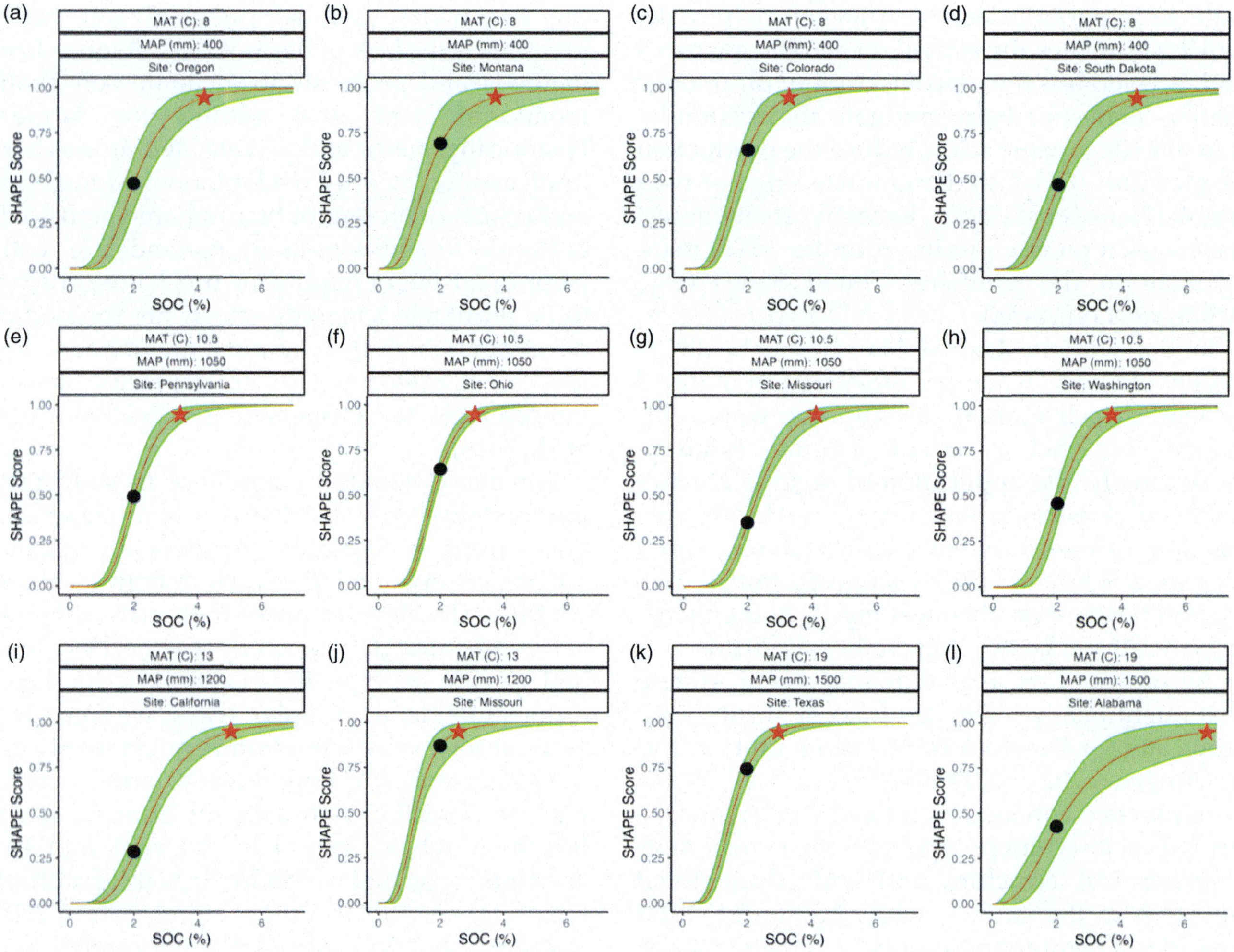

Figure 24.16. Example of SHAPEv1.0S peer-group scoring curves with scores (black dots), 95th percentile benchmarks (red stars), and uncertainty bands (green) under variable climate conditions (mean annual temperature [MAT, °C] and mean annual precipitation [in millimeters]). In these examples, soil texture group, soil suborder group, and SOC content are constant to illustrate the effect of spatial proximity. (Courtesy Veum et al., 2025).

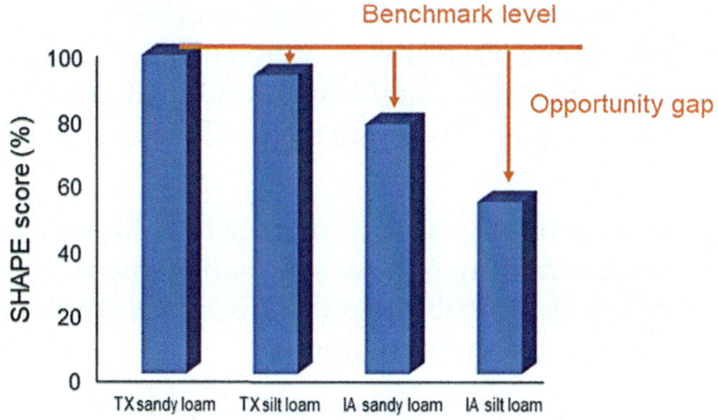

Figure 24.17. SHAPE scores for four sites with different climate and soil combinations. All sites have 2.0% SOC, yet each site has a unique SOC opportunity gap to reach its potential. (Courtesy Kristen Veum).

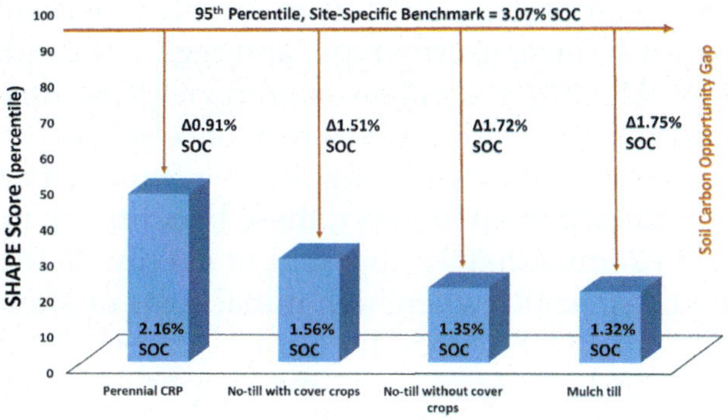

Figure 24.18. SHAPE scores for four long-term cropping systems located at the same site in Missouri, where the 95th percentile benchmark value is 3.07% SOC based on site characteristics. The SOC content varies from 1.32% to 2.16%, and the opportunity gap varies from 0.91% to 1.75% SOC as a result of management. (Courtesy Kristen Veum).

curves are available for the following soil indicators: SOC, POXC, soil respiration (4-day), total protein, aggregate stability (1–2 mm), and aggregate stability by the Cornell sprinkle method.

Examples are presented in Figure 24.17 to illustrate how SHAPE accounts for climate and soil texture. The soils from a warmer climate with less precipitation (Texas) score higher than those from a cooler climate with more precipitation (Iowa), even though the SOC content is equal (2.0%). Cooler, wetter climates tend to retain more SOC because organic matter decomposes more slowly, whereas warmer, drier climates tend to oxidize SOC more rapidly. The effect of soil texture is also evident in Figure 24.17: Coarse-textured sandy soils are not expected to retain as much SOC as finer-textured silt loam soils, so they score higher when the C content is equal.

In an example from a single site located in Missouri, USA (Veum et al., 2015), Three long-term corn–soybean management systems were assessed (Figure 24.18). The results showed that the SOC content ranged from 1.32% in a mulch-tillage system to 1.56% in a no-till, cover crop system and the opportunity gap varied from 0.91% to 1.75% SOC as a result of management. A restored perennial system that served as a reference had 2.16% SOC. All four management systems were located at the same site and shared the same inherent climate, soil texture, and soil taxonomy groupings; therefore, the benchmark value representing the SOC potential for the site was 3.07%, while each management system had a different SHAPE score and SOC opportunity gap.

Work is ongoing to expand the SHAPE tool to include additional soil health indicators, improve the underlying dataset and statistical model, and incorporate broader management and land-use categories for a more refined interpretation. More details can be found in Veum et al. (2025). The SHAPEv1.0S tool is publicly available at https://shapesoilhealth.shinyapps.io/shape_app_v1s.

Barriers to Adopting Soil Health Precision Farming Practices

For many farmers, adopting soil health practices results in increased organic matter, which enhances the soil's ability to retain water and resist erosion, providing improved resilience to extreme weather events (Clay et al., 2014). There are many barriers to the adoption of soil health practices, including technological challenges, economic constraints, cultural hurdles, and knowledge gaps. Some of these barriers may be overcome by technological advances. For example, robots are being tested to assist with pest management (Chapter 11), while machine learning algorithms are being tested to provide early warning signals of pest infestation and nutrient deficiencies (Chapter 26).

The decision to implement soil health practices is tied to changes in production costs and expected returns (Aarif K.O. et al., 2025; Getahun et al., 2024; Sharma & Shivandu, 2024). While soil health practices generate benefits, these benefits are often difficult to quantify depending on farm size, crop type, and regional conditions. Healthy soil enhances crop yields, reduces input costs, and improves resilience to extreme climatic events, delivering direct economic benefits to farmers. However, these benefits are not always immediately apparent or evenly distributed, especially when high initial costs or short-term profit pressures discourage long-term soil stewardship. Moreover, many positive outcomes of soil health, such as improved water quality, reduced erosion, and C sequestration, are externalities that extend beyond the farm, benefiting society at large. Producers frequently view profitability and stewardship as closely linked motivations, yet they may hesitate to invest in soil health when the benefits are uncertain or not directly realized by the farmer. This market failure justifies policy interventions, such as cost-sharing programs, soil health subsidies, or educational outreach to promote adoption of sustainable practices (Bagnall et al., 2020; Stevens, 2018). Programs such as the U.S. Department of Agriculture's Conservation Stewardship Program and Environmental Quality Incentives Program provide financial assistance to farmers implementing sustainable practices. Effective soil health economics must consider both the internalized returns to farmers and the broader social value of preserving soil as a public good.

Future Trends and Potential Research Areas in Precision Soil Health Management

The future of soil health management is shaped by technological innovation, interdisciplinary research, and global efforts to address climate variability. Key trends and research areas include the following:

1. Development of low-cost, high-accuracy sensors capable of real-time monitoring of soil water and nutrient concentrations (Chapter 12).
2. AI-driven models to predict soil health trends, optimize input application, and identify potential problem areas with greater precision (Chapter 27).
3. Combining remotely sensed data with Internet of Things devices to enable comprehensive, real-time monitoring of soil conditions across large areas (Chapters 11, 12, and 13).
4. Research into soil C dynamics to enhance our understanding of how soil management practices contribute to C sequestration and climate change mitigation.
5. Expansion and refinement of the SHAPE soil health interpretation tool to include additional soil health indicators, streamlined laboratory protocols, and management recommendations.
6. Development of user-friendly (e.g., farm-level) platforms that integrate soil health data with crop models, weather forecasts, and economic analysis to support informed decision-making (Sanyaolu & Sadowski, 2024; Sharma & Shivandu, 2024).

Conclusions

This chapter highlighted the foundational role of soil health in achieving sustainable agricultural productivity and ecosystem resilience. Soil health is conceptualized as the dynamic interaction of physical, chemical, and biological properties that support plant growth, enhance biodiversity, and buffer against both short- and long-term environmental stresses. The chapter also explored how precision agriculture enables site-specific soil management by accounting for the spatial variability of key soil indicators, including aggregate stability, SOC, microbial activity, and nutrient content.

A core theme of the chapter was the importance of site-specific interventions that align with soil conditions and crop needs across heterogeneous landscapes. Physical indicators, such as water-holding capacity and porosity, influence root development and water movement, chemical indicators, such as pH and CEC, affect nutrient availability, and biological indicators, such as C mineralization and microbial biodiversity, drive nutrient cycling. The chapter emphasized that soil health is both spatially variable and temporally dynamic, requiring continuous assessment using geostatistical and sensor-based approaches.

Technological innovations such as NIR spectroscopy, microbial DNA sequencing, DSS, and AI-enhanced models are improving how farmers monitor and manage soil conditions. There are many economic and practical challenges to adopting precision soil health practices, including high initial costs, knowledge gaps, and delayed returns. Nonetheless, future opportunities lie in developing novel sensors, data analysis techniques, and farmer-accessible tools, expanding farmer education, and creating policy frameworks that account for both the private and public benefits of soil stewardship.

Study Questions

1. Why is it important to link soil health indicators directly to crop yield and production in the context of precision agriculture?
   a. Because yield is the only variable influenced by soil conditions
   b. To ensure soil sampling frequency can be reduced over time
   c. To make soil health assessments economically relevant and actionable for producers
   d. To eliminate the need for biological soil testing
2. A soil test shows moderate aggregate stability but low microbial activity in a sandy loam field under continuous corn production. What combination of practices would best address these issues in the precision framework?
   a. Increase tillage frequency and synthetic N applications
   b. Introduce organic amendments and multispecies cover crops targeted to low-activity zones
   c. Apply fungicides uniformly to eliminate competing microbes
   d. Irrigate the entire field using overhead sprinkler systems
3. How do DSSs assist in soil health management?
   a. Automate pesticide spraying regardless of crop stage
   b. Monitor air temperature trends only
   c. Integrate data from sensors and models to guide site-specific decisions
   d. Replace all manual field inspections
4. Which of the following is most likely to compromise the accuracy of soil health maps generated using zone sampling?
   a. Over-sampling in eroded shoulder positions
   b. Uniform topography across the field
   c. Failure to incorporate multi-temporal yield data and topographic features
   d. Use of automated GPS guidance for sampling points
5. Define soil health and explain how its physical, chemical, and biological properties contribute to sustainable agricultural systems.
6. Why is understanding spatial variability important in soil health management? Discuss how advanced technologies, such as remote sensing and GIS, are used to assess soil health.
7. Compare the impacts of conservation tillage, crop rotation, and cover cropping on soil health.
8. Describe how precision agriculture tools such as VRT and DSS optimize soil health management. What are the economic and environmental benefits?
9. What are the major barriers to adopting precision agriculture practices for soil health? Discuss potential solutions and emerging trends that could address these challenges.

Fun Activities to Test Soil Health at Home

1. Soil Aggregate Test

Test soil aggregate stability at home. First, take two jars and shape two wire mesh baskets, one to sit about 1.5 inches below the rim of each jar. Second, fill each jar with water to within half an inch of the top. Third, place soil clods from two different sources into the baskets and gently lower them into the jars. Use a stopwatch to record how long it takes for the clods to disintegrate. The longer the time, the more stable the aggregates are.

Find directions at this link: https://www.nrcs.usda.gov/sites/default/files/2022-10/South_Dakota_Slake_Test_SD-FS-107.pdf.

2. Microbial Decomposition Test

Test microbial activity in your soil and explore soil health and the living ecosystem beneath your feet. Simply take a pair of 100% cotton underwear, bury them about 6 inches deep in your garden or yard soil, and mark the location. After about 60 days, carefully dig them up and observe the results. If your soil is biologically active, the cotton will be significantly decomposed, that is, it will be eaten away by microbes and other soil organisms. Little decomposition, on the other hand, may indicate low microbial activity.

Find directions at this link: https://www.nrcs.usda.gov/state-offices/montana/soil-your-undies-challenge#:~:text=How%20does%20it%20work%3F,deteriorated%2C%20the%20healthier%20your%20soil!

ACKNOWLEDGMENTS

This paper was funded in part by the University of Tennessee Institute of Agriculture, USDA-ARS, South Dakota State University, USDA Natural Resources Conservation Service Conservation Innovation Grants (NR233A750011G019), USDA National Institute of Food and Agriculture under the Multi-State Hatch project (KY006151-7008110-S1090), the South Dakota Corn Utilization Council, and the South Dakota Soybean Research and Promotion Council. The use of trade names is for the convenience of the reader and does not imply endorsement by the authors.

REFERENCES

Aarif K. O., M., Alam, A., & Hotak, Y. (2025). Smart sensor technologies shaping the future of precision agriculture: Recent advances and future outlooks. *Journal of Sensors*, *2025*(1), 2460098. https://doi.org/10.1155/js/2460098.

Ackerson, J. P. (2022). Precision agriculture and soil health. *Crops & Soils*, *55*(5), 35–39.

Adeli, A., Brooks, J. P., Miles, D. M., Hu, J., Huang, Y., Bheemanahalli, R., & Jenkins, J. N. (2024). Integration effects of soil amendments and cover crops on dryland no-till corn. *Agronomy Journal*, *116*(3), 1579–1589. https://doi.org/10.1002/agj2.21571.

Al-Shammary, A. A. G., Al-Shihmani, L. S. S., Fernández-Gálvez, J., & Caballero-Calvo, A. (2024). Optimizing sustainable agriculture: A comprehensive review of agronomic practices and their impacts on soil attributes. *Journal of Environmental Management*, *364*, 121487. https://doi.org/10.1016/j.jenvman.2024.121487.

Anderson, R. L. (2015). Increasing corn yield with no-till cropping systems: A case study in South Dakota. *Renewable Agriculture and Food Systems*, *31*(6), 568–573. https://doi.org/10.1017/S1742170515000435.

Andrade, M. G. D. O., Cordeiro, C. F. D. S., Calonego, J. C., Franzluebbers, A., & Rosolem, C. A. (2024). Soil amendment and N fertilization strategies to improve C sequestration and storage in soil aggregates. *Soil Science Society of America Journal*, *88*(5), 1767–1777. https://doi.org/10.1002/saj2.20722.

Ankomah, G., Amissah, S., Hollifield, S., Habteselassie, M. Y., Franklin, D. H., Snider, J. L., Kemerait, R. C., Roberts, P. M., & Sintim, H. Y. (2024). Integrating cover crops and organic amendments to mitigate the limitations of tillage on soil health and cotton productivity. *Agronomy Journal*, *116*(5), 2615–2629. https://doi.org/10.1002/agj2.21649.

Bagnall, D. K., McIntosh, W. A., Morgan, C. L. S., Woodward, R. T., Cisneros, M., Black, M., Kiella, E. M., & Ale, S. (2020). Farmers' insights on soil health indicators and adoption. *Agrosystems, Geosciences & Environment*, *3*(1), e20066. https://doi.org/10.1002/agg2.20066.

Bagnall, D. K., Rieke, E. L., Morgan, C. L. S., Liptzin, D. L., Cappellazzi, S. B., & Honeycutt, C. W. (2023). A minimum suite of soil health indicators for North American agriculture. *Soil Security*, *10*, 100084. https://doi.org/10.1016/j.soisec.2023.100084.

Bagnall, D. K., Shanahan, J. F., Flanders, A., Morgan, C. L. S., & Honeycutt, C. W. (2021). Soil health considerations for global food security. *Agronomy Journal*, *113*(6), 4581–4589. https://doi.org/10.1002/agj2.20783.

Basche, A. D., & DeLonge, M. S. (2019). Comparing infiltration rates in soils managed with conventional and alternative farming methods: A meta-analysis. *PLOS One*, *14*(9), e0215702. https://doi.org/10.1371/journal.pone.0215702.

Baumhardt, R. L., Stewart, B. A., & Sainju, U. M. (2015). North American soil degradation: Processes, practices, and mitigating strategies. *Sustainability*, *7*(3), 2936–2960. https://doi.org/10.3390/su7032936.

Blanco-Canqui, H., Ruis, S. J., Koehler-Cole, K., Elmore, R. W., Francis, C. A., Shapiro, C. A., Proctor, C. A., & Ferguson, R. B. (2023). Cover crops and soil health in rainfed and irrigated corn: What did we learn after 8 years? *Soil Science Society of America Journal*, *87*(5), 1174–1190. https://doi.org/10.1002/saj2.20566.

Blanco-Canqui, H., Ruis, S. J., Mamo, M., Shapiro, C. A., Proctor, C., Parsons, J., & Thompson, L. (2025). Interseeding cover crop into an irrigated sandy loam for 6 years: Soil, crop, and economic response. *Agronomy Journal*, *117*(1), e70013. https://doi.org/10.1002/agj2.70013.

Brus, D. J., & Heuvelink, G. B. M. (2007). Optimization of sample patterns for universal kriging of environmental variables. *Geoderma*, *138*(1–2), 86–95. https://doi.org/10.1016/j.geoderma.2006.10.016.

Cambardella, C. A., & Karlen, D. L. (1999). Spatial analysis of soil fertility parameters. *Precision Agriculture*, *1*, 5–14. https://doi.org/10.1023/A:1009925919134.

Campos, J., Gallart, M., Llop, J., Ortega, P., Salcedo, R., & Gil, E. (2020). On-farm evaluation of prescription map-based variable rate application of pesticides in vineyards. *Agronomy*, *10*(12), 102. https://doi.org/10.3390/agronomy10010102.

Carlson, G., Clay, D. E., Malo, D. D., Reese, C. L., Kerns, R., Kharel, T., Birru, G., & DeSutter, T. (2019). The management and identification of saline and sodic soils in the Northern Great Plains. In *iGrow corn: Best management practices* (pp. 32-1–32-12). South Dakota State University Extension. https://extension.sdstate.edu/sites/default/files/2019-09/S-0003-32-Corn.pdf.

Council for Agricultural Science and Technology (CAST). 2024. *Impacts of soil health practices on hydrologic processes*. Issue Paper 76. CAST. https://cast-science.org/wp-content/uploads/2024/08/IP76-Soil-Health.pdf.

Chang, J., Clay, D. E., Carson, C. G., Clay, S. A., Malo, D. D., Berg, R., Kleinjan, J., & Wiebold, W. (2003). Different techniques to identify management zones impact nitrogen and phosphorus sampling variability. *Agronomy Journal*, *95*(6), 1550–1559. https://doi.org/10.2134/agronj2003.1550.

Chang, T., Feng, G., Paul, V., Adeli, A., & Brooks, J. P. (2022). Soil health assessment methods: Progress, applications and comparison. *Advances in Agronomy*, *172*, 129–210. https://doi.org/10.1016/bs.agron.2021.10.002.

Chang, T., Feng, G., Paul, V., Adeli, A., Brooks, J. P., & Jenkins, J. N. (2023). Soil health assessment for different tillage and cropping systems to determine sustainable management practices in a humid region. *Soil and Tillage Research*, *233*, 105796. https://doi.org/10.1016/j.still.2023.105796.

Clark, J. D., Fernández, F. G., Veum, K. S., Camberato, J. J., Carter, P. R., Ferguson, R. B., Franzen, D. W., Kaiser, D. E., Kitchen, N. R., Laboski, C. A. M., Nafziger, E. D., Rosen, C. J., Sawyer, J. E., & Shanahan, J. F. (2019). Predicting economic optimal nitrogen rate with the anaerobic potentially mineralizable nitrogen test. *Agronomy Journal*, *111*(6), 3329–3338. https://doi.org/10.2134/agronj2019.03.0224.

Clay, D. E., Carlson, C. G., Clay, S. A., Chang, J., & Malo, D. D. (2005). Soil organic carbon maintenance in a corn (*Zea mays* L.) and soybean (*Glycine max* L.) as influenced by elevation zone. *Journal of Soil Water and Conservation*, *60*(6), 342–348. https://doi.org/10.1080/00224561.2005.12435825.

Clay, D. E., Chang, J., Malo, D. D., Carlson, C. G., Reese, C., Clay, S. A., Ellsbury, M., & Berg, B. (2001). Factors influencing spatial variability of apparent electrical conductivity. *Communications in Soil Science and Plant Analysis*, *32*(19–20), 2993–3008. https://doi.org/10.1081/CSS-120001102.

Clay, D. E., Clay, S. A., Reitsma, K. D., Dunn, B. H., Smart, A. J., Carlson, G. G., Horvath, D., & Stone, J. J. (2014). Does the conversion of grasslands to row crop production in semi-arid areas threaten global food supplies? *Global Food Security*, *3*(1), 22–30. https://doi.org/10.1016/j.gfs.2013.12.002.

Clay, S. A., Fiedler, D., Reese, C., & Clay, D. E. (2022). Restoring ecological function to saline–sodic soils in South Dakota with perennial grass mixtures. *Agronomy Journal*, *115*(1), 135–146. https://doi.org/10.1002/agj2.21184.

Crespo, C., O'Brien, P. L., Nunes, M. R., Ruis, S. J., Emmett, B. D., Rogovska, N., Malone, R. W., Cambardella, C., & Kovar, J. L. (2024). Contrasting soil management systems had limited effects on soil health and crop yields in a North Central US Mollisol. *Soil Science Society of America Journal*, *88*(5), 1723–1735. https://doi.org/10.1002/saj2.20716.

Crookston, B. S., Yost, M. A., Bowman, M., Veum, K., & Stevens, J. R. (2023). Microbial respiration gives early indication of soil health improvement following cover crops. *Journal of Soil and Water Conservation*, *78*(3), 272–281. https://doi.org/10.2489/jswc.2023.00015.

Derner, J. D., Smart, A. J., Toombs, T. P., Larsen, D., McCulley, R. L., Goodwin, J., Sims, S., & Roche, L. M. (2018). Soil health as a transformational change agent for US grazing lands management. *Rangeland Ecology & Management*, *71*(4), 403–408. https://doi.org/10.1016/j.rama.2018.03.007.

Dick, R. P. (1994). Soil enzyme activities as indicators of soil quality. In J. W. Doran, D. C. Coleman, D. F. Bezdicek, & B. A. Stewart (Eds.), *Defining soil quality for a sustainable environment*, SSSA Spec. Publ. No. 35 (pp. 107–124). SSSA.

Dollinger, J., & Jose, S. (2018). Agroforestry for soil health. *Agroforestry Systems*, *92*, 213–219. https://doi.org/10.1007/s10457-018-0223-9.

Dorsey, N. (2019). Understanding soil health testing. *Crops & Soils*, *52*(6), 22–24. https://doi.org/10.2134/cs2019.52.0607.

Food and Agriculture Organization of the United Nations (FAO). (2020). *Towards a definition of soil health*. ITPS Soil Letters #1. https://openknowledge.fao.org/items/c9075777-81b6-45cf-80bb-e1a8582fdf44.

Fioratti Junod, M., Reid, B., Sims, I., & Miller, A. J. (2024). Cover crops in cereal rotations: A quantitative review. *Soil and Tillage Research*, *238*, 105997. https://doi.org/10.1016/j.still.2023.105997.

Franzluebbers, A. J. (2002). Water infiltration and soil structure related to organic matter and its stratification with depth. *Soil and Tillage Research*, *66*(2), 197–205. https://doi.org/10.1016/S0167-1987(02)00027-2.

Franzluebbers, A. J. (2005). Soil organic carbon sequestration and agricultural greenhouse gas emissions in the southeastern USA. *Soil and Tillage Research*, *83*(1), 120–147. https://doi.org/10.1016/j.still.2005.02.012.

Franzluebbers, A. J., Haney, R. L., Honeycutt, C. W., Schomberg, H. H., & Hons, F. M. (2000). Flush of carbon dioxide following rewetting of dried soil relates to active organic pools. *Soil Science Society of America Journal*, *64*(2), 613–623. https://doi.org/10.2136/sssaj2000.642613x.

Franzluebbers, A. J., & Veum, K. S. (2020). Comparison of two alkali trap methods for measuring the flush of CO_2. *Agronomy Journal*, *112*(2), 1279–1286. https://doi.org/10.1002/agj2.20141.

Gergeni, T., & Scasta, J. D. (2019). Are SSURGO organic matter estimates reliable for cold arid steppes? Implications for rangeland soil health. *Arid Land Research and Management*, *33*(4), 468–475. https://doi.org/10.1080/15324982.2019.1573388.

Gamage, A., Gangahagedara, R., Subasinghe, S., Gamage, J., Guruge, C., Senaratne, S., Randika, T., Rathnayake, C., Hameed, Z., Madhujith, T., & Merah, O. (2024). Advancing sustainability: The impact of emerging technologies in agriculture. *Current Plant Biology*, *40*, 100420. https://doi.org/10.1016/j.cpb.2024.100420.

Getahun, S., Kefale, H., & Gelaye, Y. (2024). Application of precision agriculture technologies for sustainable crop production and environmental sustainability: A systematic review. *The Scientific World Journal*, *2024*(1), 2126734. https://doi.org/10.1155/2024/2126734.

Grisso, R., Alley, M., Thomason, W., Holshouser, D., & Roberson, G. T. (2011). *Precision farming tools: Variable-rate application*. Virginia Cooperative Extension, 442–505. Virginia Polytechnic Institute and State University. https://vtechworks.lib.vt.edu/server/api/core/bitstreams/9df8cff9-74f8-4826-8d5e-700e302ee2ee/content.

Guo, M. (2021). Soil health assessment and management: Recent development in science and practices. *Soil Systems*, *5*(4), 61. https://doi.org/10.3390/soilsystems5040061.

Haberern, J. (1992). A soil health index. *Journal of Soil and Water Conservation*, *47*(1), 6. https://doi.org/10.1080/00224561.1992.12456663.

Hansen, N. C., Allen, B. L., Baumhardt, R. L., & Lyon, D. J. (2012). Research achievements and adoption of no-till, dryland cropping in the semi-arid U.S. Great Plains. *Field Crops Research*, *132*, 196–203. https://doi.org/10.1016/j.fcr.2012.02.021.

Henry, C., Hardke, J., Krutz, J., Massey, J., & Reba, M. (2017). *Using alternate wetting & drying (AWD) rice flood management*. Division of Agriculture, University of Arkansas. https://www.uaex.uada.edu/farm-ranch/crops-commercial-horticulture/rice/2017%20Alternate%20Wetting%20and%20Drying%20Rice%20Management.pdf.

Hillbrand, A., Borelli, S., Conigliaro, M., & Olivier, A. (2017). *Agroforestry for landscape restoration: Exploring the potential of agroforestry to enhance the sustainability and resilience of degraded landscapes*. FAO. https://openknowledge.fao.org/server/api/core/bitstreams/8cd17c6f-051b-4105-866b-dcd344d4ba78/content.

Jagadamma, S., Singh, S., DeBruyn, J., Adotey, N., & Walker, F. (2021). What does soil health mean for Tennessee farmers? UTIA Extension W998. https://utia.tennessee.edu/publications/wp-content/uploads/sites/269/2023/10/W998.pdf.

Jenny, H. (1941). *Factors of soil formation: A system of quantitative pedology*. Dover Publications.

Joshi, D. R., Clay, D. E., Alverson, R., Clay, S. A., Westhoff, S., Johnson, J. M. F., Wang, T., & Sieverding, H. (2025). Tillage intensity reductions when combined with yield increases may slow soil carbon saturation in the central United States. *Scientific Reports*, *15*, 10697. https://doi.org/10.1038/s41598-025-95388-x.

Joshi, D. R., Sieverding, H. L., Xu, H., Kwon, H., Wang, M., Clay, S. A., Johnson, J. M., Thapa, R., Westhoff, S., & Clay, D. E. (2023). A global meta-analysis of cover crop response on soil carbon storage within a corn production system. *Agronomy Journal*, *115*(4), 1543–1556. https://doi.org/10.1002/agj2.21340.

Karlen, D. L., Goeser, N. J., Veum, K. S., & Yost, M. A. (2017). On-farm soil health evaluations: Challenges and opportunities. *Journal of Soil and Water Conservation*, *72*(2), 26A–31A. https://doi.org/10.2489/jswc.72.2.26A.

Karlen, D. L., & Stott, D. E. (1994). A framework for evaluating physical and chemical indicators of soil quality. In J. W. Doran, D. C. Coleman, D. F. Bezdicek, & B. A. Stewart (Eds.), *Defining soil quality for a sustainable environment*, SSSA Spec. Publ . No. 35 (pp. 53–72). SSSA.

Karlen, D. L., Veum, K. S., Sudduth, K. A., Obrycki, J. F., & Nunes, M. R. (2019). Soil health assessment: Past accomplishments, current activities, and future opportunities. *Soil and Tillage Research*, *195*, 104365. https://doi.org/10.1016/j.still.2019.104365.

Kelly, C., Schipanski, M. E., Tucker, A., Trujillo, W., Holman, J. D., Obour, A. K., Johnson, S. K., Brummer, J. E., Haag, L., & Fonte, S. J. (2021). Dryland cover crop soil health benefits are maintained with grazing in the U.S. High and Central Plains. *Agriculture, Ecosystems & Environment*, *313*, 107358. https://doi.org/10.1016/j.agee.2021.107358.

Khan, H., Esau, T. J., Farooque, A. A., Zaman, Q. U., Abbas, F., & Schumann, A. W. (2023). Chapter 2 - Soil spatial variability and its management with precision agriculture. In Q. Zaman (Ed.), *Precision agriculture* (pp. 19–36). Academic Press.

Lakhiar, I. A., Yan, H., Zhang, C., Wang, G., He, B., Hao, B., Han, Y., Wang, B., Bao, R., Syed, T. N., Chauhdary, J. N., & Rakibuzzaman, M. (2024). A review of precision irrigation water-saving technology under changing climate for enhancing water use efficiency, crop yield, and environmental footprints. *Agriculture*, *14*(7), 1141. https://doi.org/10.3390/agriculture14071141.

LaRose, J., & Myers, R. (2019). *Adoption of soil health systems based on data from the 2017 U.S. Census of Agriculture*. Soil Health Institute.

Lehmann, J., Bossio, D. A., Kögel-Knabner, I., & Rillig, M. C. (2020). The concept and future prospects of soil health. *Nature Reviews Earth & Environment*, *1*, 544–553. https://doi.org/10.1038/s43017-020-0080-8.

Li, C., Cano, A., Acosta-Martinez, V., Veum, K. S., & Moore-Kucera, J. (2020). A comparison between fatty acid methyl ester profiling methods (PLFA and EL-FAME) as soil health indicators. *Soil Science Society of America Journal*, *84*(4), 1153–1169. https://doi.org/10.1002/saj2.20118.

Lord, S., Veum, K. S., Sullivan, L. L., Anderson, S. H., Acosta-Martinez, V., & Clark, K. (2024). Ancient prairies as a reference for soil organic carbon content and microbial community structure. *Applied Soil Ecology*, *198*, 105355. https://doi.org/10.1016/j.apsoil.2024.105355.

Mfuka, C., Byamukama, E., & Zhang, X. (2020). Spatiotemporal characteristics of white mold and impacts on yield in soybean fields in South Dakota. *Geo-spatial Information Science*, *23*(2), 182–193. https://doi.org/10.1080/10095020.2020.1712265.

Mitchell, J. P., Shrestha, A., Horwath, W. R., Southard, R. J., Madden, N., Veenstra, J., & Munk, D. S. (2015). Tillage and cover cropping affect crop yields and soil carbon in the San Joaquin Valley, California. *Agronomy Journal*, *107*(2), 588–596. https://doi.org/10.2134/agronj14.0415.

Mizuta, K., & Grunwald, S. (2022). Reshaping how we think about soil security. *Soil Systems*, *6*(4), 74. https://doi.org/10.3390/soilsystems6040074.

Mizuta, K., Grunwald, S., Bacon, A. R., Cropper, W. P., Jr., Phillips, M. A., Moss, C. B., Gonzalez-Benecke, C. A., Markewitz, D., Clingensmith, C. M., & Xiong, X. (2022). Holistic aboveground ecological productivity efficiency modeling using data envelopment analysis in the southeastern U.S. *Science of the Total Environment*, *824*, 153802. https://doi.org/10.1016/j.scitotenv.2022.153802.

Mizuta, K., Grunwald, S., Cropper, W. P., Jr., & Bacon, A. R. (2021). Developmental history of soil concepts from a scientific perspective. *Applied Sciences, 11*(9), 4275. https://doi.org/10.3390/app11094275.

Mizuta, K., Grunwald, S., & Phillips, M. A. (2018). New soil index development and integration with econometric theory. *Soil Science Society of America Journal, 82*(5), 1017–1032. https://doi.org/10.2136/sssaj2017.11.0378.

Moebius-Clune, B. N., Moebius-Clune, D. J., Gugino, B. K., Idowu, O. J., Schindelbeck, R. R., Ristow, A. J., van Es, H. M., Thies, J. E., Shayler, H. A., McBride, M. B., Kurtz, K. S. M., Wolfe, D. W., & Abawi, G. S. (2016). *Comprehensive assessment of soil health: The Cornell Framework* (3rd ed.). Cornell University. https://www.css.cornell.edu/extension/soil-health/manual.pdf.

Moriles-Miller, J., Joshi, D. R., Reicks, G., Westhoff, S., Clay, S. A., & Clay, D. E. (2025). Delaying cover crop termination reduced corn yields in a dry spring. *Agrosystems, Geosciences, & Environment, 8*(3), e70182. https://doi.org/10.1002/agg2.70182.

Mulla, D. J., & Schepers, J. S. (1997). Key processes and properties for site-specific soil and crop management. In F. J. Pierce & E. J. Sadler (Eds.), *The state of site specific management for agriculture* (pp. 1–18). ASA, CSSA, SSSA.

Niedbała, G., Piekutowska, M., & Hara, P. (2023). New trends and challenges in precision and digital agriculture. *Agronomy, 13*(8), 2136. https://doi.org/10.3390/agronomy13082136.

Nunes, M. R., Karlen, D. L., Veum, K. S., & Moorman, T. B. (2020). A SMAF assessment of U.S. tillage and crop management strategies. *Environmental and Sustainability Indicators, 8*, 100072. https://doi.org/10.1016/j.indic.2020.100072.

Nunes, M. R., Veum, K. S., Parker, P. A., Holan, S. H., Amsili, J. P., van Es, H. M., Wills, S. A., Seybold, C. A., & Karlen, D. L. (2024). SHAPEv1.0 scoring curves and peer group benchmarks for dynamic soil health indicators. *Soil Science Society of America Journal, 88*(3), 858–875. https://doi.org/10.1002/saj2.20668.

Olness, A., & Archer, D. (2005). Effect of soil organic carbon on available water in soil. *Soil Science, 170*(2), 90–101.

Pace, R., Schiano Di Cola, V., Monti, M. M., Affinito, A., Cuomo, S., Loreto, F., & Ruocco, M. (2025). Artificial intelligence in soil microbiome analysis: A potential application in predicting and enhancing soil health—a review. *Discover Applied Sciences, 7*, 85. https://doi.org/10.1007/s42452-024-06381-4.

Pierson, F. B., & Mulla, D. J. (1990). Aggregate stability in the Palouse region of Washington: Effect of landscape position. *Soil Science Society of America Journal, 54*(5), 1407–1412. https://doi.org/10.2136/sssaj1990.03615995005400050033x.

Roper, W. R., Acosta-Martínez, V., Veum, K. S., Burgess, C. J., Moore, J. M., Manter, D. K., Stewart, C. E., Emmett, B. D., Liebig, M. A., Fischel, M. H. H., Lehman, R. M., Franco, J. G., Johnson, J. M. F., Weyers, S., Mikha, M. M., Trippe, K. M., Maul, J. E., Dungan, R. S., Gollany, H. T., . . . Reardon, C. L. (2025). Unraveling edaphic, environmental, and management drivers of soil microbial communities via ester-linked fatty acid methyl esters using a multilocation agroecosystem study. *Geoderma, 453*, 117158. https://doi.org/10.1016/j.geoderma.2024.117158.

Ruis, S. J., Kovar, J. L., Wacha, K. M., Carney, D. B., O'Brien, P. L., Delate, K., & Cambardella, C. A. (2025). Organic cropping systems enhance soil health indicators in a Mollisol of the US Corn Belt. *Soil Science Society of America Journal, 89*(1), e20795. https://doi.org/10.1002/saj2.20795.

Sanyaolu, M., & Sadowski, A. (2024). The role of precision agriculture technologies in enhancing sustainable agriculture. *Sustainability, 16*(15), 6668. https://doi.org/10.3390/su16156668.

Sharma, K., & Shivandu, S. K. (2024). Integrating artificial intelligence and Internet of Things (IoT) for enhanced crop monitoring and management in precision agriculture. *Sensors International, 5*, 100292. https://doi.org/10.1016/j.sintl.2024.100292.

Sikora, L. J., & Stott, D. E. (1997). Soil organic carbon and nitrogen. In J. W. Doran & A. J. Jones (Eds.), *Methods for assessing soil quality*, SSSA Spec. Publ. No. 49 (pp. 157–167). SSSA.

Simão, L. M., Peterson, D., Roozeboom, K. L., Rice, C. W., Du, J., Lin, X., & Lollato, R. P. (2023). Crop rotation and tillage impact yield performance of soybean, sorghum, and wheat. *Agronomy Journal, 115*(2), 658–673. https://doi.org/10.1002/agj2.21237.

Soil Survey Staff. (1999). *Soil taxonomy: A basic system of soil classification for making and interpreting soil surveys* (2nd ed.). Natural Resources Conservation Service. U.S. Department of Agriculture Handbook 436.

Stevens, A. W. (2018). Review: The economics of soil health. *Food Policy*, *80*, 1–9. https://doi.org/10.1016/j.foodpol.2018.08.005.

Stott, D. E. (2019). *Recommended soil health indicators and associated laboratory procedures*. USDA-NRCS Soil Health Technical Note No. 450-03. https://www.soils.org/files/napt/publications/method-papers/2019-nrcs-technote-450-03.pdf.

Teague, R., & Kreuter, U. (2020). Managing grazing to restore soil health, ecosystem function, and ecosystem services. *Frontiers in Sustainable Food Systems*, *4*, 534187. https://doi.org/10.3389/fsufs.2020.534187.

Tucker, C. J. (1979). Red and photographic infrared linear combinations for monitoring vegetation. *Remote Sensing of Environment*, *8*(2), 127–150. https://doi.org/10.1016/0034-4257(79)90013-0.

Veum, K. S., Goyne, K. W., Kremer, R., & Motavalli, P. P. (2012). Relationships among water stable aggregates and organic matter fractions under conservation management. *Soil Science Society of America Journal*, *76*(6), 2143–2153. https://doi.org/10.2136/sssaj2012.0089.

Veum, K. S., Kremer, R. J., Sudduth, K. A., Kitchen, N. R., Lerch, R. N., Baffaut, C., Stott, D. E., Karlen, D. L., & Sadler, E. J. (2015). Conservation effects on soil quality indicators in the Missouri Salt River Basin. *Journal of Soil and Water Conservation*, *70*(4), 232–246. https://doi.org/10.2489/jswc.70.4.232.

Veum, K. S., Lorenz, T., & Kremer, R. J. (2019). Phospholipid fatty acid profiles of soils under variable handling and storage conditions. *Agronomy Journal*, *111*(3), 1090–1096. https://doi.org/10.2134/agronj2018.09.0628.

Veum, K. S., Parker, P. A., Holan, S. H., Pais, N. V., Wills, S. A., Amsili, J. P., Nunes, M. R., van Es, H. M., Seybold, C. A., & Karlen, D. L. (2025). Spatially explicit heteroskedastic modeling for the soil health assessment protocol and evaluation version 1.0S. *Soil Science Society of America Journal*, *89*(3), e70065. https://doi.org/10.1002/saj2.70065.

Veum, K. S., Zuber, S. M., Ransom, C., Myers, R. L., Kitchen, N. R., & Anderson, S. H. (2022). Reduced tillage and rotational diversity improve soil health in Missouri. *Agronomy Journal*, *114*(5), 3027–3039. https://doi.org/10.1002/agj2.21156.

Weidhuner, A., Hanauer, A., Krausz, R., Crittenden, S. J., Gage, K., & Sadeghpour, A. (2021). Tillage impacts on soil aggregation and aggregate-associated carbon and nitrogen after 49 years. *Soil and Tillage Research*, *208*, 104878. https://doi.org/10.1016/j.still.2020.104878.

Weyers, S. L., & Spokas, K. A. (2011). Impact of biochar on earthworm populations: A review. *Applied and Environmental Soil Science*, *2011*(1), 541592. https://doi.org/10.1155/2011/541592.

Yost, M. A., Veum, K. S., Kitchen, N. R., Sawyer, J. E., Camberato, J. J., Carter, P. R., Ferguson, R. B., Fernández, F. G., Franzen, D. W., Laboski, C. A., & Nafziger, E. D. (2018). Evaluation of the Haney Soil Health Nutrient Tool for corn nitrogen recommendations across eight Midwest states. *Journal of Soil and Water Conservation*, *73*(5), 587–592. https://doi.org/10.2489/jswc.73.5.587.

Zebarth, B. J., Islam, M. M., Cambouris, A. N., Perron, I., Burton, D. L., Comeau, L.-P., & Moreau, G. (2019). Spatial variation of soil health indices in a commercial potato field in eastern Canada. *Soil Science Society of America Journal*, *83*(6), 1786–1798. https://doi.org/10.2136/sssaj2019.03.0087.

Zhang, C., & Kovacs, J. M. (2012). The application of small unmanned aerial systems for precision agriculture: A review. *Precision Agriculture*, *13*, 693–712. https://doi.org/10.1007/s11119-012-9274-5.

Zhang, Y., Lu, W., Xing, K., Guo, F., Du, Q., Zhang, X., Zhang, F., Qian, Z., & Li, F. (2025). Environmental DNA as a tool for soil health monitoring and unveiling new ecological frontiers. *Ecological Indicators*, *174*, 113438. https://doi.org/10.1016/j.ecolind.2025.113438.

Precision Conservation: A Pathway to Improved Sustainability

25

Jasmine Neupane, Gretchen Sassenrath, Deepak Joshi, Jorge Delgado, Rojina Kafle, and Dharani Suresh Babu

Chapter Purpose

Precision conservation uses geospatial technologies to improve profitability and reduce the impact on the environment. It encompasses complex, dynamic relationships among living organisms—plants, animals, microorganisms, and humans—and their physical surroundings, including soil, water, and air, within an agricultural setting. This chapter discusses the various components of precision conservation, which include (a) identifying the problem; (b) identifying potential options; (c) determining expected costs and economic assessment; (d) implementing the practice using the obtained resources; (e) assessing success, reviewing, and modifying the strategy as needed; and (f) communicating findings to the end users. In addition, this chapter provides case studies on (a) the management of salt-affected soils, (b) the replacement of nonproductive croplands with cover crops, and (c) the use of multiple approaches to solve complex water quality problems.

Introduction

Precision conservation is a set of spatial technologies and procedures used to implement conservation management practices that account for the spatial and temporal variability of the goods (products produced) and services (improvements in water quality or reduced erosion) generated by the practice across natural and agricultural systems (Berry et al., 2003). Because most agricultural management affects goods and services simultaneously, precision agriculture—or site-specific management—and environmental quality are inseparably linked. When discussing the importance of conservation, it is important to consider the impact of ecosystem services (e.g., clean air, water, soil) on our everyday lives. The consequences of failing to account for these services can severely limit our ability to produce food in the future. Unfortunately, some of the conservation practices we relied on in the past were insufficient in preserving the ecosystem goods and services produced by our soil (Sassenrath & Delgado, 2018). Therefore, precision conservation as a management strategy can be used to target and protect high-risk areas, prioritize production on land best suited for agriculture, and identify other areas that can provide ecosystem services beyond crop production, offering economic and societal benefits.

Precision conservation uses digital agricultural technologies to design and implement conservation practices that achieve the greatest ecosystem and economic benefits. While precision agriculture focuses on the 4R+ framework (*applying the right source, at the right rate, at the right time, in the right place, and in the right manner*; Khosla, 2008), any aspect of natural resource conservation can be targeted with precision conservation using the 7R framework, which includes *the right conservation practice and the right scale* in addition to 4R+ strategies (Delgado, 2016; Delgado et al., 2019). These additional measures support the idea that conservation should be targeted at a larger spatial scale and focused on stewardship. Many technologies can be used to identify and implement precision conservation. For example, archived remote sensing images can be used to track changes in soil and plant health, yield monitor data can identify unprofitable field areas, and machine learning models can forecast the impact of conservation practices on future returns on investment. These practices will become increasingly important as we attempt to meet food, energy, and environmental security issues (Sassenrath & Delgado, 2018). Precision conservation has the potential to help assess the spatial and temporal flows of agricultural and natural systems and to develop strategic control methods that managers can use to implement conservation practices within and outside the field (Berry et al., 2005). Precision conservation also provides an opportunity to address complex, multidimensional problems, including social disparities, natural resource degradation, and climate uncertainty (Delgado et. al., 2020). In this chapter, we explore (a) strategies to develop precision conservation solutions through the application of modern and emerging technologies, (b) how to construct a precision conservation project, and (c) how precision conservation has been implemented in the field.

Precision Conservation—Strategies and Technology Involved

Because precision conservation is targeted at a larger spatial scale and focuses on stewardship, we often include different site locations or landscape positions when evaluating practices. The selection of the appropriate scale depends on the problem. The strategies and technologies involved in conservation at each scale also evolve accordingly. For example, to study runoff generation and erosion risk within a single production field, the scale is smaller, so the technology used would include edge-of-field sensors, flumes, and rainfall gauges. However, to study runoff and erosion across multiple fields and topographic positions, the technology used might include LiDAR-derived digital elevation models, watershed-scale hydrologic models, and remote sensing products.

The scale of precision conservation may also extend to issues that are not necessarily generated at the field scale but nonetheless affect environmental resources and indirectly impact agricultural production systems. For example, nutrient enrichment in a watershed originates from multiple land uses beyond individual fields, yet precision conservation can target critical source areas and strategically place buffers or wetlands to protect water quality, minimize soil erosion, sustain agricultural production, and enhance biodiversity (Schulte et al., 2017).

The previous chapters in this book described how precision agriculture management influences crop production and overall profitability across the field. In this chapter, we focus more on the environmental implications and conservation aspects of agricultural production using these technologies.

Developing a Precision Conservation Plan

Precision conservation considers the long-term sustainability of a production system, addressing challenges within a specific soil type, watershed (an area of land from which water drains into a lake, river, or ocean), or landscape. It enhances long-term environmental sustainability by mitigating soil erosion, improving water quality, and supporting biodiversity beyond individual fields. By recognizing the agroecosystem as part of a larger ecological framework, we can develop strategies that not only increase agricultural efficiency but also safeguard natural resources for future generations. To be successful, we need to develop an organized plan that addresses each potential component of the system. Whether a plan is being organized within a field or within a watershed, a project should create an organizational plan that includes the following steps:

1. **Identify the problem:** This includes determining the scale and magnitude of the problem. For instance, the impact area could range from a small region within a field to a larger watershed. The conservation objective may then be identified to address the problem, such as reducing erosion or improving wildlife habitat.
2. **Identify potential options:** In this step, we need to improve our understanding of the site or region of interest. This may include determining natural soil variability and past management strategies that can be used at the study site (Chapters 5 and 6). This step also includes determining the data and technologies needed to collect information about the problem.
3. **Determine the expected costs and conduct an economic assessment:** This may include using a partial budget analysis, a capital budget analysis, or an internal rate of return to evaluate the profitability of purchasing expensive equipment. Economic analysis for the purchase and use of equipment can be found in Chapter 23.
4. **Implement the practice using the obtained resources:** In this step, the appropriate resources (tools and materials) required to address the identified problem will be acquired. Using the integrated knowledge generated from Steps 1–3, the most effective strategy will be identified and implemented at the study site.
5. **Assess success, review, and modify the strategy as needed:** Following implementation, the success of the strategy will be evaluated to determine its effectiveness in addressing the identified problem. This step may involve additional data collection, performance assessment, and adaptive decision-making to refine or modify the strategy as conditions or outcomes warrant.
6. **Communicate findings to the end users:** Finally, the results of the implemented strategies, including their effectiveness and practical implications, will be communicated to the stakeholders. Dissemination methods may include one-on-one meetings, training sessions, workshops, and other outreach activities tailored to the end users' needs.

Precision Conservation Case Studies

In this section, we analyze how three case studies put the conservation plan into action and explore the unique challenges and opportunities presented in each different scenario.

Case Study 1: Managing Salinity and Sodicity with Perennial Plants

Problem Identification

Salinity and sodicity pose significant challenges to food and environmental security and can result from ocean inundation, inadequate irrigation water to remove excess salts, and the transport of subsurface water to the soil surface. Management solutions can range from installing tile drains within a field to constructing levees for flood prevention. Each problem is unique and requires careful consideration before implementing a solution.

This case study is based on a field experiment that was conducted from 2017 to 2021 in the northern Great Plains Region of North America (Fiedler et al., 2022). In this region, many soils lie over buried marine sediments, and salts within the sediments can be transported to the soil surface by capillary action. Furthermore, salinity and sodicity problems initially start in the footslope and, within a few years, expand upslope (Figure 25.1). These areas are often intertwined with highly productive areas (Figure 25.1), and soils in the footslope can have electrical conductivity 1:1 ($EC_{1:1}$) values of >4 decisiemens per meter (dS/m) and percentage sodium (%Na) of >15%. In severe cases, the amount of salt deposited on the soil surface can be thick enough to look like snow (Figure 25.2). Prior research showed that tile drainage and the application of chemical amendments were not viable options for these soils (Birru et al., 2019; Budak et al., 2022). These techniques failed because the subsurface drainable porosity was very low, and the surface soils were supersaturated with gypsum and calcium carbonate.

Historically, gypsum, when applied to the soil, supplies calcium ions (Ca^{2+}), which replace sodium (Na) ions (Na^{+}) on soil particles. If the soil has a Na problem, applying gypsum may be beneficial, as it allows the displaced Na to be leached by percolating water. However, in the northern Great Plains, these amendments may fail if the surface soil is supersaturated with lime and gypsum, and the subsurface has low drainable porosity and a high exchangeable Na to EC ratio, as seen in the field in Figure 25.2 (Birru et al., 2019; Budak et al., 2022).

Identify Potential Options

To diagnose salt-affected soil problems, it is recommended to follow the soil sampling protocols discussed in Chapter 6 using the analytical procedures outlined by DeSutter et al. (2024). To accurately assess the extent of the salinity problem, deep soil samples should be collected and analyzed for EC and %Na (Budak et al., 2022). When evaluating the chemical analysis, it is important to consider the analytical technique. For example, the EC determined on a saturated paste (EC_e) will be about twice that of the $EC_{1:1}$ determined by mixing 10 g of soil with 10 mL of water. These differences result from different soil-to-water solution ratios and must be considered when interpreting the values. For example, to convert 3.87 dS/m ($EC_{1:1}$) to a saturated paste (EC_e), we need to multiply 3.87 by 2.26, which results in an EC_e value of 8.63 dS/m (Matthees et al., 2017). In most states, salinity and sodicity assessments are based on EC and the Na adsorption ratio (SAR) determined from a saturated paste. In general, soil with an EC_e of >4 dS/m is

Figure 25.1. Expansion of a soil salinity problem over a 6 year period. In this figure Landsat images were overlayed an elevation map (courtesy of Blake Toennies).

Figure 25.2. This image shows salt accumulation in a field located near Carpenter, Clark County, SD. The salt deposits are the result of capillary water movement from the buried marine sediments to the soil surface. Courtesy of South Dakota State University; photo taken April 22, 2021.

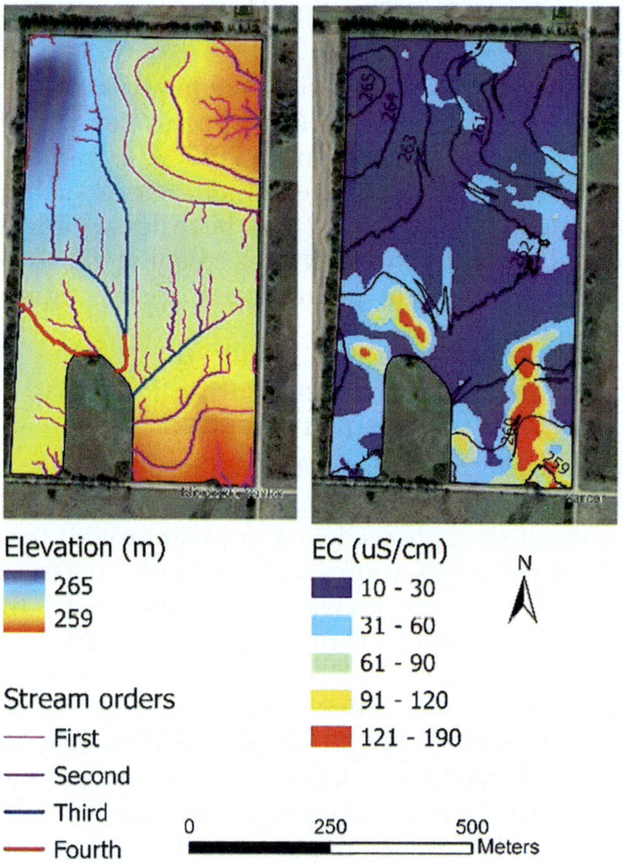

Figure 25.3. Elevation, stream order, and EC variation at the study site located in Labette County, KS.

considered saline, and soil with an SAR of >13 is considered sodic.

The extent of the problem and potential options can be assessed using remote sensing and ground-based data collection. Landsat satellite imagery overlaid with elevation data (Figure 25.3), provided spatial insights into the growing salinity problem, enabling targeted identification of affected footslope areas without the need for extensive field surveys. This was complemented by strategic soil sampling at various depths to measure EC, %Na, and drainable porosity—key data that guided the choice of management practices. Additionally, Clay et al. (2022) and Fiedler et al. (2022) have shown that a viable option for these soils is to return them to perennial plants through a process called phytoremediation.

Determine Expected Cost and Economic Assessment

The partial economic analysis conducted for this study showed that the cost of using regenerative agriculture, specifically phytoremediation, would be lower and more profitable compared with continuing conventional crop production because it reduces seed, equipment, and labor costs in the salt-affected area. Phytoremediation is useful because the deep roots of salt-tolerant plant species lower the water table and reduce the amount of salt that reaches the soil surface. This plan must also consider the costs of herbicide, perennial grass seed, and planting, which tend to be one-time investments.

The return on investment should also be determined. The yields for corn compared with the salt-tolerant perennial grass are provided in Clay et al. (2022), showing a general increase in the biomass of perennial grass compared with corn yield in the salt-affected area. Planning should also determine long-term changes in the goods and services at the site and establish the payback period. For example, planting perennial grass often creates good hunting grounds, which can be rented out for sport. Economic analysis at this site showed that economic losses could be reduced by returning the salt-affected footslope soils to perennial grasses and that there is an opportunity to use the restored lands for grazing or wildlife habitat.

Implement the Practice Using Obtained Resources

The site was seeded with perennial plant mixes after the corn harvest, in a process called dormant seeding, in December 2017. The mixtures included salt-tolerant perennial grasses such as (a) slender wheatgrass and beardless wildrye, and (b) slender wheatgrass, creeping meadow foxtail, western wheatgrass, and green wheatgrass.

Assess Success, Monitor Practices, and Modify the Strategy as Needed

By Year 2, biomass production in previously unproductive zones, such as the footslope and summit soils, showed greater plant growth from the grass mixes compared with corn; in Year 3, the perennial grasses were well established at the site. The establishment of grasses significantly improved soil health. For example, a soil analysis in Year 3 showed that restoration reduced the $EC_{1:1}$ by 0.83 ± 0.149 dS/m and the exchangeable Na by 656 ± 220 mg/kg; as a result, the soil had improved structure, drainage, and nutrient availability. This conversion of annual cropland into perennial grass areas overall resulted in reduced economic loss and opportunities to use the land for hunting or livestock grazing.

When implementing a precision conservation practice, it is important to review its impact at regular intervals. It is helpful to ask questions such as (a) Are we making a difference? and (b) Are modifications to protocols or practices needed? In this scenario, the practices implemented were beneficial both environmentally and economically, and any modifications may include adjusting seed mixes to attract specific animals for hunting or to meet the dietary requirements of livestock.

Communicate Findings to the End User

Findings from this study were communicated to end users at workshops, a regional salinity summit, and in published papers (Birru et al., 2019; Budak et al., 2022; Clay et al., 2022; Fiedler et al., 2022). Often, communicating findings leads to a better understanding of the project's impact and enables others to apply these lessons to similar issues they may be facing. For example, neighboring farmers with similar saline–sodic problems considered how they might use phytoremediation in their own salt-affected soils. Weighing the costs and benefits of removing unproductive land from annual row crop production and converting it to a perennial grass mix.

Case Study 2: Replacing Nonproductive Croplands with Cover Crops

Problem Identification

In this case study, the field under examination had undergone several agricultural intensification practices aimed at increasing the yield per unit area, such as the use of higher-yielding crop varieties, increased application of fertilizers and pesticides, and irrigation. Despite these efforts, productivity had stagnated at a level that remained unsatisfactory to the landowner. As in the first case study, some areas within this field were inherently low in productivity, contributing to reduced returns. Under these conditions, alternative management approaches, such as site-specific conservation management, offer the potential to improve economic returns by managing these underperforming areas differently from the rest of the field.

The underperforming areas were identified by analyzing production costs and the value of the products produced on the land. For example, the cost of production was $809/ac ($2,000/ha), and the products sold from that hectare provided a return of $607/ac ($1,500/ha). Typically, yields vary from year to year, with 5–10 years of data for this analysis being optimal (Luther et al., 2022; Sassenrath & Kulesza, 2017). For this case study, year-to-year variability is simplified.

The study site is a 32-ha (80-ac) field located in Labette County, KS. The field has an elevation change of 5 m, from 265 m at the northwest corner to 260 m at the northeast and southeast corners. There is also a 1.6-ha (4-ac) grassed waterway in the west-central portion of the field (Figure 25.3). Spatial differences in EC (Figure 25.3) are attributed to soil texture differences (Kulesza et al., 2024). In a typical 2-year rotation, corn is planted in March or April and harvested from August to September, wheat is planted in October and harvested in June, and soybeans are planted in June and harvested in October. Productive areas vary by crop and year (Figure 25.4).

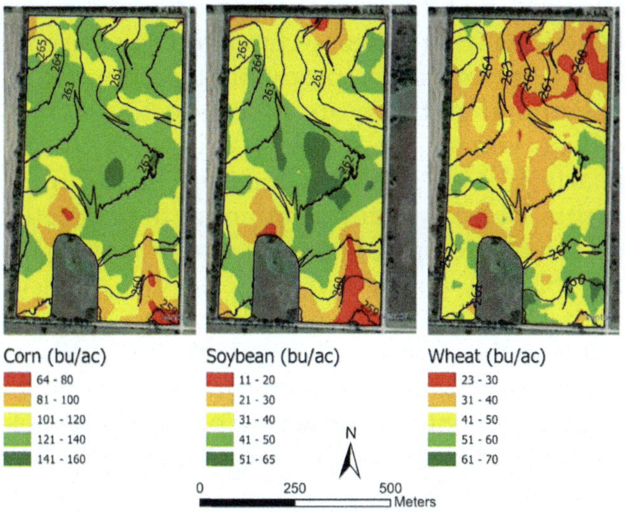

Figure 25.4. Corn, soybean, and wheat yield variation in 2015 at the study site located in Labette County, KS.

Information obtained from yield monitoring and EC, Geographic Information System (GIS), and elevation mapping was used to identify that, on average, in this field, corn and soybeans do not perform well in the high-clay soils located in the southeast corner of the field (Figures 25.3 and 25.4). Conversely, growing winter wheat produced a higher yield in that southeast corner, but a lower yield in the northeast corner.

Identify Potential Options

Two approaches have been suggested to manage nonproductive lands (Dudley & Alexander, 2017). The first of these approaches, termed "sparing," sets aside nonproductive areas for conservation practices. The second approach, termed "sharing," means that the land is used for multiple purposes, such as planting cover crops when the soil is not growing a crop. Conservation practices, such as reduced or no tillage, integrated pest management, returning the land to perennial grasses, and planting a cover crop, can be used in both sparing and sharing techniques. Several precision conservation approaches combine aspects of sparing and sharing to meet project goals

In this case study, we will evaluate precision conservation from the biodiversity or "sharing" perspective. With the introduction of commercial production systems and crop specialization, the loss of biodiversity in agricultural systems is common and can reduce many of the services provided by the land (Boetzl et al., 2021; Dainese et al., 2019; Emmerson et al., 2016; Hendrickson et al., 2008; Hsiao et al., 2025; Reidsma et al., 2006). For example, one of the primary ecosystem services affected by reduced biodiversity is soil nutrient cycling, which is disrupted by changes in the soil microbiome. The soil microbiome is a community of microorganisms that transform organic materials, such as crop residue, into available nutrients. For more information on the benefits of soil health to increased biodiversity, see Chapter 24.

In this case study, a region was established in the underperforming northeast corner of the field and was planted alternately with winter wheat (conventional production; light blue strips) or cover crops (tillage radish and oats; dark blue strips; Figure 25.5) to evaluate how improving biodiversity may increase productivity and reduce operational costs.

Determine Expected Cost and Economic Assessment

An economic analysis of the site showed that over the three crop rotations, there were financial losses in the high-clay soil in the southeast and the slightly elevated north-central portions of the field before intervention (Figure 25.6). The intervention in this context was to plant cover crops in the northeast part of the field to determine whether this would increase returns in that portion of the field. The cover crops had seed costs of \$28/ac (\$69/ha) and planting costs of \$15.11/ac

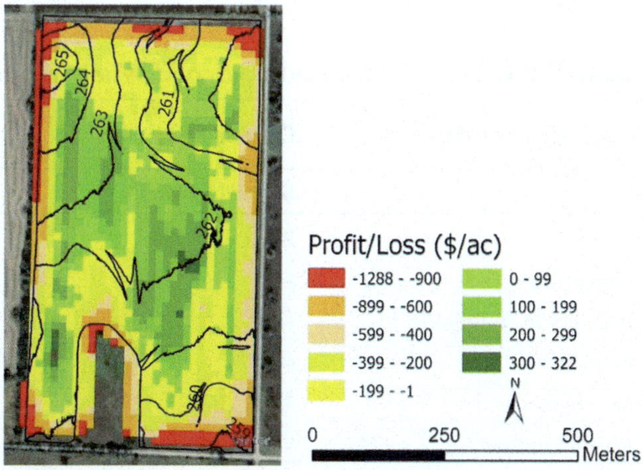

Figure 25.5. Cover crop establishment at the study site located in Kansas.

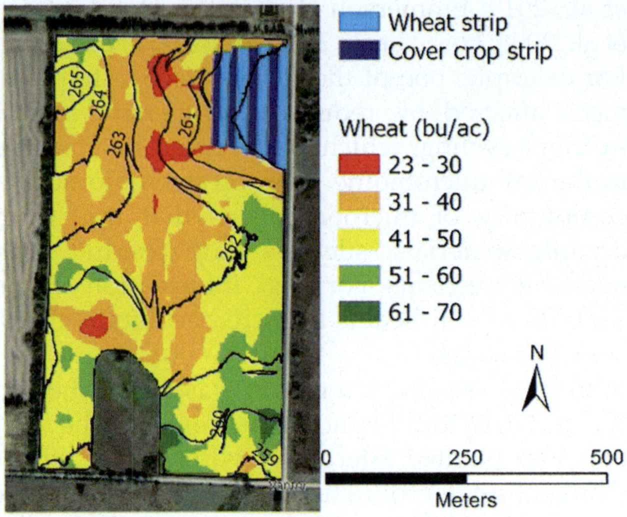

Figure 25.6. Spatial variation of profit and loss from multiple years of corn–wheat–soybean rotation in Kansas (With permission of Kansas State University).

($37.34/ha; Table 25.1). Basic crop management costs for soybean were $218.05/ac ($538.6/ha), and for wheat, the production costs were $245.03/ac ($605.3/ha).

Although this is a simplification, the production costs of both wheat and cover crops resulted in a net loss; however, the establishment of cover crops was less of a net loss at $43.11/ac (−$106.48 ha), whereas wheat was a greater loss at $62.70/ac ($164.87/ha; Table 25.2). Interestingly, the soybean crop planted after a cover crop had a greater yield of 36.4 bu/ac (2.44 Mg/ha), while the yield of soybean planted in the wheat strips was lower, at 35 bu/ac (2.35 Mg/ha). This suggests that nutrient cycling and soil health, benefiting from increased biodiversity with cover crops, are improving productivity in the area.

Estimating the costs of production for the wheat versus cover crop strips indicated a nearly $12/ac ($29/ha) increase in net return (Table 25.2, Soybean). Examining the net return on the two-crop rotation results in a $31.52/ac ($77.85/ha) net return increase by replacing winter wheat with cover crops in areas with low winter wheat productivity (Table 25.2, Wheat/Soybean). Similar changes can be made in the high–clay-content southeast portion of the field, where summer crops (soybean and corn) perform poorly.

Assess Success, Monitor Practices, and Modify the Strategy as Needed

Landowner and farmer engagement is needed to implement precision conservation practices. In these discussions, technologies such as spatial variability analysis and yield maps are important for targeting and monitoring the economic benefits of conservation practices. New technologies enable land managers to implement practices in zones with a negative return on investment, allowing these areas to be used for other purposes.

When implementing a precision conservation practice, it is important to review its impact at regular intervals. It is helpful to ask questions such as (a) Are we making a difference? and (b) Are modifications to protocols or practices needed? In this case study, it was observed that increasing biodiversity by establishing a cover crop helped improve the productivity of

Table 25.1. Cover Crop, Soybeans, and Wheat Production Costs per Acre.

Cost associated	**Cover crop**	**Soybean**	**Wheat**
	$/ac ($/ha)	**$/ac ($/ha)**	**$/ac ($/ha)**
Seed	28 (69.2)	51 (126)	21.6 (53.4)
Herbicide	–	7.76 (19.2)	4.48 (11.1)
Insecticide/fungicide	–	–	6.69 (16.5)
Fertilizer	–	15.77 (39.0)	45.39 (112.1)
Machinery	15.11 (37.3)	76.39 (188.7)	87.11 (215.2)
Miscellaneous	–	67.13 (165.8)	79.76 (197.0)
Total	43.11 (106.5)	218.05 (538.6)	245.03 (605.2)

Note: Based on K-State Extension Crop Budgets for southeast Kansas. In this table, $/ha values are in parentheses.

Table 25.2. Net Return Comparison for Wheat, Soybean, and Total Rotation for Areas Planted to Wheat or Cover Crops.

Main crop	Experimental strips	Input cost $/ac ($/ha)	Income $/ac ($/ha)	Net return $/ac ($/ha)
Wheat	Wheat	245.03 (605.22)	182.33 (450.36)	−62.70 ($164.87)
	Cover crop	43.11 (106.48)	0 (0)	−43.11 (−106.48)
Soybean	Wheat	218.05 (538.58)	298.2 (736.55)	80.15 (198.05)
	Cover crop	218.05 (538.58)	310.13 (766.02)	92.08 (227.53)
Wheat–soybean	Wheat	463.08 (1,143.81)	480.53 (1,186.91)	17.45 (43.10)
	Cover crop	261.16 (645.07)	310.13 (766.02)	48.97 (120.96)

Note: Net return is the income minus input costs. In this table, $/ha values are in parentheses.

low-yielding areas in the field. The literature also supports the idea that the establishment of a cover crop improves nutrient cycling in the following growing season (Joshi et al., 2024).

However, modifications to improve economic returns and profitability in the field may be beneficial to the farmer. While economic factors can change quickly due to shifts in input costs (e.g., seed, fertilizer, chemical expenses) and commodity market prices, additional income from these areas would improve the strategy. For example, identifying a local market for animal feed may increase profitability in underperforming areas by growing cover crops tailored to livestock dietary requirements. Additionally, seeking help from government agencies such as the USDA Natural Resources Conservation Service and nonprofit organizations can result in the area's enrollment in paid programs that create and protect wildlife habitats, waterways, and prairie strips.

Communicate Findings to the End Users

The results need to be communicated to farmers and other interested parties. For this study, the results were explained to the producers during several one-on-one meetings, along with summary reports. The goals and expectations for the land were also frequently discussed before and after implementation. This was particularly helpful for the producer to understand the concept of yield potential and how applying high inputs to a low-yielding region does not necessarily increase yield or productive capacity.

Case Study 3: Using Multiple Approaches to Solve Complex Water Quality Problems

Problem Identification

While the first two case studies focus on individual fields, Case Study 3 illustrates the broader reach of precision conservation, with impacts extending beyond agriculture to surrounding communities. The Tri-State Mining District of southeast Kansas, northeast Oklahoma, and southwest Missouri is an area rich in geological resources. Mining of minerals, including coal, uranium, phosphate, lead, and zinc, began in the 1850s and became particularly robust during the World War I and World War II eras to support munitions production. Mining waste was disposed of in piles that are clearly visible from satellite imagery (Figure 25.7). The waste material within the piles contained high concentrations of lead, zinc, arsenic, and cadmium, and it was used as foundation material in local homes (Figure 25.8). In addition to the waste materials left on the soil surface, a hydroelectric dam was opened in 1940. This dam created a 43,500-ac (17,603 ha) lake that provided 1,300 miles (2,092 km) of shoreline. However, the hydroelectric facility flooded 2,000 ac (809 ha) of land containing mining waste. These materials are a source of contamination for the water in the lake. The lake is being used for recreation, and multiple entities manage this resource.

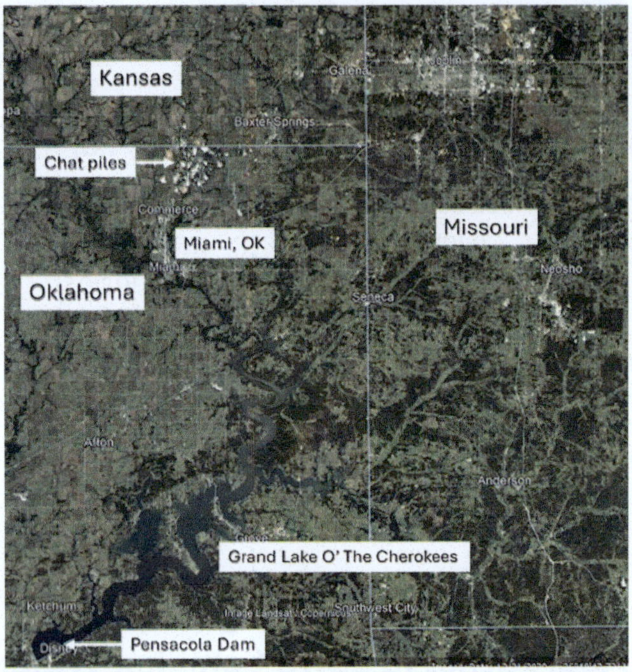

Figure 25.7. Satellite image of the Tri-State Mining District in southeast Kansas, northeast Oklahoma, and southwest Missouri, showing waste piles from mining activities present in abandoned towns of Picher, OK, and Treece, KS, as well as Baxter Springs and other towns in the area. The Neosho River flows into the Grand Lake o' the Cherokees, which is formed by the Pensacola Dam. (Google Earth images).

Figure 25.8. Orange acid runoff water from mine waste can be clearly seen at the Lytle Creek Bridge over Tar Creek. Google Earth image, 2015. 36°57′29.25″N, 94°50′41.93″W.

Identify Potential Options

The goal of this case study is to discuss potential projects and their impacts on remediating water quality in the surrounding area and mitigating health hazards. The assessment identified three potential approaches to meet the project goals. These approaches include the removal of contaminated waste materials, water filtration, and stream bank stabilization. Satellite imagery (Figures 25.7 and 25.8) provided crucial spatial insight into the challenges surrounding this project, including the ability to diagnose the problem, make decisions, and evaluate performance.

Determine the Expected Costs and Economic Assessment

The initial assessment found that the practices used in this watershed affect human health and ecosystem sustainability. Based on this assessment, the community began implementing corrective solutions using financial resources available from multiple agencies.

Implement the Practice Using the Obtained Resources

The Oklahoma Department of Environmental Quality (ODEQ), working with the U.S. Environmental Protection Agency and the Quapaw Nation, collaborated on this project. During the project, many tools discussed in this book were used, including Global Positioning System, GIS, remote sensing, and proximal sensing products.

Work on the waste piles requires careful planning to minimize the introduction of toxic metals during the removal and remediation processes. The passive water treatment systems at three locations in the region create a flow system using wind and solar power to move contaminated water over and through oxidation ponds and bioreactors to remove iron, zinc, lead, cadmium, and other contaminants from the water (Figure 25.9). Since the implementation of the water purification systems, contaminant levels in the water have decreased, and the number of fish species has increased from 8 to 15. In addition to improvements in the diversity of fish species, beavers and river otters have also returned. The group is currently working to implement additional passive treatment systems at other locations in the region. For in-stream filtration, the

Figure 25.9. Mayer Ranch passive treatment system at the Tar Creek Superfund Site near Commerce, OK. Water contaminated with toxic metal mine waste enters the system on the left. After passing through multiple stages, including an oxidative process, the water exits the system through the final polishing unit on the right.

Peoria Tribe has undertaken an effort to restore native mussel populations in the region's streams. They have built a new mussel facility at the Peoria Aquatic Fish Hatchery to repopulate the streams with the Neosho Mucket and rabbitsfoot mussels native to the region. The state-of-the-art facility simulates river water flow to support the mussels. Mussels are an important indicator species for water health, and they conduct in-stream filtration (Haygood et al., 2020). However, mussel survival is threatened by sedimentation, invasive species, and dams. Most mussels need flowing water to survive, and since they rely on fish for part of their life cycle, they also require a variety of fish species to be successful.

To reduce water sedimentation, the Cherokee Nation and the Eastern Shawnee Tribe have undertaken the restoration of two sections of Lost Creek through stream bank stabilization, building resilient creek crossings, and restoring riparian habitat. The restoration activities will address the contamination of natural resources caused by toxic materials released from historical mining operations.

Assess Success, Review, and Modify the Strategy as Needed

Because of the complexity and extent of environmental damage in the Tri-State Mining District, as well as the diversity of entities affected by the ecosystem, solutions require multiple approaches and coordinated efforts. These issues are critically important for improving the health of people in the area. Most groups are working closely with others in the area and with community members to gain support for the cleanup efforts. Because of the extent of the problem, monitoring continues, and program refinements are made as needed.

Communicate Findings to the End Users

The L.E.A.D. Agency (https://www.leadagency.org) developed a campaign to educate people on the potential toxic effects of environmental contaminants in the area and in people's homes. In addition to monitoring and measuring lead levels in homes due to heating, ventilation, and air conditioning systems and water contamination, the agency also has a strong campaign to encourage people, especially children, to get tested for high blood lead levels. Their efforts have resulted in greater recognition of the problems in the area and the identification of lead contamination and other toxic metals in children. Moreover, the efforts on this project have led to the cleaning and removal of waste materials to a stable, long-term storage site. To date, 5,500 ac (2,225 ha) of land can now be used for farming or building homes; 134 waste piles, bases, and tailings ponds have been completed, and 379 mine shafts have been encased (B. Stanila, ODEQ, personal communication). A total of 11.65 million tons (10.56 million Mg) of waste has been removed.

A survey of the area identified potentially suitable streams for the reintroduction of native mussels (M. Hamilton, personal communication). The survey revealed that drought caused 7 of the 12 potential streams to run dry at some point during the year, resulting in small, stagnant pools rather than the flowing water needed to support mussel populations. In addition to reliable streamflow, they found that streambank stabilization, willing landowners, and good fish diversity were required for mussel reintroduction. Areas with high agricultural runoff from cattle or chicken confinement operations were not suitable. Heavy equipment use in one area resulted in oil contamination. The survey identified Five Mile Creek and Sycamore Creek as two potential sites. They are implementing silo studies in these creeks where mussels have been reintroduced and will continue to grow mussel stocks for additional stream introductions in the future. In addition to helping mussels, it is anticipated that the mussels themselves will further improve stream water quality, assisting in the diversification of species within the streams. A multi-tribal group has developed the Tar Creek Apprenticeship Program (TCAP; https://www.youtube.com/watch?v=Y8fXXAu9qQc) to educate students about the area's challenges and environmental conditions. Participants learn about the restoration activities underway in the region, and work to design their own restoration proposals.

Case Study 4: Using Trail Cameras for Continuous, Non-Destructive Monitoring in Cropping Systems

Problem Identification

Monitoring plant development, phenology, and stress responses plays an important role in improving crop management and sustaining long-term productivity in an agriculture system or ecosystem. In practice, however, many traditional approaches rely on destructive sampling, repeated manual measurements, and significant labor input, which limits their temporal resolution and scalability (Tovar et al., 2018). These challenges also limit our ability to make optimal management decisions. Instead, certain precision agriculture technologies can be used as non-destructive continuous monitoring systems, improving decision making and reducing cost and labor.

These non-destructive continuous monitoring systems are particularly evident in specialty crop systems such as cranberry production but can also be used in marginal lands and natural conservation projects in which plant health needs to be consistently monitored.

Continuous imaging systems, including phenocams, have demonstrated strong potential for tracking vegetation phenology and canopy development across ecosystems (Crimmins & Crimmins, 2008; Sonnentag et al., 2012; Richardson, 2019). At the same time, high-throughput phenotyping platforms and dedicated imaging infrastructures often require substantial financial investment and technical capacity, which can restrict their use in field-based specialty crop systems (Berger et al., 2012; Pauli et al., 2016).

In this case study, the objective is different than those above; rather that assessing an isolated problem, this case study will focus on selecting equipment to help identify and address precision conservation challenges. The goal is to provide a roadmap to solve similar operational and technological challenges. Specifically, this case study evaluated the potential integration of trail cameras as a practical and affordable alternative for continuous, non-destructive plant monitoring to inform long-term resource stewardship.

Identify Potential Options

Several imaging systems were evaluated conceptually, including stationary phenocams, high-throughput phenotyping platforms, and trail cameras. Phenocams provide reliable canopy-scale monitoring but are typically deployed at broader landscape scales. High-throughput systems generate high-resolution data but require substantial financial and technical investment.

Trail cameras, which are widely used in tracking animals in wild environments and in ecological research (Newey et al., 2015), offer a more flexible and field-adapted alternative to evaluate plant health over long periods of time (Larbi et al., 2017; Morris et al., 2022), suggesting potential applications beyond wildlife monitoring and into precision agriculture systems.

Considering factors such as affordability, durability under field conditions, image resolution, and time-lapse capability, eight commercially available camera models were selected for evaluation within a cranberry production system that can be extended to other cropping systems.

Determine Expected Cost and Economic Assessment

When choosing equipment, evaluating the technology's ability to complete the job as well as the cost effectiveness of the technology is key. In this case, compared to dedicated high-throughput phenotyping systems, trail cameras offer a practical and affordable option for continuous plant monitoring. The primary costs are limited to the purchase of the cameras, batteries, and data storage, making them accessible for field-scale deployment. In contrast to labor-intensive sampling methods that require repeated manual measurements, trail cameras can operate autonomously for extended periods, reducing both labor demands and field disturbance.

From a precision conservation perspective, the ability to deploy multiple cameras across different landscape positions provides meaningful spatial coverage without requiring substantial financial investment. This expanded coverage allows researchers and producers to capture variability in canopy development and environmental conditions across the field. The resulting improvement in temporal resolution through repeated, time-lapse observations, supports more informed management decisions and facilitates targeted conservation strategies that account for spatial and temporal variability within production systems.

Implement the Practice Using Obtained Resources

Eight different trail cameras were installed at canopy height within a cranberry production field in Wisconsin and programmed to capture images at 30-minute intervals over a 15-day period. Cameras were positioned at a 45-degree diagonal angle to optimize canopy visibility and minimize shadow interference.

To ensure data consistency, image preprocessing included time-based filtering to exclude low-light and irrigation-distorted images. Exposure correction and white balance standardization were applied using image processing workflows.

Temperature data embedded within image metadata were extracted and compared with a canopy-level thermistor probe and a nearby weather station temperature probes to assess environmental sensing accuracy

Assess Success, Monitor Practices, and Modify the Strategy

Camera performance was evaluated based on image clarity (blur scores), color fidelity (color cast), noise levels, temperature measurement trends, and battery longevity. Several models demonstrated reliable image quality, strong temperature correlation with canopy-level sensors, and sustained battery performance over the deployment period (Table 25.3).

Continuous RGB imaging enabled non-destructive tracking of canopy development and environmental variability, supporting phenological monitoring without manual sampling. These findings align with prior research demonstrating the utility of digital repeat photography in vegetation monitoring (Sonnentag et al., 2012; Richardson, 2019).

Communicate Findings to End Users

Findings from this work were shared through extension discussions and stakeholder engagement activities. Early dissemination efforts have focused on communicating the practical implications of continuous imaging for specialty crop systems and gathering feedback from growers and research collaborators. By lowering economic and technical barriers to ecosystem monitoring, trail cameras provide a practical entry point into decentralized sensing networks that enhance the spatial and temporal resolution of precision conservation practices; better informing long-term resource stewardship.

Conclusions

Precision conservation, as an ecosystem management approach, builds on knowledge from precision agriculture and uses advanced geospatial data collection and analytics to conserve resources at the field, agroecosystem, and broader scales. It serves as a tool to address twenty-first century challenges. The rapid development of new information technologies and data analytics highlights issues with current management practices and the connection between production systems and natural resources. Excitingly, this has opened innovative avenues for developing decision-support tools that enhance precision conservation practices. Throughout this book, we have explored a variety of advanced technologies applied in precision agriculture management. In this chapter, we underscore the significance of using precision conservation across spatial and temporal scales. It is our intent that readers appreciate the interdependence between agricultural productivity and environmental stewardship, as well as the technological innovations that facilitate this integration. Furthermore, we aim to reinforce the notion that the responsible and judicious use of shared natural resources is a collective obligation essential to ensuring the Earth's sustainability and its capacity to support existing systems.

Table 25.3. Summary Performance Evaluation of Four Trail Camera Models for Continuous Plant Monitoring.

Performance Metric	Camera A	Camera B	Camera C	Camera D	Interpretation for Precision Monitoring
Image Sharpness	High	Moderate–High	Moderate	Moderate–High	Higher sharpness improves canopy detail detection
Color Fidelity	Moderate	Moderate	High	Low	Accurate color improves stress and phenology analysis
Noise Level	Moderate	Low	Low–Moderate	Very Low	Lower noise improves image clarity under variable light
Temperature Trend Consistency	Moderate	High	High	Moderate	Strong alignment supports microclimate monitoring
Operational Reliability	Moderate	High	High	High	Critical for long-term deployment
Battery Longevity	Moderate	High	Moderate	High	Important for remote, unattended monitoring
Overall Suitability	Moderate	High	High	Moderate	Suitable models identified for field-scale application

Study Questions

1. What is precision conservation?
2. What are the components of a conservation plan?
3. What is the 7R conservation approach?
4. Explain the concepts of sparing and sharing.
5. Why are mussels important for in-stream water filtering?
6. True/False: There is no difference between precision agriculture and precision conservation.
7. Question (short answer): How do precision agriculture and precision conservation differ in terms of primary focus and scale, and why is this distinction important for long-term sustainability?

ACKNOWLEDGMENTS

This paper was funded in part by the University of Missouri, the USDA, Kansas State University, South Dakota State University, USDA Natural Resources Conservation Service Conservation Innovation Grants (NR233A750011G019), USDA Agriculture and Food Research Initiative, the American Coalition for Ethanol, GEVO, the South Dakota Corn Utilization Council, and the South Dakota Soybean Research and Promotion Council. The use of trade names is for the convenience of the reader and does not imply endorsement by the authors.

Thank you to student reviewers Ananya Gairola from IARI and Bruno Scheffer from Purdue University for contributing to the end-of-chapter questions.

REFERENCES

Berger, B., De Regt, B., & Tester, M. (2012). High-throughput phenotyping of plant shoots. In R. Furbank & M. Tester (Eds.), *High-throughput phenotyping in plants: Methods and protocols* (pp. 9–20). Humana Press.

Berry, J. K., Delgado, J. A., Khosla, R., & Pierce, F. J. (2003). Precision conservation for environmental sustainability. *Journal of Soil and Water Conservation*, *58*(6), 332–339. https://doi.org/10.1080/00224561.2003.12457555.

Berry, J. K., Delgado, J. A., Pierce, F. J., & Khosla, R. (2005). Applying spatial analysis for precision conservation across the landscape. *Journal of Soil and Water Conservation*, *60*(6), 363–370. https://doi.org/10.1080/00224561.2005.12435828.

Birru, G. A., Clay, D. E., DeSutter, T. M., Reese, C. L., Kennedy, A. C., Clay, S. A., Bruggeman, S. A., Owen, R. K., & Malo, D. D. (2019). Chemical amendments of dryland saline–sodic soils did not enhance productivity and soil health in fields without effective drainage. *Agronomy Journal*, *111*(2), 496–508. https://doi.org/10.2134/agronj2018.04.0296.

Boetzl, F. A., Krauss, J., Heinze, J., Hoffmann, H., Juffa, J., König, S., Krimmer, E., Prante, M., Martin, E. A., Holzschuh, A., & Steffan-Dewenter, I. (2021). A multitaxa assessment of the effectiveness of agri-environmental schemes for biodiversity management. *Proceedings of the National Academy of Sciences of the United States of America*, *118*(10), e2016038118. https://doi.org/10.1073/pnas.2016038118.

Budak, M. E., Clay, D. E., Clay, S. A., Reese, C. L., Westhoff, S., Howe, L. E., Owen, R. K., Birru, G., He, Y., & Wang, Z. (2022). Increased rainfall may place saline/sodic soils on the tipping point of sustainability. *Journal of Soil and Water Conservation*, *77*(4), 418–425. https://doi.org/10.2489/jswc.2022.00131.

Clay, S. A., Fiedler, D., Reese, C., & Clay, D. E. (2022). Restoring ecological function to saline–sodic soils in South Dakota with perennial grass mixtures. *Agronomy Journal*, *115*(1), 135–146. https://doi.org/10.1002/agj2.21184.

Crimmins, M. A., & Crimmins, T. M. (2008). Monitoring plant phenology using digital repeat photography. *Environmental Management*, *41*(6), 949–958.

Dainese, M., Martin, E. A., Aizen, M. A., Albrecht, M., Bartomeus, I., Bommarco, R., Carvalheiro, L. G., Chaplin-Kramer, R., Gagic, V., Garibaldi, L. A., Ghazoul, J., Grab, H., Jonsson, M., Karp, D. S., Kennedy, C. M., Kleijn, D., Kremen, C., Landis, D. A., Letourneau, D. K., . . . Steffan-Dewenter, I. (2019). A global synthesis reveals biodiversity-mediated benefits for crop production. *Science Advances*, *5*(10), eaax0121. https://doi.org/10.1126/sciadv.aax0121.

Delgado, J. A. (2016). 4 Rs are not enough. We need 7 Rs for nutrient management and conservation to increase nutrient use efficiency and reduce off-site transport of nutrients. In R. Lal & B. A. Stewart (Eds.), *Soil specific farming: Precision agriculture* (pp. 89–126). CRC Press.

Delgado, J. A., Short, N. M., Jr., Roberts, D. P., & Vandenberg, B. (2019). Big data analysis for sustainable agriculture on a geospatial cloud framework. *Frontiers in Sustainable Food Systems*, *3*, 54. https://doi.org/10.3389/fsufs.2019.00054.

DeSutter, T., Franzen, D., He, Y., Wick, A., Lee, J., Deutsch, B., & Clay, D. (2015). Relating sodium percentage to sodium adsorption ratio and its utility in the northern Great Plains. *Soil Science Society of America Journal*, *79*(4), 1261–1264. https://doi.org/10.2136/sssaj2015.01.0010n.

DeSutter, T. M., Clay, D. E., Oster, J. D., Westhoff, S., & Alvarez-Torres, B. (2024). Laboratory methods for determining salinity and sodicity. In D. E. Clay, T. M. DeSutter, S. A. Clay, & T. Nleya (Eds.), *Salinity and sodicity: A global challenge to food security, environmental quality, and soil resilience* (pp. 31–40). ASA, CSSA, SSSA.

Dudley, N., & Alexander, S. (2017). Agriculture and biodiversity: A review. *Biodiversity*, *18*(2–3), 45–49. https://doi.org/10.1080/14888386.2017.1351892.

Emmerson, M., Morales, M. B., Oñate, J. J., Batáry, P., Berendse, F., Liira, J., Aavik, T., Guerrero, I., Bommarco, R., Eggers, S., Pärt, T., Tscharntke, T., Weisser, W., Clement, L., & Bengtsson, J. (2016). How agricultural intensification affects biodiversity and ecosystem services. *Advances in Ecological Research*, *55*, 43–97.

Fiedler, D., Clay, S. A., Westhoff, S., Reese, C. L., Bruggeman, S. A., Moriles-Miller, J., Perkins, L., Joshi, D. R., Marzano, S.-Y., & Clay, D. E. (2022). Phytoremediation and rainfall combine to improve soil and plant health in a northern Great Plains saline sodic soil. *Journal of Soil and Water Conservation*, *77*(4), 381–388. https://doi.org/10.2489/jswc.2022.00112.

Haygood, L., Roberts, K., Li, J. J., Cornell, W. C., Crawford, A., Hartwell, J., & Sharp, A. M. P. (2020). Hg, As, Cd, and Pb in water, sediment, mussel tissue and shells in the Kiamichi River Basin, southeastern Oklahoma: Sources, surface water and groundwater quality, and human health implications. In *AGU Fall Meeting Abstracts*, 2020, GH005-05.

Hendrickson, J., Sassenrath, G. F., Archer, D., Hanson, J., & Halloran, J. (2008). Interactions in integrated US agricultural systems: The past, present and future. *Renewable Agriculture and Food Systems*, *23*(4), 314–324. https://doi.org/10.1017/S1742170507001998.

Hsiao, C.-J., Mushtaq, M., Sassenrath, G. F., Zeglin, L. H., Hettiarachchi, G. M., & Rice, C. W. (2025). Long-term tillage and compost shape soil microbes under soil organic carbon equilibrium. *European Journal of Soil Science*, *76*(3), e70125. https://doi.org/10.1111/ejss.70125.

Joshi, D. R., Moriles-Miller, J., Clay, S. A., Reicks, G., Westhoff, S., & Clay, D. E. (2024). Rye cover crop had a nitrogen credit one year after termination in a frigid environment. *Journal of Soil and Water Conservation*, *79*(5), 261–267. https://doi.org/10.2489/jswc.2024.00097.

Khosla, R. (2008). The 9th International Conference on Precision Agriculture opening ceremony presentation. [Conference presentation]. July 20–23.

Knapp, M., Teder, T., Lukas, V., Štrobl, M., Knappová, J., Landis, D. A., & González, E. (2023). Ecologically-informed precision conservation: A framework for increasing biodiversity in intensively managed agricultural landscapes with minimal sacrifice in crop production. *Biological Conservation*, *288*, 110343. https://doi.org/10.1016/j.biocon.2023.110343.

Kulesza, S. E., Mathis, M. A., II, Alarcon, V. J., & Sassenrath, G. F. (2024). Critical shear stress variability in claypan soils with depth. *Journal of Soil and Water Conservation*, *79*(2), 66–77. https://doi.org/10.2489/jswc.2024.00099.

Larbi, P. A., & Green, S. (2018). Time series analysis of soybean response to varying atmospheric conditions for precision agriculture. *Precision Agriculture*, *19*(6), 1113–1126.

Luther, Z. R., Swinton, S. M., & Van Deynze, B. (2022). Potential supply of Midwest cropland for conversion to in-field prairie strips. *Land Economics*, *98*(2), 274–291. https://doi.org/10.3368/le.98.2.082020-0129R1.

Matthees, H. L., He, Y., Owen, R. K., Hopkins, D., Deutsch, B., Lee, J., Clay, D. E., Reese, C., Malo, D. D., & DeSutter, T. M. (2017). Predicting electrical conductivity of the saturation extract from a 1:1 solution to water ratio. *Communications in Soil Science and Plant Analysis*, *48*(18), 2148–2154. https://doi.org/10.1080/00103624.2017.1407780.

Morris, J. R., Petersen, S. L., Madsen, M. D., McMillan, B. R., Eggett, D. L., & Lawrence, C. R. (2022). Monitoring seedling emergence, growth, and survival using repeat high-resolution imagery. *Remote Sensing*, *14*(21), 5365.

Mulla, D. J., & Belmont, S. (2018). Identifying and characterizing ravines with GIS terrain attributes for precision conservation. In J. A. Delgado, G. F. Sassenrath, & T. Mueller (Eds.), *Precision conservation: Geospatial techniques for agricultural and natural resources conservation* (Vol. *59*, pp. 109–129). ASA, CSSA, SSSA.

Newey, S., Davidson, P., Nazir, S., Fairhurst, G., Verdicchio, F., Irvine, R. J., & van der Wal, R. (2015). Limitations of recreational camera traps for wildlife management and conservation research: A practitioner's perspective. *Ambio*, *44*(S4), 624–635.

Osipitan, O. A., Dille, J. A., Assefa, Y., & Knezevic, S. Z. (2018). Cover crop for early season weed suppression in crops: Systematic review and meta-analysis. *Agronomy Journal*, *110*(6), 2211–2221. https://doi.org/10.2134/agronj2017.12.0752.

Pauli, D., Andrade-Sanchez, P., Carmo-Silva, A. E., Gazave, E., French, A. N., Heun, J., Hunsaker, D. J., Lipka, A. E., Setter, T. L., & Gore, M. A. (2016). Field-based high-throughput plant phenotyping reveals the temporal patterns of quantitative trait loci associated with stress-responsive traits in cotton. G3: Genes, Genomes. *Genetics*, *6*(4), 865–879.

Reidsma, P., Tekelenburg, T., van den Berg, M., & Alkemade, R. (2006). Impacts of land-use change on biodiversity: An assessment of agricultural biodiversity in the European Union. *Agriculture, Ecosystems & Environment*, *114*(1), 86–102. https://doi.org/10.1016/j.agee.2005.11.026.

Richardson, A. D. (2019). Tracking seasonal rhythms of plants in diverse ecosystems with digital camera imagery. *New Phytologist.*, *222*, 1742–1750. https://doi.org/10.1111/nph.15591.

Sassenrath, G., & Kulesza, S. (2017). Measuring soil electrical conductivity to delineate zones of variability in production fields. *Kansas Agricultural Experiment Station Research Reports*, *3*(2). https://doi.org/10.4148/2378-5977.1386.

Sassenrath, G. F., & Delgado, J. A. (2018). Precision conservation and precision regulation. In J. A. Delgado, G. F. Sassenrath, & T. Mueller (Eds.), *Precision conservation: Geospatial techniques for agricultural and natural resources conservation* (Vol. *59*, pp. 361–384). ASA, CSSA, SSSA.

Schulte, L. A., Niemi, J., Helmers, M. J., Liebman, M., Arbuckle, J. G., James, D. E., Kolka, R. K., O'Neal, M. E., Tomer, M. D., Tyndall, J. C., Asbjornsen, H., Drobney, P., Neal, J., Van Ryswyk, G., & Witte, C. (2017). Prairie strips improve biodiversity and the delivery of multiple ecosystem services from corn–soybean croplands. *Proceedings of the National Academy of Sciences of the United States of America*, *114*(42), 11247–11252. https://doi.org/10.1073/pnas.1620229114.

Sonnentag, O., Hufkens, K., Teshera-Sterne, C., Young, A. M., Friedl, M., Braswell, B. H., & Richardson, A. D. (2012). Digital repeat photography for phenological research in forest ecosystems. *Agricultural and Forest Meteorology*, *152*, 159–177.

Tovar, J. C., Hoyer, J. S., Lin, A., Tielking, A., Callen, S. T., Castillo, E., Miller, M., Tessman, M., Fahlgren, N., Carrington, J. C., & Nusinow, D. A. (2018). Raspberry Pi–powered imaging for plant phenotyping. *Applications in Plant Sciences*, *6*(3), e1031.

From Control Systems to Variable Rate Seeding, Pesticide, Chemical Applicators, and Irrigation

26

Ali Mirzakhani Nafchi, Douglas Prairie, Todd Trooien, Nicholas Uilk, Xufei Yang, Mazhar Sher, Karishma Kumari, Ahmed Abdalla, and Hasan Mirzakhaninafchi

Chapter Purpose

This chapter is a revision and update of chapter 10, *Electronics and Control Systems*, from the first edition of this book, (Franzen A., 2018). It discusses how improvements in energy efficiency, artificial intelligence, and robotics have provided opportunities to reduce costs related to harvest, pest control, and planting. The goal of this chapter is to introduce students to the basics of control systems, electronics, and robotics and explain how these technologies can be used in precision nutrient and pest management.

Introduction of Electronics, Control Systems, and Robotics

In the 1990s, robotics and control systems were too costly to be commonly used in agricultural equipment. However, cost savings, along with advanced electronics that combined mobile platforms, wireless communication, mechanical actuators, control systems, visual perception, and real-time decision-making, are now making robotics a feasible option in precision agriculture. These advances became possible through improvements in electronics and advances in communication required for variable rate control of fertilizers, planting, irrigation, pest management, and manure applications. The following sections discuss these advances.

Control System Theory

Control systems rely on sensors to monitor machine operation and regulate machine function. A sensor is "a device that responds to a physical stimulus (e.g., heat, light, sound, pressure, magnetism, or motion) and transmits an impulse for measurement or control" (Mish et al., 1996). Sensors convert physical phenomena into electronic signals that are readable by a controller that processes sensor data and sends control signals to motors, valves, pumps, and meters, with specific instructions. There are several general categories of control systems including the open and closed loop systems.

Open-loop Control Systems

Figure 26.1 shows the basic structure of an open-loop control system. In this system, an input command is sent to a controller, which drives the process and produces an output. Importantly, the output is not measured or fed back to the controller, so the system cannot automatically correct for changes in load or disturbances. An example is a DC motor supplied with a variable voltage to control speed. With a constant load torque, a roughly linear relationship between motor speed and applied voltage would be expected. In practice, however, speed also depends on torque demand (the resisting twisting force). For example, a metering device on a dry-material hopper would indicate higher torque demand when the hopper is full because the material exerts greater pressure on the metering mechanism, increasing friction and resistance. As the hopper empties, resistance decreases, the motor runs faster with the same applied voltage, and the delivery rate rises. As a result, the hopper may deliver its contents at 5 kg/s when full but 8 kg/s when nearly empty, even though the control input has not changed.

Figure 26.1. Open-loop control concept used in this section: an input setting (e.g., motor voltage command) is applied to the control block to produce an output (e.g., speed/flow). Since output is not fed back, changes in load (torque demand) can change the output even when the input remains constant.

Closed-loop Control Systems

Closed-loop control systems, or feedback control systems, use sensors to monitor and adjust a device's operation. For example, they might regulate rotation speed based on the material in a hopper, requiring calibration curve adjustments. A more common approach is direct measurement feedback, where a sensor monitors meter speed and adjusts the control signal to maintain the desired rate. These systems typically measure actuator status (position, speed, angle, etc.) and use a setpoint for control (Figure 26.2). The setpoint is the desired value, such as a target speed, flow rate, pressure, position, or application rate, entered by the operator or generated by the ECU. ECUs are typically small, embedded microcontrollers that monitor sensor data. The error is the difference between the setpoint and the actual output—if they match, the error is zero.

A typical example of a closed-loop control system is cruise control in a car. When the driver sets the desired speed, the ECU monitors the speedometer sensor and adjusts the accelerator. The engine revs up on hilly roads when climbing because the ECU detects a speed drop and compensates. After cresting a hill, the ECU reduces acceleration if the speed exceeds the setpoint. On steeper hills, the transmission may be downshifted. This proportional response to error is called proportional control. There are many types of closed-loop control schemes, including Modern control, H+ control, H0 control, and Lyapunov Nonlinear control, among others. However, the afore-mentioned control schemes are well beyond the scope of this text. A common control scheme used by an ECU is proportional–integral– derivative (PID) control. A PID controller monitors the closed-loop error and makes changes to the signal

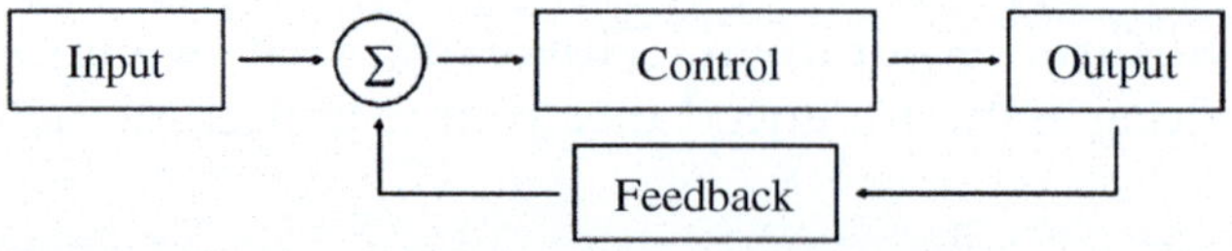

Figure 26.2. A closed-loop control system block diagram.

sent to the actuator to achieve the desired setpoint. The general equation used by the ECU for PID control is found in Eq. 26.1.

$$\text{Control Signal} = K_p \,\text{error} + K_i \int_0^i \text{error}\, dt + K_d \left[\frac{d\ \text{error}}{dt} \right] \quad (26.1)$$

$$\text{Control Signal} = K_p e(t) + K_i \int_0^t e(\tau)\, d\tau + K_d \frac{de(t)}{dt}$$

$$e(t) = \textit{desired position} - \textit{actual position}$$

K_p is the proportional gain, which determines how the controller will respond to being out of line by a distance, or position error. K_i is integral gain, which determines how the controller will respond to accumulated (past) error over time, adding corrective action when the system has been off the A–B line (or off the desired position) for a while. This action helps eliminate steady-state error (persistent offset) by "building up" control effort until the long-term error is driven toward zero. K_d is the derivative gain, which determines how the controller responds to being out of alignment with the A–B line, or angular error.

The PID control scheme can be implemented using the proportional, integral, and derivative terms in the equation above. The proportional gain term determines how much correction is applied to the signal based on the current error, while the integral gain term corrects for error over time as measured by the sum of past errors. The derivative term determines how much correction is applied based on temporal changes in the error. Derivative gain reduces the likelihood that the output will overshoot the setpoint.

A PID control scheme can be used for real-time adjustment of fertilizer application rates. For example, if soil N (nitrogen) sensors identify N levels in the field below a threshold, the system increases the fertilizer flow proportionally and continues to adjust the N levels until there is no longer a problem. The PID in this example could also prevent the over-application of N by accounting for rapid changes in the sensor readings. A real-world PID-style example is a variable-rate nitrogen setup. The goal (setpoint) is to keep soil N near a target band (e.g., 20–25 ppm): (P) if sensors read low, increase flow *in proportion* to how low it is; (I) if it stays slightly low for a long time (small but persistent deficit), slowly "ratchet up" flow until the deficit disappears; (D) if readings are dropping fast (suggests to be very low), increase sooner, but if readings rise sharply (maybe higher N), back off quickly to avoid over-application. Proportional and derivative gains are illustrated in the global positioning system (GPS)/global navigation satellite system (GNSS) auto guidance systems section later in this chapter.

Rate Control with Fast Switching Pulse Width Modulation

Many electrical actuators have input–output curves that are approximately linear: As the applied voltage increases, the actuator output (such as speed) increases in a roughly proportional way. In practice, however, generating a smoothly varying analog voltage can be expensive, inefficient, or both. A common alternative is to apply the maximum available voltage but switch it on and off very quickly. If the switching happens fast enough, the actuator behaves as though it is receiving a "blended" intermediate voltage. This technique is called pulse width modulation (PWM), and it is built into most microcontrollers. To use PWM, the microcontroller is set up to output a digital square wave on a chosen pin. Instead of outputting a constant high or low, the pin repeatedly alternates between them. High (on) is the condition in which the actuator receives the full supply voltage, and in low (off), the actuator receives 0 V (or is disconnected, depending on the driver circuit). Most microcontrollers include dedicated PWM hardware, which makes this switching accurate and efficient without tying up the processor. The duty cycle describes the fraction of each PWM cycle spent on. A higher duty cycle means the signal is on more of the time, the average power delivered is higher, and the actuator output is higher (e.g., faster motor speed). A lower duty cycle means the signal is on less of the time, average power delivered is lower, and output is lower. Therefore, duty cycle is usually expressed as a percentage (0% = always off; 50% = on half the time, off half the time; and 100% = always on). The pulse frequency is how many PWM cycles happen each second (Hz). Higher frequency means faster switching (shorter cycles), and lower frequency means slower switching (longer cycles). If the frequency is high enough, the actuator's mechanical or electrical inertia "smooths out" the rapid switching so the output appears steady. If it is too low, side effects such as visible flicker (LEDs) or audible whining/vibration (motors) may become apparent. For example, a microcontroller in a drone does not "dial" motor voltage smoothly like a wall dimmer. Instead, it sends PWM signals to an electronic speed controller, which rapidly switches power, so the motor responds as though it is getting a lower or higher effective power level (Figure 26.4). In most microcontrollers, PWM functionality is initialized using a counter with three important parameters: *bottom*, *top*, and *compare*. The timer counts in increments of one from bottom to top. Each time the counter increments, the microcontroller checks whether the counter value is greater than the compare value. If the counter value is less than the compare value, the PWM signal is set to on. If it is greater, the signal is set to off. The counter continues counting from top to bottom repeatedly, and the percentage of the time that the output signal is on versus off can be changed by resetting the compare value.

For an agronomist, technician, or farmer, it is more important to understand a few key concepts about the PWM signal. The most important concepts are duty cycle, period, frequency, and resolution. The duty cycle is the percentage of the time that the control signal is turned on in a single period. The period is the time required for the counter to count from bottom to top, which depends on the counter increment interval and the number of steps from bottom to top. For example, the period of a PWM signal for which the counter increments 1,000,000 times per second, with bottom = 0 and top = 255 (common settings), is demonstrated in the equation,

$$T = \frac{\left(\dfrac{256\text{ steps}}{\text{period}}\right)}{\left(\dfrac{1{,}000{,}000\text{ steps}}{\text{s}}\right)} = \frac{0.256\text{ ms}}{\text{period}} \qquad (26.2)$$

where, T represents the PWM period (the time for one complete PWM cycle), meaning the time it takes the timer/counter to count from bottom (0) up to top (255) and then restart (i.e., the full repeating interval of the PWM waveform). In equation 26.2, Steps represent individual counter increments (timer ticks). Each step is one count advance of the timer (e.g., 0 to 1 is one step). With top = 255 and bottom = 0, there are 256 discrete count values per PWM cycle (0 through 255). The PWM resolution is the number of distinct duty cycles the controller can represent. In this example, the counter spans 256 possible values, the frequency is 3,909 s^{-1} (= 1/0.256 ms), and the PWM resolution is the number of duty cycles that can be expressed by the controller. The counter goes from 0 to 255, resulting in 256 possible values and a resolution of 1/256 = 0.39%. Changing the compare value from 100 to 101 results in a duty cycle change from 39.22% to 39.61% (Figures 26.3 and 26.4).

To simplify the functionality of PWM systems, we will assume a period of 1 s, which is the length of time to complete a cycle. The frequency is the number of cycles per second and is generally measured in hertz. A system with a frequency of 10 Hz will be "on" 10 times per second and "off" 10 times per second. A duty cycle is used to adjust the rate, and a duty cycle of 20% would mean that it is "on" 20% (0.2 s) of the time and "off" 80% (0.8 s) of the time. If we have a frequency of 10 (10 cycles/s), it means that for each cycle, the system would be on for 0.02 s and off for 0.08 s. Modern sprayers that use PWM nozzles provide the best visual example of PWM systems (Figure 26.3).

Numbers in Electronics and Control Systems

Data storage and computations in the embedded ECU systems are slightly different from how calculations are done on paper. On a piece of paper or on a calculator, most people use a base-10 numbering system. In base-10, each digit in a number has a multiplier that ranges from zero to nine. Depending on each digit's position, or distance from the decimal point, the total value assigned is changed. The base-10 number 123.4 is equivalent to $(1 \times 10^2) + (2 \times 10^1) + (3 \times 10^0) + (4 \times 10^{-1})$, or $100 + 20 + 3 + 0.4$. The name base-10 refers to 10 being the base to the exponent for each digit, with the exponent changing depending on position relative to the decimal point. Computers, in general, store their memory and make calculations in what is known as the

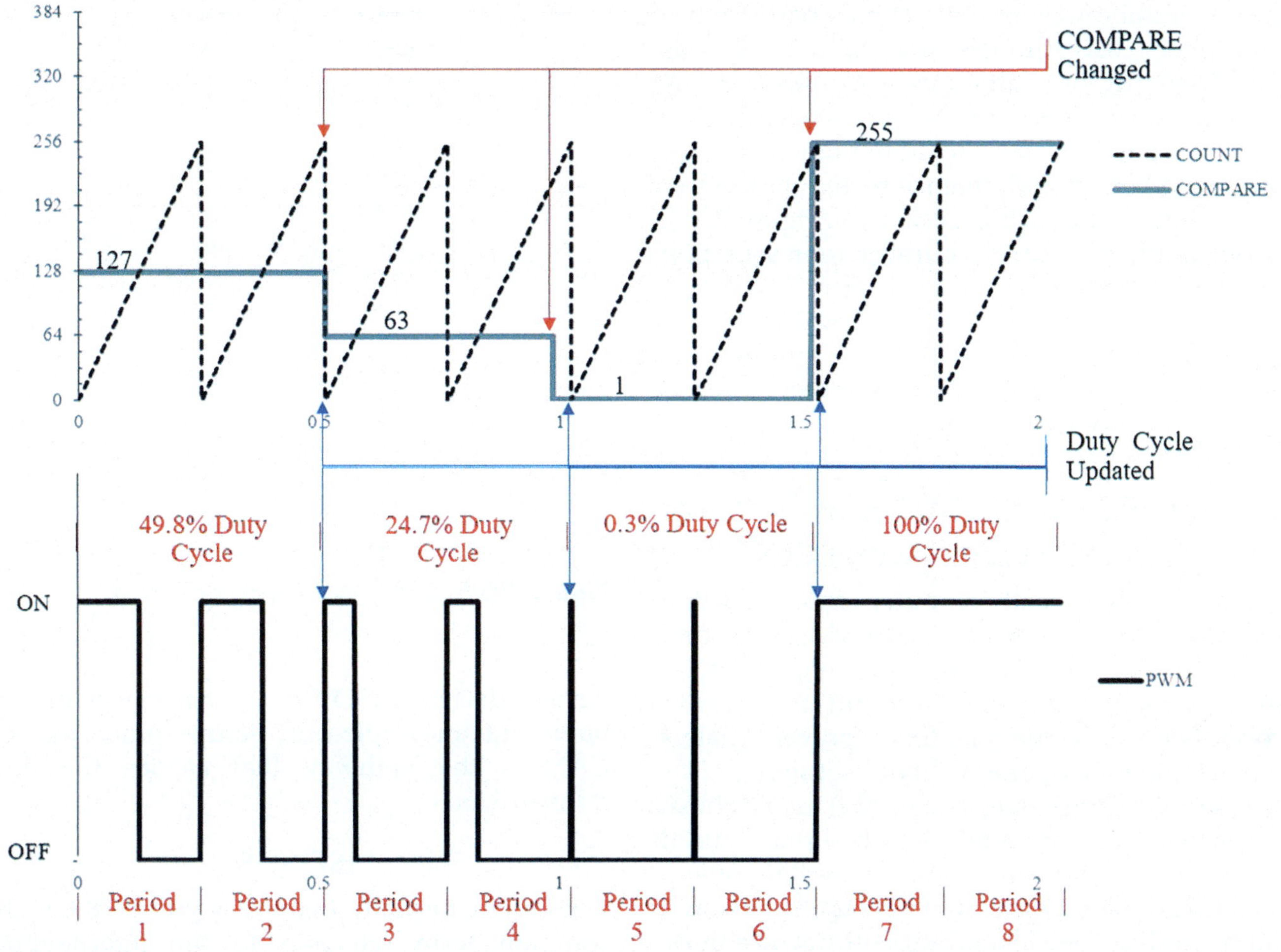

Figure 26.3. The counter and compare values for a PWM system with bottom = 0 and top = 255, 1 ms counter rate, 0.256 s period, and 3.906 Hz frequency; (b) the resulting PWM signal output. Note that this is synchronized PWM update, since the compare value is only effectively changed at the beginning of each period.

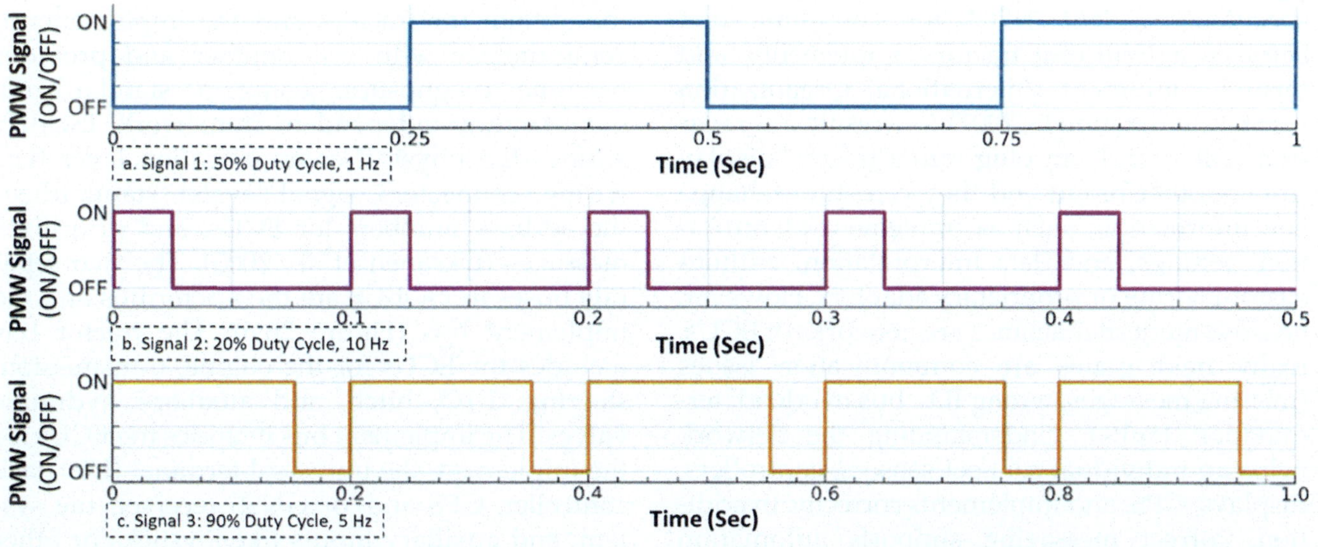

Figure 26.4. Three PWM signals: (a) 50% duty cycle, 1 Hz frequency, 1 s period; (b) 20% duty cycle, 10 Hz frequency, 100 ms period; and (c) 90% duty cycle, 5 Hz frequency, and 200 ms period (With permission of Karishma Kumari).

base-2 numbering system. While base-10 has multipliers zero through nine, base-2 only has the possibility of multipliers being zero or one because most computer memory systems have only two possible states to store numbers: on or off. The other main difference is that the base of the exponent for each digit is two (Figure 26.5). An example of a base-2 number representation might be 100,110, which is equivalent to $(1 \times 2^5) + (0 \times 2^4) + (0 \times 2^3) + (1 \times 2^2) + (1 \times 2^1) + (0 \times 2^0) = 32 + 0 + 0 + 4 + 2 + 0 = 38$ in base-10. While interesting, additional discussion on how computers count is beyond the scope of this chapter.

Binary	Decimal
0000	0
0001	1
0010	2
0011	3
0100	4
0101	5
0110	6
0111	7
1000	8
1001	9
1010	10
1011	11
1100	12
1101	13
1110	14
1111	15

Figure 26.5. Binary to decimal conversion.

Introduction to ISOBUS and/or CAN Communication with ECUs

An ECU is a small computer embedded in a machine. It reads sensors, makes decisions, and controls actuators. For example, engine ECUs manage fueling and emissions and implement/controller ECUs run planters, sprayers, balers while letting the operator change settings.

Controller area network (CAN) is a robust, message-based communication bus that is used so multiple ECUs can reliably share data over a small set of wires, especially in noisy, harsh environments like vehicles and agricultural machines. Instead of every sensor needing a dedicated wire to every device, ECUs broadcast messages (e.g., speed, PTO status, pressure), and other ECUs respond as needed. ISOBUS is the agricultural equipment standard built on top of CAN and is specified by ISO 11783 for communication between agricultural tractors, implements, and related equipment (International Organization for Standardization IS, 2007) As a result, a sprayer from one brand can plug into a tractor/display from another brand and they can still exchange key information, such as implement identification, settings, and data interpretation, without custom wiring or proprietary adapters. However, because modern machines are networks of ECUs, many field issues are communication issues (missing messages, wrong IDs, bus overload, termination faults). Understanding the network helps an individual connect sensors, controllers, displays, GPS, and implements correctly. In addition, correct messaging supports automation (section control, rate control), reduces operator workload, and prevents unsafe behavior when devices disagree. ISOBUS is what makes mixed-brand tractor/implement setups practical, and CAN is the highway that carries the data transmission.

ISOBUS

ISOBUS, formally known as ISO 11783, is the common term referring to the international standard for "tractors and machinery for agriculture and forestry serial control and communications data network" (ISO, 2007). Building on top of previous standards (SAE J1939, IEC 61162) and applying CAN 2.0B technology, ISOBUS defines the system for tractors and the implements to communicate with one another and provides machine diagnostics, using a standardized, manufacturer-independent framework. ISOBUS defines the physical communication layer (i.e., wiring, connectors, signal levels), standardizes the address numbers for ECUs, and prioritizes messages (Oksanen et al., 2005). The two separate buses in ISOBUS are the tractor bus and the implement bus (Figure 26.6). The tractor bus includes the ECUs for the engine, transmission, steering, PTO, hitch, and auxiliary hydraulic valves. The implement bus includes the ECUs for the implement, the universal terminal (UT), task controller, GPS and/or GNSS receiver, file system, and auxiliary inputs like joysticks or other electronic user interfaces. The implement bus and tractor bus are connected by a special ECU

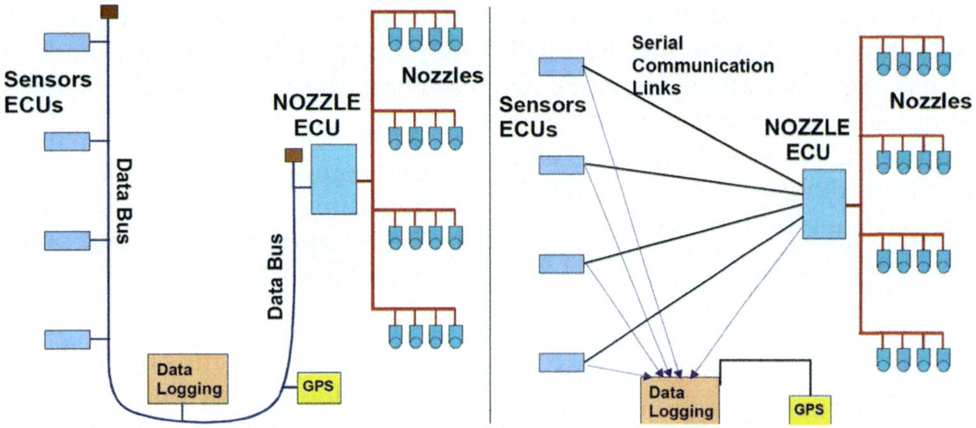

Figure 26.6. (a) A sprayer with optical canopy sensors wired in a bus topology; (b) an older, traditional serial communication sprayer wired in a star topology. In the star configuration, all ECUs that need to communicate with each other require a direct connection. With the bus configuration, all communication is shared on the same wires, simplifying the wiring enabling easier data collection.

known as the tractor ECU in a bridge configuration. The bridge configuration means that the tractor ECU listens to all communication on both the tractor and implement buses, selectively determining which messages should be relayed from tractor bus to implement bus, or vice versa.

The ISOBUS standard divides the communication network into two separate, but connected, buses. A bus topology in computer networking is a shared set of wires that multiple different ECUs can communicate through, while a star topology is a system in which each ECU on the network requires individual connections to the other ECUs. This integrated communication simplifies operations for the operator and optimizes implement performance before expanding into telemetry and data management. By using this standardized bus network, ISOBUS eliminates the need for multiple proprietary control boxes and ensures that a single UT can monitor and operate all compatible implements. This approach reduces cab clutter and simplifies the operator's experience while improving overall system interoperability. The ISOBUS standard includes defined messages for sending information concerning engine and ground speeds, transmission status, fuel economy, hitch and valve statuses, and GPS locations. Many messages have a required frequency with which they should be sent, with critical information like speed and location typically being sent more frequently (~10 times per second). Other data might be sent only once per second or when the vehicle has traveled a certain distance. The three-bit priority field for each message ensures that critical data are transmitted before less critical information. Each message on ISOBUS includes a 29-bit identifier number and up to 8 bytes (64 bits) of data. The identifier includes the ECU's address, the destination ECU address, and a number called the parameter group number that indicates what is included in the data field of that message.

ECU

There are also reserved ECU addresses that indicate that a message is intended to be broadcast or sent to all other ECUs. Typically, ECUs only listen to messages that are related to their own functionality. For example, to achieve the proper flow rate for each point in a field, a sprayer controller on the implement bus needs to know the ground speed and prescribed application rate, and it can ignore messages relating to the tractor's lighting configuration. This can be accomplished by using a filter on the ECU that separates important from non-important messages. The UT is an ECU with both a screen and user interface that allows the operator to interact with the implement, change settings, create tasks in the field, and collect data from the system (as an applied map, etc.). When an implement is connected to the bus, the ECU that controls the implement (after claiming an address) sends information needed to the interface and operator.

To ensure that equipment from different manufacturers works together seamlessly, ISOBUS products are tested at events called plug fests. At plug fests, engineers verify interoperability and compliance, improving system reliability and giving farmers the flexibility to mix and match brands without sacrificing performance.

Telemetry

Telemetry involves collecting real-time data from moving vehicles via radio or cellular signals. This information is used in tracking semitrucks in shipping, unmanned aerial vehicles (UAV) flights, or farm equipment operations. With ISOBUS, telemetry data (e.g., ground speed, application rates, fuel usage) is accessible from a common bus, enabling farm managers, dealerships, and retailers to monitor machine performance remotely. This standardization simplifies third-party telemetry integration and enhances equipment use, maintenance, and efficient decision-making. For example, large operations can analyze ISOBUS telemetry to identify their most and least productive operators. With these insights, managers can refine training, scheduling, or equipment assignments to boost efficiency. ISOBUS ensures uniform data collection, allowing tractor manufacturers and data management firms to supply, process, and interpret information seamlessly.

Telemetry is used on farms every day to monitor real-time equipment performance. For example, a grower can track the speed, fuel use, and application rates of the tractors through ISOBUS data, then adjust equipment management plans to improve efficiency and reduce costs.

Wireless Applications in Agriculture

The Internet of Things (IoT) connects sensors and actuators to the Internet and thereby facilitates data-driven decision-making and execution. IoT only needs to be connected to a network and be individually addressable. IoT technology has been widely employed across various sectors, including smart home (e.g., cell phone-controlled light bulbs), smart infrastructure (e.g., Wi-Fi-connected water meters). Agricultural IoT, or simply Ag-IoT, is the term used to describe IoT systems or technologies used in agriculture.

Since the 1980s technological advancements have made IoT systems substantially more capable, comprehensive, and complex. Despite these developments, the basic architecture of IoT remains largely unchanged and can be described as containing multiple layers that include:

- A physical layer (also known as device layer) that comprises the sensors and actuators that generate data, transfers this information to the Internet, and executes commands received from the Internet.
- A network layer (also known as transport layer) that transmits both raw and processed data, along with commands, across the Internet and other networks (e.g., Bluetooth) connected to the Internet. It interfaces with the physical layer through network devices such as gateways, routers, and concentrators. Based on the wireless communication technologies that link sensors and actuators in the physical layer to the network, communication solutions can be classified as (1) short-range, (2) cellular, and (3) long-range. It is important to note that wired communication is rarely used in real-world Ag-IoT systems owing to the high costs and installation challenges associated with cabled networks.
- An application layer that includes back-end cloud server databases and front-end dashboards for data storage, processing, visualization, and analysis. It focuses on the interaction of IoT systems, facilitating data interpretation and supporting decision-making.

Application Control Systems

The application machinery used to apply fertilizer and crop protection products is usually designed to apply the material at a desired quantity per unit area. That quantity, as specified by each product, can either be in terms of mass or volume. Regardless of the material, accurate application depends on agreement between the machine configuration, the electronics configuration and settings, and the capabilities of the control system to measure and control the application rate. Examples follow for more common application equipment in precision farming systems.

Liquid Application

Machinery designed to apply liquid materials varies from towed ground-based sprayer implements to aerial-based sprayers for orchards.

For each liquid application system, achieving the desired application rate often depends on the pump's pressure versus flow-rate curve, the concentration of active ingredient in the carrier liquid, transport delays from changes in operating condition, nozzle selection, and total achievable flow rate. One commonality that all systems share is that the flow rate must be measured or calculated accurately to achieve the desired results.

Liquid Application Goals

In liquid chemical application systems, there are many goals, including applying (1) the appropriate amount of active ingredients, (2) an appropriate amount of carrier for adequate coverage, and (3) the appropriate droplet size for good canopy coverage. In sprayers or other liquid application systems that use nozzles, the application rate, droplet size, and spray pattern for a given nozzle vary with system pressure. The application rate is determined by the concentration of active ingredients in the conveyance lines, the flow rate through the lines, and speed of vehicle travel through the field. Droplet size is determined by nozzle orifice size and the pressure drop across the nozzle.

Liquid Application Flow Rate

For a given nozzle orifice, one general equation for flow rate is:

$$Q = 60C_o A_o \sqrt{\frac{2\Delta P}{\rho}}$$

where Q is the flow rate, C_o is the orifice coefficient, A_o is the orifice area, ΔP is the pressure drop, and ρ is the liquid density. In this equation, if the nozzle and liquid remain constant, the equation simplifies to:

$$Q = K\sqrt{\Delta P}$$

where K is the nozzle-density coefficient.

This means the flow rate is determined by the pressure drop across the nozzle for any given nozzle and liquid mixture. For example, if a grower broadcasts spraying herbicide with a boom sprayer and wants to increase application rate slightly because the canopy is denser than expected, instead of changing nozzles, the operator can increase pressure. If the current pressure of 40 psi is increased to 60 psi then:

$$Q_2 / Q_1 = \sqrt{(\Delta P_2 / \Delta P_1)} = \sqrt{(60/40)} = \sqrt{1.5} \approx 1.225$$

The flow increases by 22.5%, resulting in increased application rate (gal/ac or L/ha) by about the same percentage as though the travel speed and boom configuration stayed the same.

Liquid Application Traditional Systems

Older machines regulate flow using a pressure relief valve, adjusted electronically or manually via a knob. Instead of direct measurement, they estimate flow using a calibration equation, which provides adequate application control. However, these systems struggle to maintain consistent flow, concentration, and droplet size when flow rates change, as both flow rate and droplet size depend on nozzle pressure.

Liquid Application Servo Valve

The early variable-rate liquid application systems used a flow meter sensor to measure flow rate and a radar or wheel sensor to measure ground speed. This type of control system adjusts the flow rate based on both measurements. Unlike pressure relief valves, these systems use a servo valve, which modulates flow by partially opening or closing. It offers less resistance when fully open than when partially closed (Figure 26.7). The intersection of the

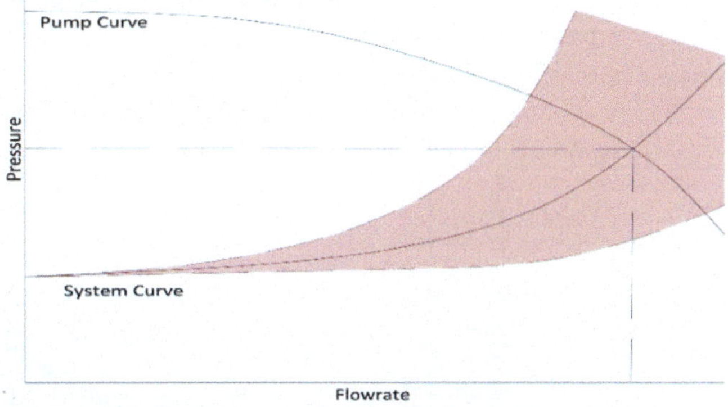

Figure 26.7. The flow rate of a centrifugal pump depends on both the pump curve and the system curve. Changes to the system configuration alter the system curve, which falls between the two red dashed lines. When a servo valve is adjusted to control flow rate, it also affects the pressure at the nozzles and can negatively affect droplet size. The red hatched area represents the possible system curves depending on servo valve setting.

pump and system curves determines the sprayer flow rate. A drawback of this setup, comprising a servo valve, flow meter, and speed sensor, is that system pressure varies with the servo valve setting and is not actively controlled, affecting droplet size.

Next Generation of Sprayers and Retrofit Kits

The next generation of sprayers and retrofit kits for variable-rate liquid applications often use (1) a variable-speed fixed-displacement chemical injection pump and (2) mixing baffles to mix pure active ingredients with a carrier liquid inside the sprayer's plumbing.

These systems also combine ground speed sensors with a pressure bypass valve to maintain constant pressure. Unlike a centrifugal pump, a fixed-displacement pump has a known or calibrated relationship between pump speed (in rpm or cycles per minute) and flow rate. The controller varies the chemical injection pump speed based on the desired application rate and ground speed to achieve variable-rate application of the active chemical. The flow rate of the carrier liquid is constant, providing constant carrier flow rate, pressure, and droplet size. However, sprayers increasingly use hydraulic PWM motors to control the centrifugal pump speed instead of a constant pump speed using a servo valve. Hydraulic PWM motors control flow with greater sensitivity and responsiveness to changing field conditions without significantly influencing system pressure and droplet characteristics.

The major drawback of this system is that the concentration of the active ingredients is varied, which can result in chemical concentrations that are not consistent with the labeled rates. Additionally, on a long spray boom, any change in the active chemical application rate requires all the pipes between the mixing baffles and the nozzles to be pumped out before the new rate of active chemical application is realized.

Nozzles farther away from the mixer have a greater tubing volume than those at the center of the boom, resulting in longer application rate transition times. This transition time, or transient, often leads to the "Christmas tree effect" since the resulting spray pattern in the field is often visible in a triangle pattern as the machine moves through the field. Some new sprayers retrofit systems combine the flow meter, servo valve, and speed meter for a carrier flow controller with the chemical injection pump for concentration control.

In a combined servo valve and chemical injection system, the controller uses the measured carrier flow rate, ground speed, and desired flow rate to control both carrier flow

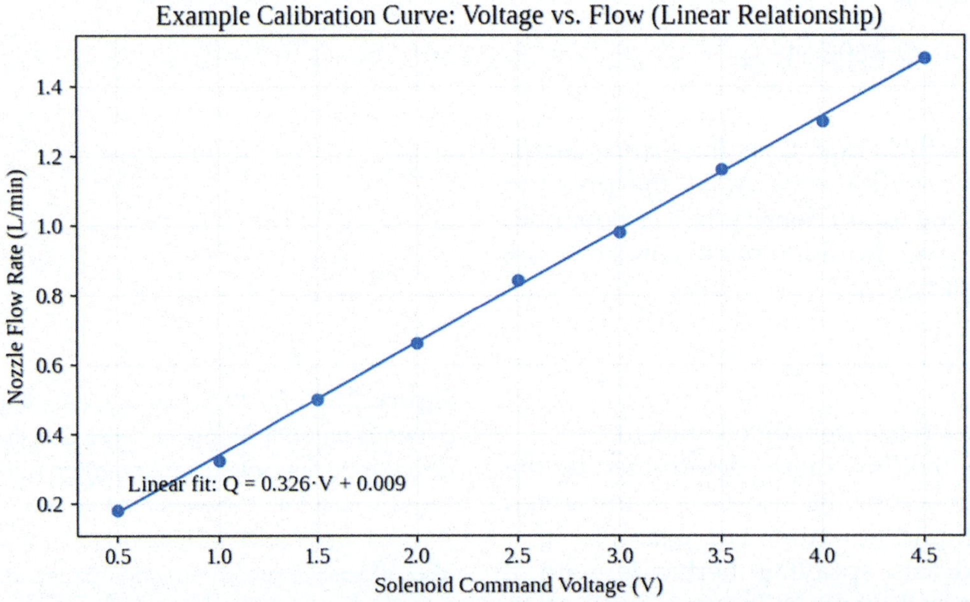

Figure 26.8. The schematic-style showing the linear relationship between solenoid command voltage and nozzle flow rate.

and concentration, assisting in providing adequate coverage without the risk of different nozzles applying different active chemical rates across the boom. The major drawback of this combined system is that different flow rates result in varying nozzle pressure and droplet sizes. The exact regulation of liquid flow via the nozzle is made possible by calibration curve studies that show a linear relationship between the voltage signals supplied to the solenoid valve and the resulting flow rate.

In addition, recirculating boom systems are used to prevent lag in chemical delivery and uneven spray transition. These systems recycle chemical mixtures through boom plumbing continuously, providing more uniform concentration at nozzles when chemicals are being applied and reducing chemical waste when startup or rate adjustments are made. A crucial design factor in these sophisticated systems is the vehicle ground speed being matched to the intended application rate to mitigate the issue of fluctuating droplet sizes and pressures.

Furthermore, integrating components such as a pressure relief valve, electronic pump, and hollow cone nozzles guarantees consistent spray distribution in various operating circumstances. A microcontroller control unit interprets sensor inputs to activate the solenoid valve, allowing for real-time chemical distribution modifications. This method not only ensures consistency in droplet size and spray deposition but also tailors the application rate to the crop's requirements.

Sprayer Boom

Many liquid application systems divide the sprayer boom into multiple sections that can be turned on and off independently, known as boom section control. For example, a 90 ft (27.4 m) spray boom might be divided into six sections, each 15 ft (4.57 m) long. In manual operation, the operator can use a section controller in the cab to turn individual sections on and off as the machine moves through a field. If a variable-rate prescription is used, the controller can automatically turn boom sections on or off based on their location in the field. These prescription maps may also define keep-out zones around water bodies or field edges to avoid unintended application in those areas.

Each boom section is controlled by its solenoid valve, which is either fully on or off. Machines with chemical injection pumps or servo valve flow control require coordination with the control system for section control to be successful. Both the main carrier pump and the chemical injection pump controllers must track how many sections are currently applying the product and adjust their flow rates accordingly. Systems that use pressure relief valves maintain constant pressure and can implement boom section control without additional controller coordination. The most advanced, and typically most expensive, combine boom section control with individual nozzle control to achieve very fine-grained control of the application rate.

Individual nozzle controller schemes typically group nozzles in pairs, with only one nozzle active at a time. Each nozzle has a small solenoid valve that is turned on or opened for short pulses. The application rate is dependent on the length of the pulse. This pulsing of valves is another example of PWM that provides variable rate control. Individual nozzle PWM differs from section PWM mainly in signal frequency. Nozzle control uses a slower signal than those used for motors or hydraulic valves. This means that the nozzle valve is allowed to open completely for each PWM cycle. To achieve even distribution, pairs of nozzles are set up (e.g., nozzles 1, 3, and 5 are respectively paired with nozzles 2, 4, and 6) so that one set of alternating nozzles on the boom (e.g., 1, 3, 5) is pulsing on, while the other (2, 4, 6) is off. The alternating pattern provides even distribution and eliminates misses during application.

Individual nozzle control improves application uniformity and precision. Because the nozzles are controlled in pairs, with only one or the other turned on at any one time, the system can maintain a constant pressure across the boom at a range of application rates. Pulse width modulation–controlled nozzles allow for compensation along the boom. As the vehicle turns, the nozzles on the outside of the turn dispense more liquid than those on the inside. The individual nozzle control systems reduce the overlap caused by multiple overlapping passes and improve coverage along field boundaries. Liquid application control systems that do not use spray nozzles are simpler because there is no need to maintain constant nozzle pressure for droplet size and flow regulation. Controllers for starter fertilizer pumps and other non–nozzle-based systems can use the same servo valve, flow meter, and ground speed sensor combination to control flow rates.

Dry Product Application

Dry materials are generally applied by a box spinner–spreader or a pneumatic spray boom. In either case, the material is loaded into a box with a conveyor belt or chain on the floor. The conveyor moves the material out of the box, and the rate can be adjusted by varying the conveyor speed. With dry material, there is no concern for active ingredient concentration versus the carrier since the carrier is simply air. However, there is a need to know the material composition to achieve desired rates. For example, if the fertilizer being applied is labeled as 20% nitrogen and the goal is to apply 50 lb/ac (55.9 kg/ha), the raw material application rate needs to be 250 lb/ac (279 kg/ha) of raw material.

New technology, such as the New Leader NL5000 G5 spinner spreader, has enabled significantly enhanced precision in dry applications. The G5 model adds 16-section swath width control, boundary and half-width spreading, and last pass control. These features allow the operator to place nutrients where they are needed, minimizing overlap and waste while maintaining consistent coverage. The separate spinner control, speed adjustments of the conveyor, and automatic feedgate function work together to provide site-specific nutrient applications.

With box spreader style machines, the conveyor drops the dry material onto spinning disks that accelerate the material to cover a larger swath width than the spreader box itself. The spread pattern for any material depends on the speed of the spinning disks, the density of the dry particles, and the particle size, while the application rate depends on the speed of the conveyor belt. Control systems for variable-rate control of box spreaders typically include speed sensors for the spinning disks and conveyor belt, servo valves to control the speed of the disks and conveyors, a ground speed sensor, and load cells to measure the weight of the material in the box. Spinning disk spreaders do not allow for high-resolution spatial control granularity. Dry box spreaders are available on the market with multiple nested boxes with their own conveyor belts to apply more than one product in a single field pass. If multiple products are applied at variable rates, each product box requires its own set of speed sensors, servo valves, and load cells. It is also important to recognize that even though each of the products is accelerated by the same spinning disks, differences in density or particle size will result in different spread patterns.

Pneumatic dry material applicators use the same type of conveyor box as the spinning disk spreader to control the application rate but use fast-moving air to accelerate the product through long spray booms. Rather than spinning disks, the material falls into a mixing manifold for each boom section, and the mixed product is then blown across the boom. A variable-rate pneumatic application control system requires the same conveyor speed sensor, servo valve controller, ground speed sensor, and load cells. It operates in much the same way as the box spreader control system. Pneumatic applicators can provide more granular spatial control than box spreaders if the air and/or product mixing manifolds are gated for each boom section. However, this practice is not common.

Boom Height Control

In rolling or hilly fields, both liquid and pneumatic dry applications can be improved by controlling the height of the booms, either to prevent damage to the boom or to improve coverage by maintaining a steady height above the crop canopy or soil surface. Most boom height control is achieved by using a hydraulic cylinder to lift and lower the ends of the boom, and an ultrasonic sensor is used to measure the distance to the canopy. Ultrasonic sensors combine a speaker and microphone into a single device. The speaker sends a signal that has a higher frequency than the human ear can hear. The microphone converts the time between the speaker emitting and the microphone receiving the reflected signal to distance based on the known speed of sound to determine the boom height. The result is that boom heights are rapidly adjusted to maintain the proper distance from the target. The boom leveling controller should be calibrated for each machine that it is installed on.

Anhydrous Ammonia Application

Anhydrous ammonia is a widely used effective fertilizer source that is injected 3–4 in. (7.6 to 10.2 cm) deep in the soil. It has several benefits including low cost, high concentration, ease of application, and effectiveness. However, there are challenging variable-rate anhydrous ammonia applications owing to the "dual-phase" nature of ammonia. Unlike water, anhydrous ammonia in the tank and in the plumbing lines exists in both liquid and vapor phases. Liquid flow meters are inaccurate because of the presence of vapor, while gas flow measurement devices do not work owing to the liquid portion. Without the ability to accurately measure the flow rate, it is difficult to control the rates. Accurate control is not possible without reliable measurement.

To overcome this difficulty, modern technologies such as Capstan's N-Ject provide novel solutions. N-Ject can be installed aftermarket on a variety of OEM toolbars, air drills/seeders, and planters, eliminating the need to manually adjust orifices for different rates or speeds. The system includes PWM solenoid valves at each NH_3 output to ensure exact flow rate control. Liquid anhydrous ammonia flows through cooling blocks and into solenoids, where pulses regulate flow to ensure precision. N-Ject has a 25 : 1 variable rate range and rapid overlap control, providing precise and equal product placement. For variable-rate anhydrous ammonia application, the product must be cooled sufficiently so that it no longer contains vapor. The amount of cooling required depends on the temperature of the ammonia and the pressure in the tank. Once cooled, the ammonia flow rate can be measured using a liquid flow meter.

The final issue that complicates variable-rate anhydrous application is that ammonia is not typically pumped; pressure in the tank forces the fertilizer out of the tank and into the soil. While ammonia pumps do exist, the caustic and corrosive nature of ammonia means that they are not often used for fertilizer applications. Additionally, the pumps only work well for liquid ammonia and would require the temperature to be held below the freezing point (−28 °F, −33.3 °C). Tru-Kare's Blackbird SR71 overcomes variable-rate anhydrous ammonia application issues by utilizing a high-pressure design that keeps the ammonia in a liquid state throughout the process. It was introduced in 2009 and is compatible with most controllers. It prevents freezing boots and lines by delivering exact streaming liquid injections through machined orifices with a difference of only 1% from shank to shank. It operates at 20–50 psi (138–345 kPa) above tank pressure and can sustain equipment ranging from 50 to 71 ft (15 to 21.6 m) broad at speeds of up to 8 mph (12.9 km/hr). Both the N-Ject and Blackbird SR71 systems represent substantial advances in variable-rate ammonia application, providing precise control, lower costs, and increased efficiency.

Manure Application

Manure is often applied to agricultural fields as a byproduct of livestock production. However, unlike commercial liquid and solid fertilizers, the concentration of nitrogen, phosphorus, potassium, and other nutrients contained within the manure are both highly valuable and variable. Owing to the high variability, manure is not generally applied in a variable-rate manner. However, it is possible to target applications to problem areas in the field. For example, more manure can be applied to soils with low soil test P values, where erosion has reduced the yield potential, or to sandy soils with low water retention.

Retrofit electronic kits are available that allow farmers to adjust manure flow rates in real-time using prescription maps or operator input. Retrofittable kits offer a cost-effective means of enhancing manure application accuracy without investing in entirely new equipment. The application rates can be based on mass or nutrient concentrations if a laboratory analysis was conducted. Retrofit kits provide a reasonable option to target manure applications to where it is most needed. However, emerging technologies like John Deere's HarvestLab 3000 use near-infrared reflectance spectroscopy to provide real-time estimates of nutrient content (e.g., nitrogen, phosphorus, potassium) during application. These systems mounted on tankers or draglines automatically adjust rates in real time to meet precise nutrient goals.

GPS-aided Guidance

GPS-aided guidance, often referred to as lightbar guidance, is a visual aid for a machine operator to reduce overlap or skips during product application or to ensure straight rows during planting. In this system, the operator defines a line that they would like to follow by pressing a button to identify the two ends of a line. This type of line is referred to as an A–B line, where point A is the first point selected and point B is the second. The lightbar has a series of LEDs or light bulbs in a horizontal line, as seen in Figure 26.9. When the vehicle is directly on the desired A–B line, the middle LED lights up, indicating that the vehicle is on the correct path. As the vehicle moves farther away from the line, the LED indicator strip lights up other indicators farther away from the center, with the vehicle's distance from the desired line proportional to how many indicators are lit. Light bar guidance saves money by ensuring that the driver maintains the proper swath spacing, which reduces fuel consumption and application overlap. While most agriculture equipment today uses auto-steering, lightbar guidance systems offer an economical approach for manually steering farm equipment not equipped with auto-steering.

Figure 26.9. The light bar–assisted guidance display. The LED indicators and map both suggest that the vehicle is to the right of the desired path. The operator should correct the course by turning the steering wheel to the left.

GPS Auto-guidance Systems

By making perfectly aligned and spaced swaths in the field, GPS auto guidance systems reduce fatigue, minimize fuel consumption, overlaps, and skips. With auto-guidance, rather than just indicating to the driver that the vehicle is out of position, the system physically controls the steering to correct the course. This can be done with a mechanical connection to the steering wheel, or with a hydraulic valve that can adjust the steering angle directly. Important parameters for auto-guidance systems include the look-ahead distance and one or two PID gain factors from the Closed-loop Control Section section of this chapter. The look-ahead distance is the point in front of the tractor that is used to calculate how the controller should steer the vehicle. The typical gain factors used in auto-guidance control systems are the proportional gain and derivative gain, which may be called "line acquisition aggressiveness" and "offline sensitivity," respectively, in a commercial product. Proportional gain determines how the controller will respond to being out of line by a distance, or position error, while derivative gain determines how the controller will respond to being out of alignment with the A–B line, or angular error. Larger gain values result in quicker steering response but might also result in an unstable system.

Implement Steering

When very small amounts of position error are allowable from one pass to another, such as in strip till and planting operations, it is possible to improve pass-to-pass repeatability by steering both the tractor and the implement. In implement steering, a second real-time kinematic (RTK) GPS receiver is placed on the implement to pinpoint its precise location. With centimeter accuracy, the implement can be moved from side-to-side using steering hitches, axles, or coulters. This ability is particularly useful on sloping or undulating terrain when using high-draft machinery like strip-till machines. Using a second GPS receiver, passive implement guiding depends on the tractor's side-to-side motion to maintain the implement's position. Although less accurate, it is an inexpensive method that works well on flat fields with low-draft tools.

Precision technologies can be used to place and record the location of a guidance line in the field. A guidance line (A–B line) can be saved for future operations, helping to reduce soil compaction and ensure accurate alignment of fall strip-tillage and spring planting.

Vision Steering

When side dressing fertilizer or controlling pests, RTK guidance may not be accurate enough to ensure that the vehicle's tires stay between the crop rows. RTK positioning uses differential corrected GPS/GNSS information to produce highly accurate location information. In vision steering, cameras can be used to steer the vehicle by communicating with the steering system in place of a GPS signal. Vision steering can be used on both mechanical drive mechanisms and hydraulic steering systems. The camera will need to identify crop rows as separate from the ground, which means that once the rows have reached full canopy, the system will not be able to steer the vehicle. One such device is the AutoTrac™ Vision system from John Deere that uses a monocular camera mounted on the front to identify maize, soybeans, and cotton early in the growing season. When the crops are at least 6 in. (15 cm) tall, the system recognizes crop rows and uses this visual cue to precisely steer the vehicle between rows. This method works especially well in fields where crops were planted without the use of guidance lines or in places with poor GPS/GNSS satellite configurations. Based on the detected crop rows, the AutoTrac Vision system provides real-time adjustment by interacting with the vehicle's mechanical or hydraulic steering system. This technique allows for up to 20% higher operating speeds than manual steering.

The system smoothly switches back to GPS-based guidance in situations in which the camera's accuracy in identifying crop rows declines. Areas with subpar crop stands or high weed pressure might be such areas. Furthermore, the AutoTrac Vision system can be combined with AutoTrac RowSense, which uses mechanical row sensors to maintain precise steering, in late-season corn applications when canopy closure restricts optical row detection. Technically, the bulk of such systems use closed-loop control algorithms, notably PID controllers, to continuously detect deviations from the row centerline. These controllers calculate exact steering adjustments, which reduce drift, overshooting, and oscillations, resulting in smoother and more stable tracking over dense or difficult terrain.

Planter Downforce

The force applied to a fixed-mount planter's row opening and closing units can cause the seed–soil contact to vary with soil moisture depending on the soil type, topography, and plant residue. A uniform downforce is desirable to ensure the best seed-to-soil contact and improve germination uniformity. Down-force control systems use load cells to measure the force on individual row units, adjusting the hydraulic or pneumatic pressure applied to each row unit to achieve the desired depth and force. A downforce system should provide sufficient downforce to maintain accurate seeding depth without compacting the sidewalls of the seed furrow, while ensuring the seed furrow is properly formed and retains its shape after the opener disks creates it, until the closing system closes the furrow. Insufficient downforce can result in air pockets, poor seed-to-soil contact, delayed germination, and reduced yield. As soil conditions vary, adjusting downforce to ensure a proper seed furrow is crucial for maximizing yield.

Recent planter downforce systems differ in terms of control systems, sensor positioning, and reactivity. Pneumatic systems use airbags and are affordable but slow to respond, whereas hydraulic systems allow for faster, more precise adjustments, appropriate for a variety of field circumstances. Sensor installations range from one per planter to one per row, with the latter providing additional control in nonlevel soils. Some systems provide both downforce and lift, giving them greater versatility for diverse soil types. Sensor position, response time, and active lifting or pushing down rows have a significant impact on seed depth uniformity and overall planting accuracy.

Seed Depth

For a seed to germinate, it must imbibe a certain amount of water, which varies depending on the type of seed. Problems can occur if field moisture is not uniform throughout the field. Again, field characteristics such as elevation, soil type and composition, residue cover, and so forth affect soil moisture. Planters are typically designed to place the seeds at a uniform depth. If the moisture content varies across the field, some seeds may be placed in dry soil, resulting in delayed germination. Some modern planters have systems that measure soil moisture and allow the row unit to adjust planting depth within a predefined range. For example, during corn planting, the operator may define the maximum depth of 3 in. (7.6 cm) and a minimum depth of 1.75 in. (4.4 cm). The planter will adjust seeding depth based on the moisture in the seed furrow, helping to ensure all seeds are placed in adequate moisture to achieve germination. It is important to note that the planter row units should be calibrated to confirm that the seeding depth is accurate. To maximize seed placement, current variable-depth planting systems combine smart soil moisture monitoring and automated depth control technologies. For example, sensors such as Precision Planting's SmartFirmer assess temperature, moisture, and organic matter in real time while maintaining a high spatial resolution. To place seeds in locations with sufficient moisture and encourage uniform germination, sensors send information to control systems to monitor depth. This includes systems, such as SmartDepth, that modify planting depth within user-defined ranges. Each row unit has electric or hydraulic actuators that enable quick responses to changing field conditions. However, some systems may also include manual control inside of the tractor cab for real-time depth adjustment. By maintaining proper ground contact in a variety of soil types and topography, integration with downforce management systems guarantees consistent seed placement.

Precision Seed Placement (Metering)

Proper seed spacing is an important factor in maximizing the potential of every seed. Skips, or seeds that should have been placed, reduce the yield potential. Doubles, or seeds placed too close together, also reduce the yield potential. Monitoring skips and doubles can be an important part of analyzing a planting operation and adjusting the equipment. Recent advances in control systems have led to improved seed metering, the uniform placement of the seed through precise measurement. For example, new metering units are capable of full singulation of soybean seeds, which was previously unattainable. Both John Deere's ExactEmerge and Precision Planting's vSet feature optical sensors and high-speed electric seed meters that track individual seed drops.

The sensors use millisecond accuracy to identify the presence, timing, and spacing of seeds. The data are sent to onboard control systems that can instantly modify row unit performance or meter speed. Even for irregularly shaped seeds like soybeans, consistent singulation is ensured by electronic control systems, which are usually led by microcontrollers or CAN-based networks. Apart from that, even at higher planting rates, seed–soil contact and spacing are made possible by the belt or brush-type systems included in high-speed planting equipment. Real-time correction with equitable seed distribution is made possible by this use of sensing, feedback, and control electronics. Seed meters with multi-hybrid and variable-rate seeding capabilities are also available to further improve planting capabilities. Variable-rate seeding uses location and speed to adjust the seeding rate. Some of the best practices for variable-rate and multi-hybrid seeding are still being developed by agronomists.

Irrigation

This section focuses on the control system aspect of variable-rate irrigation (VRI). For a broader discussion of irrigation management, including VRI, please see Chapter 16. VRI can be applied with center pivot irrigation, and it is defined as "the ability to spatially vary water application depths across a field to address specific soil, crop, and/or other conditions" (Evans et al., 2013). The simplest level of VRI can be accomplished with speed control of the center pivot. The system's speed increases in areas that require less irrigation and decreases in areas needing more. Each management zone is thus in the shape of a pie slice (Figure 26.10). This usually

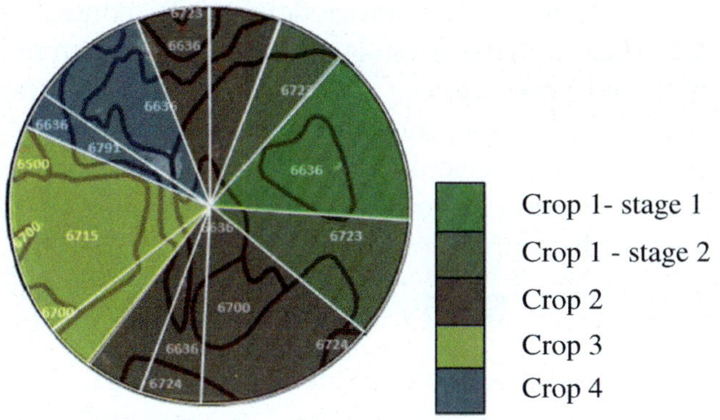

Figure 26.10. Soil map overlaid by speed-control zones for a center pivot irrigated field (Kranz et al., 2014). Courtesy of W. Kranz, University of Nebraska–Lincoln.

requires little to no additional investment in hardware, with the exception of older control panels. Many control panels are already capable of implementing this with VRI. This VRI implementation works well if potential zones are a good match to the pie slice shape.

Zone control is more complex and more flexible, using valves on individual sprinklers or banks of sprinklers to turn the sprinklers on or off (Figure 26.11). Zones can be as small as a single sprinkler, or multiple sprinklers can be connected to reduce the required valves. Size, shape, and number of zones are much more flexible when using this type of zone control.

With low soil moisture content, more irrigation is needed.

Center pivots can play another role in the implementation of precision agriculture by serving as platforms for crop and environment sensors (Bhatti et al., 2022). For example, infrared thermometers can be mounted on the center pivot to measure the crop canopy temperature of the ring covered by the sensor. Radiometers can also be mounted to measure various vegetation indices.

Subsurface Drip Irrigation

Subsurface drip irrigation can apply small amounts of water (daily or even sub-daily) while maintaining high application efficiency. Because this system can apply enough water to replace what is lost by evapotranspiration daily, yield losses due to water stress can be minimized. The combination of drip irrigation laterals mounted on a center pivot is called "precision mobile drip irrigation." It is offered in some form by various manufacturers.

Precision Drainage

When excess water is the problem, artificial subsurface drainage is an appropriate water management practice. To implement precision drainage, the drainage system must be designed, installed, and managed in zones, like a subsurface drip irrigation system. Two types of precision drainage practices that can be implemented are controlled drainage (also called drainage water management) and drainage water recycling. These drainage management practices bring additional water quality benefits.

Controlled drainage is implemented by installing a control structure near the outlet of the drainage system (Fausey et al., 2011). Within the control structure, boards are raised or lowered to raise or lower the water table in the soil profile.

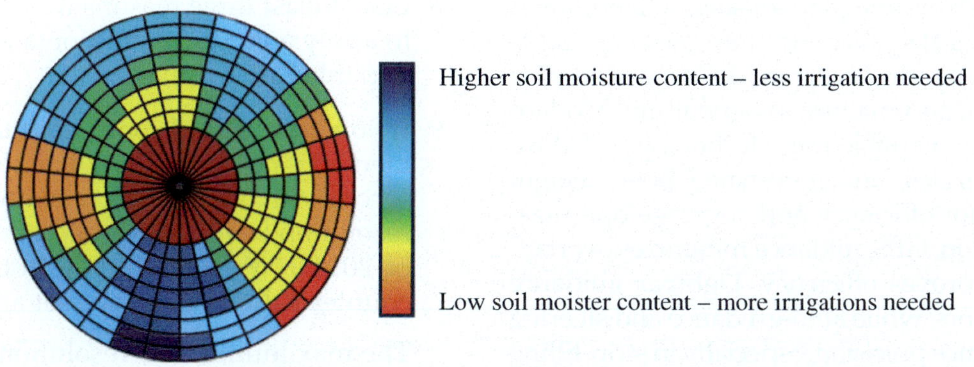

Figure 26.11. Map of center pivot irrigated fields showing 270 potential control zones (Kranz et al., 2014). Courtesy of W. Kranz, University of Nebraska–Lincoln.

Management and adjustment of the control structure can be automated (Agri Drain, 2024) to reduce the labor cost of controlled drainage. Drainage water can be captured in a reservoir or tank for later application back to the field as irrigation water.

Robotics in Agriculture

Intelligent harvesting techniques are revolutionizing agriculture by reducing human labor through the integration of advanced technology such as robotics, UAVs, smart sensors, IoT devices, and computer vision techniques. These intelligent systems have many advantages over traditional methods, including reduced labor costs, increased crop output accuracy, and more cost-effective production. Owing to the practical requirements for labor-saving and efficient agricultural production, the categories of agricultural robots have been continuously expanding, and their application scenarios have become more diversified. The integration of agronomic and biomimetic principles to improve operational efficiency is one of the key problems facing agricultural robotics.

Mobile robots with vision systems are now used for weed detection and selective weeding, automated harvesting, and crop monitoring. They are also reducing labor needs and improving accuracy. Advanced sensors are now used for monitoring soil moisture, nutrient levels, plant health, and weather conditions. This has enabled the use of localized and automatic monitoring/control systems as well as real-time data collection that can be connected to various farm equipment and sensors.

Chapter Summary

Control systems integrate sensors, actuators, and wireless networks enabling informed decision-making and ensuring accurate delivery of fertilizers, pesticides, and manure in agriculture. Modern sprayers use variable-rate technology, PWM-controlled nozzles, and automated boom height adjustments for efficiency and coverage optimization. In addition, GPS guidance minimizes overlapping and improves efficiency. Lightbar guidance assists operators, while auto-guidance and steering systems enhance precision, especially in strip-tilling and spraying. Precision planters optimize seed placement with downforce control and variable-rate seeding. VRI and subsurface drip irrigation can improve water efficiency, while controlled drainage enhances soil moisture management. Artificial intelligence, robotics, UAVs, computer vision, and autonomous vehicles are increasingly being used in agricultural monitoring, application, and harvesting systems and are expected to play a larger role in precision agriculture.

Study Questions

1. What types of electrical signals would you expect to see as the output of a sensor?

2. Current signals (current proportional to the measured value). In open-loop control, why is it not possible to be certain that the speed or rate of an actuator is at its desired value?

3. In a closed-loop control system, what is the setpoint? How is it related to the error?

4. Thought experiment: Open-loop control using feedback is never guaranteed to exactly meet the setpoint if only proportional gain is applied to the error. Which gain, integral or derivative, would need to be added to the system to ensure the setpoint is reached?

5. What are the benefits and drawbacks of liquid sprayer system with a servo valve to control flow rate?

6. Why is it possible to control flow rate on a conventional nozzle-based liquid application system only by changing the pressure relief valve?

7. Why is manure less likely than granular or liquid-based fertilizers to be applied in a variable-rate manner?

8. List at least three reasons why it is difficult to apply anhydrous ammonia fertilizer in a variable-rate manner.

9. If you had a PWM controller that incremented its counter 16 million times per second, what would the maximum PWM resolution be if you wanted the frequency to be 20 kHz? What would the bottom and top counter values need to be set at?

   The maximum PWM resolution is 800 steps, or approximately 9.64 bits. The bottom counter value should be 0, and the top counter value should be 799.

10. How are lightbar guidance and auto-steer guidance similar? What is the major difference? How could the two technologies improve efficiency in a tillage operation?
11. What are at least two major benefits of ISOBUS using a bus network topology instead of a star topology?
12. How does implement steering improve the potential accuracy of product and seed placement?
13. How can improved seed placement and seed meter monitoring improve yields?
14. What is the universal terminal in an ISOBUS system?
15. What are the advantages of filtering messages in an ISOBUS system?

ACKNOWLEDGMENTS

Support for this document was provided by South Dakota State University, Precision Farming Systems community in the American Society of Agronomy, International Society of Precision Agriculture, the USDA-AFRI Higher Education Grant (2014-04572), and USDA National Institute of Food and Agriculture (Multistate Project S1090) Program under (7008110). The use of trade names is for the convenience of the reader and does not imply endorsement by the authors.

REFERENCES

Bhatti, S., Heeren, D. M., O'Shaughnessy, S. A., Evett, S. R., Maguire, M. S., Kashyap, S. P., & Neale, C. M. U. (2022). Comparison of stationary and mobile canopy sensing systems for maize and soybean in Nebraska, USA. *Applied Engineering in Agriculture*, *38*, 331–342. https://doi.org/10.13031/aea.14945.

Agri Drain. (2024). *Valves & gates*. Retrieved May 22, 2025, from https://www.agridrain.com/shop/c133/smart-drainage-system%C2%AE.

Evans, R. G., LaRue, J., Stone, L. C., & King, B. A. (2013). Adoption of site-specific variable-rate sprinkler irrigation systems. *Irrigation Science*, *31*, 871–887. https://doi.org/10.1007/s00271-012-0365-x.

Fausey, N., Williams, M., & King, K. (2011). *Using inline water level control devices for improved drainage water management*. https://drainagesolutionsinc.com/uploads/AGRI-DRAIN-USDA-ARS-%20upright.pdf.

Franzen, A. (2018). Electronics and control systems. In D. K. Shannon, D. E. Clay, & N. R. Kitchen (Eds.), *Precision Agriculture Basics*. ASA, CSSA, SSSA.

ISO. (2007). *ISO 11783. Tractors and machinery for agriculture and forestry—Serial control and communications data network. Part 10: Task controller and management information system data interchange: 100*. American National Standards Institute. https://www.iso.org/standard/61581.html?utm_source.

Kranz, W. L., Irmak, S., Martin, D. L., Shaver, T. M., & van Donk, S. J. (2014). *Variable rate application of irrigation water with center pivots*. https://extensionpubs.unl.edu/publication/ec2000/2014/pdf/view/ec2000-2014.pdf.

Mish, F. C., Morse, J. M., & Gillman, E. W. (1996). Sensor. In *Merriam-Webster's collegiate dictionary* (10th ed.). Merriam-Webster Incorporated.

Nise, N. S. (2011). *Control systems engineering*. John Wiley & Sons, Inc.

Oksanen, T., Öhman, M., Miettinen, M., & Visala, A. (2005). ISO 11783–Standard and its implementation. *IFAC Proceedings Volumes*, *38*, 69–74.

Ullah, M. H., Gunawan, T. S., Raihan, S. M., & Muhida, R. (2012). Development of a Electrically Inspired Low Emission Microcontroller Based Hybrid Vehicle. *American Journal of Applied Sciences*, *9*(10), 1729–1735. https://doi.org/10.3844/ajassp.2012.1729.1735.

Machine Learning and Artificial Intelligence-Driven Solutions in Precision Agriculture

27

Deepak R. Joshi, Skye Brugler, David Mulla, Jitender Rathore, Ahmed Harb Rabia, and Pappu Yadav

Chapter Purpose

This is a new chapter that was not included in *Precision Agriculture Basics*, 1st ed. Machine learning (ML), a subset of artificial intelligence, is transforming precision agriculture by improving the performance of computers without the need for additional programming. There are many different ML programs, each with distinct strengths and weaknesses. The purpose of this chapter is to provide an overview of ML and to discuss the strengths and weaknesses of the various approaches used to develop ML algorithms for precision farming techniques.

Introduction to Machine Learning and Artificial Intelligence

Over the past several thousand years, innovations have transformed food production. The Norfolk four-source crop rotation that was implemented in England in the early eighteenth century replaced the two- or three-field rotation of (1) wheat or rye; (2) barley, oat, pea, or bean; and (3) fallow with a four-course rotation of (1) wheat, (2) turnip, (3) barley, and (4) clover/rye grass. This replaced the fallow system and introduced a grazing crop to improve the quality of their animal products. This change was pivotal for the British Agricultural and Industrial Revolutions and demonstrates that change in agriculture has been happening for centuries. Numerous other changes affecting yields, labor requirements, and profitability have occurred since the eighteenth century. For example, the tractor auto-steering technologies that became commercially available in 1999 reduced driver fatigue, skips, and overlaps. However, growers continue to face challenges, including low profit margins, limited labor availability, and difficulties in collecting and using information from their farms. Artificial intelligence (AI)–based tools can help address these needs.

Artificial intelligence is a nonhuman intelligence capable of performing tasks, solving problems, and reasoning within a rule-based system. Artificial intelligence is often confused with machine learning (ML), but there are important distinctions. While AI can perform many functions independently, it is rigid and rule based. For example, if AI were tasked with solving the maze in Figure 27.1, it would need hard-coded rules, such as complex *if–then* statements, to find the solution. Artificial intelligence addresses the issue by using the rules provided by the human who developed the program. For example, the rule might be to search for an opening from left to right, and *if* there is an opening, *then* to evaluate the depth of the opening. This basic process might then be repeated multiple times until the maze is completed (Mehra, 2024).

Machine learning, a subset of AI, provides computers with greater flexibility and adaptability in problem-solving (Figure 27.2). For example, a system that uses both AI and ML can train a computer to navigate a novel maze by teaching it how to solve other mazes. Instead of providing a set of rules that must be systematically followed, ML uses essentially the same learning process as human brains to develop an understanding of the world. For example, human infants are born unable to speak, but as they practice and learn, they start to identify objects around them and eventually speak their native language fluently. When the computer functions like a brain, the learning process accelerates, enabling the system to train and test itself with numerous data points. In agriculture, ML can be used with a variety of collected data, including imagery data from satellites or unmanned aerial vehicles (UAVs), proximal soil or plant sensors, yield monitors, and more. These examples encompass a large volume of data points that may be gathered over time, referred to as "Big Data," and which may be either organized and structured or disorganized and unstructured.

Deep Learning (DL) is a subset of ML under the umbrella of AI. Deep Learning aims to build a neural network architecture that resembles the human brain. Deep Learning neural networks (also called deep neural networks; DNNs) often have layers that make them extremely powerful problem-solving tools, but they require a vast amount of data to be useful and efficient. Deep Learning algorithms have demonstrated strong capabilities in tasks such as object detection, image segmentation, and image classification. In agricultural applications, these algorithms are often integrated with sensing technologies that produce large-scale, intensive datasets for characterizing target objects. Examples may include, but are not limited to, identifying types of weeds or their

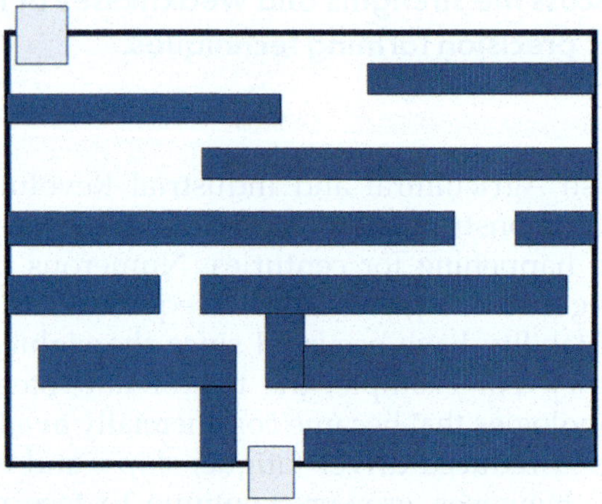

Figure 27.1. In a maze, the pathway can be discovered by trial and error. Machine learning programs use a similar approach. (Courtesy Skye Brugler).

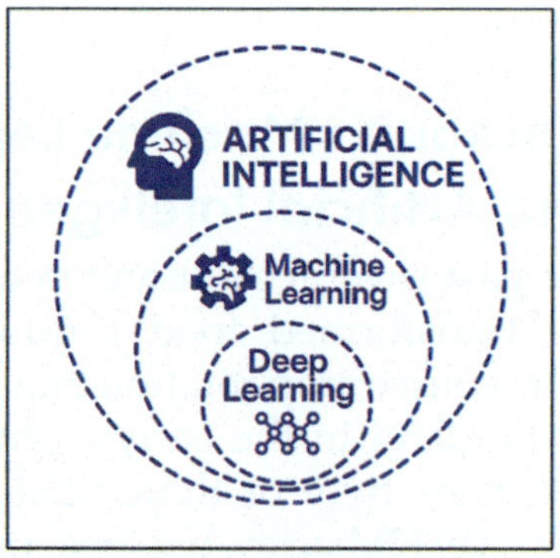

Figure 27.2. Machine learning and DL are subsets of AI that can make systems more flexible and adaptable in problem-solving. (Courtesy Deepak R. Joshi, Created using ChatGPT).

abundance among crops, detecting plant diseases, and diagnosing water or nutrient deficiencies (Wang et al., 2025). A specialized branch of DL, known as generative AI (GenAI), refers to systems that not only learn from the input data but also generate new content, ideas, or solutions, often producing creative and adaptive outputs. Models like Generative Adversarial Networks and transformer-based architectures such as Generative Pre-trained Transformer (GPT), which are examples of Large Language Models (LLMs), go beyond traditional pattern recognition. They can synthesize text, code, images, audio, and even videos, allowing for applications across various fields ranging from agriculture and medicine to art and education. Unlike rule-based AI, regenerative AI models learn representations from data and use probabilistic reasoning to produce outputs that are novel yet contextually relevant. This capability makes them especially valuable for decision support, simulation, and predictive modeling in precision agriculture. The number of applications for these highly computational algorithms is rapidly increasing (e.g., Guler et al., 2024).

Many agronomists use AI/ML/DL algorithms to help them make decisions. For example, numerous smartphone apps can accurately identify weeds, diseases, or pests from an image collected while scouting a field. This real-time information can be combined with the experience of a producer, certified crop advisor, or agronomist to identify the magnitude and economic impact of the problem and to make quick decisions that can reduce yield loss.

As these technologies become increasingly integrated into our daily lives, it is essential to understand the terms and associated definitions of modality, labeling, feature space, feature selection, training, testing, segmentation, attention, encoder, and downstream tasks. Definitions for these terms are provided below:

- **Modality.** Modality refers to the type of data that an AI model can process. Common types of modalities include text, image, video, sensor, and audio data, as well as synthetic data from crop simulation model predictions. Multimodal models can simultaneously incorporate multiple types of data.

- **Labeling.** Labeling is the process by which a human assigns a meaningful identifier to data for supervised learning purposes. An example of labeling is assigning land cover categories (labels)—such as urban, forest, cropland, grassland, and water—to different pixels in a remotely sensed image.

- **Feature space.** Feature space is a multi-dimensional space in which each dimension represents a feature (or attribute) used to describe data points. Every data point in a dataset is represented as a vector within this space. For example, satellite remote sensing images with 30-m pixels that cover an area of 100 by 100 pixels and are captured every 2 weeks for 5 months in red, green, and blue (RGB) and near-infrared (NIR) spectral bands can be characterized by a feature space for the dataset (100, 100, 10, 4), which represents x, y, time, and spectral dimensions. The structure of feature space affects the ability of the models to learn patterns.

- **Feature selection.** Feature selection is the process of choosing the most relevant and useful features (i.e., variables, wavelengths, or inputs) from your dataset to build a model. This process improves model accuracy, reduces memory requirements, and helps reduce the risk of overfitting that can arise from having too many variables. Feature selection methods involve assessing and ranking the correlation of each independent variable with the dependent variable, using recursive feature selection to evaluate how combinations of variables perform, or using the least absolute shrinkage and selection operator (LASSO) regression method to find complementary wavelengths in a remote sensing image that account for most of the variability in the dependent variable.

- **Training.** Training is the process of teaching a ML model to make predictions or decisions by feeding it data and adjusting its parameters to minimize errors. For example, a neural network model can be trained on a large dataset of weather, soil, and crop yield data to learn how to predict spatial and temporal variability in crop yield.

- **Testing.** Testing is the evaluation of a trained model on data that was not used in training to assess the accuracy of the model. For example, a neural network model could be trained on 75% of the crop yield data collected between 2011 and 2020. Once the model is trained, it will test itself on the remaining 25% of the data it did not see to evaluate the accuracy of its predictions.

- **Segmentation.** Segmentation is the process of using computer vision to divide data into meaningful classes. For example, computer vision can be applied to a drone image of a farmer's field to distinguish areas with crop vegetation in shadow, crop vegetation in sunlight, bare soil, and field-access roads.

- **Attention.** Attention refers to the ability of models to focus on a specific subset of the dataset by assigning weights to different input elements. For example, satellite images of agricultural fields can be assigned a greater weight during crop growth periods when nitrogen (N) stress is important, while images taken during periods of cloud cover can be assigned a lower weight.

- **Encoder.** The encoder is a component of AI models that processes input data into a contextualized representation (vector) that reflects the data's meaning. For example, transformers are a type of DNN that can contextualize information. They can be applied to satellite imagery of agricultural fields to identify the key spectral information that distinguishes cornfields from soybean fields.

- **Downstream tasks.** Downstream tasks are the real-world outputs where a pretrained model is used for a specific purpose. These tasks come after (or downstream of) the initial model training. For example, a Vision Foundation Model trained on very large datasets of general images, rather than specifically agricultural ones, can be fine-tuned to predict different types of agricultural crops. The application process of AI and ML in precision agriculture begins with data collection, followed by preprocessing and cleaning to ensure data quality.

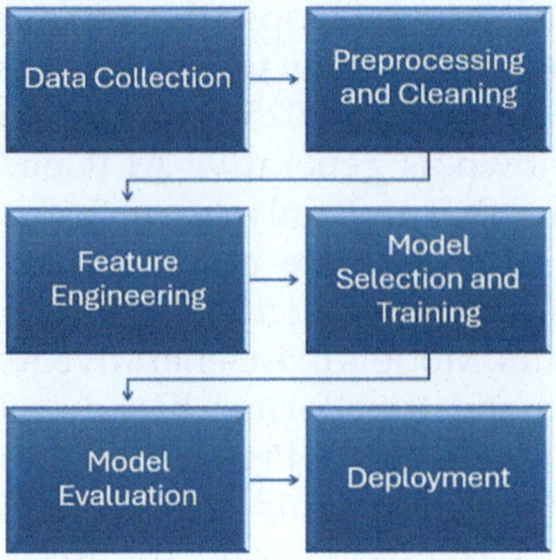

Figure 27.3. General workflow diagram illustrating the key steps of applying AI/ML in precision agriculture.

In addition to these key terms used in AI/ML, this chapter presents a general AI/ML algorithm workflow, as outlined in Figure 27.3.

Data Collection

The workflow for implementing AI/ML starts with collecting the data needed to solve a specific problem. Data can include photos, satellite imagery, sensor data, soil test values, soil texture, crop yield, weather, crop disease, weeds, or insects. However, to be efficient, only appropriate information should be collected. Relevant information can be identified in an aligned study using various approaches, such as regression and correlation analysis. For example, if the purpose is to develop a yield model and yield is highly correlated to soil moisture, then soil moisture data should be collected along with yield data. Similarly, if the purpose is to develop a weed identification model, reflectance information from areas containing weeds (green) and from areas without weeds (bare soil) is needed.

Preprocessing and Cleaning

The collected data need to be inspected and cleaned to remove outliers and checked for alignment with other datasets (Chapters 8 and 9). For example, when a yield monitor slows down, yield can increase. This incorrect data must be removed from the dataset (Chapter 20). Preprocessing for an AI/ML model may also include normalizing, standardizing, scaling, or transforming features in the dataset. Features are measurable input variables that are used to make predictions. For example, if a model is trained to predict crop yield, the dataset may include features such as soil N content, soil moisture, crop biomass, rainfall, average temperature, and vegetation indices from satellite imagery.

A key part of data preprocessing is feature scaling. It is important to ensure that all features used in the prediction are considered appropriately by the model. For example, soil moisture measured in cubic meters per cubic meter will have small numbers, while biomass values may be large. If not appropriately scaled, the biomass feature will be inherently more important to the model than the soil moisture feature, potentially skewing the prediction results. There are a variety of scaling and transformation techniques, such as standardization, minimum–maximum scaling, and log transformation that could be explored further, but that is beyond the scope of this chapter.

Feature Selection and Engineering

Researchers often work with a large dataset with many attributes, known as features. However, including redundant or irrelevant features can complicate a model's use or even reduce its accuracy. Therefore, selecting the "best" features is an important step in model building. There are a variety of feature selection strategies, with the most common being principal component analysis (PCA), correlation coefficients, collinearity diagnostics, and variance-based methods. For example, a correlation matrix can be used to show features that are either related or not related to the variable that the model needs to predict.

The purpose of feature selection is to choose the features that are most likely to help the model identify patterns and make predictions. For example, a multispectral satellite image includes RGB and NIR wavelengths that can be used as features. These features can be combined to create a new metric known as the Normalized Difference Vegetation Index (NDVI), which can predict yield more accurately than either feature alone (Chapter 14).

A key difference between conventional ML and DL is that DL models can automatically perform more complex patterns that traditional methods and human experts may miss (Sarker, 2021).

Model Selection

In agriculture, supervised ML models are used most often. Supervised ML models involve training a model using labeled input data that is paired with corresponding output data. These models use labeled training data to identify the relationship between the input and output data. A supervised model compares the actual answers to the predicted ones to determine error. For example, a supervised learning model would be trained using labeled yield data along with crop reflectance measurements to predict crop yield. Unsupervised models are not often used in agriculture. Although they are extremely powerful, these models do not evaluate their predictions on labeled datasets and are often used for pattern and structure discovery through association and clustering. These types of models are interesting, but their discussion is outside the scope of this chapter.

There are many different types of supervised ML models. The most commonly used types include decision tree, random forest (RF), support vector machine (SVM), DNN, and fully convolutional network (FCN) models. The decision tree model consists of a single tree with multiple branches, with each branch representing a decision that is passed down to the next branch until a final prediction is reached. To understand how a decision tree works, consider the task of classifying seeds in a bowl that contains wheat, corn, and rice. The first branch may classify each seed by size. Therefore, any seed with a diameter larger than 4 mm will be classified as corn. This leaves the corn as correctly classified, but the model still classifies wheat and rice as the same type of seed. This decision is then passed on to the next branch, which will classify the seeds further, but now using color as the feature. This will ultimately separate the brown wheat seed from the yellow rice seed.

The RF model uses the same concept as the decision tree; however, it uses multiple trees that assess the features in a different order (Figure 27.4). The RF model gets its name from

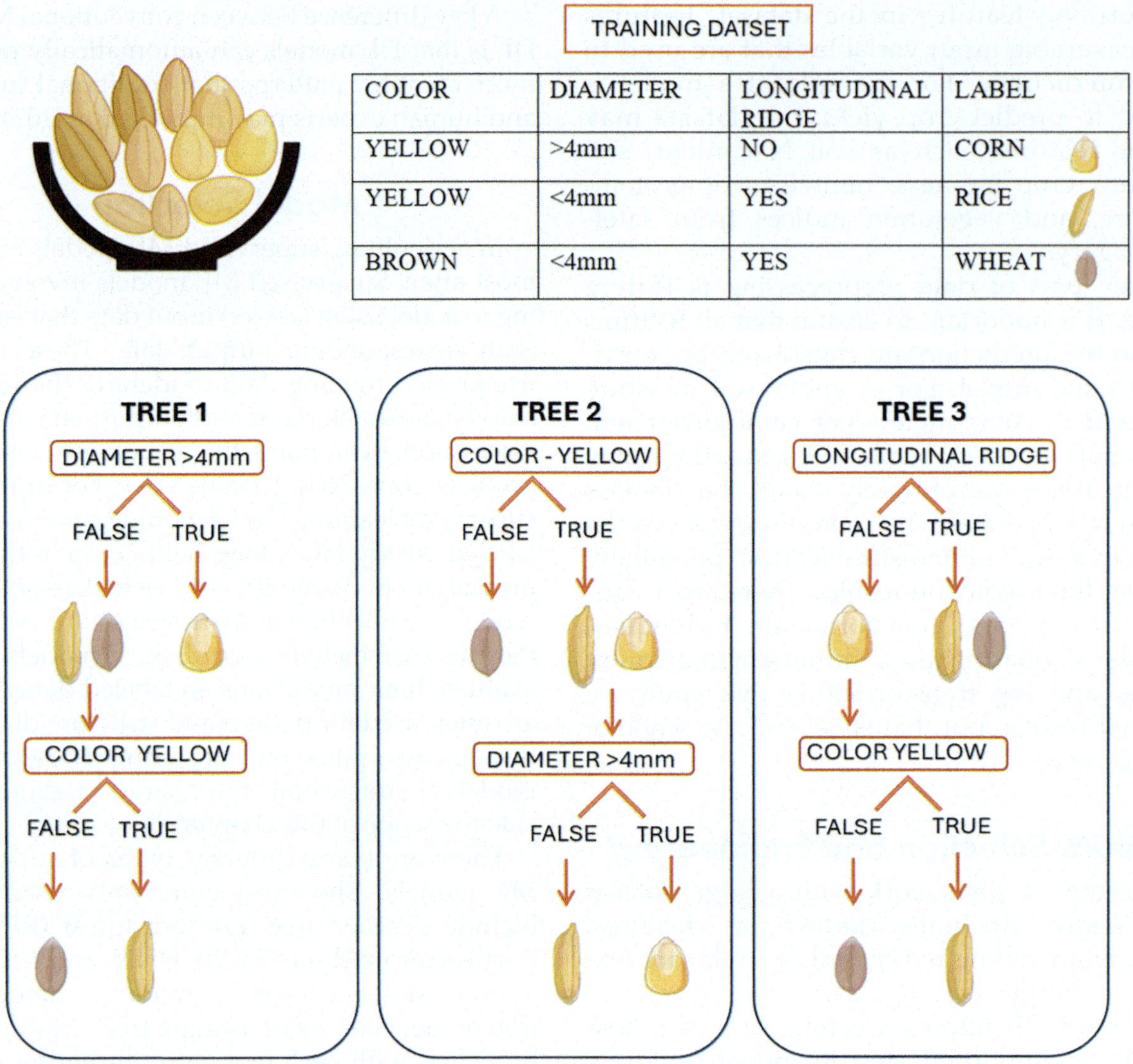

Figure 27.4. Random Forest made up of three trees, demonstrating a classification problem involving wheat, corn, and rice seeds using color, diameter, and longitudinal ridge as the three features to perform classification. (Image created by BioRender, courtesy Sravanthi Bachina).

the random selection and order of features in each tree. This process, called bootstrapping, makes RF a robust model. Returning to the seed example, consider a RF of three trees. One may classify the seeds first by size and then by color, while the second tree may classify them by color first and then by size. On the other hand, the third tree may use the longitudinal ridge feature before considering the color feature, completely ignoring the size feature. The decision trees will then vote on the final prediction. In highly complex problems, using a different order and different features can change the prediction, giving the model a broader perspective. This has made RF a popular model in agriculture, especially in yield prediction (Joshi et al., 2024) and land use classification, both of which require complex pattern recognition (Nguyen et al., 2020).

Support vector machine models are used in classification problems, while Support Vector Regression (SVR) models are used in regression (Lachgar et al., 2024; Ribeiro et al., 2024). The SVM model constructs a hyperplane, essentially dividing a line between data points. The two data points closest to either side of the dividing line are called support vectors, hence the model's name. The SVM aims to optimize the margin between the support vectors, while the SVR aims to find a hyperplane that fits as many data points as possible within a specified margin of error, which essentially involves finding the shortest width in the boundary line margin (Figure 27.5).

A DNN is a fully connected neural network that consists of an input layer, one or more hidden layers, and an output layer. The number of features in a dataset makes up the number of neurons in the network. A DNN works through forward and backward propagation (Figure 27.6). Forward propagation begins with data moving through the network with randomly assigned parameters, producing a result. The model then assesses the error (used to assess the model in regression) or the uncertainty (used to assess the model in classification) to update the weights and biases of its parameters. The process of making these alterations is called backward propagation. The model repeatedly conducts forward and backward propagation to reduce the error or uncertainty until further reduction is not possible. A FCN is similar to a DNN; however, it does not contain neurons, meaning the dataset can be made up solely of unfiltered and unaltered images, making it primarily a segmentation method used in mapping.

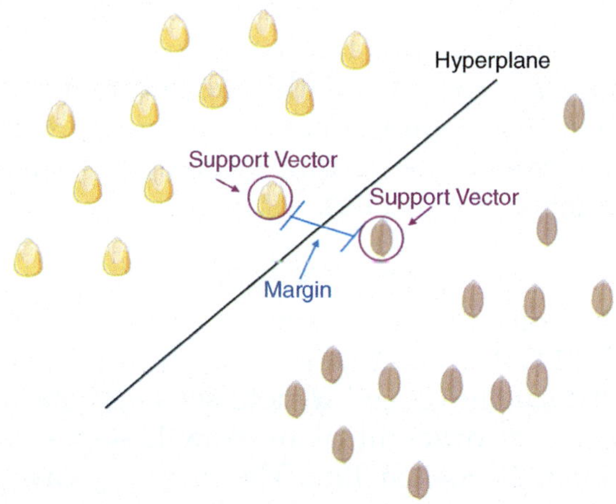

Figure 27.5. Support Vector Machine demonstrates a classification problem between wheat and corn seeds by drawing a hyperplane with as small a margin as possible between the two closest data points, also called support vectors. (Image created by BioRender, courtesy Sravanthi Bachina).

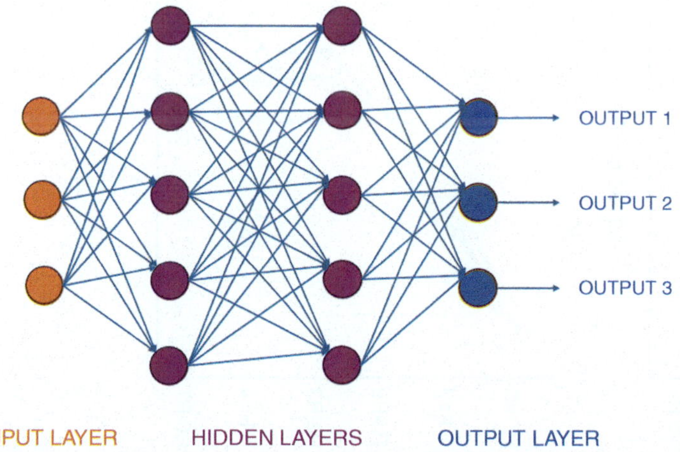

Figure 27.6. Deep Neural Network with two hidden layers demonstrates how the network is connected through forward and backward propagation to improve accuracy and complex pattern recognition. (Image courtesy Sravanthi Bachina).

Model Training and Evaluation

Once the model has been selected, it must be given a dataset to train on and learn from. There are various methods for training models. Two common methods include (a) manually splitting the data, and (b) using cross-validation. Manually splitting the data typically involves a 70/30 split in which 70% of the dataset is used for training and the remaining 30% is set aside for testing or validation. Cross-validation is commonly used in training because it can directly estimate a model's predictive performance while also training. Leave-one-out cross-validation (LOOCV) and *k*-fold cross-validation allow a model to be trained and tested on all available data in a structured manner. In LOOCV, the model is trained on all but one data point, using the excluded point for testing. This process repeats until every data point has been used for testing. *k*-Fold cross-validation generalizes this approach by splitting the data into *k* parts, training on $k - 1$ parts, and testing on the remaining one, cycling through all partitions.

Once the model is trained, it is tested to evaluate its performance, essentially determining how accurate and reliable the predictions are. There are many approaches to assessing model accuracy. For example, common statistics are used to determine the mean square error (MSE), mean absolute error (MSE), the R^2 value, and the number of false positives and false negatives.

Model Deployment

Once a model has been evaluated and shown to perform well, it can be applied to real-world problems. Below are some examples of how models built using similar approaches can be applied in practical applications.

Crop Monitoring and Yield Prediction

Crop monitoring and end-of-year yield prediction involve observing and assessing crop health and growth throughout the season. Traditionally, crop monitoring was conducted by walking through various parts of the field. However, such methods are time consuming, labor intensive, and often fail to provide full field coverage. For example, the corn plant population per 0.001 ac is estimated by counting the number of plants within 17 ft 5 in. (5.31 m) of the rows, which are separated by 30 in. (76 cm). However, advances in remote sensing now make it possible to conduct this calculation using high-resolution UAV images (Acharya et al., 2021; Kharel et al., 2025).

Reflectance data can also be analyzed using AI models to detect early signs of stress, disease, pest infestation, drought, and nutrient deficiencies more accurately and efficiently. For instance, vegetation indices like the NDVI and Enhanced Vegetation Index (EVI) use reflectance data to assess plant health, while tree-based models (such as decision trees and RF) can use remote sensing, weather, topographic, and soil nutrient data to predict crop yields (Khan et al., 2022).

Case Studies

The case studies presented here illustrate practical applications of ML in precision agriculture; they do not constitute a comprehensive catalog of all approaches or AI-based technologies. As science, innovation, and technology continue to rapidly advance, we anticipate substantial progress in the coming decades, including further improvements to existing technologies and the emergence of new ones.

Yield Prediction

The following case study is based on Joshi et al. (2024). In that study, high-resolution (3-m [9.8 ft]) satellite-based remote sensing data were collected at different plant growth stages using the PlanetScope satellite. The imagery was used to derive 10 distinct vegetation indices across 24,282 grid cells. These vegetation indices were selected as features to be used by five ML models

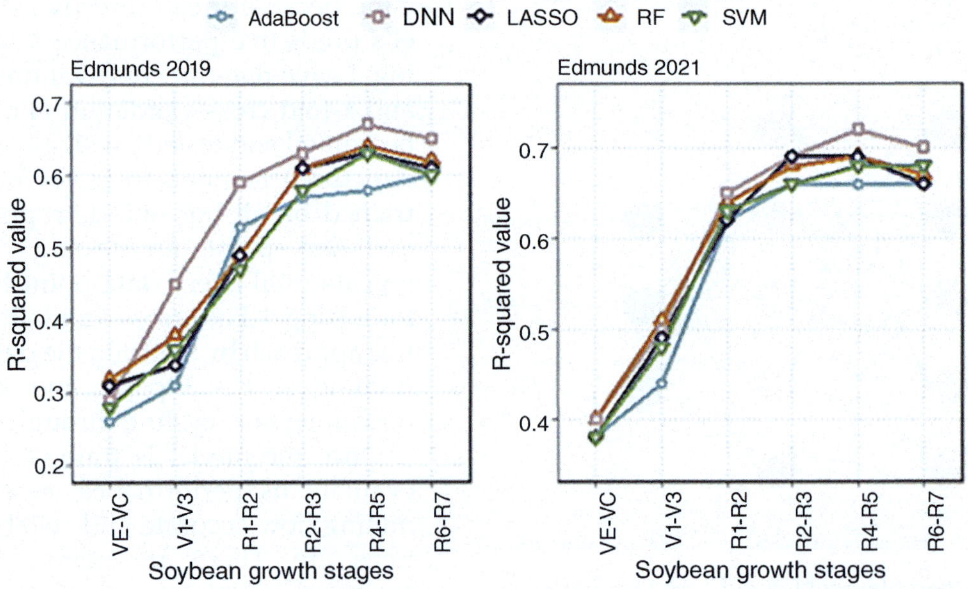

Figure 27.7. Proportion of soybean yield explained by five models at the Edmunds field site of South Dakota in 2019 and 2021. In this analysis, yield prediction was performed at various times corresponding to different soybean growth stages and included early germination; early, medium, and vegetative; and early reproductive to fully matured (VE/VC, V1/V2, R1/R2, R4/R5, and R6/R7) growth stages. (Joshi et al., 2024).

(AdaBoost, LASSO, RF, SVM, and DNN) to predict soybean yields at different growth stages in a 145-ac (58.7-ha) farm field in 2019 and 2021 (Figure 27.7). The models were trained on data collected from the current and previous years of the study, which included different vegetative and reproductive growth stages.

Of the five models tested, the DNN model had the highest accuracy, and its accuracy increased from approximately 30% ($R^2 = 0.30$) at the vegetative growth stage (VE) to approximately 70% ($R^2 = 0.72$) at reproductive stage 4 (R4) to R5. This analysis showed that including data from different stages of growth and using ML models improved yield estimates.

Greenhouse Gas Prediction

The following case study investigated the influence of cover crops on soil health and N fertilizer efficiency. The research was reported by Joshi et al. (2022). The rate of decomposition of the cover crop biomass and the rate of the gaseous loss of N were measured six times daily for 3 years. Soil moisture, soil temperature, air temperature, microbial biomass, inorganic N, and rainfall were also quantified, along with CO_2 and N_2O emissions. Models, including SVM, RF, and DNN, were used to predict gaseous losses from the soil and from the decomposition of the cover crop residue. This analysis showed that air temperature had the strongest relationship ($r = 0.67$) with CO_2 emissions, while CO_2 emissions had the strongest relationship ($r = 0.44$) with N_2O emissions. Contrary to expectations, soil moisture and rainfall were poorly correlated with N_2O emissions. This case study demonstrates that ML models can be used to help explain complex biological, chemical, and physical relationships in the soil. The findings from this study indicate that these relationships are more complex and site-specific than previously thought.

Weed, Pest, Disease, and Pathogen Detection

A variety of studies have evaluated the usefulness of ML models in detecting weeds, pests, and diseases in crop fields. In the past, pest detection using statistically based models was difficult because of changes in the reflectance signature as the plant matured. Now, with AI-powered image recognition and predictive analytics, researchers can overcome this barrier (Ajayi & Ashi, 2023; Hennessy et al., 2022; Sree et al., 2024). For example, Azghadi et al. (2025) implemented a robotic-based AI/ML system that identified weeds in real time and reduced herbicide use by up to 65% in sugarcane. This work trained a model to identify weeds using machine vision camera imagery data as the training data. Similarly, by using this type of imagery as training data for an AI/ML model, researchers were able to detect citrus black spot, citrus canker, and aphid infestations in wheat fields. Both studies achieved over 90% classification accuracy, changing the way researchers and farmers prevent widespread pest and disease issues (Yadav, Burks, et al., 2022; Yadav, Thomasson, et al., 2022; Figure 27.8).

Computer vision models have also been used for similar goals. For example, Yadav et al. (2023), used a computer vision–based DL model called YOLOv6 (you only look once) to detect volunteer cotton plants in cornfields with an accuracy of up to 98%. Machidon et al. (2025) had similar results and reported that a ML model called SqueezeSlimUNet accurately identified weedy areas based on UAV information.

Challenges and Limitations

Machine learning can be used for many different purposes, ranging from driving a tractor to identifying pests to processing data. Machine

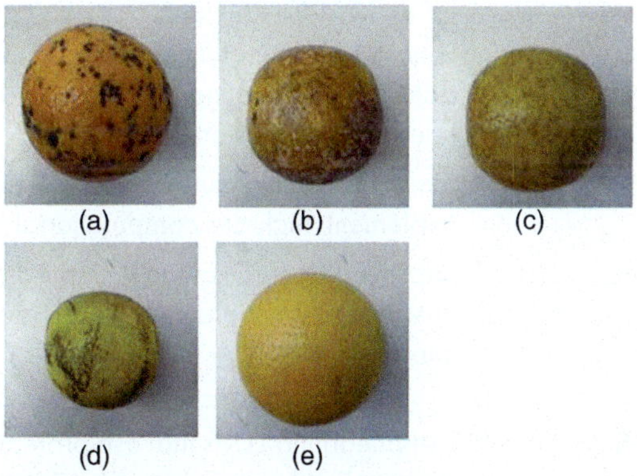

Figure 27.8. Red, green, and blue color images of five different citrus peel conditions: (a) citrus black spot, (b) greasy spot, (c) melanose, (d) wind scar, and (e) marketable peel of orange fruits. These images were scanned using a HyperSpectral Imaging (HSI) system, and a CNN-based ML model was subsequently used to classify them (Yadav et al., 2024).

learning provides a new approach for processing data for decision-making. In general, predictions improve as the amount of data increases. Despite its many benefits, adoption has been relatively slow due to the complexity of model design and implementation. While many commonly used coding programs, such as Python and R, have built-in libraries to use conventional ML models like RF and SVM, more complex models that use neural or convolutional networks require considerable skill and time investment. Additionally, the user must weigh the strengths and weaknesses of a model before implementation, meaning the user must have background knowledge of the data being used, as well as an understanding of the model being implemented (Table 27.1).

The gap in college-level agriculture education that trains students in the use of these models is shrinking. Training in the statistical and data science elements of AI and ML, along with coding, has mostly been taught through computer science and statistics courses and has not been geared toward agriculture-based problems. However, universities like Cornell University, Purdue University, the University of Illinois, the University of Nebraska, South Dakota State University, New Mexico State University, and others have begun offering minors in precision agriculture and digital agriculture. These programs aim to improve student competency in both computer science and agriculture by focusing on topics such as AI, ML, data science/analytics, computer vision, precision

Table 27.1. Comparison of the strengths and weaknesses of different AI approaches

Approach	Strengths	Weaknesses
ANN	Versatile and simplistic. Also includes classification, regression, and pattern recognition.	Limited ability to capture complex patterns. Lack of memory awareness.
DNN	Hierarchical representation learning includes automatic feature learning.	Prone to overfitting and requires large amounts of training data, which can also be computationally intensive.
CNN	Spatial hierarchical learning using image processing and computer vision with parameter sharing. This is efficient for image classification.	Translation invariance. Limited sequential processing.
RNN	Sequential modeling able to retain memory of previous inputs. Includes variable-length inputs and uses natural language processing.	Vanishing gradient memory. Poor long-term memory.
LSTM	Long-term dependences with memory as sequential data. Includes gating mechanisms and selective retention or omission of memory.	Complex and harder to train and understand. Computational intensity limits large-scale applications.
GRU	Simplified architecture, which is easier to train and implement. Includes computational efficiency.	Limited representational power and limited fine-scale control.
KGML	Improved generalization with predictions at unsampled locations with a high level of interpretability.	Includes the assumption of known physics, which increases uncertainties and limits the model's flexibility. The model cannot capture complex, nonlinear relationships in data.
GenAI	Deep learning LLMs that use prompts to generate personalized answers by mimicking patterns and structure of data.	Answers must be constrained based on context and are limited by biases in training data.
FM	Broad, general-purpose DL models that are versatile enough to be used for a wide range of tasks, domains, and outputs.	Requires pretraining using large databases, followed by fine-tuning for specific applications.

Note: ANN = artificial neural network; CNN = convolutional neural network; DL = deep learning; DNN = deep neural network; FM = foundation model; GenAI = generative artificial intelligence; GRU = gated recurrent unit; KGML = knowledge graph machine learning; LLMs = large language models; LSTM = long short-term memory; RNN = recurrent neural network.

agriculture, agronomy, soil science, animal science, and environmental science. At both the undergraduate and graduate levels, there is a pressing need for greater collaboration and cross-disciplinary educational exchanges between agricultural scientists (e.g., agronomists, soil scientists, biosystems engineers) and computer scientists.

Some challenges associated with data also increase the barrier to adoption among researchers. Agricultural data are classified as either public or private. Public data are available across a range of spatial and temporal scales. These include annual summaries of crop-specific land cover (from the USDA National Agricultural Statistics Service Cropland Data Layer [NASS CDL]), static soil databases from the USDA Natural Resources Conservation Service (NRCS) Soil Survey Geographic Database (SSURGO), and 5-year county-level survey data from the USDA NASS Census of Agriculture. Public satellite remote sensing data are available from multiple sources, including Landsat 7, 8, and 9, as well as Sentinel 1 and 2. These satellites have revisit frequencies ranging from 5 days for Sentinel 2, to 6 days for Sentinel 1, to 16 days for Landsat 7. More frequent revisits by the satellites result in more images being captured. Spatial resolution ranges from 10 m for Sentinel 2, to 20 m for Sentinel 1, to 30 m for Landsat 7. Landsat 8 and 9 operate in tandem, providing 8-day revisit intervals at a spatial resolution of 30 m. Data Fusion, the interaction of several datasets from different sources and with different spatial and temporal scales, presents an ongoing challenge that has exciting potential to change how environmental and agricultural data is analyzed and interpreted (Singh et al., 2025).

Private sources of data include data held by individual landowners or land managers and agribusiness companies, along with commercial satellite remote sensing data. Data from individuals and agribusiness companies include detailed information about intensity, timing, and sequencing of specific management practices, which are often missing from AI training and testing procedures that rely on public data sources. Commercial satellites such as WorldView and Planet Dove offer near-daily revisit frequencies with spatial resolutions of 1–3 m. Acquiring this type of data is difficult and costly. Additionally, it is challenging to directly assess subsurface soil properties collected as private and public satellite remote sensing data.

Data privacy is also a major concern for holders of private agricultural data, leading some to hesitate to adopt AI/ML-integrated systems. A broader discussion of data management and data privacy is provided in Chapter 22. However, it is important to note that new innovations in cloud-based decentralized data storage techniques may reduce concerns about data leaks. A process known as federated learning allows the model to train on data that is inaccessible to humans due to the absence of a central storage location. Federated learning is emerging as a viable approach to ensuring data privacy while allowing data to be used confidentially for AI training and testing (Dembani et al., 2025).

Future Directions and Innovations

Artificial intelligence and ML are being used to address challenges faced in agriculture today in 2025 through in-season yield predictions, improved fertilizer recommendations, and the identification of weeds, insects, and diseases. Many of these AI applications have been implemented in the field through the development of advanced robotics and autonomous systems. For example, AI-powered "green-on-brown" and "green-on-green" weed detection systems have been integrated into tractors. These systems use computer vision models to distinguish weeds either on bare soil or among green crop plants. The resulting weed pressure maps, created using AI-based advanced image processing, facilitate site-specific herbicide application, thereby reducing the overuse of agricultural chemicals by targeting only the weeds in the field. Similarly, robotic arms trained with AI models are used for harvesting crops and picking fruit. In addition to this, many mobile apps built with AI/ML algorithms can be quickly downloaded by a user to help make decisions based on weather prediction, identification of weeds, diseases, and more.

Although numerous AI-based applications have already been developed and implemented in agriculture, the global adoption and integration of these technologies into agriculture are still expanding (Chapter 2). Many areas within the agricultural domain remain underexplored in terms of AI applications. For instance, there is the concept of digital twin technology. This technology could theoretically use AI to simulate a physical farming system and evaluate the impact of different decisions before implementation. This could aid producers and agricultural professionals in making the most optimal site-specific decisions for the land. This is an emerging and promising direction in agricultural innovation. Similarly, Real-Time Edge AI is an emerging area in agricultural technology that enables the implementation of management decisions and the collection of information directly in the field via a smartphone or tablet. Such technology allows farmers to collect and use data from drones, yield monitors, ground-based moisture and temperature sensors, and other sources to analyze data in real time. The ability to act on data in real time enhances the speed, precision, and efficiency of farm operations, making Edge AI a transformative tool in precision agriculture. Besides these emerging technologies, autonomous agricultural robots for planting, weeding, nutrient management, and crop harvesting continue to advance. Moreover, new forms of AI analysis techniques are being developed and are continuously emerging. For example, as of 2025, GenAI and LLMs have been developed and used in chatbots, copilots, and virtual assistants to assist in many agronomic decisions, including providing real-time recommendations for crop management and enhancing farmer–advisor interactions through natural language communication.

Conclusions

Artificial intelligence is transforming agriculture by enabling real-time monitoring, precise soil management, early pest detection, accurate yield prediction, and autonomous machinery. Conventional AI/ML models and neural-network–based models are increasingly being used to predict spatial and temporal variability in soil characteristics and crop stresses, aiding farmers in their decision-making on irrigation and weed management. These advancements improve efficiency, reduce costs, and promote sustainable farming. As AI/ML continues to develop, its role in agriculture will expand, making food production more resilient and productive. Challenges to adopting AI in precision agriculture are gradually being overcome. Data scarcity issues are being addressed through new techniques in data sharing and storage. Additionally, computer scientists are working collaboratively with precision agriculture scientists to develop AI/ML-guided models that incorporate sensor data, soil and crop sampling data, yield data, topographic data, remote sensing data, and weather data. Educational initiatives focusing on digital agriculture are emerging at many universities, helping to train the next generation of scientists skilled in developing and using AI/ML-guided decision tools for precision agriculture.

Study Questions

1 What is artificial intelligence, and how is it different from machine learning and deep learning in the context of agriculture?

2 What are some examples of AI-powered technologies currently used in farming systems?

3 How can AI support site-specific nutrient, weed, and irrigation management?

4 How can AI improve decision-making in agricultural systems?

5 What are the main challenges of implementing AI technologies on farms?

ACKNOWLEDGMENTS

This paper was funded in part by National Science Foundation grants 2202706 and 2026431, South Dakota State University, USDA Natural Resources Conservation Service Conservation Innovation Grants (NR233A750011G019), the USDA Agriculture and Food Research Initiative, the American Coalition for Ethanol, GEVO, the South Dakota Corn Utilization Council and the South Dakota Soybean Research and Promotion Council, Kansas State University, USDA National Institute of Food and Agriculture under the AI-LEAF project (grant 2023-67021-39829) and the Multi-State Hatch Project (project number KS24-1883-S1090), the Kansas Department of Commerce, North Dakota State University, and the USDA-ARS under agreement number 58-6064-3-011. The use of trade names is for the convenience of the reader and does not imply endorsement by the authors.

REFERENCES

Acharya, B. S., Bhandari, M., Bandini, F., Pizarro, A., Perks, M., Joshi, D. R., Wang, S., Dogwiler, T., Ray, R. L., Kharel, G., & Sharma, S. (2021). Unmanned aerial vehicles in hydrology and water management: Applications, challenges, and perspectives. *Water Resources Research*, *57*(11), Article e2021WR029925. https://doi.org/10.1029/2021WR029925.

Ajayi, O. G., & Ashi, J. (2023). Effect of varying training epochs of a Faster Region-Based Convolutional Neural Network on the Accuracy of an Automatic Weed Classification Scheme. *Smart Agricultural Technology*, *3*, 100128. https://doi.org/10.1016/j.atech.2022.100128.

Azghadi, M. R., Olsen, A., Wood, J., Saleh, A., Calvert, B., Granshaw, T., Fillols, E., & Philippa, B. (2025). Precision robotic spot-spraying: Reducing herbicide use and enhancing environmental outcomes in sugarcane. *Computers and Electronics in Agriculture*, *235*, 110365. https://doi.org/10.1016/j.compag.2025.110365.

Dembani, R., Karvelas, I., Akbar, N. A., Rizou, S., Tegolo, D., & Fountas, S. (2025). Agricultural data privacy and federated learning: A review of challenges and opportunities. *Computers and Electronics in Agriculture*, *232*, 110048. https://doi.org/10.1016/j.compag.2025.110048.

Guler, N., Kirshner, S. N., & Vidgen, R. (2024). A literature review of artificial intelligence research in business and management using machine learning and ChatGPT. *Data and Information Management*, *8*(3), 100076. https://doi.org/10.1016/j.dim.2024.100076.

Hennessy, P. J., Esau, T. J., Schumann, A. W., Zaman, Q. U., Corscadden, K. W., & Farooque, A. A. (2022). Evaluation of cameras and image distance for CNN-based weed detection in wild blueberry. *Smart Agricultural Technology*, *2*, 100030. https://doi.org/10.1016/j.atech.2021.100030.

Joshi, D. R., Clay, D. E., Clay, S. A., Moriles-Miller, J., Daigh, A. L. M., Reicks, G., & Westhoff, S. (2022). Quantification and machine learning based N_2O–N and CO_2–C emissions predictions from a decomposing rye cover crop. *Agronomy Journal*, *116*(3), 795–809. https://doi.org/10.1002/agj2.21185.

Joshi, D. R., Clay, S. A., Sharma, P., Rekabdarkolaee, H. M., Kharel, T., Rizzo, D. M., Thapa, R., & Clay, D. E. (2024). Artificial intelligence and satellite-based remote sensing can be used to predict soybean (*Glycine max*) yield. *Agronomy Journal*, *116*(3), 917–930. https://doi.org/10.1002/agj2.21473.

Khan, S. N., Li, D., & Maimaitijiang, M. (2022). A geographically weighted random forest approach to predict corn yield in the US Corn Belt. *Remote Sensing*, *14*(12), 2843. https://doi.org/10.3390/rs14122843.

Kharel, T. P., Tyler, H. L., Mubvumba, P., Huang, Y., Bhandari, A. B., Fletcher, R. S., Anapalli, S., Joshi, D. R., Mengistu, A., Birru, G., Adhikari, K., Dhakal, M., Maskey, M. L., Reddy, K. N., & Clay, D. E. (2025). Machine learning on multispectral imagery to estimate nutrient yield of mixed-species cover crops. *Agricultural & Environmental Letters*, *10*(1), e70009. https://doi.org/10.1002/ael2.70009.

Lachgar, A., Mulla, D. J., & Adamchuk, V. (2024). Implementation of proximal and remote soil sensing, data fusion and machine learning to improve phosphorus spatial prediction for farms in Ontario, Canada. *Agronomy*, *14*(4), 693. https://doi.org/10.3390/agronomy14040693.

Machidon, A. L., Krašovec, A., Pejović, V., & Machidon, O. M. (2025). SqueezeSlimU-Net: An adaptive and efficient segmentation architecture for real-time UAV weed detection. *IEEE Journal of Selected Topics in Applied Earth Observations and Remote Sensing*, *18*, 5749–5764. https://doi.org/10.1109/JSTARS.2025.3536175.

Mehra, A. (2024). Hybrid AI models: Integrating symbolic reasoning with deep learning for complex decision-making. *Journal of Emerging Technologies and Innovative Research*, *11*(8), f693–f705. https://www.jetir.org/papers/JETIR2408685.pdf.

Nguyen, L. H., Joshi, D. R., Clay, D. E., & Henebry, G. M. (2020). Characterizing land cover/land use from multiple years of Landsat and MODIS time series: A novel approach using land surface phenology modeling and random forest classifier. *Remote Sensing of Environment*, *238*, 111017. https://doi.org/10.1016/j.rse.2018.12.016.

Ribeiro, A. V., Lacerda, L. N., Windmuller-Campione, M. A., Cira, T. M., Marston, Z. P. D., Alves, T. M., Hodgson, E. W., MacRae, I. V., Mulla, D. J., & Koch, R. L. (2024). Economic-threshold-based classification of soybean aphid, *Aphis glycines*, infestations in commercial soybean fields using Sentinel-2 satellite data. *Crop Protection*, *177*, 106557. https://doi.org/10.1016/j.cropro.2023.106557.

Sarker, I. H. (2021). Deep learning: A comprehensive overview on techniques, taxonomy, applications and research directions. *SN Computer Science*, *2*(6), 420. https://doi.org/10.1007/s42979-021-00815-1.

Singh, R., Kaur, S., Joshi, D. R., Iboyi, J., Dar, E., Sharma, L. K., & Singh, H. (2025). Estimating cotton biomass and nitrogen content using high-resolution satellite and UAV data fusion with machine learning. *Smart Agricultural Technology*, *12*, 101191. https://doi.org/10.1016/j.atech.2025.101191.

Sree, V. P., Davuluri, A. J., & Polepalli, S. (2024). Classification and identification of weeds using machine learning classifiers. *Informatica*, *48*(4). https://informatica.si/index.php/informatica/article/view/5132.

Wang, S., Xu, D., Liang, H., Bai, Y., Li, X., Zhou, J., Su, C., & Wei, W. (2025). Advances in deep learning applications for plant disease and pest detection: A review. *Remote Sensing*, *17*(4), 698. https://doi.org/10.3390/rs17040698.

Yadav, P. K., Burks, T., Frederick, Q., Qin, J., Kim, M., & Ritenour, M. A. (2022). Citrus disease detection using convolution neural network generated features and Softmax classifier on hyperspectral image data. *Frontiers in Plant Science*, *13*, 1043712. https://doi.org/10.3389/fpls.2022.1043712.

Yadav, P. K., Burks, T., Qin, J., Kim, M., Frederick, Q., Dewdney, M. M., & Ritenour, M. A. (2023, June 13). *Citrus disease classification with convolution neural network generated features and machine learning classifiers on hyperspectral image data* [Paper presentation]. Proceedings of the Society of Photo-Optical Instrumentation Engineers (SPIE) 12539, *Autonomous air and ground sensing systems for agricultural optimization and phenotyping VIII*, 1253902. https://doi.org/10.1117/12.2665768.

Yadav, P. K., Burks, T., Qin, J., Kim, M., Frederick, Q., Dewdney, M. M., & Ritenour, M. A. (2024). Automated classification of citrus disease on fruits and leaves using convolutional neural network generated features from hyperspectral images and machine learning classifiers. *Journal of Applied Remote Sensing*, *18*(1), 014512. https://www.spiedigitallibrary.org/journalArticle/Download?urlId=10.1117%2F1.JRS.18.014512.

Yadav, P. K., Thomasson, J. A., Searcy, S. W., Hardin, R. G., Braga-Neto, U., Popescu, S. C., Martin, D. E., Rodriguez, R., Meza, K., Enciso, J., Diaz, J. S., & Wang, T. (2022). Assessing the performance of YOLOv5 algorithm for detecting volunteer cotton plants in corn fields at three different growth stages. *Artificial Intelligence in Agriculture*, *6*, 292–303. https://doi.org/10.1016/j.aiia.2022.11.005.

An Introduction to Using Life Cycle Analysis as an Agricultural Sustainability Assessment Tool

28

Heidi Sieverding, Nowrin Mostafa Tuly,
Skye Brugler, and James J. Stone

Chapter Purpose

Life cycle analysis (LCA) is a method for evaluating the environmental impact of a product from its creation to its destination. The practice of conducting LCAs can be highly variable, but the analyses are becoming increasingly important in agronomy and agri-business. Greater understanding and awareness of LCAs are needed. The chapter goal is to ensure that individuals in agri-business and those affected by it are aware of the basics of how LCAs work and how to use them to increase profitability and efficiency in their operations.

Definition of a Life Cycle Analysis

Life cycle analysis or assessment (LCA) is a method used to calculate the environmental impact of a product or process. The terms analysis and assessment are used interchangeably within LCAs. Typically, an LCA is done to give consumers a better idea about the long-term benefits and disadvantages for nature and people related to the manufacture and use of products. In addition, a wide variety of businesses worldwide use LCAs to make decisions on which products to buy. LCA results can have an impact on national and international trade, including the activities of manufacturing firms (automotive, electronics, chemicals etc.), retail companies, fuel/energy industries, and agriculture.

Within the context of an LCA, the terms reuse, redistribute, and remanufacture (Figure 28.1) are used in estimating the benefits of using tools or

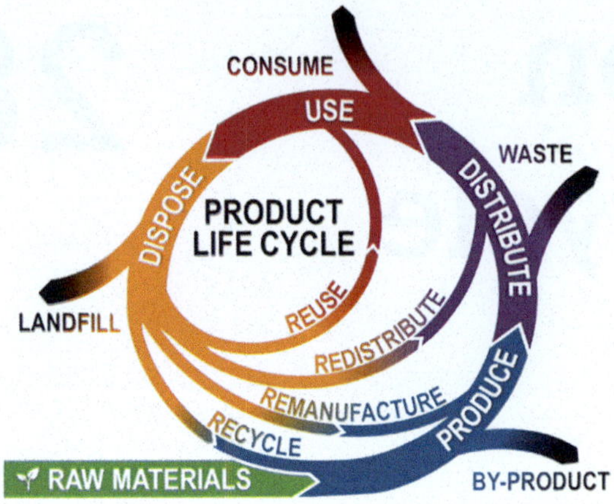

Figure 28.1. Conceptual diagram showing factors considered in a LCA and circular bioeconomy.

equipment many times, purchasing them secondhand, making repairs, or adapting them for other tasks. Farmers frequently follow these practice in their operations, and agri-businesses and agronomists have begun to use a simplified form of LCA to improve their decision-making. These simplified analyses often focus primarily on the production of raw materials (crops) before they leave the farm. A commonly used term to describe a simplified LCA is a "footprint," which reflects the impact of one factor, such as water use, on nature owing to how crops are grown. Generally, if a farmer is focusing on a single aspect of soil health, resource, or concern, then the impact of that aspect would be considered a footprint. However, if the farmer is evaluating more than one aspect at once and considering their collective impacts to make better decisions from a systems viewpoint, then they are doing an LCA. For example, the use of tools such as Field to Market's Fieldprint Platform® (https://fieldtomarket.org/our-programs/fieldprint-platform/) or the USDA's COMET-Planner (https://comet-planner.com/) can help farmers understand the amount of nitrogen fertilizer being lost to the atmosphere as N_2O or which actions to take to improve soil carbon. As greater soil organic matter and more efficient nitrogen use result in higher productivity, high losses due to erosion or application of nitrogen just before rain or freezing weather can result in lower yield. Additionally, Fieldprint Platform and other LCA-based tools (Figure 28.2) help farmers better understand and document the amount of fuel being used by equipment, which can help them reduce fuel consumption and thus crop production costs. Similarly, fertilizer and irrigation are expensive. Footprints can help farmers save resources by identifying ways to keep nitrogen and water where plants can use it. As a result, farmers can improve profitability over time and protect resources for the future.

Life cycle assessments can get very complex and integrate even more categories, such as land use, nutrient runoff, or human health impacts, into calculations. A professional LCA should follow ISO (i.e., ISO 14040, 14044, 14046) or Environmental Product Declaration standards

Figure 28.2. Online tools can be used by farmers to calculate their footprint based on their main crop and farming practices. COMET-Planner can show how much soil organic matter is stored each year, while Field to Market's Fieldprint Platform can assist farmers in managing and documenting additional factors such as nitrogen or water use efficiency.

to ensure that the analysis is creating value for both the user and consumer through the accounting practices used to assess the systemic life cycle impacts. The most accepted broad categories of LCAs are consequential and attributional LCAs (Weidema et al., 2018). Consequential LCAs are system-based models that are often national in scale and include many indirect effects. These LCAs address not only the inputs and outputs of a system, but also the consequences of subsequent use and production. Many of the factors in a consequential LCA are not within a farmer's or manufacturer's control. Consequential LCAs can be useful in policy development; for example, they can be used to evaluate the effect of biofuel and bioenergy policies supporting the introduction of more industrial crops into food crop rotations (Styles et al., 2015). Attributional LCAs are detailed, primarily process specific, and frequently only consider direct effects. Attributional LCA can be useful to companies for evaluating and improving production processes and factors such as supply chains, material sourcing, and marketing. This type of LCA can assist in comparing products or processes and answering questions, such as which biorefinery has the lowest water use per liter (gallon) of ethanol produced. Attributional LCAs focus on the parts of the system that people and manufacturers can control. Most agricultural LCAs are focused on improving or evaluating a specific farm product or source relative to past practices or competitors and can help with long-term planning. However, most LCAs are not user friendly and are outside of the scope of this chapter. For more information on this topic, see Weidema et al. (2018), Finkbeiner (2014), and the United Nations' Life Cycle Initiative online tutorials (https://www.lifecycleinitiative.org/).

Analytical Steps

Life cycle analyses are designed to be circular, so they can support continuous improvement. They involve an accounting process that follows many of the same basic rules of economic accounting practices such as balancing a checkbook. The four analytical steps (Figure 28.3) in a typical LCA study include the following:

1. The scope, which encompasses the system boundary and level of detail of an LCA, depends on the subject and the intended use of the study. Within the scoping step of an LCA, the functional unit (what will be the basis for measurement such as per bushel or per acre), the system (what will be analyzed such as a field, farm, or region), and the system boundary (limits of the analysis, what will be included or excluded) will be determined. A system boundary that includes production of a raw material such as corn or a specific product such as ethanol is called cradle-to-gate. A system boundary that includes production of raw material through use or consumption and disposal, such as the combustion of ethanol, is called a cradle-to-grave system. A system boundary that includes a cyclic process is called cradle-to-cradle. A good example of a

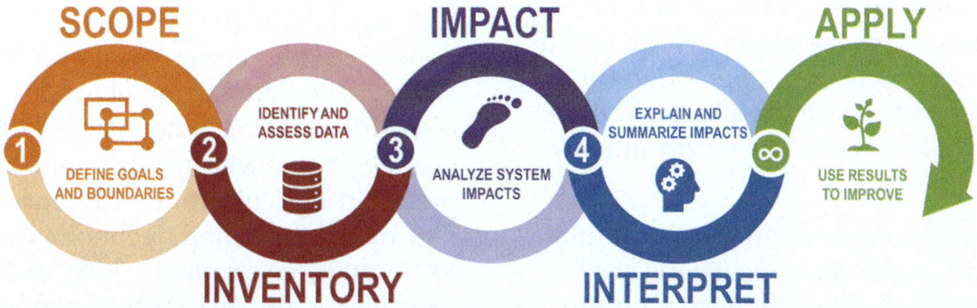

Figure 28.3. Flow diagram showing the primary steps of an LCA study. An LCA is an iterative process. For example, if data cannot be found to document a targeted system boundary, then the scope of the project needs to be re-evaluated and redefined before any impacts can be analyzed. Once an LCA is completed, the results are used to identify areas where processes could be improved or inputs could be changed to help reduce environmental impacts.

cradle-to-cradle system is soil formation: Soil is used to grow a plant, the plant dies, and the decayed matter builds the soil. If the analogy of an economic account process is used, this stage is the point at which a person is doing something like calculating their personal budget or a budget for their farm business, determining last year's net or gross income, or planning for the coming month.

2. The life cycle inventory (LCI) analysis phase involves collecting and checking data on the product or process needed to meet the goals of the defined study. Within the LCI step of an LCA, the inputs and outputs are documented and evaluated. This process entails accounting for all raw materials, processing steps, and sources used to produce a crop or bioproduct and all products, emissions, and waste created. Within the economic analogy, this step would involve gathering receipts and total sales, balancing checking accounts, checking market values, summarizing assets, estimating depreciation, and looking up the amount of money in savings.
3. The life cycle impact assessment (LCIA) phase calculates the costs and benefits following one of the many methods available (such as the EPA's TRACI) to better understand the net impact on nature and people. Within the LCIA step, a scientific process is used to summarize the environmental impact. Within the economic analogy, this step is the point at which a person balances their income with their expenses, calculates their taxes, and develops a budget forecast for the coming year. Just as there are many ways for a person to file their taxes and balance their budget, there are many ways to calculate environmental impacts.
4. The life cycle interpretation phase summarizes and discusses the conclusions and recommendations in accordance with the goal and scope definition (DIN EN ISO 14040:2009-11). Within the economic analogy, this step is where a person estimates their profits and losses and decides if they can afford to purchase new equipment or if they need to make changes to their expenses within the coming year.

Use of LCAs by Buyers and Sellers of Agricultural Products

The global trade of commodity crops and bioproducts is complex, and buyers want assurance that the products they are buying were grown or made following best practices. LCAs are used to document best practices. Farming has always required an understanding that the entire operation fits within a certain market or supply chain. Most farmers modify their management practices to meet expected market requirements. For example, farmers selling to the organic food markets adapt their management strategies to avoid synthetic fertilizers and pesticides. Although this change may reduce yield relative to conventional strategies, selling organic fruits and vegetables yields much higher profits for some farmers owing consumers' willingness to pay premium prices for such produce. Buyers considering LCAs are no different. Companies operating internationally or within specific markets often use LCAs to decide which source they should use to buy the products they need. Some countries and regions, such as Europe, have rules that require ecolabeling (tags that document where and how crops were grown and the environmental impact of their production) before being able to sell products for a variety of uses. Crops that have many market uses, such as corn, which can be used to produce ethanol or provide feed for livestock, need different LCA models depending on use. For example, the use of LCAs within biofuel-related policies, rules, and regulations has increased owing to growing demands for agricultural products (Bhat & Wiley, 2017; Sonnemann et al., 2018; Mohammadi & Saddler, 2025). Since about 2015, substantial changes have occurred in crop production (e.g., precision agriculture, no-till, agrochemicals, biologicals), biorefining technologies (e.g., milling techniques, co-products, heat exchangers, water recycling), and distribution (e.g., supply chains, storage, safety, traceability, transport efficiency). Therefore, to keep up with the market demand, farmers must demonstrate they are meeting production quality expectations and are better than their competitors. In some markets, an LCA is an important marketing tool for selling crops at premium pricing.

It is important to note that farmers may already be meeting or exceeding market demands but are not given the credit because of a lack of supporting data. This situation is especially true within international markets. To address this issue, the U.S. Grains Council developed the Corn Sustainability Assurance Protocol. The use of precision agriculture (Figure 28.4) to document the work farmers have done to adopt soil health practices is the key to ensuring that farmers have good Records of Sustainability (ROS). To make the models more manageable, many agricultural LCAs focus on a single crop, component, impact, or region; as a result, many inventories and datasets insufficiently represent crop production system complexities. Consequently, these analyses may not reflect all agroecosystem dynamics. These planned or unplanned oversights can alter important findings. The way to address these oversights is to provide detailed information on crop production. For example, a new analyst asked to quickly ecolabel a single cornbread muffin for direct online sales through Amazon might produce an evaluation that includes transport from where the muffin is baked in Tennessee to storage in the Amazon distribution center in Denver, Colorado, and finally, delivery to the home of a consumer in Oregon. However, if the analyst does not account for the fact that delivery trucks carry more than one muffin at a time, they would mistakenly overestimate the product impact. Because such mistakes are difficult to find by casual inspection, many regulators, companies, and academia use peer review and verification processes. When issues are identified, well-constructed arguments using analyses of data that are representative, reproducible, and transparent are necessary. Because agricultural products are used by so many industries, which affects the price farmers are given for their crop, care must be taken when using LCA model results to evaluate whether they are representative of the system(s) and product(s) being analyzed.

Figure 28.4. Simplified process diagram showing the key components of the Corn Sustainability Assurance Protocol (CSAP) accessible at https://www.sustainablecornexports.org/.

Precision farming produces a wealth of data, potentially exceeding a terabyte of data per unit of area annually on some farms (Nelson, 2024). These data can be used to build better and fairer LCA models because they include information that changes over time or based on location, and updates can then be added to the LCAs. Additionally, the ability of LCA practitioners to use these datasets has improved with advancements in artificial intelligence to analyze complex system dynamics and trade-offs, confirming for many farmers that they have improved soil and environmental quality on their lands. For example, variable rate maps can show the efforts made by a farmer to significantly reduce the application of nitrogen fertilizer near water sources. As another example, records of reduced tillage passes and soil sample tests can show the growth of organic matter in the field. These records can generate ecosystem service credits, ecolabels, or ROS for the farmer to sell their products at a premium. A well-built and fair LCA model using accurate, representative, and tested data can help farmers answer important questions, identify knowledge gaps, evaluate efficiency, document changes, and compare options. For example, an LCA might evaluate how precision nitrogen applications influence fertilizer losses under different moisture regimes and in different soil textures. An LCA might also enable calculating how the addition of a cover crop can influence organic matter decomposition and thus improve soil moisture and long-term crop production. How, when, where, and by whom precision data are used to enhance models will likely continue to rapidly evolve, but the expectation is that these datasets will improve the quality and accuracy of models, allowing farmers to negotiate premium prices for their crops in national and international markets.

Tracking Resource Use and System Changes Using LCAs

An LCA can evaluate individual components of a system, such as field fuel and energy use, or the system in its entirety. Traditionally, LCAs have focused on national or global averages; however, interest has increased for creating regional or smaller-scale evaluations to explore and address geographic and farm-specific differences in production. Crop production LCAs can vary depending on the production methods, equipment, yields, and analytical assumptions made in model creation. One of the largest modeling differences is whether direct or indirect impacts are included in an LCA. Direct impacts are decisions that are made by farmers, while indirect impacts are the system-wide factors associated with the production and use of crops that are beyond a farmer's control. For example, a farmer's decision to buy fertilizer has a direct impact on their crop yield and profitability. If all farmers in a region are buying more fertilizer at the same time, indirect impacts are a supply shortage and an increased in the price of fertilizer. These types of market impacts are unintended, but they can have a big effect on how farmers operate both locally and internationally. Therefore, we will go into detail on what direct and indirect effects are to help clarify how these model inputs are used within LCAs and in turn how they change how the results of an LCA can be used.

Direct Resource Use and System Impacts

Products, processes, and other things that can be measured and controlled are considered direct resource uses and impacts. Field inputs, including activities and soil amendments such as water, seed, fertilizer, fuel, electricity, pesticide, and land use, as well as tillage practices, can be used in an LCA, and the effects of these practices can be estimated with confidence (Figure 28.5; Kebreab et al., 2016). This confidence rewards farmers using management strategies that will improve productivity while also reducing environmental impacts. An example of

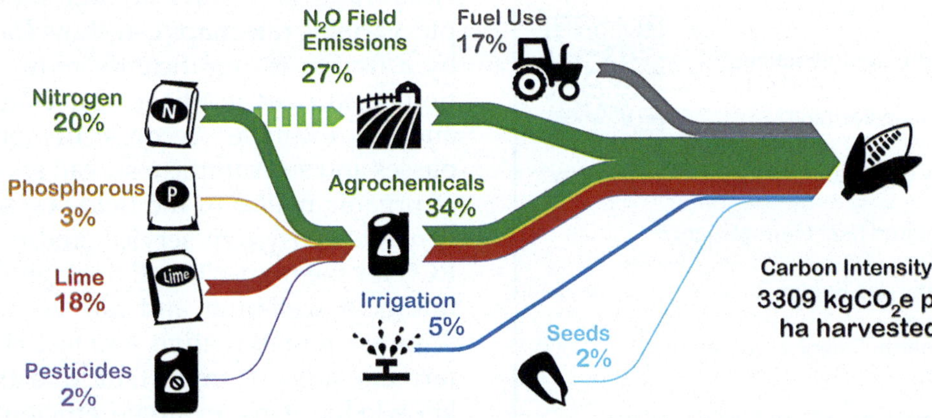

Figure 28.5. Field gate US national average emissions for corn production of approximately 0.35 kg CO_2e kg^{-1} (9.0 kg CO_2e bu^{-1}) of corn grain from Pelton (2019) shown as a Sankey diagram. Adapted from Pelton (2019), found that corn production impacts ranged from 0.20 to 0.70 kg CO_2e kg^{-1} (4.9–17.7 kg CO_2e bu^{-1}) of corn grain across the United States. Nitrogen fertilizer manufacturing and N_2O emissions associated with use affect the carbon intensity (CI) of corn.

a simple way to reduce environmental impacts is switching to pesticides with a targeted, single mode of action to avoid damage to important pollinator species. Tillage practices have a direct impact on crop LCAs because they affect fuel use, but calculating the impact of soil disturbance is more challenging. For no-till fields, the soil carbon disturbed by past tillage will slowly be redistributed within the soil profile as natural soil horizons are re-established (Cusser et al., 2020; Mondal et al., 2020). Within a natural soil system, organic matter is concentrated at or near the surface of the soil and decreases with depth as the soil system transitions to mineral soils and parent rock materials. When soil carbon sequestration is considered in LCAs, well-replicated long-term soil equivalent mass basis measurements and models are needed to represent changes to the soil carbon pools (Paustian et al., 2019). Neither models nor measurements alone are adequate to represent the variability and dynamics of natural systems. Models need measurements for calibration and validation, while measurements need models to account for spatial and temporal variations. Together, they can improve the accuracy of life cycle inventories and model simulations.

Indirect Resource Use and System Impacts

Indirect impacts, such as land use (the conversion of land into agriculture), ecosystem services (environmental stewardship to maintain clean air and water), and biodiversity (diverse root systems including cover cropping and crop rotations), are often recognized and included in governmental policies and life cycle accounting (Figure 28.6). Although indirect impacts are important, there is currently no standardized method to evaluate their effects. Despite the lack of standardization, indirect benefits can help demonstrate system improvements. For example, when reducing fertilizer input using variable rate technologies, a farmer can use a simplified LCA such as the Cover Crop Nitrogen Calculator (https://covercrop-ncalc.org/) to identify how much nitrogen is being released into the soil and how much is being lost to denitrification (N_2O) or through runoff or leaching (NO_2^-). From this quick estimation, management strategies such as using cover crops or change in fertilizer timing and placement can be implemented to avoid further profit losses. If a single farmer reduces nitrate runoff into a river, the impact is typically not large. However, if all farmers within a

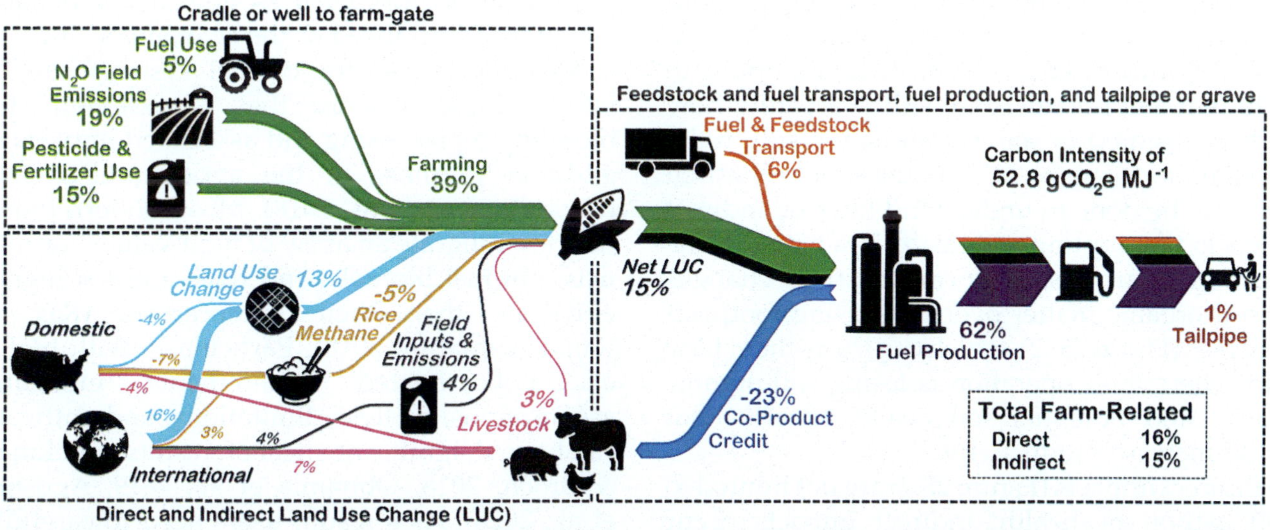

Figure 28.6. Sankey diagram of US average corn-grain ethanol CI contributions with indirect effects based on the ICF 2014 survey and consequential LCA scenario results. In this visualization, livestock feed co-product credits are applied to farm-related emissions to represent net farm-related impacts, and domestic and international livestock, rice methane, and land use change indirect impacts were assumed to be additive.

watershed are ensuring that the fertilizer they apply is not lost, their collective action can reduce algal blooms and help improve water quality within rivers.

Indirect Impacts to Agroecosystems

Agroecosystems are complicated and the interactions between soils, plants, insects, animals, weather, and management practices require both effort to understand and careful decision-making to avoid unintended consequences. Indirect impacts affect the relative contribution of life cycle sources and sinks and are important to note when using and interpreting these models. In the future, it is likely that more indirect effects of agriculture will be recognized as more accurate, peer-reviewed, and reproducible data become available. For example, an easy indicator of improved soil and environmental quality is increased soil organic matter. Higher levels of soil organic matter are known to increase soil health and provide a variety of benefits such as soil structure and aggregation, drainage, aeration, and nutrient cycling and availability. The complexities of soil carbon dynamics require careful evaluation and consideration of sequestration rates and benefits. As soils aggrade (e.g., build or increase) and accumulate organic matter, the densities and thicknesses of soil horizons change. Accumulation rates vary with both climate and agronomic management system (Lal, 2019; Joshi et al., 2023). It is important to note that soil scientists are still evaluating how carbon is stored in soil, which may result in the reevaluation of many LCA models. More research needs to be done to understand the true indirect impacts of adopting soil health practices. In the meantime, farmers today can still track changes in soil organic matter over time using footprint analyses (Figure 20.2), which can show them how and where their operation is losing soil organic matter and potential soil health benefits that could increase productivity.

Indirect impacts (Figure 28.7) are not limited to soil carbon or health; indirect impacts occur across many natural and artificial systems. The United Nations has established an independent intergovernmental body to document and evaluate human impacts on biodiversity and ecosystem services (Intergovernmental science-policy Platform on Biodiversity and Ecosystem Services [IPBES]). Such an evaluation requires large amounts of data that can be supplied by farmers using precision agriculture technologies. The use of these indirect effects in LCAs globally would significantly benefit farmers who have put the time and energy into increasing soil and environmental health on their farms, while also improving their productivity, using precision agriculture techniques. The IPBES has identified key indirect effects caused by invasive species, pollution, and other challenges to nature on global ecosystems and food security with the goal of supporting policy changes (IPBES, 2019).

LCAs and Biofuel

Crops are part of an agroecological system that, if well designed, provides beneficial biofeedback and efficient use of natural resources to maximize production, while reducing harm to nature. This context is especially considered in LCAs that focus primarily on crops grown for biofuel production. Typically, the selection of functional units used in LCAs is closely related to the evaluation's goals. For biofuel-related LCAs, an energy unit such as megajoules or British thermal units is typically used because energy units facilitate comparing LCA results with those from fossil fuel or other biofuel types. In other sectors, such as food, the functional units can be based on production factors such as how many bushels per acre are produced or the economic value.

The boundaries for biofuel systems can be more complex because they must account for biorefinery processing and associated impacts in addition to those of the agricultural system (Thornley & Adams, 2018). Most modern biorefineries integrate an array of high-value bioproducts, not just biofuels, and have a much higher efficiency than their predecessors (Ray & Ramachandaran, 2018). Particularly with biofuels, policy-related requirements, modeling assumptions, analysis limitations, and software can have a significant effect on results (Hoekman & Broch, 2018, Obnamia et al., 2019, Agostini et al., 2020). As a result, users need to carefully consider the context, usefulness, and applicability of these complex analyses. Many biofuel life cycle accounting policies and studies try to evaluate the changes that biofuel production has on food production systems. To evaluate impacts on production systems, most analyses use some

Invasive Species: *Salt cedar (Tamarix sp.) was imported in the early 1900s for use as wind breaks to reduce soil erosion on US croplands but at a substantial cost to native plants, water resources, and soil salinity. Deliberate introductions of non-native plants, insects, and animals have created negative agroecosystem impacts and costly, ongoing remediation efforts. Image is from [Montana Weed Control Association](#)* (MWCA 2025).

The Intergovernmental Science-Policy Platform on Biodiversity and Ecosystem Services (IPBES) estimates that the financial cost of invasive species has quadrupled every decade since 1970. The majority (92%) of that cost is due to degradation of ecosystem services providing essential resources to people and harming their quality of life (IPBES 2023).

Insect Control: *Insecticides are used to protect plants and seeds from insect damage. Insects can also transmit plant pathogens and diseases which can cause yield losses. Neonicotinoids are broad-spectrum insecticides used by agricultural producers and pet owners (neonicotinoids are common in pet flea, tick, and lice treatments). The term neonicotinoids refer to a family of chemicals including acetamiprid, clothianidin, dinotefuran, imidacloprid, and thiamethoxam. Even small quantities used in seed coatings, are harmful to bees and other essential pollinators (IPBES 2016). The unintentional effect broad-spectrum insecticide use has on agroecosystems is a growing concern {Guzman, 2024 #142}.*

Harmful Algal Blooms: *Stormwater and irrigation run-off from fields or drain tile can remove valuable nutrients from fields and cause eutrophication (deoxygenated water), commonly resulting in harmful algal blooms is downgradient lakes and rivers. Nitrate (NO_3^-) presence in drinking water can cause severe health impacts, particularly to small children who can develop life-threatening blue-baby syndrome within hours or days of exposure. Excess nutrients increase the growth of microscopic organisms in surface water, creating a high biological oxygen demand (BOD). Over time, all the oxygen in the water is used by the algae, leaving not enough oxygen in the water to support fish or other aquatic creatures. Algal blooms can release toxic compounds into surface water and/or cause dead zones.*

Figure 28.7. Case studies on indirect effects considered in life cycle assessments (Leslie Thomas Stayner).

metric to first assess the status quo, such as a reference system (Hoekman & Broch, 2018), business-as-usual scenario, or baseline, and then quantify changes that biofuel production creates. As a result, most biofuel-related system boundaries (Figure 28.6) also include quantifications that involve product substitution or displacement (Vadenbo et al., 2017). Inclusion of these factors are critical to properly applying more advanced allocation methods capable of addressing biofuel production system complexities. The nuances of these concepts can be hard to grasp, but it is important to understand the core principles. As an illustration, think of a recipe in an old cookbook that calls for lard. In LCA terms, the recipe would be the business-as-usual scenario or point of reference used to evaluate change. However, it is now known that eating lard is not healthy. So, to adapt the recipe, the lard could be replaced by an omega-6 rich oil, which represents a healthier alternative to support dietary changes that can be made to improve overall health. The reasons for, cumulative changes of, and effects on a person's heart health can be complex. Life cycle analyses for biofuels work the same way, documenting each good thing that can be added or each bad thing that can be removed and how the system responds. Over time, such information will lead to better understanding of long-term health and the ability to make better decisions in the future.

It is important for LCAs to be fair and representative analyses of both what can be controlled and measured (direct) and what can only be estimated because it is difficult to measure or it is a systemic effect that is not fully understood (indirect). Only through this process can we understand how to make decisions that help ensure a better future for all people.

How to Start a Footprint Analysis or LCA

Life cycle analyses are tools that can be and are used by people in nearly every profession. Subject matter expertise is important in explaining and interpreting results. As a result, the best and most useful analyses are conducted by teams with a diverse set of skills and backgrounds that enable them to tackle complex problems. Adding subject matter expertise to an LCA project can be a good introduction to the steps and processes that a person needs to learn in order to conduct their own analyses or lead an LCA team in the future.

For most agronomy-related projects, the initial and most important steps are to determine what questions to answer and what goals to target. Once these steps are accomplished, it is necessary to get a handle on the scope of work. Is the primary need a basic or general understanding of only one metric, such as water, energy, or carbon? If not, then an online search for a footprint calculator may provide the answers needed. If a project- or process-specific understanding or an evaluation of trade-offs is the aim, then more in-depth analyses, resources, and tools are likely needed. Whether a person is planning to do the analysis on their own on their farm or will be working with an LCA professional or team, they will likely need to gather data and build an inventory. They will need to determine if they have the experimental or statistical data to answer these questions. If they do not, they will need to find out if it is available from the past work of others in publications or LCA inventories. It is important to develop a good plan and ensure that resources are available prior to starting an LCA.

Summary

As agricultural production continues to evolve and advance, LCAs will likely play a key role in documenting, tracking, and marketing improvements. It is important to be aware of the uses, limitations, and benefits that LCAs provide. Numerous public resources, such as the United Nations Life Cycle Initiative (http://lifecycleinitiative.org), are available to the agricultural community to learn more about LCAs. Agricultural products are critical resources that form the foundation of civilization. As such, the impact of agricultural products and their production systems are important to understand and evaluate. LCAs can be used to identify production strengths and weaknesses, which in turn, can be used to improve agroecosystem sustainability. To learn more about how to conduct an agricultural LCA, please see Sieverding et al., (2020).

Study Questions

Question 1 (Multiple Choice)

1. Which of the following best distinguishes a footprint analysis from a full life cycle assessment (LCA)?
   a. Footprints consider indirect impacts, while LCAs do not.
   b. Footprints focus on a single impact, while LCAs integrate multiple impacts across a system.
   c. LCAs are only used in industrial sectors, not agriculture.
   d. Footprints require ISO standards, while LCAs do not.

Question 2 (True/False)

2. Attributional LCAs primarily focus on system-wide policy consequences that are outside a farmer's control.

Question 3 (Short Answer)

3. Why are precision agriculture data valuable for improving agricultural LCAs?

Question 4 (Short Answer)

4. Name three of the benefits that high-quality data and precision agriculture provide to agriculture LCAs?

Question 5 (Multiple Choice)

5. Which of the following is a major limitation of including indirect impacts in LCAs?
   a. They are irrelevant to agriculture.
   b. They are too easy to quantify.

c. There is a lack of standardized methods for evaluating them.
d. They only apply to biofuel systems.

ACKNOWLEDGMENTS

This chapter was partially funded by the South Dakota School of Mines and Technology, South Dakota State University, South Dakota Corn Utilization Council, and the South Dakota Soybean Research and Promotion Council. The use of trade names is for the convenience of the reader and does not imply endorsement by the authors. We thank the following student reviewers for contributing to the questions at the end of the chapter: Riya Jain and Blake Toennies from South Dakota State University, Clair Spickermann from Utah State University, Egondu Umeobi from The Ohio State University, and Acharya Bhawana from University of California Riverside.

REFERENCES

Agostini, A., Giuntoli, J., Marelli, L., & Amaducci, S. (2020). Flaws in the interpretation phase of bioenergy LCA fuel the debate and mislead policymakers. *The International Journal of Life Cycle Assessment*, *25*(1), 17–35. https://doi.org/10.1007/s11367-019-01654-2.

Bhat, R., & Wiley, J. (2017). *Sustainability challenges in the agrofood sector*. John Wiley & Sons.

Cusser, S., Bahlai, C., Swinton, S. M., Robertson, G. P., & Haddad, N. M. (2020). Long-term research avoids spurious and misleading trends in sustainability attributes of no-till. *Global Change Biology*, *26*(6), 3715–3725. https://doi.org/10.1111/gcb.15080.

Environmental Protection Agency. (2018, July 5). *Harmful algal blooms (HABs) in water bodies*. https://www.epa.gov/habs.

Finkbeiner, M. (2014). Product environmental footprint—Breakthrough or breakdown for policy implementation of life cycle assessment? *International Journal of Life Cycle Assessment*, *19*, 266–271. https://doi.org/10.1007/s11367-013-0678-x.

ICF. (2014) Waste, Residue and By-Product Definitions for the California Low Carbon Fuel Standard Final Report, December 2015, ICF International, San Francisco, CA, USA. 71 pages. https://theicct.org/wp-content/uploads/2021/06/ICF_LCFS_Biofuel_Categorization_Final_Report_011816-1.pdf.

IPBES. (2019). In E. S. Brondizio, J. Settele, S. Díaz, & H. T. Ngo (Eds.), *Global assessment report on biodiversity and ecosystem services of the Intergovernmental Science-Policy Platform on Biodiversity and Ecosystem Services*. IPBES secretariat, . 1148 pages. https://doi.org/10.5281/zenodo.3831673.

Joshi, D. R., Sieverding, H. L., Xu, H., Kwon, H., Wang, M., Clay, S. A., Johnson, J. M., Thapa, R., Westhoff, S., & Clay, D. E. (2023). A global meta-analysis of cover crop response on soil carbon storage within a corn production system. *Agronomy Journal*, *115*, 1543–1556. https://doi.org/10.1002/agj2.21340.

Kebreab, E., Liedke, A., Caro, D., Deimling, S., Binder, M., & Finkbeiner, M. (2016). Environmental impact of using specialty feed ingredients in swine and poultry production: A life cycle assessment. *Journal of Animal Science*, *94*(6), 2664–2681. https://doi.org/10.2527/jas.2015-9036.

Kent Hoekman, S., & Broch, A. (2018). Environmental implications of higher ethanol production and use in the U.S.: A literature review. Part II – Biodiversity, land use change, GHG emissions, and sustainability. *Renewable and Sustainable Energy Reviews*, *81*, Part 2, 3159–3177. ISSN 1364-0321, https://doi.org/10.1016/j.rser.2017.05.052.

Lal, R. (2019). *Management of carbon sequestration in soil*. CRC Press eBooks. https://doi.org/10.1201/9781351074254.

Mohammadi, H., & Saddler, J. (2025). Biofuel policies used by IEA Bioenergy Task 39 countries: The transition to using the carbon intensity (CI) of biofuels to set targets. *Biofuels, Bioproducts & Biorefining*, *19*(3), 929–941. https://doi.org/10.1002/bbb.2732.

Mondal, S., Chakraborty, D., Bandyopadhyay, K., Aggarwal, P., & Rana, D. S. (2020). A global analysis of the impact of zero-tillage on soil physical condition, organic carbon content, and plant root response. *Land Degradation & Development*, *31*(5), 557–567. https://doi.org/10.1002/ldr.3470.

Nelson, B. (2024). *Precision agriculture and data transparency*. Federal Communications Commission Precision Agriculture Connectivity Task Force.

Obnamia, J. A., Dias, G. M., MacLean, H. L., & Saville, B. A. (2019). Comparison of U.S. Midwest corn stover ethanol greenhouse gas emissions from GREET and GHGenius. *Applied Energy*, *235*, 591–601. https://doi.org/10.1016/j.apenergy.2018.10.091.

Paustian, K., Collier, S., Baldock, J., Burgess, R., Creque, J., DeLonge, M., Dungait, J., Ellert, B.,

Frank, S., Goddard, T., Govaerts, B., Grundy, M., Henning, M., Izaurralde, R. C., Madaras, M., McConkey, B., Porzig, E., Rice, C., Searle, R., . . . Jahn, M. (2019). Quantifying carbon for agricultural soil management: From the current status toward a global soil information system. *Carbon Management*, *10*(6), 567–587. https://doi.org/10.1080/17583004.2019.1633231.

Pelton, R. (2019). Spatial greenhouse gas emissions from US county corn production. *International Journal of Life Cycle Assessment*, *24*(1), 12–25. https://doi.org/10.1007/s11367-018-1506-0.

Ray, R., & Ramachandaran, S. (2018). *Bioethanol production from food crops*. Elsevier.

Sieverding, H. L., Kebreab, E., Johnson, J. M. F., Xu, H., Wang, M., Del Grosso, S. J., Bruggeman, S., Stewart, C. E., Westhoff, S., Ristau, J., Kumar, S., & Stone, J. J. (2020). A life cycle analysis (LCA) primer for the agricultural community. *Agronomy Journal*, *112*(5), 3788–3807. https://doi.org/10.1002/agj2.20279.

Sonnemann, G., Gemechu, E. D., Sala, S., Schau, E. M., Allacker, K., Pant, R., Adibi, N., & Valdivia, S. (2018). Life cycle thinking and the use of LCA in policies around the world. In M. Hauschild, R. Rosenbaum, & S. Olsen (Eds.), *Life cycle assessment* (Vol. 429, pp. –463). Springer. https://doi.org/10.1007/978-3-319-56475-3_18.

Stayner, L. T., Almberg, K., Jones, R., Graber, J., Pedersen, M., & Turyk, M. (2017). Atrazine and nitrate in drinking water and the risk of preterm delivery and low birth weight in four Midwestern states. *Environmental Research*, *152*, 294–303. https://doi.org/10.1016/j.envres.2016.10.022.

Styles, D., Gibbons, J., Williams, A. P., Dauber, J., Stichnothe, H., Urban, B., Chadwick, D. R., & Jones, D. L. (2015). Consequential life cycle assessment of biogas, biofuel and biomass energy options within an arable crop rotation. *GCB Bioenergy*, *7*(6), 1305–1320. https://doi.org/0.1111/gcbb.12246.

Thornley, P., & Adams, P. (2018). *Greenhouse gas balances of bioenergy systems*. Elsevier.

Vadenbo, C., Hellweg, S., & Astrup, T. F. (2017). Let's be clear(er) about substitution: A reporting framework to account for product displacement in life cycle assessment. *Journal of Industrial Ecology*, *21*, 1078–1089. https://doi.org/10.1111/jiec.12519.

Weidema, B. P., Pizzol, M., Schmidt, J., & Thoma, G. (2018). Attributional or consequential life cycle assessment: A matter of social responsibility. *Journal of Cleaner Production*, *174*, 305–314. ISSN 0959-6526, https://doi.org/10.1016/j.jclepro.2017.10.340.

Precision Agriculture Basics for Major Agronomic Crops Worldwide

29

Unius Arinaitwe, Thandiwe Nleya, Dwarika Bhattarai, Rojina Kafle, Shailesh Pandit, Janet Moriles-Miller, Abraham Hangamaisho, Rafael Otto, Guilherme Martineli Sanches, Prakriti Sharma, and Varshith Kommineni

Chapter Purpose

This is a new chapter that was not contained within the first edition of *Precision Agriculture Basics*. This chapter discusses precision farming opportunities for important global food crops. In precision farming, cultivar selection is a key strategy to improve yield, and understanding crop selection options may also enhance biodiversity and strengthen crop rotations. It is important to understand the basics associated with the different crops. In this chapter, students will be introduced to the climate requirements and production differences among corn, rice, wheat, oat, barley, millet, sorghum, teff, soybean, cotton, sunflower, rapeseed/canola, potato, chickpea, lentil, field pea, common bean, sugarcane, banana, and grape.

Introduction to Precision Agronomic Crop Management

From the earliest days of human settlement, fertile, well-drained land with access to water has been the foundation of agriculture. The crops that were grown depended on the local climate, soil conditions, and community food preferences (Figure 29.1). These differences resulted in a wide variety of foods being produced worldwide. However, independent of a society's preferences, the failure to adopt a sustainable system resulted in societal decline. Precision agriculture can help societies reduce this risk.

Precision agriculture is often discussed in terms of technologies and tools, including proximal and remote sensors, robots, Global Positioning System (GPS)–enabled equipment, weather stations, machine learning, geographic information systems (GIS), and variable rate application technologies. However, precision agriculture is more than just equipment;

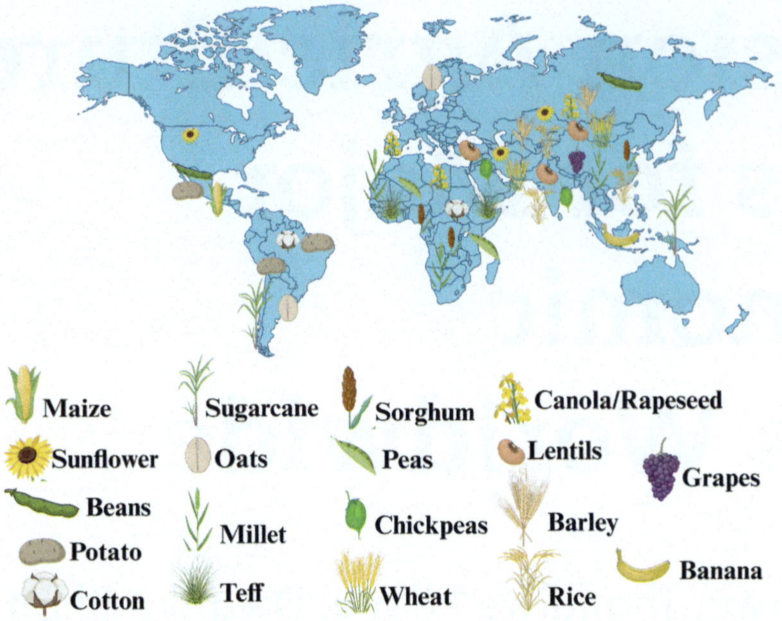

Figure 29.1. World map showing the general origin locations of key crops. Courtesy of Sravanthi Bachina and Skye Brugler.

it involves using information to match decisions to specific problems. This involves increasing root and crop diversity as a soil health practice and understanding how plants respond to stress, as well as how to use this information to make informed decisions. Informed decisions start with understanding the characteristics of the crop plants (Table 29.1).

Plant Growth Characteristics: Cool- Versus Warm-season and C3 Versus C4 Photosynthesis

Cool-season plants grow better at cool temperatures than warm-season plants (Table 29.1). Cool-season plants generally bloom from spring to early summer, whereas warm-season plants bloom from summer to early fall. This is why cool-season plants generally complete their life cycles before warm-season plants.

A second important plant growth characteristic is how plants convert light energy into stored chemical energy. This process, called photosynthesis, converts CO_2, water, and light into oxygen and glucose. There are three types of photosynthesis: C4, C3, and Crassulacean acid metabolism (CAM). Plants with C4 and CAM photosynthesis generally have higher water use efficiency than plants with C3 photosynthesis. Both the plants' growth characteristics and the type of photosynthesis influence water use. For example, warm-season plants that use C3 photosynthesis, such as cotton, generally require more water than warm-season plants that use C4 photosynthesis, such as corn. Due to these differences, there are varying opportunities to use these plants in precision agriculture. To simplify the discussion, crops are categorized according to their temperature adaptation (warm- or cool-season) and the type of photosynthesis they use.

Corn: Warm-season, C4 Photosynthesis

Corn (*Zea mays* L.) is a warm-season plant that was domesticated nearly 9,000 years ago in Central America (Mamani-Pati et al., 2011; Matsouka et al., 2002; van Heerwaarden et al., 2010). Corn can be grown at latitudes from 50°N to 50°S, at elevations ranging from sea level to over 3,000 m (9,843 ft). Currently, it has many uses, including use as livestock feed, as a feedstock for ethanol, and for human consumption (e.g., tortillas, tamales, cornmeal, cooking oils). The United States leads global

Table 29.1. The List of Crops Discussed Within this Chapter.

Crop	Characteristics			
	Optimal growing season	Photosynthetic pathway	Salt tolerance	Drought tolerance
Maize	Warm	C4	Sensitive to moderate	Moderate
Rice	Warm	C3	Sensitive	Sensitive
Wheat	Cool	C3	Moderate	Sensitive
Oat	Cool	C3	Moderate	Sensitive
Barley	Cool	C3	Moderate to tolerant	Sensitive
Millet	Warm	C4	Moderate to tolerant	Tolerant
Sorghum	Warm	C4	Tolerant	Tolerant
Teff	Warm	C4	Moderate	Tolerant
Soybean	Warm	C3	Moderate	Sensitive
Cotton	Warm	C3	Moderate to tolerant	Moderate
Sunflower	Warm	C3	Moderate	Moderate
Rapeseed/Canola	Cool	C3	Moderate to tolerant	Moderate
Potato	Cool	C3	Moderate	Sensitive
Chickpea	Cool	C3	Moderate	Tolerant
Lentil	Cool	C3	Sensitive	Sensitive
Field pea	Cool	C3	Moderate	Sensitive
Common bean	Warm	C3	Sensitive	Sensitive
Sugarcane	Perennial, warm	C4	Moderate	Tolerant
Banana	Perennial, warm	C3	Sensitive	Sensitive
Grape	Cool	C3	Sensitive	Tolerant

Note: Information on photosynthetic pathways, salt and drought tolerance of different crops discussed in this chapter.

production with approximately 390 million metric tons (MT) annually, followed by China with 280 million MT.

Depending on the variety, corn requires a certain number of growing degree days (GDDs) to progress through its vegetative (emergence to tasseling) and reproductive (silking, blister, milk dough, and maturity) growth stages. In the Northern Hemisphere, this typically lasts from early May to mid-October. Corn, as a C4 plant, can tolerate very warm temperatures but requires around 1 cm (0.40 inch) of water per day, with its maximum requirement during the reproductive stages (Kimball et al., 2019). However, while corn is grown in semi-arid conditions, it is less economical to grow it in drier areas or in locations with average summer temperatures at or above 30 °C (86 °F). Corn is moderately sensitive to salt during its later growth stages but highly sensitive during the seeding and emergence stages. Once electrical conductivity levels (EC_e) reach a threshold above 2 decisiemens (dS)/m, a 12% yield decrease can be expected.

Corn is harvested at physiological maturity, which is indicated by the formation of a black layer at the base of the kernel. Harvesting at the correct grain moisture level is critical to minimize losses and maintain grain quality. For grain harvest using a combine, an optimal moisture range of 20%–25% is recommended. Harvesting grain at moisture levels below 20% can increase field losses due to stalk lodging, ear drop, and kernel shattering. Conversely, moisture levels above 25% may result in kernel damage, as the kernels are more prone to rupture during separation from the cob. After harvest, grain should be dried to approximately 13%–15.5% moisture to ensure safe, long-term storage (Carlson & Reese, 2016).

Precision agriculture tools commonly used for corn growth include variable seeding and fertilizer rates (Ferguson et al., 2017; Jiang et al., 2025; Mahlein, 2016; Singh et al., 2024), as well as equipment that improves plant spacing uniformity and nutrient use efficiency (Bruggeman et al., 2017). Earlier chapters (Chapters 7–9, 11–16) discussed a variety of remote, proximal and software-based techniques that can be used on corn as precision agriculture tools, including the use of satellite imagery, creating prescription maps with GIS software and techniques, and using variable rate applications for both fertilizer and irrigation systems.

Millet: Warm-season, C4 Photosynthesis

Millet refers to a diverse group of small-seeded grass grain crops within the family Poaceae, including pearl millet (*Pennisetum glaucum* L.), foxtail millet (*Setaria italica* L.), proso millet (*Panicum miliaceum* L.), finger millet (*Eleusine coracana* L.), and fonio (*Digitaria exilis*). Millet species are cultivated across various regions based on local adaptability, demand, and agronomic conditions (Lemus, 2015). With origins in Africa (East and North Africa) and Asia (India), millet is highly valued for its drought resilience. Millet thrives in arid and semi-arid environments where other grains struggle, and it is an important food source in many tropical countries.

All millet species are C4 warm-season plants with low-input requirements (Pardo & VanBuren, 2021). They are produced in tropical, subtropical, and temperate climates up to altitudes of 2,700 m (9,000 ft) above sea level (Pardo & VanBuren, 2021). The optimal growing temperatures range from 10 to 35 °C (50 to 95 °F), with some cultivars adapted to temperatures as high as 45 °C (113 °F). Millet species are moderately tolerant to mild salinity but do not produce high yields under these conditions (Table 29.1). Millet is especially important agriculturally given that many regions with salt-affected soils, such as South Asia, produce 41% of the world's millet.

Most millet species require between 200 and 600 mm (7.9–23.6 inches) of rainfall; however, finger millet and pearl millet can withstand drier conditions. Millet species are susceptible to multiple pathogens and pests, including blast disease (*Magnaporthe oryzae*), thrips (family Thripidae), aphids (family Aphididae), and stemborers (*Busseola fusca*). For these, early detection and timely intervention are essential to minimize yield loss. Tools, such as unmanned aerial vehicles (UAVs) equipped with high-resolution cameras and multispectral sensors, can aid in the early identification of diseases and nutrient deficiencies. For more information about how remote sensing imagery can be used to detect pests and diseases, see Chapters 17 and 19.

Sorghum: Warm-season, C4 Photosynthesis

Sorghum [*Sorghum bicolor* (L.) Moench] is a versatile cereal grain belonging to the family Poaceae and is grown in tropical, subtropical, and temperate regions at altitudes as high as 2,000 m (6,600 ft) above sea level (Wortmann & Eledu, 1999). Sorghum has its origin in eastern Africa and is adapted to low-input dryland cropping systems. Around the world, it is cultivated for food, fodder, and industrial purposes.

Sorghum is a warm-season crop that uses C4 photosynthesis. It needs a base temperature of 7–10 °C (45–50 °F) for seedling emergence, but requires 27–30 °C (80–86 °F) for better yields. Sorghum can be grown on a variety of soils, except flooded soils, and some cultivars can tolerate moderate salinity (Kausar et al., 2012). All sorghum cultivars are very sensitive to weed competition during their early stages of growth (the first 30–40 days after planting). A fine seedbed is preferable for better emergence and improved weed control. A per-hectare seeding rate of 6–12 kg/ha (5.4–10.8 lb/ha) and row spacing between 45 and 90 cm (17.7–35 inches) are optimal but can be adjusted according to moisture availability and production goals. Seeds emerge best when placed at depths ranging from 1.25 to 3 cm (0.5 to 1.2 inches).

Variety selection depends on the intended use, suitability to agronomic conditions, and tolerance of biotic and abiotic stresses. Due to its lower water requirement (200–600 mm; 7.9–23.6 inches), it is preferred over corn, wheat, sunflower, and soybean in hot, water-limited conditions. Due to its high tolerance to a wide range of stressors and substantial genetic diversity, sorghum is a key crop for precision molecular breeding programs, enabling the development of varieties tailored to specific stresses and capable of producing profitable yields on marginal lands. For example, genes associated with stay-green traits have been identified that improve photosynthetic capacity under post-flowering drought conditions. Additionally, genes associated with the phytohormone auxin and its transporter proteins have been identified as key regulators of dwarfism in sorghum. Manipulation of these genes can enhance the plant's ability to allocate resources toward grain yield rather than stem elongation, thereby reducing lodging and improving overall yield (Hao et al., 2021).

Sorghum is also affected by many insect pests, including sorghum midge [*Contarinia sorghicola*], stem borers (*B. fusca*) and related species, and aphids (family Aphididae), as well as diseases such as anthracnose (*Colletotrichum graminicola*) and grain mold (*Fusarium* spp. and *Curvularia* spp.). Pest and disease problems can be reduced by using various proximal and remote sensing techniques or computer vision models incorporated into smartphone applications. For example, a web-based application has been developed that uses computer vision and machine learning to detect and quantify winged adult aphids on sorghum plants (Grijalva et al., 2024).

Teff: Warm-season, C4 Photosynthesis

Teff [*Eragrostis tef* (Zuccagni) Trotter] is a small-seeded, warm-season C4 grass native to northern Africa, where it is widely cultivated for its nutritious grain and forage value. Although production is concentrated in Ethiopia, teff has gained increasing attention in the United States and other regions due to its role in gluten-free markets and its potential as a drought-tolerant alternative forage crop. Teff performs best in well-drained, friable soils with good tilth, as it is highly sensitive to soil crusting and waterlogging, both of which can severely limit seedling emergence and stand establishment.

Teff is broadcast-seeded in subsistence farming systems and drill-seeded in mechanized farming systems. Like other cereals, teff responds well to split-applied nitrogen (N), with an initial application at planting and a second application during the panicle initiation stage to promote tillering and grain development. Teff is drought tolerant, making it suitable for dryland farming systems. However, it is highly sensitive to water deficits during germination and the period between panicle initiation and flowering. Teff's moderate tolerance to soil salinity, combined with its drought tolerance, makes it well suited for precision farming in arid and semi-arid environments (Table 29.1).

Teff requires a 30- to 40-day weed-free period after seedling establishment because its seedlings are slow-growing, weak competitors early in the season. During this time, the crop is also vulnerable to insect damage, especially aphids (*Rhopalosiphum padi* and *Melanaphis sacchari*) and cutworms (*Agrotis* spp.), as well as to diseases such as ergot (*Claviceps purpurea*) and leaf spot (*Helminthosporium* spp.) when the season is humid. Disease and insect pressure influence variety selection, crop rotation, and other management strategies. Regular field scouting, combined with UAV-based imagery, aids in the timely detection and management of plant stress. Further agronomic insights and region-specific recommendations are detailed on the Ethiopian Institute of Agricultural Research (EIAR) website (http://www.eiar.gov.et).

Sugarcane: Warm-season, C4 Photosynthesis

Sugarcane (*Saccharum* spp.) is a crop of global importance, cultivated to produce sugar, ethanol, and bioelectricity. Modern sugarcane cultivars are hybrids of the *Saccharum officinarum* L. and *S. spontaneum* L. species. Sugarcane is a warm-season plant that belongs to the C4 photosynthesis group and is managed as a semi-perennial crop, with a productive cycle that can last from 5 to 7 years, depending on the cultivar and management conditions (Table 29.1). The crop is highly adapted to humid tropical lowlands (such as Brazil, Indonesia, and Thailand) and to areas with dry seasons (such as Australia, India, and South Africa). The largest sugarcane producer in the world is Brazil. In Brazil, and, more recently, in India, sugarcane is also a feedstock for ethanol production. Usually, sugarcane produces more ethanol per unit area than corn-based ethanol and requires less N fertilizer (Otto et al., 2022).

Sugarcane production varies globally. Small-scale manual cultivation dominates in India and China, while in Brazil and Australia, large-scale mechanized operations are favored. Planting uses stalk cuttings called "seed-cane" and is done manually on small farms and mechanically on large ones. High yields require intensive management, including soil tillage, fertilization, control of weeds, insect pests, and diseases, as well as manual or mechanical harvesting. Precision farming techniques are being used to reduce yield losses from water stress, salinity, and nutrient stress (Vinayaka et al., 2025). Monitoring soil fertility through soil EC_e to create management zones, variable rate application of soil amendments, and measurement of planting failures by UAV imagery are precision agriculture techniques being widely adopted for modern sugarcane cultivation (Sanches et al., 2021).

Rice: Warm-season, C3 Crops

Rice (*Oryza sativa* L.) originated in the Yangtze River basin in China, with domestication occurring as early as 9,000 years ago. Rice thrives in a wide range of climates. Based on agronomic adaptation, rice production is categorized into upland and lowland (paddy) rice. Globally, about 700 – 800 million metric tonnes of rice are produced annually. Of this total, 90-96% is paddy (lowland) rice and remainder, upland rice. Upland rice is grown in a dryland environment (with no standing water). In Asia, upland rice is grown at elevations of up to 2,000 m (6,500 ft) above sea level. Paddy rice is grown in flooded fields primarily in Asia and dominates global production. Other important rice-producing regions include Latin America, the United States, and West Africa.

Upland and paddy rice have different challenges and opportunities for implementing precision agriculture. Challenges for upland rice cultivation include declining fertility and increasing pest pressure (Goud et al., 2022). Challenges for paddy rice cultivation include synchronizing flowering and pollination with ambient temperature and moisture conditions.

Because rice is a C3 plant, it is sensitive to extreme heat and drought. Some precision agriculture techniques to improve yield in this case include a "rice planning calendar," which can be used (Mishra et al., 2021) in combination with planting optimization, weather forecasting, and remote sensing to make site-specific and time-sensitive decisions. Paddy rice also requires large volumes of water, ranging from 1,200 to 1,500 mm/ha (47–59 inches/ha) for each wetting and drying cycle (Lampayan et al., 2015).

Precision agriculture techniques to optimize yield for upland and paddy rice include using paddy rice planters equipped with a GPS-guided Laser Land Leveling system. This improves the precision and efficiency of transplanting and reduces labor requirements (Akdemir, 2025). Fertilizer management using variable rate and prescription maps has also been used to apply less N closer to specific stages, particularly basal, tillering, and panicle initiation.

Precision molecular breeding of rice is an important strategy for developing cultivars that maintain high yields under abiotic stress conditions. Many regions where rice is a primary food staple are affected by saline soils, to which rice is extremely sensitive. Therefore, researchers are actively breeding salt-tolerant rice varieties that can grow in saline–alkali soils with salt concentrations of up to 0.3%. These cultivars represent an important precision approach for farmers in salt-affected regions of Bangladesh, China, and Thailand (Qin et al., 2020).

Soybean: Warm-season, C3 Photosynthesis

Soybean [*Glycine max* (L.) Merr.] is a warm-season C3 plant that was first domesticated in northeastern China. In China, it has been an important food source for several thousand years (Kim et al., 2012); however, it was not cultivated in Europe and North America until the eighteenth century. Soybean is an oilseed and a major global source of protein, widely consumed, and a major commodity traded in world markets (Wilcox, 2004). Argentina, Brazil, and the United States are the largest exporters of soybean products to many countries, including China, India, and Iran (Wilcox, 2004).

Soybean, as a legume, can fix atmospheric N into a plant-usable form through a symbiotic relationship with specific bacteria known as rhizobia. This unique trait has made soybean a key component of many crop rotations, as it helps diversify rooting systems and reduces the need for inorganic N fertilizers. As a result, soybean plays an important role in improving sustainable land management and soil fertility. Soybean has been a major focus of many breeding programs aimed at developing cultivars with broad adaptability. For example, soybean varieties have been bred to grow at many different latitudes.

Choosing the correct hybrid variety is a critical aspect of precision agriculture. This is especially important for soybean, which can be either determinate or indeterminate. Determinate varieties cease stem growth once flowering begins, whereas indeterminate varieties continue growing during part of their reproductive period. For instance, cultivating soybean varieties with shorter growing seasons may be particularly well suited to regions with limited rainfall.

Canopy imaging technologies, including thermal, multispectral, and hyperspectral sensors, are often used to detect drought, salt, weed, or disease stress symptoms before visual signs emerge or before the critical weed-free period of 30–45 days after planting. One important cause of soybean yield loss to watch out for is soybean cyst nematode (*Heterodera glycines* Ichinohe), which can be detected through soil sampling and/or remote sensing to identify affected areas. Global Positioning System–enabled precision planters are used to support optimal seed placement and spacing, and variable rate seeding can be used to match population density to soil productivity zones.

Cotton: Warm-season C3, Photosynthesis

Cotton (*Gossypium* spp.) is a very important fiber crop that was domesticated over 5,000 years ago at multiple locations (Table 29.1). It is a perennial shrub in the family Malvaceae that can be grown in tropical and subtropical climates. In the United States, it is grown as an annual. Cotton exhibits indeterminate growth, meaning it produces vegetative and reproductive structures concurrently. Of the several species, *Gossypium hirsutum* L. (upland cotton) accounts for 90% of global production due to its high yield and adaptability.

Cotton thrives in well-drained sandy loam soils and warm, humid climates with extended frost-free periods. Optimal planting temperatures range from 21 to 37 °C (70 to 99 °F), with tolerance for brief periods of temperatures of up to 43 °C (109 °F). Nitrogen levels must be consistently monitored, as inadequate N leads to stunted growth, reduced leaf area, delayed flowering, and poor boll development. However, excessive N can prolong vegetative growth at the expense of reproductive development, increasing the risk of lodging and pest pressure, reducing fiber quality, and raising the likelihood of frost damage before maturity.

Cotton is a drought-sensitive plant that can recover from short droughts; however, frost damage can reduce both yield and crop quality. The complex physiological link between boll size, lint fiber length, and seed size requires balancing the source–sink distribution of resources, a process further complicated by the plant's indeterminate growth habit. These three traits are metabolically interconnected and often represent trade-offs based on plant genetics, management, and environmental conditions. For example, planting cotton in dry soil exposes it to insect damage, reducing stands; however, it can increase its branching to compensate for the lost stands when soil moisture, N, and other nutrients are not limited.

Remote and proximal sensors can be used to track nutrient stress and detect diseases such as Fusarium wilt (*Fusarium oxysporum*) and Verticillium wilt (*Verticillium dahliae*). Moisture sensors that detect boll moisture content are used to determine harvesting time, minimizing yield losses. Additionally, N and moisture demands can be predicted through UAV-based Normalized Difference Vegetation Index (NDVI) mapping, tissue testing, soil moisture sensing, and machine learning models to optimize input use and improve yield and profitability (Jia et al., 2025).

Sunflower: Warm-season, C3 Photosynthesis

Cultivated sunflower (*Helianthus annuus* L.) is native to North America. However, commercialization began in Russia in the 1800s. Sunflower is an important oilseed crop for human food and livestock feed. Two main types of sunflower are cultivated. Oilseed sunflower is primarily grown for its seeds, which are used to extract edible oil ideal for cooking, baking, and industrial oil production. In contrast, confectionery sunflower produces kernels that are consumed directly in a variety of ways, including as snacks, in salads, in baking, and in products such as sunflower butter. Non-oilseed varieties are also commonly used as bird feed. The leading producers are Russia, Ukraine, the European Union, Argentina, China, Kazakhstan, Turkey, Moldova, South Africa, and the United States.

Sunflower, although a temperate crop, can grow and perform well under a wide range of climatic and soil conditions and is moderately tolerant of salt-affected soils. It thrives in well-drained sandy soil that warms up quickly. Sunflower crops are widely cultivated under rainfed conditions. Compared with other annual crops, sunflower can withstand short droughts because the plants have a deep taproot that can extract soil moisture from deeper layers of the soil profile (Hussain et al., 2018). In drier regions, no-till practices are recommended to conserve soil moisture. However, under such conditions, increasing the seeding rate by 20% can mitigate the reduced emergence caused by poor soil-to-seed contact. By adjusting its seed and head size, sunflower can adapt to various plant populations and spacing. Balanced fertilizer application at planting supports healthy growth and yield, but excessive N levels reduce seed oil content (Kafle, 2024). Integrated pest and disease management enhances plant health and limits yield loss. Harvest timing depends on the end use: oilseed sunflower should be harvested at 10%–13% moisture, while birdseed or snack seed should be harvested at less than 7% moisture, when head backs are fully brown and seeds are mature.

Precision tools that measure general plant health, including NDVI and the Green Normalized Difference Vegetation Index (GNDVI), can be used to scout for yield-limiting factors. The GNDVI is especially useful for this crop because this vegetative index is slightly more sensitive to variations in canopy color and density, which are important in later stages due to the yellow petals (ray florets) of the sunflower changing how light is reflected (Shao et al., 2023). This data can be correlated with in-season N status to inform management, including the use of variable rate prescription maps (Shanmugapriya et al., 2019).

These proximal and remote sensors can also be used to diagnose diseases and insect damage (Xu et al., 2019). Ercan et al. (2024) reported that PlanetScope satellite imagery can provide useful information for detecting damage caused by the beet webworm moth (*Loxostege sticticalis*) in sunflower fields. Early detection has been shown to reduce yield loss and help maintain oilseed quality.

Pulse Crops: (Chickpea, Lentil, and Field Pea [Cool-season] and Common Bean [Warm-season]), C3 Photosynthesis

Pulse crops are N-fixing plants (leguminous) that produce edible dry seeds with high protein and fiber content and low oil fat content. Most pulse crops are native to the Middle East, with lentil and pea being native specifically to the Fertile Crescent region. The common bean is native to Central and South America. The common bean, unlike chickpea, lentil, and field pea, is a warm-season crop. This suggests that this group of crops has a wide diversity of management opportunities. Pulse crops are widely cultivated globally, with production occurring in nearly every hemisphere. The Northern Hemisphere is a leading producer, with India alone accounting for approximately 25% of total production (Singh et al., 2016).

Pulse crops prefer well-drained, loamy soils with a pH above 6.0. Field selection should avoid saline or poorly drained areas due to pulse crops' high susceptibility to root rot pathogens under these conditions. A firm, fine, and weed-free seedbed supports consistent emergence and root establishment. Where moisture conservation is critical, minimal tillage systems paired with crop residue retention are preferred to increase soil water retention. Remote sensing and yield maps from previous seasons can help delineate productive zones and inform variable rate tillage or targeted soil amendment applications. Topography-based zone mapping also aids in predicting drainage challenges and guides field layout decisions. Seeding dates depend on geographic location. For example, in temperate North America, seeding occurs from early to mid-spring (April–May), while in many tropical regions, planting may align with the start of the rainy season (e.g., October–November). Planting dates also depend on the regional climate and the specific crop's sensitivity to frost. Lentil and field pea can tolerate light spring frosts, while chickpea and common bean are sensitive to frost.

The bacteria used by pulse crops to fix N are host specific, meaning they are most effective when they can form a relationship with their preferred plant species. For example, field pea is commonly associated with *Rhizobium leguminosarum* biovar *viciae*, a strain that can also nodulate lentil. Use of the correct inoculation species, like crop selection, is an important aspect of precision agriculture that does not necessarily rely on advanced technologies; instead it uses our understanding and knowledge of the plant to improve yield.

Pulses generally compete poorly with weeds and are susceptible to a broad spectrum of fungal and viral pathogens, such as Ascochyta blight (*Ascochyta* spp.), Rhizoctonia root rot (*Rhizoctonia solani*), Pythium damping-off (*Pythium* spp.), Fusarium wilt (*F. oxysporum*), and white mold (*Sclerotinia sclerotiorum*), especially under conditions of high humidity, continuous pulse cropping, and poor residue management. Precision agriculture management—including crop rotation (3 or more years), certified disease-free seed, and residue decomposition through conservation tillage—reduces the spread of pathogens and their carryover into the next growing season.

Pulse crop productivity can also be enhanced by combining advanced precision agriculture tools with traditional agronomic practices. For instance, precision planters equipped with variable rate seeding can optimize plant population density, while UAV-based thermal imaging and NDVI mapping enable early detection of nutrient deficiencies and identification of stress zones before visual symptoms appear. Precision-guided pesticide applications, based on UAV scouting or in-field sensors, can reduce blanket spraying and help preserve beneficial insects. Successful production of these legume crops hinges on site-specific management practices, informed input decisions, and timely interventions, all of which can be optimized through the use of precision agriculture tools.

Banana: Warm-season, C3 Photosynthesis

Banana (*Musa* spp.) comprises several species, including *Musa acuminata* Colla, *Musa balbisiana* Colla, and their hybrids, which constitute the edible banana cultivated worldwide. Banana originated in Southeast Asia and the South Pacific, with domestication occurring approximately 7,000–10,000 years ago. Early domestication of *Musa banksii* in New Guinea involved the selection of seedless forms and led to the development of triploid cultivars through natural hybridization. Currently, banana is grown in tropical and subtropical regions between 30N and 30S latitudes, at elevations of up to 2,000 m (6,500 ft), across more than 130 countries (Perrier et al., 2011; Soluri, 2003; Volkaert, 2018). Global banana production was estimated at 114.92 million Mg in 2017 (Evans et al., 2020). Ecuador, the Philippines, and Costa Rica collectively supply nearly 50% of banana to the global market, while the United States, Germany, Western Europe, and Russia represent major importing regions (Evans et al., 2020). The banana growth cycle ranges from 9 to 18 months, depending on the cultivar and environmental conditions.

Globally, cultivar selection is based on adaptability, growth traits, and purpose (fresh consumption, cooking, processing, brewing, baking, and aesthetic value). All species are herbaceous perennials with a pseudostem and reproduce by suckers. Once the plant matures, it produces a flower with both male and female inflorescences. The banana fruit develops from part of the female inflorescence and can weigh variably from 5 to over 60 kg/bunch (11–132 lb/bunch), depending on the variety (Hauser & Van Asten, 2008; Wairegi et al., 2009), with some weighing up to 100 kg/bunch (220 lb/bunch). Most commercially cultivated varieties are parthenocarpic and sterile due to their triploid nature.

Banana is a high-yielding crop; however, its productivity can be significantly reduced by numerous pests and diseases. Major constraints include plant-parasitic nematodes (*Radopholus similis* and *Pratylenchus coffeae*); insect pests, such as the banana weevil (*Cosmopolites sordidus*); and numerous diseases, including Panama disease (caused by *F. oxysporum* f. sp. *cubense*), black Sigatoka (caused by *Mycosphaerella fijiensis*; Bouwmeester et al., 2016), and banana bacterial wilt (caused by *Xanthomonas vasicola* pv. *musacearum*). Yields can also be reduced by drought, poor soil fertility, and extreme temperatures. On average, a banana plant requires approximately 0.5–0.7 cm of water per day (0.2–0.28 inches/day). However, excessive water (flooding) can reduce photosynthetic efficiency and fruit quality, as well as increase the chances of stem and root lodging. Nutrient demands—particularly for N and potassium—are high, and responses to fertilization vary based on soil conditions and weather.

Precision agriculture can enhance banana input optimization and stabilize yields. However, the application of these tools can vary by banana production region, depending on the specific production challenges encountered. For example, tropical regions are characterized by high temperatures, humidity, and rainfall, which favor high disease pressure (including Fusarium wilt and black Sigatoka), as well as waterlogging. Disease surveillance with artificial intelligence (AI) and UAVs for early detection and targeted pesticide application, as seen in studies covering Africa, Asia, and Latin America (Selvaraj et al., 2019, 2020), can reduce disease impacts on yield. In subtropical regions, frost damage and irrigation inefficiencies are addressed with UAV-based microclimate monitoring and precision spraying (Al-Najadi et al., 2025). Seasonal temperature fluctuations are managed by adjusting planting schedules using soil moisture sensors for real-time data (Vellidis et al., 2008), and weather variability is countered with AI-driven yield prediction and precision irrigation based on soil moisture (Ramezanpour & Farajpour, 2022). These region-specific technologies demonstrate the adaptability of precision agriculture to unique environmental constraints.

Wheat: Cool-season, C3 Photosynthesis

Wheat (*Triticum aestivum* L.) is a cool-season, C3 crop that can be planted either in the fall prior to the first frost (winter wheat) or in the spring (spring wheat). Wheat is one of the oldest domesticated cereal crops, with origins dating back more than 10,000 years to the Fertile Crescent region of the Middle East. Even though it is moderately tolerant to salinity and sensitive to drought, it thrives in a wide range of temperate climates around the world (Table 29.1). Wheat is cultivated across latitudes from approximately 60°N to 45°S and at elevations ranging from near sea level to over 3,000 m (9,800 ft). The top two producers of wheat are China (138 million MT) and India (110 million MT). Wheat is a global staple and one of the most widely consumed crops as a primary source of carbohydrates. The crop provides over 20% of global human dietary calories and protein (Shiferaw et al., 2013).

Wheat development proceeds through distinct phenological stages, including germination, tillering, stem elongation, heading, flowering (anthesis), and grain filling. Among these stages, flowering and early grain filling represent the periods most sensitive to abiotic stresses such as heat and drought. Exposure to high temperatures or water stress during these stages can reduce pollen viability, increase spikelet sterility, shorten the grain filling duration, and ultimately result in substantial yield losses (Farooq et al., 2014).

Wheat performs best under cool temperatures (15–22 °C or 59-72 °F), with full sunlight, and in well-drained loam soils with a pH of 6.0–7.5, combined with moderate, well-distributed rainfall. The crop exhibits moderate tolerance to drought, cold temperatures (particularly in winter wheat), and salinity; however, it is highly sensitive to heat stress, waterlogging, and severe moisture deficits during critical reproductive stages, particularly flowering and grain filling. Wheat has a fibrous root system that can access soil moisture from deeper layers, contributing to its moderate drought tolerance. Root growth and function, however, are constrained by soil compaction, poor drainage, and subsoil chemical constraints, which can restrict water and nutrient uptake and ultimately reduce crop performance (Kirkegaard et al., 2007).

Nitrogen availability plays a central role in wheat yield formation by influencing tiller initiation and survival, canopy development, and grain protein concentration. Insufficient N supply during early vegetative growth restricts tiller production and biomass accumulation, while N deficiency during the reproductive and grain-filling stages reduces kernel weight and grain protein content. Conversely, excessive N application increases the risk of lodging and environmental N losses, emphasizing the need for optimized N management strategies (Cassman et al., 2002).

Precision agriculture has become increasingly important in wheat production for improving disease detection, nutrient management, and overall input-use efficiency. A major area of research involves the application of machine learning and remote sensing techniques for the early identification and classification of economically important wheat diseases, including leaf rust (*Puccinia triticina*), stripe (yellow) rust (*P. striiformis* f. sp. *tritici*), stem rust (*P. graminis* f. sp. *tritici*), and Septoria leaf blotch (*Zymoseptoria tritici*; Singh et al., 2023, 2025). These diseases exhibit strong spatial and regional variability in occurrence and severity, making traditional uniform management strategies less effective. Integration of multispectral imagery, canopy reflectance indices, and machine learning-based pattern recognition enables timely, site-specific disease detection and targeted management, reducing yield losses and unnecessary fungicide applications (Chapters 17, 18, and 19 provide more details; Guebsi et al., 2024; Getahun et al., 2024). In addition to disease monitoring, precision agriculture tools are increasingly used to optimize N management in wheat by matching nutrient supply with crop demand, thereby improving N-use efficiency while minimizing lodging risk and environmental N losses (Raun et al., 2002).

Oat: Cool-season, C3 Photosynthesis

Archeological evidence suggests that oat (*Avena sativa* L.) is an ancient crop, with cultivation beginning in the Jordan Valley over 11,400 years ago. Approximately 25.13 million MT of oat are produced globally in over 70 countries. Because oat is sensitive to drought conditions, production is concentrated in temperate regions between 45 and 65°N latitude. Spring varieties dominate in northern climates, whereas winter oat prevails in Mediterranean areas. Oat is used for animal feed (48%–70%) and in human food applications (25%–30%). Traditional products such as rolled oats remain important, while emerging applications, including oat-based meat alternatives, have become popular.

Oat plants are generally hardy and thrive in challenging growing conditions, such as acidic soil (pH range of 4.5–8.6), cold temperatures, and soil with high clay content (often a difficult soil composition for other cereals to grow in). However, as a C3 plant, it is sensitive to temperatures above 28 °C (82 °F), becoming less efficient at carbon fixation when it needs to conserve water. This adaptability, however, also makes oat more sensitive to drought, as it has a higher water requirement than most other cereal crops (Hakala et al., 2020).

Precision agriculture technologies, specifically variable rate applications and remote sensing to evaluate general plant health (using NDVI and other indices), are key for oat cultivation. This is because N fertilizer rates that are too high can reduce test weight and milling quality. Prescription maps and variable rate applications can help keep soil N levels between 33 and 39 kg/ha (30–35 lb N/ac). This also keeps disease levels in check, as higher soil N rates can increase the risk of several diseases in oat, including loose smut (*Ustilago nuda*), barley yellow dwarf virus (BYDV), and Septoria leaf blotch (*Septoria passerinii*). To reduce diseases, resistant varieties are commonly cultivated in rotation. The NDVI can be used to assess crop health, detect diseases, and support variable rate fungicide application. Oat can also serve as a tool for weed suppression without herbicide use, as its vigorous vegetative growth naturally suppresses competing weed species.

Barley: Cool-season, C3 Photosynthesis

Barley (*Hordeum vulgare* L.), like oat, also has ancient origins in the Jordan Valley. Barley is among the first domesticated cereals, originating 10,000 years ago from wild *H. vulgare* ssp. *spontaneum* in the Fertile Crescent (Haas et al., 2019). Genetic evidence suggests that there were multiple domestication events from the eastern Mediterranean to Afghanistan, as well as in Central Asia (Morrell & Clegg, 2007). Barley is cultivated on an estimated 47 million ha globally, producing more than 14 million MT of grain. Barley can be grown at latitudes from 60°N to 45°S, as well as in high-altitude regions, such as Tibet and the Andes. Barley has been used primarily as animal feed and in malting and brewing applications; however, increasing interest in functional foods has created new market opportunities due to its relatively high β-glucan content (3%–7%).

Barley has a strong tolerance to salt-affected soils, meaning it can be grown in degraded soils and used in phytoremediation to rebuild soil health. For barley used in malt production, monitoring soil N using precision agriculture technologies is key. Sensor technologies that can monitor plant health and provide guidance on soil N levels, such as NDVI and chlorophyll meters, including GreenSeeker and SPAD-502 m, are key (Chapter 13). Nitrogen rates that are too low risk causing the protein concentration to fall below 12.5%, resulting in reduced grain quality and overall lower selling prices. Fertilizer rates of 80–120 kg N/ha (71–107 lb N/ac) for spring barley and 120–180 kg N/ha (107–161 lb N/ac) for winter types are usually adequate.

Rapeseed/Canola: Cool-season, C3 Photosynthesis

Rapeseed (*Brassica napus* L.), also known as oilseed rape and canola, is a member of the family Brassicaceae. Originally domesticated in Europe, Asia, and Africa, rapeseed is used to produce edible oil and protein-rich meals for livestock feed. Canola denotes a group of rapeseed cultivars bred for lower levels of erucic acid and glucosinolates, as found in the species *B. napus*, *B. rapa* ssp. *rapa* L., and *B. juncea* (Arinaitwe et al., 2023). These crops thrive in temperate and subtemperate climates and have demonstrated adaptability to a range of rainfall and soil conditions, making them key components of diversified farming systems.

Rapeseed is a cool-season, long-day annual that prefers temperatures between 10 and 25 °C (50–77 °F) during vegetative growth, with flowering and seed development requiring temperatures under 30 °C (86 °F) to avoid heat stress and yield loss (Diepenbrock, 2000). Over time, breeding gave rise to two main types of rapeseed: winter types sown in the fall and harvested the following summer, and spring types planted in spring and harvested in late summer or early fall. While the crop is moderately drought tolerant, consistent soil moisture during flowering and pod filling is essential for optimal yield. Ideal rainfall for canola ranges between 400 and 800 mm per season (15.7–31.4 inches), and supplemental irrigation is often used in semi-arid zones. The crop prefers well-drained loamy soils over clay loam soils and thrives in a pH range of 5.5–7.5. *B. juncea* cultivars offer improved tolerance to more alkaline conditions, making them suitable for marginal lands (Nleya et al., 2025).

Brassica spp. establish poorly in no-till production systems. In high-residue environments, precision planters or no-till drills equipped with residue-cutting coulters, press wheels, and seed firmers should be used to ensure consistent depth placement and seedling emergence. Proper fertilization regimes, guided by soil testing and yield targets, improve nutrient efficiency and lessen disease and insect pest pressure, especially in warm, moist environments. Key pathogens include *Leptosphaeria maculans* (blackleg), *S. sclerotiorum* (white mold), and *Alternaria brassicae* (leaf spot), while major insect pests include flea beetles (*Phyllotreta* spp.), diamondback moths (*Plutella xylostella*), and aphids. Timely monitoring and integrated pest management strategies are essential to mitigate losses. At maturity, rapeseed pod shattering is a major problem in many species (Arinaitwe et al., 2023); therefore, harvesting rapeseed requires precise timing to avoid pod shattering and seed losses. Physiological maturity is indicated by a uniform change in pod color from green to brown and a target seed moisture content of 8%–10%. Precision harvesting tools, including near-infrared (NIR) sensors and onboard moisture meters, help determine optimal harvest timing and reduce post-harvest losses. Swathing prior to combine harvesting is common in Canada to allow for uniform ripening under variable weather and soil nutrient conditions.

Other opportunities to improve rapeseed performance with precision tools during planting and in-season nutrient monitoring include using UAVs equipped with multispectral imaging and machine learning models. These can detect the onset of disease, assess canopy health, and optimize input usage. Real-time monitoring supports early interventions, especially during critical stages such as flowering and pod development.

Potato: Cool-season, C3 Photosynthesis

Potato (*Solanum tuberosum* L.) is a shallow-rooted tuber crop that is sensitive to pH, salinity, nutrient stress, and drought (Table 29.1). It is often grown successfully in loamy and sandy loam soils that are rich in organic matter and well-drained. It is a temperate crop with high nutrient demand that grows well under a wide range of climatic conditions. Potato originated in the Andes Mountains (modern-day Peru and northwestern Bolivia). At present, more than 5,000 different varieties of potato are cultivated worldwide, with China and India being the largest producers. In addition to being a global food staple, potato is fermented into alcohol and used in the production of cosmetics.

Disease prevention in potato production faces unique challenges, as many common diseases—such as late blight (*Phytophthora infestans*), early blight (*Alternaria solani*), and bacterial wilt (*Ralstonia solanacearum*)—spread easily by insects, such as aphids, and through generational transmission, where a tuber seed carrying the disease may pass it on to new growth. Because many symptoms are extremely subtle and have a limited window during which the crop can be saved, specialized precision agriculture equipment has been developed. For example, a robotic system called AgriScout, developed in Canada, can traverse the raised ridges of these fields using a red–green–blue (RGB) camera and a GPS system. These robots capture high-quality images that can be evaluated by computer vision models in real time (Singh et al., 2025). These models detect symptoms of infection early and generate infestation maps. These maps can be used to address the specific problem site. Avoiding the spread of the disease reduces the need for chemical amendments (Singh et al., 2025).

Grape (*Vitis* spp.): Cool-season, C3 Photosynthesis

Grapevine (*Vitis vinifera* L.) is among the most valuable fruit crops in the world and is cultivated for winemaking, table grapes, raisins, and juice. Members of the family Vitaceae are remarkably adaptable, thriving in temperate, subtropical, and even some tropical climates. The species is highly diverse, with over 10,000 recognized cultivars (Lopez-Ruiz, 2024). Archeological and genetic evidence indicate that grapevine domestication began in a region that spans present-day Iran to the Black Sea 4,000 to 6,000 years ago (Chataigner, 1995; McGovern et al., 2003; Zohary & Hopf, 1988). From this primary center, cultivated vines were disseminated to the Middle East and Europe, which subsequently became significant secondary centers of domestication (Arroyo-García et al., 2006). Grapevine is drought tolerant; however, a reliable water supply, abundant sunlight, and sufficient chilling hours during dormancy are all essential for successful production (Charalampopoulos et al., 2024).

With the emergence of advanced agricultural technologies, viticulture has been gradually transitioning to mechanized approaches for operations such as planting, pruning, weeding, spraying, fertilizer application, irrigation, and harvesting. Specifically, precision viticulture applications that have been widely adopted in recent years include smart irrigation with crop and soil sensors, mechanized pruning, and harvesting equipment that uses computer vision combined with NIR spectroscopy for ripeness assessments (Passias et al., 2024). Additional advances include integrating remote sensing platforms and computer-based decision-making technologies to facilitate precise operational management through information systems, mobile applications, decision support systems, and weather-based advisories (e.g., 300viti; AgTech: Vineyard; PureSense; Ranch system; e-vineyard, AgriSense; Khadatkar et al., 2025). Recent research has emphasized the integration of AI with sensor-based robotics for real-time crop canopy management, pest and disease surveillance, and early detection of biotic and abiotic stressors (Giovos et al., 2021; Leolini et al., 2023; Sharma et al., 2024). Moreover, these AI-based predictive analytics are used to forecast weather patterns and assist vineyard managers in preparing for extreme events and determining the optimal harvest time based on grape maturity and environmental factors. The adoption of precision viticulture technologies ensures sustainable and resilient vineyard operations by effectively navigating the complexities of modern grape cultivation.

Conclusion

In this chapter, we explored a variety of common crops grown throughout the world and the different ways precision agriculture can be used to improve crop production. Many aspects of precision agriculture involve advanced technologies and equipment, such as satellite imagery, UAVs, robots, and proximal sensors, which have been explored in other chapters of this book. However, precision agriculture also uses site-specific knowledge—including information on a region's climate, environmental conditions, soil type, and topography—to make optimal management decisions such as crop selection. The goal of this chapter was to demonstrate how this knowledge has shaped where and how crops are grown across the Earth, and how precision agriculture is continuing to improve production.

Study Questions

1. Where do you plant sorghum as opposed to corn?
2. What is a pulse, and why is it important in rotation?
3. When might you plant potato in tropical versus temperate climates?
4. For which climatic region is cotton best suited?
5. What is the difference between rice grown in a paddy and rice grown in upland areas?
6. What makes a plant well suited for precision technologies?
7. Which plants are best suited for salt-affected soil?
8. What is the importance of measuring planting failures in sugarcane fields using UAV imagery?
9. What are the main species of banana cultivated globally, and why are they preferred?
10. What are the reproductive structures of banana plants? Which part of the flower is of economic importance?
11. Why are most commercial banana varieties seedless?
12. Name the top banana-importing countries and the main importing region.
13. What are the key environmental requirements for successful grape production, and where are the prime viticultural zones typically located?
14. What are some examples of precision viticulture technologies, and how do they contribute to sustainable vineyard operations?

ACKNOWLEDGMENTS

Support for this document was provided by South Dakota State University, the Precision Farming Systems community in the American Society of Agronomy, the International Society of Precision Agriculture, the USDA National Institute of Food and Agriculture (Multistate Project S1090) Program under grant number 7008110, and the USDA Agriculture and Food Research Initiative Higher Education Grant (2014-04572). The use of trade names is for the convenience of the reader and does not imply endorsement by the author.

REFERENCES

Akdemir, B. (2025). Determination of GPS-based land leveling system performance in land conditions. *Tarım Makinaları Bilimi Dergisi*, *21*(1), 51–64. https://izlik.org/JA62WA24SF.

Al-Najadi, R., Al-Mulla, Y., Al-Abri, I., & Al-Sadi, A. M. (2025). Effectiveness of drone-based thermal sensors in optimizing controlled environment agriculture performance under arid conditions. *Scientific Reports*, *15*, 9042. https://doi.org/10.1038/s41598-025-94432-0.

Arinaitwe, U., Clay, S. A., & Nleya, T. (2023). Growth, yield, and yield stability of canola in the Northern Great Plains of the United States. *Agronomy Journal*, *115*, 744–758.

Arroyo-García, R., Ruiz-García, L., Bolling, L., Ocete, R., López, M. A., Arnold, C., Ergul, A., Söylemezoğlu, G., Uzun, H. I., Cabello, F., Ibáñez, J., Aradhya, M. K., Atanassov, A., Atnassov, I., Balint, S., Cenis, J. L., Costantini, L., Goris-Lavets, S., Grando, M. S., . . . Martinez-Zapater, J. M. (2006). Multiple origins of cultivated grapevine (*Vitis vinifera* L. ssp. *sativa*) based on chloroplast DNA polymorphisms. *Molecular Ecology*, *15*(12), 3707–3714. https://doi.org/10.1111/j.1365-294X.2006.03049.x.

Bouwmeester, H., Heuvelink, G. B. M., & Stoorvogel, J. J. (2016). Mapping crop diseases using survey data: The case of bacterial wilt in bananas in the East African highlands. *European Journal of Agronomy*, *74*, 173–184. https://doi.org/10.1016/j.eja.2015.12.013.

Bruggeman, S. A., Clay, S. A., Reese, C. L., & Carlson, C. G. (2017). Mathematics associated with seed emergence, plant population, stand uniformity, and harvest losses. In D. E. Clay, S. A. Clay, & S. A. Bruggeman (Eds.), *Practical mathematics for precision farming* (pp. 201–207). ASA, CSSA, SSSA.

Byrne, P. (2023). Teff (*Eragrostis tef*). In K. Chen & P. Byrne (Eds.), *Understudied indigenous crops*. Colorado State University. https://colostate.pressbooks.pub/understudiedindigenouscrops/chapter/teff/.

Carlson, C. G., & Reese, C. L. (2016). Grain marketing – Understanding corn moisture content, shrinkage and drying. In D. E. Clay, C. G. Carlson, S. A. Clay, & E. Byamukama (Eds.), *iGrow Corn: Best Management Practices*. South Dakota State University. https://extension.sdstate.edu/sites/default/files/2019-09/S-0003-35-Corn.pdf.

Cassman, K. G., Dobermann, A., & Walters, D. T. (2002). Agroecosystems, nitrogen-use efficiency, and nitrogen management. *Ambio*, *31*(2), 132–140. https://doi.org/10.1579/0044-7447-31.2.132.

Charalampopoulos, I., Polychroni, I., Droulia, F., & Nastos, P. T. (2024). The spatiotemporal evolution of the growing degree days agroclimatic index for viticulture over the northern Mediterranean Basin. *Atmosphere*, *15*(4), 485. https://doi.org/10.3390/atmos15040485.

Chataigner, C. (1995). *La Transcaucasie au Néolithique et au Chalocolithique*. BAR International Series No. 624. BAR Publishing.

Diepenbrock, W. (2000). Yield analysis of winter oilseed rape (*Brassica napus* L.): A review. *Field Crops Research*, *67*(1), 35–49. https://doi.org/10.1016/S0378-4290(00)00082-4.

Ercan, B. S., Maden, B., Kara, S., Sunar, F., Aysal, T., Ozkaya, N., & Saglam, O. (2024). A comparative study of satellite image resolutions for detecting pest damage in sunflower fields. *The International Archives of the Photogrammetry, Remote Sensing and Spatial Information Sciences*, *48*, 133–139. https://doi.org/10.5194/isprs-archives-XLVIII-4-W9-2024-133-2024.

Evans, E. A., Ballen, F. H., & Siddiq, M. (2020). Banana production, global trade, consumption trends, postharvest handling, and processing. In M. Siddiq, J. Ahmed, & M. G. Lobo (Eds.), *Handbook of banana production, postharvest science, processing technology, and nutrition* (pp. 1–18). John Wiley & Sons. https://doi.org/10.1002/9781119528265.ch1.

Farooq, M., Hussain, M., & Siddique, K. H. M. (2014). Drought stress in wheat during flowering and grain-filling periods. *Critical Reviews in Plant Sciences*, *33*(4), 331–349. http://dx.doi.org/10.1080/07352689.2014.875291.

Ferguson, R. B., Luck, J. D., & Stevens, R. (2017). Developing prescriptive soil nutrient maps. In D. E. Clay, S. A. Clay, & S. A. Bruggeman (Eds.), *Practical mathematics for precision farming* (pp. 149–166). ASA, CSSA, SSSA.

Food and Agriculture Organization of the United Nations (FAOSTAT). (2015) *FAO Global Statistical Yearbook, FAO Regional Statistical*. https://www.fao.org/faostat/en/#data/QCL.

Getahun, S., Kefale, H., & Gelaye, Y. (2024). Application of precision agriculture technologies for sustainable crop production and environmental sustainability: A systematic review. *The Scientific World Journal, 2024*(1), 2126734. https://doi.org/10.1155/2024/2126734.

Giovos, R., Tassopoulos, D., Kalivas, D., Lougkos, N., & Priovolou, A. (2021). Remote sensing vegetation indices in viticulture: A critical review. *Agriculture, 11*, 457.

Goud, R. B., Tripathi, R., Guru, P. K., Mohanty, S., Kumar, A., Khanam, R., Munda, S., Vijayakumarr, S., Debnath, M., Sivashankari, M., Kumar, K., Mohapatra, S. D., & Nayak, A. J. (2022). Advanced techniques for precision farming in rice. In P. Bhattacharyya, K. Chakraborty, K. A. Molla, A. Poonam, D. Bhaduri, R. P. Sah, S. Paul, P. S. Hanjagi, G. Basana-Gowda, & P. Swain (Eds.), *Climate resilient technologies for rice based production systems in Eastern India* (pp. 224–247). ICAR-National Rice Research Institute.

Graham, C., & Carlson, G. (2019). Chapter 8: Corn Seeding Rates in South Dakota. In *iGrow corn: Best management practices*. South Dakota Extension. https://extension.sdstate.edu/sites/default/files/2019-09/S-0003-08-Corn.pdf.

Grijalva, I., Kang, Q., Flippo, D., Sharda, A., & McCornack, B. (2024). Unconventional strategies for aphid management in sorghum. *Insects, 15*(7), 475. http://dx.doi.org/10.3390/insects15070475.

Guebsi, R., Mami, S., & Chokmani, K. (2024). Drones in precision agriculture: A comprehensive review of applications, technologies, and challenges. *Drones, 8*(11), 686. https://doi.org/10.3390/drones8110686.

Haas, M., Schreiber, M., & Mascher, M. (2019). Domestication and crop evolution of wheat and barley: Genes, genomics, and future directions. *Journal of Integrative Plant Biology, 61*(3), 204–225. https://doi.org/10.1111/jipb.12737.

Hakala, K., Jauhiainen, L., Rajala, A. A., Jalli, M., Kujala, M., & Laine, A. (2020). Different responses to weather events may change the cultivation balance of spring barley and oats in the future. *Field Crops Research, 259*, 107956. https://doi.org/10.1016/j.fcr.2020.107956.

Hao, H., Li, Z., Leng, C., Lu, C., Luo, H., Liu, Y., Wu, X., Liu, Z., Shang, L., & Jing, H.-C. (2021). Sorghum breeding in genomic era: Opportunities and challenges. *Theoretical and Applied Genetics, 134*, 1899–1924. https://doi.org/10.1007/s00122-021-03789-z.

Hauser, S., & van Asten, P. (2008). Methodological considerations on banana (*Musa* spp.) yield determinations. In *IV International Symposium on Banana: International Conference on Banana and Plantain in Africa: Harnessing International Partnerships to Increase Research Impact. 879* (pp. 53–56).

Hussain, M., Farooq, S., Hasan, W., Ul-Allah, S., Tanveer, M., Farooq, M., & Nawaz, A. (2018). Drought stress in sunflower: Physiological effects and its management through breeding and agronomic alternatives. *Agricultural Water Management, 201*, 152–166. https://doi.org/10.1016/j.agwat.2018.01.028.

Jia, Y., Li, Y., He, J., Biswas, A., Siddique, K. H. M., Hou, Z., Luo, H., Wang, C., & Xie, X. (2025). Enhancing precision nutrient management for cotton cultivation in arid environments using remote sensing techniques. *Field Crops Research, 321*, 109689. https://doi.org/10.1016/j.fcr.2024.109689.

Jiang, Y., Li, H., Ma, W., Yu, W., Chen, J., Gao, Y., Qi, G., Yin, M., Kang, Y., Ma, Y., Wang, J., & Xu, L. (2025). A meta-analysis of the effects of nitrogen fertilizer application on maize (*Zea mays* L.) yield in Northwest China. *Frontiers in Plant Science, 15*, 1485237. https://doi.org/10.3389/fpls.2024.1485237.

Kafle, R. (2024). *Nitrogen fertilizer placement and rate impacts sunflower seed yield, oil and protein content.* [Master's thesis, South Dakota State University]. *Electronic Theses and Dissertations*. 1352. https://openprairie.sdstate.edu/etd2/1352.

Kausar, A., Ashraf, M. Y., Ali, I., Niaz, M., & Abbass, Q. (2012). Evaluation of sorghum varieties/lines for salt tolerance using physiological indices as screening tool. *Pakistan Journal of Botany, 44*(1), 47–52. https://www.pakbs.org/pjbot/PDFs/44(1)/07.pdf.

Khadatkar, A., Sawant, C. P., Thorat, D., Gupta, A., Jadhav, S., Gawande, D., & Magar, A. P. (2025). A comprehensive review on grapes (*Vitis* spp.) cultivation and its crop management. *Discover Agriculture, 3*, 9. https://doi.org/10.1007/s44279-025-00162-2.

Khan, T. N., Timmerman-Vaughan, G. M., Rubiales, D., Warkentin, T. D., Siddique, K. H. M., Erskine, W., & Barbetti, M. J. (2013). *Didymella pinodes* and its management in field pea: Challenges and opportunities. *Field Crops Research, 148*, 61–77. https://doi.org/10.1016/j.fcr.2013.04.003.

Kim, M. Y., Van, K., Kang, Y. J., Kim, K. H., & Lee, S.-H. (2012). Tracing soybean domestication history: From nucleotide to genome. *Breeding Science*, *61*(5), 445–452. https://doi.org/10.1270/jsbbs.61.445.

Kimball, B. A., Boote, K. J., Hatfield, J. L., Ahuja, L. R., Stockle, C., Archontoulis, S., Baron, C., Basso, B., Bertuzzi, P., Constantin, J., Deryng, D., Dumont, B., Durand, J.-L., Ewert, F., Gaiser, T., Gayler, S., Hoffmann, M. P., Jiang, Q., Kim, S.-H., . . . Williams, K. (2019). Simulation of maize evapotranspiration: An inter-comparison among 29 maize models. *Agricultural and Forest Meteorology*, *271*, 264–284. https://doi.org/10.1016/j.agrformet.2019.02.037.

Kirkegaard, J. A., Lilley, J. M., Howe, G. N., & Graham, J. M. (2007). Impact of subsoil water use on wheat yield. *Australian Journal of Agricultural Research*, *58*(4), 303–315. https://doi.org/10.1071/AR06285.

Lampayan, R. M., Rejesus, R. M., Singleton, G. R., & Bouman, B. A. M. (2015). Adoption and economics of alternate wetting and drying water management for irrigated lowland rice. *Field Crops Research*, *170*, 95–108. https://doi.org/10.1016/j.fcr.2014.10.013.

Lemus, R. (2015). Pearl millet: Pinpoint forage for summer grazing. *Forage News*, *8*(4), 1–2. Mississippi State University Extension Service. https://extension.msstate.edu/sites/default/files/newsletter/201504.pdf.

Leolini, L., Bregaglio, S., Ginaldi, F., Costafreda-Aumedes, S., Di Gennaro, S. F., Matese, A., Maselli, F., Caruso, G., Palai, G., Bajocco, S., Bindi, M., & Moriondo, M. (2023). Use of remote sensing-derived fPAR data in a grapevine simulation model for estimating vine biomass accumulation and yield variability at sub-field level. *Precision Agriculture*, *24*(2), 705–726. https://doi.org/10.1007/s11119-022-09970-8.

Lopez-Ruiz, F. J. (2024). The grape crop. In R. P. Oliver (Ed.), *Agrios' plant pathology* (6th ed., pp. 715–721). Elsevier.

Mahlein, A.-K. (2016). Plant disease detection by imaging sensors—Parallels and specific demands for precision agriculture and plant phenotyping. *Plant Disease*, *100*(2), 241–251. https://doi.org/10.1094/PDIS-03-15-0340-FE.

Mamani-Pati, F., Clay, D. E., & Smeltekop, H. (2011). Geospatial management of Andean technologies by the Inca Empire. In D. E. Clay & J. F. Shanahan (Eds.), *GIS applications in agriculture* (pp. 255–265). CRC Press.

Matsouka, Y., Vigouroux, Y., Goodman, M. M., Sanchez G, J., Buckler, E., & Doebley, J. (2002). A single domestication for maize is shown by multilocus microsatellite genotyping. *Proceedings of the National Academy of Sciences of the United States of America*, *99*(9), 6080–6084. https://doi.org/10.1073/pnas.052125199.

McGovern, P. E., Fleming, S. J., & Katz, S. H. (2003). *The origins and ancient history of wine: Food and nutrition in history and anthropology*. Routledge.

Mishra, B., Busetto, L., Boschetti, M., Laborte, A., & Nelson, A. (2021). RICA: A rice crop calendar for Asia based on MODIS multi year data. *International Journal of Applied Earth Observation and Geoinformation*, *103*, 102471. https://doi.org/10.1016/j.jag.2021.102471.

Morrell, P. L., & Clegg, M. T. (2007). Genetic evidence for a second domestication of barley (*Hordeum vulgare*) east of the Fertile Crescent. *Proceedings of the National Academy of Sciences of the United States of America*, *104*(9), 3289–3294. https://doi.org/10.1073/pnas.0611377104.

Nleya, T., Clay, S. A., & Arinaitwe, U. (2025). Poor emergence of *Brassica* species in saline–sodic soil is improved by biochar addition. *Agronomy*, *15*(4), 811. https://doi.org/10.3390/agronomy15040811

Otto, R., Ferraz-Almeida, R., Sanches, G. M., Lisboa, I. P., & Cherubin, M. R. (2022). Nitrogen fertilizer consumption and nitrous oxide emissions associated with ethanol production—A national-scale comparison between Brazilian sugarcane and corn in the United States. *Journal of Cleaner Production*, *350*, 131482. https://doi.org/10.1016/j.jclepro.2022.131482.

Pardo, J., & VanBuren, R. (2021). Evolutionary innovations driving abiotic stress tolerance in C_4 grasses and cereals. *The Plant Cell*, *33*(11), 3391–3401. https://doi.org/10.1093/plcell/koab205.

Passias, A., Kleitsiotis, G., Tompris, I., Stavroulakis, E., Tsipas, E., Tsakalos, K.-A., Rallis, K., Fyrigos, I.-A., Pantazi, X. E., & Sirakoulis, G. C. (2024). A holistic approach to grapevine cultivation with precision viticulture. In *2024 IEEE 2nd Conference on AgriFood Electronics (CAFE)* (pp. 1–5). IEEE. https://doi.org/10.1109/CAFE63183.2024.11069354.

Perrier, X., De Langhe, E., Donohue, M., Lentfer, C., Vrydaghs, L., Bakry, F., Carreel, F., Hippolyte, I., Horry, J.-P., Jenny, C., Lebot, V., Risterucci, A.-M., Tomekpe, K., Doutrelepont, H., Ball, T., Manwaring, J., de Maret, P., & Denham, T. (2011). Multidisciplinary perspectives on banana (*Musa* spp.) domestication. *Proceedings of the National Academy of Sciences of the United States of America*, *108*(28), 11311–11318. https://doi.org/10.1073/pnas.1102001108.

Qin, H., Li, Y., & Huang, R. (2020). Advances and challenges in the breeding of salt-tolerant rice. *International Journal of Molecular Sciences*, *21*(21), 8385. https://doi.org/10.3390/ijms21218385.

Ramezanpour, M. R., & Farajpour, M. (2022). Application of artificial neural networks and genetic algorithm to predict and optimize greenhouse banana fruit yield through nitrogen, potassium and magnesium. *PLoS One*, *17*(2), e0264040. https://doi.org/10.1371/journal.pone.0264040.

Raun, W. R., Solie, J. B., Johnson, G. V., Stone, M. L., Mullen, R. W., Freeman, K. W., Thomason, W. E., & Lukina, E. V. (2002). Improving nitrogen use efficiency in cereal grain production with optical sensing and variable rate application. *Agronomy Journal*, *94*(4), 815–820. https://doi.org/10.2134/agronj2002.8150.

Sanches, G. M., Magalhães, P. S. G., Kolln, O. T., Otto, R., Rodrigues, F., Jr., Cardoso, T. F., Chagas, M. F., & Franco, H. C. J. (2021). Agronomic, economic, and environmental assessment of site-specific fertilizer management of Brazilian sugarcane fields. *Geoderma Regional*, *24*, e00360. https://doi.org/10.1016/j.geodrs.2021.e00360.

Selvaraj, M. G., Vergara, A., Montenegro, F., Ruiz, H. A., Safari, N., Raymaekers, D., Ocimati, W., Ntamwira, J., Tits, L., Omondi, A. B., & Blomme, G. (2020). Detection of banana plants and their major diseases through aerial images and machine learning methods: A case study in DR Congo and Republic of Benin. *ISPRS Journal of Photogrammetry and Remote Sensing*, *169*, 110–124. https://doi.org/10.1016/j.isprsjprs.2020.08.025.

Selvaraj, M. G., Vergara, A., Ruiz, H., Safari, N., Elayabalan, S., Ocimati, W., & Blomme, G. (2019). AI-powered banana diseases and pest detection. *Plant Methods*, *15*, 92. https://doi.org/10.1186/s13007-019-0475-z.

Shanmugapriya, P., Rathika, S., Ramesh, T., & Janaki, P. (2019). Applications of remote sensing in agriculture—A review. *International Journal of Current Microbiology and Applied Sciences*, *8*(1), 2270–2283. https://doi.org/10.20546/ijcmas.2019.801.238.

Shao, C., Shuai, Y., Wu, H., Deng, X., Zhang, X., & Xu, A. (2023). Development of a spectral index for the detection of yellow-flowering vegetation. *Remote Sensing*, *15*(7), 1725. https://doi.org/10.3390/rs15071725.

Sharma, A., Ceasar, S. A., Pandey, H., Devadas, V. S., Kesavan, A. K., Heisnam, P., Vashishth, A., Mishra, V., & Mall, A. K. (2025). Millets: Nutrient-rich and climate-resilient crops for sustainable agriculture and diverse culinary applications. *Journal of Food Composition and Analysis*, *137* (Pt B), 106984. https://doi.org/10.1016/j.jfca.2024.106984.

Sharma, P., Villegas-Diaz, R., & Fennell, A. (2024). Predicting grapevine physiological parameters using hyperspectral remote sensing integrated with hybrid convolutional neural network and ensemble stacked regression. *Remote Sensing*, *16*(14), 2626. https://doi.org/10.3390/rs16142626.

Shiferaw, B., Smale, M., Braun, H.-J., Duveiller, E., Reynolds, M., & Muricho, G. (2013). Crops that feed the world 10. Past successes and future challenges to the role played by wheat in global food security. *Food Security*, *5*, 291–317. https://doi.org/10.1007/s12571-013-0263-y.

Singh, A., Rudnick, D., Snow, D. D., Proctor, C., Puntel, L., & Iqbal, J. (2024). Impact of split nitrogen applications on nitrate leaching and maize yield in irrigated loamy sand soils of northeast Nebraska. *Agrosystems, Geosciences & Environment*, *7*(3), e20554. https://doi.org/10.1002/agg2.20554.

Singh, C., Randhawa, G. S., Farooque, A. A., Gill, Y. S., Fraser, A., Lokesh Kumar, K. M., Barrett, R., & Al-Mughrabi, K. (2025). AgriScout: AI-powered robot for precise detection of PVY-infected potato plants. *Computers and Electronics in Agriculture*, *238*, 110781. https://doi.org/10.1016/j.compag.2025.110781.

Singh, J., Chhabra, B., Raza, A., Yang, S. H., & Sandhu, K. S. (2023). Important wheat diseases in the US and their management in the 21st century. *Frontiers in Plant Science*, *13*, 1010191. https://doi.org/10.3389/fpls.2022.1010191.

Singh, P., Shahi, B., & Singh, K. M. (2016). Trends of pulses production, consumption and import in India: Current scenario and strategies. *SSRN Electronic Journal*. https://doi.org/10.2139/ssrn.2780102.

Soare, E., Bold, N., Stoicea, P., David, L., Dobre, C.-A., & Firățoiu, A.-R. (2023). Survey on the worldwide barley production and trade. *Bulgarian Journal of Agricultural Science*, *29*(Suppl 1), 119–124. https://www.agrojournal.org/29/01-14s.pdf.

Soluri, J. (2003). Banana cultures: Linking the production and consumption of export bananas, 1800–1980. In S. Striffler & M. Moberg (Eds.), *Banana wars: Power, production, and history in the Americas* (pp. 48–79). Duke University Press.

van Heerwaarden, J., Doebley, J., Briggs, W. H., Glaubitz, J. C., Goodman, M. M., de Jesus Sanches Gonzalez, J., & Ross-Ibarra, J. (2010). Genetic signals of origin, spread, and introgression in a large sample of maize landraces. *Proceedings of the National Academy of Sciences of the United States of America*, *108*(3), 1088–1092. https://doi.org/10.1073/pnas.1013011108.

Vellidis, G., Tucker, M., Perry, C., Kvien, C., & Bednarz, C. (2008). A real-time wireless smart sensor array for scheduling irrigation. *Computers and Electronics in Agriculture*, *61*(1), 44–50. https://doi.org/10.1016/j.compag.2007.05.009.

Vinayaka, Prasad, P. R. C., Avinash, G., Amaresh, Kumar, R. A., Murali, P., Palaniswami, C., & Govindaraj, P. (2025). Harnessing AI and remote sensing for precision sugar cane farming: Tackling water stress, salinity, and nitrogen challenges. *Frontiers in Agronomy*, *7*, 1681294. https://doi.org/10.3389/fagro.2025.1681294.

Volkaert, H. (2018). The origin, domestication and dispersal of bananas. In G. H. J. Kema & A. Drenth (Eds.), *Achieving sustainable cultivation of bananas: Volume 1. Cultivation techniques* (pp. 3–20). Burleigh Dodds Science Publishing.

Volkova, E., & Smolyaninova, N. (2024). Analysis of world trends in soybean production. In *IX International Scientific Conference on Agricultural Science 2024 "Current State, Problems and Prospects for the Development of Agricultural Science"* (Vol. *141*, pp. 01026). EDP Sciences.

Wairegi, L. W. I., van Asten, P. J. A., Tenywa, M., & Bekunda, M. (2009). Quantifying bunch weights of the East African Highland bananas (*Musa* spp. AAA-EA) using non-destructive field observations. *Scientia Horticulturae*, *121*(1), 63–72. https://doi.org/10.1016/j.scienta.2009.01.005.

Wilcox, J. R. (2004). World distribution and trade of soybean. In R. M. Shibles, J. E. Harper, R. F. Wilson, & R. C. Shoemaker (Eds.), *Soybeans: Improvement, production, and uses* (3rd ed., Vol. *16*, pp. 1–14). ASA, CSSA, SSSA.

Wortmann, C. S., & Eledu, C. S. (1999). *Uganda's agroecological zones: A guide for planners and policy makers*. Centro Internacional de Agricultura Tropical (CIAT).

Xu, J.-L., Gobrecht, A., Héran, D., Gorretta, N., Coque, M., Gowen, A. A., Bendoula, R., & Sun, D.-W. (2019). A polarized hyperspectral imaging system for *in vivo* detection: Multiple applications in sunflower leaf analysis. *Computers and Electronics in Agriculture*, *158*, 258–270. https://doi.org/10.1016/j.compag.2019.02.008.

Zhang, J., Wang, J., Zhu, C., Singh, R. P., & Chen, W. (2024). Chickpea: Its origin, distribution, nutrition, benefits, breeding, and symbiotic relationship with *Mesorhizobium* species. *Plants*, *13*(3), 429. https://doi.org/10.3390/plants13030429.

Zohary, D., & Hopf, M. (1988). *Domestication of plants in the Old World: The origin and spread of cultivated plants in West Asia, Europe and the Nile Valley*. Clarendon Press.

Instructor's Guide to Teaching with *Precision Agriculture Basics*

30

Katsutoshi Mizuta, Skye Brugler, Sravanthi Bachina, Sarah E. LaRose, and Nick Schiltz

Chapter Purpose

The *Precision Agriculture Basics* manual serves as an educational resource for students at all levels, Certified Crop Advisers (CCAs), Certified Professional Soil Scientists, and anyone wishing to learn more about digital agriculture. This chapter is intended to assist instructors and extension educators in a variety of settings, from teaching a full-length course on digital agriculture to organizing a workshop on precision agriculture technologies. This chapter also provides an overview of course planning using a "backward" design, instructional strategies to enhance student engagement, adapt content to various academic levels, and improve learning outcomes using supplementary teaching tools. With the addition of new chapters, revisions of the chapters from the first edition, and new videos, the second edition now contains more content than can typically be covered in a standard 15-week semester. This allows extension educators and instructors to selectively tailor their lessons to the specific needs of their students.

Backward Design

Before selecting the order of your course topics, it is important to (a) identify the goals for your course, (b) what students should be able to accomplish when they complete the class, and (c) the key takeaways or big ideas and concepts that they should grasp as a result of your course. Identifying these with precision (no pun intended) before creating your assignments, assessments, and course activities will ensure that all your learning activities are aligned with your desired outcomes. This approach to instructional design is called a backward design or "understanding by design" because this method intentionally focuses on ensuring students develop deep comprehension or understanding of the key ideas in the course (Wiggins & McTighe, 2005). The backward design is based on three steps: (1) identifying learning objectives, (2) planning assessments, and (3) planning learning activities. It is important to note that these steps should all align with each other.

The learning objectives planned in Step 1 should be assessed by the assignments or assessments in Step 2, and the planned learning activities in Step 3 should help the learners demonstrate their new skills in Step 2. Using a backward design provides a mechanism for aligning daily lesson plans with the course goals. This approach is focused on what truly matters.

Suggested Chapter Groupings

You can design your course flexibly by selecting chapters from each of the seven thematic sections:

1. Characterizing and managing system variability
2. Geographic information systems in precision agriculture
3. Proximal and remote sensing technologies for precision agriculture
4. Pest and disease management
5. On-farm data collection, management, and utilization
6. Precision agriculture as conservation
7. Process and trends in precision agriculture.

This modular structure allows you to tailor the curriculum to your specific learning objectives, whether you are focusing on foundational knowledge, advanced sensing technologies, crop management strategies, or data-driven decision-making. By mixing and matching chapters across sections, you can create a well-rounded course that aligns with the needs of your students or research focus. Instructors are encouraged to design their syllabi by selecting chapters that align with their students' academic levels, professional aspirations, and the overall structure of the course. This flexibility supports the development of both introductory and advanced tracks within a precision agriculture curriculum. If additional information is needed, do not hesitate to contact the chapter authors.

Instructional Resources and Pedagogical Tools

Each chapter begins with the purpose and concludes with sample questions that can be used to assess students' comprehension, stimulate classroom discussion, or serve as the basis for quizzes. The textbook also provides access to real-world datasets, such as those presented in Chapter 31, which can be used for hands-on exercises in data analysis and interpretation. These resources are particularly valuable for reinforcing lecture content, supporting student projects, and laying the groundwork for future research. They are designed to be adaptable, allowing instructors to integrate them into weekly lesson plans or use them as the foundation for independent student projects.

To further support instruction, accompanying videos have been developed for each chapter. These materials provide ready-to-use content that enhances instructional flexibility and promotes student engagement. A link to the repository containing these resources should be provided to ensure easy access.

Experiential Learning, Active Learning, and Student Engagement

Kolb (1984) argued that learning is the result of experiences. Experiences can be "hands-on," listening to a story, or watching a thought-provoking video. Regardless of the type of experience, it is important that the instructor provide opportunities for the learner to reflect on that experience, consider what the experience means in the bigger picture, and then tweak their approach to similar future experiences. For example, the instructor might set up a demonstration where students participate in precision and non-precision field scouting for pests. Students could be prompted to reflect on the pros and cons of their experiences and consider techniques to improve future scouting.

Using *Experiential Learning Theory* (Kolb, 1984) as a guide, instructors can build courses that create active learning environments that promote student engagement. According to Felder and Brent (2016), active learning is anything in a class session that students are called on to do other than watching, listening, and taking notes. This means planning activities that require students to *do* something other than passively listen to information. Some examples might include asking students to discuss a topic in small groups, analyze data, or diagnose why a piece of equipment is malfunctioning. The following section provides examples of active learning strategies that could be integrated into a workshop or classroom.

Examples of Active Learning Strategies in Precision Agriculture

Here are eight examples of active learning strategies:

- **Peer Teaching.** Instructors organize students into small groups of two to three members. Each group can be assigned—or allowed to choose—a chapter from the textbook to summarize and present to the class. This model encourages collaboration, enhances presentation skills, and deepens students' understanding of the material through active engagement.
- **Keyword Challenges.** In this approach, key terms are omitted from selected passages, and students are required to identify and insert the correct terms based on the context and corresponding textbook sections in assignments and examinations. This technique not only assesses comprehension but also encourages close reading and critical engagement with the text.
- **Case Studies.** Incorporating real-life case studies from peer-reviewed research and documented field trials can enhance the learning experience. Case studies offer empirical evidence that aligns with the chapter objectives, enabling students to bridge theoretical knowledge with practical applications. By analyzing real-world scenarios, students develop critical thinking skills, engage in hands-on research, and gain insights into the types of problems that they will face in the future.
- **Scenarios.** Very similar to case studies, but not as in-depth, a scenario puts the learner into the shoes of the decision-maker and asks them to make predictions, analyze situations, or ask further questions. Scenarios are typically only a few sentences to a paragraph or two long, whereas case studies are often much more in-depth and might be several pages long or be introduced through a video clip or farm visit.
- **Guest Speakers/Collaborators.** Inviting chapter contributors or local stakeholders—such as growers or crop consultants—to share their experiences can further ground the material in practical reality. Hearing first-hand accounts of how decisions are made in response to dynamic, day-to-day challenges (e.g., weather variability, labor and equipment availability, logistical constraints) helps students contextualize their learning. These interactions also highlight the unpredictable nature of farming.
- **Group or independent study.** The instructor separates the materials into small sections that can be explored through group or independent study. This is a common approach in laboratory practicals, which require identification of tools, parts, or specimens, or can allow students to extend their understanding of the content through the analysis of similar problems.
- **Simulations.** These are immersive learning activities that place the learner into a situation that closely mimics real life. There are increasingly more online computer-based simulations available to replicate decision-making on the farm or in the research lab, such as the online simulation game *Farmers 2050*.
- **Role-Play.** Closely aligned with simulations, role-play activities can place learners into situations where they can practice certain real-life skills in a controlled environment. For example, you might have students practice to be service technicians for farm equipment, while you or a guest serves as their supervisor or customer. The role-play simulates a real-life interaction between the role that the

student is playing and the problem. In role-playing, students learn how others respond to a common problem.

Integrating Course Language and Structure into Your Syllabus

To support instructors in course planning, this section provides sample language and structural guidance that can be adapted for use in syllabi or course outlines when teaching with *Precision Agriculture Basics*. Students entering a precision agriculture course often come from diverse academic disciplines and career trajectories. Some may be encountering the subject for the first time, while others may be building on prior coursework or field experience. A welcoming and inclusive course introduction can help set the tone for collaborative learning. For example:

> Welcome to (YOUR COURSE NUMBER)! This (#)-credit course includes one interactive lecture each week (delivered in person on Tuesdays with laboratory sessions on Thursdays). Whether this is your only exposure to precision agriculture or the beginning of a deeper academic journey, this course is designed to help you develop practical strategies to optimize agricultural and environmental outcomes through data-driven decision-making. The knowledge and skills you gain here will be valuable whether you plan to become a grower, crop consultant, GIS specialist, data scientist, or natural resource manager.

As part of setting the tone for the course, developing a syllabus that promotes a growth mindset and communicates a sense of belongingness can lead to increased student engagement and accomplishment with the course material (Taylor et al., 2019). A useful tool when updating or creating your course syllabus is the Social Justice Syllabus Design Tool (Taylor et al., 2019), which can help the instructor reflect on how best to build community, communicate the learning process, and build positive relationships.

One of the ways to build community with your students through the syllabus is to clearly convey how the course content is relevant to their lives and future careers, or how what they learn in class can be used to improve the lives of individuals or communities. A course in precision agriculture easily lends itself to practical connections in these ways.

Course Description

The following description can be adapted to reflect the course's alignment with the textbook:

> Precision agriculture is rapidly evolving in response to technological innovation and environmental challenges. It introduces a data-intensive, spatially explicit approach to optimizing inputs and practices for improved productivity, sustainability, and environmental stewardship. This course explores the principles and applications of precision agriculture through lectures, case studies, group discussions, and hands-on computer-based exercises. Students will gain practical experience in spatial analysis and develop a foundational understanding of the technologies and strategies that support agronomic innovations in managed natural resource systems.

Learning Objectives

The textbook supports a range of learning outcomes that can be tailored to the course level and focus. The following objectives reflect the core competencies students are expected to develop.

1. Define and discuss key concepts in precision management, including the following:
   a. Spatial variability in soil, water, and crop systems
   b. Core technologies such as Geographic Information Systems (GIS), Global Positioning Systems (GPS), Digital Elevation Models (DEMs), proximal and remote sensing tools, variable rate application machinery, and yield monitoring systems
   c. Practical applications of spatial data to enhance soil, water, and crop management
   d. Broader considerations that include environmental sustainability, socioeconomic impacts, and the role of digital marketing in precision agriculture
2. Compare the opportunities and limitations of precision management across geographic regions and agricultural systems

3. Develop practical skills to generate prescription maps tailored to specific management goals (e.g., fertilizer application).

Continued Growth as an Instructor

This chapter just barely scratches the surface of instructional design and how you can approach the design of your precision agriculture course or program. If you are new to teaching, or looking to brush up on your skills, we recommend exploring the professional development options available at your institution. You might have a center for instructional excellence that provides faculty, staff, and graduate students with support in revising or designing courses, or they might offer a professional development series on the basics of teaching. There might also be study groups within your college, department, profession, or local technical school that focus on the scholarship of teaching and learning, which could be a great resource. If you have an agricultural education program at your school, reach out to the teacher educators for ways you might partner with them not only to enhance your own professional growth as an educator, but also to provide future secondary agriculture teachers the opportunity to benefit from your technical expertise!

At the end of this chapter is a reference list that also serves as a resource list for you as you continue to expand your expertise as a precision agricultural educator.

Summary and Conclusion

By leveraging the book's structure, resources, and pedagogical tools, instructors can create impactful learning experiences that prepare students for the evolving demands of agriculture and natural resource management. The following chapter, *Chapter 31: Real-World Data Applications*, builds on this instructional foundation by introducing datasets and case studies that support hands-on learning and applied research in precision agriculture.

Study Questions

Understanding Key Concepts

1. What is the purpose of the *Precision Agriculture Basics*, and who is its intended audience?
2. Explain the concept of *backward design* in instructional planning. What are its three main steps?
3. Of the seven thematic sections of the textbook, identify how they can be used to customize the course for agronomy majors and for data science students?
4. What types of instructional resources are provided with each chapter to support teaching and learning?
5. What are some examples of *active learning strategies* mentioned in the chapter? Choose two and explain how they promote student engagement.

ACKNOWLEDGMENTS

Support for this document was provided by the Precision Farming Systems community in the American Society of Agronomy, South Dakota State University, USDA Natural Resources Conservation Service Conservation Innovation Grants (NR233A750011G019), USDA National Institute of Food and Agriculture under the Multi-State Hatch project accession no. 7008110, American Coalition for Ethanol, the South Dakota Corn Utilization Council and the South Dakota Soybean Research and Promotion Council. The use of trade names is for the convenience of the reader and does not imply endorsement by the authors.

REFERENCES

Felder, R. M., & Brent, R. (2016). *Teaching and learning STEM: A practical guide*. John Wiley & Sons.

Kolb, D. A. (1984). *Experiential learning: Experience as the source of learning and development*. Prentice Hall.

Taylor, S. D., Veri, M. J., Eliason, M., Hermoso, J. C. R., Bolter, N. D., & Van Olphen, J. E. (2019). The social justice syllabus design tool: A first step in doing social justice pedagogy. *Journal Committed to Social Change on Race and Ethnicity*, *5*(2), 133–166. https://doi.org/10.15763/issn.2642-2387.2019.5.2.132-166.

Wiggins, G. P., & McTighe, J. (2005). *Understanding by design* (2nd ed.). Association for Supervision & Curriculum Development.

Precision Agriculture Datasets

31

Liujun (Johnny) Li, Sravanthi Bachina, Viacheslav Adamchuk, David Bullock, Qianqian Du, Ivan Grijalva, Saurav Upadhyaya, Brian McCornack, and Brennan Lewis

Chapter Purpose

This chapter builds on Chapter 22 of the first edition of *Precision Agriculture Basics* (Fulton & Port, 2018) and *Practical Mathematics for Precision Farming* (Clay et al., 2018). The goals of this chapter are to provide (a) example datasets, (b) data or links to data that can be used in classrooms or workshops for learners with a range of skills, and (c) a discussion of different software options to develop technical skills in handling precision agriculture data. The datasets can be used by early learners who want to investigate how to create field trials and management zones and to understand the basic concepts in multilayer analysis of field-growth conditions. Advanced learners are provided with datasets for machine learning modeling using digital imagery. Readers will critically assess data quality, apply FAIR principles, use analytical tools, and recognize ethical considerations in data usage.

An Introduction to Geospatial Data

Geospatial data is information linked to a geographic location. Geospatial data typically combines location with attribute and temporal information. Geospatial data can be stored in many different formats and may have been developed using various quality control criteria; thus, understanding the limitations and accuracy of geospatial data is a key component in developing algorithms that improve management. For example, if the data were collected using inappropriate standards and quality control techniques, it may have little value in precision farming. Therefore, careful selection of input data sources and the most reliable methods for data processing, modeling, and interpretation is essential to achieve positive net revenue from adopting site-specific crop management. To meet these goals, data should be Findable, Accessible, Interoperable, and Reusable (FAIR).

Data curation in precision agriculture is critical for ensuring accuracy, reliability, and suitability for decision-making. Data curation is important for data you have collected or obtained from other sources, and it often involves collecting, cleaning, and annotating the data with metadata in accordance with FAIR principles. The curation process for data from other sources includes (a) ensuring that sufficient datasets are available in data repositories, (b) using the digital object identifiers (DOIs) to verify data authenticity and traceability, and (c) assessing the completeness, consistency, and accuracy of the datasets. In addition, the dataset should contain (a) detailed metadata schemas describing the data sources, collection methods, units, coordinate systems, and any transformations applied; and (b) notes specifying limitations, assumptions, and potential sources of bias. Within the dataset, data types (e.g., yield, soil, climate) should be clearly identified, and standardized tags and keywords should be assigned to facilitate data retrieval and interoperability.

Datasets should identify any ethical considerations, such as privacy concerns, data provider requests, potential biases, and suggestions for the practical application of the data. For example, they can be used in the classroom to create maps of yield, soil properties, or nutrient prescriptions. They also have potential usefulness in advanced research and in developing decision support tools. By adhering to data curation methodologies, precision agriculture datasets become valuable and reliable resources that support both academic excellence and practical advancements in sustainable farming.

Data Types, Standards, and FAIR Data Principles

There are many types of data, including tabular, raster, vector, and time series. For additional information on these data types, see Chapters 8 and 9. Standards can be divided into several groups, including Open Geospatial Consortium (OGC) Standards (GeoTIFF, GML, WMS/WFS), ISO 19115 (metadata for geospatial data), International Consortium for Agricultural Systems Applications (ICASA), Data Standards (agricultural field experimentation and crop modeling), AgGateway, and AgXML (interoperability in farm management systems).

When collecting, processing, and making information available to others, the FAIR data principles should be followed. The F in the FAIR acronym is for "Findable" and refers to persistent identifiers (DOIs, handles) indexed in agricultural data repositories, such as Ag DataCommons and GARDIAN. The A is for "Accessible" and refers to open licenses (CC BY, CC0) and machine-readable formats with application programming interfaces (APIs) or portals. The I is for "Interoperable" and refers to shared vocabularies/ontologies (Agrovoc, ENVO) and standardized formats (Geographic Tagged Image File Format [GeoTIFF], Network Common Data Form [NetCDF], Comma-separated Values [CSV]). Finally, the R is for "Reusable" and refers to clear documentation, usage rights/citations, and Machine Learning–Geographic Information Systems (ML/GIS)–friendly formats. Additional information on FAIR is available in Chapter 22 and Wilkinson et al. (2016).

Open-source Categorized Dataset Overview

There are many different data sources for soil, climate, remote sensing, and yield information. Information on such datasets is available in Tables 31.1–31.5. When using these datasets, it is important to understand the benefits and limitations associated with each information layer. For example, it may be different to relate county, state, and national yield databases to explain that occur within a field, local weather may be estimated from a coarse grid of weather stations, and soil information may be estimated based on a coarse grid of soil samples collected at some point in the past.

Table 31.1. Soil Characteristics.

Dataset	Source	Description	Format	Link
SoilGrids	ISRIC	Global soil properties (pH, OC, sand/silt/clay) at 250m resolution	GeoTIFF	https://soilgrids.org
Web Soil Survey	NRCS (USA)	County- and field-level soil maps at 10-30m resolution	Shapefile, PDF	https://websoilsurvey.sc.egov.usda.gov

Table 31.2. Selected Climate and Hydrology Source Links.

Dataset	Source	Description	Format	Link
NASA POWER	Global	Agro-climatic variables (temperature, solar, wind)	CSV, API	https://power.larc.nasa.gov
NOAA NCDC	USA	Precipitation, temperature extremes	CSV, NetCDF	https://www.ncdc.noaa.gov/cdo-web
CHIRPS	Global (Africa/Asia focus)	Precipitation estimates from satellite and weather station data	GeoTIFF	https://www.chc.ucsb.edu/data/chirps
GRACE	Global	Soil moisture and hydro anomalies	NetCDF	https://grace.jpl.nasa.gov/

Abbreviations: CHIRPS, Climate Hazards Group InfraRed Precipitation with Statin data; GRACE, Gravity Recovery & Climate Experiment; NASA POWER, National Aeronautics and Space Administration Prediction of Worldwide Energy Resources; NOAA NCDC, National Oceanic and Atmospheric Administration National Centers for Environmental Information.

Table 31.3. Selected Remote Sensing and Imagery Links.

Dataset	Source	Description	Format	Link
Skysat	Planet Lab	Multispectral imagery 0.5–3 m	GeoTIFF	https://www.planet.com/products/high-resolution-satellite-imagery/
Sentinel-2	ESA	Multispectral imagery, 10–60 m	GeoTIFF	https://eos.com/landviewer/
Landsat	USGS	Multispectral, historical archive	GeoTIFF	https://earthexplorer.usgs.gov
Radiant Earth MLHub	Global	ML-ready satellite data	TIFF, CSV	https://source.coop
UAV-based datasets	Various	High-res agriculture field imagery	JPG, TIFF	https://www.plant-phenotyping.org

Abbreviations: ESA, European Space Agency; USGS, U.S. Geological Survey.

Table 31.4. Selected Yield Data Source Links.

Dataset	Source	Description	Format	Link
USDA NASS	USA	National yield statistics	CSV, API	https://www.nass.usda.gov
ICASA	Global	Field-scale crop modeling data	Text, CSV	https://github.com/DSSAT/ICASA-Dictionary
AgMIP	Global	Multilocation agronomic trials	ZIP, JSON	https://agmip.org

Abbreviations: AgMIP, Agricultural Model Intercomparison and Improvement Project; NASS, National Agricultural Statistics Service.

Case Study Dataset for Early Learners

This dataset has been used in several studies, including Chang et al. (2003, 2004, 2005) and Clay et al. (1997). It contains yield-monitoring and soil-test data collected over a 20-year period. The yield data were cleaned and compared with hand-collected information, and the site's rotation was corn followed by soybean. The field size was 65 ha, and soil samples were collected on a 26 × 26 point grid and analyzed at a commercial laboratory. This dataset can be used by readers to calculate management zones, as well as means

Table 31.5. Selected Links for Machine Learning–ready Datasets.

Dataset	Source	Description	Format	Link
DeepWeeds	Australia	UAV imagery for weed detection	PNG, labels	https://github.com/AlexOlsen/DeepWeeds
Agriculture-Vision dataset	USA	Large-scale aerial agricultural image dataset		https://www.agriculture-vision.com/agriculture-vision-2021/dataset-2021
PIC4SeR	Synthetic	Labeled datasets for precision agriculture robot navigation simulation	Images, 3D models	https://sites.google.com/view/datasetpic4ser/home-page?utm_source=chatgpt.com
AgML	Project AgML	Crop type, segmentation, time-series	PNG, CSV, JSON	https://github.com/Project-AgML/AgML

Abbreviation: 3D, three-dimensional.

and medians. See Clay et al. (2018) for additional information on these calculations. The data layer and description are summarized in Table 31.6. This dataset can be accessed at Final_Version_Moody_Combined_with_GPS.xlsx.

Table 31.6. Moody Field Dataset Overview.

Data layer	Year	Data description
Moody 1995 Soil Test Info 0–6 in	1995	It contains longitude, altitude, elevation (m), Olsen P (ppm), K (ppm), pH (1:1), zinc (ppm), NO_3 (ppm), NH_4 (ppm).
Moody 1997 0–6 in pre-plant	1995	It contains longitude, altitude, elevation (m), Olsen P (ppm), K (ppm), pH (1:1), zinc (ppm).
2000 0–6 in pre-plant	2019	It contains longitude, altitude, elevation (m).
Moody Oct_20_2000 EM38	2018	It contains longitude, altitude, elevation (m), EC shallow (dS/m), EC deep (dS/m).
Moody_Corn_(bu_ac)_15.5_1995	2017	It contains longitude, altitude, elevation (m), Corn dry yield (bu/ac) at 15.5% moisture.
Soybeans_(bu_ac)_13_1996	2014	It contains longitude, altitude, elevation (m), Soybean dry yield (bu/ac) at 13% moisture.
Moody_Corn_(bu_ac)_15.5_1997	2020	It contains longitude, altitude, elevation (m), Corn dry yield (bu/ac) at 15.5% moisture.
Soybeans_(bu_ac)_13_1998	2014	It contains longitude, altitude, elevation (m), Soybean dry yield (bu/ac) at 13% moisture.
Moody_Corn_(bu_ac)_15.5_1999	2013	It contains longitude, altitude, elevation (m), Corn dry yield (bu/ac) at 15.5% moisture.

Abbreviation: EC, electrical conductivity.

Case Study for Intermediate Learners

This dataset is characterized as Field 26 at the Macdonald Campus Farm, McGill University (Sainte-Anne-de-Bellevue, Quebec, Canada). The crop rotation includes corn (grain or silage), soybean, and an alfalfa/grass mix; the field size is 11 ha (27 ac) (Adamchuk et al., 2020). The data types and formats are shown in Table 31.7. This list includes a range of formats typically used in precision agriculture. The link to the other training dataset is https://sites.google.com/site/macgeodata. The data types and formats are listed in Table 31.7.

The potential exercises include the following (Adamchuk et al., 2020):

1. Create a field boundary using the Geographic Identifier(GID) data layer. This exercise allows users to use Quantum GIS (QGIS) and plugins and convert GPS tracking points from a text file into a field boundary polygon.
2. Create arbitrary sampling points. This exercise involves manually adding sampling locations to a map using Google Maps and coordinate digitization tools.

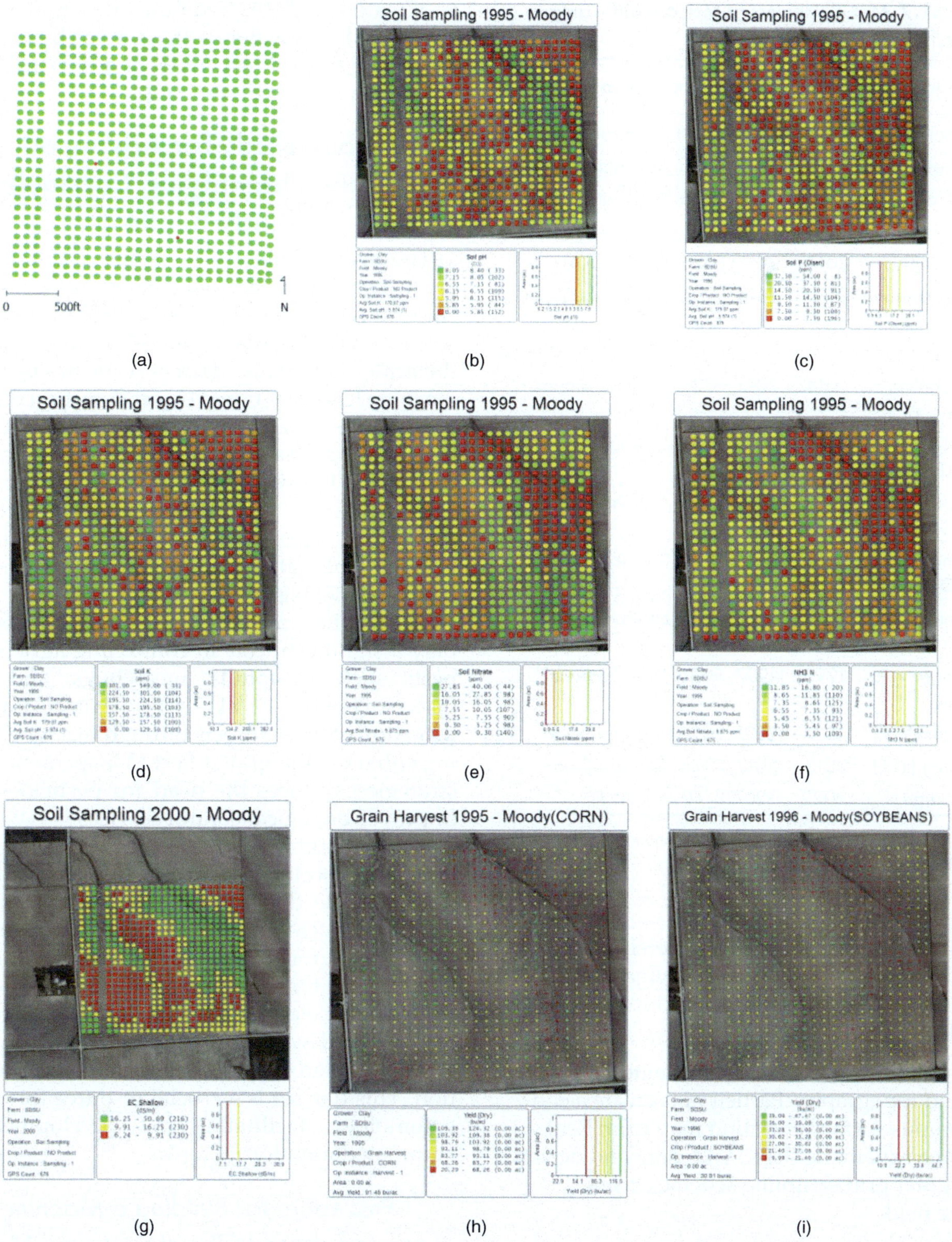

Figure 31.1. Datasets from the Moody field in South Dakota (a) 26 × 26 point-sampling scheme, (b) soil pH from 1995 preplant top 6-inch soil sampling, (c) soil Olsen P (ppm) from 1995 preplant top 6-inch soil sampling, (d) soil K (ppm) from 1995 preplant top 6-inch soil sampling, (e) soil nitrate NO_3-N (ppm) from 1995 preplant top 6-inch soil sampling, (f) soil ammonia nitrogen (NH_3-N; ppm) measurement from 1995 preplant top 6-inch soil sampling, (g) shallow electrical conductivity (EC) measurement 2000 EM mapping, (h) 1995 corn yield map, and (i) 1996 soybean yield map.

Table 31.7. McGill University Macdonald Campus Research Farm F026 Dataset.

Data layer	Year	Data description
Field boundary	2025	2025_F026_Boundary.kml
Field boundary	2010	2010_F026_Boundary_shp.zip
Yield (corn)	2019	2019_F026_YieldCorn.txt
Yield (soybean)	2018	2018_F026_YieldSoybeans.txt
Yield (corn)	2017	2017_F026_YieldCorn.txt
Yield (soybean)	2014	2014_F026_YieldSoybeans.txt
Soil sampling analysis	2020	2020_F026_SoilSamgling.txt
Soil sampling analysis	2014	2014_F026_SoilSamgling.txt
Soil EC_a (DUALEM)	2013	2013_F026_SoilOgtix.txt
Gamma-ray (Soil Optix)	2024	2024_F026_SoilOptix.txt
Gamma-ray (Soil Optix)	2013	2013_F026_SoilOptix.txt
Soil series survey	1971	1971_F026_SoilSurvey.kml
Drainage installed	1997	1997_F026_Drainage.kmz
Elevation (RTK)	2021	2021_F026_Togograghy.txt
Elevation (RTK)	2013	2013_F026_Togograghy.txt
UAV image	2022	220718_F026_UAV.tif

Abbreviations: EC_a, soil apparent electrical conductivity; RTK, Real-Time Kinematic.

3. Create a yield map. In this exercise, users import yield monitor data to create a yield raster map using inverse distance weighting (IDW) interpolation,
4. Establish a grid sampling scheme over the field topography. Participants create a continuous surface elevation map from Real-Time Kinematic (RTK) GPS data and design a regular grid sampling scheme across the field.
5. Create a soil property map. This exercise involves creating a continuous surface soil property map through kriging interpolation using soil analysis point data.
6. Create K_2O, P_2O_5, and lime prescription maps. Users generate these maps by reclassifying the soil test values and filtering the resulting data (Figure 31.2).

Case Study for Advanced Learners: Identifying Coccinellids (Lady Beetles)

The primary objectives of this case study are (a) to provide a comprehensive dataset of coccinellids, commonly known as lady beetles, captured in wheat fields (Figure 31.3a and b); and to illustrate how this dataset can be used for machine learning–based object detection models. See Chapter 17 and Grijalva et al. (2024a, 2024b) for more information. Each image has an associated label file containing bounding boxes and class labels for coccinellids, enabling efficient training of object detection models. The dataset is accessible at https://data.mendeley.com/datasets/j9735xjspw/1.

Coccinellids are vital beneficial insects that play a significant role in controlling agricultural pests such as aphids, mealybugs, and scale insects Hewlett et al. (2019). Automating the detection of these insects through machine learning enhances Integrated Pest Management (IPM) strategies, reduces the need for pesticides, and improves biological control services.

The dataset consists of 2,133 high-quality images of coccinellids on wheat heads, collected using mobile devices. Each image is standardized to 640 × 640 pixels in the red–green–blue (RGB) color format and is provided in the You Only Look Once (YOLO) format with corresponding labeled files. The images were captured in replicated wheat subplots, focusing on clear visibility of lady beetles to ensure accurate detection. This dataset is a crucial resource for digital entomology, facilitating the development of automated monitoring tools.

Procedure for Building a Machine Learning–based Object Detection Model

Download the dataset that includes the images and their corresponding YOLO annotation files. Check the annotations and fine-tune them using

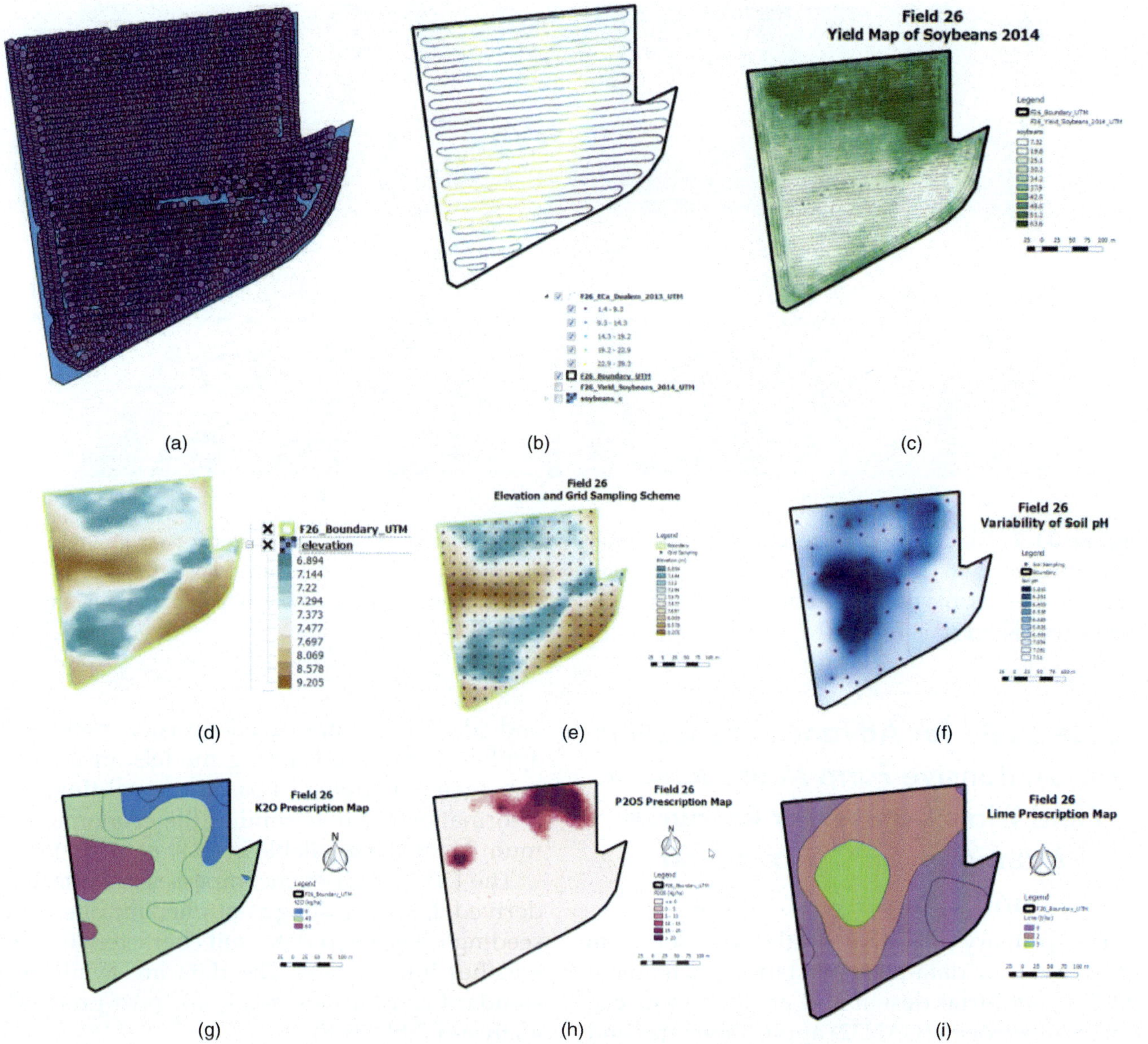

Figure 31.2. Hands-on lab exercise based on the McGill University Macdonald Campus Research Farm F026 dataset (a) field boundary and raw yield data map (data source: 2014_F026_YieldSoybeans, (b) soil apparent electrical conductivity (EC_a) map of Field 26 in 2013, (c) 2014 soybean yield map, (d) 2021 elevation map of Field 26, (e) grid sampling map, (f) soil pH map, (g) K_2O prescription map, (h) P_2O_5 precision map, and (i) lime prescription map.

Roboflow. This platform offers user-friendly annotation features, including bounding boxes and polygons for precise object detection. Build the model on a local PC. A detailed guide and code repository are available on GitHub (https://github.com/SauravUpadhyaya/wheat-coccinellid-detection-project). Users can follow the instructions to set up the environment in Visual Studio Code and train the YOLOv8m model on a local PC equipped with a compatible graphics processing unit(GPU) for optimal performance.

Models can also be developed on Google Colab. For users preferring a cloud-based solution, the project is also available on Google Colab (https://github.com/SauravUpadhyaya/wheat-coccinellid-detection-colab). This option provides an accessible platform for model training without requiring high-end local hardware.

(a) (b)

Figure 31.3. Object detection of coccinellids on wheat heads: (a) image example, (b) detection result. These results indicate a strong balance between precision and recall, confirming the model's reliability in detecting coccinellids. Example detection results, including original images and model outputs with highlighted bounding boxes, are illustrated in Figure 31.1.

Case study for Advanced Learners: Data-intensive Farm Management Field Trials Dataset for Resource Management Optimization and Economy Analysis

The primary objective of this dataset is to showcase how data-intensive farm management (DIFM) field-trial design and analysis tools can optimize nitrogen (UAN32) application rates and seeding rates, thereby enhancing crop yield and farm profitability while minimizing environmental impacts. By leveraging precision agriculture techniques, the dataset provides comprehensive insights into the effects of variable nitrogen and seeding rates on yield and economic returns. Employing the DIFM on-farm precision implementation tool facilitates data-driven decision-making to optimize nitrogen application through (a) optimizing nitrogen application rates by determining both uniform and site-specific rates to maximize yield and profitability; (b) enhancing site-specific management; (c) promoting the use of field variability data, such as soil type, slope, and elevation to refine nutrient strategies; and (d) using data-driven decision making by applying machine learning models for adaptive management strategies based on real-time data. Information for determining the economic optimum N rates is available in Clay et al. (2018).

The DIFM dataset encompasses raw machine-derived data, including as-applied nitrogen rates, seeding rates, yield data, soil characteristics, and weather information. The data are available in standardized formats, ready for post-processing analysis (Table 31.8).

The dataset pertains to an 80-ac field ("L" field; 32.4 ha) with plots ranging from 204 to 250 ft (62.2-76.2 m) in length and 60 ft (18.3 m) in width. The planter width is 60 ft (18.3 m). The harvester width is 30 ft (9.14 m). The experimental design is based on a Latin square trial. The targeted seeding rates are 24,000 seeds/ac (59,300 seeds/ha), 30,000 seeds/ac (74,100 seeds/ha), 36,000 (99,000 seeds/ha), 40,000 seeds/ac (98,800 seeds/ha), and 44,000 seeds/ac (108,700 seeds/ha) based on the producer's status quo strategy of 36,000 seeds/ac). The price of corn is \$5.00/bu (\$197/Mg) at 15.5% moisture, and the price of seed is \$4.00/1,000 seeds. Liquid fertilizer was applied at a rate between 24 and 31 gal/ac (224–299 L/ha)

Table 31.8. Data Intensive Farm Management (DIFM) Data Overview and Description.

Data layer	Year	Data description
boundary.shp	2021	Field boundary file
ab-lines-farmer-n-2021	2021	AB line for planting
ssurgo.shp	2021	Complete SSURGO soil data
ec.shp	2021	EM38 EC mapping
trial-deisgn-n.shp	2021	Designed nitrogen rate field trial design
as-applied-n.shp	2021	Actual nitrogen rate from machine
trial-deisgn-seedrate.shp	2021	Designed seeding rate field trial design
as-applied-seedrate.shp	2021	Actual seeding rate from machine
Sentinel2-07_04_2021.shp	2021	10-m satellite remote sensing imagery for in-season monitoring
yield.shp	2021	Corn yield data in bu/ac at 15.5% moisture

Abbreviation: SSURGO, Soil Survey Geographic Database.

of UAN32 (32-0-0). The density of UAN32 was 11.06 lb/gal.

The variable-rate planter and yield monitor provided raw as-planted and harvest data. An initial cleaning removed observations with extreme yield or as-planted rates ("outliers") from the raw data. Points were also removed from the headlands, where the data are less reliable due to differences in variables such as sun exposure, changes in driving speed, and potential application overlaps. The yield points were grouped into polygons based on the distances between points, swath width, and the headings recorded in the raw yield data. Subplots were created by grouping contiguous yield polygons with similar seed rates into sets of four. Details of the data processing are documented by Edge et al. (2024).

Common Software Platform for Precision Agriculture Data Analysis

Precision agriculture thrives on the integration of spatial, temporal, and agronomic data. A variety of software platforms—ranging from open-source GIS tools to commercial farm management systems—support the analysis and visualization of these datasets. Below is a detailed overview of key tools, followed by a comparative analysis. Selected commonly used GIS programs are discussed in Table 31.9. Additional information on GIS is provided in Chapters 8 and 9.

The choice of software in precision agriculture depends on the user's goals, technical expertise, and budget. Open-source tools, such as QGIS, offer flexibility and cost savings, while commercial platforms, such as Ag Leader SMS and FieldView, provide streamlined, agriculture-specific workflows. Climate Engine adds a valuable dimension by integrating environmental data into decision-making. A hybrid approach—leveraging multiple tools—often yields the most comprehensive insights.

Table 31.9. Comparative Summary of Common Software Platforms.

Software	Type	Strengths	Limitations	Best for
QGIS	Open-source	Free, flexible, plugin-rich	Steep learning curve, limited agricultural support	Researchers, GIS-savvy agronomists
ArcGIS	Commercial	Advanced analytics, enterprise-ready	Expensive, complex	Large farms, consultants
Ag leader SMS	Commercial	Equipment integration, agronomic tools	Proprietary, costly	Precision agriculture professionals
FieldView	Commercial	Real-time data, user-friendly	Subscription, limited analytics	Farmers, crop advisors
Climate engine	Web-based	Satellite/climate data, no installation needed	Internet-dependent, less field-specific	Researchers, extension agents

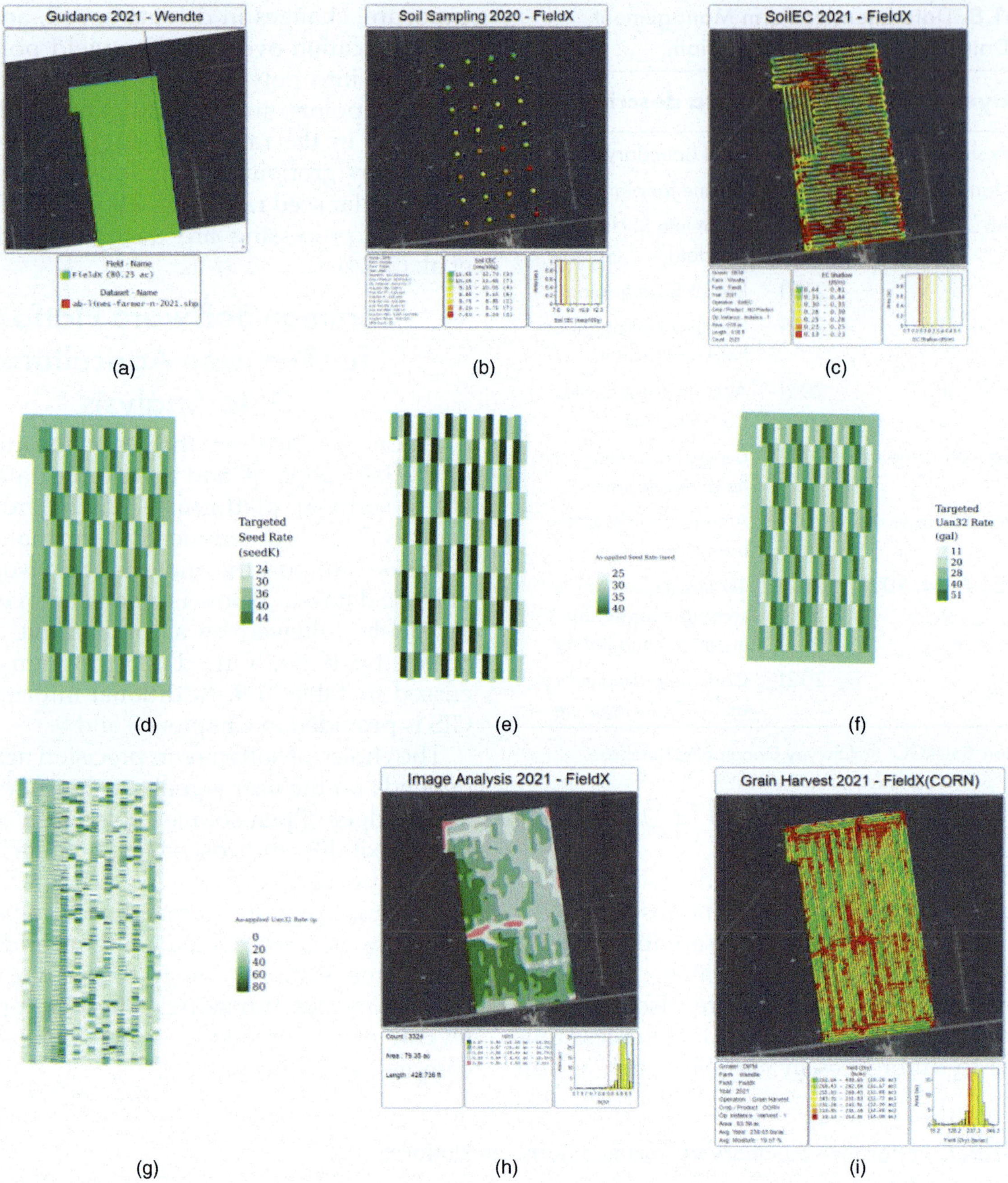

Figure 31.4. Hands-on lab exercise based on the DIFM on-farm field trial dataset: (a) field boundary and AB line, (b). Soil Survey Geographic Database (SSURGO) soil EC data, (c) EC mapping in shallow soil, (d) nitrogen trial design, (e) as-applied nitrogen map, (f) seed rate trial design, (g) as-applied seed rate map, (h) 10 m resolution remote sensing imagery from Sentinel 2, and (i) corn harvested yield map 2021.

Conclusion and Future Outlook

This chapter provided an introduction to geospatial data, FAIR principles, links to datasets, and case studies for early, intermediate, and advanced learners, as well as software commonly used to process datasets. It offered tips for merging tabular and spatial datasets and included classroom exercises. The chapter also featured QGIS lab exercise, Google Colab to support practical learning. In conclusion, the chapter emphasized the growing need for standardized datasets and the promotion of global data sharing. It highlighted the increasing role of synthetic datasets and simulations in modeling rare events.

ACKNOWLEDGMENTS

The authors express their gratitude to Jiyul Chang for sharing the data in Section 31.4; Macdonald Campus Farm, McGill University (Sainte-Anne-de-Bellevue, Quebec, Canada) for sharing their datasets. This chapter has been approved by the Director of the Louisiana Agricultural Experiment Station as manuscript no. 2025-234-40431. We express sincere thanks to the Data-intensive Farm Management (DIFM) project (USDA-NRCS award no. NR213A7500013G021), as well as to Brittani Edge and Scott Wahl for their data preparation and cleaning, coding development and statistical analysis. We extend special acknowledgment to Wendte Farms Ltd. for generously sharing the complete DIFM on-farm trial datasets. The authors express their gratitude to the institutions and individuals who contributed to the dataset collection, model development, and manuscript preparation.

REFERENCES

Adamchuk, V. I., Huang, H. H., & Marmette, M.-C. (2020). *Tutorial on application of QGIS software for precision agriculture*. https://adamchukpa.mcgill.ca/qgis.

Chang, J., Clay, D. E., Carlson, C. G., Reese, C. L., Clay, S. A., & Ellsbury, M. M. (2004). Defining yield goals and management zones to minimize nitrogen and phosphorus fertilizer recommendations errors. *Agronomy Journal*, *96*(1), 825–831. https://doi.org/10.2134/agronj2004.0825.

Chang, J., Clay, D. E., Dalsted, K., Clay, S. A., & O'Neill, M. (2003). Corn (*Zea mays* L.) Yield predictions using multispectral and multidate reflectance. *Agronomy Journal*, *95*(6), 1447–1453. https://doi.org/10.2134/agronj2003.1447.

Chang, J., Clay, S. A., Clay, D. E., Aaron, D., Helder, H., & Dalsted, K. (2005). Clouds influence precision and accuracy of ground-based spectroradiometers. *Communication in Soil Science and Plant Analysis*, *36*(13–14), 1799–1807. https://doi.org/10.1081/CSS-200062449.

Clay, D. E., Chang, J., Clay, S. A., Ellsbury, M., Carson, C. G., Malo, D. D., Woodson, D., & DeSutter, T. (1997). Field scale variability of nitrogen and $\delta^{15}N$ in soil and plants. *Communication in Soil Science and Plant Analysis*, *28*(17–18), 1513–1527. https://doi.org/10.1080/00103629709369893.

Clay, D. E., Clay, S. A., & Bruggeman, S. A. (2018). *Practical mathematics for precision farming*. ASA, CSA, SSSA.

Edge, B., Mieno, T., & Bullock, D. S. (2024). *Processing of on-farm precision experiment data in the DIFM project*. OSF Preprints n9rgu, Center for Open Science. https://osf.io/preprints/osf/n9rgu_v1.

Fulton, J. P., & Port, K. (2018). Precision agriculture data management. In D. K. Shannon, D. E. Clay, & N. R. Kitchen (Eds.), *Precision agricultural basics* (pp. 169–187). ASA, CSSA, SSSA.

Grijalva, I., Clark, N., Hamilton, E., Orpin, C., Perez, C., Schaefer, J., Vogts, K., & McCornack, B. (2024a). Comprehensive wheat coccinellid detection dataset: Essential resource for digital entomology. *Data in Brief*, *55*, 110585. https://doi.org/10.1016/j.dib.2024.110585.

Grijalva, I., Clark, N., Hamilton, E., Orpin, C., Perez, C., Schaefer, J., Vogts, K., & McCornack, B. (2024b). Wheat_coccinellid, *Mendeley Data, V1*. https://doi.org/10.17632/j9735xjspw.1.

Hewlett, J. A., Szczepaniec, A., & Eubanks, M. D. (2019). The effects of sugarcane aphid density in sorghum on predation by lady beetles and lacewings. *Biological Control*, *129*, 171–177. https://doi.org/10.1016/j.biocontrol.2018.10.015.

Wilkinson, M. D., Dumontier, M., Jan Aalbersberg, I., Appleton, G., Axton, M., Baak, A., Blomberg, N., Boiten, J.-W., da Silva Santos, L. B., Bourne, P. E., Bouwman, J., Brookes, A. J., Clark, T., Crosas, M., Dillo, I., Dumon, O., Edmunds, S., Evelo, C. T., Finkers, R., . . . Mons, B. (2016). The FAIR guiding principles for scientific data management and stewardship. *Scientific Data*, *3*, 160018. https://doi.org/10.1038/sdata.2016.18.

Soil and Plant Calculations

32

Brennan Lewis, Dwarika Bhattarai, Skye Brugler, Shailesh Pandit, Nowrin Mostafa Tuly, Janet Moriles-Miller, Sravanthi Bachina, Norah Ochonye, Varshith Kommineni, Rojina Kafle, Graig Reicks, and Deepak Joshi

Chapter Purpose

This chapter builds on *Practical Mathematics for Precision Farming* (Clay et al., 2018) and contains some of the most useful calculations used in agronomy. This chapter contains selected calculations related to fertilizer rates, estimating cation exchange capacity, estimating yields at different moisture contents, estimating carbon and nitrogen mineralization, determining bulk density, calculating plant-available water, calculating economic optimum fertilizer and seeding rates, using remote sensing to estimate crop responses, and sprayer calculations.

Soil Test and Nitrogen, Phosphorus, and Potassium Levels

After a soil sample is collected (Chapter 6), it is delivered to an appropriate laboratory in accordance with the soil testing lab's instructions. It is important to ensure that the laboratory conducts the state-specific chemical analysis prior to submitting the sample for analysis (Lyons et al., 2023). Another important thing to remember is that different laboratories may use slightly different soil extractant methods, which can influence the resulting values. In addition to soil nutrient levels, the soil test reports provide information on soil texture, soil pH, electrical conductivity (EC), and an estimate of the soil's cation exchange capacity (CEC; Table 32.1).

In most soil reports, soil nutrient values are expressed in parts per million, indicating the number of units of interest per 1,000,000 units of sample. Parts per million can also be reported as milligrams per kilogram or pounds per million pounds. For example, if a soil sample contains 20 ppm nitrate-N (NO_3-N), it means it has 20 lb NO_3-N/1,000,000 lb of soil.

When converting one unit to another, it is important to multiply by "1." For example, 1,000 g/1 kg = 1 because 1 kg contains 1,000 g.

Table 32.1. Hypothetical Soil Test Laboratory Report from a Submitted Sample.

Sample ID	Depth	SOM	EC or soluble salts (1:1)	Nitrate-N	P	K	Ca	Mg	Na	Sum of cation
	inches	%	mmho/cm	lb/ac	ppm	ppm	ppm	ppm	ppm	meq/100 g
2250	0–6	4.6	0.53	47	21	1,066	2,261	218	30	17.1
2250	6–24	1.9	0.85	20	17	621				

Note: For this analysis, soil samples were collected from two depths (0–6 and 6–24 inches). Each sample consisted of a composite collected from a whole field, point, grid point, grid cell, or management zone. In this case, each composite sample consisted of 10 individual cores that were combined to produce one sample. The findings may be reported in different units.
Abbreviations: Ca, calcium; ID, identifier; K, potassium; Mg, magnesium; Na, sodium; N, nitrogen; P, phosphorus; SOM, soil organic matter.

QUESTION 32.1

A soil sample report indicates that the 0- to 6- and 6- to 24-inch soil depths contain 12 and 16 ppm NO_3-N, respectively. How many pounds of NO_3-N are contained in this soil? Assume that each 6 inches of soil contains 2 million lb and that 18 inches contains 6 million lb of soil.

Answer:

$$\frac{12\,lb}{1{,}000{,}000\,lb\,soil} \times \frac{2{,}000{,}000\,lb}{6\,inches} = \frac{24\,lb}{6\,inches}$$

$$\frac{16\,lb}{1{,}000{,}000\,lb\,soil} \times \frac{6{,}000{,}000\,lb}{18\,inches} = \frac{96\,lb}{18\,inches}$$

Adding the NO_3-N content in these two depths results in 120 lb.

It is also important to note that when written as NO_3-N, only the nitrogen (N) is considered. When written as NO_3, the entire molecule is considered. Nitrate is converted to NO_3-N by multiplying NO_3 by 14/62, where 14 is the molecular weight of N, and 62 is the molecular weight of NO_3. The reported N value does not include soil ammonium (NH_4^+).

To convert 10 ppm NO_3 to NO_3-N, write the parts-per-million value as milligrams per kilogram and use the following calculation,

$$\frac{10\,mg\,NO_3}{kg\,soil} \times \frac{g\,NO_3}{1{,}000\,mg} \times \frac{1\,mole\,NO_3}{62\,g\,NO_3} \times \frac{1\,mole\,NO_3-N}{1\,mole\,NO_3} \times \frac{14\,g\,N}{1\,mole\,NO_3-N} \times \frac{1{,}000\,mg}{kg\,soil} = 2.26\,ppm\,NO_3{-}N$$

In this unit conversion, 10 ppm was multiplied by a series of "1"s. For example, 1 mol NO_3/62 g NO_3 = 1.

Most soil samples are not analyzed for total organic N; however, total organic N can be estimated from the soil organic carbon (SOC) content, which is obtained by multiplying soil organic matter (SOM) by 0.58. The total N is then estimated by multiplying SOC by 0.10. The amount of N that may be mineralized is about 2% of the total organic N value (Joshi et al., 2025). For example, if the soil contains 2% SOM, then the surface SOC is approximately 40,000 lb SOC/ac (i.e., $0.02 \times 2{,}000{,}000$ lb/ac). This calculation assumes that 6 inches of soil contains 2,000,000 lb of soil/ac. If 2% of the carbon is mineralized annually, then multiply 4,000 lb of soil organic N by 0.02 to determine that the amount of mineralized N is approximately 80 lb N/ac (40,000 lb SOC/ac × 0.02 g mineralized N/lb of soil organic N) from the surface 6 inches (Joshi et al., 2025).

One common use of soil test results is to determine the amount of fertilizer to apply. The nutrient recommendation is determined using the appropriate algorithm. For example, if the N algorithm is N rate = 1.2 × yield goal NO_3-N, then the NO_3-N used in the equation is determined by the laboratory. Based on the nutrient recommendations, the amount of fertilizer is determined. The fertilizer application rate is then calculated based on the fertilizer grade. Fertilizer grades are expressed as the percentage of N, P_2O_5, and K_2O contained in the fertilizer. For example, a fertilizer with a grade of 20-5-10 contains 20% N, 5% P_2O_5, and 10% K_2O. However, fertilizers can also contain secondary nutrients. For example, ammonium sulfate [$(NH_4)_2SO_4$] has a fertilizer grade of 21-0-0-24 and contains 21% N and 24% S.

QUESTION 32.2

The surface soil (0–6 inches) contains 5% organic matter. Estimate the amount of N that will be available during the growing season if 2% of the N in the organic matter is mineralized. Assume the soil weighs 2,000,000 lb.

Answer:

$$\frac{0.05\,lb\,organic\ matter}{lb\,soil} \times \frac{0.58\,lb\,C}{1\,lb\,organic\ matter} \times \frac{2{,}000{,}000\,lb\,soil}{ac} = \frac{58{,}000\,lb\,C}{ac}$$

Assuming that every 10 lb of C in the soil contains 1 lb of total N, 58,000 lb C/ac soil will contain 5,800 lb total N.

Assuming that 2% of the N in the soil organic matter is mineralized, then the N available during the growing season is 0.02 × 5,800 = 116 lb N.

QUESTION 32.3

How much monoammonium phosphate (MAP; 11-52-0) and anhydrous ammonia (82-0-0) should be applied if the recommendation is 80 lb P_2O_5 and 90 lb N/ac?

Answer:

$$\frac{80\,lb\,P_2O_5}{ac} \times \frac{1\,lb\,MAP}{0.52\,lb\,P_2O_5} = \frac{153.8\,lb\,MAP}{ac}$$

$$\frac{153.8\,lb\,MAP}{1\,ac} \times \frac{0.11\,lb\,N}{1\,lb_{MAP}} = \frac{16.9\,lb\,N}{ac}$$

$$90\,lb\,N - 16.9\,lb\,N = 73.1\,lb\,N$$

$$\frac{73.1\,lb\,N}{1\,ac} \times \frac{1\,lb\,Anhydrous\ ammonia}{0.82\,lb\,N} = \frac{89.1\,lb\,Anhydrous\ ammonia}{1\,ac}$$

Hence, 153.8 lb of MAP and 89.1 lb of anhydrous ammonia should be applied per acre.

Soil pH, and EC

Important information about nutrient availability, as well as salt concentration in the soil, is also provided in soil test results, such as soil pH and EC tests. These parameters can help in making important management decisions. pH is the negative logarithm of the hydrogen ion concentration ($pH = -\log[H^+]$), conversely, pOH is the negative logarithm of the hydroxyl ion concentration ($pOH = -\log[OH-]$). The availability of many nutrients depends on pH, and different plants perform better at different pH values. For example, corn and soybean are well suited to soils with a pH of 6.5, whereas blueberries prefer a lower pH. The pH also influences the solubility of many pesticides, which is determined by their pKa values. The pKa can be calculated as pKa = −log(Ka), where Ka is the dissociation constant. When the Ka value is very small, we can use the equation $pH = \frac{pKa + pHA}{2}$, with the pHA as the pH of the total weak acid solution calculated as pHA = −log(HA). Understanding how the pH of the soil interacts with the pKa of an agrichemical, such as N fertilizer, can be used to reduce losses. For example, the pKa of NH_4^+ is 9.26. This means that if urea fertilizer is applied to an extremely alkaline soil with a pH higher than 9.26, the expected hydrolysis reaction of the urea fertilizer into the product NH_4^+ ($2NH_3 + CO_2 + H^+ \rightarrow NH_4^+$) will not occur, and the soil will only contain NH_3. This will significantly increase the soil's volatilization risk.

QUESTION 32.4

What is the pH of a 0.038M nitric acid (HNO_3) solution?

Answer:

Nitric acid is a strong acid. It completely dissociates. So, pH = −log [0.038M] = 1.4.

QUESTION 32.5

The EC value of a soil measured in a 1:1 solution is 2 dS/m. Would the EC of a saturated paste, EC_e, be higher or lower than 2 dS/m? What is the $EC_{1:1}$ value expressed as EC_e, and can this soil be classified as saline–sodic?

Answer:

The EC_e value compared with the $EC_{1:1}$ will be higher due to the higher soil-to-water ratio used in the $EC_{1:1}$ method. To convert $EC_{1:1}$ into EC_e multiply $EC_{1:1}$ by 2.14. The value of EC_e is 4.28 dS/m, which indicates the soil may be classified as saline–sodic.

Soil EC depends on the methods used to determine it. In general, increasing the amount of water used in the extraction reduces the EC. If the water-to-soil ratio used in the analysis is 10 g water to 10 g soil, then EC is reported at $EC_{1:1}$. If EC is reported for a saturated paste, then the value is reported at EC_e. Increasing the EC reduces the germination of many plants, and soils with an EC_e of greater than 4 dS/m are considered saline. Additional information on saline and sodic soils is available in Clay et al. (2024).

The Soil CEC

Cation exchange capacity is an estimate of the total number of negative charges that are balanced by cations. A cation is a positively charged ion that can be monovalent (e.g., K^+, Na^+, NH_4^+, H^+) with a single positive charge, whereas divalent cations (e.g., Ca^{+2}, Mg^{+2}) have two positive charges. Thus, a monovalent cation balances a single negative charge, and a divalent cation balances two negative charges. The CEC units are centimoles of charge per kilogram soil ($cmol_c$/kg

Table 32.2. The Approximate Cation Exchange Capacity for Different Components of the Soil.

Soil phase	Approximate CEC, cmolc/kg
Organic matter	200
Clays	
Vermiculite	150
Smectite	100
Mica	30
Kaolinite	9
Gibbsite	4

soil, or meq/100 g soil). The CEC provides information about the soil's ability to retain positive charges. The CEC is a function of organic matter content and the type of clay. On average, SOM contains 200 $cmol_c$/kg (Table 32.2).

QUESTION 32.6

Estimate the soil CEC if the soil contains 30% smectite and 3% organic matter.

Answer:
Smectite = 100 cmolc/kg × 0.30 = 30 cmolc/kg
Organic matter = 200 cmolc/kg × 0.03 = 6 cmolc/kg
CEC = 36 cmolc/kg

QUESTION 32.7

What are the CEC and percent base saturation (%BS) of a soil if the soil's negative charges are balanced by 14.7, 0.2, 3.1, 1.2, and 3.6 cmolc/kg soil of Ca^{2+}, Mg^{2+}, K^+, Na^+, and H^+, respectively?

Answer:

1. CEC is the sum of cations extracted from the soil.

$$CEC = 14.7 + 0.2 + 3.1 + 1.2 + 3.6 = \frac{22.8\, cmol_c}{kg\ soil}$$

2. The BS is given by [100(basic cations/CEC)], where the basic cations are all cations except H^+ and Al^+. Use CEC to solve for %BS using only the basic cations.

$$\%BS = \frac{14.7 + 0.2 + 3.1 + 1.2}{22.8} \times 100 = 84\%$$

Economic Optimum N Rate

The economic optimum N rate (EONR) is the point on a response curve where the instantaneous return on investment equals the instantaneous investment. Details of the calculation are available in Clay et al. (2024). One approach to solving this problem is to use a second-order polynomial equation to define the relationship between inputs and outputs. The second-order equation can be expressed as yield = $a + b \times N^1 + c \times N^2$, where N is the N or seeding rate. The EONR is determined using the equation,

$$EONR = \left(\frac{\frac{Cost\ of\ N\$/lb}{Price\ of\ corn\$/bu} - b}{2c} \right)$$

where b and c are defined in the polynomial equation. In this equation, the cost of N/lb N and the price of corn per bushel are self-defined. This identical equation can be used for seeding rates.

QUESTION 32.8

Assume the following estimated quadratic equation values for a corn return function: b = 0.73 and c = (−0.0032), the selling price of corn is \$3.85/bu, and the purchase price of N is \$0.60/lb.

Answer:

$$EONR = \left(\frac{\frac{Cost\ of\ N\$/lb}{Price\ of\ corn\$/bu} - b}{2c} \right)$$

$$= \left(\frac{\frac{\$0.60/lb}{\$3.85/bu} - 0.73}{2(-0.0032)} \right) = 89.7\, lb/ac$$

QUESTION 32.9

What is the EONR if the cost of N increases to \$0.65 per pound?

Answer:

$$EONR = \left(\frac{\frac{Cost\ of\ N\$/lb}{Price\ of\ corn\$/bu} - b}{2c} \right) = \left(\frac{\frac{\$0.65/lb}{\$3.85/bu} - 0.73}{2(-0.0032)} \right) = 87.7\ lb/ac$$

QUESTION 32.10

What is the EONR for the following quadratic model?

$$Yield = 106 + 0.933x - 0.00296x^2$$

where N is the N rate in pounds per acre? In this calculation, N is selling for \$0.60 per pound, and corn is worth \$3.85 per bushel.

Answer:

$$EONR = \left(\frac{\frac{Cost\ of\ N\$/lb}{Price\ of\ corn\$/bu} - b}{2c} \right) = \left(\frac{\frac{\$0.60/lb}{\$3.85/bu} - 0.933}{2(-0.00296)} \right) = 131.3\,lb/ac$$

Bulk Density

Bulk density is used to measure the amount of soil per unit volume and is defined as the dry weight of soil per unit volume. Therefore, bulk density is the dry soil weight divided by the volume of the soil. Measuring bulk density requires taking a soil sample that occupies a known volume. This is often done with a core sampler. Once the soil is collected, it is removed from the core sampler and placed into an oven to dry at 105 °C for approximately 48 hours. The total volume of soil is the sum of the volume occupied by solids, water, and air. The total volume (VT) is defined by the equation, VT = $V_{solids} + V_{water} + V_{air.}$ The volume occupied by solids is used to determine the particle density and is defined as the ratio of the weight of the dry soil to the volume of the solids. Porosity is 100 times the ratio between the volume of air (V_{air}) and VT. Percentage porosity (%porosity or ϕ) is calculated using the equation,

$$\%Porosity = 1 - \frac{Bulk\ density}{Particle\ density} \times 100$$

In productive agriculture, %porosity generally ranges from 45% to 55%, and particle density is typically 2.65 g/cm^3. The bulk density is used to convert soil test values in parts per million to pounds per acre.

QUESTION 32.11

Calculate the dry bulk density of a soil sample taken with a core sampler that has an inner diameter of 5.1 cm and a length of 7.62 cm. The dry weight is 220 g.

Answer:

$$\text{Dry bulk density}\left(\rho_b\right)=\text{Dry soil weight}(g)/\text{Soil Volume}\left(cm^3\right)$$

Core sampler is a cylinder. So, soil volume = $\pi r^2 h$

$$\text{Core diameter}=5.1\,cm;\ \text{core radius}=5.1/2=2.55\,cm$$

$$\text{Dry bulk density}=(2.55)^2\times(3.14)\times(7.62)=155.6\,cm^3=\frac{220}{155.6}=\frac{1.41\,g}{cm^3}$$

QUESTION 32.12

Calculate the pounds of soil per acre contained in the surface 15 cm of soil. In this calculation, the bulk density is 1.07 g/cm^3. [1 ha = 10,000 m^2 = 2.471 ac; 1 kg = 2.205 lb]

Answer:

$$\text{Amount of soil}(kg/ha)=\text{bulk density}\left(g/cm^3\right)\times\text{soil depth}(cm)\times\text{conversion factors}$$

$$\text{Kg of soil per ha}=\frac{1.07\,g}{cm^3}\times 15\,cm\times\frac{10{,}000\,cm^2}{1\,m^2}\times\frac{10{,}000\,m^2}{ha}\times\frac{1\,kg}{1{,}000\,g}=\frac{1{,}605{,}000\,kg}{ha}$$

$$\text{Pounds of soil per acre}=\frac{1{,}605{,}000\,kg}{ha}\times\frac{1\,ha}{2.471\,ac}\times\frac{2.205\,lb}{kg}=\frac{1{,}432{,}223\,lb}{ac}$$

QUESTION 32.13

If a 35-cm^3 volume soil sample has an oven-dried weight of 48 g, what is its bulk density?

Answer:

$$\frac{48\,g}{35\,cm^3}=\frac{1.37\,g}{cm^3}$$

QUESTION 32.14

If a soil has a bulk density (ρ_b) of 1.31 g/cm^3 and a particle density (ρ_s) of 2.76 g/cm^3, what is its %porosity (ϕ)?

Answer:

$$\phi=\left[1-\frac{P_b}{P_s}\right]100\%,\quad=\left[1-\left(\frac{1.31\,g}{cm^3}\right)\left(\frac{cm^3}{2.76\,g}\right)\right]100\%,\quad=[1-0.475]100\%,\quad=52.5\%$$

Soil Water and Plant-available Water

Soil is very much like a sponge in a bucket of water. If a sponge is placed in a bucket of water, its pores will fill with water (Table 32.3). When the sponge is removed, free-draining water will run out of the pores. Once the water has stopped draining from the sponge, it will still be damp. In this example, the water remaining in the sponge represents the upper limit of plant-available water (PAW), or field capacity. If plants were growing in this sponge, they would remove a portion but not all of this water. The point at which the plants have removed all the water they can represents the permanent wilting point. The difference in water content between field capacity and the permanent wilting point is the PAW.

The gravimetric water content is the water contained in the soil per unit weight of soil, and the volumetric water content is the water contained in the soil per unit volume of soil. The volumetric water content is equal to the gravimetric water content multiplied by the bulk density.

Table 32.3. Table Relating the Soil Characteristics at the Saturation Point, Field Capacity (or PAW), and Permanent Wilting Point.

Water potential	Description	Characteristics
0 to −⅓ bars	0 bar is when all the pores are filled with water, and −1/3 bar is at field capacity	All pores are filled with water (saturation) Free drainage of water
−⅓ to −15 bars	Plant-available water is the difference between field capacity and the permanent wilting point	The amount of water remaining after free drainage The water can be absorbed by plant roots
−15 bars	Permanent wilting point	Point where plants will not recover from water stress

Additional information is available in Ostrem et al. (2016).

QUESTION 32.15

Calculate the volume of water (V_w) in a field-moist sample that has a total weight (M_{total}) of 250 g and a weight after oven-drying (M_s) of 184 g. The temperature of the field-moist sample is 20 °C.

Answer:
To find V_w, we first find the weight of the water (M_w) using the difference between M_{total} and M_s, and then convert grams to cubic centimeters (i.e., volume).

$$250\,g - 184\,g = 66\,g \times \frac{1cm^3}{1g} = 66\,cm^3$$

QUESTION 32.16

How many cubic centimeters of water are in the surface 15 cm if the volumetric water content is 32.7%?

Answer:
(0.327 cm^3 water/cm^3 soil) × 15 cm soil = 4.9 cm^3 water/cm^2 soil. This value represents the amount of water per unit area of soil.

QUESTION 32.17

Convert 1 ac-inch to gal. For this calculation, it is important to know that 7.481 gal are contained in 1 ft^3 and 1 ac is 43,560 ft^2.

Answer:

$$1\,ac-inch\ water \times \frac{1\,ft}{12\,inches} \times \frac{43{,}560\,ft^2}{1\,ac} \times \frac{7.481\,gal}{1\,ft^3} = 27{,}156\,gal\,of\ water$$

QUESTION 32.18

How much PAW is in the surface 15 cm of soil that has a gravimetric soil moisture content of 0.35, a permanent wilting point of 0.14, and a bulk density of 1.32 g/cm^3?

Answer:

$$\left[15\,cm \times (0.35-0.14)\right] \times 1.32\,g/cm^3 = 4.16\,cm\,PAW$$

QUESTION 32.19

Calculate the total amount of available water (in mm) across all depths using the following data.

Depth (cm)	Inches in segment	water % at − ⅓ bar	Water % at −15 bar	Bulk density
0–12	12	25	8	1.2 g/cm^3
12–36	24	30	12	1.3 g/cm^3

Answer:

Amount of PAW = (soil depth) (bulk density) (available water)

For the 0–12 and 12–36 inches,

$$(12\,inches)(1.2\,g/cm^3)(0.25-0.08) = 2.45\,inches\,of\ water$$

$$(24\,inches)(1.3\,g/cm^3)(0.30-0.12) = 5.63\,inches\,of\ water$$

Total water is 2.45 *inches* + 5.63 *inches* = 8.08 *inches PAW*

QUESTION 32.20

How long will it take to apply 1 inch of water to a 30 ft × 50 ft garden with a garden hose that delivers 10 gal/min? For this calculation, it is important to know that 7.481 gal are contained in 1 ft^3.

Answer:
Calculate the volume of water needed:

$$\text{Water needed} = (30\ ft) \times (50\ ft) \times \left(1 \times \frac{1\ ft}{12\ inches}\right) = 125\ ft^3$$

Calculate the number of gallons of water needed:

$$\text{Gallons of water} = 125\ ft^3 \times \frac{7.481\ gal}{1\ ft^3} = 935.125\ gal\ of\ water$$

Calculate the time it will take to water:

$$\text{Length of time} = (935\ gal) \times \left(\frac{1\ gal}{10\ min}\right) = 93.5\ min = 1\ hour\ 34\ min$$

Economic Optimum Seeding Rate

Determining the economic optimum seeding rate (EOSR) increases the likelihood of profiting from grain production. Planting at too high a seeding rate can reduce yields by causing adjacent plants to compete with each other for nutrients, sunlight, and water. Inversely, planting at too low a rate can open the field to weeds. The seeding rate is influenced by a multitude of factors that a producer must consider, including plant stand, row spacing, germination rate, expected seedling mortality rate, grain prices, and planter calibration, among many other considerations. The first step is to identify the desired population. This may come from past years' population at harvest, and the percentage emergence (%emergence) of planted seeds.

The %emergence can be estimated based on past years, or it can be estimated as between 90% and 95%. It can also be calculated if the plant population after emergence and the seeding rate are known. Percentage seed emergence is the ratio of plant emergence to the seeding rate. The emergence of germinated seeds can also be calculated using the percentage seed emergence and the percentage seed germination rate, as provided on the seed tag or by the seller. The percentage seed purity (%purity) is the actual percentage of crop seed in a bag of seed. These percentages can be used to calculate the expected live plant population using the equation,

$$\text{Expected live plant population} = \text{Seeding rate} \times \frac{\%purity}{100} \times \frac{\%emergence\ rate}{100}$$

QUESTION 32.21

A corn producers' group requested that a research agronomist determine the EOSR for corn in their region. The agronomist experimented with four different plant populations, 20,000, 26,000, 32,000, and 36,000 plants/ac and obtained yields of 155, 158, 162, and 169 bu/ac, respectively. How is this information used to determine the EOSR?

Answer:
This is solved using the same approach as in Question 32.8, with seeding rates substituted for N rates. In MS Excel, a second-order polynomial equation can be created using the following steps.

Step 1. Open a blank sheet in MS Excel and tabulate the populations and yield data as shown in the table below.

	A	B
1	Populations (×1,000)	Yield (bu/ac)
2	20	155
3	26	158
4	32	162
5	36	169

Step 2. Select cells A6 and B6. This can be done by pressing the "Ctrl" key and selecting those cells. Alternatively, click on cell A6, press the "Shift" key, and press the "Right Arrow" key on the keyboard to select cell B6.

Step 3. Type the formula "=LINEST(B2:B5, A2:A5^{1,2})" (do not use " " in the formula) into the selected cells and press the "Enter" key. This gives the following equation.

$$Yield\left(bushels\ per\ acre\right) = 173.94 - 1.92\left(seed\right) + 0.049\left(seed\right)^2$$

If the cost of the seed (P_s) is $240 per 80,000 kernels (a 50-lb bag), and the corn selling price (P_c) is $4.00 per bushel, what is the EOSR per acre, using the equation above?

The EOSR can be calculated using the following equation.

$$EOSR = \left(\frac{\frac{\$/1,000\,seeds}{Price\ of\ corn\,\$/bu} - b}{2c} \right)$$

In the equation, b and c represent the coefficients of the nonlinear relationship between plant population and yield. Since the second-order polynomial equation was developed using ×1,000 seeds, the cost of 1,000 seeds should be calculated.

So, the cost of seeds in $/1,000 seeds = $240/80 = $3 per 1,000 seeds, and the corn selling price (P_c) = $4 per bushel.

From the second-order polynomial equation, coefficient $b = -1.92$ and coefficient $c = 0.049$.

Calculate the EOSR using the seeding rate equation,

$$EOSR = \left(\frac{\frac{\$3}{\$4/bu} - 1.92}{2 \times 0.049} \right) = 27.24\,seeds\ rounded,\ to\,27.30\,seeds$$

Given that the unit used was 1,000 seeds, the EOSR is 27,300 seeds/ac or 67,400 seeds/ha (1 ha = 2.47 ac).

QUESTION 32.22

A farmer decided to plant corn based on the research results. However, the planter was not calibrated correctly, and seeding occurred at 25,000 seeds/ac. What is the expected yield loss compared with the EOSR?

Answer:

$$Yield\left(bushels\ per\ acre\right) = 173.94 - 1.92\left(seed\right) + 0.049\left(seed\right)^2$$

$$Farmer - planted\ corn\ yield = 173.94 - 1.92(25) + 0.049\left(25^2\right) = 156.57\ bu\ /\ ac$$

$$EOSR\ corn\ yield = 173.94 - 1.92(27.3) + 0.049\left(27.3^2\right) = 158.04\ bu\ /\ ac$$

$$Expected\ yield\ loss = 158.04 - 156.57 = 1.47\ bu\ /\ ac$$

QUESTION 32.23

If the emergence rate is 90%, the seed purity is 98%, and the seeding rate is 27,300 plants/ac, what is the expected final population?

Answer:

$$Expected\ live\ plant\ population = Seeding\ rate \times \frac{\%purity}{100} \times \frac{\%emergency\ rate}{100}$$

$$Expected\ live\ plant\ population = 27{,}300 \times \frac{90\%}{100} \times \frac{98\%}{100} = 24{,}100\ plants\ /\ ac$$

Using Remote Sensing to Identify Yield Variation

Global Navigation Satellite Systems (GNSS) are satellite-based technologies that provide accurate positioning and timing information anywhere on Earth. Major GNSS include the U.S. Global Positioning System (GPS), Russia's GLONASS, Europe's Galileo, and China's BeiDou. These systems allow users to calculate precise geographic coordinates (latitude and longitude) by receiving signals from multiple satellites. Latitude and longitude are fundamental to GNSS, as they define specific locations on the Earth's surface, enabling consistent spatial referencing across different platforms and datasets. This positional accuracy is crucial for remote sensing applications, where spatial data must align correctly.

There are different base systems in which latitude and longitude are recorded: the decimal degree (DD) system and the degree, minutes, and seconds (DMS) system. The DD format denotes latitude and longitude as a decimal fraction written as a single number with a decimal point. The degree, minute, and second base is divided into three parts: the degree, minute (1/60°), and second (1/3,600°) while also noting the cardinal direction. To convert DMS into DD, use the equation below, along with the cardinal direction to determine whether to place a positive or negative sign in front of the final number:

$$Decimal\,degrees\left(DD\right) = Degrees + \frac{Minutes}{60} + \frac{Seconds}{3{,}600}$$

Cardinal Directions:

Latitude North: Positive → Latitude South: Negative

Longitude East: Positive → Longitude West: Negative

QUESTION 32.24

A farmer decided to sample the field for soil phosphorus (P) testing and sectioned it into three different zones based on yield differences. He collected yield data from his yield monitor in DD units, but when he collected soil samples, the GPS coordinate system used by the GPS software used DMS units.

Convert the DMS units to DD units to ensure that each point from the soil P test matches up with the yield monitor data.

Zone	Point number	DMS	DD	P ppm	Yield bu/ac
1	1	44°18′21.70″N, 96°40′11.86″W		1	170
1	2	44°18′15.53″N, 96°40′03.40″W		5	174
1	3	44°18′16.66″N, 96°40′05.46″W		4	172
2	4	44°18′18.51″N, 96°40′18.63″W		8	150
2	5	44°18′27.05″N, 96°40′08.19″W		9	155
2	6	44°18′19.63″N, 96°40′15.28″W		10	157
3	7	44°18′27.73″N, 96°40′17.44″W		18	180
3	8	44°18′26.64″N, 96°40′19.50″W		19	184
3	9	44°18′28.84″N, 96°40′18.60″W		21	179

Answer:

$$Decimal\ Degrees(DD) = Degrees + \frac{Minutes}{60} + \frac{Seconds}{3{,}600}$$

$$Decimal\ Degrees(DD) = 44 + \frac{18}{60} + \frac{21.70}{3{,}600} = 44.30602778$$

$$Decimal\ Degrees(DD) = 96 + \frac{40}{60} + \frac{11.86}{3{,}600}$$

Zone	Point number	DMS	DD	P ppm	Yield bu/ac
1	1	44°18′21.70″N, 96°40′11.86″W	44.305972, −96.669961	1	170
1	2	44°18′15.53″N, 96°40′03.40″W	44.304314, −96.66761	5	174
1	3	44°18′16.66″N, 96°40′05.46″W	44.304626, −96.668183	4	172
2	4	44°18′18.51″N, 96°40′18.63″W	44.305142, −96.671842	8	150
2	5	44°18′27.05″N, 96°40′08.19″W	44.307514, −96.668942	9	155
2	6	44°18′19.63″N, 96°40′15.28″W	44.305453, −96.670911	10	157
3	7	44°18′27.73″N, 96°40′17.44″W	44.307703, −96.671511	18	180
3	8	44°18′26.64″N, 96°40′19.50″W	44.307400, −96.67208	19	184
3	9	44°18′28.84″N, 96°40′18.60″W	44.30801, −96.671833	21	179

(continued)

Identify the P ppm and yield corresponding to the sampling point numbers, and draw a management zone map. Use the map drawn below.

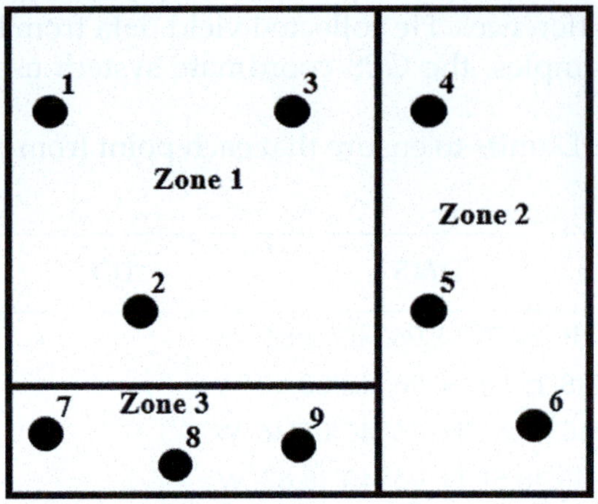

Answer:

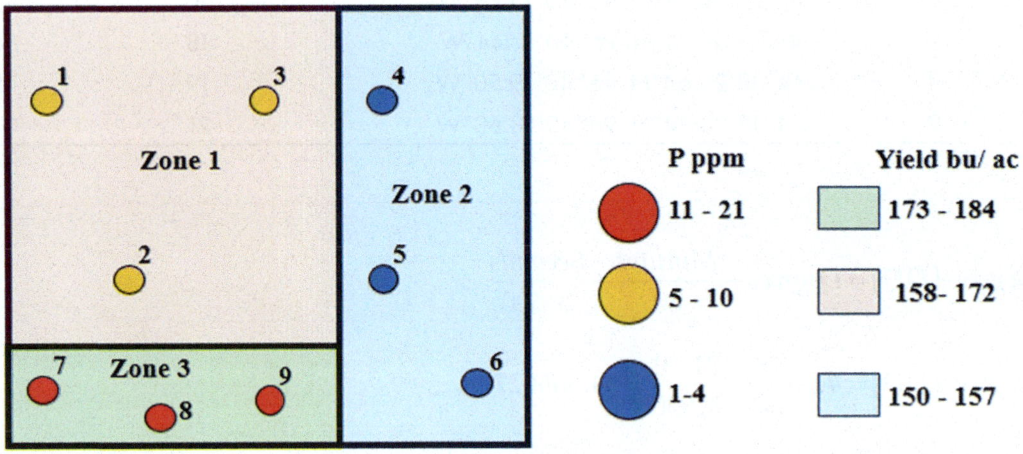

Based on your management zone map, does P ppm appear to impact yield? Which zone is most likely to have been the area where livestock was kept during homesteading?

Answer:
Yes, the high yields seem to correspond to the high P ppm from the P soil test, suggesting this is affecting yield. Zone 3 appears to have the highest P ppm, most likely because it was formerly a livestock area, leaving it with high legacy P levels.

Remote sensing data uses GNSS-based coordinates to assess vegetation health over a particular field or location. Remote sensing can be performed by satellites and drones equipped with multispectral sensors that capture the visible red–green–blue (RGB) spectrum, as well as near-infrared (NIR). For more information, see Chapters 13 and 14. This allows the calculation of indices such as the Normalized Difference Vegetation Index: $NDVI = (NIR - Red)/(NIR + Red)$.

Plant health can vary across a field due to a variety of factors, such as soil texture, slope, pests, disease, nutrients, soil moisture, and soil temperature. Using NDVI values from drone or satellite imagery is useful for identifying variability and reducing stress before yield loss occurs. Remote sensing data are used in combination with ground truth data, such as the average and standard deviation of yield.

QUESTION 32.25

A research station conducted an experiment over a 3-year period (2019–2021) to study how N application affects plant health. The experimental site was divided into six plots, with three plots receiving an N rate (pounds N per ac) of 0N and the other three receiving 200N. The researchers recorded yield in bushels per acre after harvest and used a multispectral handheld proximal sensor midseason to quantify the red and NIR wavelengths for each plot. Calculate the NDVI values for each plot in each year.

2019

Plot number	N rate	Red	NIR	NDVI	Yield bu/ac
1	0	0.15	0.65		120
2	0	0.20	0.7		110
6	0	0.14	0.69		115
3	200	0.12	0.68		150
4	200	0.18	0.72		140
5	200	0.05	0.7		160

2020

Plot number	N rate	Red	NIR	NDVI	Yield bu/ac
2	0	0.17	0.5		110
4	0	0.18	0.5		100
6	0	0.15	0.61		105
1	200	0.15	0.61		130
3	200	0.15	0.63		140
5	200	0.10	0.8		150

2021

Plot number	N rate	Red	NIR	NDVI	Yield bu/ac
1	0	0.14	0.64		115
3	0	0.11	0.67		125
4	0	0.17	0.71		120
2	200	0.10	0.6		140
5	200	0.04	0.6		165
6	200	0.11	0.64		133

(continued)

Calculate the NDVI values for 2019, 2020, and 2021.

Answer:

The NDVI values are calculated using the equation: NDVI = (NIR – Red)/(NIR + Red).

2019

Plot number	N rate	Red	NIR	NDVI	Yield bu/ac
1	0	0.15	0.65	0.63	120
2	0	0.20	0.7	0.56	110
6	0	0.14	0.69	0.66	115
3	200	0.12	0.68	0.70	150
4	200	0.18	0.72	0.60	140
5	200	0.05	0.7	0.87	160

2020

Plot number	N rate	Red	NIR	NDVI	Yield bu/ac
2	0	0.17	0.5	0.49	110
4	0	0.18	0.5	0.47	100
6	0	0.15	0.61	0.60	105
1	200	0.15	0.61	0.61	130
3	200	0.15	0.63	0.62	140
5	200	0.10	0.8	0.78	150

2021

Plot number	N rate	Red	NIR	NDVI	Yield bu/ac
1	0	0.14	0.64	0.64	115
3	0	0.11	0.67	0.72	125
4	0	0.17	0.71	0.61	120
2	200	0.10	0.6	0.71	140
5	200	0.04	0.6	0.88	165
6	200	0.11	0.64	0.71	133

Next, calculate the average and standard deviation for the 0N treatment and the 200N treatment for each year.

Year	N rate	Average yield bu/ac	Standard deviation
2019	0N		
	200N		
2020	0N		
	200N		
2021	0N		
	200N		

Answer:

The mean is the average of the values within a treatment; it is calculated by summing the values and dividing by the number of values summed. A standard deviation is the square root of the summed squared differences between the treatment mean and the individual values. This squared sum difference is divided by the number of measurements (n) minus 1. See Chapter 21 for additional information.

Year	N rate	Average yield bu/ac	Standard deviation
2019	0N	115	±5
	200N	150	±10
2020	0N	105	±5
	200N	140	±10
2021	0N	120	±5
	200N	146	±17

QUESTION 32.26

In the 2019 experiment, the 0N rate resulted in an average yield of 115 bu/ac, while the 200N rate resulted in 150 bu/ac. Use the NDVI values calculated above for each plot in 2019 to create an NDVI map showing the variation in plant health due to differences in N application. How does NDVI vary with yield?

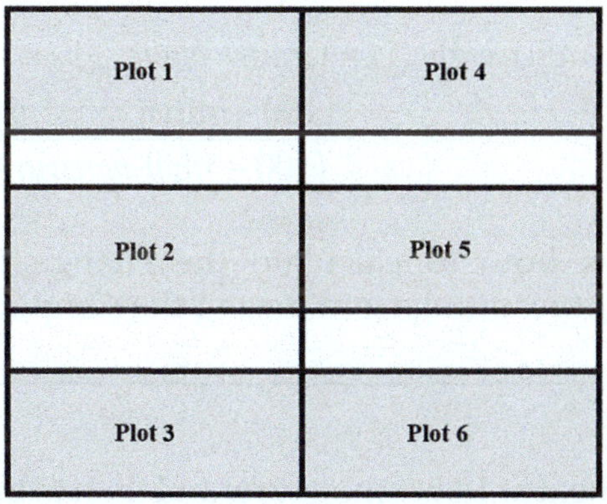

Answer:

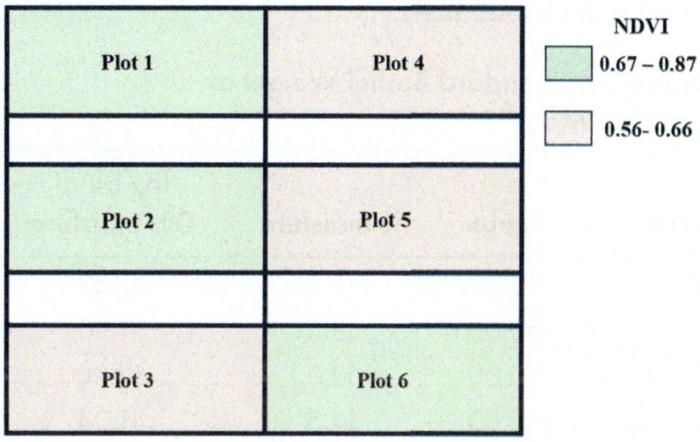

Plots 3, 4, and 5 have lower NDVI values than plots 1, 2, and 3, suggesting that NDVI varies based on yield. Additionally, the N rate plays a role in yield variation and plant health.

QUESTION 32.27

If the NIR, green, red edge, and red reflectance values are 0.55, 0.10, 0.20, and 0.05, respectively, determine the Green Normalized Difference Vegetation Index (GNDVI), Normalized Difference Red Edge Index (NDREI), and Modified Soil-Adjusted Vegetation Index (MSAVI). Briefly explain the scenarios in which these indices are used.

(continued)

Answer:

$$GNDVI = \left(\frac{NIR - Green}{NIR + Green}\right) = \left(\frac{0.55 - 0.10}{0.55 + 0.10}\right) = 0.69$$

The GNDVI can be used to detect early signs of crop water stress or N deficiency.

$$NDRE = \left(\frac{NIR - Red\,Edge}{NIR + Red\,Edge}\right) = \left(\frac{0.55 - 0.20}{0.55 + 0.20}\right) = 0.47$$

The NDRE index can be used to monitor crops with dense canopies or during the mid- to late-growth stages.

$$MSAVI = \frac{2 \cdot NIR + 1 - \sqrt{(2 \cdot NIR + 1)^2 - 8(NIR - Red)}}{2} = \frac{2 \cdot 0.55 + 1 - \sqrt{(2 \cdot 0.55 + 1)^2 - 8(0.55 - 0.05)}}{2} = 0.73$$

The MSAVI is suitable for areas with sparse vegetation or a high soil background reflectance. This index adjusts for soil brightness.

Grain Moisture

Grain yields are typically reported on a wet weight basis, as defined by international conventional use (Table 32.4).

Table 32.4. Standard Bushel Weight at Indicated Moisture.

Grain	lb/bu	% moisture	lb/bu at 0% moisture
Corn	56.00	15.5	47.32
Wheat	60.00	13.5	51.90
Soybean	60.00	13.0	52.20
Barley	48.00	14.5	41.04
Oats	32.00	14.0	27.52
Rye	56.00	14.0	48.16

In addition to international marketing, grain moisture content is used to assess the selling value of the crop and to determine whether drying is needed. When grain is harvested, it is rarely at the required standard moisture content. If the moisture content is greater than required, drying may be necessary; if it is below the required moisture, drying may not be needed (Table 32.4). The following equation

$$Grain\,weight_{final} = Grain\,weight_{initial} \cdot \frac{(100 - initial\ moisture\%)}{(100 - final\ moisture\%)}$$

shows how to calculate the change in grain weight resulting from drying the sample.

QUESTION 32.28

Convert 5,000 lb of wheat at 15% moisture to pounds of wheat at 13.5% moisture, and then to bushels.

Answer:

$$Grain\,weight_{final} = Grain\,weight_{initial} \cdot \frac{(100 - initial\ moisture\%)}{(100 - final\ moisture\%)}$$

$$Grain\,weight_{final} = 5{,}000 \cdot \frac{(100 - 15\%)}{(100 - 13.5\%)} = 4{,}913\,lb$$

$$4{,}913\,lb\ grain \times \frac{1\,bu}{60\,lb} = 814.9\,bu$$

Wheat production in the United States has an additional standard besides grain moisture. Wheat price at the elevator is also based on percentage protein (%protein), with the standard calculated at 12% moisture, rather than the wheat standard bushel measurement of 13.5%. The percentage N (%N) from the laboratory analysis, at 0% moisture, can be converted to %protein by multiplying the %N by 6.25. The protein content will need to be converted to 12% moisture. This can be done using following the equations,

$$\%Dry\ weight = 100\% \times \frac{(Wet\ weight - Dry\ weight)}{Dry\ weight}$$

After the percentage dry weight is determined, the wet weight basis needs to be calculated. This is done using the following equation,

$$\%wet\ protein = \%dry\ protein \times \frac{100 - wetcrop\ moisture\%}{100}$$

The %protein of wheat is used at the grain elevator to determine whether the grain receives a premium or is docked for unfavorable protein levels. The premiums are determined by market conditions and the quality of the grain and its protein content.

QUESTION 32.29

A sample of harvested soybean was placed in a dryer to measure its constant weight. The grains were ground and analyzed for %N. If the grain's %N was 7.5%, calculate the protein content at 13% moisture.

Answer:

$$\%dry\ protein = 7.5\% \times 6.25 = 46.88\%$$

$$\%wet\ protein = 46.88\% \times \frac{100 - 13\%}{100} = 41\%$$

Growing Degree Days

The combination of temperature and time, expressed as heat units, is used to understand the growth and development of plants and other organisms. The accumulation of these heat units over time is referred to as Growing Degree Days (GDDs). Growing Degree Days are used for many purposes, including hybrid selection, irrigation scheduling, pesticide application timing, and determining optimal scouting periods. Converting between Celsius (°C) and Fahrenheit (°F) is important when calculating GDDs. The equation for converting between Celsius (°C) and Fahrenheit (°F) is

$$°F = \frac{9°F}{5°C}(°C) + 32°F \text{ and } °C = (°F - 32°F) \cdot \frac{5°C}{9°F}.$$

QUESTION 32.30

Convert 15 °C to °F.

Answer:

$$\frac{9°F}{5°C}(15°C) + 32°F = 59°F$$

QUESTION 32.31

Convert 77 °F to °C.

Answer:

$$(77°F - 32°F) \cdot \frac{5°C}{9°F} = 25°C$$

GDDs are calculated based on the daily minimum temperature (**T_{min}**), the daily maximum temperature (**T_{max}**), and the base temperature (**T_{base}**), which is the lowest temperature at which the organism being measured can grow. Each day is calculated separately and summed to yield the accumulated GDDs. This equation is

$$GDDs = \frac{T_{max} + T_{min}}{2} - T_{base}$$

where T_{base} is defined for each organism (Table 32.5).

When the T_{min} value is below the T_{base} value, the T_{base} value should replace the T_{min} value. Some crops and pests also have a T_{max} cap due to reduced growth from excessive heat. For example, corn has a T_{base} value of 50 °F and a T_{max} value of 86 °F.

Table 32.5. Commonly Used Base Values (T_{base}) When Calculating Heat Units.

Crop	T_{base} (°F)	Weeds	T_{base} (°F)	Insects	T_{base} (°F)
Corn	50[a]	Barnyard grass	50	European corn borer	54
Wheat	Range from 32 to 40	Yellow foxtail	55	Alfalfa weevil	48
Soybean	50	Common cocklebur	40	Western corn rootworm	53
Sunflower	44	Sowthistle	47	Gypsy moth	50
		Wild buckwheat	40	Stalk borer	41

[a] T_{max} value for corn is 86 °F.

QUESTION 32.32

Calculate the GDDs for Days 1 and 2 for corn.

	Maximum temperature (T_{max})	Minimum temperature (T_{min})
	°F	°F
Day 1	75	48
Day 2	90	77

Answer:

$$\text{GDDs for Day 1} = \frac{75+50}{2} - 50 = 12.5$$

Please note: T_{min} was changed from 48 °F to 50 °F because 48 is below the T_{base} for corn, which is 50 °F.

$$\text{GDDs for Day 2} = \frac{86+77}{2} - 50 = 31.5$$

Please note: T_{max} was changed from 90 °F to 86 °F because 90 is above the T_{max} for corn, which is 86 °F.

QUESTION 32.33

Calculate the accumulated GDDs for the 2 days above for corn.

Answer:

GDDs for Day 1 = 12.5

GDDs for Day 2 = 31.5

12.5 + 31.5 = 44 Accumulated GDD, calculated with T_{base} as 50 °F.

QUESTION 32.34

Calculate the accumulated GDDs for corn over the 2 days in °C. (You must first convert the temperatures of each day to °C to begin.) Be sure to also calculate T_{base} and T_{max} in °C.

Answer:

$$T_{base} \text{ for corn } ^\circ C = (50\,^\circ F - 32\,^\circ F) \cdot \frac{5^\circ C}{9^\circ F} = 10\,^\circ C$$

$$T_{max} \text{ for corn } ^\circ C = (86\,^\circ F - 32\,^\circ F) \cdot \frac{5^\circ C}{9^\circ F} = 30\,^\circ C$$

	Maximum temperature (T_{max})	Minimum temperature (T_{min})
	°F	**°F**
Day 1	$^\circ C = (75\,^\circ F - 32\,^\circ F) \cdot \frac{5^\circ C}{9^\circ F} = 23.9\,^\circ C$	$^\circ C = (48\,^\circ F - 32\,^\circ F) \cdot \frac{5^\circ C}{9^\circ F} = 8.9\,^\circ C$
Day 2	$^\circ C = (90\,^\circ F - 32\,^\circ F) \cdot \frac{5^\circ C}{9^\circ F} = 32.2\,^\circ C$	$^\circ C = (77\,^\circ F - 32\,^\circ F) \cdot \frac{5^\circ C}{9^\circ F} = 25\,^\circ C$

$$\text{GDDs for Day 1} = \frac{23.9 + 10}{2} - 10 = 7.0$$

Please note: T_{min} was changed from 8.9 °C to 10 °C because 8.9 °C is below the T_{base} for corn, which is 10 °C.

$$\text{GDDs for Day 2} = \frac{30 + 25}{2} - 10 = 17.5$$

Please note: T_{max} was changed from 32.2 °C to 30 °C because 32.2 °C is above the T_{max} for corn, which is 30 °C.

7.0 + 17.5 = 24.5 Accumulated GDDs, calculated with T_{base} as 10 °C.

QUESTION 32.35

Using the data from the table at the beginning of this section (Question 32.32), calculate the accumulated GDDs for the western corn rootworm from the temperatures given below.

	Maximum temperature (T_{max})	Minimum temperature (T_{min})
	°F	**°F**
Day 1	75	48
Day 2	90	77

(continued)

Answer:

$$\text{GDDs for Day 1} = \frac{75+53}{2} - 53 = 11$$

Please note: T_{min} was changed from 48 °F to 53 °F because 48 is below the T_{base} for western corn rootworm, which is 53 °F.

$$\text{GDDs for Day 2} = \frac{90+77}{2} - 53 = 30.5$$

Please note that western corn rootworm does not have a reported maximum temperature.

11 + 30.5 = 41.5 Accumulated GDDs, calculated with T_{base} as 53 °F.

QUESTION 32.36

If western corn rootworm eggs require 360 GDDs to hatch, would you need to start scouting the field for this pest according to the accumulated GDDs calculated in Question 32.35?

Answer:
No, the accumulated GDDs from Question 32.33 are 41.5, which is not close to 360 GDDs. If the farmer is concerned about this pest they should wait for another 318.5 GDDs before scouting to achieve optimal results.

Yield Calculations and Estimations

During the growing season, yield estimates are useful for postharvest planning. Estimating corn and soybean yields requires first estimating the plant population in 1/1,000th of an acre. To do this, identify a defined length of measurement, for example, 20 ft in a row, and for the row spacing, 30-in row spaces. Count the number of plants in the defined area and estimate the number of plants per acre using the equation,

$$\textit{Plants per acre} = \frac{\textit{Number of plants in defined length}}{\textit{Number of feet in defined length} \times \textit{row inch spacing}} \times \frac{12\,\textit{inches}}{1\,\textit{ft}} \times \frac{43{,}560\,\textit{ft}^2}{1\,\textit{ac}}$$

Use Table 32.6 to determine the defined area. Count the number of plants along the length of the row, then multiply the final number by 1,000 to determine the number of plants per acre.

To estimate yields, the yield per plant is needed. For soybean, the number of beans per plant must be estimated. If it is assumed that one soybean plant contains 60 beans and a bushel contains 215,000 beans, then the soybean yield can be estimated. In corn, we can assume that one plant will produce a single ear, and we can use the weight of the grains from that ear to calculate the estimated yield. The equation would look like the following example:

$$\textit{Bushels per acre corn} = \frac{\textit{Number plants}}{\textit{ac}} \times \frac{1\,\textit{ear}}{1\,\textit{plant}} \times \frac{0.28\,\textit{lb}}{1\,\textit{ear}} \times \frac{1\,\textit{bu}}{56\,\textit{lb}}$$

In corn, the yield potential (bu/ac) is most accurately estimated between reproductive growth stages R4 and R6. This estimate can be improved by replacing the expected weight of an ear with the number of kernels per ear and the weight of each kernel. The number of kernels per ear is estimated by multiplying the number of rows on an ear by the number of kernels per row.

Table 32.6. Length of Row Relative to Row Width Required for 1/1,000 of an Acre.

Row spacing in inches	Row length required for 1/1,000 ac-inch-ft
7	74 ft 8 inches
10	52 ft 4 inches
15	34 ft 10 inches
20	26 ft 2 inches
22	23 ft 8 inches
28	18 ft 8 inches
30	17 ft 5 inches
32	16 ft 4 inches
34	15 ft 5 inches
36	14 ft 6 inches
38	13 ft 9 inches

QUESTION 32.37

If 30 corn plants are counted in a measured 20 ft of row and the rows are spaced 30 in apart, what is the estimated plant population in bushels per acre? Assuming one ear has a kernel weight of 0.28 lb and each plant has one ear.

Answer:

$$\frac{30\,plants\ counted}{20\,ft \times 30\,inches} \times \frac{12\,inches}{1\,ft} \times \frac{43{,}560\,ft^2}{1\,ac} \times \frac{1\,ear}{1\,plant} \times \frac{0.28\,lb}{1\,ear} \times \frac{1\,bu}{56\,lb} = 131\,bu / ac\,corn$$

QUESTION 32.38

If 30 soybean plants are counted in a measured 20 ft of row and the rows are spaced 15 inches apart, what is the estimated plant population in bushels per acre? Assuming one plant has 60 beans and 215,000 beans make up one bushel (60 lb).

Answer:

$$\frac{30\,plants\ counted}{20\,ft \times 15\,inches} \times \frac{12\,inches}{1\,ft} \times \frac{43{,}560\,ft^2}{1\,ac} \times \frac{60\,beans}{1\,plant} \times \frac{60\,lb}{215{,}000\,beans} \times \frac{1\,bu}{60\,lb} = 15\,bu / ac\,soybean$$

QUESTION 32.39

Calculate the bushels per acre of corn if the number of kernels is 20,400,000 kernels/ac.

Answer:

$$\frac{20{,}400{,}000\,kernels}{ac} \times \frac{1\,bu}{90{,}000\,kernels} = 227\,bu / ac\,corn$$

QUESTION 32.40

Calculate the bushels per acre of soybean if the number of soybean plants counted in 1/1,000th of an acre is 275 plants. The average number of pods per plant is 26 pods, and the average number of beans per pod is 2.5 beans. Assume 1 bu contains 215,000 beans.

Answer:

$$Plants\ per\ acre = \frac{Number\ of\ plants\ counted\ in\ sampling\ area \times 1{,}000}{ac}$$

$$\frac{275\ plants \times 1{,}000}{ac} = 275{,}000\ plants/ac$$

$$Bushels\ per\ acre = \frac{Number\ of\ plants}{1\ ac} \times \frac{Number\ of\ pods}{1\ plant} \times \frac{Number\ of\ beans}{1\ pod} \times \frac{1\ bu}{215{,}000\ beans}$$

$$\frac{275{,}000\ plants}{1\ ac} \times \frac{26\ pods}{1\ plant} \times \frac{2.5\ beans}{1\ pod} \times \frac{1\ bu}{215{,}000\ beans} = 83\ bu\ soybean/ac$$

Harvest Loss

Prior to and during harvest, grain can be lost in many ways, including lodging or ear snapping off the plant (preharvest loss), header loss, ear or pod loss, and separation loss. Preharvest loss is estimated by determining the number of ears or kernels on the ground prior to harvesting. This is determined by following a similar approach to that used for estimating yields (Section 32.11).

Header loss occurs when the combine header fails to harvest the grain, and instead allows it to fall to the ground. To measure header loss, begin harvesting, then stop in the row and back up to where the chaff has been spread. Consider three or four areas, each measuring 1 ft^2, and see how many seeds have fallen. Do the same behind the combine and compare the losses. This helps determine whether the harvest loss is due to the header. If the losses from the header are significantly higher than those from preharvest losses, then consult the operator's manual or a dealer to learn how to adjust the machinery to reduce the losses that are being incurred. After adjustments, if a significant loss persists, it may be wise to consider an aftermarket header to help reduce it. It would be best to compare the yearly cost of a new header with the cost of the grain loss to determine whether the purchase would be the best financial move.

When walking down the harvested rows, whole ears of corn, pods of soybean, or various other grain units may be found. This loss is called ear or pod loss. This can occur when the grain ear, pod, or head is knocked off the plant as the header hits it, causing the whole unit to fall to the ground before it can be collected by the header. Separation loss occurs when shelled grain remains in the crop residue that is discharged out the back of the combine. If the losses are greater than your expectations, check the operator's manual or contact the dealer to determine what corrective adjustments should be made.

Table 32.7. Seeds on the Ground in a 1-ft^2 Area that Represents 1 bu of Grain Lost.

Crop	Seeds/bu
Corn	2
Soybean	5
Wheat	22

QUESTION 32.41

What is the yield loss if 8 corn kernels are found in a 1-ft² area behind the combine?

Answer:

$$\left(\frac{8\,kernels}{1\,ft^2}\right)\left(\frac{4,3560\,ft^2}{1\,ac}\right)\left(\frac{1\,bu}{80,000\,kernels}\right) = 4.4\,bu\,/\,ac\,yield\ loss$$

Sprayer Calibration

In modern agriculture, chemical pesticide applications may be required to control pests that could pose economic or agronomic issues. It is important to ensure that pesticides are applied at the correct label rates. This is because improper use of these agronomic tools can lead to severe consequences, including fines and possible jail time. To ensure that these chemicals are applied correctly, sprayers need to be calibrated, and the correct amount of chemical must be calculated. Factors that affect the calibration of a sprayer include boom length, boom output, nozzle selection, applicator speed, and the area to be sprayed.

QUESTION 32.42

What is the output (gal/min) of a sprayer that has an average nozzle output of 25 fl oz in 30 s when there are 30 nozzles on the boom that are spaced every 30 inches (30 inches on the center)?

Answer:

$$Sprayer\ output\ per\ nozzle = \left(\frac{25\ oz}{30\,s}\cdot\frac{1\,gal}{128\,oz}\cdot\frac{60\,s}{1\min}\right) = 0.39\,gal\,/\min/\,nozzle$$

$$Sprayer\ output\ over\ the\ boom = 0.39\ gal/\ min\ /nozzle\cdot 30\ nozzles = 11.7\ gal/min$$

The speed at which an operator moves the spray equipment will influence the amount of chemical applied to a given area. By setting up a course with a known length and measuring how long it takes to complete the course, the implement's speed can be calculated. It is important to note that speed should be calculated on a course that is similar to the conditions in the field.

QUESTION 32.43

The course length is 300 ft, and the time to drive the course is 30 s. Calculate the tractor's speed in feet per minute and in miles per hour.

Answer:

$$Speed = \left(\frac{length\ of\ course}{time\ over\ course}\cdot Conversion\right)$$

$$Speed = \left(\frac{300\,ft}{30\,s}\cdot\frac{60\,s}{1\min}\right) = 600\,ft\,/\min$$

$$Speed = \left(\frac{300\,ft}{30\,s}\cdot\frac{1\,mile}{5,280\,ft}\cdot\frac{60\,s}{\min}\cdot\frac{60\min}{hour}\right) = 6.8\,miles\,/\,hour$$

QUESTION 32.44

Calculate the number of acres sprayed in 5 min if the sprayer has 40 nozzles on a 30-inch center spacing and the tractor's speed is 430 ft/min. [1 ac = 43,560 ft²]

Answer:

$$Area = length \cdot boom\ width$$

$$Boom\ width = \left(\frac{30\,inches}{1\,nozzle} \cdot \frac{1\,ft}{12\,inches}\right) \cdot 40\,nozzles = 100\,ft$$

$$Area\ covered/\mathrm{min} = \left(\frac{430\,ft}{1\,\mathrm{min}} \cdot 100\,ft\right) = 43{,}000\ ft^2/\mathrm{min} = 0.987\,\mathrm{ac/min}$$

$$\text{Area covered in } 5\,\text{min} = 0.987\,\text{ac/min} \times 5\,\text{min} = 4.936\,\text{ac}$$

Once the area is calculated for a unit of time, we can find the sprayer's output. Use the equation, $Output/area = \left(\frac{output}{\mathrm{min}} \cdot \frac{\mathrm{min}}{area}\right)$, for this calculation. Alternatively, if the nozzle output, speed, nozzle spacing, and number of nozzles are known, then the sprayer output can be calculated using the equation, $gpa = \left(\frac{gpm \cdot 5{,}940}{mph \cdot w}\right)$, where gpa is gallons per acre, gpm is gallons per minute per nozzle, mph is miles per hour of the applicator, w is nozzle spacing in inches, and 5,940 is a conversion constant used for calculations using inches, miles, minutes, hours, and acres.

QUESTION 32.45

If the sprayer output is 18.2 gal/min and the area is sprayed at a rate of 0.852 gal/min, what is the output in gallons per acre?

Answer:

$$Sprayer\ output = \left(\frac{18.2\,gal}{1\,\mathrm{min}} \cdot \frac{1\,\mathrm{min}}{0.852\,ac}\right) = 21.36\,gal/ac$$

After calibrating the sprayer and determining its output, the correct amount of pesticide to add to the tank needs to be calculated. Pesticide labels are required by law to contain information such as the appropriate application rates. For liquid pesticides, application rates will be expressed in ounces, pints, or quarts per acre, and dry-formulation pesticides in ounces or pounds. To calculate the amount of pesticide needed, multiply the label recommendation by the number of acres.

QUESTION 32.46

A spray tank holds 800 gal, the intended application rate of solution is 12 gal/ac, and the pesticide product rate is 2.0 pt/ac.

a. How many acres can be treated with a single tank of solution?
b. How many gallons of pesticide need to be added to the tank to get the correct application rate?

Answer:

$$\left(\frac{800\,gal}{tank}\cdot\frac{1\,ac}{12\,gal}\right)=66.7\,ac\,/\,tank$$

$$\left(\frac{2\,pt}{ac}\cdot\frac{66.7\,ac}{tank}\right)=\frac{200\,pt}{tank}=\left(\frac{200.1\,pt}{tank}\cdot\frac{1\,gal}{8\,pt}\right)=25.01\,gal\,pesticide\,/\,tank$$

QUESTION 32.47

Calculate the output of a dry spreader in pounds per acre if the spreader travels 500 ft and has an output of 8 oz with 16 inches of coverage.

Answer:

$$Area\ covered=(500\ ft)\cdot(16\ inches)\cdot\left(\frac{1\ ft}{12\ inches}\right)\cdot\left(\frac{1\ ac}{43{,}560\ ft^2}\right)=0.015\,ac$$

$$Pounds\ of\ product\ spread=(8\,oz)\cdot\left(\frac{1\,lb}{16\,oz}\right)=0.5\,lb$$

$$Output\,lb\,/\,ac=\left(\frac{0.5\,lb}{0.015\,ac}\right)=33.3\,lb\,/\,ac$$

Other common additives to in-tank mixes are surfactants, adjuvants, and crop oils. These additives are used to increase the effectiveness of the pesticides being applied. Labels for these products generally provide a volume-per-volume format for adding them to the tank mix (water + pesticide + additive). In the volume-per-volume format, the additive added to the tank must be based on the amount of liquid in the tank, not on the application rate. Often, entire areas cannot be sprayed using only one tank load of chemicals, and it is important that partial tank mixes do not contain excess solution that cannot be sprayed.

QUESTION 32.48

A spray tank holds 800 gal of solution. The intended application rate is 16 gal/ac. A crop oil concentrate is recommended at 0.25% (v/v), and the pesticide is applied at 14 fl oz/ac. There are 80 ac being treated. Answer the following questions based on this information.

a. How many acres can be treated with one tank of solution?
b. How many pints of pesticide and crop oil concentrate need to be added to achieve the correct application rate?
c. How many acres will be treated with the second tank of solution?
d. What is the total amount of solution in gallons needed to treat the area?
e. How many pints of pesticide and crop oil concentrate should be added to the second tank?

Answer:

$$\left(\frac{800\,gal}{Tank\ load}\right)\cdot\left(\frac{1\,ac}{16\,gal}\right)=50\,ac\,/\,1\,tank\ load$$

$$\left(\frac{14\,fl\,oz}{1\,ac}\right)\cdot\left(\frac{50\,ac}{Tank}\right)\cdot\left(\frac{1\,gal}{128\,fl\,oz}\right)\cdot\left(\frac{8\,pt}{1\,gal}\right)=44\,pt\ of\ pesticide\,/\,tank\ load$$

$$\left(\frac{800\,gal}{Tank}\right)\cdot\left(\frac{0.25\,gal}{100\,gal}\right)\cdot\left(\frac{8\,pt}{1\,gal}\right)=16\,pt\ of\ crop\ oil\ concentrate\,/\,tank\ load$$

$$80\,ac-50\,ac=30\,ac\ remaining\ to\ be\ treated$$

$$30\,ac\cdot\left(\frac{16\,gal}{1\,ac}\right)=480\,gal\ of\ solution\ needed\ for\ Tank\ 2$$

$$\left(\frac{14\,fl\,oz}{ac}\right)\cdot\left(\frac{1\,gal}{128\,fl\,oz}\right)\cdot\left(\frac{8\,pt}{1\,gal}\right)\cdot 30\,ac=26.24\,pt\ of\ pesticide\ needed\ for\ Tank\ 2$$

$$\left(\frac{480\,gal}{Tank}\right)\cdot\left(\frac{0.25\,gal}{100\,gal}\right)\cdot\left(\frac{8\,pt}{1\,gal}\right)=9.6\,pt\ of\ crop\ oil\ concentrate\ needed\ for\ Tank\ 2$$

The cost of the treatment must also be determined. Pesticides can come in multiple forms, either as single-active ingredient (a.i.) products or as mixtures containing two or more active ingredients. The question that needs to be asked is whether it is more economically advantageous to purchase a pesticide individually or as a Premix. A premix may fulfill one pesticide requirement but not contain enough of another, and so a separate pesticide may need to be purchased to apply the required rate.

QUESTION 32.49

Product A (6.4 lb a.i./gal) may be applied at 1.0–1.5 pt/ac and tank-mixed with Product B (4.2 lb a.i./gal) at 0.6–1.3 pt/ac. A premix, Product C, can be applied at 1–2 pt/ac and contains 5.5 lb/gal a.i. of Product A and 3.8 lb a.i. of Product B. Based on the product guide and the level of weed pressure in the field, the recommendation is to apply 1.3 pt/ac of Product A and 0.8 pt/ac of Product B.

Product A, $100/gal, containing 6.4 lb a.i./gal

Product B, $110/gal, containing 4.2 lb a.i./gal

Premix C, $105/gal, containing 5.5 lb a.i./gal of Product A and 3.8 lb a.i./gal of Product B

Should the chemicals be purchased as a premix or separately and mixed in the tank to achieve the lowest cost?

Answer:
Determine the cost of the tank mix by calculating the amount of each product needed and the total cost of the tank mix.

Product A:

$$\left(\frac{1.3\,pt}{1\,ac}\right)\cdot\left(\frac{6.4\,lb\,a.i.}{1\,gal}\right)\cdot\left(\frac{1\,gal}{8\,pt}\right)=1.04\,lb\,a.i./\,ac$$

$$\left(\frac{1.3\,pt}{1\,ac}\right)\cdot\left(\frac{1\,gal}{8\,pt}\right)\cdot\left(\frac{\$100}{1\,gal}\right)=\$16.25\,/\,ac$$

Product B:

$$\left(\frac{0.8\,pt}{1\,ac}\right)\cdot\left(\frac{4.2\,lb\,a.i.}{1\,gal}\right)\cdot\left(\frac{1\,gal}{8\,pt}\right)=0.42\,lb\,a.i./\,ac$$

$$\left(\frac{0.8\,pt}{1\,ac}\right)\cdot\left(\frac{1\,gal}{8\,pt}\right)\cdot\left(\frac{\$110}{1\,gal}\right)=\$11.00\,/\,ac$$

$$Total\ cost\ per\ acre=\$16.25+\$11.00=\$27.25$$

Determine the cost of the premix based on the desired amount of active ingredient that needs to be applied. Start by calculating the amount of Premix C that is required to provide the recommended amount of Product A.

$$Premix\,C\ needed\ to\ equal\ Product\ A=\left(\frac{1.04\,lb\,a.i.A}{1\,ac}\right)\cdot\left(\frac{1\,gal\,Premix\,C}{5.5\,lb\,a.i.}\right)=\frac{0.19\,gal\,Premix\,C}{1\,ac}$$

Calculate the amount of Premix C required to provide the correct amount of Product B.

$$Premix\,C\ needed\ to\ equal\ Product\,B=\left(\frac{0.42\,lb\,a.i.B}{1\,ac}\right)\cdot\left(\frac{1\,gal\,Premix\,C}{3.8\,lb\,a.i.}\right)=\frac{0.11\,gal\,Premix\,C}{1\,ac}$$

(continued)

As we see there is a difference between the amount of Product A and Product B that is needed, Premix C should be applied according to the lower requirements between Product A and Product B. In this example, we would use the requirement for Product B. Then, more Product A can be added to reach the desired rate of the active ingredient. To determine how much Product A is needed, the amount of Product A applied with Premix C must be calculated, and then cost calculations can be made.

$$Pounds\ of\ a.i.A\ applied\ from\ Premix\,C = \left(\frac{0.11\,gal\ Premix\,C}{1\,ac}\right) \cdot \left(\frac{5.5\,lb\,a.i.A}{1\,gal\ Premix\ C}\right) = \frac{0.605\,lb\,a.i.A}{1\,ac}$$

$$Additional\ Product\ A\ needed = 1.04 - 0.605 = 0.435\,lb$$

$$Gallons\ of\ Product\ A\ needed = \left(\frac{0.435\,lb\,a.i.A}{1\,ac}\right) \cdot \left(\frac{1\,gal\ Product\ A}{6.4\,lb\,a.i.}\right) = \frac{0.07\,gal\ Product\ A}{1\,ac}$$

$$Cost\ of\ Premix\,C = \left(\frac{0.11\,gal\ Premix\,C}{1\,ac}\right) \cdot \left(\frac{\$105}{1\,gal}\right) = \frac{\$11.55}{ac}$$

$$Cost\ of\ Product\ A = \left(\frac{0.07\,gal\ Product\ A}{1\,ac}\right) \cdot \left(\frac{\$100}{1\,gal}\right) = \frac{\$7}{ac}$$

$$Total\ cost\ of\ Premix\,C\ plus\ Product\ A = \$11.55 + \$7 = \$18.55.$$

In this situation, the use of Premix C and Product A is more cost-efficient than purchasing and applying products A and B separately.

ACKNOWLEDGMENTS

This paper was partially funded by South Dakota State University, the USDA Natural Resources Conservation Service Conservation Innovation Grants (NR233A750011G019), the USDA Agriculture and Food Research Initiative, American Coalition for Ethanol, GEVO, the South Dakota Corn Utilization Council, and the South Dakota Soybean Research and Promotion Council. The use of trade names is for the convenience of the reader and does not imply endorsement by the authors.

REFERENCES

Clay, D. E., Clay, S. A., & Bruggeman, S. A. (2018). *Practical mathematics for precision farming*. ASA, CSA, SSSA.

Clay, D. E., DeSutter, T., Clay, S. A., & Nleya, T. (2024). *Salinity and sodicity: A global challenge to food security, environmental quality, and soil resilience*. ASA, CSA, SSSA.

Joshi, D. R., Clay, D. E., Alverson, R., Clay, S. A., Westhoff, S., Johnson, J. M. F., Wang, T., & Sieverding, H. (2025). Tillage intensity reductions when combined with yield increases may slow soil carbon saturation in the central United States. *Scientific Reports*, *15*, 10697.

Lyons, S. E., Clark, J. D., Osmond, J. D., Parvel, M. R. P., Pearse, A. W., Slaton, N. A., & Spargo, T. T. (2023). Current status of US soil test phosphorus and potassium recommendations for analytical methods. *Soil Science Society of America Journal*, *87*(4), 985–998. https://doi.org/10.1002/saj2.20536.

Ostrem, D., Trooien, T., & Hay, C. (2016). Irrigating corn in South Dakota. In D. E. Clay, C. G. Carlson, S. A. Clay, & E. Byamukama (Eds.), *iGrow Corn: Best Management Practices*. South Dakota State University. https://extension.sdstate.edu/sites/default/files/2019-09/S-0003-33-Corn.pdf.

Glossary

A

4R (also known as 4Rplus or 5R) – A nutrient stewardship framework promoting the Right Source, Right Rate, Right Time, and Right Place of nutrient application. Some versions add a fifth "R" Right Manner for 4Rplus or 5R approaches.

abiotic stressors – Non-living stress factors (e.g., temperature, drought, salinity).

Absolute accuracy – A receiver's closeness to a true geographic reference; critical for aligning multiple machines and datasets.

Acoustic sensors – Use sound waves to measure properties of solids, liquids, or gases.

Action threshold – The pest population level at which control measures must be taken.

Active acidity (pH) – The concentration of free hydrogen ions (H^+) in soil solution, measured in a 1:1 soil-to-water mixture.

Active sensor – Emit their own energy (at specific wavelengths) to illuminate a target and measure the reflected signal that include RADAR and LiDAR sensors.

Active sensors – Sensors that generate their own light source, allowing consistent measurements regardless of ambient lighting. They reduce the influence of sunlight or shadows on data quality.

Aggregate stability – The soil's ability to resist breaking apart when exposed to water, wind, or tillage. It indicates the structural integrity of a soil.

Agricultural IoT (Ag-IoT) – IoT technologies applied to agriculture for monitoring and decision-making.

Allocative efficiency – Directing inputs to areas where they provide the highest economic return.

alternative hypothesis – Opposes the null, stating that observed differences are due to the treatment effect.

Anthropogenic – Human-made or caused by human activity.

AOI (Area of Interest) – The user-defined geographic area for soil data, which can be selected by address, county, soil survey area, latitude/longitude, legal land description, or manual selection.

Application layer IoT – The software layer that stores, processes, visualizes, and analyzes IoT data, often using cloud servers and dashboards.

Associations – Areas where two or more soil types occur together but are more spatially distinct than in a complex.

Atmospheric windows – Specific wavelength ranges in the electromagnetic spectrum where atmospheric the atmosphere is more transmissive and does not significantly absorb the radiation that passes through it.

Attention – A mechanism that directs a model to focus on specific parts of the input by assigning different weights to them.

Attribute data – non-spatial information linked to spatial features, in formats such as text, numbers, and dates.

Automatic section control (ASC) – A technology that automatically turns application sections on or off, or adjusts duty cycle, using look-ahead distance to account for vehicle speed and system response time, reducing overlapping and skips across the field.

Available water capacity – The amount of water a soil can store for plant use, measured as the difference between field capacity (–10 to –33 kPa) and the wilting point (–1,500 kPa).

B

Big data – Large volumes of structured or unstructured data collected over time.

Big data – refers to large volumes of structured or unstructured data gathered over time.

Biochemical sensors – Use biological components (enzymes, antibodies, nucleic acids) to selectively detect target molecules.

Biological factors of variability – Variability caused by living organisms (e.g., pests, diseases, weeds, microbes).

biotic stressors – Living stress factors that negatively affect plants.

blocking – Grouping experimental units to reduce spatial variability due to soil types, field topography, management history, or other field characteristics.

boom section control – Divides a sprayer boom into multiple sections that can be turned on/off independently to avoid overlap and reduce waste.

buffer – Untreated areas between plots to prevent cross-contamination.

Bulk density – The dry weight of soil per unit volume.

C

Calibration – Establishing a relationship between sensor output and true reference values using standards.

Calibration – Process of adjusting sensor accuracy by harvesting and weighing multiple loads of grain, then recording the results in the yield monitor. Calibration should be performed on uniform, level field areas, with each load representing a distinct yield condition. The moisture sensor also requires calibration to obtain accurate estimates by collecting samples and comparing the sensor output to a certified moisture sensor device.

Capital budgeting – A process for evaluating long-term investments by calculating metrics such as net present value (NPV), internal rate of return (IRR), and payback period.

Cartesian (X, Y, Z) position correction – Improves calculated positions by applying corrections directly in a 3D coordinate system.

Cartographic – relating to maps and the process of making them.

Categorical data – data representing categories with a natural order (ordinal) or without a natural order (nominal), such as soil quality ratings, land cover types, or soil units, often mapped with graduated (ordinal) or unique (nominal) colors.

categorical variables – Variables with a finite number of distinct groups.

Cell/pixel depth – The number of possible values a raster cell/pixel can store, which determines how precisely differences in brightness/value are recorded. Referred to as radiometric resolution in remote sensing.

Clock imprecision – Errors caused by limited precision and imperfect synchronization of satellite and receiver clocks.

Closed-loop control system – Also called a feedback control system; uses sensors to monitor and adjust device operation to reach a target state.

Complexes – Areas where two or more soil types are closely intermixed and cannot be separated at the map scale.

Composite sampling – Combines multiple soil cores into one sample; may be done for an entire field or specific locations.

confounded data – Results influenced by an unaccounted variable, making it difficult to isolate the treatment effect.

Consociations – Areas where one dominant soil type occupies most of the map unit.

Contact temperature sensors – Devices that measure temperature by direct contact, often used in soil probes.

Continuous data – data representing phenomena that change gradually across space, such as elevation, temperature, or rainfall, expressed as ratio (with a true zero) or interval (without a true zero) data.

continuous variables – Numeric variables with an infinite range of possible values.

Continuum – A continuous sequence where adjacent elements appear similar, but differences become more distinct with greater separation.

Control systems – Use sensors to monitor machine operations and allow a computer to regulate function automatically.

CORS program – A U.S. program certifying base stations used for GNSS corrections, managed by the National Geodetic Survey.

Cost-Benefit Ratio (CBRA) – An indicator of how effectively money is invested, expressed as return per dollar spent.

CRISPR – (Clustered Regularly Interspaced Short Palindromic Repeats) A genome-editing tool that allows precise targeting and modification of DNA sequences.

Crop Water Stress Index (CWSI) – Temperature-based index comparing canopy temperature to air temperature, normalized against well-watered and stressed conditions.

Crop Water Stress Index (CWSI) – The Crop Water Stress Index (CWSI) is a remote sensing metric used to detect plant water stress based on canopy temperature, air temperature, and vapor pressure deficit (VPD).

D

Data fusion – Combining datasets from different sources, scales, or times into a single, integrated dataset.

Decision Support Systems (DSS) – Tools that help farmers make informed crop management decisions to protect plant health and improve yields.

Decision tree model – A tree-like model where each branch represents a decision, leading to a final prediction.

Deep Neural Network (DNN) – A fully connected neural network with input, hidden, and output layers. The number of input features determines the number of neurons in the input layer.

Deep percolation losses – Water moving below the root zone.

Dendrometers – Sensors that track trunk diameter changes to indicate plant water stress.

Dense till – Glacially deposited soil material that can be compact and restrictive.

Derivative gain – Determines how strongly a controller reacts to angular error (heading misalignment).

Dielectric constant – A measure of a material's ability to store electrical energy in an electric field.

Dielectric constant – A measure of a material's ability to store electrical energy in an electric field.

Differential GNSS (DGNSS) – Increases accuracy by using corrections from a stationary base station at a known location.

Digital Elevation Models (DEM) – a digital model of the Earth's "bare" land surface, without trees or buildings

Dipolar polarization – The rotation of molecules with permanent dipoles (like water) in response to an electric field.

Discount rate – A percentage used to convert future cash flows into their present value, reflecting the time value of money.

Disease vectors – Insects (or other organisms) that transmit pathogens (viruses, bacteria, or parasites), increasing their spread.

Down-force control systems – Use load cells to measure force on row units and adjust hydraulic/pneumatic pressure to maintain target depth and force.

Downstream tasks – Real-world applications where a pre-trained model is used for specific purposes after initial training.

Drift – Drift is the gradual change in thermal sensor measurements over time, often causing temperature differences between consecutive UAV images.

duty cycle – The percentage of time a control signal is ON during one cycle (e.g., 20% duty cycle = ON 20% of the time, OFF 80%).

E

Economic efficiency – Achieving the highest profit per unit of investment by combining technical and allocative efficiency.

Economic threshold – The pest population level at which expected yield loss will exceed the cost of control, making treatment economically justified.

Ecosystem services – Benefits provided by nature, directly or indirectly, such as clean air and water.

Electrochemical sensors – Sensors that convert plant-related chemical signals—such as hydrogen peroxide—into electrical signals. They help monitor biochemical processes in real time.

Electromagnetic (EM) induction and electrical resistivity sensors – EM induction instruments measure soil electrical conductivity (EC) without direct soil contact. An alternating current in a transmitter coil generates a primary electromagnetic field, inducing eddy currents in the soil that create a secondary field measured by a receiver. The response is influenced by soil moisture, clay content, salinity, and temperature.

Electronic Control Unit (ECU) – A compact, embedded microcontroller that processes sensor data, performs calculations, and communicates with other ECUs.

Electronic control unit (ECU) – The central controller that integrates prescription maps or sensor algorithms with RTK positioning, IMU-derived orientation, and machine speed to generate control signals that achieve the desired application rate.

Emissivity – The ratio of radiation emitted by a surface to that of a perfect blackbody at the same temperature.

Emitted energy – Radiation released directly from an object (e.g., thermal energy), measurable at any time.

Encoder – A model component that converts input data into a compact, meaningful representation (vector).

Ephemeris variation – Errors due to differences between a satellite's predicted and actual position.

Ephemeris – The predicted orbital location of a satellite.

Exchangeable sodium percentage (ESP) – The proportion of sodium ions on soil exchange sites, calculated as (meq Na / CEC) × 100.

F

factors – The experimental variables, such as fertilizer rates, pesticides, cover crops, or tillage types.

False-color composite (FCC) image – Combines any combination of visible, near-infrared, middle-infrared and/or thermal-infrared bands to visualize regions that are not visible observable and can be graphically presented in any color combination.

Feature class – a vector layer with multiple features of the same geometric type (point, line, or polygon) in an area stored in the same spatial reference.

Feature dataset – multiple vector layers of different geometric types (point, line, and polygon) in the same area and stored in the same spatial reference.

Feature selection – Choosing the most relevant features (variables, wavelengths, or inputs) from a dataset to build a model.

Feature space – A multi-dimensional space where each dimension represents a feature (attribute) describing a data point.

Features – The attributes or variables used to describe data points.

Federated learning – A training approach where the model learns from decentralized data stored locally (not centrally) to preserve privacy.

Field systematic variability – Patterned within-field differences in soil properties caused by management practices like tillage, irrigation, or fertilization.

Five soil-forming factors – Introduced by Hans Jenny (1941): climate, organisms, relief (topography), parent material, and time.

fixed effects – Treatment effects that apply only to the specific conditions studied or the average effect.

Fluorescence – Light emitted by a molecule after it absorbs energy and its electrons return to a lower energy state. This emitted light can indicate plant stress or biochemical activity.

Fragipans – Naturally occurring brittle subsurface layers that limit drainage and root penetration.

Fully Convolutional Network (FCN) – A neural network without fully connected layers, designed for tasks like image segmentation and mapping.

Function arguments – Inputs that provide required or optional information to a function.

Function – A command in R that performs a specific task (e.g., mean()).

Furrow irrigation – Applies water through shallow, evenly spaced furrows in the soil.

G

Geodatabase – an all-inclusive data structure used by ArcGIS to store geospatial data, including both raster and vector data.

Geoidal height – The predicted elevation of a point on the earth's surface relative to a mathematical ellipsoidal model of the earth.

Geometric analysis – Capturing object surfaces as images (pixels) or by scanning (lasers, line cameras).

Geoprocessing – the use of tools to transform geospatial data into new output data containing useful information.

Geospatial data – data representing objects or phenomena tied to specific geographic locations on the Earth's surface.

GPS spoofing – A cyberattack that transmits fake GNSS signals to mislead receivers into reporting false positions or times.

Grain moisture sensors – Capacitance-based sensors that estimate grain moisture by measuring dielectric properties, which reflect the grain's ability to store energy in an electric field.

Green-on-brown (GOB) – Refers to weed detection where green plants are identified against bare soil or residue, typically in burndown or pre-emergence applications.

Green-on-green (GOG) – Refers to advanced weed detection systems that distinguish green weeds from green crop plants, often using AI and ML, and typically used in postemergence applications.

Growing Degree Days (GDD) – A measure of heat accumulation (temperature × time) used to predict plant and insect development stages.

Gypsum block – A soil moisture sensor made of a gypsum block containing two embedded electrodes. As the block equilibrates with surrounding soil moisture, its electrical resistance changes, allowing soil moisture to be indirectly estimated.

H

Hardpans – Compacted soil layers that restrict root growth and water movement.

Hydraulic systems – Provide faster, more precise adjustments—better suited for variable field conditions.

Hyperspectral camera – Capture hundreds to thousands of narrow spectral bands across narrow ranges of wavelengths spanning from 250–15,000 nm in the electromagnetic spectrum. These are passive sensors.

I

Image texture analysis – Examining spatial patterns within objects to describe surface characteristics.

Imagery – raster data in the form of satellite images, aerial photos, or scanned maps.

inoculum – Pathogenic material introduced into soil or plants to initiate infection.

Internal Rate of Return (IRR) – The discount rate at which an investment's NPV equals zero, representing the break-even rate of return over time.

International GNSS Service (IGS) – An organization that develops GNSS operational and transmission standards.

Ionospheric interference – Position errors caused by GNSS signals passing through charged atmospheric particles.

Ion-selective electrodes – Sensors that measure the activity of specific ions in a solution using membranes that are selective to particular ions.

Irrigation trigger (refill point) – The soil water level at which irrigation is applied to avoid plant stress.

J

Join – Combines two tables by matching rows on a common field.

Join – temporarily combines two attribute tables into one by matching rows based on a shared field (key).

K

K-fold cross-validation – Splits data into k parts, training on k-1 parts and testing on the remaining one, cycling through all parts.

L

Labeling – Assigning meaningful identifiers (labels) to data so it can be used for supervised learning.

Land capability classes – A classification system that rates soils for suitability in land use activities, identifying major limitations (e.g., erosion, wetness, texture, salinity, pH, thickness, climate) with maps and descriptions.

Laser-Induced Breakdown Spectroscopy (LIBS) – A rapid analytical technique that uses a high-energy laser pulse to ablate a tiny amount of tissue and form a micro-plasma. As the plasma cools, it emits element-specific wavelengths that reveal the sample's elemental composition.

Leave-One-Out Cross-Validation (LOOCV) – A model validation method where all data points except one are used for training, and the remaining point is used for testing. This repeats until all points have been tested.

lightbar guidance – GPS-aided visual guidance that helps operators maintain straight paths and minimize overlap/skips during application or planting.

linear mixed effects model (LMM) – A statistical approach that accounts for both fixed and random effects: correlations within experimental units, strips or fields.

Local calibration – Collecting reference samples and laboratory measurements to adjust and validate sensor readings.

Long-term accuracy – Accuracy measured over extended time, usually about three times worse than pass-to-pass accuracy.

Low Elevation Spray Application (LESA) – Sprays water from 3–5 ft (0.9–1.5 m) above the soil surface.

Low Energy Precision Application (LEPA) – Applies water at low pressure directly onto the soil surface.

M

Management allowable depletion (MAD) – The fraction of PAW that can be depleted before irrigation is needed.

Management-induced variability – Long-term variability resulting from human actions (e.g., tillage, fertilization, irrigation).

Map elements (map surrounds) – components of a map layout that define and provide additional information about a map, such as a legend, scale bar, etc.

Map scale – shows the quantitative relationship between a distance on the map and the corresponding distance on the ground.

map unit (MU) – A symbol or identifier on a soil survey map representing an area with soils of similar characteristics.

map-based cloning – A method for identifying pathogen genotypes linked to resistance, used to study and dissect resistance mechanisms.

Mass flow rate (kg/s) – The rate at which material moves along a temperature or pressure gradient.

Maximum daily shrinkage (MDS) – The daily difference between maximum and minimum trunk diameter.

Measured force – The force detected by an impact plate of clean grain flow, which depends on the mass and velocity (i.e. momentum) of the bulk material.

mechanical damage – Plant tissue injury that allows organisms to infect through natural openings or wounds.

Mechanical sensors – Detect position, motion, force, mass, pressure, stress, or density.

Microbolometers – Microbolometers are temperature-sensitive resistors, commonly made of vanadium oxide (VOx) or amorphous silicon (a-Si), used in uncooled thermal sensors.

Microwave mass-flow sensors – A non-contact, non-intrusive device that uses low-power microwave energy to detect and quantify the flow of material (such as cotton) through a harvester. It operates on a principle similar to radar, often utilizing the Doppler effect to measure material flow., [JF1]

Mobile drip irrigation (MDI) – Attaches drip lines beneath pivot drops to dispense water directly on the soil surface.

Modality – The type or form of data that an AI model can process (e.g., text, images, sound).

Mode of action – The overall manner in which a herbicide affects a susceptible plant at the tissue or cellular level, resulting in suppression or control.

Molecular signatures – Distinct spectral patterns produced by the vibration of chemical bonds when they interact with specific wavelengths of light. Different bonds (e.g., C–H, N–H, O–H) produce unique signatures that allow spectroscopic techniques to distinguish among chemical compounds.

Multipath errors – Occur when signals bounce off objects, causing the receiver to receive the same signal at different times.

Multispectral camera – Capture 3–10 bands of the electromagnetic spectrum, including RGB and non-visible wavelengths such as Red-Edge and near-infrared (NIR). These are passive sensors.

N

Negative buffer – Sinks a feature inward; often used to clean yield data near field edges.

nematodes – Microscopic roundworms that attack plant roots, disrupting water and nutrient uptake.

Net Present Value (NPV) – The difference between the present value of benefits and the present value of costs, used to determine whether an investment is financially worthwhile.

Net profit analysis – Accounts for all costs and revenues; often simplified to revenue minus costs.

Network layer IoT – The transport layer that transmits raw/processed data and commands over networks (e.g., Internet, Bluetooth).

Neutron probes – Instruments with neutron emitters used for precise soil water content measurements.

Nodes and Vertices – in vector data, lines are made of points. The start and end points are called nodes, while the points in between are vertices.

Noncontact temperature sensors – Measure thermal radiation in the infrared and microwave range (2–14 μm) without touching the object.

Normalized Difference Red Edge (NDRE) index – A vegetation index that substitutes the red-edge band for the red band in the equation: (NIR – Red Edge) / (NIR + Red Edge).

Normalized Difference Vegetation Index (NDVI) – A widely used vegetation index based on reflectance in the red and NIR regions, using the equation: (NIR – Red) / (NIR + Red) .

NTRIP (Network Transport of RTCM via Internet Protocol) – A standard protocol (Network Transport of RTCM via Internet Protocol) for sending GNSS correction data over the internet.

null hypothesis – States that no differences exist among treatments.

Nutrient use efficiency metrics – Measure how much applied nutrient is utilized by crops; higher values mean lower losses.

O

Object – In a vector data model, an object represents a real-world thing or event.

Objects – Named variables in R that store information for later use.

Observation Space Representation (OSR) – Uses multiple reference stations to measure satellite errors, which are applied directly to rover calculations.

Open-loop control system – A DC motor supplied with variable voltage to control speed, without feedback.

Optical emission – Light emitted by an excited sample, forming a spectrum that reveals its composition or structure.

Optical sensing – Collects reflected solar radiation from visible (0.4–0.7 μm), near-infrared (0.7–1.4 μm), and shortwave infrared (1.4–3 μm) wavelengths.

outliers – Data points that are far from the average or the majority of the data compared to the rest of the dataset.

P

Package – A collection of related functions bundled for use.

Partial budgeting analysis – A simple decision-making tool that evaluates the financial impact of a practice change by comparing added costs/returns and reduced costs/returns.

Partial factor productivity (PFP) – Yield divided by fertilizer application rate.

Passive sensors – Use naturally existing energy (e.g., sunlight) to detect signals.

Pass-to-pass accuracy – Short-term accuracy (often 15 minutes) measuring precision between consecutive passes.

Penetrometers – Rod- or cone-shaped instruments that are pushed into the soil to directly measure soil resistance to penetration, which is related to compaction. When properly calibrated for soil moisture, penetrometer data can be used to estimate bulk density.

period – The time required for a counter to count from its minimum (BOTTOM) to maximum (TOP), determined by step size and increment speed.

Permanent wilting point – Soil water content (volumetrically defined) at –15 bar, when plants can no longer extract water.

Photoelectric effect – The conversion of incoming photons into an electrical current or voltage proportional to the number of photons.

Photoelectric sensors – Devices that estimate the volumetric flow of clean grain by detecting interruptions in a light beam between a light emitter and a receiving sensor. The magnitude of light interruptions is proportional to the volume of grain passing through the clean grain elevator per unit of time (m3/s or similar).

Physical factors of variability – Abiotic influences on crop performance, such as weather, water, temperature, sunlight, wind, and soil properties.

Physical layer IoT – The device layer containing sensors and actuators that collect data, send it to the Internet, and execute commands.

PID control – A controller that monitors the error between setpoint and actual value, then adjusts actuator signals to reach the setpoint.

plant pathogens – Biotic (living) stressors that reduce plant health and productivity.

Pneumatic dry material applicators – Use conveyor boxes like spinning disk spreaders but accelerate product with air through long booms.

Pneumatic systems – Use airbags for adjustments—affordable but slower.

Porosity – The amount of pore space in soil that allows air and water movement.

Precise Point Positioning (PPP) – A correction method using global base station data to correct satellite clock, ephemeris, and bias errors.

precision drainage – Designs and manages drainage systems in zones, similar to SDI, for optimal water management.

Precision Integrated Pest Management (PIPM) – Integration of site-specific data, sensing technologies, and advanced analytics into pest management strategies that uses proper pest identification, knowledge of life cycles, crop susceptibility, treatment effectiveness, and environmental factors. Strategies focus on prevention, action thresholds, and economic injury levels, tailored to differences in management, climate, history, and soils.

Precision irrigation – Uses soil, canopy, and atmospheric sensors and improved irrigation infrastructure to optimize water use.

Precision Mobile Drip Irrigation – Combines drip irrigation laterals with a center pivot system for targeted water delivery.

Present Value (PV) – The current worth of a future amount of money, calculated using a discount rate.

Pressure chamber – Measures plant water potential by applying pressure to a leaf until sap emerges.

Projected coordinate system – a reference system using linear units to define locations on a 2D surface, as derived from a map projection.

Proportional gain – Determines how strongly a controller reacts to position error (distance off-line).

Proportional–Integral–Derivative (PID) control equation – A mathematical feedback-based control mechanism for continuous regulation.

Proximal plant sensing – Detects heat, light, or electrochemical signals generated within about 2 meters of a plant. These sensors measure specific plant or environmental parameters to support management decisions.

Proximal soil sensors – Detect heat, light, and/or electrochemical signals generated by physical, chemical, or biological processes occurring within approximately 2 meters of the soil surface.

Pulse Width Modulation (PWM) – Rapidly switches a voltage on and off to control power delivery.

Appears continuous because switching is very fast. Built into most microcontrollers.

PWM frequency – The reciprocal of the PWM period, measured in Hertz (cycles per second). Determines how many PWM cycles occur per second.

Q

Query – a request to select a subset of data based on specific criteria, often used to explore spatial data.

R

Radiation sensors – A device that uses a radiation source and a detector to measure crop flow during harvesting. The material stream passes between the source and the sensor, with the amount of radiation reaching the sensor being inversely proportional to the grain flow rate; less radiation indicates a higher flow. Their use is prohibited in the U.S. because of the radioactive source required.

Radiation – High-energy particles or waves (>10eV) that can ionize atoms and break chemical bonds.

Radio sensing – Functions like radar, detecting soil moisture and field structure through clouds and darkness.

Radioactive radiation – Emission of α-particles, β-particles, γ-rays, X-rays, neutrons, and others.

Radiometric calibration – Adjusts waveband intensities measured under varying environmental conditions (e.g., atmospheric conditions) to standardized values.

Raman spectroscopy – A proximal sensing method that uses a laser to scatter light off molecules. Each molecule's unique vibrational energy shift creates a "fingerprint" that can be used to identify and quantify compounds in plant tissue.

random effects – Treatment effects that represent a broader population or distribution.

Random forest (RF) – An ensemble of multiple decision trees that evaluate features in different orders to improve accuracy.

Random sampling – Each point in a field has an equal chance of selection, ensuring unbiased soil sampling.

randomized complete block design (RCBD) – A design where each block (replication) contains all treatments, arranged randomly.

Raster – geospatial data that represents continuous surfaces using a grid of cells.

Readily available water (RAW) – The portion of soil water above the MAD threshold.

Real-Time Kinematic (RTK) GNSS – A high-accuracy DGNSS method that uses carrier signal phase information from satellite transmissions to calculate corrections.

Real-Time Kinematic (RTK) positioning – Uses differential GPS/GNSS corrections to achieve highly accurate location data.

Records – the rows in an attribute table that store different attributes of respective features.

Reflected energy – Electromagnetic radiation that is are redirected from ithe surface of a target that can be detected by a remote sensing instrument.

Relate – Links two attribute tables without merging them, allowing information from either to be accessed separately, while carrying over queries from one to the other.

Relative performance – A receiver's ability to produce consistent positions compared to its own earlier measurements.

Relative radiation – A measure of the intensity of a specific wavelength as it moves through plant tissue. It helps assess how leaves interact with or modify incoming light.

replication – The number of times each treatment is repeated; used to capture variation and enable statistical analysis.

research hypothesis – A statement predicting what researchers or farmers expect to discover.

Reserve acidity – The hydrogen ions bound to soil particles, measured using a $CaCl_2$ solution; indicates long-term pH buffering and lime needs.

Residual parent materials – Soils formed in place from weathered bedrock

Resonance – Occurs when the energy of incoming light matches the difference between quantized energy states in a material.

response variable – The factor measured to show the effect of the treatment.

return-on-investment (ROI) – The percentage yield gain relative to input cost per acre.

Row control module (RCM) – A system that maintains the target seeding rate for each individual row by adjusting seed plate speed based on real-time ground speed and sensor feedback on seed singulation and spacing.

S

Satellite-Based Augmentation Systems (SBAS) – Wide-coverage correction systems using geostationary satellites to transmit correction data, less precise than local base stations.

Sector control variable rate irrigation (VRI) – An irrigation control approach that changes pivot speed by sector during rotation, allowing different water application rates within a field based on spatial variability.

seed metering – The precise, uniform placement of seeds at a controlled rate.

Segmentation – Dividing data (often images) into meaningful regions or classes, typically using computer vision.

Semantic alignment – Defining and standardizing object names, measurement units, and data relationships across multiple data sources.

Sensor fusion – Combines data from multiple sensors to improve spatial, spectral, or temporal resolution.

sensor – A device that responds to a physical stimulus (e.g., heat, light, sound, pressure, motion) and transmits a signal for measurement or control.

servo valve – A valve that modulates flow by partially opening or closing, offering variable resistance.

setpoint – The desired value, rate, or position specified by an operator or ECU.

Shape analysis – Extracting and analyzing object outlines.

Shapefile – the most common vector data format, storing geometry (.shp) and attributes (.dbf) in separate files. Other simple vector data formats include KML and GeoJSON.

signal – The measurable treatment effect, distinct from random variation (noise).

Site-Specific Weed Management (SSWM) – The use of targeted weed control technologies to improve efficiency and precision compared with broadcast herbicide applications or field-wide tillage.

SOC maintenance requirement – The amount of fresh biomass that must be added to maintain current levels of soil organic carbon (SOC).

Soil Adjusted Vegetation Index (SAVI) – A vegetation index designed to reduce soil brightness effects on canopy reflectance in red and NIR regions.

Soil health indicators – Physical, biological, or chemical measurements used to assess soil condition and compare soil health across sites and time.

Soil organic matter (SOM) – Decomposed plant and animal material that stores nutrients, water, and energy for soil organisms.

Spatial alignment – Matching observations from different sources to the same locations.

Spatial calibration – Aligns sensor data with real-world coordinates using GPS and reference points (e.g., ground control points).

Spectral profile (or spectral curve) – A chart showing an object's reflectance or response strength across multiple wavelengths. Also referred to as 'spectra'.

spectral wavelengths – Bands of the electromagnetic spectrum, both visible and invisible, ranging from about 100 nm to 12,000 nm.

split-planter trials – Comparisons set up by applying different treatments with separate planter sections in one pass.

split-plot – An experimental design that has at least two treatment types, one is difficult to change/apply (whole plots) and the other is easy to change (split plots) during an experiment.

Staggered grid start – Introduces random offsets in grid rows to reduce bias and improve sampling accuracy.

State Space Representation (SSR) – Separates satellite errors (clock, ephemeris, atmosphere, biases) and sends these components to the rover for corrections.

STEP-AWBH (phonetically "step-up") - A space-time model (Grunwald et al., 2011) that explains within-field variability using eight factors: Soil (S), Topography (T), Ecology (E), Parent material (P), Atmosphere (A), Water (W), Biotic (B), and Human (H).

Stereovision - Generating 3D data using multiple images or laser-based distance measurements (e.g., TOF cameras).

Stop-and-go method - A sampling approach in which detailed measurements are collected at fixed, predetermined locations before moving to the next sampling point.

Sub-meter receivers - GNSS devices with accuracy under one meter, often using SBAS; common in phones, yield monitors, and basic guidance.

Subsurface Drip Irrigation (SDI) - Applies small, frequent amounts of water below the soil surface with high efficiency.

Subsurface drip irrigation (SDI) - Applies water directly into the root zone via buried tubing (drip tape).

Sudden Death Syndrome (SDS) - A soybean disease caused by the soil-borne fungus Fusarium virguliforme, characterized by leaf yellowing (chlorosis), browning (necrosis), and premature defoliation.

Super-resolution - Enhances the resolution of a dataset (e.g., satellite imagery) using algorithms trained with higher-resolution reference data.

Supervised ML model - A model trained on labeled input-output pairs to learn relationships between inputs and outputs.

Support Vector Machine (SVM) - A classification model that finds the optimal dividing line (hyperplane) between data points, using the closest points (support vectors) to define the boundary.

Support Vector Regression (SVR) - Similar to SVM but used for regression, fitting a line or hyperplane to predict continuous values.

Surface drip irrigation - Delivers small amounts of water directly to the soil surface near roots, reducing evaporation, runoff, and percolation losses.

Surge irrigation - Furrow irrigation method with cyclical on/off water delivery to improve uniformity and reduce losses.

Symmetrical join - Used when all layers have the same extent and resolution.

Synchronized spoofing - A spoofing method where counterfeit signals are aligned with real ones, making detection very difficult.

Synthetic Aperture Radar (SAR) - Synthetic aperture radar (SAR) is an active remote sensing system that emits energy pulses and measures the reflected signal from the surface.

Systematic unaligned grid - Uses a fixed grid layout, but selects a random point within each grid cell for sampling.

T

Tail-water runoff - Water lost as it drains off the field surface.

target population - The broader group of fields or conditions to which results are meant to apply.

Technical efficiency - Using inputs (e.g., seed, fertilizer, water) in the most productive way possible to maximize output.

Temporal alignment - Matching observations to the same time points.

Terahertz (THz) spectroscopy - A non-destructive imaging and sensing method using radiation in the 0.1–10 THz range, between microwaves and infrared. It can reveal structural and water-related properties in plant tissues.

Terrestrial RF link - Radio frequency communication from land-based transmitters used to send corrections from base stations to rovers.

Testing - Evaluating a trained model on data not used during training to measure its accuracy.

Thematic data - data that represents a theme or class, often (but not exclusively) in the form of categories, such as land use, soil type, or crop type.

Time Domain Reflectometry (TDR) - Instruments that use two or three parallel metal rods inserted into the soil to measure the soil's dielectric constant. Precisely timed electromagnetic pulses are sent along the rods, and the travel time of the pulse depends on the surrounding soil's dielectric properties. Because water has a much higher dielectric constant than soil minerals, TDR measurements are commonly used to estimate soil water content.

Timing code – A component of the data transmitted by GNSS satellites that is utilized to measure the travel time of a signal from the satellite to the receiver.

Torque – A twisting or rotational force.

Training – Teaching a machine learning model to make predictions by feeding it data and adjusting parameters to minimize error.

Transmittance – The amount of light that passes through a leaf or other material. It helps indicate internal structure or chemical composition.

Transpiration – The movement of water from plant roots to leaves, where it evaporates into the atmosphere. It is a key process in plant cooling and water transport.

Transpiration – Water loss through stomata during gas exchange.

Transported parent materials – Soils formed from materials moved by water (alluvium), gravity (colluvium), wind (aeolian deposits), or ice (glacial till/outwash).

treatment design – A map showing how treatments are assigned to experimental units or plots.

treatment levels – The categories or values within each factor.

treatment – A variable of interest manipulated by the experimenter, usually based on the research hypothesis or prior knowledge.

trial protocol – A set of written instructions (usually 1–2 pages) for carrying out the experiment.

Trilateration – A mathematical method where a receiver calculates its position using distances from multiple satellites.

True-color composite (TCC) image – Displays images using the blue, green, and red bands, matching the real-world colors perceived by human vision. Also known as 'natural color' image.

U

Undifferentiated – Areas containing two or more soil types that cannot be individually mapped or distinguished with accuracy.

Unsupervised model – A model trained without labeled data, used to discover patterns, clusters, or structures in data.

Urban canyon – Areas between tall buildings where satellite visibility is blocked or reduced.

User interface (UI) – Typically referred to as an in-cab display or mobile application on a smartphone/tablet that provides real-time data, enables user input for field management, and records georeferenced information for post-harvest analysis. It serves as the link between the harvester operator and the sensors, integrating peripheral components into a working system for operation and monitoring.

V

variability – Differences in experimental results caused by error, spatial variation, environment, equipment, or human factors.

Variable rate technologies (VRT) – Systems that use spatial information about field variability to automatically adjust input application rates (e.g., seed, fertilizer, or chemicals) across a field.

Variable-Rate Irrigation (VRI) – Adjusts water application depths across a field to match soil, crop, or environmental variability.

Variable-Rate Irrigation (VRI) – Adjusts water application depths across a field to match soil, crop, or environmental variability.

Vector – geospatial data that represents discrete objects with points, lines, and polygons.

Virtual reference – A statistical method that creates a reference value for calibration and crop response analysis without requiring a physical reference strip in the field.

viruses – Obligate parasites that require living plant cells to replicate, often causing systemic infections throughout the plant.

W

Water use efficiency – Ratio of crop yield produced to water used.

Weather - Atmospheric conditions at a given location and time, including temperature, precipitation, wind, cloud cover, and air pressure.

Web Soil Survey - A free online platform (launched in 2005) that provides soils data for the U.S., including maps, reports, and analyses for agriculture, engineering, land management, and soil health.

Z

Zonal statistics - Summarizing pixel values (mean, min, max, etc.) within a polygon.

Zone Control VRI Systems - Divide fields into irrigation zones managed by valves along the pivot for precise control.

Zone control - Uses valves to independently control water flow to individual sprinklers or zones for precise irrigation.

Zone control - Uses valves to independently control water flow to individual sprinklers or zones for precise irrigation.

Zone sampling (stratified random sampling) - Divides the field into zones based on spatial variability, then samples within each zone to capture targeted differences.

Study Questions Answer Key

Chapter 2

1. **How has agricultural innovation changed productivity?**
Agricultural innovation has transformed productivity through successive waves of technological and methodological advancement. Mechanization replaced labor-intensive practices with efficient machinery (for example, the combine harvester), which dramatically increased work output. Genetic improvements have developed crop varieties with substantially higher yields and better disease resistance. Modern corn hybrids yield more than what was possible a century ago, while improved livestock genetics have increased efficiency of meat and milk production. Chemical innovations revolutionized soil fertility management and pest control through synthetic fertilizers and targeted pesticides. Similarly, the digital revolution introduced data-driven precision agriculture. GPS-guided equipment reduces input waste, while VRT enables site-specific application based on field variability rather than uniform treatment. These innovations collectively contributed to the increased crop productivity post the 1960s.

2. **What are the potential data sources in precision agriculture?**
Precision agriculture integrates multiple data sources to inform farm management decisions. Satellite imagery provides broad field perspectives on crop development and stress patterns, identifying anomalies before they become visible to the naked eye. Drones equipped with specialized cameras offer on-demand, high-resolution field scanning, detecting plant health variations with great precision. Ground-based sensor networks continuously monitor soil moisture, nutrient levels, and local weather conditions, transmitting real-time data for irrigation and input decisions. Farm equipment serves as mobile data collectors through GPS-enabled tractors and harvesters with yield monitors, generating detailed maps of field operations and crop responses across different zones. Weather stations and climate forecasts help with timing decisions for planting, spraying, and harvesting. Historical farm records establish valuable baselines, while public databases provide reference information on soil types, topography, and regional growing conditions. When properly integrated, these diverse data streams enable evidence-based decisions tailored to specific field conditions, optimizing input use and maximizing yield while reducing environmental impact.

3. **What are the barriers to precision agriculture adoption?**
Despite the benefits, there are several barriers that hinder precision agriculture adoption. Financial constraints represent the primary obstacle, with substantial upfront investments required for specialized equipment, sensors, and software. Evaluating the return-on-investment parameters, affording the precision agriculture technology especially becomes challenging for the small- and medium-scale farmers. Technical complexity is another substantial hurdle. Modern precision agriculture systems require digital literacy and specialized knowledge for data collection, troubleshooting, and translating analytics into practical field decisions. Without adequate training support, even purchased technologies may remain underutilized. Infrastructure limitations in rural areas compound difficulties, as reliable broadband internet access remains inadequate across many agricultural regions. This connectivity gap prevents seamless technology integration. Another barrier to adoption is a lack of trust of precision agriculture technologies and agricultural corporations. The advanced data-driven technologies majorly rely on farm-based data, which creates suspicions around transparency and control of data.

4. **Why is privacy important to farmers?**
Privacy is important to farmers because their data represents both intellectual property and competitive advantage. Farm data includes local management practices (such as optimal plating dates, input rates, and equipment settings) developed through years of experience. Field-level yield performance data directly impacts land valuation, rental negotiations, and financing opportunities. When this information becomes accessible to competitors or market speculators, it can weaken a farmer's bargaining position and potentially increase land costs in high-productive areas. Sometimes farmers also express concerns about a third party's access to their operational information, as these data are often collected without consent or compensation.

5. **Why are interdisciplinary studies important?**
Interdisciplinary approaches are important because agricultural systems operate at the intersection of biological systems, technological applications, economic situations, and social contexts. Combining expertise from different disciplines, such as agronomy, engineering, economics, ecology, and social sciences, develops comprehensive solutions that address this complexity. Complex issues, such as climate adaptation and water and nutrient management, require multifaceted approaches that no single discipline can completely address. For example, soil management practices differ across states due to differences in climatic conditions, resource availability, topography, and local knowledge, as well as the nature of the crops grown and the scale of farming. Therefore, scientific evidence and solutions for a problem in one region cannot be replicated in another region without accounting for the social and economic factors. This can only be achieved by involving expertise from different disciplines in approaching a farming issue.

6. **What is co-creation?**
Co-creation is a collaborative approach to agricultural innovation that directly involves farmers as active participants rather than passive recipients. This methodology recognizes that effective solutions emerge when scientific expertise combines with practical farming knowledge using a participatory approach. This approach acknowledges and values farmers' practical expertise and local knowledge and uses it, along with other experts from universities, industry, and government bodies, to identify and solve problems in real-world contexts. Farmers provide critical assessment and feedback on the practicality of the solution in terms of integration with existing systems, economic feasibility, and the refinement of innovations before widespread introduction.

Chapter 3

Activity #1: You are an extension professional in South Dakota tasked with creating a workshop to help farmers understand how to use Web Soil Survey data for decision-making. Discuss how you would create and implement this workshop.

Example answer:

1. Identify the farmer's long-term goals and explain how using the Web Soil Survey can help achieve them.
    - Potential farmer responses may include increased productivity, profitability, and sustainability on their operations. Their responses may be about something in general, or they may have a specific area of interest in their field they want information about.

2. Obtain data from previous years using the Web Soil Survey.
    - This may include an in-person activity in a computer lab in which farmers are guided by an instructor and provided with written materials to access the NRCS Web Soil Survey, select their farm field as the area of interest (AOI), and use the "Soil Data Explorer" tab. They will select the "Soil Properties and Qualities" section and identify dominant soil types and variable soil conditions.
    - The farmers may identify that the variable soil conditions necessitate site-specific nutrient and irrigation

management or that variable rate application can be used to optimize crop performance.

3. Discuss on-farm strip trial approaches with the farmers to demonstrate how they may scale up an implementation plan.
   - Farmer responses may include learning how to apply the newly identified management practices as a strip trial rather than across the entire field. The farmer may learn to collect and analyze data from this experiment, and an instructor-guided discussion may include how to determine whether the treatment was statistically significant and if it can then be used to accomplish the farmer's overall goal.

Activity #2: Nitrogen-rate recommendations using the corn nitrogen rate calculator. Instructions: You are an extension professional in Iowa, and you are tasked with creating a workshop that helps farmers calculate the optimal nitrogen application rate for their field. Discuss how you would create and implement this workshop.

Example answer:

1. Identify the farmer's long-term goals regarding nitrogen rate applications.
   - Potential farmer responses may include that they want to remain profitable while also reducing nitrogen application. This may also include a discussion between the farmers and instructors about how precision nitrogen applications can improve profitability and environmental outcomes, as well as the potential risks of overapplication and nitrogen runoff. A discussion of precision agriculture technologies, such as remote sensing and variable rate applicators, may also be helpful.
2. Obtain data from previous years using soil test data and crop rotation information.
   - This may include encouraging the farmers to bring in old soil test results and previous crop rotation records, as well as previous records on nitrogen application rates.
   - This may also include accessing tools such as the Iowa State University's Corn Nitrogen Rate Calculator tool at https://www.cornnratecalc.org
3. Discuss on-farm strip trial approaches with the farmers to demonstrate how they can scale up an implementation plan using the information gathered from the Corn Nitrogen Rate Calculator tool.
   - Farmer responses may include learning how to apply the information generated from the new research as a strip trial before adopting it across the entire field. The farmer may have an opportunity to collaborate with extension specialists to learn data collection and analysis from such experiments. This could be complemented with an instructor guided discussion that may include how to determine whether the treatment was statistically significant and if it can then be used to accomplish the farmer's overall goal.

Chapter 4

1. **Explain how variable cultivar seeding is accomplished.**
   A multi-cultivar planter is used containing separate seed boxes for the different cultivars.
2. **There are two types of lag time in a flow rate control routine. (1) What are they, and (2) how can variable rate applicators control for them?**
   (1) Control system response time and dynamic stabilization. (2) These can be controlled by variable rate applicators using look-ahead distance, which initiates a rate change in advance or by auto-calibrating controllers that estimate lag time and can dynamically adjust the look-ahead distance.
3. **How can automatic section control (ASC) in variable application systems improve efficiency?**
   ASC can automatically turn application sections ON and OFF based on position, coverage maps, and field conditions. This

improves resource use efficiency, enhancing environmental sustainability and reducing input costs.

4. **How does pulse-width modulation (PWM) control spray application rates?**
   PWM controls spray application rates by opening and closing the valves at specific cycles independent of system pressure. This allows for more precise adjustment of the rate without affecting the overall system pressure.

5. **Why must constant pressure be maintained in a PWM system?**
   Maintaining constant pressure is important to ensure uniform droplet size. This improves the precision and consistency of the spray application.

6. **Why are flow-based control systems easier to manage than pressure-based control systems in liquid applications?**
   Flow-based systems are easier to manage because of the volumetric flow. This flow is linear with valve position, allowing for simpler control over the flow.

7. **What are some differences between the electric drive seed meter and the hydraulic drive system for variable rate seeding? How can they help improve performance?**
   Electric drive seed meters are less expensive, require less maintenance, and have fewer moving parts than hydraulic drives. They can help improve performance by giving each row independent control, given that each row's speed is individually controlled based on real-time speed. They also allow contour farming and point-row shutoff.

8. **How does the controller in a hydraulic-driven dry fertilizer application system use conveyor speed calibration to maintain the target application rate?**
   The controller uses calibration curves along with input from radar or GNSS. These curves correlate the PWM% to the rpm, which allows the conveyor to adjust its speed based on radar or GNSS input.

9. **What is the main difference between sector control and zone control when using a variable rate irrigation system?**
   Sector control changes the pivot's rotational speed creating a pie-shaped zone. In this case, the water application depth is influenced by speed. Zone control changes the water application depth based on the pivot length.

10. **How does having either too much or too little water influence yield, and how can variable rate application help with this?**
    Water stress limits the plant's growth and development, ultimately reducing productivity. Too little water can reduce nutrient uptake and the plant's ability to fight pests and diseases or resist weed pressure. Too much water can risk nutrient loss, especially N, through leaching and denitrification, thereby reducing productivity. Variable rate application can improve management strategies by matching water application to plant needs, reducing operational costs, improving yield, and minimizing resource use.

11. **What role can machine learning play in variable rate agriculture?**
    Machine learning processes large, high-resolution datasets from sensors, satellites, UAVs, and soil tests to generate site-specific prescriptions. It helps identify yield-limiting factors and provides actionable recommendations for nutrient, seed, and pesticide management.

12. **Why is response or lag time critical in variable rate application systems?**
    Lag time impacts the accuracy of input placement. If application rates are not adjusted quickly enough when transitioning between zones, inputs may be over- or underapplied. Controllers use look-ahead functions and auto-calibration to minimize these errors.

13. **How does automatic section control (ASC) help reduce environmental impact?**
    ASC prevents double application in overlap zones and avoids applying chemicals to sensitive areas, such as waterways or buffer

strips. This reduces input waste and minimizes off-target environmental contamination.

14. **What is the difference between flow- and pressure-based control in liquid application?**
Flow-based systems measure and control the actual product volume applied, making them more precise and easier to calibrate. Pressure-based systems are more complex because nozzle flow varies with the square root of pressure, making control less direct.

15. **Why are electric drive seed meters better suited for contour farming than hydraulic or ground-drive systems?**
Electric drives independently control each row's seeding rate, allowing compensation for turns and irregular field shapes. This prevents over- or underseeding on inside versus outside planter rows during curves.

16. **How does pulse-width modulation (PWM) improve droplet size control in spraying?**
PWM separates pressure control from flow rate control. Pressure remains constant to maintain consistent droplet size, while flow is adjusted by changing the nozzle duty cycles. This ensures consistent spray quality at varying speeds.

17. **What are the two main types of variable rate irrigation (VRI), and how do they differ?**
Sector (speed) control adjusts the pivot rotation speed to vary water depth in pie-shaped zones. Zone control adjusts sprinkler flow directly, either in banks or per-sprinkler, providing finer control of water distribution.

18. **How do UAVs enhance N management using the Nitrogen-Rich Biosensor Spot (NRBS) method?**
UAVs capture multispectral imagery to compare crop growth in N-rich spots with surrounding areas. Algorithms use these differences to identify N deficiencies and generate zone-specific fertilizer prescriptions.

19. **What limitations do UAV-based direct application systems have?**
UAVs have limited payload capacity and flight duration, reducing their efficiency for large-scale operations. They are best suited for spot treatments, cover crop seeding, or small fields.

20. **What role are UAV swarms expected to play in agriculture by 2050?**
UAV swarms will likely serve as autonomous farm partners, capable of continuous monitoring, scouting, and applying inputs. Enabled by AI and machine-to-machine communication, they will operate collaboratively with ground robots and large equipment.

Chapter 5

1. **Describe how soil erosion creates variability in crop grain yield.**
Soil erosion removes highly productive surface soil from the soil profile. When it comes to growing crops, the original topsoil (or A horizon) is the most productive. For a given location the topsoil typically has the highest concentration of soil organic matter, which helps with soil aggregation, nutrient cycling, water retention, and more. All these properties are important for healthy crop growth. And when soil organic matter is lost, then the function of the soil for growing crops is diminished. Both water and wind erosion strip this most valuable topsoil with its organic matter and move it elsewhere.

2. **What technology enabled inexpensive and rapid mapping of spatial variability?**
GNSS. GPS. GIS.

3. **What does it mean for some forms of variability to be derived from an anthropogenic source?**
Human-made differences. "Anthro" means coming from humans or human activity. So, any "management" activity on fields that creates variability is anthropogenic. Examples include tillage, livestock activity

or manure applications, synthetic fertilizers or lime, earth work like terraces or pond building, and crop types and rotations selections.

4. **Provide an example of how pests create variability and then identify a management practice you might use to address that variability.**
Pests cause biotic stress in crops, are often not uniformly distributed across the field, and can be influenced by weather variability. Because of the variability in types and severity of specific biotic stresses due to pests, yield variations are not uniform across a field. One example of a management practice is planting varieties resistant to a specific pathogen (e.g., nematodes) in problematic areas, with other varieties planted across the rest of the field.

5. **What does the term "EST" in the expression G × EST × M represent?**
E_{ST} represents environment (E) as impacted by spatial (S) and temporal (T) factors.

6. **Name two examples of abiotic factors that can cause within-field variability.**
Examples include weather, water, temperature, sunlight, wind speed, and the physical properties of soil.

7. **Name two examples of biotic factors that can cause within-field variability.**
Examples include viruses, fungi, and bacteria (when unhealthy for crops, they are called pathogens), unwanted plants (weeds), beneficial and pest insects, nematodes, and vertebrate animals (ground squirrels, mice, birds, deer, etc.).

8. **What types of agronomic knowledge are needed to optimize precision farming possibilities?**
Knowledge about soils, how they impact crop growth and how they vary across a landscape, crop growth and development, soil fertility and crop nutrient needs, pest management principles, and more.

Chapter 6

1. **Which soil-forming factor physically contributes to the development of unstable zones that affect yield over time?**
Relief/topography.

2. **Name soil-forming factors other than topography that might influence soil nutrient variability.**
Climate, biota/organisms, parent materials, time, and human factors.

3. **Fields where high rates of phosphate and potash fertilizer were applied following a "build soil test" program would benefit from a higher intensity grid soil sampling method (more than one sample per 2.5 ac) next time, especially if 90–95% of the soil test values in a field exceed the critical level. (T/F).**
False. (This is because the soil nutrient levels are already well above the critical threshold, making additional sampling less likely to provide a positive return on investment.)

4. **Name four possible variables that can help delineate management zones.**
Spatial yield trends, temporal stability, soil electrical conductivity, elevation. . . etc.

5. **Compare sampling strategies discussed in this chapter and address their pros and cons from the aspects of bias, cost, and feasibility.**
Open-discussion. (Consider how the methods mentioned in this chapter differ in their ability to capture variability, their implementation costs, and their practicality in various field conditions.)

6. **What are the 4Rs, and what might be considered the 5th R or the "plus" in 4Rplus?**
Time, source, rate, and place. Manner/technology would be considered the 5th R.

7. **Discuss the pros and cons of the common assumption that soil mass per acre at a 6-in. depth is equivalent to 2,000,000 lb (when the actual soil bulk density is 1.25 g/cm³.**

$$6\,\text{in.} \times \frac{2.54\,\text{cm}}{1\,\text{in.}} \times \frac{1.25\,\text{g}}{\text{cm}^3} \times \frac{10{,}000\,\text{cm}^2}{\text{m}^2} \times \frac{10{,}000\,\text{m}^2}{\text{ha}} \times \frac{1\,\text{ha}}{2.471\,\text{acre}} \times \frac{1\,\text{kg}}{1{,}000\,\text{g}} \times \frac{2.205\,\text{lb}}{\text{kg}}$$

The value of this calculation is 1.7 million lb of soil/ac in the surface 6 in. Applying 2 million lb of soil/ac × 6 in. will overestimate residual nutrients in the soil since it assumes more soil mass than is actually present. Bulk density varies by soil texture; sandy soils generally have higher bulk density than silty soils.

8. **Grid sampling is the most cost-effective method for detecting soil nutrient variability in highly heterogeneous fields. (T/F).**
False

9. **Soil compaction from heavy machinery can lead to nutrient stratification and reduced root penetration by plants. (T/F).**
True

10. **Which soil property is most affected by parent material?**
 a. Soil aggregate stability
 b. Soil temperature
 c. Soil texture and mineral composition
 (C is the correct answer)
 d. Soil microbial activity

11. **What is the main advantage of zone sampling over grid sampling?**
 a. It is cheaper and requires fewer samples
 (A is the correct answer)
 b. It provides a higher resolution soil nutrient map
 c. It does not require prior knowledge of the field
 d. It is the most unbiased method available

Chapter 7

1. **What is the minimum number of satellite signals needed for a GNSS receiver to compute a 3D position? Explain.**
Four satellites are the minimum required for complete position determination, but most receivers will use more than four satellite signals simultaneously if they are available. Some receivers will acquire a fix with only three satellites, but it will only be a 2D fix.

2. **Use internet resources to explore the status of the GPS constellation and report the total number of satellites currently functional in the constellation, how many have L2C capabilities, and how many broadcast the L5 codes.**
The goal of this question is exploration; please seek out this information through the internet or your local library. If needed, some resources (links below) can be helpful: https://www.gps.gov/systems/gps/space and https://www.navcen.uscg.gov/?Do=constellationStatus

3. **Use internet resources to explore the CORS stations in your state. Print a state map showing all the locations. For the one closest to your location, find out where it is and when it was established.**
The goal of this question is exploration; please seek out this information through the internet or your local library. If needed, a resource (link below) can be helpful for more information: https://www.ngs.noaa.gov/CORS

4. **A farmer wants to purchase an RTK GNSS system for a farm. The base station antenna will be placed on top of a centrally located grain bin. If the accuracy of the RTK GNSS system is 1.5 cm + 4 ppm, what is the accuracy at a field located 35 km away?**
The 4-ppm accuracy value means that for every 1,000,000 m from the base station, the accuracy decreases by 4 m.
 35 km × (1, 000 m/km) = 35, 000 m of baseline separation

→Degradation = 35, 000 m × (4 m/1, 000, 000 m) = 0.14 m = 14 cm

→Accuracy = 1.5 + 14 = 15.5 cm

5. **For a particular date and location, use a constellation modeling tool such as the one found at http://gnssplanning.com to plot the availability of GPS and GNSS satellites and the DOP values. Are there particular times of the day that would be really good or potentially bad for GNSS work?**
Because most GNSS satellites are moving relative to the Earth, the configuration of the satellite constellation overhead at any point in time can vary significantly. The configuration of the satellites at any given point in time is quantified by the Dilution of Precision (DOP). Dilution of Precision can be predicted using orbital models. Some users needing the utmost precision will plan to do field work during the times of the day when the satellites will be in the best configuration. A number of software tools are available to help users plan for variations in satellite constellations.

6. **Explain how the absolute and relative accuracy of GNSS receivers would affect their use in machine guidance applications.**
A receiver with good relative accuracy would exhibit consistent position determination relative to earlier position solutions from the same receiver. That receiver might be very adequate for providing position solutions for vehicle guidance systems if the end result were simply to reduce the number of overlaps and skips in a single field operation. If, however, it would be desirable to use multiple machines in the same field or to overlay data from one field operation with another, the lack of absolute accuracy could cause significant spatial offsets between the field operations. The absolute accuracy of a receiver is a measure of the position deviation from some established reference frame. A high absolute accuracy would ensure that multiple machines working in the same field would be able to follow the same path, and that multiple data layers would overlay correctly without geographic offset.

7. **Explain the difference between short-term and long-term accuracy and how these two types of accuracy relate to GNSS position quality.**
A critical aspect of GNSS position quality is the distinction between short-term and long-term accuracy. Because most GNSS satellite constellations are constantly changing, GNSS position measurements tend to drift over time. Consequently, position measurements taken over a shorter time interval will tend to be more precise than those collected over extended periods. Because of this, many manufacturers will quote a short-term accuracy performance parameter often called "pass-to-pass" accuracy. This is typically based on a 15-min relative accuracy performance. For many receivers, the long-term accuracy is approximately three times larger than the short-term pass-to-pass accuracy.

8. **A farmer has a field that is bordered on one side by a large manufacturing facility in a three-story metal-clad building. Why might their GNSS equipment perform poorly near that edge of the field?**
Multipath errors occur when the same radio signal is received at two different times. This will happen when the radio signal bounces off some object. In this case, a GNSS transmission could come straight to the receiver and also bounce off the metal building, causing the same signal to arrive at the receiver a short time after the initial receipt. Multipath errors can cause significant problems with GNSS receivers especially when operating near objects that reflect radio waves, such as metal buildings and bodies of water. Placing the receiver antenna near any metallic objects could increase the chances of multipath. Manufacturers effectively use choke rings in antennae along with other hardware and software filtering techniques to minimize the effects of multipath errors. However, the user should still be careful to scrutinize GNSS information when operating in places where multipath is likely to occur.

9. **A GPS satellite is at an altitude of 20,100 km directly above a receiver. Calculate how many wavelengths of the L1 and L2 GPS frequencies there would be between the satellite and the receiver.**
The L1 band operates at 1,575.42 MHz and the L2 band operates at 1,227.60 MHz. One hertz (Hz) means that an event repeats once per second. In the case of radio waves, one hertz (Hz) is interpreted as a wave per second (wave/s).
    i. The speed of light is 299,792,458 m/s, which is approximately equal to 300,000,000 m/s.
    ii. Speed of light (C) = wavelength × frequency.
    iii. 1 MHz = 1,000,000 Hz = 1,000,000 wave/s.
    iv. L1 wavelength = (300,000,000 m/s)/(1,575,000,000 wave/s) = 0.19 m/wave.
        1. Number of wavelengths = 20,100,000 m/0.19 m/wave = 105,525,000 waves.
    v. L2 wavelength = (300,000,000 m/s)/(1,227,000,000 wave/s) = 0.2445 m/wave.
        1. Number of wavelengths = 20,100,000 m/0.2445 m/wave = 8,220,900 waves.

10. **A farmer in Canada is getting ready to plant a field that is perfectly square and nearly completely flat using a GNSS-based autosteer system. In which direction (N-S or E-W) should they plant the field to get the best guidance accuracy? Explain your reasoning.**
Satellites do not cross the poles of the Earth. In Canada, which is relatively close to the North Pole, there will be fewer satellites visible to the north. For the best accuracy based on satellite configuration, it is desirable to have satellites spread evenly in all directions from the receiver. When traveling N-S in Canada, there is a good chance that satellites will be evenly distributed to the left and right of the receiver, giving good accuracy in the lateral direction, which would mean good guidance accuracy. However, when traveling from east to west, there will be fewer satellites on the receiver's right side, which could reduce lateral guidance accuracy. Therefore, from a purely guidance perspective, the receiver would be expected to perform better with N-S rows than with E-W rows.

Chapter 8

These questions are designed to help readers review the key concepts presented in the chapter. Most answers are found in the respective sections of the chapter. Certain open-ended questions have answers that are user-specific.

1. **Describe GIS functions and workflow.**
Acquire, store/manage, display, analyze/transform, present/share.

2. **What are the different types of GIS software?**
QGIS, ArcGIS Pro, etc.

3. **Imagine you are tasked with creating a GIS-based decision support system for a precision farming operation. Which GIS programs would you choose to work with and why?**
Open questions. See characteristics of the different programs in Table 8.1.

4. **Describe the characteristics of raster and vector data.**
Raster: pixels; vector: points, lines, and polygons.

5. **What is a good approach for sharing shapefiles?**
Compress to a Zip file.

6. **How is attribute data managed in GIS?**
A matrix or a spreadsheet with rows (records) and columns (fields). Relational database structure.

7. **Define and describe geographic coordinate systems and horizontal datum.**
See the section "Geographic Coordinate Systems."

8. **What datum is loaded onto most GPS systems?**
WGS84.

9. **What datum is commonly used in the GNSS in your country?**
Open question.

10. **What are projected coordinate systems? Can you give some examples and describe them? Which system is commonly used in precision agriculture?**
See details in the section "Projected Coordinate Systems."

11. **What are some of the typical map symbology types and attribute data types with respect to measurement scales for all data types (i.e., nominal, ordinal, interval, and ratio data)?**
See details in Table 8.4.

12. **What are the essential elements to include when making a map?**
Title, legend, north arrow, data sources, scale bar, etc.

Chapter 9

1. **Georeferenced points are essential for creating accurate boundaries in precision agriculture.**
True

2. **Vector-based operations in GIS are used to analyze and manipulate raster data.**
(False)

3. **What type of data is typically represented in raster format in GIS?**
A. Points
B. Lines
C. Polygons
D. Continuous surfaces (**D is the correct answer**)

4. **How would you create a workflow for creating a field boundary (polygon) based on GPS locations collected by your mobile device (e.g., smartphone)?**
(1) Walk the field using your smartphone GPS to put down GPS points.
(2) Export the GPS points to a GIS software, such as ArcGIS or QGIS.
(3) Digitize a polygon by connecting the points in the GIS software.
(4) Save or export it as a shapefile.

5. **Should you use a projected coordinate system commonly used in precision agriculture, such as UTM, to produce maps using functions that work based on latitude and longitude instead of a distance-based system?**
(False)

6. **Several examples of GIS processing tools are provided within this chapter. Find and research two GIS tools not listed and describe their characteristics and functions.**
QGIS plugins like FieldImageR, Remote Analysis of Vegetation Indices (RAVI), Precision Agriculture Tools (PAT).

7. **List four examples of spatial data characterized as vector data used in precision agriculture.**
(1) Field boundaries (polygon)
(2) Soil sampling points (points)
(3) Plot boundaries (polygon)
(4) Crop rows (lines)
(5) Yield monitor data (points)

8. **List four examples of spatial data characterized as raster data used in precision agriculture.**
(1) Cleaned and interpolated yield map
(2) NDVI or another vegetative index from a UAV image
(3) NDVI or another vegetative index from a satellite image
(4) Topographical elevation map (a DEM)
(5) Interpolated gridded soil data

9. **A grower asks you to perform gridded soil sampling on one of her corn fields. Describe the sequence of steps required to accomplish them, from the initial data needed, tools used, all the way to delivering an interpolated map.**
(1) Collect field boundary data—GPS around perimeter OR digitize with an image basemap.
(2) With the boundary, create a grid of a predefined cell size.
(3) Extract the centroid or corner of the grid cells to represent the sampling point.
(4) Sample at the GPS-extracted points in the field and send the soil samples to a soil laboratory for analysis.

(5) Connect the soil data report from the laboratory with the sampling locations (through a common column in both data, such as sampleID or locationID).
(6) Input the combined data into a GIS software, such as QGIS or ArcGIS.
(7) Clean and interpolate the data (Kriging or IDW).
(8) Generate an interpolated map and export it.

10. A grower asks you to create management zones on one of his cotton fields. List at least five geospatial data layers that could be used for this purpose.
(1) Yield map
(2) Soil properties layer
(3) Soil electrical conductivity layer
(4) DEM elevation layer
(5) Remote sensing vegetative index layer, such as NDVI
(6) Hail or pest damage layer

11. Following from the situation in Question 10, if you were using R as GIS, list the functions and their arguments required to import a vector geospatial data, set its native CRS (if missing), and transform its CRS to a different one.
Import a vector: read_sf("path/to/file.shp").
Set native CRS: st_set_crs(x = objectname, value = nativeEPSGcode).
Transform to another CRS: st_transform(x = objecname, value = newEPSGcode).

12. Should precision agriculture prioritize environmental sustainability over maximizing crop yields? (Debate)
Debate

Yes. Answer: Protecting environmental sustainability protects the resources humans rely on to grow food, such as the soil, water, and air, ensuring long-term productivity.

No. Answer: It is economically crucial that farmers maximize their yields to stay profitable. There is a significant demand for food as the human population is growing, and farmers must prioritize feeding as many people as possible with the land they have.

Additional thoughts to debate and/or consider:

- The connection/ balance between profitability, resource use, and food security globally
- The profitability and the challenge of maintaining profitability for small farms vs. large farms (how might this factor into the farmer's ability to remain sustainable?)
- What programs or resources might be available to small farmers to help financially support a transition to more sustainable resource use?
- How can understanding and analyzing geospatial data improve a farmer's ability to be both sustainable and maximize yields?

Chapter 10

1. What is the Web Soil Survey (WSS), and who provides the materials, knowledge, and information within the WSS?
The WSS is an interactive online tool that can be used to create maps, provide fundamental soils knowledge, and improve management decisions. The USDA NRCS maintains it. The soils data comes from the National Cooperative Soil Survey (NCSS), collected from federal, state, and local agencies, universities, and other partners.

2. Provide three reasons why a farmer might use the WSS in their farming practice.
Some reasons a farmer might use the WSS in their farming practice are to

- Determine soil suitability before planting specific crops
- Evaluate drainage class for tile drainage/irrigation decisions/compaction/and so on
- Identify highly erodible areas
- Understand available water capacity with the field soil types
- Assist with improving fertilizer applications
- Estimate SOM/carbon/soil health levels for sustainability or carbon credit programs

- Help in cover crop selection
- Guide precision agriculture systems
- Plan grazing rotations or forage production based on soil productivity
- Understand soil limitations for structures
- Understand soil compaction potential
- Assist in manure or compost applications
- Support on-farm research trials
- Help meet compliance or USDA documentation
- Track historical soil data and/or monitor changes after management transformations.

3. **You are advising a producer to use cover crops within the crop rotation. How could the WSS be used to support your recommendations related to soil suitability, drainage, or erosion risk?**
   WSS can support cover crop recommendations in several ways:
   - The WSS provides data on soil texture, depth, and rooting restrictions, which helps determine whether certain cover crops will survive.
   - Drainage Class identifies the water speed movement in the soil, which impacts the types of cover crops selected.
   - Erosion Risk indicates how susceptible soils are to wind and water erosion.
   - By using WSS, a person can select a cover crop mix to match the field's conditions criteria and circumstances, and hence, improve the likelihood of success and improving soil health.

4. **Within the WSS, create an area of interest (AOI) from your selected field. See references for step-by-step guidance (Malo, 2008, 2016a, 2016b), as well as Section 10.2.**
   a. Identify the MU.
   b. While looking at the MUs, how are they different from one another?
   c. Create a Crop Productivity Index (CPI) Map from the AOI-selected field.
      i. Identify and explain the trends being displayed with the CPI map.
      ii. How will these trends impact the Best Management Practices within the farming practice decisions?
   d. Create a Suitability and Limitations Report from the AOI-selected field.
      i. What are the limitations for crop production?
      ii. What are the limitations for rangeland production?
   e. Create reports for bulk density, available water capacity/storage, organic matter (OM), and compaction risk.
      i. Explain how bulk density, available water storage, and OM are impacting the soil health in this field.
   f. From these reports, what management changes would you suggest the producer make to increase profitability while also improving soil health?
   g. How would these reports impact the crop rotation selection?

5. **If the soil depth is 0–15 cm, the bulk density is 1.26 g/cm³, and the SOM is 4%, how much SOC is contained in the surface 15 cm?**

$$15\,\text{cm} \times \frac{1.26\,\text{g}}{\text{cm}^3} \times \frac{10{,}000\,\text{cm}^2}{\text{m}^2} \times \frac{10{,}000\,\text{m}^2}{\text{ha}}$$
$$\times \frac{1\,\text{kg}}{1{,}000\,\text{g}} \times \frac{0.04\,\text{kh SOM}}{\text{kg soil}} \times \frac{0.58\,\text{kg SOC}}{1\,\text{kg SOM}}$$
$$= \frac{43{,}848\,\text{kg SOC}}{\text{ha}\,15\,\text{cm}}$$

Chapter 11

1. **What are the main components of a proximal sensing system?**
   Sensor, recording unit, GNSS.

2. **List at least four types of proximal sensors.**
   EM meter, soil moisture sensor, soil temperature sensor, cone penetrometer.

3. **What are some of the types of sensors your smartphone has?**
   GNSS, camera, compass, acceleration sensor, temperature sensor, among others.

4. **Compare proximal and remote sensing in terms of spatial sampling elements, including their advantages and disadvantages.** Remote sensing typically offers complete coverage of the area with a fixed sampling interval (pixels). Proximal sensing provides approximate measurements (within a point) that must be interpolated. The sample support of optical remote sensing (pixel) is a "flat tile" that mainly represents the visible surface, while some proximal soil sensors look deeper into the ground. Deployment of mobile proximal sensors is often limited by field conditions and growth stage, whereas optical remote sensing is restricted by cloud cover.

5. **Compare the support (footprint) of the antimony electrode with that of the electrical conductivity sensor shown in Figure 11.4.** The tip of the antimony electrode has a diameter of about 5 mm, and it senses the immediate surface. The electrical conductivity sensor covers a half-sphere area about 1.4 m wide and 0.7 m deep.

6. **True or false: Neutron radiation is commonly used in agricultural field sensors because of its precision and ease of handling.** False. It is not commonly used due to the dangers associated with *n*-rays

7. **Which region of the electromagnetic spectrum is primarily responsible for exciting electrons to higher energy states in atoms or molecules?**
   A. Infrared
   B. Microwave
   C. Visible and ultraviolet **(C is the correct answer)**
   D. Radio

8. **True or false: X-ray fluorescence spectroscopy (XRF) is used to measure soil water content.** False. XRF is used to detect elemental concentrations, not water content.

9. **Match each sensor characteristic with its correct description:**

Characteristic	Description
Selectivity	A. Ability to maintain consistent performance over time
Stability	B. Ability to detect a specific analyte without interference
Resolution	C. Smallest detectable change in the measured variable
Uncertainty	D. Degree of accuracy/ correctness in sensor output

Selectivity → B
Stability → A
Resolution → C
Uncertainty → D

10. **True or false: Smart sensors are capable of real-time data processing and decision-making without the need for post-processing.** True.

11. **Why is resonance important in the absorption of electromagnetic radiation by materials?** It enables selective and efficient absorption of electromagnetic radiation to nondestructively detect specific elements of soil and plants.

12. **What is the principle behind the photoelectric effect used in optical sensors?** The photoelectric effect is the conversion of incident light (photons) into an electrical signal (current or voltage) proportional to the number of photons detected.

13. **Discuss the advantages and risks of using radioactive radiation in agricultural sensing. How do safety concerns affect the choice of technology in the field?** Radioactive radiation is useful in detecting specific elements or moisture levels with high precision. However, because of its ionizing nature, it poses health risks that require strict safety protocols. As a result, safer alternatives, such as passive γ-ray detectors or non-ionizing sensors, are often preferred for field applications.

14. **How does reflectance differ from transmittance in terms of light interaction with a surface?**
Reflectance is the portion of light that bounces off the surface, whereas transmittance is the portion that passes through the material.

15. **What is the difference between physical and chemical sensors?**
Chemical sensors require direct interaction between the object and the sensing element, whereas physical sensors do not.

16. **What is the advantage of chemical and biological sensors as compared with physical sensors?**
Chemical and biological sensors can be more selective.

17. **What is calibration, and why is it necessary?**
Calibration relates the sensor reading to the properties of interest. The calibration mode is typically a mathematical function or a machine learning model. Calibration measurements of objects with known properties are required to tune the calibration model. Calibration is necessary because sensor readings often do not exactly represent the properties of interest. Furthermore, environmental conditions can affect the measurements. Thus, calibration must be updated when moving to different environments, when ambient conditions change, or when the sensors show drift due to wear.

18. **What are some advantages to fusing data from multiple sensors?**
Improvements in accuracy, enlargement of footprint, obtaining complementary information, alignment of scales, and generation of time series.

19. **How can data collected at different points within a field be fused?**
By interpolation or by attribution (e.g., if management zones are used, data points can be labeled with zone IDs).

Chapter 12

1. **Compare deployment methods: Analyze advantages and limitations of stop-and-go, static, and on-the-go measurement approaches with specific examples.**
Most sensor systems can be categorized as stop-and-go, static, and on-the-go. Stop-and-go sensors collect information at a predefined location and then travel to the next point of investigation. Static chambers measure at a fixed point, and on-the-go sensors measure as the sensor moves over the field

2. **Explain apparent conductivity: Why do electromagnetic sensors measure "apparent" rather than "true" electrical conductivity, and why is calibration essential?**
In some systems, current is injected through two electrodes while potential difference is measured between two additional electrodes. The spacing and arrangement of these electrodes determine the depth of investigation, allowing for flexible measurement depths. When using this system, the soil becomes part of the electromagnetic circuit and the changing local conditions immediately affect the signal recorded by a data logger. For example, changes in soil texture or soil moisture will influenced the measured values. Apparent EC references the use of a sensor to assess EC, whereas EC references the actual EC in the laboratory.

   Soil salinity mapping can also use electromagnetic (EM) mapping. EM sensors can be used for on-the-go, stop-and-go, and static measurements. These sophisticated instruments generate a primary electromagnetic field that induces eddy currents in the soil, which in turn generate a secondary magnetic field. The ratio of these fields provides a measure proportional to soil EC_a. Contemporary EMI instruments incorporate multiple receiver coils at different orientations and separations from the transmitter, enabling simultaneous measurement at different soil depths.

3. **Discussion question: Evaluate future potential: Assess the statement "Proximal soil sensing will eventually eliminate laboratory analysis discussing strengths and limitations."**

Multiple Choice Questions

1. **Primary advantage of proximal sensing:**
   a. Can be as accurate as lab analysis
   b. **Real-time field measurements without sample preparation**
   c. Costs nothing after purchase
   d. Works identically in all conditions

2. **Electromagnetic sensors measure:**
   a. Soil pH precisely
   b. Specific nutrients directly
   c. **Apparent electrical conductivity**
   d. Biological activity levels

3. **"Apparent" conductivity means:**
   a. Sensors lack precision
   b. **It is related to the electrical conductivity**
   c. True conductivity unmeasurable
   d. Intentionally imprecise design

4. **Most chemically specific sensors:**
   a. Electromagnetic induction
   b. Ground penetrating radar
   c. **Ion-selective electrodes**
   d. Mechanical penetrometers

5. **Management zone delineation:**
   a. Creates legal boundaries
   b. **Divides fields by similar soil properties for management**
   c. Determines equipment size
   d. Establishes crop rotations

Answer Key: 1-b, 2-c, 3-b, 4-c, 5-b

Chapter 13

1. **State some examples of active and passive sensors.**

   Active sensors include The Crop Circle ACS-435, SPAD meter, and GreenSeeker. Passive sensors are OceanOptics, FieldSpec, etc. Measurements depend on ambient light.

2. **Explain how a transmittance-based chlorophyll meter differs in function from reflectance-based sensors such as GreenSeeker?**

   Chlorophyll meters measure light that passes through the leaf (transmittance) to estimate chlorophyll content, while GreenSeeker measures reflected light to calculate NDVI. In both cases, light absorbance is measured so the two technologies are correlated as well. In the chlorophyll meter, assuming no reflectance, absorbance =
   1 – transmittance. In the GreenSeeker, assuming no transmittance, absorbance =
   1 - reflectance.

3. **How can THz spectroscopy distinguish between healthy and drought-stressed leaves?**

   THz radiation is highly sensitive to water; differences in hydration alter absorption and refractive index, which can be measured to assess stress.

4. **In what way does LiDAR outperform RGB imaging for 3D canopy reconstruction in high-density crop systems?**

   LiDAR captures fine structural detail through precise distance measurements, even in complex or shaded canopies, while RGB is limited by surface visibility.

5. **Name one of the plants' signaling molecules detected by electrochemical sensors.**

   Hydrogen peroxide

6. **Which of the following is a key advantage of proximal sensing over satellite sensing?**
   a. Lower cost
   b. **High spatial resolution**
   c. Larger area coverage
   d. Wide spectral coverage

7. **What is the principle behind Raman spectroscopy?**
   a. Emission of fluorescence per millisecond
   b. Reflectance of light energy per millisecond
   c. **Inelastic scattering of monochromatic light per millisecond**
   d. Elastic scattering of light per millisecond

8. **What is the ratio used to calculate nitrogen sufficiency index (NSI)?**

$$NSI = \frac{N\text{-}Limited\ zone}{Well\text{-}fertilized\ zone}$$

Chapter 14

1. **What portion of the electromagnetic spectrum constitutes the visible range (visible to the human eye)?**

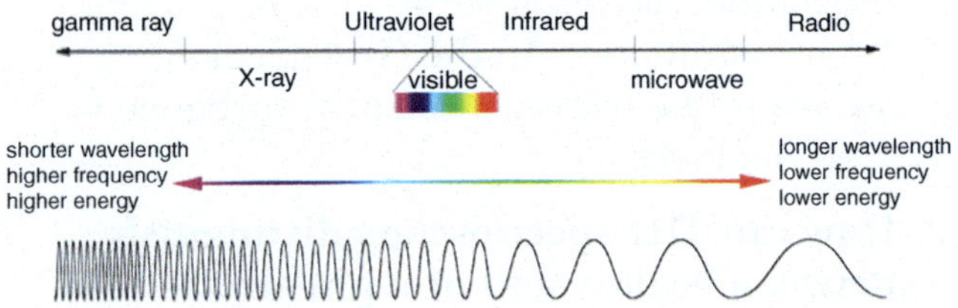

Source: NASA GSFC.

The visible portion of the spectrum ranges from 400 to 700 nm.

2. **Provide a definition of "vegetation index."**
A vegetation index is a spectral transformation of at least two bands that are designed to improve our understanding of the plant system. Indices allow for reliable spatial and temporal intercomparisons of terrestrial photosynthetic activity and canopy structural variations.

3. **Differentiate "remote" sensing from "proximal" sensing.**
Remote sensing is the act of monitoring an object without direct contact. In proximal sensing the sensors are placed very close to the target. They could be in physical contact with the target.

4. **What is the term for a sensor that produces its own energy for sensing?**
An **active** sensor

5. **What is the principal wavelength range for thermal infrared sensing?**
There are specific "atmospheric windows" where remote sensing is possible, but, unfortunately, there are also many areas of the spectrum where remote sensing is not possible. An example is the thermal infrared range, where it is only possible to do remote sensing using detectors sensitive to energy with wavelengths between 3000 and 5000 nm, and 8000 and 14.2600 nm; thus, these two wavelength regions constitute atmospheric windows for thermal infrared.

6. **Which remote sensing platform is best suited to the coverage of many square miles on a weekly basis?**
See Table 14.1.

7. **Which remote sensing platform can be readily deployed on short notice for crop scouting?**
UAV.

Chapter 15

1. **Why are stomata important for plant water management?**
When stomata close, transpiration is reduced, increasing leaf temperature.

2. **What are the limitations of thermal scanners placed on a UAV?**
Because thermal scanners placed on a UAV are generally not cooled, they can be subject to temperature drift. This drift increases uncertainty in the data.

3. **What are the limitations of thermal scanners placed on space-based platforms?**
They may not have the spatial and temporal resolution needed for precision irrigation.

4. **What are the benefits of SAR?**
This information has the potential to help farmers and water resource managers make more informed decisions throughout the growing season. For example, the OpenET and the Farm and Ranch Management Support (FARMS) systems can be used to generate field-specific ET reports. The OpenET datasets are also available via an application programming interface.

5. **What are the disadvantages of SAR?**
   1. Underestimation of ET can occur during the winter and spring.
   2. Lower accuracy is possible in semi-arid climates with higher cloud cover compared with Mediterranean climates with less cloud cover.

3. Combining multiple models together generally increases accuracy over individual models; however, these individual models can produce vastly different answers than when combined.
4. Estimating ET for orchards and vineyards is still challenging and uncertain because of the complex canopy structures and greater shading.

Chapter 16

1. In fine sandy soils, typical plant-available water (PAW) per foot of depth is closest to

a. 0.2–0.4 inch/ft
b. 0.7–1.0 inch/ft **(B is the correct answer)**
c. 1.8–2.6 inches/ft
d. 3.0–3.5 inches/ft

2. Management allowable depletion (MAD) represents

a. total soil water at wilting point
b. fraction of plant-available water (PAW) that can be depleted before irrigation **(B is the correct answer)**
c. water lost during deep percolation
d. difference between RAW and field capacity

3. Which soil moisture sensing technology measures the travel time of an electromagnetic pulse along a probe?

a. Gypsum block
b. Tensiometer
c. Capacitance sensor
d. Time-domain reflectometry (TDR) **(D is the correct answer)**

4. Thermal infrared canopy sensors can detect water stress primarily because stressed plants

a. emit more long-wave radiation due to higher leaf temperature **(A is the correct answer)**
b. display lower chlorophyll content
c. increase reflectance in the near-infrared band
d. exhibit reduced stomatal conductance causing canopy cooling

5. Low Energy Precision Application (LEPA) sprinkler packages are characterized by

a. high-pressure operation with fine-mist nozzles
b. water applied 10 ft above the canopy
c. very low-pressure heads wetting a narrow soil band **(C is the correct answer)**
d. buried emitters delivering sub-surface water

6. Computerized Hole Selection (CHS) in poly-tube irrigation adjusts hole diameters to compensate for

a. variable soil texture
b. declining pressure along tubing length **(B is the correct answer)**
c. changes in water salinity
d. slope-induced runoff at the tail-end of furrows

7. Subsurface drip irrigation typically achieves field-level application efficiencies

a. below 60%
b. around 70%
c. above 95% **(C is the correct answer)**
d. exactly 100%

8. Variable rate irrigation (VRI) by speed control modifies application depth by

a. changing nozzle pressure while pivot speed stays constant
b. altering travel speed of the machine across the field **(B is the correct answer)**
c. opening and closing individual tower valves only
d. substituting sprinkler heads with drip tapes

9. Smartphone irrigation scheduling "apps" generally need which minimum inputs from the user?

a. soil sand–silt–clay percentages only
b. local weather station ownership details
c. crop type, location, planting date, and system characteristics **(C is the correct answer)**
d. satellite image mosaics at 10-cm resolution

10. **In the two-step FAO-56 approach widely used for irrigation scheduling, which prerequisite variable must be estimated before calculating crop evapotranspiration (ET_c)?**
   a. Soil matric potential
   b. Reference evapotranspiration (ET_o or ET_r) from weather data **(B is the correct answer)**
   c. Canopy temperature
   d. Soil electrical conductivity

Chapter 17

1. **What is IPM?**
   Integrated Pest Management (IPM) is a management-based strategy that focuses on long-term pest prevention that uses biocontrol, chemical management, habitat manipulation, modification of cultural practices, and pest-resistant varieties.
2. **Why is insect biology important to understand?**
   Insect biology is important because it provides the foundation for understanding insect behavior, life cycles, and vulnerability.
3. **What are some examples of scouting techniques used to determine insect populations (See Figure 17.1)?**
   There are many different types of scouting, three techniques are shown above with different sampling patterns.
4. **How could new technologies be used to inform insect management decisions?**
   New technologies can be used to assist in pest identification and the use of models to match problems to potential solutions. In addition, machine learning models generate intelligent predictions by finding complex patterns in large datasets. These algorithms can be extremely accurate because they learn and modify themselves as data are collected.
5. **Why is machine learning important in IPM?**
   Machine learning is important in IPM because it can be used to help identify the pest and match remediation techniques to problems.
6. **What is a CNN?**
   Convolutional neural network (CNN) is a machine learning model used for image identification.
7. **What is the difference between object classification and object detection?**
   Object classification mainly focuses on determining an assigned identification value (ID). Research has been conducted to improve classification efficiency (Arroyo, 2021; Li et al., 2021). For example, Spiesman et al. (2021) investigated an approach where the area of interest was identified, followed by classification.

   Object detection using machine learning models generally involves identifying the location of the object, and subsequently assigning it to a classification category (Wang et al., 2022). For example, Vega et al. (2021) used the weighted Hausdorff distance as a loss function in combination with the convolutional neural network (CNN) machine learning model to detect beetles on sensor images, with a mean accuracy of 94.3%. Grijalva et al. (2023a, 2023b) used this same CNN machine learning model to detect aphid individuals on sorghum plants.

Chapter 18

1. **What is a weed?**
   A weed is a plant that is growing where it is not desired.
2. **Why are weed populations spatially and temporally variable?**
   Within a field many factors influence weed variability. Some of these factors include variable soil moisture and temperatures as well as weed seeds germinating at different times.
3. **Why does SSWM reduce herbicide application rates?**
   There are portions of the field where herbicides are not applied.

4. **What are the risks of not applying a preemergence herbicide to a portion of the field?**
   Not applying a preemergence herbicide to small area of a field can result in a high germination of weed seeds in that area.
5. **What is green-on-brown?**
   Live vegetation vs soil
6. **What is green-on-green?**
   Crop plant vs weed
7. **Explain the process of creating a weed spray map.**
   See section Example of Prescription Map Development Using Pix4DFields
8. **Explain the benefits and barriers to adopting SSWM.**
   See section Considerations for Adoption

Chapter 19

1. **Define plant disease.**
   Plant disease is a condition in a plant caused by biotic stressors (e.g., fungi, bacteria, viruses, nematodes) that disrupts normal physiological functions and results in symptoms like wilting, chlorosis, necrosis, stunted growth, or reduced yield.
2. **Explain how precision agriculture technologies contribute to the early identification of plant diseases.**
   Precision agriculture helps in disease diagnosis by leveraging technologies such as Global Position System mapping, remote sensing, drones, and soil and crop sensors to monitor field conditions in real time and detect early changes in plant health, such as shifts in chlorophyll content or canopy temperature, that often precede visible symptoms. If the stress is suspected to be caused by a plant disease, confirmation requires sending samples to a plant disease diagnostic lab, where accurate identification methods are used. These may include microscopic examination of pathogens, molecular diagnostic tools, and consultation with plant pathology experts to ensure a correct diagnosis and inform appropriate management strategies.
3. **Provide a successful example where precision spraying technologies have led to a significant reduction in pesticide use.**
   A good example is the management of cotton root rot, caused by *Phymatotrichopsis omnivora*. High-resolution satellite imagery was used to develop prescription maps that identified diseased areas, enabling targeted application of foliar fungicides only where needed. This site-specific approach significantly reduced the number of fungicide applications and lowered production costs (Yang et al., 2017).

Chapter 20

1. **What sensors need to be calibrated when preparing to harvest a field?**
   The moisture and mass or volume sensors need to be calibrated, along with moisture sensor. The accuracy of the calculated yield is dependent on the calibration of the moisture and mass or volume sensors, as well as the accuracy of the DGPS system receiver.
2. **What factors influence calibration constants?**
   Systematic errors are produced by using old calibration data. With time, calibrations can change due to machine wear or sensor output drift. To minimize these errors, routine calibration is recommended.
3. **You harvest 50,000 lb of corn from a field at 22% moisture. How many pounds and bushels of corn do you have at 15.5% moisture?**

$$\frac{(100\% - 22\%)}{(100\% - 15.5\%)} \times 50{,}000\,\text{lb} = 46{,}153\,\text{lb corn}$$

$$46{,}153\,\text{lb corn} \times \frac{\text{bu}}{56\,\text{lb corn}} = 824.2\,\text{bushels}$$

4. **What will happen to the yield monitor data when the combine slows down? What happens when the combine speeds up?**
   When a yield monitor slows down, yields increase, and when it speeds up, yields decrease.

5. **Why does yield monitor data need to be cleaned?**
Yield monitors create files containing large amounts of spatial information, including the type of monitor, longitude, latitude, field name, object ID, track, swath width, distance, elevation, header switch status, GPS differential status, time, *Y* and *X* offset, material flow rate, harvester speed, productivity (in acres per hour), grain moisture content, type of crop harvested, and date. However, before using the data, erroneous data must be removed. Erroneous data is collected when the combine slows down or speeds up, as well as in the end rows. For example, if 10 bu of grain move from the grain auger through the clean grain elevator, and the combine covers 0.1 acre, then the yield is 100 bu/acre; however, if the same amount of grain is transferred but the combine slows down and covers only 0.005 ac (0.002 ha), then the calculated yield would be 2,000 bu/acre.

6. **True/False: Mass flow sensors on combines must be calibrated for each crop type to ensure accurate yield estimates.**

   True. Mass flow sensors must be calibrated for each crop type.

7. **True/False: Grain moisture sensors estimate moisture content by measuring the dielectric properties of the grain.**

   True. Grain moisture sensors use dielectric properties to estimate moisture.

8. **True/False: The use of radiation sensors for yield monitoring is permitted in the United States.**

   False. Radiation sensors are prohibited in the United States due to the radioactive source.

9. **True/False: Yield monitor data does not need to be cleaned before being used for management decisions.**

   False. Yield monitor data must be cleaned before use.

10. **True/False: AI algorithms in modern combines can automatically adjust ground speed and harvest settings in real time.**

    True. AI can automatically adjust ground speed and harvest settings.

Fill in the Blank

11. **The process of adjusting sensor accuracy by harvesting and weighing multiple loads of grain is called __________.**
Calibration

12. **The __________ sensor is typically mounted at the top of the clean grain elevator and measures the force of grain impacting a plate.**
Impact plate

13. **__________ correction is a technique used to improve the accuracy of GNSS or GPS measurements by comparing data from a mobile receiver with a stationary base station.**
Differential

14. **The __________ provides a link between the farmer and the sensors and is commonly referred to as the in-cab display.**
User interface (UI)

15. **Imagery from __________ and drones is improving our understanding of yield estimates by providing high-resolution aerial perspectives.**
UAVs

Multiple Choice

16. **Which of the following is NOT a component of a typical grain yield monitoring system?**
    a. Mass flow sensor
    b. Moisture sensor
    c. GNSS or GPS receiver
    d. **Soil pH sensor**

17. **What is the main reason for cleaning yield monitor data before analysis?**
    a. To reduce file size
    b. **To remove erroneous data points caused by speed changes, partial swaths, or GPS errors**
    c. To improve combine speed
    d. To calibrate the moisture sensor

18. **What technology is used in John Deere's Predictive Ground Speed Automation to generate predictive field maps?**
    a. Infrared sensors
    b. **Forward-facing stereo cameras and preharvest satellite imagery**
    c. Soil moisture probes
    d. Weigh wagons

19. **Which sensor is used to estimate the volumetric flow of clean grain by detecting interruptions in a light beam?**
    a. Microwave sensor
    b. **Photoelectric sensor**
    c. Impact plate sensor
    d. Load cell

20. **What happens to yield monitor data when the combine slows down?**
    a. Yield estimates decrease
    b. **Yield estimates increase**
    c. Yield estimates remain unchanged
    d. Data is deleted

Chapter 21

1. **List two benefits of on-farm replicated strip trials as a research tool in agronomic studies, specifically for farmers, agronomists, and researchers.**
   On-farm replicated strip trials enable farmers to (a) test new products, technologies, and management practices on their farms; and (b) develop site-specific recommendations tailored to their local conditions, improving long-term profitability and environmental stewardship.

2. **List key differences between on-farm trials conducted by farmers and small-plot controlled field experiments carried out by university researchers and graduate students.**
   The main differences between on-farm trials and small-plot controlled field experiments are the size (5–10 ft wide and 20–50 ft long for small plots vs. field scale or field length for on-farm trials) of experimental units or plots and the size of the equipment used in applying treatments and managing crops. Farmers use their standard equipment for conducting on-farm trials, making these studies more practical and relevant to them.

3. **What aspects of on-farm experiments require the most attention?**
   Planning on-farm studies together with farmers is the first step for successful on-farm experiments. Any mistake during the trial execution can lead to the failure of the study, resulting in the loss of data and research resources. Effective planning and communication between participating farmers and researchers are critical to avoid costly mistakes.

4. **On-farm strip trials fall into the category of "learning by doing." List the roles of modern technologies, the internet, and social media in on-farm research.**
   University extension services and private industry offer numerous online resources to educate farmers and agronomists about the fundamentals of on-farm research. Additionally, there are public and commercial tools and digital platforms designed to assist with planning, executing on-farm trials, processing and analyzing data, and providing insights and recommendations.

5. **What new technologies may be helpful for conducting on-farm research in the near future?**
   Some new technologies, including cloud- and AI-based digital data collection and analytics platforms, UAVs, autonomous tractors, and small robotic vehicles, are expected to be used in planning and executing on-farm trials, as well as in data collection, data analyses and decision-making.

6. **Why is a research hypothesis needed, and what are its key elements?**
   A research hypothesis, which is a critical part of a well-designed on-farm study, specifies the research question that guides the entire research process to ensure clarity, scientific integrity, and the development of reliable management decisions. The two key elements of a research hypothesis are the null

hypothesis and the alternate hypothesis. The null hypothesis states there are no differences among the treatments, while the alternative hypothesis asserts that at least one treatment pair has significant differences.

7. **List key weather and environmental variables needed to interpret results from on-farm trials studying foliar fungicide applications on soybean.**
   Examples are daily maximum and minimum temperatures, monthly rainfall before planting or during the growing season, humidity, leaf wetness, and the presence and severity of diseases.

8. **Describe the role of aerial imagery in cleaning yield data from on-farm trials.**
   Aerial imagery is helpful for identifying areas within fields affected by treatment application errors, flooding, hail, wind, soil compaction, equipment damage, and other factors.

9. **What can go wrong during the planning of an effective on-farm trial?**
   The key to avoiding many mistakes in the beginning or planning stage of on-farm trials is to establish strong communication between farmers and researchers and to develop a reliable and practical trial execution protocol. Moreover, failing to identify the important agronomic, economic, or environmental research questions for farmers can diminish their interest in on-farm research. Focusing on overly complex research questions can also lead to potential errors and loss of data.

10. **What common considerations should be taken into account when using farmers' equipment, such as planters or sprayers, to conduct on-farm research?**
    Treatment prescriptions can be developed for variable-rate or fixed-rate seeding trials, as well as for variable-rate variety or hybrid trials, if planters are equipped with electric or hydraulic drive capabilities. Another factor to consider is whether planters have individual seed boxes, two bulk seed tanks, or planter section shutoffs. Treatment prescriptions can be uploaded to execute on-farm sprayer trials. Sprayer trials can be conducted along and across crop rows. Considering potential drift is essential when planning sprayer trials. Guidance or AB lines can be used to create treatment maps that spatially match planting, fertilization, spraying and harvest patterns within fields.

11. **Describe how publicly available tools can be used to select on-farm research locations. Explain three external factors to consider when selecting a research site.**
    Depending on on-study objectives, data on (1) climate and weather extremes, (2) soil and field topography, and (3) current and historical field management can be used to select on-farm research locations. For example, if one of the study objectives is to quantify the potential interaction between field topography and yield response to increasing seeding rates of corn, DEM data can be obtained from public or private sources, or from soil apparent electrical conductivity maps generated by EC field surveys.

12. **Explain the difference between "signal" and "noise" in the context of data analysis.**
    The distinction between "signal" and "noise" helps in data interpretation and decision-making. The "signal" is meaningful information in the data collected or effect we are trying to measure. The "noise" is unwanted variation, whether random, or systematic, that interferes with the meaningful information. Often, soil reflectance can be considered as noise when measuring the reflectance of crop canopies using remote sensing.

13. **Give examples of metadata for on-farm research. What is the role of metadata?**
    Metadata is information about the data, treatments, environmental conditions, equipment calibration, and all details relevant to the study. Metadata is critical in data analy-

ses, result interpretations, and building databases for predictive modeling. For example, early season rainfall, soil characteristics (texture, organic matter, drainage), and previous crops can help explain yield potential yield responses to additional side dress N applications on corn.

14. **List key advantages of joining an on-farm research network.**
Key advantages of joining an on-farm research network include increasing the ability to combine datasets from multiple locations, improved data summarization, greater statistical power and decision-making, shared experiences and knowledge with other participants, and the potential to attract sponsors and new partners.

Chapter 22

1. **Identify at least three types of georeferenced data commonly collected at the field level in precision agriculture today.**
Possible answers include crop yield data, soil properties (e.g., moisture, nutrients, organic matter), elevation or topography data, soil electrical conductivity, pest and disease scouting reports, as-applied planting and fertilizer data, and remote sensing imagery.

2. **Data records can be categorized as either __________ or __________.**
Qualitative or quantitative.

3. **Provide four ways in which precision agriculture tools and data analytics can directly benefit farmers or agricultural consultants in current practice.**
(1) Optimizing the placement and timing of inputs (fertilizers, pesticides, irrigation). (2) Improving field-level decision-making and efficiency. (3) Enhancing sustainability and promoting environmental stewardship. (4) Facilitating detailed documentation, verification, and benchmarking to improve management practices.

4. **In which decade were computer-generated maps and the Global Positioning System (GPS) first widely integrated into agricultural practices?**
In the early 1990s.

5. **Which two key technological advances have driven the transition toward modern "digital agriculture"?**
Machine connectivity and the adoption of consumer mobile devices.

6. **Define the type of field boundary most commonly used in precision agriculture data management and explain what it represents.**
An operational boundary, representing the actual area managed, cultivated, or actively farmed within a field.

7. **List the four primary classifications of agricultural data and provide a contemporary example for each.**
(1) Agronomic Data: Yield maps, soil nutrient analyses. (2) Machine Data: Equipment performance metrics (fuel usage, engine load, speed). (3) Production Data: Weather records, farm notes, planting/application dates. (4) Remote sensing imagery: Normalized Difference Vegetation Index images or multispectral imagery.

8. **Identify four practical applications of remote sensing imagery in modern precision agriculture.**
(1) Identifying variability in crops and soil. (2) Improving crop scouting efficiency. (3) Determining fertilizer and pesticide needs. (4) Evaluating the effectiveness of treatments and assessing crop health and potential yield.

9. **Explain the primary purpose of conducting multiyear data analyses in the context of precision agriculture.**
Evaluating whether adopting site-specific management practices offers economic benefits and sustains profitability over time.

10. Describe the purpose of data mining.
Data mining, particularly through machine learning, is used for predictive modeling, enabling farmers to forecast crop yields and disease outbreaks, and to optimize management strategies by recognizing patterns in large datasets.

11. Discuss the importance of data compatibility and explain why data compatibility is slow.
Data compatibility (interoperability) refers to the ability of different systems or software to seamlessly communicate, exchange, and use information without extensive manual intervention. Its significance lies in improving the efficiency, accuracy, and integration of precision agriculture technologies from various manufacturers, ultimately promoting widespread adoption.

12. List and briefly define each component of digital agriculture as it is currently understood.
(1) Financial analysis: The use of farm cost-related data (costs, revenues, cash flow, risk, and return on investment) to guide operational and capital decisions, optimize resource allocation, and evaluate technology and infrastructure investments under real-world constraints. (2) Prescriptive Agriculture: Data-driven decision support for input rate recommendations and site-specific applications—what input to apply, where, when, and how much—using agronomic and predictive models, along with analytics to improve profitability, efficiency, and sustainability. (3) Big Data: Using large, high-volume, and high-variety datasets generated from farm operations (e.g., sensors, machinery, satellites, weather, soils, markets) that require advanced storage and analytics to extract insights, make predictions, and support better decisions and financial analysis.

13. Identify two advantages and two disadvantages associated with the practice of data sharing in agriculture today.
a. Advantages include the following:
- Reducing the duplication of data collection efforts.
- Enabling comprehensive analyses, innovation, benchmarking, and quicker decision-making.

b. Disadvantages include the following:
- Increasing the risk of data misinterpretation or misuse due to complexity or poor quality.
- Increasing the risk of the potential unauthorized use of data leading to competitive disadvantages or privacy concerns.

Chapter 23

1. Distinguish between technical, allocative, and economic efficiency in precision agriculture, and explain how each contributes to farm performance.
Each contributes to farm performance in its own way.
- Technical efficiency describes the ability to achieve the maximum output from a given input. Precision agriculture technologies enhance this by reducing overlapping operations (e.g., Global Positioning System–guided auto-steering) and applying inputs more accurately. It answers the question: Are you using your input without waste?
- Allocative efficiency involves directing resources to where they generate the highest returns. With soil and yield maps, farmers can target inputs to the most productive zones, optimizing resource allocation. It answers the question: Are you spending your money on the right inputs in the right amounts?
- Economic efficiency combines both technical and allocative efficiencies, ensuring the highest profit for each dollar spent. Precision agriculture tools improve profitability by increasing input efficiency and boosting yields. It answers the question: Is a farm achieving the highest possible profit

by combining efficient use of inputs (technical efficiency) and the optimal allocation of resources based on costs and returns (allocative efficiency)?

2. **Why do information-intensive technologies, such as yield monitors and soil sampling, have relatively lower adoption rates than automated technologies?**
Information-intensive tools require advanced data interpretation skills and seamless integration into decision-making processes. Unlike automation tools (like Global Positioning System–guided steering), which offer immediate time and labor savings, information-based tools often involve
   - Data overload without clear decision support
   - The need for technical training or software to interpret results
   - Delayed benefits, since farmers may have to test, analyze, and adapt before results are visible.

3. **What are the main risks that precision agriculture technologies can help farmers manage?**
Precision agriculture helps mitigate the following major risks:
   - Production risk—By using tools such as remote sensors, drones, and Variable Rate Technology, farmers can respond promptly to pests, drought, or nutrient deficiencies, thereby reducing yield losses.
   - Market/price risk—Providing more accurate yield forecasts and cost assessments, precision agriculture helps farmers plan harvest timing, input purchases, and marketing strategies more effectively.
   - Legal/compliance risk—Global Positioning System–tracked application logs and georeferenced data help document compliance with pesticide use regulations and conservation programs.
   - Safety risk—Automation and remote sensing reduce direct exposure to hazardous chemicals and stressful labor, protecting health and safety.

4. **A farmer is considering adopting an integrated Real-Time Kinematic (RTK) auto-steering system to reduce chemical input use and enhance crop yield over a period of 5 years. The initial investment cost is $18,000, with no salvage value at the end of its useful life. The farmer incurs an annual operating cost of $1,500 for software installation. Each year, this technology generates two types of economic benefits: (a) input cost savings of $3,000 from reduced pesticide and labor use and (b) additional revenue from yield improvement that amounts to $4,500.**

Assuming an annual discount rate of 8%,

A. Calculate total benefit, total cost, and net benefit.
Annual Net Benefit = Annual Benefits − Annual Operating Cost

$$\begin{aligned} &= (\text{chemical and labor savings} \\ &\quad + \text{yield improvement}) \\ &\quad - \text{software installation cost} \\ &= \$7{,}500 - \$1{,}500 = \$6{,}000. \end{aligned}$$

Table 23.A1. A list of cost and benefit items

Cost/benefit item	Amount
Initial investment cost	$18,000
Software installation cost	$1,500
Annual chemical and labor savings	$3,000
Annual yield improvement	$4,500

B. Calculate the net present value (NPV).

The NPV is calculated using the formula,

$$NPV_T = \sum_{t=1}^{T} \frac{C_t}{(1+r)^t} - C_0$$

Initial investment (C_0) = $18,000
Net cash flow = $6,000
Discount rate = 8%.

Table A23.2. Net cash flows and present values for the first 5 years

Year	Net cash flow (C_t)	Discount factor (8%) $1/(1+r)^t$	Present value (PV_t)	Cumulative PV (PV_T)
1	$6,000	0.926	$5,555	$5,555
2	$6,000	0.857	$5,144	$10,699
3	$6,000	0.794	$4,763	$15,462
4	$6,000	0.735	$4,410	$19,872
5	$6,000	0.681	$4,084	$23,956

For Years 1 and 2, cumulative PV can be calculated as

$$PV_1 = (\$6,000)\left[\frac{1}{(1+0.08)^1}\right] = \$5,555$$

$$PV_2 = \$5,555 + \$5,144 = \$10,699$$

NPV can be calculated as

$$NPV_5 = \frac{C_1}{(1+r)^1} + \frac{C_2}{(1+r)^2} + \frac{C_3}{(1+r)^3} + \frac{C_4}{(1+r)^4} + \frac{C_5}{(1+r)^5} - C_0$$
$$= (\$5,555 + \$5,144 + \$4,763 + \$4,410 + \$4,084) - \$6,000$$
$$= \$23,956 - \$18,000 = \$5,956$$

Since the NPV is positive, the investment is financially viable.

C. Based on your result in Part B, should the farmer invest in this technology? Provide an economic justification.

A positive NPV means that the present value of future cash inflows (benefits) exceeds the present value of the investment cost, accounting for the time value of money. Therefore, the farmer should invest in this technology.

D. Calculate the cost–benefit ratio (CBR).

$$CBR = \text{Total Present Value of Benefits / Initial Investment} = PV_T / C_0$$
$$= \$23,956 / \$18,000 = 1.33.$$

This means the investment returns $1.33 for every $1 spent.

E. Find out the payback period of this investment.

To determine how long it takes for the investment in auto-steering to pay for itself, we refer to the cumulative present values in Table A23.2. The initial investment of $18,000 is recovered between Years 3 and 4. By the end of Year 3, the cumulative present value of benefits stands at $15,462, leaving $2,538 still to be recovered.

In Year 4, the present value of the benefit is $4,410, which is more than enough to cover this shortfall. To find out how far into Year 4 the break-even point occurs, we compare the remaining $2,538 to the Year 4 benefit:

$$\$2,538 / \$4,410 = 0.58.$$

This means that about 58% of the fourth year's benefit is required to fully repay the investment. Since 58% of a year corresponds to roughly 7 months, the break-even point is reached about 7 months into Year 4. This translates to a payback period of approximately **3 years and 7 months.**

F. Suppose input savings decline to $2,000 annually due to pesticide price change. Recalculate the NPV and reassess the investment decision.
The annual benefits will drop from $6,000 to $5,000.

The NPV will drop from $5,956 to $1,963.

CBR will also decrease, from 1.33 to 1.11.

The investment remains profitable, but the margin of profit is much narrower.

Chapter 24

1. **Why is it important to link soil health indicators directly to crop yield and production in the context of precision agriculture?**
   a. Because yield is the only variable influenced by soil conditions
   b. To ensure soil sampling frequency can be reduced over time
   c. To make soil health assessments economically relevant and actionable for producers **(C is the correct answer)**
   d. To eliminate the need for biological soil testing

2. **A soil test shows moderate aggregate stability but low microbial activity in a sandy loam field under continuous corn production. What combination of practices would best address these issues in the precision framework?**
   a. Increase tillage frequency and synthetic N applications
   b. Introduce organic amendments and multi-species cover crops targeted to low-activity zones **(B is the correct answer)**
   c. Apply fungicides uniformly to eliminate competing microbes
   d. Irrigate the entire field using overhead sprinkler systems

3. **How do DSSs assist in soil health management?**
   a. Automate pesticide spraying regardless of crop stage
   b. Monitor air temperature trends only
   c. Integrate data from sensors and models to guide site-specific decisions **(C is the correct answer)**
   d. Replace all manual field inspections

4. **Which of the following is most likely to compromise the accuracy of soil health maps generated using zone sampling?**
   a. Over-sampling in eroded shoulder positions
   b. Uniform topography across the field
   c. Failure to incorporate multi-temporal yield data and topographic features **(C is the correct answer)**
   d. Use of automated GPS guidance for sampling points

5. **Define soil health and explain how its physical, chemical, and biological properties contribute to sustainable agricultural systems.**
   Soil health is the continued capacity of soil to function as a living ecosystem. Physical properties (like aggregate stability) support water retention, chemical properties (like pH) dictate nutrient availability, and biological properties (like microbial activity) drive nutrient cycling and disease suppression.

6. **Why is understanding spatial variability important in soil health management? Discuss how advanced technologies, such as remote sensing and GIS, are used to assess soil health.**
   Soils naturally vary across a landscape. Mapping this variability allows farmers to apply inputs only where needed, avoiding uniform over-application on healthy areas or under-application on degraded spots. Remote sensing provides aerial imagery to detect crop stress and moisture variations, while GIS layer this data to create accurate, interactive management zones across fields.

7. **Compare the impacts of conservation tillage, crop rotation, and cover cropping on soil health.**
   Conservation tillage protects soil structure and prevents erosion, crop rotation breaks pest cycles and diversifies nutrient demands, and cover cropping continuously adds soil organic matter while preventing nutrient runoff.

8. **Describe how precision agriculture tools such as VRT and DSS optimize soil health management. What are the economic and environmental benefits?**

VRT equipment applies fertilizers and water at exact localized rates based on soil maps, while DSS analyzes collected data to recommend the most effective practices for those applications. Economically, precision tools reduce wasted input costs and optimize yields. Environmentally, they drastically reduce nutrient runoff, minimize greenhouse gas emissions, and prevent overall soil degradation.

9. **What are the major barriers to adopting precision agriculture practices for soil health? Discuss potential solutions and emerging trends that could address these challenges.**
   Major barriers to adopting precision soil health practices include high initial equipment costs, delayed financial returns, and a lack of standardized, affordable testing protocols. To overcome these hurdles, emerging solutions focus on providing financial subsidies to farmers and developing unified, regionally calibrated evaluation frameworks like the SHAPE tool. Furthermore, future technological trends are driving the creation of low-cost, real-time field sensors and AI-powered decision support systems to make complex soil data more accessible and actionable.

Chapter 25

1. **What is precision conservation?**
   Precision conservation is a set of spatial technologies and procedures used to implement conservation management practices that account for the spatial and temporal variability of the goods and services produced by these practices across natural and agricultural systems.
2. **What are the components of a conservation plan?**
   The components a conservation plan include (a) identifying the problem; (b) identifying potential options; (c) determining the expected costs and economic assessment; (d) implementing the practice using the obtained resources; (e) assessing success, reviewing, and modifying the strategy as needed; and (f) communicating findings to the end users
3. **What is the 7R conservation approach?**
   The 7R framework includes applying the right fertilizer, at the right rate, using the right method, with the right conservation practice, in the right place, at the right scale, and at the right time.
4. **Explain the concepts of sparing and sharing.**
   "Sparing" sets aside nonproductive areas for conservation practices. The second approach, termed "sharing," involves using the land for multiple purposes. This revises agricultural management approaches.
5. **Why are mussels important for in-stream water filtering?**
   Mussels are an important indicator species for water health and they conduct in-stream filtration which contributes to the conservation of water quality.
6. **True/False: There is no difference between precision agriculture and precision conservation.**
   False. Precision agriculture focuses on site-specific strategies to optimize productivity and resource use at the field or farm level, addressing localized challenges such as soil fertility, irrigation, pest management, and resource utilization. In contrast, precision conservation considers the long-term sustainability of a production system, tackling challenges within a specific soil, watershed, or landscape. It enhances long-term environmental sustainability by mitigating soil erosion, improving water quality, and supporting biodiversity beyond individual fields.
7. **Question (short answer): How do precision agriculture and precision conservation differ in terms of primary focus and scale, and why is this distinction important for long-term sustainability?**

Precision agriculture focuses on site-specific management to optimize crop productivity and input use efficiency at the field or farm level (e.g., fertilizer, irrigation, pest control). Precision conservation, in contrast, focuses on long-term environmental sustainability by targeting soil, water, biodiversity, and ecosystem services at broader scales such as fields, watersheds, and landscapes. This distinction is important because it emphasizes that sustaining production requires managing not only yield but also the underlying natural resources and ecosystem processes that support agriculture over time.

Chapter 26

1. **What types of electrical signals would you expect to see as the output of a sensor?**
   a. Analog signals (continuous signals that vary smoothly over a range)
   b. Digital signals (discrete signals, high/low or 1/0).
   c. Frequency signals (output varies with the measured variable)
   d. Pulse-width modulated signals (output is in the width of digital pulses)

2. **Current signals (current proportional to the measured value). In open-loop control, why is it not possible to be certain that the speed or rate of an actuator is at its desired value?**
   In open-loop control, there is no feedback to ensure that the actuator is moving at the proper speed or rate.

3. **In a closed-loop control system, what is the setpoint? How is it related to the error?**
   The desired value of the system output, such as speed, location, or flow rate, is known as the setpoint. The difference between the measured output and the setpoint is known as the error. To determine the error, the control system continuously compares the measured value to the setpoint. To shift the process variable closer to the setpoint and lower the error, the controller uses the error signal to modify the actuator's output.

4. **Thought experiment: Open-loop control using feedback is never guaranteed to exactly meet the setpoint if only proportional gain is applied to the error. Which gain, integral or derivative, would need to be added to the system to ensure the setpoint is reached?**
   Integral gain should be applied to make sure the system reaches and stays at the setpoint. To ensure that the output finally reaches the setpoint, the integral term steadily collects the error over time and removes steady-state error.

5. **What are the benefits and drawbacks of liquid sprayer system with a servo valve to control flow rate?**

   Benefits:
   - Enables precise control of flow rate
   - Minimizes over-application and waste
   - Reduces the need for constant manual adjustments
   - Is suitable for large-scale operations

   Drawbacks:
   - Requires more complex installation, calibration, and maintenance
   - May react more slowly relative to PWM systems
   - Increases the overall cost because of associated electronic components

6. **Why is it possible to control flow rate on a conventional nozzle-based liquid application system only by changing the pressure relief valve?**
   Since conventional nozzle systems depend on a constant orifice size, pressure changes are the only means to alter flow. By regulating system pressure, a pressure relief valve indirectly modifies flow rate. However, this modification changes the spray pattern and droplet size, which could compromise the quality of the application.

7. **Why is manure less likely than granular or liquid-based fertilizers to be applied in a variable-rate manner?**
   Because the nutritional content varies according to the animal, food, handling, and

storage. It is challenging to detect nutrients in real time, and the technology available today is usually costly and sophisticated. The viscosity and particle concentration make it difficult to assess and control flow, especially in comparison with granular or liquid fertilizers. Furthermore, the application is limited in time, usually only available during pre-planting or post-harvest intervals, which makes thorough planning difficult.

8. **List at least three reasons why it is difficult to apply anhydrous ammonia fertilizer in a variable-rate manner.**
   Although anhydrous ammonia has been a widely used nitrogen fertilizer source, the variable-rate application is more challenging to implement owing to the dual-phase nature of ammonia. Unlike water, anhydrous ammonia in the tank and in plumbing lines exists in both liquid and vapor phases. Liquid flow meters are inaccurate because of the presence of vapor, while gas flow measurement devices do not work owing to the liquid portion. Without the ability to accurately measure the flow rate, it is difficult to control the rates. Remember, you cannot control it if you cannot measure it.

9. **If you had a PWM controller that incremented its counter 16 million times per second, what would the maximum PWM resolution be if you wanted the frequency to be 20 kHz? What would the bottom and top counter values need to be set at?**
   The bottom counter value should be 0 and the top counter = • Frequency = 20,000 Hz = 1/20,000 s

   Timer frequency = 1/16,000,000 s

   Max counter value = **800 counts**

   The maximum PWM resolution is 800 steps, or approximately 9.64 bits. The bottom counter value should be 0, and the top counter value should be 799.

10. **How are lightbar guidance and auto-steer guidance similar? What is the major difference? How could the two technologies improve efficiency in a tillage operation?**
    Although they both use GPS to help drivers stay on an established path in the field, lightbar guidance and auto-steer guidance work differently to guide the vehicle. While auto-steer systems steer the tractor or other equipment automatically, lightbar systems give the operator visual indications. By decreasing overlap, accelerating, and maybe lengthening the workday, both can increase tillage operations' efficiency.

11. **What are at least two major benefits of ISOBUS using a bus network topology instead of a star topology?**
    Compared with a star network, in which every device needs a separate dedicated connection, ISOBUS simplifies the wiring and lowers the overall cost of implementation by using a single communication line (the bus) to which all devices are attached. A more controllable and effective network is also facilitated by this shared communication channel.

12. **How does implement steering improve the potential accuracy of product and seed placement?**
    Particularly on hills or curves, implement steering makes sure the implement and tractor travel in the same direction. This guarantees more uniform product and seed placement, enhances row alignment, and minimizes skips or overlaps.

13. **How can improved seed placement and seed meter monitoring improve yields?**
    Through improving the seeding procedure, which results in less waste and more consistent plant stands. While monitoring systems detect and handle problems like missed seeds, doubles, or skips, all of which can affect plant emergence and total yield,

accurate seed placement ensures improved seed-to-soil contact.

14. **What is the universal terminal in an ISOBUS system?**
A universal terminal is an in-cab display that acts as a user interface for all ISOBUS-compliant devices.

15. **What are the advantages of filtering messages in an ISOBUS system?**
   A. Enables users to concentrate on particular message types or data sources, which simplifies analysis and troubleshooting.
   B. Improves privacy by restricting access to unnecessary data or hiding important information.
   C. Decreases the amount of data, improving the efficiency and focus of analysis.
   D. Enhances system security by making it possible to prevent the processing or presentation of particular communications.

Chapter 27

1. **What is artificial intelligence, and how is it different from machine learning and deep learning in the context of agriculture?**
Artificial Intelligence (AI) is a broad computer-based field where machines or computers mimic human intelligence to perform tasks like decision-making and problem-solving. Machine Learning (ML) is a subset of AI that uses data to train models for specific predictions, such as crop yield or disease detection. Deep Learning (DL) is a type of ML that uses complex neural networks, often for analyzing images or large datasets. In agriculture, AI helps with tasks like weather prediction, ML powers models for yield forecasting, and DL enables image-based applications like weed detection.

2. **What are some examples of AI-powered technologies currently used in farming systems?**
Some examples of AI-powered technologies in farming include chatbots for crop advisory purposes, drones for field scouting and weed detection, and sensors for real-time monitoring of soil moisture, nutrients, and crop health.

3. **How can AI support site-specific nutrient, weed, and irrigation management?**
AI can analyze data from sensors, satellite imagery, and weather to optimize nutrient, weed, and irrigation management at specific locations within a field. This helps farmers apply inputs precisely where and when they are needed, improving efficiency and sustainability.

4. **How can AI improve decision-making in agricultural systems?**
AI improves decision-making in agriculture by analyzing large, complex datasets from weather, soil, and crop sensors to generate insights. These insights help farmers make better choices on planting, fertilizing, irrigation, pest control, and harvesting, ultimately enhancing productivity and sustainability.

5. **What are the main challenges of implementing AI technologies on farms?**
The main challenges of implementing AI on farms are limited access to high-quality data, a lack of technical expertise among farmers, the high costs of advanced equipment, connectivity issues in rural areas, and concerns regarding data privacy and ownership.

Chapter 28

Question 1 (Multiple Choice)

1. **Which of the following best distinguishes a footprint analysis from a full life cycle assessment (LCA)?**
   a. Footprints consider indirect impacts, while LCAs do not.
   b. Footprints focus on a single impact, while LCAs integrate multiple impacts across a system. **(B is the correct answer)**
   c. LCAs are only used in industrial sectors, not agriculture.
   d. Footprints require ISO standards, while LCAs do not.

Question 2 (True/False)

2. **Attributional LCAs primarily focus on system-wide policy consequences that are outside a farmer's control.**
   (False)
   (The statement describes consequential LCAs; attributional LCAs focus on direct, controllable processes.)

3. **Why are precision agriculture data valuable for improving agricultural LCAs?**
   Precision agriculture data provide time- and space-specific information that improves the accuracy, representativeness, and transparency of LCA inventories, allowing better evaluation of management practices and environmental impacts.

4. **Name three of the benefits that high-quality data and precision agriculture provide to agriculture LCAs?**
   1. Improved accuracy and representativ eness
   2. Better evaluation of management practices and trade-offs
   3. Documentation and verification of sustainability improvements

Question 5 (Multiple Choice)

5. **Which of the following is a major limitation of including indirect impacts in LCAs?**
   a. They are irrelevant to agriculture.
   b. They are too easy to quantify.
   c. There is a lack of standardized methods for evaluating them. **(C is the correct answer)**
   d. They only apply to biofuel systems.

Chapter 29

1. **Where do you plant sorghum as opposed to corn?**
   Sorghum is more drought tolerant than corn and has higher water use efficiency than corn. At low water levels, it produces more biomass than corn. Therefore, sorghum is planted in drier areas than corn.

2. **What is a pulse, and why is it important in rotation?**
   Pulses are the edible seeds of plants in the legume family, such as lentil, pea, chickpea, and dry bean. The name is limited to legume crops that are harvested as dry grain. Pulses are important in crop rotation because they fix atmospheric nitrogen through symbiosis with rhizobia, reducing the fertilizer needs of subsequent crops.

3. **When might you plant potato in tropical versus temperate climates?**
   Potato is a cool-season crop that is better suited to cooler environments than those found in the tropics. However, in warm environments, it can be grown in highland areas. Research is being conducted to develop cultivars suitable for tropical conditions.

4. **For which climatic region is cotton best suited?**
   Cotton grows best in warm humid environments. It needs abundant sunlight and a long frost-free period.

5. **What is the difference between rice grown in a paddy and rice grown in upland areas?**
   These are two different methods of rice cultivation. Paddy rice is grown in flooded fields (often known as paddies), while upland rice is grown in non-flooded conditions. Flooding is used to help control weeds and to provide adequate water for growing plants.

6. **What makes a plant well suited for precision technologies?**
   Plants most suitable for precision farming are those that have needs that vary across a landscape.

7. **Which plants are best suited for salt-affected soil?**
   More than 20% of the world's soils have high salt concentrations. Salt stress increases

intracellular osmotic pressure and adversely affects germination, growth, and development. Most of the plants discussed in this chapter are not salt tolerant. Of those discussed, barley is an annual plant that has high salt tolerance. Other plants discussed in Chapter 25 are AC Saltlander (green wheatgrass), western wheatgrass, and slender wheatgrass.

8. **What is the importance of measuring planting failures in sugarcane fields using UAV imagery?**
   Sugarcane is a semi-perennial crop. Failures in the establishment of a given field will compromise yield formation not only in the first cycle, but also in sequential crop cycles. Monitoring planting failures and replanting zones with intense failures will allow for achieving a high yield in the first cycle and maintaining high yields throughout the entire life of the field.

9. **What are the main species of banana cultivated globally, and why are they preferred?**
   The main species are *M. acuminata*, *M. balbisiana*, and their hybrids. They are preferred due to their adaptability, growth traits, and suitability for various purposes, such as fresh consumption, cooking, brewing, baking, and aesthetics.

10. **What are the reproductive structures of banana plants? Which part of the flower is of economic importance?**
    Banana plants produce both male and female inflorescences. The female inflorescence is economically important because it develops into the seedless fruit we eat.

11. **Why are most commercial banana varieties seedless?**
    Most commercial banana varieties are seedless due to their triploid nature, which causes sterility and prevents seed formation.

12. **Name the top banana-importing countries and themain importing region.**
    The top importers are the United States, Germany, and Russia, while the main importing region is Western Europe. .

13. **What are the key environmental requirements for successful grape production, and where are the prime viticultural zones typically located?**
    - A reliable water supply, abundant sunlight, and sufficient chilling hours during dormancy are all essential for the successful production of grape, which is characterized by uniform bud break and fruit set.
    - *Vitis vinifera* requires 1,390–2,220 growing degree days (GDDs) above 10 °C (50 F) for optimal development.
    - In both hemispheres, prime viticultural zones are typically situated between 30° and 50° latitude.

14. **What are some examples of precision viticulture technologies, and how do they contribute to sustainable vineyard operations?**
    - Key precision viticulture applications include smart irrigation using crop and soil sensors, as well as mechanized pruning and harvesting employing computer vision and near-infrared spectroscopy for ripeness assessment.
    - Remote sensing platforms and computer-based technologies facilitate precise operational management and decision-making through information systems, mobile applications, decision support systems, and weather-based advisories.

Chapter 30

Understanding Key Concepts

1. **What is the purpose of the *Precision Agriculture Basics*, and who is its intended audience?**
   The manual is designed for students at all levels, CCAs, Certified Professional Soil Scientists, and anyone interested in precision agriculture. This chapter is especially useful for instructors and extension educators teaching precision agriculture concepts or organizing workshops.

2. **Explain the concept of *backward design* in instructional planning. What are its three main steps?**
   The general concept is to align learning objectives, assessments, and activities in course design. If learning objectives are not aligned with assessments or activities, students may miss key concepts or feel confused. The three main steps include (1) identifying learning objectives, (2) planning assessments to measure those objectives, and (3) designing activities that help students achieve them.
3. **Of the seven thematic sections of the textbook, identify how they can be used to customize the course for agronomy majors and for data science students?**
   For agronomy majors: focus on soil variability, pest management, and sensing technologies. For data science students: emphasize GIS, remote sensing, and data management.
4. **What types of instructional resources are provided with each chapter to support teaching and learning?**
   Each chapter includes sample questions, PowerPoint slides, videos, and access to real-world datasets. These save prep time and help instructors deliver engaging lessons.
5. **What are some examples of *active learning strategies* mentioned in the chapter? Choose two and explain how they promote student engagement.**
   Peer Teaching: Students present chapters to the class, reinforcing their understanding; Keyword Challenges: Great for quizzes—students fill in missing terms based on context.

Wiley The manufacturer's authorized representative according to the EU General Product Safety Regulation is Wiley-VCH GmbH, Boschstr. 12, 69469 Weinheim, Germany, e-mail: Product_Safety@wiley.com.

Printed and bound by CPI Group (UK) Ltd, Croydon, CR0 4YY
02/10/2026
02227445-0001